Supramolecular Chemistry in Corrosion and Biofouling Protection

Supramolecular Chemistry in Corrosion and Biofouling Protection

Edited by
Viswanathan S. Saji

CRC Press
Taylor & Francis Group
Boca Raton London New York

CRC Press is an imprint of the
Taylor & Francis Group, an **informa** business

First edition published 2022
by CRC Press
6000 Broken Sound Parkway NW, Suite 300, Boca Raton, FL 33487-2742

and by CRC Press
2 Park Square, Milton Park, Abingdon, Oxon, OX14 4RN

CRC Press is an imprint of Taylor & Francis Group, LLC

Library of Congress Cataloging-in-Publication Data
Names: Saji, Viswanathan S., editor.
Title: Supramolecular chemistry in corrosion and biofouling protection /
edited by Viswanathan S. Saji.
Description: First edition. | Boca Raton : CRC Press, 2022. |
Includes bibliographical references and index.
Identifiers: LCCN 2021034124 (print) | LCCN 2021034125 (ebook) |
ISBN 9780367769024 (hbk) | ISBN 9780367769628 (pbk) | ISBN 9781003169130 (ebk)
Subjects: LCSH: Corrosion resistant materials. | Protective coatings. |
Supramolecular chemistry. | Fouling--Prevention. | Macromolecules.
Classification: LCC TA418.75 .S885 2022 (print) | LCC TA418.75 (ebook) |
DDC 620.1/1223--dc23
LC record available at https://lccn.loc.gov/2021034124
LC ebook record available at https://lccn.loc.gov/2021034125

ISBN: 978-0-367-76902-4 (hbk)
ISBN: 978-0-367-76962-8 (pbk)
ISBN: 978-1-003-16913-0 (ebk)

DOI: 10.1201/9781003169130

Typeset in Times
by codeMantra

Contents

SECTION I   Supramolecules

SECTION II   Applications in Corrosion Protection
   A   Protective Coatings (Chapters 8 to 14)
   B   Corrosion Inhibitors (Chapters 15 to 19)

SECTION III   Applications in Biofouling Protection

Preface

Supramolecular chemistry, "the chemistry beyond the molecule," is a fascinating realm of modern science. The design of novel supramolecular structures, surfaces, and techniques are at the forefront of research in different application areas, including corrosion and biofouling protection. The book's primary objective is to provide accurate descriptions of the type and fundamentals of supramolecules and assemblies to researchers in the broad domain of corrosion and biofouling. The first part of the book (Part I) is dedicated to the basics of supramolecular chemistry with this objective. Here, the chapters are arranged in the following sequence: fundamentals of supramolecules (Chapter 1), supramolecular host–guest inclusion complexes (Chapter 2), and supramolecular assemblies (Chapters 3–7). In Part II, we have attempted to present several supramolecular chemistry applications in corrosion protection. This part has two sections; the first section deals with surface coatings, and the second section, the corrosion inhibitors. Part III of the book deals with supramolecular applications in biofouling protection.

Part I starts with a chapter (Chapter 1) on supramolecules. The chapter provides clear descriptions of different supramolecular building blocks and supramolecular assemblies. Chapter 2 details fundamental aspects of host–guest inclusion complexes. The chapter also presents representative examples of the use of host–guest inclusion complexes in corrosion and biofouling prevention. Chapter 3 focuses on supramolecular organic polymers. The chapter provides the basics of preparation and characterization of supramolecular polymers, their mechanism and the pathway of supramolecular polymerization. A description of the application of supramolecular polymer networks as self-healing materials is also provided. Chapter 4 gives a good account of the fundamentals and types of supramolecular self-healing hydrogels. These hydrogels have several potential biomedical and industrial applications, including corrosion and biofouling protection. Chapter 5 presents an interesting overview of supramolecular aspects in coordination polymers. The chapter provides basic details of the type, classification, and structure of coordination polymers and an overview of covalent and non-covalent interactions in the supramolecular architecture. Chapter 6 focuses on metal-organic frameworks. The chapter explains supramolecular host–guest chemistry in metal-organic frameworks along with different examples. The last chapter of the section (Chapter 7) describes synthesis and applications of coordination-driven metallo-assemblies. A detailed discussion on the rational design and synthesis of two and three-dimensional metallo-macrocycles and metallo-cages is provided. In addition, important applications of metallo-assemblies are also described.

Section A of Part II of the book contains chapters focusing on supramolecular chemistry-based self-healing (extrinsic and intrinsic) anti-corrosion coatings. The section starts with an informative chapter (Chapter 8) on self-healing mechanisms in smart protective coatings. The self-healing properties of both dynamic covalent chemistry and supramolecular chemistry are explained. The second chapter (Chapter 9) concentrates on polymeric nanocapsules as nanocarriers in surface coatings. Chapter 10 describes different applications of cyclodextrins in surface coatings, whereas Chapter 11 presents a broad outlook on mesoporous silica, their synthesis, encapsulation methods, release kinetics, characterization, and self-healing properties. Chapter 12 provides an account of metal-organic frameworks based anti-corrosion coatings. Chapter 13 describes the performance and mechanisms of graphene-based anti-corrosion coatings. The last chapter of part IIA (Chapter 14) focuses on intrinsic self-healing in polyurethane coatings.

Chapters 15–19 of Section B of Part II present supramolecules based corrosion inhibitors. The section begins with a concise description of supramolecular polyurethane corrosion inhibitors (Chapter 15). The next chapter (Chapter 16) explains four supramolecules, crown ethers, dendrimers, cyclodextrins, and calixarenes and their derivatives as corrosion inhibitors. The chapter also provides good descriptions of these supramolecules' physical, chemical, structural, and adsorption properties and solution chemistry. Chapter 17 gives an account of coordination polymer inhibitors, emphasizing Schiff's base polymer complexes, whereas Chapter 18 presents different metal–organic frameworks based corrosion inhibitors reported for steel, copper, and magnesium. The last chapter of the section (Chapter 19) describes self-assembled monolayers in corrosion inhibition.

Part III of the book starts with a chapter (Chapter 20) on supramolecular polymers and gels in anti-biofouling applications. The chapter explains various anti-biofouling systems utilizing different non-covalent interactions and chemical modifications. The next chapter (Chapter 21) provides an account of cyclodextrins in anti-biofouling coatings. Cyclodextrin-functionalized polymer-based coatings and their potential applications in health care, membrane separation, and marine industries are presented. Chapter 22 gives an interesting description of mesoporous silica-based systems in anti-biofouling applications. Different approaches concerning biocide encapsulation and immobilization are discussed. Chapter 23 details metal-organic frameworks and their nanocomposites in anti-biofouling approaches. Antibacterial mechanisms and recent developments are described. The next chapter (Chapter 24) describes the author's pilot plant study on an eco-friendly anti-biofouling and anti-corrosion coating with incorporated nanocontainers. The last chapter of the book (Chapter 28) accounts for supramolecular surface modifications for titanium implants.

This book results from our genuine effort to extend the supramolecular concept and approaches to the domain of corrosion and biofouling prevention. We have tried to bring together different sections of supramolecular chemistry under one roof. We hope that the present book will be a handy reference tool for students and researchers working in different areas of supramolecular chemistry, corrosion, and biofouling.

Viswanathan S. Saji

Acknowledgment

We are thankful to all the authors for their valuable contributions to this book. We would also like to express our gratitude to all the sources and publishers granting us the copyright permissions for reproducing illustrations. We acknowledge King Fahd University of Petroleum and Minerals, Saudi Arabia, for the moral support provided. Our sincere thanks to the CRC team for evolving this book into its final shape.

Editor

Viswanathan S. Saji is a Research Scientist III/Assistant Professor at the Interdisciplinary Research Center for Advanced Materials, King Fahd University of Petroleum and Minerals (KFUPM), Saudi Arabia. He received M.Sc. (1997), M.Phil. (1999) and Ph.D. (2003) degrees from the University of Kerala, India. He was a Research Associate at Indian Institute of Technology (IIT) Bombay (2004–2005) & Indian Institute of Science (IISc) Bangalore (2005–2007), Postdoctoral Researcher at Yonsei University (2007–2008) & Sunchon National University (2009), Senior Research Scientist at Ulsan National Institute of Science and Technology (UNIST) (2009–2010), Research Professor at Chosun University (2008–2009) & Korea University (2010–2013), and Endeavour Research Fellow at University of Adelaide (2014). He has authored 75 journal publications, contributed 8 books and 15 book chapters. He has more than 50 international/national conference presentations and invited talks. His research interest lies in electrochemistry, corrosion science, and smart materials.

Contributors

Bimalendu Adhikari
Department of Chemistry
National Institute of Technology Rourkela
Rourkela, India

Khamdam Akbarov
Faculty of Chemistry
National University of Uzbekistan
Tashkent, Uzbekistan

Idalina Vieira Aoki
Chemical Engineering Department
Polytechnic School of the University
    of São Paulo
São Paulo, Brazil

Vasudeva Rao Bakuru
Inorganic and Analytical Chemistry
    Department, School of Chemistry
Andhra University
Visakhapatnam, India

Rakesh C. Barik
Corrosion and Materials Protection Division
CSIR-Central Electrochemical Research
    Institute
Karaikudi, India
Academy of Scientific and Innovative
    Research (AcSIR)
Ghaziabad, India

Elyor Berdimurodov
Faculty of Natural Sciences
Karshi State University
Karshi, Uzbekistan

Achikanath C. Bhasikuttan
Radiation & Photochemistry Division
Bhabha Atomic Research Centre
Mumbai, India
Homi Bhabha National Institute
Mumbai, India

Ik Sung Cho
Department of Bioengineering
University of Illinois at Chicago
Chicago, Illinois, USA

Bhawna Chugh
Department of Chemistry
Netaji Subhas University of Technology
Dwarka, India

Vitalis I. Chukwuike
Corrosion and Materials Protection Division
CSIR-Central Electrochemical
    Research Institute
Karaikudi, India
Academy of Scientific and Innovative Research
    (AcSIR)
Ghaziabad, India

Peter J. Cragg
School of Pharmacy and Biomolecular
Sciences, University of Brighton,
    Brighton, UK

Brunela Pereira da Silva
Chemical Engineering Department
Polytechnic School of the University
    of São Paulo
São Paulo, Brazil

Viviane Dalmoro
Instituto Federal de Educação Ciência e
    Tecnologia Sul-rio-grandense,Camaquã /
    RS, Brazil

Marilyn Esclance DMello
Materials Science Division
Poornaprajna Institute of Scientific Research
Bengaluru, India
Graduate Studies
Manipal Academy of Higher Education
Manipal, India

João Henrique Z. dos Santos
Department of Chemistry,
Universidade Federal do Rio Grande do Sul,
Porto Alegre/RS, Brazil

Ubong Eduok
College of Engineering
University of Saskatchewan
Saskatoon, Canada

Drishya Elizebath
Photosciences and Photonics Section,
    Chemical Sciences and Technology
    Division
CSIR-National Institute for Interdisciplinary
    Science and Technology (CSIR-NIIST)
Thiruvananthapuram, India

Academy of Scientific and Innovative
Research (AcSIR)
Ghaziabad, India

Jesus Marino Falcón-Roque
Environmental Engineering
    Department
Universidad San Ignacio de Loyola
Lima, Peru

Karan Gulati
School of Dentistry
University of Queensland
Queensland, Australia

Lei Guo
School of Materials and Chemical Engineering
Tongren University
Tongren, China

Bashirul Haq
Department of Petroleum Engineering
King Fahd University of Petroleum
    and Minerals
Dhahran, Saudi Arabia

Tracey Jackson
Technology, Chemicals and Industrial
    Services, Oilfield Services
Baker Hughes Company, Inc.
Sugar Land, Texas, USA

Suresh Babu Kalidindi
Inorganic and Analytical Chemistry
    Department, School of Chemistry
Andhra University
Visakhapatnam, India

Abduvali Kholikov
Faculty of Chemistry
National University
    of Uzbekistan
Tashkent, Uzbekistan

George Kordas
Peter the Great St. Petersburg Polytechnic
    University
St. Petersburg, Russian Federation

T.V. Krishna Mohan
Water and Steam Chemistry Division,
    Chemistry Group
BARC Facilities
Kalpakkam, India

Dusan Losic
School of Chemical Engineering
    and Advanced Materials
The University of Adelaide
Adelaide, Australia
ARC Hub for Graphene Enabled Industry
    Transformation
The University of Adelaide
Adelaide, Australia

A. Madhan Kumar
Interdisciplinary Research Center
    for Advanced Materials
King Fahd University of Petroleum
    and Minerals
Dhahran, Saudi Arabia

Suman Mandal
Department of Chemistry
Indian Institute of Technology
    Hyderabad
Sangareddy, India

Bobbie M. McVey
School of Pharmacy and Biomolecular
    Sciences
University of Brighton
Brighton, UK

Carlos Menendez
Technology, Chemicals and Industrial
    Services, Oilfield Services
Baker Hughes Company, Inc.
Sugar Land, Texas, USA

Rakesh K. Mishra
Department of Chemistry
National Institute of Technology Uttarakhand (NITUK),
Srinagar (Garhwal), Uttarakhand, India

Binduja Mohan
Discipline of Chemistry
Indian Institute of Technology
Palakkad, India

Indrajit Mohanta
Department of Chemistry
National Institute of Technology Rourkela
Rourkela, India

Jyotirmayee Mohanty
Radiation & Photochemistry Division
Bhabha Atomic Research Centre
Mumbai, India

Homi Bhabha National Institute
Mumbai, India

Muthukumar Nagu
Research and Development Center
Saudi Aramco, Dhahran, Saudi Arabia

Muneshwar Nandeshwar
Department of Chemistry
Indian Institute of Technology Hyderabad
Sangareddy, India

Md Julker Nine
School of Chemical Engineering
and Advanced Materials
The University of Adelaide
Adelaide, Australia

ARC Hub for Graphene Enabled Industry
Transformation
The University of Adelaide
Adelaide, Australia

Tooru Ooya
Department of Chemical Science and
Engineering, Graduate School of
Engineering
Kobe University
Kobe, Japan

Akhil Padmakumar
Photosciences and Photonics Section, Chemical
Sciences and Technology Division
CSIR-National Institute for Interdisciplinary
Science and Technology (CSIR-NIIST)
Thiruvananthapuram, India

Academy of Scientific and Innovative
Research (AcSIR)
Ghaziabad, India

Balaram Pani
Department of Chemistry, Bhaskaracharya
College of Applied Science
University of Delhi
Dwarka, India

Ganesan Prabusankar
Department of Chemistry
Indian Institute of Technology Hyderabad
Sangareddy, India

Vakayil K. Praveen
Photosciences and Photonics Section,
Chemical Sciences and Technology
Division
CSIR-National Institute for Interdisciplinary
Science and Technology (CSIR-NIIST)
Thiruvananthapuram, India

Academy of Scientific and Innovative
Research (AcSIR)
Ghaziabad, India

Mohammad Mizanur Rahman
Interdisciplinary Research Center for
Advanced Materials
King Fahd University of Petroleum and
Minerals
Dhahran, Saudi Arabia

T. P. D. Rajan
Material Science and Technology
Division
CSIR-NIIST
Trivandrum, India

Puspalata Rajesh
Water and Steam Chemistry Division,
    Chemistry Group
BARC Facilities
Kalpakkam, India

Sunder Ramachandran
Technology, Chemicals and Industrial Services,
    Oilfield Services
Baker Hughes Company, Inc.
Sugar Land, Texas, USA

Silvia Rosane S. Rodrigues
Department of Chemistry,
Universidade Federal do Rio Grande do Sul,
Porto Alegre/RS, Brazil

Nihar Sahu
Department of Chemistry
National Institute of Technology Rourkela
Rourkela, India

Viswanathan S. Saji
Interdisciplinary Research Center
    for Advanced Materials
King Fahd University of Petroleum and
    Minerals
Dhahran, Saudi Arabia

Sankarasekaran Shanmugaraju
Discipline of Chemistry
Indian Institute of Technology
Palakkad, India

Sheetal
Department of Chemistry
Netaji Subhas University of Technology
Dwarka, India

Ashish Kumar Singh
Department of Applied Science
Bharati Vidyapeeth's College of Engineering
New Delhi, India

Akhilesh K. Srivastava
Department of Chemistry
Deen Dayal Upadhyay Gorakhpur University
Gorakhpur, India

Alicja Stankiewicz
coat-it sp. z o. o.
Puławy, Poland

Kalaivanan Subramaniyam
Department of Chemistry
Indian Institute of Technology Hyderabad
Sangareddy, India

Sanjeeve Thakur
Department of Chemistry
Netaji Subhas University of Technology
Dwarka, India

Sarah B. Ulaeto
Department of Chemical Sciences
Rhema University
Aba, Nigeria

Sabari Veerapathiran
Department of Chemistry
Indian Institute of Technology Hyderabad
Sangareddy, India

Brian D. Wagner
Department of Chemistry
University of Prince Edward Island
Charlottetown, Canada

Mengyue Zhu
School of Materials Science and Engineering
East China Jiaotong University
Nanchang, China

Katarzyna Zielińska
coat-it sp. z o. o.
Puławy, Poland

Section I

Supramolecules

1 Supramolecules

Bobbie M. McVey and Peter J. Cragg
University of Brighton

CONTENTS

1.1 INTRODUCTION

Supramolecules, structures that are literally "beyond molecules," have their origins in descriptive biology where proteins and plant fibers were examined for "the presence of supra-molecular discrete and discontinuous units" (Baas-Becking and Galliher 1931). Observations of structures in the cytoplasm noted that they had "been considered until recently to be devoid of structure at the supramolecular level of organization" (Palade 1955) and, in later work, enzymes had been described as having "supramolecular organization" (Mitchell 1961). The term "supramolecular" itself had been defined over a century ago as referring to systems "composed of an aggregation of molecules; of greater complexity than the molecule" (Century Dictionary 1909). Later, it was defined as: "The transition between molecular structure and morphology is approached by what we may call 'supra-molecular biology'" (Luria 1970).

The concept entered the chemical lexicon in the 1930s when Wolf and co-authors investigated the physical properties of liquids and reported that certain molecules formed strong associations with others of the same type (Wolf et al. 1937). The simplest of these was acetic acid, which was believed to form dimers held together through strong hydrogen bonds, a concept popularized by Pauling, and called by Wolf an "*übermolecül.*" This idea was greatly expanded upon by Lehn who, in his 1987 Nobel Prize lecture, defined supramolecular as: "the chemistry beyond the molecule bearing on the organized entities of higher complexity that result from the association of two or

DOI: 10.1201/9781003169130-2

more chemical species held together by intermolecular forces" (Lehn 1988). Elsewhere, Lehn has likened molecules to words with their atoms, or letters, linked irreversibly, and supramolecules to sentences in which the words, though not their spelling, may adopt different orders to give different meanings or functions. As Lewis Carroll's Hatter points out, "I see what I eat" is not the same thing as "I eat what I see" (Carroll 1866), and the same can be said to be true for the effect of molecular ordering in supramolecular systems.

Supramolecules form through complementary, and usually reversible, interactions between their component species. These range from individually weak hydrogen bonds and hydrophobic interactions to metal–ligand coordinate bonds and disulfide bonds. Molecules with numerous opportunities to form such interactions result in properties such as gelation, whereas reversibility ensures allows them to reform broken bonds to give self-healing polymers. Where supramolecules are built upon a surface, carefully designed complementary functionality allows the order of these self-assembling layers to be controlled.

One further aspect of supramolecule formation also has profoundly useful effects; the formation of host–guest complexes (Fieser and Fieser 1959). Previous examples may result in complex polymer-like behavior; however, discrete binding between a host, with converging bonding sites, and a guest, with diverging binding sites, can also occur. Where this is coupled to a chromophore, fluorophore or redox-active group, the host can act as a sensor and disclose the presence of the guest through a change in color, fluorescence, or electrochemical potential.

1.2 SUPRAMOLECULAR BUILDING BLOCKS

1.2.1 Non-Cyclic Molecules

1.2.1.1 Polyethers

A series of ten non-cyclic polyethers with different aromatic termini, including the podand in Figure 1.1 were synthesized, and their complexation thermodynamics was studied by Vögtle (Tümmler et al. 1979). They acted as neutral ionophores and formed stable 1:1 complexes with alkali metals, with stability constants ranging between 1 and $10^4 \, M^{-1}$. The stability of the complexes depended on factors such as the number of coordinating atoms, cation size, donor strength, and the rigidity of the aromatic end groups. Thermodynamic studies showed complex formation was driven enthalpically, with an unfavorable decrease in entropy caused by forming a pseudo-circular conformation with much lower flexibility than the free ligand. While these derivatives are excellent at binding alkali metals, they lack the ability to discriminate between cations due to their flexibility when compared with crown ethers, which have a defined cavity size which will be complementary to certain metal ions. An advantage of these acyclic compounds over macrocycles is that the cyclization reaction, which often leads to a low yield, is omitted. In addition, they can be easily tailored to many applications by changing the end groups and they are readily soluble in most organic solvents.

1.2.1.2 Dendrimers

Dendrimers are molecules that branch from a central core. Named after the Greek *dendron*, meaning tree, they have been of continued academic and industrial interest since the pioneering work of Tomalia, Newcome and Fréchet (Tomalia et al. 1985; Newkome et al. 1985; Hawker and Fréchet 1990). Each addition of a branch generates a higher "generation" of dendrimer leading to well-ordered three-dimensional molecules with relative molecular masses usually between 5 and 500 kDa; Figure 1.1 shows an example of a first-generation poly(amidoamine), or PAMAM, dendrimer. Functional terminal groups at the outer surface of the dendrimer can be selected for an intended application. Possibilities for using these "molecular sponges" to absorb and release therapeutic compounds were clear from the outset, and they are now used in the controlled release of non-steroidal anti-inflammatory compounds, anticancer drugs such as doxorubicin, paclitaxel and methotrexate, anti-HIV retrovirals, and glaucoma treatments (Madaan et al. 2014).

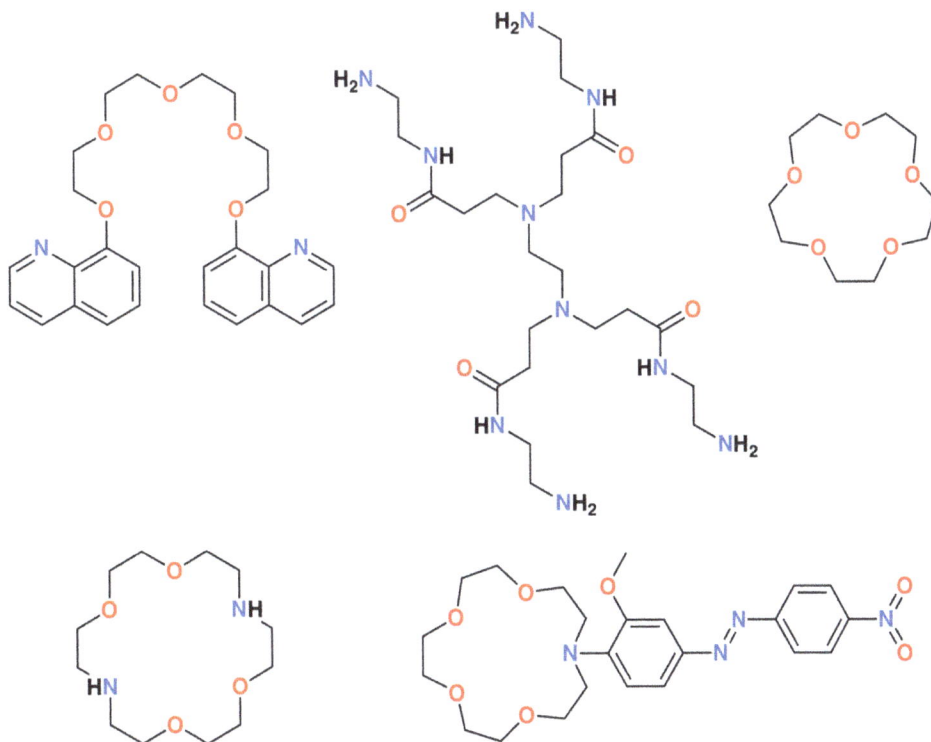

FIGURE 1.1 Flexible supramolecule components: (left to right, top) podand, dendrimer, crown ether, (bottom) 1,10-diazacrown ether, and lariat ether.

1.2.1.3 DNA

Although usually associated with replication, the base-specific interactions between strands of DNA facilitate their use in molecular design. Extensive work by Seeman has demonstrated how mismatched terminal sequences can be used as "sticky ends" to recognize other strands (Seeman 1982). Careful consideration of matched and mismatched regions allows DNA to be used to build frameworks much like nanosized wicker baskets or boxes, which could be used to store guest molecules (Chen and Seeman 1991).

1.2.2 MACROCYCLES

1.2.2.1 Crown Ethers

Crown ethers are cyclic polyethers made up of oxygen atoms bridged by ethylene groups. The name "crown" is reflective of the shape made when a crown ether forms an inclusion complex with a metal ion through ion–dipole interactions between the negatively charged oxygen atoms in the ring and the cation. The structural nomenclature N-crown-M is derived from the number of atoms in the ring N, and the number of those which are oxygen atoms M, for example, 18-crown-6. The structure in Figure 1.1 is therefore 15-crown-5.

Early work by Nobel laureate Pedersen demonstrated the formation of stable 1:1 complexes of crown ethers with alkali and alkali earth metals (Pedersen 1967). The stabilities of these complexes were influenced by factors such as cavity size, the number and basicity of oxygens in the ring, the oxygens' coplanarity and symmetry, steric hindrance, and electrical charge on the cation. He also found that the most easily formed rings were composed of five or six oxygen atoms each separated by two carbon atoms.

It was soon realized that some or all of the oxygen could be replaced by other main group elements such as sulfur and, in particular, nitrogen. The latter gave rise to the azacrown ethers. Following from Pedersen's original nomenclature, the non-carbon atom is included to give, for example, thia-15-crown-5 or, where two or more heteroatoms are present, their relative positions in the macrocyclic ring are given as in 1,10-diaza-18-crown-6 (Figure 1.1). Replacing oxygen with the softer Lewis base sulfur gives the thiacrowns an affinity for a number of transition metals, whereas incorporating nitrogen enables side chains to be introduced to give the lariat ethers (Schultz et al. 1985). Linking a fluorophore, electroactive substituent or chromophore, as shown in Figure 1.1, to the crown allows guest binding to be detected through a change in the spectroscopic absorption maximum or redox potential, thus turning the crowns into sensors.

1.2.2.2 Calixarenes and Related Macrocycles

Calixarenes are cyclic oligomers formed through the condensation reaction of p-tert-butylphenol and formaldehyde, which were first synthesized by Baeyer in 1872 (Baeyer 1872a). The name was coined by Gutsche and originates from two Greek root words: *calix* which means chalice, reflecting the wide upper rim and narrow lower rim of the macrocycle, and *arene* indicating the presence of aromatic rings (Gutsche and Muthukrishnan 1978). The bracketed number in calix[n] arene nomenclature reflects the number of aromatic units. Thus, Figure 1.2 shows 4-tert-butyl-calix[4]arene as an example. Calixarenes of varying compositions can be obtained by varying reaction conditions such as reactant ratios and the use of different bases or solvents. The basic structure comprises hydrophobic tertiary butyl groups on the upper rim, a hydrophobic cavity for

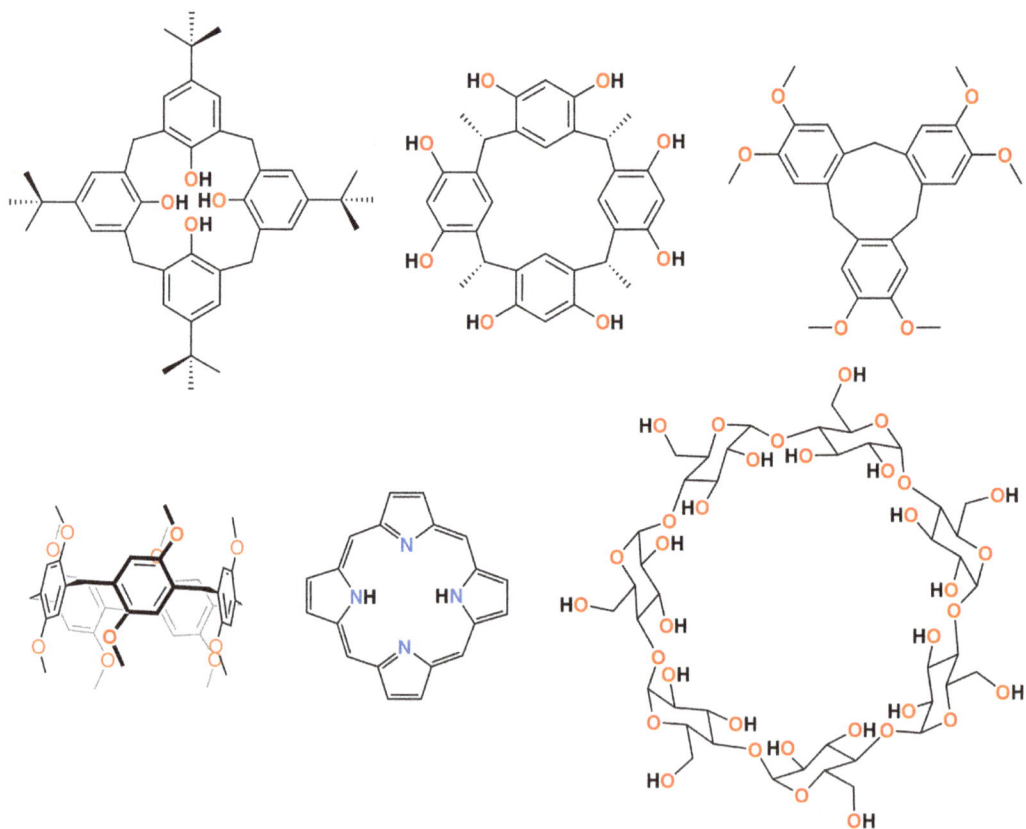

FIGURE 1.2 Rigid supramolecule components: (left to right, top) calixarene, resorcinarene, cyclotriveratrylene, (bottom) pillarene, porphyrin, and cyclodextrin.

small organic molecules, and hydroxyl groups on the lower rim for adding functional components through hydroxyalkylation reactions.

Resorcinarenes are close cousins of calixarenes with an analogous nomenclature and similar chalice-like structures as illustrated in Figure 1.2. They are formed through the condensation of resorcinol and aldehydes in acid solution. The first resorcinarene was synthesized in 1872 by Baeyer who obtained a red product from the condensation reaction of benzaldehyde and resorcinol in concentrated sulfuric acid; however, he did not have the analytical tools available at the time to fully characterize them (Baeyer 1872b). Later, Högberg later investigated similar products and discovered the two stereoisomeric resorcinarene products resulting from a resorcinol-acetaldehyde condensation (Högberg 1980).

The reaction of veratrole and formaldehyde produces the cyclotrimer, cyclotriveratrylene (CTV). Originally reported in 1915 (Robinson 1915), and believed to be a dimer, it was not until 1965 that X-ray crystallography revealed it to be the trimer shown in Figure 1.2 (Lindsey 1965). Despite the lack of accessible central cavity, CTV can be functionalized and used to synthesize dimeric capsules known as cavitands, which are capable of encapsulating guest solvents (Canceill et al. 1984).

In 2008, Ogoshi reported a new class of macrocycle, the pillar[n]arenes (Ogoshi et al. 2008). Pillar[n]arenes, or pillarenes, are almost structural analogs of calixarenes and are formed by the condensation of 1,4-dimethoxybenzene and paraformaldehyde in the presence of an appropriate Lewis acid. They are also named analogously to the calix[n]arenes, with the bracketed number indicating the number of units that make up the macrocycle. The example in Figure 1.2 is dimethoxypillar[5]arene. While methylene groups link calixarene units in the meta-position, pillarene units are linked at the para-positions, forming a more symmetric topology made up of pillar-like units. An advantage to this increased symmetry is the ability of the macrocycle to form some inclusion complexes without modification, although the pillarenes may be easily functionalized by cleaving the methyl ether bonds and reacting at the resulting hydroxyl groups. Alternatively, other dialkoxybenzene derivatives can be cyclized to introduce functionality as in Ogoshi's phenylethynylpillar[5] arene (Ogoshi et al. 2009).

1.2.2.3 Porphyrins

Porphyrins are macrocycles formed of four modified pyrrole units linked by a methine bridge at the α-carbons as shown in Figure 1.2. Due to the large conjugated system formed of 18 out of 26 π-electrons in a planar, continuous electron cycle, porphyrins absorb light strongly in the visible region. An example of a naturally occurring porphyrin is heme, the pigment in red blood cells, essential in the actions of hemoglobin.

1.2.2.4 Cyclodextrins

Cyclodextrins (CDs) are macrocycles formed of glucose units joined through α-1,4-glycosidic bonds. They are products of enzymatic reactions with starch, α-amylase, and a glycosyltransferase enzyme, with different compositions resulting in varying ratios of differently sized cyclic products, which can be made on a large scale at a low cost. These α-, β-, and γ-CDs, comprising six, seven, or eight units, respectively, are separated based on their relative solubilities. The least water-soluble, β-CD (Figure 1.2), crystallizes from an aqueous solution, whereas the smaller α-CD and larger γ-CD require selective complexation and membrane filtration to be isolated. The secondary hydroxyl groups at the upper rim and primary hydroxyl groups at the lower rim provide aqueous solubility and the hydrophobic central cavity provides a chiral environment for guest encapsulation. Consequently, CDs have found extensive use as delivery agents for poorly soluble drugs and as extractants in air fresheners where volatile odoriferous molecules are extracted from the air. Their inherent chirality, coupled with hydrogen bonding potential, has been exploited in technologies such as the separation of chiral compounds using CD-coated columns for HPLC. The CDs can be easily functionalized, either through all accessible hydroxyl groups or through just one, such that solubility or guest binding properties can be modified.

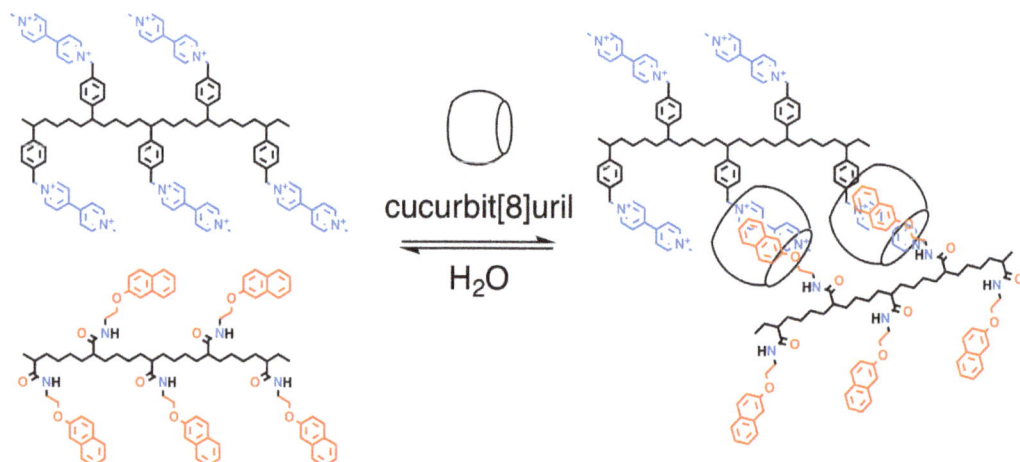

FIGURE 1.3   Supramolecular polymerization.

1.3   SUPRAMOLECULAR ASSEMBLIES

1.3.1   SUPRAMOLECULAR POLYMERS

Conventional polymers are formed when monomers react to form covalent bonds, but if, instead of covalent bonds, reversible interactions hold the structure together, then a supramolecular polymer result (Brunsveld et al. 2001). Complementary hydrogen bonding involving alternating donors, such as amines, and acceptors, such as pyridines or carbonyls, allows very strong and specific polymeric assembly to occur (Kotera et al. 1994), whereas the alternating amide interactions in cyclic peptides result in stacking and nanotube formation (Ghadiri et al. 1993). Other interactions, such as enforced π-stacking of naphthyl and quaternarized bipyridyl groups inside cucurbit[8]uril macrocycles (Figure 1.3), can be used to prepare di-block co-polymers (Rauwald and Scherman 2008) which can stabilize waterlogged archaeological artefacts (Walsh et al. 2014).

1.3.2   SUPRAMOLECULAR GELS

Gelation occurs when molecules crosslink, usually in the presence of a large excess of solvent, and expand in volume (Sangeetha and Maitra 2005). In supramolecular gels, the intermolecular interactions are those associated with supramolecule formation. While hydrogen bonding is the most obvious driving force, other interactions, such as the addition of metal cations to change molecular conformation and induce gelation, exist. What distinguishes these gels from supramolecular polymers is their ability to form expanded structures when they trap solvent or gas. Hydrogels are those that form in water, while xerogels are those with a gaseous second phase. For gels, and specifically hydrogels, to form, other functional groups capable of binding solvent molecules, such as urea moieties (Figure 1.4), amides or alcohols, are also present. The small amount of gelator present is able to trap the solvent and provide an extended porous surface for solvent binding.

1.3.3   COORDINATION POLYMERS AND METAL ORGANIC FRAMEWORKS

The term "coordination-polymère" was coined by Shibata in 1916 to describe cobalt complexes linked through ambidentate ligands (Shibata 1916). Subsequent usage covers any solid where bridging ligands impart a polymeric repeat pattern, generally in two or three dimensions. Consequently, metal–organic frameworks (MOFs), where the key "framing" ligands are organic, are a subset of coordination polymers. MOFs are materials made up of two main structural components; metal

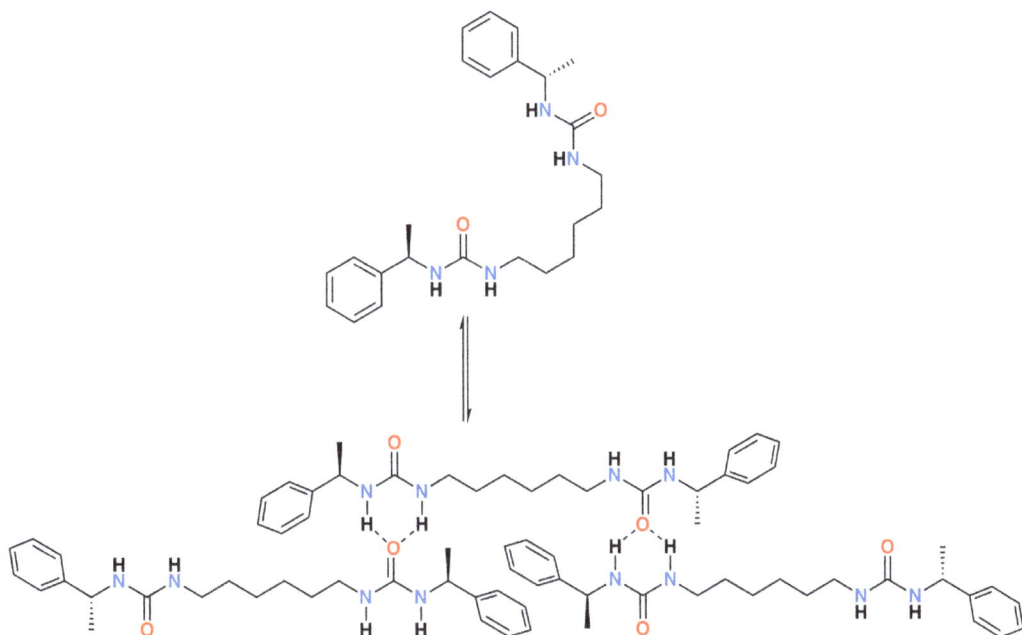

FIGURE 1.4   A supramolecular gel.

ions (or metal ion clusters) and organic linker molecules which can form a diverse range of one-, two- or three-dimensional network structures through molecular self-assembly (Perry et al. 2009). Figure 1.5 shows how a linear ditopic ligand can combine with a metal exhibiting a square planar geometric preference to generate a two-dimensional grid. In the context of supramolecular design, these ligand and metal components may also be referred to as supramolecular building blocks. Having metallic nodes separated by organic linkers results in unique material properties such as high surface area and high porosity, making them ideal for the inclusion of small molecules.

Pore size can be easily adapted by changing the length of the linker molecules and the topology, structure and functionality can also be controlled through careful selection of each structural component while the overall geometry of the network is largely determined by the coordination modes and geometry of the nodes. The pore size of MOFs can be used to make them function as selective

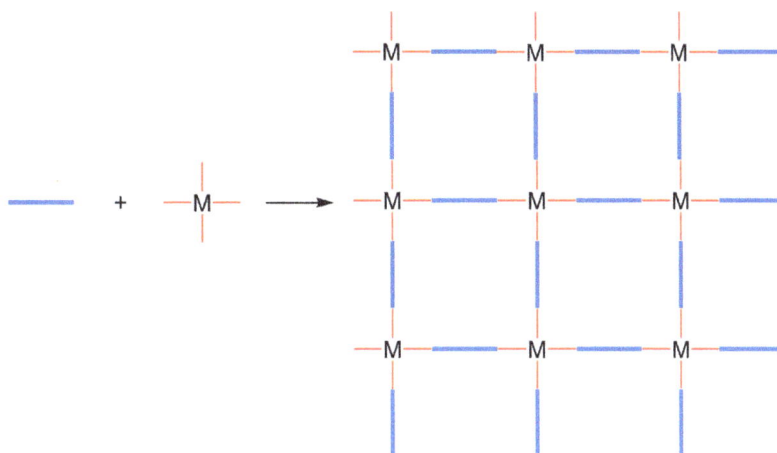

FIGURE 1.5   Constructing a metal–organic framework.

sponges for gases and small organic molecules with the rate of release controlled by heat or pressure changes (Li et al. 2009).

Perhaps the most ingenious application is to trap flexible organic molecules inside the crystalline matrix (Figure 1.6). Fujita demonstrated that it was possible to take single crystals of MOFs, determine their solid-state structure, then soak them in solutions of the desired guest and re-determine the X-ray structure (Hoshino et al. 2016). Subtraction of the electron density due to the framework reveals the density due to the guest and allows its structure to be calculated. There are drawbacks with this approach as the guests will not be uniformly distributed within the crystalline network, which can render the crystallographic solution ambiguous, but, in general, it is an excellent method by which the structures of rare natural products or complex drug molecules can be elucidated.

1.3.4 NANOCONTAINERS

Encapsulation of molecules has long held a fascination in the field of supramolecular chemistry whether the interior space is only large enough for a solvent molecule or is capable of facilitating an "impossible" reaction. At the smallest end of the scale are the cryptates developed by Lehn (Dietrich et al. 1969) to bind alkali metals and which, subsequently, allowed Dye to isolate alkali metal anions (Tehan et al. 1974). These molecules were prepared by linking the two nitrogen atoms in a diazacrown to give a relatively flexible "three-dimensional" crown ether (Figure 1.7). Similar examples include Sargeson's ethylenediamine cobalt complexes which, when reacted with formaldehyde and ammonia or nitromethane, form sepulchrates and sarcophagenes (Figure 1.7), respectively (Creaser et al. 1977). More rigid examples include Schiff bases formed from tris(2-aminoethyl)amine and aromatic dialdehydes (Arthurs et al. 2001).

Continuing with the early trend in nomenclature to describe encapsulating macrocycles in descriptive terms, Cram linked two resorcinarene derivatives to form a carcerand in which to imprison guest molecules (Cram et al. 1985). It was noted that the restrictive interior of the carcerands could protect guests from the effects of external gases or solvents. This was demonstrated by the cyclization of butadiene inside a carcerand, resulting in the stabilization of a molecule that is quite simply too reactive to prepare under any other conditions (Cram et al. 1991).

One question that arose concerned the small volumes that these compounds can enclose and asked how much of that space could be occupied by guest molecules. This was answered by Rebek who prepared self-assembling capsules, such as those in Figure 1.8, to trap a number of guest molecules (Mecozzi and Rebek 1998). Several dimeric capsules were investigated computationally to

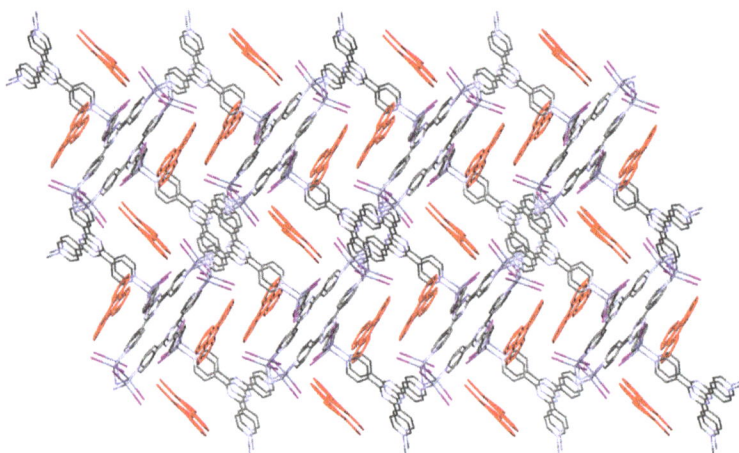

FIGURE 1.6 Guest (shown in red) inclusion inside a metal–organic framework (Cambridge Crystallographic Data Centre, CCDC 241417).

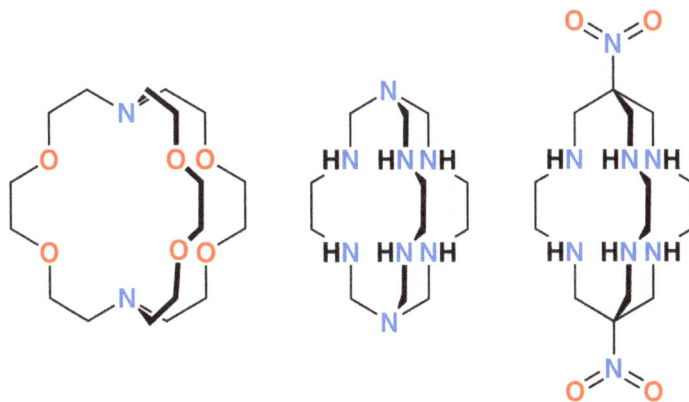

FIGURE 1.7   Cryptate, sepulchrate and sarcophagene.

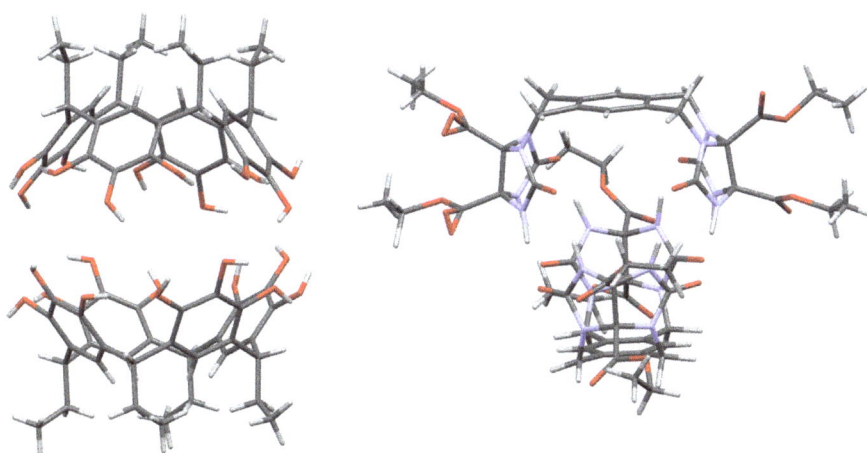

FIGURE 1.8   Hydrogen bonded molecular capsules (CCDC 131412 and 220346).

determine an internal volume and then assembled in solution with a number of guest molecules. Evidence from relative integration of host and guest molecules indicated that the most efficient packing occurred when the guest occupied 55% of the available space within the capsule.

Coordination chemistry can also be used to create capsules through careful choice of rigid ligand and metal center. Using principles now much more widely understood through their application to the construction of MOFs, Raymond showed that four tin(IV) or titanium(IV) atoms could bind to four tripodal ligands and adopt a tetrahedral geometry although the central cavity appeared too small to contain a guest (Brückner et al. 1998).

A similar approach taken by Nitschke, in which six rigid, twisted bidentate ligands assemble around four iron centers to form a tetrahedron with enough room to bind a tetrahedral P_4 molecule as shown by its X-ray crystal structure in Figure 1.9 (Mal et al. 2009). The encapsulated air-sensitive white phosphorus molecule becomes air-stable and, as the complex, water-soluble. The spontaneously combustible guest can then be hydrolyzed under controlled conditions to the far safer phosphoric acid.

Careful consideration of a rigid ligand's geometry and the preferred coordination mode of transition metals has allowed Fujita to create a number of capsules including one with a $M_{24}L_{48}$ formula and an inner diameter of 3.6 nm (Sun et al. 2010). Evidence of the capsule's formation came initially from mass spectrometry but was proved by X-ray crystallography (Figure 1.9).

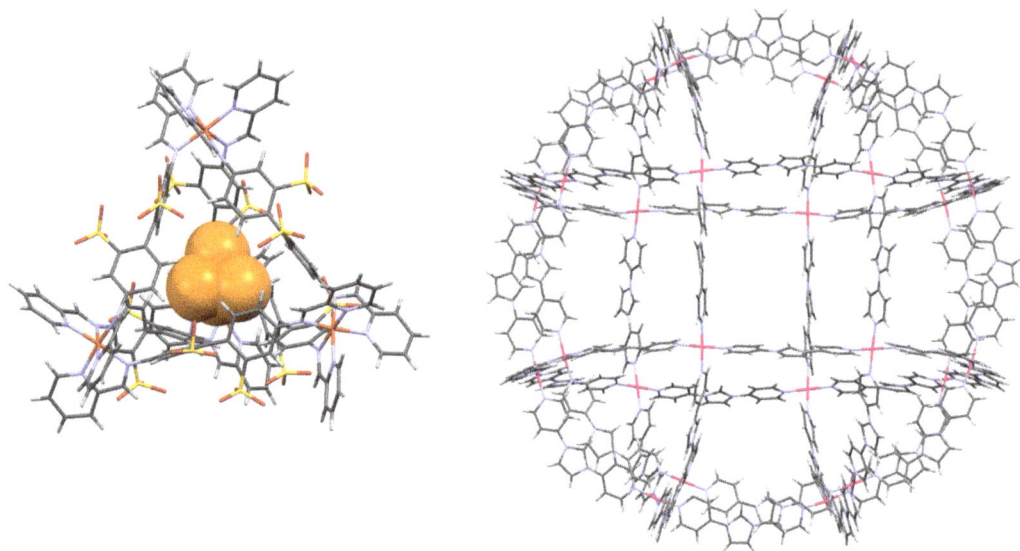

FIGURE 1.9   Tetrahedral and spherical capsules (CCDC 727817 and 860617).

1.3.5 SUPRAMOLECULAR MACHINES

Supramolecular chemists have exploited reversible interactions in many ways, from selective detection of small molecules to the creation of nanospheres, but the greatest challenge is surely to create analogues of simple machines on the molecular scale. When interlocked molecules, whether the interpenetrating catenands or shuttle-like rotaxanes (Figure 1.10), incorporate recognition motifs then the movement of one molecule between them becomes possible with the appropriate external stimulus. In doing so, mechanical movement becomes an on-off switch or a binary 0 and 1. Stoddart has used this phenomenon to create "nanoelevators"; rotaxanes linked through an aromatic moiety that rise and fall in concert (Badjic et al. 2004). The same system could be used as a piston, to pump drugs as the rotaxane reaches one of its termini, but perhaps the most interesting application is in molecular computing where electronic components, cross-linked by an array of rotaxanes, can move between "on" and "off" states (Collier et al. 1999). Molecules become detached over time so using an array, such as 10×10, allows for some redundancy so that, as long as a majority of the shuttles are in one of the two positions, the correct 0 or 1 will be recorded.

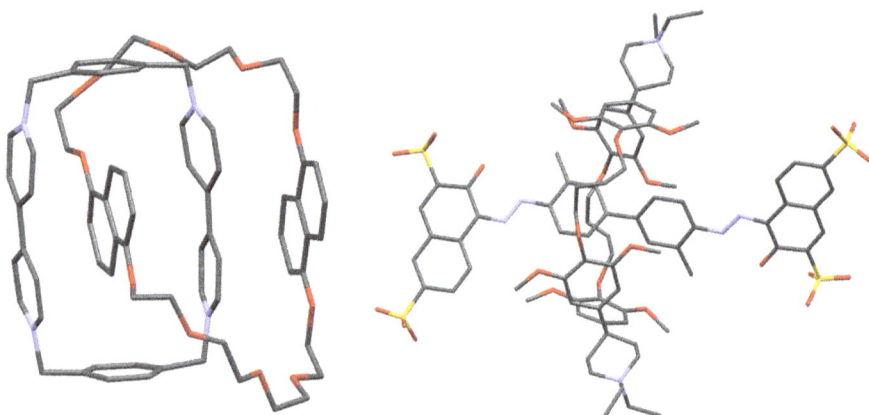

FIGURE 1.10   Catenane (left) and rotaxane (right) (CCDC 1197478 and 104321).

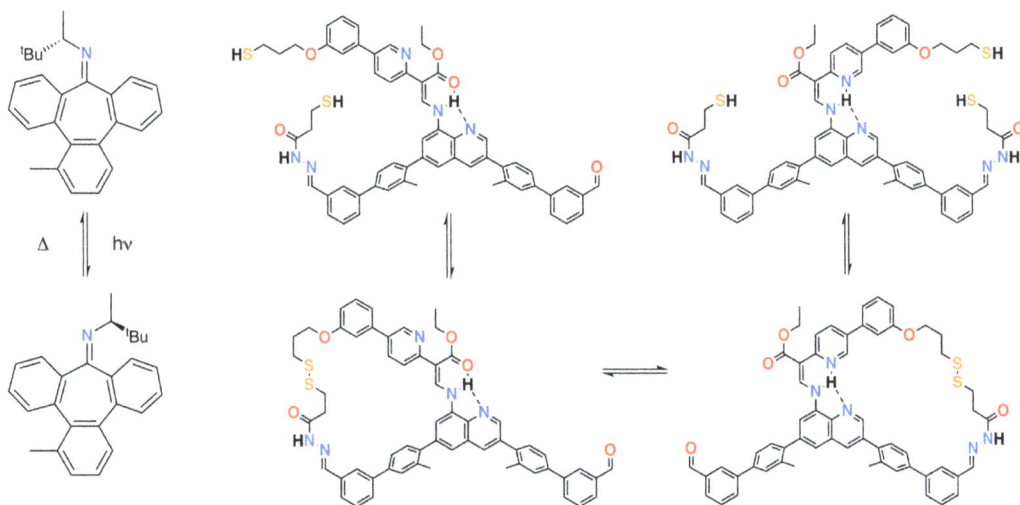

FIGURE 1.11 Molecular machines: a rotor (left) and nanoassembler (right).

Simple molecular ratchets, where part of a molecule can be made to rotate while the remainder acts as the stator in a motor (Figure 1.11), were first demonstrated by Kelly (Kelly et al. 1999) and Feringa (Feringa et al. 1999). The rotor may be attached by a double bond, such as an imine (Greb and Lehn 2014), that can be partially broken for rotation to occur or may interact through remote hydrogen or disulfide bonding while rotating around a single bond. The former method has been used by Tour to power "nanocars" by illuminating a paddle-like rotor which responds to incoming photons (Morin et al. 2006). The latter has been used by Leigh to create a "nanoassembler" (Figure 1.11) which can be programmed to prepare one specific enantiomer resulting from a multi-step synthesis with four possible products (Kassem et al. 2017).

1.4 HOST–GUEST COMPLEXES

Many supramolecules are complexes between a host, often a macrocycle with convergent binding sites, and a guest, a smaller molecule with divergent binding sites. At its simplest, this could be an alkali metal cation and a macrocycle as in Bell's torand complex shown in Figure 1.12 (Bell et al.

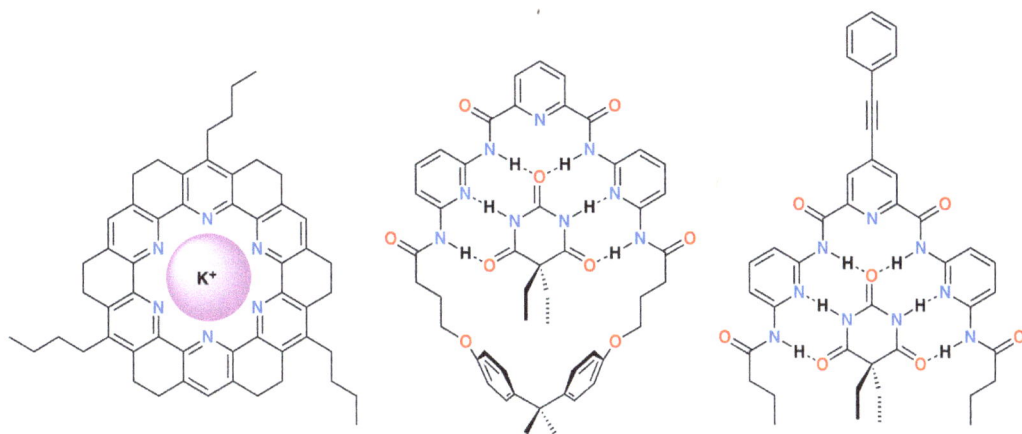

FIGURE 1.12 Host–guest complexes: Bell's [K·torand]$^+$ complex (left), Hamilton's barbiturate receptor (center) and Kondo's fluorescent barbiturate sensor (right).

1992). Complexation occurs through mutually beneficial, or complementary, interactions such as hydrogen bond donors aligned with acceptors, stacking of hydrophobic aromatic rings or simple electrostatic attraction. With careful design, made easier by powerful computational chemistry programs to model the systems, it is possible to design hosts with high specificity for their chosen guests. An excellent example of this can be seen in the work of Hamilton (Chang and Hamilton 1988) who used complementary arrays of hydrogen bond donors and acceptors to bind barbituric acid derivatives (Figure 1.12). The incorporation of sterically hindering moieties resulted in high specificity for barbital over phenobarbital. To demonstrate how this could be used to detect illegal substances, Kondo incorporated a phenylethynyl group that signals barbiturate binding by switching on fluorescence (Kondo et al. 2017). The molecule was fashioned as a molecular cleft that allows any barbiturate to bind (Figure 1.12) which, while it gives reduced selectivity for barbital, allows a broad group of chemically related barbituric acid derivatives to be detected by a single sensor. Host–guest complexes are discussed in Chapter 2.

1.5 CONCLUSIONS AND PERSPECTIVES

Supramolecules, structures that are literally "beyond molecules," have developed over the past few decades from chemical curiosities to complex systems with well-understood properties and a wide range of applications. Supramolecules range from macrocyclic complexes of guest molecules, which reveal the nature of these guests, to complex MOFs that restrict guest molecule motion to facilitate the determination of their structures. Dendrimers are capable of releasing life-saving drugs in response to external stimuli and nanoassemblers also fall into the same category. In every case, the properties and behaviors of the supramolecules relate to weak, reversible interactions between the molecular components which can be harnessed by the supramolecular chemist. Once the fundamental chemistry behind supramolecule formation had been understood – through the work of Nobel laureates Pedersen, Lehn, Cram, Feringa, Sauvage, Stoddart, and many other chemists – it can be used in the design of functional supramolecules. Since there is a wide range of data presently available and advances in computational chemistry, complex supramolecular systems can now be designed, modeled, and created for a range of applications.

Supramolecular chemistry has come a long way from the early days of Baeyer's isolation of cyclic products from experiments on phenol-formaldehyde resins and Pedersen's serendipitous discovery of alkali metal-binding crown ethers. Research is directed outward from the field to impact drug delivery, the detection of illicit substances, preserving ancient artifacts, self-healing materials, and inhibition of corrosion processes.

ACKNOWLEDGMENT

Crystallographic data were obtained from the Cambridge Structural Database. Deposition numbers are given in the figure captions.

REFERENCES

Arthurs, M., McKee, V., Nelson, J., and Town, R. M. 2001. Chemistry in cages: dinucleating azacryptand hosts and their cation and anion cryptates. *J. Chem. Ed.* 78:1269–1272.
Baas-Becking, L. G. M., and Galliher, E. W. 1931. Wall structure and mineralization in corraline algae. *J. Phys. Chem.* 35:467–479.
Badjic, J. D., Balzani, V., Credi, A., and Stoddart J. F. 2004. A molecular elevator. *Science* 303:1845–1849.
Baeyer, A. 1872a. Ueber die Verbindungen der Aldehyde mit den Phenolen. *Ber. Dtsch. Chem. Ges.* 5:280–282.
Baeyer, A. 1872b. Ueber die Verbindung der Aldehyde mit den Phenolen. *Ber. Dtsch. Chem. Ges.* 5:25–31.
Bell, T. W., Cragg, P. J., Drew, M. G. B., Firestone, A., and Kwok, D.-I. A. 1992. Conformational preference of the torand ligand in its complexes with potassium and rubidium picrate. *Angew. Chem. Int. Ed.* 31:345–347.

Brückner, C., Powers, R. E., and Raymond, K. N. 1998. Symmetry-driven rational design of a tetrahedral supramolecular Ti_4L_4 cluster. *Angew. Chem. Int. Ed.* 37:1837–1839.

Brunsveld, L., Folmer, B. J. B., Meijer, E. W., and Sijbesma R. P. 2001. Supramolecular polymers. *Chem. Rev.* 101:4071–4098.

Canceill, J., Lacombe, L., and Collet, A. 1984. Synthèse de nouveaux cavitands possédant une structure bis-cyclovératrylénique. Application á la complexation sélective du dichlorométhane en présence de chloroform. *C. R. Acad. Sc. Paris, Ser. II* 298:39–42.

Carroll, L. 1866. *Alice's Adventures in Wonderland*. London: Macmillan and Co., p. 98.

Century Dictionary, Supplement Vol XII. 1909. New York: The Century Company, p. 1301.

Chang, S. K., and Hamilton, A. D. 1988. Molecular recognition of biologically interesting substrates: synthesis of an artificial receptor for barbiturates employing six hydrogen bonds. *J. Am. Chem. Soc.* 110:1318–1319.

Chen, J., and Seeman, N. C. 1991. Synthesis from DNA of a molecule with the connectivity of a cube. *Nature* 350:631–633.

Collier, C. P., Wong, E. W., Belohradsky, M., Raymo, F. M., Stoddart, J. F., Kuekes, P. J., Williams, R. S., and Heath, J. R. 1999. Electronically configurable molecular-based logic gates. *Science* 285:391–394.

Cram, D. J., Karbach, S., Kim, Y. H., Baczynskyj, L., and Kallemeyn, G. W. 1985. Shell closure of two cavitands forms carcerand complexes with components of the medium as permanent guests. *J. Am. Chem. Soc.* 107:2575–2576.

Cram, D. J., Tanner, M. E., and Thomas, R. 1991. The taming of cyclobutadiene. *Angew. Chem. Int. Ed.* 30:1024–1027.

Creaser, I. I., Harrowfield, J. M., Herlt, A. J., Sargeson, A. M., Springborg, J., Geue, R. J., and Snow, M. R. 1977. Sepulchrate: a macrobicyclic nitrogen cage for metal ions. *J. Am. Chem. Soc.* 99:3181–3182.

Dietrich, B., Lehn, J.-M., and Sauvage, J.-P. 1969. Les cryptates. *Tet. Lett.* 10:2889–2893.

Feringa, B. L., Koumura, N., Zijlstra, R. W. J., Van Delden, R. A., and Harada, N. 1999. Light-driven mono-directional molecular rotor. *Nature* 401:152–155.

Fieser, L. E., and Fieser, M. 1959. *Steroids*. London: Chapman & Hall Ltd., p. 58.

Ghadiri, M. R., Granja, J. R., Milligan, R. A., McRee, D. E., and Khazanovich, N. 1993. Self-assembling organic nanotubes based on a cyclic peptide architecture. *Nature* 366:324–327.

Greb, L., and Lehn, J.-M. 2014. Light-driven molecular motors: imines as four-step or two-step unidirectional rotors. *J. Am. Chem. Soc.* 136:13114–13117.

Gutsche, C. D., and Muthukrishnan, R. 1978. Calixarenes. 1. Analysis of the product mixtures produced by the base-catalyzed condensation of formaldehyde with para-substituted phenols. *J. Org. Chem.* 43:4905–4906.

Hawker, C. J., and Fréchet, J. M. 1990. Preparation of polymers with controlled molecular architecture. A new convergent approach to dendritic macromolecules. *J. Am. Chem. Soc.* 112:7638–7647.

Högberg, A. G. S. 1980. Two stereoisomeric macrocyclic resorcinol-acetaldehyde condensation products. *J. Org. Chem.* 45:4498–4500.

Hoshino, M., Khutia, A., Xing, H. Z., Inokuma, Y., and Fujita, M. 2016. The crystalline sponge method updated. *IUCrJ* 3:139–151.

Kassem, S., Lee, A. T. L., Leigh, D. A., Marcos, V., Palmer, L. I., and Pisano, P. 2017. Stereodivergent synthesis with a programmable molecular machine. *Nature* 549:374–378.

Kelly, T. R., De Silva, H., and Silva, R. A. 1999. Unidirectional rotary motion in a molecular system. *Nature* 401:150–152.

Kondo, S., Endo, K., Iioka, J., Sato, K., and Matsuta, Y. 2017. UV–vis and fluorescence detection by receptors based on an isophthalamide bearing a phenylethynyl group. *Tet. Lett.* 58:4115–4118.

Kotera, M., Lehn, J.-M., and Vigneron, J.-P. 1994. Self-assembled supramolecular rigid rods. *J. Chem. Soc. Chem. Commun.* 197–199.

Lehn, J.-M. 1988. Supramolecular chemistry – scope and perspectives. Molecules, supermolecules and molecular devices. *Angew. Chem. Int. Ed.* 27:89–112.

Li, J. R., Kuppler, R. J., and Zhou, H. C. 2009. Selective gas adsorption and separation in metal-organic frameworks. *Chem. Soc. Rev.* 38:1477–1504.

Lindsey, A. S. 1965. The structure of cyclotriveratrylene (10,15-dihydro-2,3,7,8,12,13-hexamethoxy-5H-tribenzo[a, d, g]cyclononene) and related compounds. *J. Chem. Soc.* 1685–1692.

Luria, S. E. 1970. Molecular biology: past, present and future. *BioScience* 20:1289–1296.

Madaan, M., Kumar, S., Poonia, N., Lather, V., and Pandita, D. 2014. Dendrimers in drug delivery and targeting: drug-dendrimer interactions and toxicity issues. *J. Pharm. Bioallied. Sci.* 6:139–150.

Mal, P., Breiner, B., Rissanen, K., and Nitschke, J. R. 2009. White phosphorus is air-stable within a self-assembled tetrahedral capsule. *Science* 324:1697–1699.

Mecozzi, S., and Rebek, Jr., J. 1998. The 55% solution: a formula for molecular recognition in the liquid state. *Chem. Eur. J.* 4:1016–1022.

Mitchell, P. 1961. Coupling of phosphorylation to electron and hydrogen transfer by a chemi-osmotic type of mechanism. *Nature* 191:144–148.

Morin, J.-F., Shirai, Y., and Tour, J. M. 2006. En route to a motorized nanocar. *Org. Lett.* 8:1713–1716.

Newkome, G. R., Yao, Z., Baker, G. R., and Gupta, V. K. 1985. Micelles. Part 1. Cascade molecules: a new approach to micelles. A [27]-arborol. *J. Org. Chem.* 50:2003–2004.

Ogoshi, T., Kanai, S., Fujinami, S., Yamagishi, T., and Nakamoto, Y. 2008. *para*-Bridged symmetrical pillar[5] arenes: their Lewis acid catalyzed synthesis and host–guest property. *J. Am. Chem. Soc.* 130:5022–5023.

Ogoshi, T., Umeda, K., Yamagishi, T., and Nakamoto, Y. 2009. Through-space π-delocalized pillar[5]arene. *Chem. Commun.* 4874–4876.

Palade, G. E. 1955. A small particulate component of the cytoplasm. *J. Biophys. Biochem. Cytol.* 1:59–68.

Pedersen, C. J. 1967. Cyclic polyethers and their complexes with metal salts. *J. Am. Chem. Soc.* 89:7017–7036.

Perry IV, J. J., Perman, J. A., and Zaworotko, M. J. 2009. Design and synthesis of metal–organic frameworks using metal–organic polyhedra as supermolecular building blocks. *Chem. Soc. Rev.* 38: 1400–1417.

Rauwald, U., and Scherman, O. A. 2008. Supramolecular block copolymers with cucurbit[8]uril in water. *Angew. Chem. Int. Ed.* 47:3950–3953.

Robinson, G. M. 1915. A reaction of homopiperonyl and of homoveratryl alcohols, *J. Chem. Soc. Trans.* 107:267–276.

Sangeetha, N. M., and Maitra, U. 2005. Supramolecular gels: functions and uses. *Chem. Soc. Rev.* 34:821–836.

Schultz, R. A., White, B. D., Dishong, D. M., Arnold, K. A., and Gokel, G. W. 1985. 12-, 15-, and 18-Membered-ring nitrogen-pivot lariat ethers: syntheses, properties, and sodium and ammonium cation binding properties. *J. Am. Chem. Soc.* 107:6659–6668.

Seeman, N. C. 1982. Nucleic acid junctions and lattices. *J. Theor. Biol.* 99:237–247.

Shibata, Y. 1916. Recherches sur les spectres d'absorption des ammine-complexes métalliques. *J. Coll. Sci. Imp. Univ. Tokyo* 37:8.

Sun, Q.-F., Iwasa, J., Ogawa, D., Ishido, Y., Sato, S., Ozeki, T., Sei, Y., Yamaguchi, K., and Fujita, M. 2010. Self-assembled $M_{24}L_{48}$ polyhedra and their sharp structural switch upon subtle ligand variation. *Science* 328:1144–1147.

Tehan, F. J., Barnett, B. L., and Dye, J. L. 1974. Alkali anions - preparation and crystal-structure of a compound which contains cryptated sodium cation and sodium anion. *J. Am. Chem. Soc.* 96:7203–7208.

Tomalia, D. A., Baker, H., Dewald, J., Hall, M., Kallos, G., Martin, S., Roeck, J., Ryder, J., and Smith, P. 1985. A new class of polymers: starburst-dendritic macromolecules. *Polym. J.* 17:117–132.

Tümmler, B., Maass, G., Vögtle, F., Sieger, H., Heimann, U., and Weber, E. 1979. Open-chain polyethers. Influence of aromatic donor end groups on thermodynamics and kinetics of alkali metal ion complex formation. *J. Am. Chem. Soc.* 101:2588–2589.

Walsh, Z., Janeček, E.-R., Hodgkinson, J. T., Sedlmair, J., Koutsioubas, A., Spring, D. R., Welch, M., Hirschmugl, C. J., Toprakcioglu, C., Nitschke, J. R., Jones, M., and Scherman, O. A. 2014. Multifunctional supramolecular polymer networks as next-generation consolidants for archaeological wood conservation. *Proc. Natl. Acad. Sci.* 111:17743–17748.

Wolf, K. L., Frahm, H., and Harms, H. 1937. The state of arrangement of molecules in liquids. *Z. Phys. Abt. B* 36:237–287.

2 Host–Guest Inclusion Complexes

Brian D. Wagner
University of Prince Edward Island

CONTENTS

2.1 INTRODUCTION

Supramolecular chemistry, or "chemistry beyond the molecule" (Lehn 1988), is the study of chemical structures resulting from the interaction of two or more participating molecules held together only by intermolecular noncovalent attractions (Beer et al. 1999; Schneider and Yatsimirsky 2000; Cragg 2005; Steed 2009). Supramolecular host–guest inclusion refers to a specific type of supramolecular process, in which a small guest molecule is included within the internal cavity of a cagelike, hollow host molecule (Wagner 2020; Steed and Atwood 2009). Supramolecular approaches including host–guest inclusion have found a significant application to the prevention of corrosion and biofouling (Zheludkevich 2012; Saji 2012; Saji and Cook 2019). A recent review article provides a comprehensive examination of the concepts and applications of supramolecular chemistry in this area (Saji and Cook 2019). In this chapter, the basic principles and theory behind the formation of host–guest inclusion complexes in solution are reviewed, in order to provide a firm basis of understanding of the underlying chemistry involved. In addition, some representative examples of the application of host–guest inclusion complexation to anti-corrosion and anti-biofouling are briefly presented and discussed, in order to provide some context for the role of host–guest complexation in this area.

DOI: 10.1201/9781003169130-3

This chapter focuses on host–guest chemistry involving discrete molecular hosts in solution, and so will not, for example, cover in detail applications of metal-organic frameworks (MOFs), polymeric materials, porous nanomaterials and other such non-discrete systems (although a few representative such examples are briefly discussed). However, this chapter is not meant to be a comprehensive review of such applications of molecular hosts, but rather to provide an overview of the breadth of types of host molecules that have been used for this purpose, which includes cyclodextrins, cucurbiturils, calixarenes, crown ethers and dendrimers.

Chapter 1 of this book provided a general overview and discussion of *Supramolecules* and supramolecular chemistry. Host–guest inclusion complexation in solution represents the simplest example of supramolecular chemistry in solution and is discussed in detail in this chapter, with a few illustrative examples of applications presented, as mentioned above. In terms of host encapsulation to form host–guest inclusion complexes, by far the most commonly applied molecular hosts are cyclodextrins (CDs) (Szejtli 1998), which are discussed in Section 2.2.1. Details of the applications of CDs in surface coatings for corrosion and biofouling protection are presented in Chapters 10 and 21 of this book.

Before discussing host–guest complexation and molecular hosts in detail in the following sections, a brief overview of the phenomenon of host–guest inclusion in solution (Wagner 2020) is presented here. In general, host–guest complexation occurs when a guest molecule is encapsulated within the internal cavity of a large, cagelike host molecule. This process is illustrated in Figure 2.1, where the guest is represented by naphthalene as a typical aromatic guest molecule, and the host is represented in a cartoon form as a molecular bucket, representative of a CD host.

A key aspect of supramolecular host–guest inclusion is that it occurs via *self-assembly* (Biedermann 2017). In other words, the guest and host molecules in solution will spontaneously form the host–guest complex upon mixing. This makes the preparation of such complexes relatively simple and straightforward, as long as there is a good match between the size, shape, and properties of the guest of interest and the host cavity. The stoichiometry of host:guest complexation is another essential aspect of the phenomenon, and the nature of the complex(es) formed for a specific host–guest pair. The stoichiometry is defined as the host:guest ratio in the formed complex. In the simplest case, 1:1 complexes form if one guest molecule is included within the cavity of a single host molecule, as illustrated in Figure 2.1. However, other stoichiometries can also occur, most commonly 2:1, in which one guest (often relatively elongated) is encapsulated at each of its two ends by a host, and 1:2, in which two guests (possibly as a dimer) are encapsulated in a single host cavity. These various stoichiometry complexes (including 2:2 in which a pair of guests is encapsulated by a pair of hosts) are shown in Figure 2.2, again for naphthalene as a representative guest and bucket-shaped hosts.

One fundamental question that needs to be addressed is: why do host–guest inclusion complexes form? In other words, why does a solvated guest which is free in solution undergo inclusion within a host cavity, self-assembling into a host–guest complex? This important question is addressed in Section 2.3, in terms of the thermodynamics, dynamics, and driving forces for inclusion. The simple answer is that the resulting solvated host–guest complex is thermodynamically more stable than the separate solvated guest and host. The stability of the complex can be quantified by the equilibrium

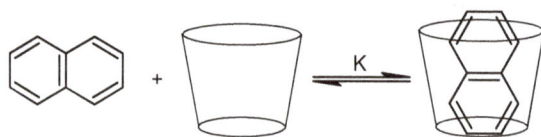

FIGURE 2.1 Cartoon illustration of the formation of a host–guest inclusion complex for a bucket-shaped host and naphthalene as a representative aromatic guest.

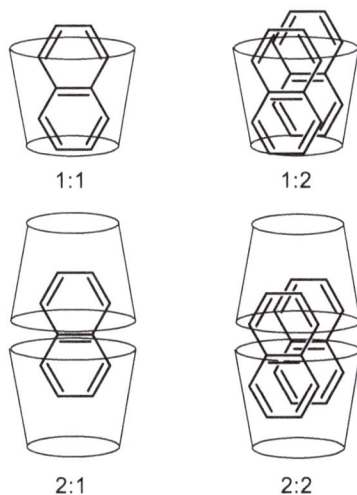

FIGURE 2.2 Illustration of the most common stoichiometries of host–guest inclusion complexes for a bucket-shaped host and naphthalene as a representative aromatic guest.

constant K (referred to herein as the binding constant), as indicated in Figure 2.1 for the indicated equilibrium for the case of 1:1 complexation. In this case, the binding constant K is defined as:

$$K = \frac{[H:G]}{[H][G]} \tag{2.1}$$

As is discussed in Section 2.3, thermodynamically, K is related to the change in Gibbs energy upon inclusion, and strong binding constants are favoured by large decreases in the enthalpy and large increase in the entropy of the complex relative to the free host and guest. The stronger the intermolecular forces between the host and guest, the more highly negative the enthalpy change will be (more stable complex), resulting in a higher magnitude of K.

The inclusion of a guest molecule within a host cavity can result in significant changes in its physicochemical properties. In some cases, these changes result in improved properties for the application of the guest. Such properties can include solubility, stability, activity, electrochemical properties and optical properties. Depending on the value of the binding constant K, the host inclusion can also be used to control the release of the guest into the system of interest. It is the potential to control and improve guest properties and activity, as well as the potential to control the guest's release, which has made supramolecular host–guest inclusion of interest in applications involving anti-corrosion and anti-biofouling agents as guests.

2.2 COMMON HOST MOLECULES

There are a number of families of macrocyclic molecules with internal cavities, which have been utilised in the formation of host–guest inclusion complexes in solution. By far, the three most widely used families of molecular hosts are the CDs, followed by cucurbiturils and calixarenes. Each of these types of host molecules has been applied to host–guest enhancement of anti-corrosion and biofouling prevention and is discussed below. In addition, a few other types of hosts have also been used in such applications, including crown ethers and dendrimers; these are briefly discussed as well. In general, all such molecular hosts have some common attributes: they have well-defined internal cavities with relatively nonpolar interiors, they are sufficiently water-soluble to be able to be applied in an aqueous solution, and they are capable of undergoing intermolecular interactions with

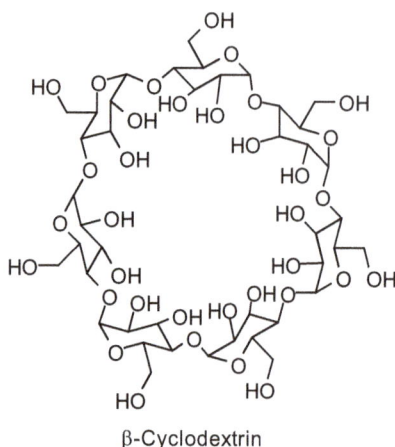

β-Cyclodextrin

FIGURE 2.3   Chemical structure of the macrocyclic host compound β-cyclodextrin.

target guest molecules. Each type of host molecule has unique structural properties, interactions, and binding affinities for particular types of guests, which is discussed here.

2.2.1   CYCLODEXTRINS

Cyclodextrins are macrocyclic oligomers of glucopyranose, formed by an enzymatic action on glucose (Szejtli 1998; Hashidzume et al. 2017). CDs are commonly found in three sizes, referred to as α-CD (6 glucopyranose units), β-CD (7 glucopyranose units) and γ-CD (8 glucopyranose units). The chemical structure of β-CD is shown in Figure 2.3. Numerous reviews have been published on CD physicochemical and host properties (Szejtli 1998; Hashidzume et al. 2017) as well as their applications in the pharmaceutical and other industries (Kurkov and Loftsson 2013; Cheirsilip 2016; Sharma and Baldi 2016; Amiri and Rahimi 2017; Iacovino 2017).

In an aqueous solution, intramolecular hydrogen bonding around the secondary (top rim) and primary (bottom rim) hydroxyl groups results in the CD molecule taking on a bucket-like shape, with a well-defined internal cavity with larger (top) and smaller (bottom) cavity openings, as is illustrated in Figure 2.4. CDs are sometimes referred to as *molecular buckets*. Different sizes of CDs available allow for three different cavity sizes to be used as hosts, for matching to the size of the guest of interest. As shown in Figure 2.4, the upper rim of the CD cavity has an opening size of 5.7, 7.8 and 9.5 Å, respectively, for α, β and γ. The interior of the CD cavity is relatively nonpolar, compared with the aqueous solution, with a micropolarity within the cavity similar to that of an ethanol solvent (Heredia et al. 1985). This difference in polarity within the CD cavity relative to the

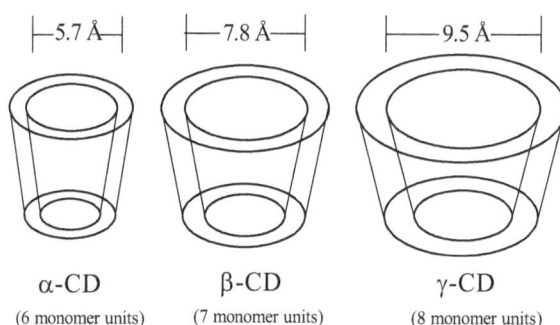

FIGURE 2.4   Depiction of the relative cavity dimensions of α-, β- and γ-cyclodextrins.

bulk aqueous solution provides a major driving force for the inclusion of nonpolar guests, namely the hydrophobic effect, which is discussed in Section 2.3.1. Further details on the size and physical properties of the three main CDs can be found in Szejtli (1998).

CDs have found widespread usage as the most popular molecular host for forming host–guest inclusion complexes in an aqueous solution for a wide variety of reasons, including their commercial availability, relatively low cost, well-defined relatively nonpolar cavities, which can accommodate a wide variety of guests, high aqueous solubility, ease of chemical modification and nontoxic nature. In terms of room temperature aqueous solubility, α- and γ-CD are the most soluble of the native (non-modified) CDs (14.5 and 23.2 g/100 mL, respectively), whereas β-CD is significantly less soluble (1.85 g/100 mL) (Szejtli 1998). This relative solubility can be explained by differences in solvation and intramolecular hydrogen bonding in aqueous solution (Szejtli 1998). Despite its lower solubility, β-CD is often the host of choice, as its cavity size is a good match for many molecular guests of interest. In addition, β-CD (as well as α- and γ-CD) can be chemically modified to improve its properties, including increasing its aqueous solubility. One commonly used modified β-CD is hydroxypropyl-β-CD (HP-β-CD), in which some of the 21 hydroxyl protons have been replaced by $CH_2CHOHCH_3$ groups; an example of the use of HP-β-CD in an anti-corrosion application is discussed in Section 2.5.1 as well as numerous applications involving the unmodified parent α-, β- and γ-CD.

2.2.2 Cucurbiturils

Cucurbit[n]urils (CB[n]) are macrocyclic oligomers of glycoluril, joined together by pairs of methylene bridges (Lagona et al. 2005), as shown in Figure 2.5. These pairs of bridges result in a highly rigid structure of cucurbit[n]urils, compared, for example, to CDs and calixarenes (see Section 2.2.3), in which adjacent the monomers are attached by one bridge only. For CB[n] up to $n=8$, a highly symmetric structure about the central axis is obtained, giving a unique, spherical internal cavity quite different from that of CDs. In addition, the opposing carbonyl groups in the monomer units result in the presence of n pairs of carbonyl groups lining the opposing cavity rims of the host molecule. These carbonyls point slightly inward towards the central axis, resulting in portal openings that are smaller than the diameter of the cavity. This results in what is referred to as constrictive host–guest binding, in which the cavity entrance is smaller than the cavity itself. Furthermore, the highly electronegative carbonyl oxygens provide the opportunity for strong dipole–dipole interactions with polar guests and most importantly, electrostatic interactions with cationic guests. CB[n] are particular good hosts for binding cations, either metal or molecular (Lagona et al. 2005). These carbonyl portals are another unique feature of cucurbit[n]urils, making them uniquely useful as hosts, and as alternatives to CDs, for certain guests. Yet another unique aspect of CB[n] as hosts is the microenvironment within the cavity. The polarity within the CB[n] cavity is similar

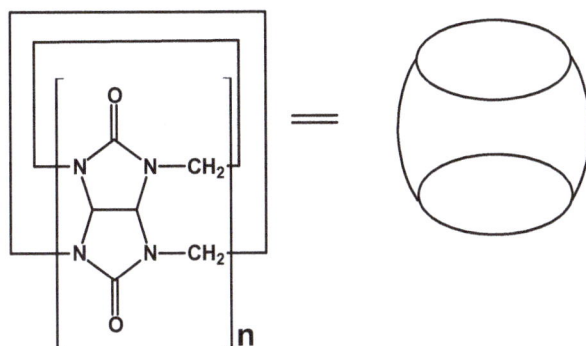

FIGURE 2.5   General chemical structure of the cucurbit[n]uril family of macrocyclic host compounds.

to that of n-octanol solvent (Nau et al. 2011) and hence similar to that of a CD cavity. However, it is the polarizability within a CB[n] cavity that is so unique; it is similar to that of a vacuum, as opposed to that of a solvent (Nau et al. 2011). This has significant implications on the host properties of CB[n], and their binding abilities for different types of guests (Nau et al. 2011), which are different from those of CDs.

The first member of the family prepared and characterised was the six-membered macrocycle CB[6], first prepared in 1905 (Behrend 1905) but not characterised (including a full crystal structure revealing its spherical nature and unique, well-defined cavity) in 1981 (Freeman et al. 1981). They named this cyclic molecule cucurbituril, after the Latin word for pumpkin, which they felt it resembled in shape. The utility of cucurbit[n]urils was vastly expanded in 2000 when Kim et al. reported the synthesis and characterisation of three additional members of the cucurbituril family, CB[5], CB[7] and CB[8] (Kim et al. 2000). Since then, even larger cucurbiturils have been made and characterised, including CB[10] (Liu 2005a). One limiting factor in the use of CB[n] as hosts are their relatively low aqueous solubility, on the order of 20–30 mM in the case of CB[5] and CB[7], and 0.018 and <0.01 mM in the case of CB[6] and CB[8], respectively (Lagona et al. 2005). This low solubility can, however, be significantly increased in acidic or salt solutions. The most commonly utilised cucurbit[n]uril is cucurbit[7]uril, which has a favourable cavity size and has (along with CB[5], the cavity size of which is too small for most guests of interest) the highest aqueous solubility of the cucurbituril family (Lagona et al. 2005). This highly useful CB[7] host has a cavity diameter of CB[7] 7.3 Å (similar to that of β-CD) and a portal diameter of 5.4 Å. The chemical structure of CB[7], including its highly symmetrical spherical shape, and the carbonyl-lined portals, is shown in Figure 2.6.

The interesting and distinctive cavity properties, the carbonyl-lined portals, and the rigidity of cucurbit[n]urils (at least up to $n = 8$) have given these compounds unique host properties (Lagona et al. 2005; Liu 2005b; Nau et al. 2011; Assaf and Nau 2015) and applications (Kim et al. 2007; Das et al. 2019). As discussed in Section 2.5.2, CB[n] hosts have been applied to corrosion and biofouling inhibition applications.

2.2.3 CALIXARENES

Calixarenes are macrocyclic oligomers of *p*-substituted phenols, with adjacent monomers attached by single methylene bridges, oriented *meta* to each other on the phenol ring (Shinkai 1986; Pochini and Ungara 1996; Reinhoudt 2016). The general structure of a calix[n]arene, with *n meta*-methylene-attached phenol units, is shown in Figure 2.7 (with R' often = H). The cyclic nature of calixarenes results in the presence of an internal cavity, which makes calixarenes good potential hosts for a variety of molecular guests. The name calixarene derives from the Latin word for chalice, or cup, which this host can resemble in certain conformations.

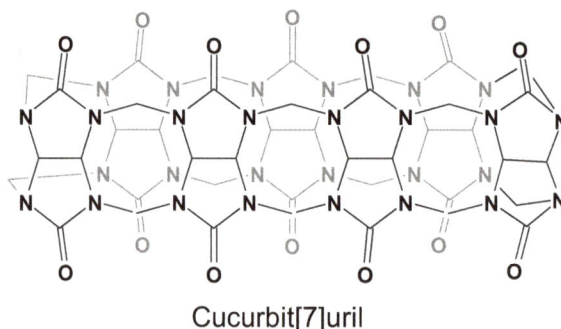

Cucurbit[7]uril

FIGURE 2.6   Chemical structure and shape of the macrocyclic host compound cucurbit[7]uril.

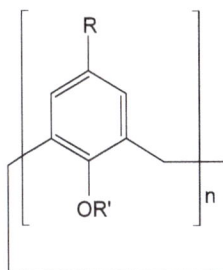

FIGURE 2.7   General chemical structure of the calix[n]arene family of macrocyclic host compounds ($R=H$ or an alkyl or other substituent; R' typically$=H$).

The single methylene bridges between adjacent phenol rings, and the lack of coordinated hydrogen bonding between adjacent monomers as in the case of CDs, makes calixarenes highly flexible and rather floppy. Calixarene cavities are less well-defined than those in the case of CDs and in particular compared to cucurbit[n]urils. There are various conformers possible for calix[n]arene hosts in the solution. For example, for calix[4]arene, two prominent orientations are the cone orientation, in which all phenols are pointing in the same direction, and 1,3-alternate, where opposite phenols are pointing in the same direction, but adjacent phenols are pointing in *opposite* directions. This results in the ability of calixarenes to adopt different cavity shapes in response to a specific guest. In addition, the presence of the aromatic rings in the host walls allows for host–guest interactions which are not available in CDs and cucurbiturils, such as π–π interactions, making calixarenes excellent hosts for aromatic guests, for example. The presence of the π-electrons in the cavity walls contributes to the unique host properties of calixarenes.

Calixarenes themselves are practically water-insoluble. However, calixarenes can be derivatised to yield water-soluble host molecules, typically by the addition of anionic or cationic substituent R groups in the para position as shown in Figure 2.7. One common, commercially available example of water-soluble calixarenes is the p-sulfonatocalix[n]arenes, in which the anionic alkylsulfonate groups give the molecules significant aqueous solubility, allowing for their application in water. These types of water-soluble calixarenes have been applied to corrosion and anti-biofouling applications in acidic solution, which are described in Section 2.5.3.

2.2.4   OTHER MOLECULAR HOSTS AND HOST MATERIALS

CDs, cucurbiturils and calixarenes represent the most commonly used molecular hosts in reported applications to enhance the performance of anti-corrosion and anti-biofouling agents. However, there have been additional such studies involving other types of hosts, including crown ethers, dendrimers and the broad category of nanocontainers. These studies themselves will be described in Section 2.5.4, but a brief description of these hosts is given here. There are also a wide array of other types of molecular hosts such as pillar[n]arenes and bambus[n]urils which are available for host–guest inclusion (Wagner 2020), which may show useful applications to corrosion and biofouling protection in the future.

Crown ethers (Gokel 2017) are macrocyclic carbon rings containing oxygen atoms, i.e. cyclic polyethers. The first crown ether identified and used to coordinate metal ions was dibenzo-18-crown-6 in 1961 (Izatt 2007); its structure is shown in Figure 2.8. Crown ethers can tightly bind metal cations, and their applications as hosts tend to be limited to such metal cations as guests. An example of the use of crown ether in an anti-corrosion application is described in Section 2.5.4. Another metal-complexing molecular host of interest here is tin phthalocyanin "nanocaps", which have been investigated for their corrosion inhibiting activity (Beltrán et al. 2005).

Dendrimers are highly branched oligomeric/polymeric molecules in which added monomer units branch out from a central core (Zheng and Zimmerman 1997; Abbasi et al. 2014). The size of

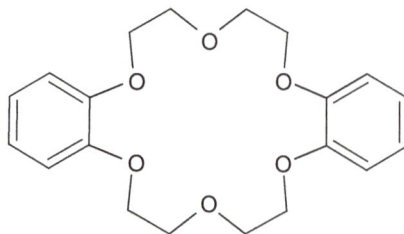

FIGURE 2.8   Chemical structure of the crown ether dibenzo-18-crown 6.

the dendrimer, or a number of branching events, is referred to as the generation. Each subsequent generation results from the addition of new branched units, or dendrons, to each end group of the previous generation, quickly increasing the number of end groups. Depictions of the structures of typical dendrimer molecular can be found (Abbasi et al. 2014). The end groups can be chemically altered to provide the desired functionality, for example, with fluorescent probes to impart fluorescent properties to the dendrimer. Large dendrimers tend to have an overall spherical shape, but with large pockets or cavities in between the separate branches. It is the presence of these pockets that allows dendrimers to act as hosts to small guest molecules in solution. Unlike all of the molecular hosts considered thus far, each host dendrimer can have multiple host cavities, allowing for a large number of guests to be included within a single host. Dendrimers can have significant aqueous solubility, depending on the polarity of the end groups. One common type of water-soluble dendrimer which has seen significant applications is polyimidoamide (PAMAM) dendrimers (Esfand and Tomalia 2001). There have been a few applications of water-soluble dendrimers to the enhancement of anti-corrosion agents, which are described in Section 2.5.4.

In addition to the molecular hosts in solution described thus far, there have been a number of studies of the use of other types of host materials in anti-corrosion applications. One example is polyurea (Gite et al. 2015) and other microcapsules, which are significantly larger than the molecular hosts described above. However, the most predominant category of host materials applied to anti-corrosion is the broad class of materials referred to collectively as *nanocontainers*. Nanocontainers, in general, are defined as materials with interior cavities larger than 1 nm in at least one dimension (Zhao et al. 2013; Nguyen and Assadi 2018; Nguyen-Tri et al. 2020). Although the term "nanocontainer" has sometimes been applied to molecular hosts such as CDs or their inclusion complexes (Amiri and Rahimi 2014, 2015, 2016), for this chapter, nanocontainers will be used to refer to nanoscale host materials as opposed to discrete molecular hosts, which are being considered separately herein as specific families of host molecules, and which are the main focus of this chapter. Typical nanocontainer materials often have sizes on the order of 100 nm or more, and such materials can be dispersed in certain specific solvents. These types of host materials include nanotubes, nanoshells and nanobottles (Zhao et al. 2013) and can be based on organic or inorganic materials or a hybrid of both. Further details on the chemical structures, cavities, active ingredient loading, and other properties of nanocontainer materials can be found elsewhere (Zhao et al. 2013; Nguyen and Assadi 2018; Nguyen-Tri et al. 2020). A number of anti-corrosion studies involving host materials that broadly fit into this category of nanocontainers have been reported. Although the focus of this chapter is on molecular hosts forming host–guest complexes, a few representative examples are described in Section 2.5.4.

In addition, a number of other types of solid host materials have also been applied in this area, including zeolites, nanocomposites, coordination polymers and MOFs. The use of such solid host materials is covered in detail elsewhere in this book, in Chapters 5 and 6 for coordination polymers and MOFs, respectively. A few examples involving such solid materials are briefly mentioned in this chapter, in Section 2.5.4.

2.3 FORMATION AND THERMODYNAMICS OF HOST–GUEST INCLUSION COMPLEXES

As mentioned above, host–guest complexation in solution occurs via self-assembly – mixtures of the host and guest dissolved in solution will spontaneously form inclusion complexes as a result of the guest molecule moving into the host cavity, becoming either fully encapsulated within the cavity or partially included, with part of the guest remaining exposed to solvent. For this to happen, there must be a negative Gibbs energy change for inclusion, i.e. $\Delta_{inc}G < 0$. The detailed mechanism for this phenomenon, however, can be relatively complicated, involving solvent molecules, for example, and vary for different types of hosts and guests. In this section, the mechanisms for inclusion, as well as the driving forces and detailed thermodynamics involved for different types of molecular hosts, are discussed.

2.3.1 Mechanisms and Driving Forces

Host–guest inclusion complexes self-assemble in solution, whereby the guest spontaneously enters the internal cavity of the host to form the complex. As indicated in Figure 2.1, this is a reversible, equilibrium phenomenon, and the guest typically can leave the cavity as well. Overall, the rate of entry into the cavity must exceed the rate of exit (or egress) for a substantial equilibrium population of the complex to be formed. Measurement of the rate of entry and exit is the realm of chemical dynamics, which has been well applied to host–guest inclusion (Bohne 2006, 2014; Douhal 2004). However, for this chapter, this process will be considered from the overall equilibrium point of view, in terms of the thermodynamics involved. Before considering these thermodynamic factors in detail, it is informative to consider both the detailed mechanisms for the complexation, as well as the driving forces for inclusion, such as the forces of attraction between the host and guest, which drive the inclusion process forward. In an aqueous solution, the water solvent molecules play a significant role in both the mechanism of host–guest complexation and the driving forces for inclusion. The precise mechanism depends on the nature and properties of the host itself, as well as the guest, and in particular on the nature of the host cavity in terms of shape, cavity opening, internal cavity polarity and other host properties (Detellier 1996; Pluth and Raymond 2007; Schneider 2009; Biedermann 2017). For this reason, the mechanism for inclusion of a guest will only be considered for CDs and cucurbiturils as a representative of different types of hosts.

In the case of CDs, Szejtli proposed a general mechanism for the binding of guests within their internal cavity (Szejtli 1996). This mechanism directly involves participation by water (solvent) molecules in the inclusion process. In the first step of this inclusion mechanism, water molecules leave the CD cavity and are re-absorbed into the bulk solvent, which may result in a change in shape of the host molecule. Concomitantly, the solvated "free" guest loses its solvation shell, becoming essentially a gas-phase molecule, and the released solvent molecules also become re-absorbed by the bulk solvent. The un-solvated guest now enters the empty host cavity, forming the complex stabilised by host–guest interactive forces. Finally, water molecules re-solvate any exposed parts of the guest, yielding a fully solvated, stable CD-guest conclusion complex. These changes in host–guest interactions and solvation result in changes in the enthalpy and entropy for the process, which are discussed in Section 2.3.2.

Nau et al. have proposed sets of detailed mechanisms for the binding of guests by cucurbiturils (Márquez et al. 2004). As opposed to CDs, cucurbiturils are constrictive hosts, with cavity openings (the opposing carbonyl-lined portals) significantly smaller than the cavity diameter. In addition, the presence of the highly electronegative carbonyl rims and the low polarizability and polarity within the cavity also impact the mechanism by which guests become included. They proposed two sets of mechanisms, one for neutral guests and one for cationic guests, in particular organic ammonium species. The details can be found in Márqez et al. (2004), including schematic illustrations of these two sets of mechanisms. To summarise briefly, the inclusion of neutral guests occurs in a series of equilibria, involving not only the host and neutral guest but also any metal cations or protons (in the case of acidic medium) binding the carbonyl rims, which need to dissociate from the CB rims before the guest can enter the cavity. Overall, the inclusion of a neutral guest in a CB[n] cavity involves a total of six different equilibrium constants (Márquez et al. 2004). The case of an organic ammonium cationic guest requires a different mechanism with an additional layer of complexity and a new mechanism, in which the ammonium moiety first binds to the carbonyl portals (after dissociation of any metal cations or protons from the portal), after which the entire molecule can undergo a "flip" to allow the remainder of the guest to enter the cavity, with the ammonium group still associated with the portal carbonyl oxygens. In either case, the inclusion of a guest into a CB[n] cavity in an acidic or salt solution involves a competitive binding process, in which the metal cation or proton must be displaced before guest binding can occur.

The possible driving forces for inclusion in terms of guest–host interactions are many and varied and can include, for example, van der Waals forces (Mettry and Hooley 2017), hydrogen bonding, electrostatic interactions (when ions are involved as host or guest), the hydrophobic effect, and the displacement of cavity solvent molecules. For complexation by calixarene hosts, the presence of the aromatic cavity walls allows for additional interactions between the host and guest including π–π interactions (Fagnani et al. 2017) and CH-π hydrogen bonds (Nishio et al. 2009), which can also contribute to the overall driving forces. A specific host–guest inclusion can involve a combination of a number of these potential driving forces, depending on the nature of the host, guest and solvent.

In an aqueous solution, the hydrophobic effect (Schneider 2009; Biedermann et al. 2014) is often the major contributor to the binding of hydrophobic guests within the relatively nonpolar internal cavities of these various hosts. In general, the hydrophobic effect refers to the collected set of driving forces that are responsible for a hydrophobic guest to become encapsulated within a relatively nonpolar host cavity. The hydrophobic effect is based on the tendency for hydrophobic ("water-fearing") molecules to avoid water and is responsible for the low solubility of such molecules in an aqueous solution. One way that such hydrophobic molecules can avoid water is to become included within a host cavity. This results in the removal of both the cavity waters and the waters of solvation of the guest and results in only the solvation of the entire complex (which is similar to that of the host itself), minimising the interaction of water with the hydrophobic guest. This effect applies to all hosts with relatively nonpolar internal cavities.

In the case of CDs, Liu and Guo concluded that the main driving forces involved are electrostatic, van der Waals interactions, hydrogen bonding and the hydrophobic effect (Liu and Guo 2002). For cucurbiturils, it has been shown that the release of high-energy water from the host cavity is an essential driving force, which is mainly responsible for the large binding constant (high host affinity) reported for many, particularly cationic, guests (Biedermann et al. 2012). These authors re-considered the hydrophobic effect in the case of cucurbituril host binding in terms of the expulsion of high-energy water from the cucurbituril cavity (Biedermann et al. 2014). In calixarenes, in addition to all of these driving forces, π-π interactions also play an important role in the case of aromatic guests.

2.3.2 THERMODYNAMICS OF INCLUSION

To consider the strength of host–guest binding, as indicated by the binding constant K, it is essential to consider and understand the thermodynamics of the inclusion process (Izatt and Oscarson 2017). As mentioned above, the strength of a host–guest inclusion complexation depends on the Gibbs energy of the resulting solvated complex compared to the free, solvated host and guest; this value will be referred to as ΔG_{inc}. The value of the binding constant K (e.g. for the simple 1:1 host:guest inclusion illustrated in Figure 2.1 and defined in eq. (2.1) is given by:

$$K = e - \Delta G_{inc} / RT \qquad (2.2)$$

The more negative the ΔG_{inc}, the larger the binding constant (the stronger the binding). It is informative to consider the enthalpic (ΔH_{inc}) and entropic (ΔS_{inc}) contributes to ΔG_{inc}, remembering that by definition $\Delta G = \Delta H - T\Delta S$. Thus, the more negative the ΔH_{inc}, and the more positive the ΔS_{inc}, the greater the enthalpic and entropic contributions to the stability of the host–guest complex, respectively.

ΔH_{inc} is the change in enthalpy of the complex compared to the free solvated host and guest and is a measure of the energy stabilisation of the complexation. Thus, ΔH_{inc} comes from the intermolecular interactions between the host and guest, such as van der Waals interactions, hydrogen bonding, π–π interactions, and so on. The greater the intermolecular attractions between host and guest, the more negative will be ΔH_{inc}. In general, host–guest complexes are more stable than their free solvated components, so ΔH_{inc} is typically negative, and there is generally a significant enthalpic component to complexation. The entropy term is more difficult to predict. In the gas phase, the formation of a host–guest complex will typically have a negative entropy change, as it involves the combination of two free species into one. In addition, the inclusion of a guest may involve hindering of intramolecular rotations and vibrations, which also reduces motional freedom and hence entropy. In solution, however, the expulsion of multiple solvent (such as water) molecules to be replaced by a guest molecule can result in a significant increase in entropy. Thus, values of ΔS_{inc} can be positive or negative depending on the specific host–guest pair. This is clearly seen in compilations of thermodynamic data for CD complexation, for example, for which ΔH_{inc} are almost always negative and ΔS_{inc} are positive in some cases and negative in others (Rekharsky and Inoue 1998). For more details on the role that entropy plays in molecular self-assembly, Nash has written an excellent recent article that explores this topic in detail (Nash 2017).

The values of ΔH_{inc} and ΔS_{inc} for a specific host–guest complex can be determined experimentally, either directly using calorimetry, or indirectly by determined the values of K as a function of temperature, using a spectroscopic method such as nuclear magnetic resonance (NMR) for example. In the latter approach, a van't Hoff plot of $\ln K$ vs. $1/T$ is constructed and fit to a linear equation; the slope will equal $-\Delta H_{inc}/R$, and the y-intercept will equal $\Delta S_{inc}/R$. An overview of experimental methods for studying host–guest inclusion complexes is presented in the next section.

2.4 EXPERIMENTAL STUDY OF HOST–GUEST INCLUSION COMPLEXATION

This section provides a brief discussion of the experimental methods commonly used to study host–guest inclusion in solution, with an emphasis on how the binding constant K can be obtained from experimental data. This will put into context the experimental studies of the applications of molecular hosts to corrosion and antifouling prevention addressed in Section 2.5.

2.4.1 SPECTROSCOPIC METHODS

Spectroscopic methods involve the measurement of electromagnetic radiation emitted or absorbed by molecules upon transitions between quantum mechanical states. They are well-suited for the characterisation of host–guest inclusion complexes in solution. To apply such methods to this phenomenon, some spectroscopic parameter of the guest, host or both must change measurably upon formation of the inclusion complex. The most commonly utilised spectroscopic methods for studying host–guest complexation, in general, are UV–vis absorption, fluorescence, Fourier-transform Infrared (FTIR), and NMR. These experimental methods are briefly discussed as they apply to this area of study.

UV–vis absorption (Johnston and Wagner 1996) and fluorescence (Johnston and Wagner 1996; Wagner 2003, 2006) spectroscopy both involve transitions between electronic energy levels of a molecule, the energy gaps of which correspond to light in the UV (200–400 nm) or visible (400–750 nm) wavelength range. Upon UV–vis absorption, molecules are excited to a higher energy level, whereas in the case of fluorescence spectroscopy molecules in higher excited states relax back to the ground state via the emission of a photon of light. In the case of both of these spectroscopies inclusion within the host results in significant changes in the guest spectrum, which allows for the study of the inclusion process. In the case of UV–vis absorption, inclusion-induced changes in the molar absorptivity of the guest at a specific wavelength are measured; these changes are relatively small in most cases, but quantifiable. In the case of fluorescence, much larger inclusion-induced changes in the intensity of the emission are possible. Furthermore, fluorescence is the most highly sensitive of the common spectroscopy techniques, requiring only small amounts of the compounds of interest and also showing extreme sensitivity of the measured emission intensity to local environment for many fluorescent molecules. This means that large changes in fluorescence signal upon inclusion can often be observed, making for highly accurate determination of binding constants. For these reasons, fluorescence spectroscopy has been highly utilised in the experimental study of host–guest complexation (Wagner 2003, 2006). However, it requires that either the guest or the host (or both) be significantly fluorescent and exhibit significant polarity sensitivity. All three of the most common hosts, CDs, cucurbiturils and calixarenes, are non-fluorescent, so the guest must be fluorescent in these cases. In the case of the corrosion prevention studies which are the focus of this chapter, corrosion inhibiting agents as guests are also typically non-fluorescent, so fluorescence has not been much used for studying these inclusion complexes.

FTIR is another absorption method, in this case involving transition between vibrational energy levels of molecules, corresponding to electromagnetic radiation in the infrared region of the spectrum, typically 200–4000 cm^{-1} in wavenumber. As in the case of UV–vis absorption, inclusion-induced changes in the IR absorption spectrum are typically relatively small (e.g. compared to fluorescence), but can be utilised. Furthermore, changes in specific bands can indicate the nature of the host–guest complex geometry, by indicating which guest functional groups are most impacted by inclusion.

NMR (Riek et al. 2002) is overall the most commonly applied spectroscopic technique for studying host–guest inclusion. Although smaller than in the case of fluorescence, inclusion-induced changes to the chemical shift of proton bands in ^{1}H NMR spectra can be significant and well quantifiable, allowing for accurate determination of the binding constant from NMR titration data (see below). In addition, NMR studies reveal much more information about the nature of the complex (Mugridge et al. 2011), and how the guest is oriented within the cavity, as each NMR band corresponds to a specific proton or sets of protons on the host and guest. 2D NMR techniques can also be used to determine the specific interactions between parts of the host and guest to determine the nature of the complex.

For all of these optical spectroscopic methods of studying host–guest inclusion, the all-important value of the binding constant K can be obtained through titration studies, in which the spectrum of the guest is measured as a function of added host concentration (or less commonly, vice versa). Various fit equations are available to fit this titration data and extract the numerical value of the binding constant K, depending on the complex stoichiometry and the type of experiment (absorption, fluorescence, NMR, etc.) (Wagner 2020). These equations can be found in the literature, for example for 1:1 complexation for fluorescence-based titration data (Muñoz de la Peña et al. 1993). The data in these cases are generally fit to the appropriate equation via nonlinear least squares. However, a linear approach based on the Benesi-Hildebrand method (Benesi 1949) can also be employed, but in general, this is less accurate than the nonlinear least-squares approach (Thordarson 2011; Wagner 2020). Details on the extraction of binding constant from experimental data, and errors involved, can be found in the literature (Hirose 2001; Thordarson 2011; Wagner 2020).

2.4.2 OTHER EXPERIMENTAL METHODS

There are many non-spectroscopic experimental methods, which have been applied to study host–guest inclusion complexation. Three of particular relevance to this chapter are electrochemistry, diffraction methods, and thermal methods; each of these three is briefly discussed here, as they have all been applied to host–guest studies involving corrosion and biofouling prevention.

Various types of electrochemical techniques have been used to study host–guest complexation (Kaifer and Gómez-Kaifer 1999). One of the most important electrochemical techniques is cyclic voltammetry (Elgashi et al. 2018), which can be used to determine the oxidation/reduction potential of a compound in solution. For an electrochemically active guest molecule, encapsulation within a host cavity can have significant, measurable effects on its oxidation/reduction potential, allowing for the study of its complexation using cyclic voltammetry. Another important electrochemical technique that has found particular application to corrosion studies is electrochemical impedance spectroscopy (EIS), which uses a potentiostat/galvanostat apparatus to measure the AC impedance of a material.

X-ray diffraction (XRD) methods (Zolotoyabko 2014) can be used to determine the structure of host–guest complexes in the solid state (Barbour 2017; Rissanen 2017). Although the structure of the complex in solution may differ from that in the solid state, this method does allow for a fundamental understanding of the exact nature of the inclusion complex and its geometry. Another useful XRD method for host or host–guest complex materials is powder XRD, which analyses solid samples as powders, so does not require single crystals. This method does not allow for the solution of the exact structure of the material but does allow for the determination of the degree of crystallinity, the spacing of layers in the case of layered materials and other useful bulk material information.

Thermal methods (White 1996; Nassimbeni and Báthori 2017) involve the study of the behaviour of materials upon the addition of heat. Of most importance are calorimetric methods (Arena and Sgarlatai 2017), which measure the heat released or absorbed during a reaction, or in this case, upon the inclusion of a guest within a host. Adiabatic calorimetric experiments are ones in which samples are analysed in an insulated reaction container, so that all heat changes can be accurately measured. Such methods can be used to directly and accurately determine the thermodynamics of inclusion, including ΔH_{inc} and the heat capacity of the complex material. A modern calorimetric technique which is often applied to supramolecular host–guest inclusion systems is isothermal titration calorimetry (ITC) (Bertaut and Landy 2014), in which host and guest solutions are mixed incrementally in a flow system, and changes in temperature recorded. These methods represent an accurate alternative to the use of a van't Hoff plot with temperature-dependent K data obtained via a spectroscopy-based method, as discussed above. Also of importance is thermal gravimetric analysis (TGA), in which the mass of a sample is measured as it is heated, to determine its decomposition, and differential scanning calorimetry (DSC), which measures the difference in heat required to increase the temperature of a sample of interest relative to that of a reference material. Both of these latter thermal analysis techniques, for which commercial, bench-top instruments are readily available, have been applied in studies of interest in this chapter, discussed in Section 2.5.

Many other experimental techniques have also been used to study host–guest complexes, including mass spectrometry (Brodbelt and Dearden 1996), and in particular electrospray ionisation mass spectrometry (Schalley et al. 1999; Banerjee and Mazumdar 2012), and chromatographic techniques (Buha et al. 2012; Kalchenko et al. 2017). In addition to experimental techniques, computational approaches have also been applied to host–guest inclusion (Wipff 1994).

2.5 EXAMPLES OF HOST–GUEST INCLUSION COMPLEXES IN CORROSION AND BIOFOULING PROTECTION

As mentioned in the Introduction, Saji has provided a comprehensive recent review of the application of supramolecular concepts to corrosion and biofouling prevention (Saji 2019). Some particularly interesting recent examples of such applications involving host–guest inclusion complexes are briefly described here.

2.5.1 Cyclodextrin Hosts

CDs have been used to form host–guest inclusion complexes with various anti-corrosion agents as guests in aqueous solution. The effect of CD inclusion on the anti-corrosion properties and efficiencies of these corrosion inhibiting molecules has been the focus of the research into their applications. Figure 2.9 shows the chemical structure of some of these corrosion inhibitor molecular guests, which are discussed in this section.

Leclercq et al. (2012, 2016) have reported on the effects of adding CDs to solutions of didecyladimethylammonium chloride (DiC$_{10}$, Figure 2.9a). DiC$_{10}$ is a disinfectant and biocide which has both antifungal (Leclercq et al. 2012) and antiviral (Leclercq et al. 2016) properties. It is used, for example, to disinfect, clean and remove fungal colonies which can cause biofouling of equipment. However, self-aggregation can result in inhibition of its biocidal activity. The addition of CDs was proposed as a way to prevent self-aggregation, by encapsulating the individual DiC$_{10}$ molecules as guests, thus physically preventing them from forming aggregates. These researchers found that while α- and γ-CD did not affect the bioactivity of DiC$_{10}$, the addition of β-CD caused a synergistic effect which increased the antifungal activity of the guest compound. They proposed a mechanism in which the host–guest complex of the guest in its cationic ammonium ion form diffuses through the aqueous solution, then absorbs onto the membrane surface whereupon it dissociates releasing the guest, the ammonium cation moiety of which inserts into the fungus membrane, causing cell lysis (Leclercq 2012). This is an excellent, illustrative example of both the guest protection and guest transport and release abilities of CD hosts, and a clear demonstration of the utility of these highly useful and available host molecules and their host–guest complexes. They also demonstrated that CD hosts significantly boost the virucidal activity of this same DiC$_{10}$ compound, yielding widely applicable, eco-friendly biocidal formulations (Leclercq et al. 2016). In this case, γ-CD was found to give the largest increase in biocidal activity.

Another study of the inclusion of a corrosion inhibitor as a guest within a CD host in an aqueous solution was recently reported (Dehghani et al. 2020). This study investigated the effects of β-CD host–guest inclusion on the corrosion inhibitor cerium acetylacetonate (CeA) in aqueous saline solution using both experimental and computational methods. They found that the CD-CeA inclusion complex showed significant inhibition of the corrosion of metal substrates in an aqueous saline solution. For example, they found an 82% corrosion inhibition after 48 h of silane-coated metal exposure to the saline solution, and that the CD host–guest complex enhanced the protective silane coating.

Amiri and Rahimi took an interesting approach to using CDs, as components in self-protective/self-healing hybrid nanocomposite coatings (Amiri 2014, 2015, 2016). These coatings contain the corrosion inhibitor agents MBT (Figure 2.9b) or 2MBI (Figure 2.9c). The idea is that these coatings will release the inhibitor guests only when demanded, by the onset of corrosion. They

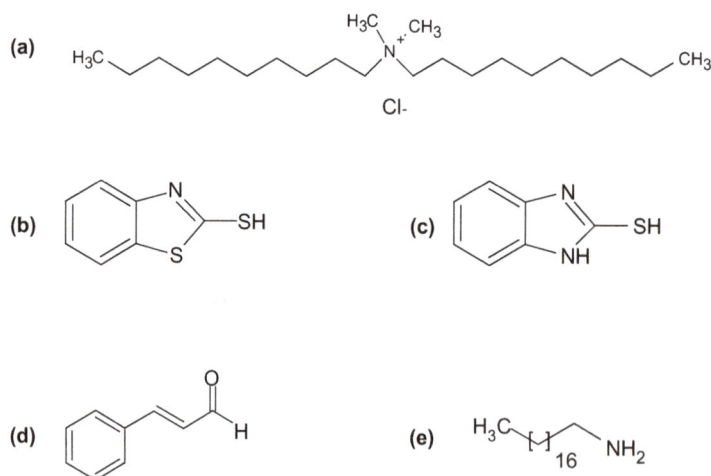

FIGURE 2.9   Corrosion inhibitor compounds included as guests within cyclodextrin hosts in the application studies discussed in Section 2.5.1. (a) Di-n-decyldimethylammonium chloride (DiC$_{10}$); (b) 2-mercaptobenzothiazole (MBT); (c) 2-mercaptobenzoimidazole (MBI); (d) trans-cinnamaldehyde; (e) octadecylamine.

investigated a range of such CD-based coatings, including ones based on α-CD (Amiri 2014), β-CD (Amiri 2015) and γ-CD (Amiri 2016) encapsulating either MBT or MBI. They found, for example, that the coating based on the largest CD, γ-CD, provided the best anti-corrosion protection as tested on scratched samples in a sea-spray medium. They also showed that the guest corrosion inhibitors could slowly diffuse out of the coating material, providing long-term corrosion protection.

Other researchers have also incorporated β-CD into the preparation of smart, self-protective anti-corrosion coatings, including coatings based on ferrocenyl polymethacrylate and other polymers (Chuo et al. 2016), carbon nanotubes (He et al. 2016) and graphene (Liu 2018).

Polymers containing β-CD have also been used to encapsulate and deliver corrosion inhibitor compounds for the protection of carbon steel in an acidic solution (He et al. 2014; Ma et al. 2019). He and co-workers prepared a supramolecular polymer containing β-CD linked to the polymer backbone that exhibited anti-corrosion properties for carbon still in acidic aqueous solution (He et al. 2014). They found that the CD-containing polymer itself acted as a mixed-type (as opposed to anodic- or cathodic-type) inhibitor (based on electrochemical methods), and showed that the inhibition efficiency increased with polymer concentration, indicating its role as the inhibitor. In the recent work reported by Ma et al., once again a polymer was prepared to incorporate β-CD, but in this case, the polymer was used as a host material to encapsulate a corrosion inhibitor as a guest (Ma et al. 2019). The inhibitor guest studied was *trans*-cinnamaldehyde (Figure 2.9d). Also different with this polymer as compared to that described above is that the CD host was incorporated into the polymer chain, not attached as a tethered group; this was done using crosslinking of glutaraldehyde with β-CD, creating what the authors refer to as a SCDP: a soluble CD polymer. They applied this inhibitor-containing SCDP to the corrosion protection of mild steel in 3.5% NaCl aqueous solution. Electrochemical measurements showed that the polymer containing *trans*-cinnamaldehyde could suppress both cathodic and anodic reactions, and found a maximum anti-corrosion efficiency of 92.2% at a temperature of 30°C. Surface analysis was used to show that the *trans*-cinnamaldehyde was released from the polymer assembly and adsorbed onto the steel surface, providing the observed corrosion inhibition.

In addition to the native (unmodified) CDs used in the studies described above, modified CDs have also been used in corrosion protection studies. For example, Fan et al. (2014) examined the impact of HP-β-CD on the hydrophobic corrosion inhibitor compound octadecylamine (ODA, Figure 2.9e). They found that HP-β-CD formed a highly soluble 2:1 host:guest inclusion complex with ODA, which served to solubilise the ODA (which is otherwise insoluble in aqueous solution), and that tests showed that this complex had a corrosion efficiency in simulated industrial boiler condensate water medium of 95%, much higher than that of the guest itself, which is nearly insoluble in aqueous solution.

Chen and Fu used an interesting two-pronged supramolecular host approach to an anti-corrosion application. They used both α-CD and cucurbit[6]uril as caps in the synthesis of hollow mesoporous silica nanoparticles (HMSNs), preparing acid- or alkaline-responsive "supramolecular nanocontainers", respectively (Chen 2012a). Both of these materials were used to encapsulate the corrosion inhibitor benzotriazole (BTA, Figure 2.10a). They then used both of these host–guest complexes as dopants in sol–gel materials as intelligent anti-corrosion coatings for the protection of metal alloys in an aqueous solution. They showed that this interesting CD-CB hybrid material offered significant long-term corrosion protection relative to the un-doped sol–gel coating. This is an interesting use of two molecular hosts, combining their differing host properties to achieve a material with much better anti-corrosion properties than could be achieved with a material containing only one or the other of these hosts.

These few selected examples clearly show that CDs as molecular hosts have found widespread application in the area of supramolecular corrosion and biofouling protection, and their host properties have been utilised in various ways, either directly as free hosts in aqueous solution, as covalently linked units in polymeric materials (as tethered groups or as part of the chain itself), incorporated with other types of materials, such as carbon nanotubes graphene, or as integral components in complex smart coatings.

2.5.2 Cucurbituril Hosts

Compared to CDs, cucurbiturils have found much less application in this area. However, there have been some such applications, a few examples of which will be briefly presented here. As mentioned above, cucurbit[6]uril was used in conjunction with α-CD to develop a complex smart anti-corrosion coating material, with the CB[6] providing an alkaline-responsive corrosion inhibitor delivering nanocapsules (Chen 2012a). This same group also looked at related pH-responsive "nanovalves" based on hollow silica spheres and CB[6] to obtain the controlled release of the same corrosion inhibitor BTA (Figure 2.10a) (Chen 2012b). They monitored the release of BTA using UV–vis absorption spectroscopy and showed that in neutral solution, the inhibitor was practically unreleased, but in basic solution, the BTA inhibitor guests were rapidly released, with a rate which increased with increasing pH. This demonstrated pH-dependent release showed that a smart CB-based anti-corrosion material was indeed obtained.

These same researchers also used cucurbit[7]uril in a similar application with mesoporous silica, this time for the controlled release of caffeine (Figure 2.10b) as a corrosion inhibitor (Fu et al. 2013). As can be seen in Figure 2.10, comparing the size of caffeine (Figure 2.10b) to that of BTA (Figure 2.10a), caffeine is significantly larger, requiring the use of the larger cucurbituril CB[7], as opposed to CB[6] that was used as described above for the inhibitor BTA. This is reflective of the general principle of the importance of the size match between the host cavity and the guest. Again, using a wide variety of experimental techniques, the authors were able to show the successful preparation of a smart container material, that in this case could release inhibitor under both acidic and alkaline pH conditions (compared to the CB[6]-based material which released inhibitor only in alkaline conditions). Full details on the material synthesis, structure and properties can be found in Fu et al. (2013).

In another study, Berdimurodov et al. investigated the use of cucurbiturils themselves as corrosion inhibitors for steel in acidic solution (Berdimurodov et al. 2016). They found that cucurbiturils showed corrosion protection as mixed-type inhibitors, with both anodic and cathodic activity. They used both SEM to examine the steel surfaces, and computational approaches, to demonstrate the corrosion inhibition properties of cucurbiturils. Interestingly, these authors used a synthesised mix

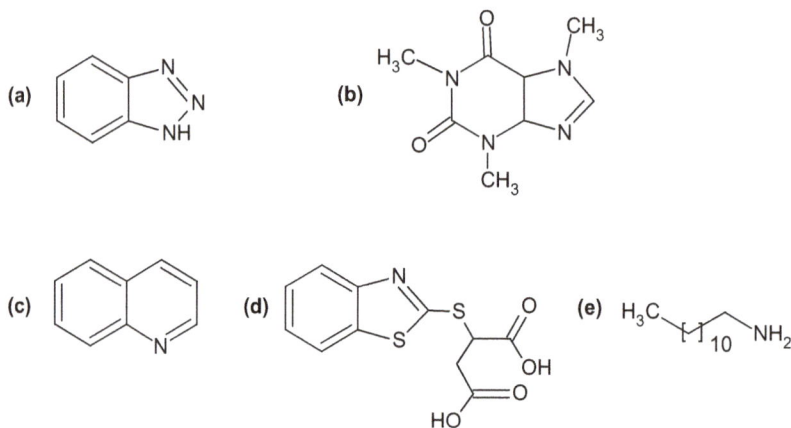

FIGURE 2.10 Corrosion inhibitor compounds included as guests within the host materials in the application studies discussed in Sections 2.5.2–2.5.4. (a) benzotriazole (BTA); (b) caffeine; (c) quinoline; (d) 2-(benzothiazol-2-ylsulfanyl)-succinic acid; (e) dodecylamine.

of cucurbit[n]urils, with varying values of n. They did not report on the relative composition of the mixture in terms of cucurbituril host size.

2.5.3 CALIXARENE HOSTS

Unlike in the case of CD hosts as described in Section 2.5.1, but similar to a study of cucurbiturils described in Section 2.5.2, applications of *calixarene* hosts in this area have included investigations the anti-corrosion properties of specially designed calixarene hosts themselves, not just their encapsulation of corrosion inhibitors as guests. This effect arises due to the ability of calixarenes to adsorb onto metal surfaces such as steel (Kaddouri et al. 2008). Kaddouri, Vicens et al. (Benabdellah et al. 2007; Kaddouri et al. 2008, 2013) have investigated the corrosion inhibition activity of some novel calixarene derivatives. By substituting groups with corrosion inhibitor properties, these specialised calixarenes can adsorb onto surfaces to provide effective corrosion protection. For example, they have investigated a set of imidazolylethylamidocarbonyl derivatives calix[4]arene (Kaddouri et al. 2008). Referring to general calixarene structure shown in Figure 2.7, these derivatives have $n=4$, $R={}^t$butyl, and 1–4 of the R' groups as imidazolylethylamidocarbonyl group (see Kaddouri et al. 2008 for the detailed structure). Imidazoles are well-known corrosion inhibitors (e.g. MBI). They used various experimental techniques to determine the inhibition efficiency of these calix[4]arene derivatives, which showed maximum inhibition efficiencies of between 94% and 100% for mild steel in 1 M HCl solution, at a calixarene concentration of 10^{-4} M. They also showed good inhibition results with much larger water-soluble sulfonated calix[8]arenes under similar conditions (Kaddouri et al. 2013).

As in the case of CDs and cucurbiturils, calixarenes have also been used to prepare smart anti-biofouling coatings. For example, Troian-Gautier et al. reported on the preparation of robust monolayers of ethylene glycol-derivatised calix[4]arenes, which could be directly grafted onto metal and semiconductor surfaces via their tetradiazonium salts (Blond et al. 2018). These were found to be effective anti-biofouling coatings for gold and germanium surfaces, reducing surface absorption of bovine serum albumin proteins by 85% for coated versus uncoated germanium for example. Most recently, this same group published a review article on the use of such calixarenes with diazonium groups for a variety of specific surface applications (Troian-Gautier et al. 2020).

2.5.4 OTHER HOSTS

In addition to the "big three" family of molecular hosts discussed above, a number of other types of molecular hosts have also been applied to supramolecular-enhanced corrosion and biofouling. Fouda and co-workers investigated the corrosion inhibition properties of the *crown ether* Dibenzo-18-crown-6 (Figure 2.8) for protection of stainless steel in an acidic aqueous solution (Fouda et al. 2010). They found significant concentration-dependent inhibition of the corrosion of steel under these conditions, the efficiency of which decreased with increasing temperature. They attributed the corrosion inhibition property of this crown ether to its adsorption on the steel surface, as evidenced by SEM and other surface analysis techniques.

Dendrimers have also been applied in this area; Verma et al. have provided an informative recent review (Verma et al. 2018). Their group reported on the corrosion inhibitor properties of two different ammonia-core dendrimers (Verma et al. 2016) and used a wide variety of experimental techniques to show that both dendrimers acted as efficient corrosion inhibitors for steel in acidic aqueous solution. The inhibition efficiency increased with increasing dendrimer concentration, with a maximum efficiency obtained at a 50 ppm concentration. As in the case of crown ether, the inhibition arose from the adsorption of the dendrimers on the steel surface, providing a protective coating. Zhang and co-workers combined PAMAM dendrimers with sodium silicate and observed a synergistic corrosion inhibition effect for carbon steel in soft water (Zhang et al. 2015). Co-adsorption of the two components was observed, yielding an efficient mixed inhibitor that provided significant protection of the steel from corrosion in this aqueous medium.

Another interesting molecular host that has shown corrosion inhibition activity is *nanocap*-shaped tin phthalocynanines, mentioned briefly in Section 2.2.4 (Beltrán et al. 2005). These molecules are π-electron-rich, and show square-antiprismatic coordination of the tin atoms, properties which give these compounds anti-corrosion abilities. Specific versions of these compounds, the structures of which are shown in Beltrán et al. (2005), showed significant corrosion inhibitor performance for carbon steel in sour brine medium (a common aqueous environment in the petroleum industry), as shown through extensive electrochemical analysis methods.

A variety of microcapsules and *nanocapsules* have also been applied as hosts for corrosion prevention. For example, polyurea-based microcapsules have been used as micro-reservoirs to contain and release the corrosion-inhibiting compound quinoline (Figure 2.10c) (Gite et al. 2015). The release of the inhibitor was monitored by mass loss and UV–vis spectroscopy experimental studies, and the inhibition of the corrosion of mild steel by this system was investigated. The use of both micro- and nanocapsules in self-healing, smart anti-corrosion coatings was reviewed in 2015; in these applications, the capsules were used as a coating layer or shell in a controlled manner (Wei et al. 2015).

As mentioned in Section 2.2.4, the term "nanocontainers" has been rather broadly and indiscriminately applied to a wide variety of host molecules and materials, including even CD and cucurbiturils. However, technically the term refers to hosts with internal cavities that are 1 nm or greater in diameter in at least one direction. Even with this more focused definition, there has been a wide array of *nanocontainer* materials used in corrosion prevention applications. Since the focus of this chapter has been on molecular hosts as already covered, just a brief overview of some of the anti-corrosion applications involving nanocontainer materials will be presented here. Shchukin and co-workers have published a series of papers on various types of nanocontainer host materials which they used to entrap corrosion inhibitors. For example, they used polyelectrolyte-modified halloysite nanotubes, nanoparticles and nanocapsules to incorporate the well-utilised corrosion inhibitor BTA (Shchukin and Möhwald 2007), which has already been discussed as a guest in various hosts above. These materials were found to release the corrosion inhibitors under both basic and acidic conditions. They demonstrated the versatility of these different types of nanocontainers, with different types working best for specific anti-corrosion applications. They also applied these nanocontainers impregnated with BTA within sol–gel coatings to provide corrosion protection for aluminium surfaces (Zheludkevich 2007), and used BTA in monodisperse polymeric core-shell nanocontainer-based self-healing anti-corrosion coatings (Li et al. 2014). In addition, they used similar nanocontainer-based approaches to develop anti-corrosion coatings and materials based on other corrosion inhibiting compounds, including MBT (Figure 2.9b, also discussed previously) (Borisova 2012, 2013) and 2-(benzothiazol-2-ylsulfanyl)-succinic acid, the structure of which is shown in Figure 2.10d (Skorb et al. 2009).

Some other interesting studies of nanocontainer-based host materials for corrosion protection include the use of silica-based nanocontainers loaded with MBT applied to metal surfaces in saline environments (Maia et al. 2012), and silica nanoparticles encapsulating the corrosion inhibitor dodecylamine (Figure 2.10e), again for use in saline solution (Falcón 2014, 2015). In the latter case, for example, efficient release of this inhibitor was found under highly acidic conditions (pH = 2), as determined by a variety of experimental methods (Falcón 2014).

In terms of solid-state host materials, *zeolites* have been loaded with corrosion-inhibiting materials to improve the inhibition efficiency. For example, Estevão and Nascimento used zeolites to reduce the volatilisation rate of some volatile corrosion inhibitors (VCIs), thereby increasing their corrosion inhibition efficiencies by reducing their rate of loss from the system of interest (Estevão and Nascimento 2001). Most recently, Lv et al. used zeolites loaded with fluorescent supramolecular "on-of" probes to develop sensors for corrosion detection (Lv et al. 2021). Chitosan-zinc oxide nanocomposites have been used as coatings for the prevention of marine biofouling (Al-Naamani et al. 2017). Layered MOFs have been applied as protective films for the prevention of corrosion of copper (Fernando et al. 2012).

Finally, a few other miscellaneous hosts or host materials which have been used recently in corrosion prevention studies, but which do not fit into any of the categories above, include biomimetic anchors, which have been used in host–guest antifouling for titanium (Cai et al. 2016); dicarboxylate bola-amphiphiles, which have been applied to the inhibition of corrosion of steel (Schmelter et al. 2017); and the use of supramolecular ureidopyridinone-based materials as anti-biofouling coatings (Goor et al. 2017).

2.6   CONCLUSIONS AND PERSPECTIVES

Host–guest complexation has found significant and advantageous applications to the prevention of corrosion and biofouling of materials and infrastructure. Particularly impressive and striking is the wide range of types of molecular hosts and hosts materials that have been utilised in this area of research and practical applications. CDs are the molecular host that has been most widely utilised, typically for the formation of inclusion complexes of inhibitor compounds as guests. The inclusion of corrosion inhibitors into the CD cavity has been shown in a wide range of studies to improve the inhibition efficiency of the anti-corrosion agents. Cucurbituril hosts have also been occasionally used for the same purpose, binding corrosion inhibitors for subsequent release, and have also been investigated as corrosion inhibitors themselves. In the case of calixarenes, these hosts have also been investigated as corrosion inhibitors, both directly by the covalent attachment of inhibitor groups, and in the preparation of smart corrosion inhibiting coatings. Crown ethers and dendrimers have both been applied as molecular hosts, and provide corrosion protection via adsorption to target surfaces. Microcapsules and nanocap-shaped hosts and other unique hosts and host materials have also been applied to corrosion prevention. Nanocontainer host materials have been widely applied, particularly in the development of smart coatings for the release of corrosion inhibitors when conditions require them, for example at high or low pH.

The biggest advantage of the application of host–guest chemistry in this area is the opportunity for controlled release of corrosion inhibitor guests from the host–guest inclusion complex in an aqueous solution. This can be a slow release, extending the timescale of the corrosion protection available, or more interestingly, triggered release when conditions warrant it, for example when the aqueous environment becomes highly acidic, or highly alkaline. This allows for the development of smart coatings for example, which respond to conditions and release corrosion protection when needed. A number of such host–guest inclusion-based smart anti-corrosion coating technologies were presented and briefly discussed, involving for example cucurbituril hosts and nanocontainer materials. This is an area that will continue to develop and grow, as the protection offered by such smart materials is much superior to that of the simple addition of corrosion inhibitors.

Thus, host–guest inclusion complexation has proven to be a valuable tool and approach in the prevention of the corrosion and biofouling of materials and surfaces. With the continual development of new and exciting families and types of host molecules and materials yet to be explored for this purpose, and continued improvements in and new ways to utilise existing hosts and host materials, the future is bright for these types of applications. The applications of supramolecular host–guest chemistry to the prevention of corrosion will continue to be a leading-edge relevant research area in this economically and society-important field of study and technology.

REFERENCES

Abbasi, E., Aval, S. F., Akbarzadeh, A., Milani, M., Nasrabadi, H. T., Joo, S. W., Hanifehpour, Y., Nejati-Koshki, K., and Pashaie-Asl, R. 2014. Dendrimers: synthesis, applications and properties. *Nano Res. Lett.* 9:247–256.
Al-Naamani, L., Dobretsov, S., Dutta, J., and Burgess, J. D. 2017. Chitosan-zinc oxide nanocomposite coatings for the prevention of marine biofouling. *Chemosphere* 168:408–417.

Amiri, S., and Rahimi, A. 2014. Preparation of supramolecular corrosion-inhibiting nanocontainers for self-protective hybrid nanocomposite coatings. *J. Polym. Res.* 21:566.

Amiri, S., and Rahimi, A. 2015. Synthesis and characterization of supramolecular corrosion inhibitor nano-containers for anticorrosion hybrid nanocomposite coatings. *J. Polym. Res.* 22: 66.

Amiri, S., and Rahimi, A. 2016. Anticorrosion behavior of cyclodextrins/inhibitor nanocapsule-based self-healing coatings. *J. Coat. Technol. Res.* 13:1095–1102.

Amiri, S., and Amiri, S. 2017. *Cyclodextrins: Properties and Industrial Applications.* Bridgewater, NJ: Wiley, ISBN: 978-1-119–24752–4.

Arena, G., and Sgarlata, C. 2017. Modern calorimetry: an invaluable tool in supramolecular chemistry, Chapter 11 In *Comprehensive Supramolecular Chemistry II, Volume 2: Experimental and Computational Methods in Supramolecular Chemistry*, Atwood, J. L., (Ed.), Amsterdam: Elsevier.

Assaf, K. I., and Nau, W. M. 2015. Cucurbiturils: from synthesis to high-affinity binding and catalysis. *Chem. Soc. Rev.* 44:394–418.

Banerjee, S., and Mazumdar, S. 2012. Electrospray ionization mass spectrometry: a technique to access the information beyond the molecular weight of the analyte. *Int. J. Anal. Chem.* 2012:282574.

Barbour, L. J. 2017. Single-crystal X-ray diffraction. Chapter 3 In *Comprehensive Supramolecular Chemistry II, Volume 2: Experimental and Computational Methods in Supramolecular Chemistry*, Atwood, J. L., (Ed.), Amsterdam: Elsevier.

Beer, P. D., Gale, P. A., and Smith, D. K. 1999. *Supramolecular chemistry.* Oxford: Oxford University Press.

Behrend, R., Meyer, E., and Rusche, F. 1905. Ueber cindensationsproducte aud glycoluril und formaldehyde. *Libigs Ann. Chem.* 339:1–37.

Beltrán, H. I., Esquivel, R., Liozada-Cassou, M., Dominiguez-Aguilar, M. A., Sosa-Sánchez, A., Sosa-Sánchez, J. L., Höpfl, H., Baraba, V., Luna-García, R., Farfán, N., and Zambudio-Rivera, L. S. 2005. Nanocap-shaped tin phthalocyanines: Synthesis, characterization and corrosion inhibition activity. *Chem. Eur. J.* 11:2705–2715.

Benabdellah, M., Souane, R., Cheriaa, N., Abidi, R., Hammouti, B., and Vicens, J. 2007. Synthesis of calix-arene derivatives and their anticorrosive effect on steel in 1M HCl. *Pig. Res. Tech.* 36(6):373–381.

Benesi, H.A., and Hildebrand, J.H. 1949. A spectrophotometric investigation of the interaction of iodine with aromatic hydrocarbons. *J. Am. Chem. Soc.* 71:2703–2707.

Berdimurodov, E., Wang, J., Kholikov, A., Akbarov, K., Burikhonov, B., and Umirov, N. 2016. Investigation of a new corrosion inhibitor cucurbiturils for mild steel in 10% acidic medium. *Adv. Eng. Forum* 18:21–38.

Bertaut, E., and Landy, D. 2014. Improving ITC studies of cyclodextrin inclusion compounds by global analysis of conventional and non-conventional experiments. *Beilstein J. Org. Chem.* 10:2630–2641.

Biedermann, F. 2017. Self-assembly in aqueous media. Chapter 11 In *Comprehensive Supramolecular Chemistry II, Volume 1: General Principles of Supramolecular Chemistry and Molecular Recognition*, Atwood, J. L. (Ed.), Amsterdam: Elsevier.

Biedermann, F., Nau, W. M., and Schneider, H.-J. 2014. The hydrophobic effect revisited – studies with supramolecular complexes imply high-energy water as a noncovalent driving force. *Angew. Chem. Int. Ed.* 53:11158–11171.

Biedermann, F., Uzunova, V. D., Scherman, O. A., Nau, W. M., and De Simone, A. 2012. Release of high-energy water as an essential driving force for the high-affinity binding of cucurbit[n]urils. *J. Am. Chem. Soc.* 134:15318–15323.

Blond, P., Mattiuzzi, A., Valkenier, H., Troian-Gautier, L., Bergamini, J.-F., Doneux, T., Goormaghtigh, E., Raussens, V., and Jabin, I. 2018. Grafting of oligo(ethylene glycol)-functionalized calix[4]arene-tetradiazonium slats for antifouling germanium and gold surfaces. *Langmuir* 34:6021–6027.

Bohne, C. 2006. Supramolecular dynamics studied using photophysics. *Langmuir* 22:9100–9111.

Bohne, C. 2014. Supramolecular dynamics. *Chem. Soc. Rev.* 43:4037–4050.

Borisova, D., Möhwald, H., and Shchukin, D. G. 2012. Influence of embedded nanocontainers on the efficiency of active anticorrosive coatings for aluminum alloys part I: Influence of nanocontainer concentration. *ACS Appl. Mater. Interfaces* 4:2931–2939.

Borisova, D., Möhwald, H., Shchukin, D.G. 2013. Influence of embedded nanocontainers on the efficiency of active anticorrosive coatings for aluminum alloys part II: Influence of nanocontainer position. *ACS Appl. Mater. Interfaces* 5:80–87.

Brodbelt, J. S., and Dearden, D. V. 1996. Mass spectrometry. Chapter 14 In *Comprehensive Supramolecular Chemistry, Volume 8, Physical Methods in Supramolecular Chemistry*, Davies, J. E. D. and Ripmeester, J. A. (Eds.), New York: Pergamon.

Buha, S. M., Baxi, G. A., and Shrivastav, P. S. 2012. Liquid chromatography study on atenolol-b-cyclodextrin inclusion complex. *ISRN Anal. Chem.* 2012:423572.

Cai, X. Y., Li, N. N., Chen, J. C., Kang, E.-T., and Xu, L. Q. 2016. Biomimetic anchors applied to the host-guest antifouling functionalization of titanium substrates. *J. Coll. Interface Sci.* 475:8–16.

Cheirsilp, B., and Rakmai, J. 2016. Inclusion complex formation of cyclodextrin with its guest and their applications. *Biol. Eng. Med.* 2:1–6.

Chen, T., and Fu, J. 2012a. An intelligent anticorrosion coating based on pH-responsive supramolecular nano-containers. *Biotechnology* 23:505705.

Chen, T., and Fu, J. 2012b. pH-responsive nanovalves based on hollow mesoporous silica spheres for controlled release of corrosion inhibitor. *Nanotechnology* 23:235605.

Chuo, T.-W., Yeh, J. M., and Liu, Y. M. 2016. A reactive blend of electroactive polymers exhibiting synergistic effects on self-healing and anticorrosion properties. *RSC Adv.* 6:55593–55598.

Cragg, P. J. 2005. *A Practical Guide to Supramolecular Chemistry.* Chichester, UK: John Wiley & Sons.

Das, D., Assaf, K. I., and Nau, W. M. 2019. Applications of cucurbiturils in medicinal chemistry and chemical biology. *Front. Chem.* 13:619.

Dehghani, A., Bahlakeh, G., and Ramezanzadeh, B. 2020. Construction of a sustainable/controlled-release nano-container of non-toxic corrosion inhibitors for the water-based siliconized film: estimating the host-guest interactions/desorption of inclusion complexes of cerium acetylacetonate (CeA) with beta-cyclodextrin (β-CD) via detailed electronic/atomic-scale computer modeling and experimental methods. *J. Hazard. Mat.* 399:123046.

Detellier, C. 1996. Complexation mechanisms. Chapter 9 In *Comprehensive Supramolecular Chemistry, Volume 1, Molecular Recognition: Receptors for Cationic Guests*, Gokel, G. W. (Ed.), New York: Pergamon.

Douhal, A. 2004. Ultrafast guest dynamics in cyclodextrin nanocavities. *Chem. Soc. Rev.* 104:1955–1976.

Elgrishi, N., Rountree, K. J., McCarthy, B. D., Rountree, E. S., Eisenhart, T. T., and Dempsey, J. L. 2018. A practical beginner's guide to cyclic voltammetry. *J. Chem. Educ.* 95:197–206.

Esfand, R., and Tomalia, D. A. 2001. Poly(amidoamine) (PAMAM) dendrimers: from biomomicry to drug delivery and biomedical applications. *Drug Discovery Today* 6:427–436.

Estevão, L. R. M., and Nascimento, R. S. V. 2001. Modifications in the volatilization rate of volatile corrosion inhibitors by means of host–guest systems. *Corros. Sci.* 43:1133–1153.

Fagnani, D. E., Sotuyo, A., and Castellano, R. K. 2017. π-π interactions. Chapter 6 In *Comprehensive Supramolecular Chemistry II, Volume 1: General Principles of Supramolecular Chemistry and Molecular Recognition*, Atwood, J. L. (Ed.), Amsterdam: Elsevier.

Falcón, J. M., Batista, F. F., and Aoki, I. V. 2014. Encapsulation of dodecylamine corrosion inhibitor on silica nanoparticles. *Electrochim. Acta* 124:109–118.

Falcón, J. M., Sawczen, T., and Aoki, I. V. 2015. Dodecylamine-loaded halloysite nanocontainers for active anticorrosion coatings. *Front. Mater.* 2:104–111.

Fan, B., Wie, G., Zhang, Z., and Qiao, N. 2014. Preparation of supramolecular corrosion inhibitor based on hydroxypropyl-b-cyclodextrin/octadecylamine and its anti-corrosion properties in the simulated condensate water. *Anti-Corros. Meth. Mater.* 61:104–111.

Fernando, I. R., Jianrattanasawat, S., Daskalakis, N., Demadis, K. D., and Mezei, G. 2012. Mapping the supramolecular chemistry of pyrazole-4-sulfonate: layered inorganic–organic networks with Zn^{2+}, Cd^{2+}, Ag^+, Na^+ and NH_4^+, and their use in copper anticorrosion protective films. *Cryst. Eng. Commun.* 14:908–919.

Fouda, A. S., Abdallah, M., Al-Ashrey, S. M., and Abdel-Fattah, A. A. 2010. Some crown ethers as inhibitors for corrosion of stainless steel type 430 in aqueous solutions. *Desalination* 250:538–543.

Freeman, W. A., Mock, W. L., and Shih, N. Y. 1981. Cucurbituril. *J. Am. Chem. Soc.* 103:7367–7368.

Fu, J., Chen, T., Wang, M., Yang, N., Li, S., Wang, Y., and Liu, X. 2013. Acid and alkaline dual stimuli-responsive mechanized hollow mesoporous silica nanoparticles as smart nanocontainers for intelligent anticorrosion coatings. *ACS Nano* 7:11397–11408.

Gite, V. V., Tatiya, P. D., Marathe, R. J., Mahulikar, P. P., and Hundiwale, D. G. 2015. Microencapsulation of quinoline as a corrosion inhibitor in polyurea microcapsules for application in anticorrosive PU coatings. *Prog. Org. Coat.* 83:11–18.

Gokel, G. 2017. Cation binding by crown ethers. Chapter 9 In *Comprehensive Supramolecular Chemistry II, Volume 1: General Principles of Supramolecular Chemistry and Molecular Recognition*, Atwood, J. L. (Ed.), Amsterdam: Elsevier.

Goor, O. J. G. M., Brounsa, J. E. P., and Dankers, P. Y. W. 2017. Introduction of anti-fouling coatings at the surface of supramolecular elastomeric materials via post-modification of reactive supramolecular additives. *Polym. Chem.* 8:5228–5238.

Hashidzume, A., Takashima, Y., Yamaguchi, H., and Harada, A. 2017. Cyclodextrin. Chapter 12 In *Comprehensive Supramolecular Chemistry II, Volume 1: General Principles of Supramolecular Chemistry and Molecular Recognition*, Atwood, J. L. (Ed.), Amsterdam: Elsevier.

He, Y., Yang, Q., and Xu, Z. 2014. A supramolecular polymer containing β-cyclodextrin as corrosion inhibitor for carbon steel in acidic medium. *Russ. J. Appl. Chem.* 87:1936–1942.

He, Y., Zhang, C., Wu, D., and Xu Z. 2016. Fabrication study of a new anticorrosion coating based on supramolecular nanocontainers. *Synth. Met.* 212:186–194.

Heredia, A., Requena, G., and Garcia Sánchez, F. 1985. An approach for the estimation of the polarity of the b-cyclodextrin internal cavity. *J. Chem. Soc. Chem. Commun.* 1814–1815.

Hirose, K. 2001. A practical guide for the determination of binding constants. *J. Inclus. Phenom. Macro. Chem.* 39:193–209.

Iacovino, R., Caso, J. V., Di Donato, C., Malgieri, G., Palmeiri, M., Russo, L., and Isernia, C. 2017. Cyclodextrins as complexing agents: preparation and applications. *Curr. Org. Chem.* 21:162–176.

Izatt, R. M. 2007. Charles J. Pedersen: innovator in macrocyclic chemistry and co-recipient of the 1987 Nobel Prize in Chemistry. *Chem. Soc. Rev.* 36:143–147.

Izatt, R. M., and Oscarson, J. L. 2017. Binding constants and related thermodynamic quantities: Significance to supramolecular chemistry. Chapter 8 In *Comprehensive Supramolecular Chemistry II, Volume 1: General Principles of Supramolecular Chemistry and Molecular Recognition*, Atwood, J. L. (Ed), Amsterdam: Elsevier.

Johnston, L. J., and Wagner, B. D. 1996. Electronic absorption and luminescence. In *Comprehensive Supramolecular Chemistry Volume 8: Physical Methods in Supramolecular Chemistry*, Ripmeester, J. E. D. (Ed.), pp. 537–566. Oxford: Pergamon.

Kaddouri, M., Cheriaa, N., Souane, R., BouKlah, M., Aouniti, A., Abidi, R., Hammouti, B., and Vicens J. 2008. Novel calixarene derivatives as inhibitors of mild C-38 steel corrosion in 1 M HCl. *J. Appl. Electrochem.* 38:1253–1258.

Kaddouri, M., Rekhab, S., Bouklah, M., Hammouti, B., Aouniti, A., and Kabouche Z. 2013. Experimental study of inhibition of corrosion of mild steel in 1 M HCl solution by two newly synthesized calixarene derivatives. *Res. Chem. Intermed.* 39:3649–3667.

Kaifer, A. E., and Kaifer-Gómez, M. 1999. *Supramolecular Electrochemistry*. Weinheim: Wiley-VCH.

Kalchenko, O., Lipkowski, J., and Kalchenko, V. 2017. Chromatography in supramolecular and analytical chemistry of calixarenes. Chapter 12 In *Comprehensive Supramolecular Chemistry II, Volume 2: Experimental and Computational Methods in Supramolecular Chemistry*, Atwood, J. L. (Ed.), Amsterdam: Elsevier.

Kim, J., Jung, I. S., Kim, S.-Y. Lee, E., Kang, J.-K., Sakamoto, S., Yamaguchi, K., and Kim, K. 2000. New cucurbituril homologues: Synthesis, isolation, characterisation and X-ray crystal structures of cucurbit[n] uril (n = 5, 7, and 8). *J. Am. Chem. Soc.* 122:540–541.

Kim, K., Selvapalam, N., Ko, Y. H., Park, K. M., Kim, D., and Kim, J. 2007. Functionalized cucurbiturils and their applications. *Chem. Soc. Rev.* 36:267–279.

Kurkov, S. V., and Loftsson, T. 2013. Cyclodextrins. *Int. J. Pharmaceut.* 453:167–180.

Lagona, J., Mukhopadhyay, P., Chakrabarti, S., and Isaacs, L. 2005. The cucurbit[n]uril family. *Angew. Chem. Int. Ed.* 44:4844–4870.

Leclercq, L., Dewilde, A., Aubry, J. M., and Nardello-Rataj, V. 2016. Supramolecular assistance between cyclodextrins and didecyldimethylammonium chloride against enveloped viruses: Toward eco-biocidal formulations. *Int. J. Pharm.* 512:273–281.

Leclercq, L., Lubart, Q., Dewilde, A., Aubry, J. M., and Nardello-Rataj, V. 2012. Supramolecular effects on the antifungal activity of cyclodextrin/di-n-decyldimethylammonium chloride mixtures. *Eur. J. Pharm. Sci.* 46:336–345.

Lehn, J.-M. 1988. Supramolecular chemistry – scope and perspectives. Molecules, supermolecules, and molecular devices (nobel lecture). *Angew. Chem. Int. Ed.* 27:89–112.

Li, G. L., Schenderlein, M., Men, Y., Möhwald, H., and Shchukin, D. G. 2014. Monodisperse polymeric core–shell nanocontainers for organic self-healing anticorrosion coatings. *Adv. Mat. Interf.* 1:1300019.

Liu, C., Zhao, H., Hou, P., Qian, B., Wang, X., Guo, C., Wang, L. 2018. Efficient graphene/cyclodextrin-based nanocontainer: synthesis and host–guest inclusion for self-healing anticorrosion application. *ACS Appl. Mater. Interf.* 10(42):36229–36239.

Liu, L. and Guo, Q.-X. 2002. The driving forces in the inclusion complexation of cyclodextrins. *J. Inclus. Phenom. Macro. Chem.* 42:1–14.

Liu, S., Ruspic, C., Mukhopadhyay, P., Chakrabarti, S., Zavalij, P. Y., and Isaacs, L. 2005b. The cucurbit[n]uril family: prime components for self-sorting systems. *J. Am. Chem. Soc.* 127:15959–15967.

Liu, S., Zavalij, P. Y., and Isaacs, L. 2005a. Curcurbit[10]uril. *J. Am. Chem. Soc.* 127:16798–16799.

Lv, J., Yue, Q.-X., Ding, R., Han, Q., Liu, X., Liu, J.-L., Yu, H.-J., An, K., Yu, H.-B., and Zhao, X.-D. 2021. Construction of zeolite-loaded fluorescent supramolecular on-off probes for corrosion detection based on a cation exchange mechanism. *Nanomaterials* 11(1):169.

Ma, Y., Fan, B., Zhou, T., Hao, H., Yang, B., and Sun, H. 2019. Molecular assembly between weak crosslinking cyclodextrin polymer and trans-cinnemaldehyde for corrosion inhibition towards mild steel in 3.5% NaCl solution: experimental and theoretical studies. *Polymers* 11:635.

Maia, F., Tedim, J., Lisenkov, A. D., Salak, A. N., Zheludkevich, M. L., Ferreira, M. G. S. 2012. Silica nanocomposites for active corrosion protection. *Nanoscale* 4:1287–1298.

Márquez, C., Hudgins, R. R., and Nau, W. M. 2004. Mechanism of host-guest complexations by cucurbituril. *J. Am. Chem. Soc.* 126:5806–5816.

Mettry, M., and Hooley, R. J. 2017. Receptors based on van der Waals forces. Chapter 4 In *Comprehensive Supramolecular Chemistry II, Volume 1: General Principles of Supramolecular Chemistry and Molecular Recognition*, Atwood, J. L. (Ed.), Amsterdam: Elsevier.

Mugridge, J. S., Bergman, R. G., and Raymond, K. N. 2011. ^{1}H NMR chemical shift calculation as a probe of supramolecular host-guest geometry. *J. Am. Chem. Soc.* 133:11205–11212.

Muñoz de la Peña, A., Salinas, F., Gómez, M. J., Acedo, M. I., and Sánchez Peña, M. 1993. Absorptiometric and spectrofluorimetric study of the inclusion complexes of 2-naphthyloxyacetic acid and 1-naphthylacetic acid with β-cyclodextrin in aqueous solution. *J. Inclus. Phenom. Mol. Recog. Chem.* 15:131–143.

Nash, T. 2017. The role of entropy in molecular self-assembly. *J. Nanomed. Res.* 5(4):00126.

Nassimbeni, L. R., and Báthori, N. B. 2017. Thermal analysis, Chapter 2 in *Comprehensive Supramolecular Chemistry II, Volume 2: Experimental and Computational Methods in Supramolecular Chemistry*, Atwood, J. L. (Ed.), Amsterdam: Elsevier.

Nau, W. M., Florea, M., and Assaf, K. I. 2011. Deep inside cucurbiturils: physical properties and volumes of their inner cavity determine the hydrophobic driving force for host-guest complexation. *Isr. J. Chem.* 51:559–577.

Nguyen, T. A., and Assadi, A. A. 2018. Smart nanocontainers: preparation, loading/release processes and applications. *Kenkyu J. Nanotech. Nanosci.* 4:S1-1-6.

Nguyen-Tri, P., Do, T.-O., Nguyen, T. A., Le, V. T., and Assadi, A. A. 2020. Nanocontainer: an introduction. Chapter 1 In *Smart Nanocontainers*, Nguyen-Tri, P., Do, T.-O., Nguyen, T. A., (Eds.), Amsterdam: Elsevier.

Nishio, M., Umezawa, Y., Honda, K., Tsuboyama, S., and Suezawa, H. 2009. CH/π hydrogen bonds in organic and organometallic chemistry. *CrystEngComm* 11:1757–1788.

Pluth, M. D., and Raymond, K. N. 2007. Reversible guest exchange mechanisms in supramolecular host-guest assemblies. *Chem. Soc. Rev.* 36:161–171.

Pochini, A., and Ungaro, R. 1996. Calixarenes and related hosts, Chapter 4 In *Comprehensive Supramolecular Chemistry, Volume 2, Molecular Recognition: Receptors for Molecular Guests*, Vögtle, F. (Ed.), New York: Pergamon.

Reinhoudt, S. 2016. Rocco Ungaroa, 40 years of calixarene chemistry. *Supramol. Chem.* 28:342–350.

Rekharsky, M. V., and Inoue, Y. 1998. Complexation thermodynamics of cyclodextrins. *Chem. Rev.* 98:1875–1917.

Riek, R., Fiaux, J., Bertelsen, E. B., Horwich, A. L., and Wüthrich, K. 2002. Solution NMR techniques for large molecular and supramolecular structures. *J. Am. Chem. Soc.* 124:12144–12153.

Rissanen, K. 2017. Crystallography of encapsulated molecules. *Chem. Soc. Rev.* 46:2638–2648.

Saji, V. S. 2019. Supramolecular concepts and approaches in corrosion and biofouling prevention. *Corros. Rev.* 37(3):187–230.

Saji, V. S., and Cook, R. (Eds.) 2012. *Corrosion Protection and Control Using Nanomaterials*. Amsterdam: Elsevier, ISBN: 978-1-84569-949-9.

Schalley, C. A., Rivera, J. M., Martin, T., Santamaria, J., Siuzdak, G., and Rebek, Jr., J. 1999. Structural examination of supramolecular architectures by electrospray ionization mass spectrometry. *Eur. J. Org. Chem.* 1999:1325–1331.

Schmelter, D., Langry, A., Koenig, A., Keil, P., Leroux, F., and Hinze-Bruening, H. 2017. Inhibition of steel corrosion and alkaline zinc oxide dissolution by dicarboxylate bola-amphiphiles: Self-assembly supersedes host-guest conception. *Sci. Rep.* 7:2785.

Schneider, H.-J. 2009. Binding mechanisms in supramolecular complexes. *Angew. Chem. Int. Ed.* 48:3924–3977.

Schneider, H.-J., and Yatsimirsky, A. 2000. *Principles and Methods in Supramolecular Chemistry*. Chichester, UK: John Wiley & Sons.

Sharma, N., and Baldi, A. 2016. Exploring versatile applications of cyclodextrins: an overview. *Drug Deliv.* 23:729–747.

Shchukin, D. G., and Möhwald, H. 2007. Surface-engineered nanocontainers for entrapment of corrosion inhibitors. *Adv. Funct. Mater.* 17:1451–1458.

Shinkai, S. 1986. Calixarenes as new functional host molecules. *Pure Appl. Chem.* 58:1523–1528.

Skorb, E. V., Fix, D., Andreeva, D. V., Möhwald, H. and Shchukin, D. G. 2009. Surface-modified mesoporous SiO_2 containers for Corrosion protection. *Adv. Funct. Mater.* 19:2373–2379.

Steed, J. W., and Atwood, J. L. 2009. *Supramolecular Chemistry.* 2nd Ed., Bridgewater, NJ: Wiley.

Szejtli, J. 1996. Inclusion of guest molecules, selectivity and molecular recognition by cyclodextrins. Chapter 5 In *Comprehensive Supramolecular Chemistry, Volume 3, Cyclodextrins*, Szejtli, J., Osa, T. (Eds.), New York: Pergamon.

Szejtli, J. 1998. Introduction and general overview of cyclodextrin chemistry. *Chem. Rev.* 98:1743–1753.

Thordarson, P. 2011. Determining association constants from titration experiments in supramolecular chemistry. *Chem. Soc. Rev.* 40:1305–1323.

Troian-Gautier, L., Mattiuzzi, A., Reinaud, O., Lagrost, C., and Jabin, I. 2020. Use of calixarenes bearing diazonium groups for the development of robust monolayers with unique tailored properties. *Org. Biomol. Chem.* 18: 3624–3637.

Verma, C., Ebenso, E. E., Vishal, Y., and Quraishi, M. A. 2016. Dendrimers: a new class of corrosion inhibitors for mild steel in 1 M HCl: experimental and quantum chemical studies. *J. Mol. Liq.* 234:1282–1293.

Verma, C., Ebenso, E. E., and Quraishi, M. A. 2018. Dendrimers as a novel class of polymeric corrosion inhibitors. *Int. J. Corros. Scale Inhibit.* 7:593–608.

Wagner, B. D. 2003. Fluorescence studies of supramolecular host-guest inclusion complexes. In *Handbook of Photochemistry and Photobiology Volume 3: Supramolecular Photochemistry*, Nalwa, H. S. (Ed.), pp. 1–57. Stevenson Ranch, CA: American Scientific Publishers.

Wagner, B. D. 2006. The effects of cyclodextrins on guest fluorescence, Chapter 2 In *Cyclodextrin Materials Photochemistry, Photophysics and Photobiology*, Douhal, A. (Ed.), Amsterdam: Elsevier B.D.

Wagner, B. D. 2020. *Host-Guest Chemistry: Supramolecular Inclusion in Solution.* Berlin: De Gruyter.

Wei, H., Wang, Y., Guo, J., Shen, N. Z., Jiang, D., Xhang, X., Yan, X., Zhu, J., Wang, Q., Shao, L., Lin, H., Wei, S., and Guo, Z. 2015. Advanced micro/nanocapsules for self-healing smart anticorrosion coatings. *J. Mater. Chem. A* 3:469–480.

White, M. A. 1996. Thermal analysis and calorimetry methods. Chapter 4 In *Comprehensive Supramolecular Chemistry, Volume 8, Physical Methods in Supramolecular Chemistry*, Davies, J. E. D. and Ripmeester, J. A. (Eds.), New York: Pergamon.

Wipff, G. (Ed.) 1994. *Computational Approaches in Supramolecular Chemistry.* Berlin: Springer.

Zeng F., and Zimmerman, S. C. 1997. Dendrimers in supramolecular chemistry: From molecular recognition to self-assembly. *Chem. Rev.* 97:1681–1712.

Zhang, B., He, C., Chen, X., Tian, Z., and Li, F. 2015. The synergistic effect of polyamidoamine dendrimers and sodium silicate on the corrosion of carbon steel in soft water. *Corros. Sci.* 90:585–596.

Zhao, X., Meng, G. Han, F., Li, X., Chen, B., Xu, Q., Zhu, X., Chu, Z., Kong, M., and Huang, Q. 2013. Nanocontainers made of various materials with tunable shape and size. *Sci. Rep.* 3:2238.

Zheludkevich, M.L., Shchukin, D.G., Yasakau, K.A., Möhwald, H., Ferreira, M.G.S. 2007. Anticorrosion coatings with self-healing effect based on nanocontainers impregnated with corrosion inhibitor. *Chem. Mater.* 19:402–411.

Zheludkevich, M. L., Tedim, J., and Ferreira, M. G. S. 2012. Smart coatings for active corrosion protection based on multi-functional micro and nanocontainers. *Electrochim. Acta* 82:314–323.

Zolotoyabko, E. 2014. *Basic Concepts of X-Ray Diffraction.* New York: John Wiley & Sons.

3 Supramolecular Polymers

Nihar Sahu and Indrajit Mohanta
National Institute of Technology Rourkela

Akhilesh K. Srivastava
Deen Dayal Upadhyay Gorakhpur University

Bimalendu Adhikari
National Institute of Technology Rourkela

CONTENTS

3.1 INTRODUCTION

Supramolecular chemistry is the chemistry "beyond the molecule" and "the science of non-covalent intermolecular interactions" (Lehn 1988). Conventional molecular chemistry is based on covalent bonds, whereas supramolecular chemistry addresses molecular assemblies based on weak noncovalent interactions that include hydrogen-bonding, ion–ion interaction, ion–dipole, dipole–dipole, π–π interactions, van der Waals forces, hydrophobic forces, and metal coordination. The use of these noncovalent interactions provides an opportunity to synthesize large, complex structures by the assembly of small building blocks and the resulting supramolecular structures exhibit interesting properties and functions. The supramolecular systems typically exhibit dynamic and stimuli-responsive behavior as they are made of small building blocks, which are associated with weak noncovalent interactions (Brunsveld et al. 2001; Hoeben et al. 2005). The advantages of noncovalent interactions are their reversible nature and hence the supramolecular assemblies respond toward different external stimuli (Segarra-Maset et al. 2013; Ma and Tian 2014; Zhang et al. 2014) such as temperature, concentration, polarity of the medium, pH, and so on. Biological systems are often the biggest inspiration for supramolecular chemistry as the structure and function of many biological systems, for instance, DNA is based on noncovalent forces.

Supramolecular polymer (SP) arises from the combination of polymer science and supramolecular chemistry. SPs are an important class of nanomaterials with fascinating properties and functions (Brunsveld et al. 2001; Hoeben et al. 2005; De Greef and Meijer 2008; Aida et al. 2012; Yang et al. 2015; Yagai et al. 2019; Wehner and Würthner 2020). The term "supramolecular polymers" can be defined as molecular assemblies where the small molecular units are associated by some directional noncovalent forces, such as hydrogen bonds, metal–ligand coordination, π–π stacking, leading to a high degree of internal order and reversibility in the resulting one-dimensional array of molecules (aggregates). The strength, directionality, and reversibility of these noncovalent forces are important for the construction of an array of molecules, categorized as SPs that not only preserve their polymeric properties in solution but also endow them with dynamic properties such as stimuli-responsiveness, environmental adaptation, and self-healing. However, to avoid confusion, the readers are highly recommended to read the following well-accepted definitions of SPs: (i) "Supramolecular polymers are defined as polymeric arrays of monomeric units that are brought together by reversible and highly directional secondary interactions, resulting in polymeric properties in dilute and concentrated solutions, as well as in the bulk. The monomeric units of the SPs themselves do not possess a repetition of chemical fragments. The directionality and strength of the supramolecular bonding are important features of systems that can be regarded as polymers and that behave according to well-established theories of polymer physics." provided by Prof. E. W. Meijer in 2001 (Brunsveld et al. 2001). (ii) "Supramolecular polymers are polymeric arrays of monomeric units held together by reversible and directional noncovalent interactions such as hydrogen bonds, π–π interactions, and metal–ligand binding, and so on. The directionality and strength of the interactions are precisely tuned so that the resulting array of molecules behaves as a polymer; the resulting materials, therefore, maintain their polymeric properties in solution." provided by Prof. Thorfinnur Gunnlaugsson in 2017 (Savyasachi et al. 2017). (iii) "Supramolecular polymers as arrays of monomeric units that are interconnected by directional secondary interactions such as hydrogen bonds, metal–ligand coordination, π–π stacking or combinations of them. Accordingly, SPs are one-dimensional and typically consist of a single chain, thereby being distinguished from nanocrystalline or liquid-crystalline three-dimensional materials to afford the characteristic properties of macromolecules in dilute solution (for example, increased viscosity), for SPs well- designed receptor pairs are needed which combine binding strength and directionality" provided very recently by Prof. F. Würthner (Wehner and Würthner 2020).

Directional noncovalent interactions such as hydrogen bonds (Armstrong and Buggy 2005; Adhikari, Lin, et al. 2017), metal–ligand coordination (Dobrawa and Würthner 2005; Yang et al. 2015; Winter and Schubert 2016) or π–π stacking (Brunsveld et al. 2001; Hoeben et al. 2005; Chen et al. 2009; Babu, Praveen, and Ajayaghosh 2014) are essential for the formation of SPs. However, in this chapter, we discuss organic SP where building blocks are usually small organic molecules, which are interconnected by hydrogen bonds, π–π stacking, or host–guest interactions (Zheng et al. 2012; Ma and Tian 2014; Zhang et al. 2014; Yang et al. 2015) mostly. Apart from these main driving forces, metal–ligand coordination as a highly directional force is also commonly exploited for the synthesis of metal-based supramolecular systems, namely, metallo-SP (Winter and Schubert 2016) and coordination polymer (Dobrawa and Würthner 2005), which are out of scope for this chapter. Herein, we have discussed the fundamentals, preparation methods, characterization techniques, and properties of SP (Seiffert and Sprakel 2012; Yang et al. 2015; Ghislaine and Meijer 2019; Thompson and Korley 2020). First, we have highlighted the design principle of monomers with a variety of noncovalent interactions such as multiple-hydrogen-bonding, host–guest interactions, and aromatic π–π stacking employed as driving forces for the construction of SPs. We have also delineated the SPs based on the energy profile of the product, like thermodynamic, kinetic, and out of equilibrium as well as the mechanisms for supramolecular polymerization, including isodesmic and cooperative methods (Matern et al. 2019; Hartlieb et al. 2020; Wehner and Würthner 2020). Finally, we have outlined the properties of SP networks and their self-repairing ability (De Greef et al. 2009; Krieg et al. 2016; Lutz et al. 2016).

3.2   METHOD OF PREPARATION AND CHARACTERIZATION OF SPs

SPs are prepared by following the self-assembly approach. The self-assembly of small organic molecules in non-aqueous solvents is generally an enthalpy-driven exothermic process (Syamala and Würthner 2020), and hence, an enhancement in supramolecular polymerization is observed at a lower temperature (Sorrenti et al. 2017). Subsequently, cooling down a molecularly dissolved hot solution of the supramolecular building blocks to induce aggregation became a common methodology in supramolecular polymerization (Noro et al. 2008; Adhikari, Yamada, et al. 2017). Another common method for inducing supramolecular polymerization is to dissolve the monomers in a good solvent initially, followed by the addition of a bad solvent to tune the polarity of medium that imposes self-assembly of the monomers (Adhikari et al. 2019). Self-assembly can also be induced by the alteration of physical parameters like pH of the medium to protonate/deprotonate the building blocks or by the addition of inorganic salts to shield the aggregate charges and for multicomponent self-assembly (Petka et al. 1998; Paulusse, Van Beek, and Sijbesma 2007; Lewis and Dell 2016).

Covalent bond formation in the case of conventional polymerization typically occurs under kinetic control where the energy barrier for the backward reaction, described as depolymerization reaction, is often much larger compared to the forward reaction. As a consequence, the molecular weight of the polymer is not altered by dilution or heating of the system. In sharp contrast, the extent of a noncovalent reaction for supramolecular polymerization is directly linked to thermodynamic forces like temperature and concentration of the system, and hence, depolymerization may occur by the application of dilution or heating. Hence, the change in temperature, concentration, and environmental conditions can considerably affect the molecular weight of SPs, which makes their characterization difficult, especially the parameters like average molar mass and polydispersity (Yang et al. 2015). Moreover, as a result of well-controlled polymerization and depolymerization, SPs are easily constructed and destructed, which makes their synthesis, degradation, recovery, and recycling much easier.

Supramolecular polymerization is a process of self-assembly with a certain thermodynamic equilibrium constant. The average molar mass of the product is difficult to measure experimentally for SPs, and it can only be obtained through some theoretical models. Like conventional polymers, viscometry is a useful means to characterize the SPs where the intrinsic viscosity is enhanced with the increasing molecular weight of the polymer chain. The degree of polymerization (DP) for an isodesmic supramolecular polymerization can be determined as $DP \approx (K_a C)^{1/2}$, where K_a is the equilibrium constant and C is the total concentration of monomer (Brunsveld et al. 2001). Size exclusion chromatography (SEC), specifically gel permeation chromatography (GPC), is commonly employed to characterize conventional polymers. However, SEC is not suitable for most of the SPs as degradation happens in the column packing, followed by nonspecific adsorption. In the case of some rigid SPs, for instance, those involving metal coordination or multiple-hydrogen-bonding with a high binding constant, SEC has been used with modest success (Meier et al. 2006; Chiper et al. 2007). Like covalent polymers, light scattering experiments, namely static light scattering (SLS) and dynamic light scattering (DLS), are valuable methods for characterizing SPs. DLS is often utilized to determine the size distribution of the SP chains. Moreover, small-angle X-ray scattering (SAXS) and small-angle neutron scattering (SANS) study is an important tool to obtain information such as the size and shape of the SPs (Hollamby et al. 2016).

3.3   VARIOUS NONCOVALENT INTERACTIONS FOR THE DESIGN OF SPs

A number of noncovalent interactions, namely, multiple-hydrogen-bonding, metal coordination, aromatic π–π stacking, and host–guest interactions work as driving forces for the construction of SPs. The chemical structure of a supramolecular building block directly accounts for the type and strength of association, specificity in interactions, and architecture of the association like chain extension and lateral stacking (Li et al. 2012). Such parameters are programmed into the supramolecular building blocks during the synthesis of SP.

3.3.1 HYDROGEN BONDS

The hydrogen bond is a particular kind of dipole–dipole interaction or electrostatic attraction force between an electronegative atom (N/O/F) and a hydrogen atom (H) bonded to another electronegative atom (N/O/F). Hydrogen bonds are typically expressed as D-H·A where both D and A are electronegative atoms. The electronegative atom attached to hydrogen is the donor (D) and the other electronegative atom containing a lone pair that interacts with hydrogen noncovalently is the acceptor (A). H-bonding is considered as an electrostatic attraction along with some degree of orbital overlap between non-bonding electrons of the acceptor (A) atom and D-H bonding electrons that introduces directionality. The H-bonding is considered the most important noncovalent interaction for the fabrication of SP due to its suitable strength (4–60 kJ/mol) and high degree of directionality. The facile design and specificity of hydrogen bonds also play a pivotal role in its great usefulness for designing SPs. In biology, H-bondings play an important role in the stabilization of secondary structures of proteins such as α-helices and β-sheets (Li et al. 2012). In the construction of supramolecular architecture, amide, urethane, and urea moieties containing specified hydrogen-bonding donor (N–H) and hydrogen-bonding acceptor (C=O) are mostly utilized for the construction of hydrogen-bonding arrays toward fabrication of SPs (Brunsveld et al. 2001). In general, a single hydrogen bond is not well enough to support SPs. However, the overall strength and directionality of the interactions between two molecules can be dramatically enhanced by the help of multiple hydrogen-bonding arrays. Figure 3.1a represents typical examples of different multiple-hydrogen-bonding arrays (Yang et al. 2015). The binding constant or dimerization constant (K_a) of

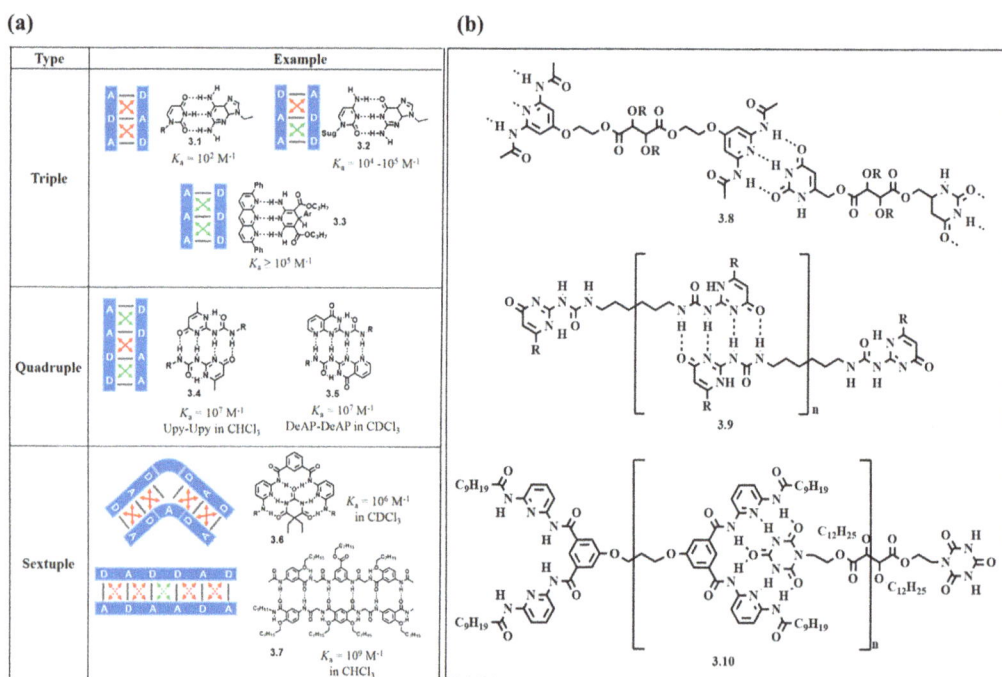

FIGURE 3.1 (a) Examples (3.1–3.7) of molecules having multiple-hydrogen-bonding units and their corresponding binding strength. The binding constant (K_a) of the motif increases with an increasing number of hydrogen-bonding sites and it is also dependent on the arrangement of the donor–acceptor atoms with respect to each other in the arrays. In the schematic representation, double-headed red arrows indicate repulsive secondary electrostatic interactions and double-headed green arrows indicate attractive secondary electrostatic interactions. (b) Chemical structures of monomers (3.8–3.10) that form linear SPs by multiple hydrogen-bonding interactions between the designed arrays. ((a) Reproduced with permission from Yang et al. (2015) © 2015 American Chemical Society.)

the multiple-hydrogen-bonding motif increases with an increasing number of hydrogen-bonding in the arrays. Hence, triple H-bonds forming motif 3.1/3.2/3.3 show a lower K_a compared to 3.4/3.5, which is capable of forming quadruple H-bonds. Similarly, 3.4/3.5 has lower binding strength with respect to 3.6/3.7 featuring sextuple-hydrogen-bonding arrays. Apart from the number of hydrogen bonding, the K_a between two arrays greatly depends on the arrangement of the donor and acceptor atoms in relation to each other in the arrays (Figure 3.1a) (Wilson 2007). In the first row of Figure 3.1a, the binding strength among three different motifs featuring triple hydrogen-bonding arrays is compared. Although 3.1, 3.2, and 3.3 have an equal number of hydrogen bonds, they have different arrangements or order of the donor and acceptor in the arrays and the observed K_a is reported to be in the order of 3.1 < 3.2 < 3.3 (Jorgensen and Pranata 1990; Murray and Zimmerman 1992). This difference is due to the occurrence of secondary electrostatic interactions along with the hydrogen-bonding arrays where the partial positive and negative charges developed on the donor (H) atoms and acceptor (O/N) atoms, respectively, repel adjacent groups with the same partial electrostatic charge. In a supramolecular array where these adjacent electrostatic secondary charges are attractive, K_a values are greater than the systems where the secondary charges repel each other. Although each of the three moieties (3.1, 3.2, and 3.3) has three attractive primary electrostatic interactions for triple H-bonds, they have a different number of additional attractive/repulsive secondary electrostatic interactions. 3.1 with ADA-DAD arrays produces two pairs of repulsive secondary electrostatic interactions, whereas 3.2 with DAA-AAD arrays possesses one pair of repulsive and one pair of attractive secondary electrostatic interactions and 3.3 with AAA-DDD arrays has two pairs of attractive secondary electrostatic interactions. Such secondary electrostatic interactions of hydrogen-bonding arrays are very useful approaches in supramolecular material design with tunable binding strength (Thompson and Korley 2020).

In 1997, Meijer and co-workers invented self-complementary quadruple hydrogen-bonding unit 3.4 (Figure 3.1a), 2-ureido-4[1H]pyrimidinone (UPy) featuring a unique donor-donor-acceptor-acceptor (DDAA) array (Sijbesma et al. 1997; Folmer et al. 2000). The presence of four hydrogen-bonding sites and the favorable cross interactions between the two central donor and acceptor sites in its AADD–DDAA arrays result in strong dimerization ($K_a \sim 10^7$ M^{-1} in chloroform) (Folmer et al. 2000). Like 3.4, Zimmerman et al. reported a DeAP unit, 3.5, that contains a self-complementary AADD array (Figure 3.1a) (Corbin and Zimmerman 1998). Moreover, sextuple-hydrogen-bonding arrays between Hamilton wedge (diaminopyridine substituted isophthalamide) and barbiturates 3.6 (Figure 3.1a) have also been recently reported. The K_a for this hetero complementary recognition pair is up to 1.37×10^6 M^{-1} in CDCl$_3$ (Chang and Hamilton 1988). Another very stable sextuple-hydrogen-bonding array, 3.7 is reported with a $K_a \approx 10^9$ M^{-1} in CDCl$_3$. The presence of intramolecular hydrogen bonding in 3.7 plays a key role to maintain an organized planar arrangement of the donor and acceptor units that are required for the formation of intermolecular sextuple H-bonds between a pair of 3.7 (Zeng et al. 2000; M. Li et al. 2006).

The use of multiple hydrogen bonds in the formation of SPs was demonstrated by Lehn et al. in 1990. Two bifunctional monomers, namely diamidopyridines- and uracil-based compounds, were mixed to form linear polymeric chains 3.8 as a result of the assembly through triple hydrogen-bonding interactions between the two complementary hydrogen-bond forming monomers (Figure 3.1b) (Fouquey et al. 1990). The motif UPy has been widely used in the design of SPs due to its facile synthesis along with strong self-complementary interactions (Hirschberg et al. 1999; Ten Cate and Sijbesma 2002; Sijbesma and Meijer 2003). The Meijer group constructed a famous linear SP 3.9 from a bifunctional monomer containing two UPy end groups (Figure 3.1b) (Sijbesma et al. 1997). The molecular mass of SPs can be altered by changing the solvent and concentration. The incorporation of monofunctional units resulted in a decrease in the solution viscosity and the DP. The calculated DP for the pure monomer was about 700 at 40 mM in CHCl$_3$. Barbiturate/cyanurate-isophthaloyldiamidopyridine sextuple H-bonding arrays were exploited to form a series of SPs 3.10 (Figure 3.1b) (Kolomiets et al. 2006; Berl et al. 2002). Rigid fibers of SP were observed in toluene, while gel-like assemblies were reported in decane.

3.3.2 π–π INTERACTIONS

Although the formation of SPs is mostly driven by hydrogen bonding and metal–ligand interactions, π–π stacking has also contributed significantly to the production of SPs. Usually, π–π stacking interactions act as a secondary force to reinforce and stabilize SPs formed by other stronger noncovalent interactions. π–π interactions represent a special case of van der Waals forces between large π-conjugated surfaces (Hartlieb et al. 2020). Although there is some debate about the actual nature of its contributing forces, π–π interaction is realized as an interaction between a π-electron cloud and the σ-scaffold of aromatic molecules (Hunter and Sanders 1990). For large aromatic π-surfaces, the molecules either assume a face-to-face orientation either with a large offset (called J-aggregates) or with an overlapping but twisted state (called H-aggregates) to ensure an ideal interaction.

As π–π-interactions are comparatively weak, generally multiple fused benzene rings are needed to form stable stacks in solution. In polar or in very apolar solvents, the π–π stacking can be reinforced considerably by solvophobic interactions, which are generally stronger than π–π interaction. In water, the presence of a hydrophobic effect along with π–π interactions leads to relatively stable assemblies of columnar stacks (Hoeben et al. 2005). Unlike hydrophobic interactions between long alkyl chains (with higher degrees of conformational freedom), π–π interactions of large π-conjugated systems (which are rigid) are associated with a strong directional component which is responsible for the formation of one-dimensional SP (Martinez and Iverson 2012; Krieg et al. 2016). This electronic coupling between the delocalized π-electrons of molecules makes π–π systems highly interesting candidates for making functional SPs for light-harvesting and solar energy conversion. Self-assembly study of π-conjugated molecules often gets easier as the systems absorb or emit light in the UV/vis region and various spectroscopic techniques can be used for the quantitative investigation of their supramolecular polymerization process (Ghosh et al. 2008). The self-assembly of π-surface molecules in various solvents has been well investigated and reviewed in several reports (Elemans et al. 2003; Zhang et al. 2007; Greenland et al. 2008; Babu, Praveen, and Ajayaghosh 2014). Examples of large π-conjugated molecules, including perylene and other functional aromatic molecules, are discussed in terms of their ability to form SPs. In the absence of additional secondary forces, π–π stacking usually leads to the formation of liquid-crystalline material and short SP of loose stacks. In general, long SPs are obtained from π-conjugated molecules with additional strong intermolecular interactions.

3.3.2.1 Only π–π Interactions

Triphenylenes (3.11a–c) are among the initial π-conjugated molecules that were studied in the context of supramolecular association (Figure 3.2). Alkoxy substituted triphenylenes were found to exhibit liquid-crystalline properties (Brunsveld et al. 2001). The triphenylenes possess a relatively small aromatic core and supramolecular polymerization is observed by SANS. The DP was found to be smaller at low concentrations, and it gradually increased and showed rodlike polymers at higher concentrations of 10^{-3} M. However, the molecules are expected to be loosely stacked as the intercore distance was determined to be ~6 Å higher than the molecular stacking distance of ~3.5 Å in the liquid-crystalline state. Such loose stacking likely arises due to the nonavailability of secondary interactions other than the π–π stacking.

Phthalocyanines have a significantly larger π-conjugated core compared to triphenylenes, giving rise to stronger intermolecular π–π interaction (Van Nostrum and Nolte 1996) and their optical and electrical properties can also be modified by the integration of a metal into the core. The aggregation of phthalocyanines has been studied intensively in various solvents. The molecules stack cooperatively in which dimerization is the favored step and further oligomerization only happens at higher concentrations (Shankar, Jha, and Vasudevan 1993; Schutte, Sluyters-Rehbach, and Sluyters 1993). However, larger columnar aggregates composed of 2–6 molecules are observed on the Langmuir-Blodgett films of phthalocyanine 3.12 (Figure 3.2) (Schutte, Sluyters-Rehbach, and Sluyters 1993).

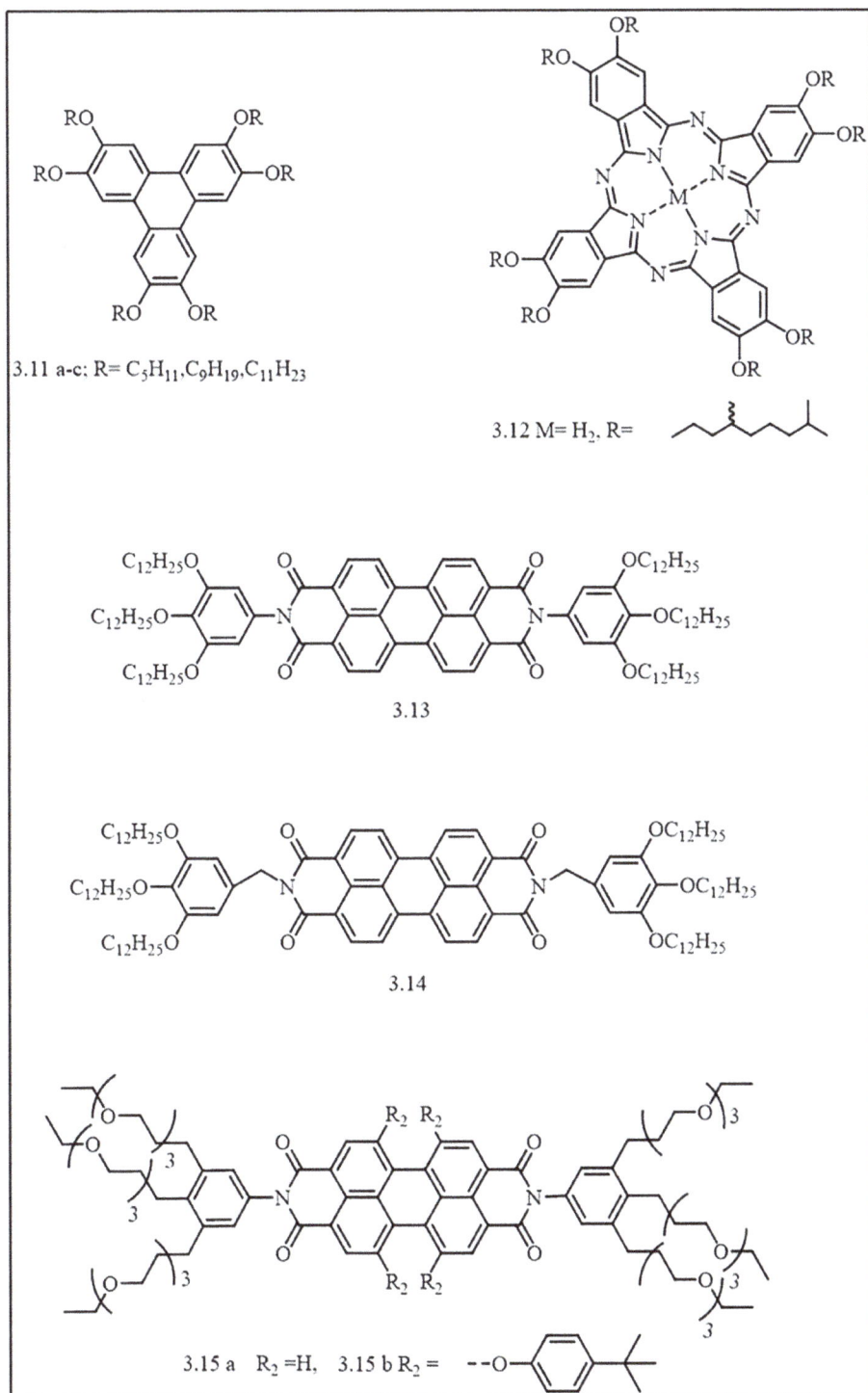

FIGURE 3.2 Chemical structures of large aromatic π-molecules capable of self-assembly by only π–π interactions. Triphenylenes (3.11a–c) that display loosely stacked liquid-crystalline property, Phthalocyanine (3.12) with the ability to form columnar aggregates, perylene diimides (3.13, 3.14) that undergo aggregation in organic solvents, substituted perylene diimide 3.15a&b forms aqueous supramolecular copolymer and gels upon mixing.

Perylene diimides (PDIs) are another promising building block for functional supramolecular polymeric systems as the π-conjugated core of PDI shows strong π–π stacking interaction, strong absorption in the visible region, high fluorescence quantum yield, and, more importantly, for its n-type organic semiconducting properties (Würthner 2004; Langhals 2005; Görl et al. 2012; S. Chen et al. 2015; Würthner et al. 2016; Krieg et al. 2019). The Würthner group reported that the solubility of PDIs increased in organic solvents by substituting 3,4,5-tridodecyloxyphenyl groups at the imide nitrogens in molecules 3.13 and 3.14 (Figure 3.2) (Würthner et al. 2001). The difference between 3.13 and 3.14 is the alkyl spacer $-CH_2$ present between tridodecyloxyphenyl and the imide N atoms in 3.14. Hypochromicity (decreasing absorption coefficients) for 3.13 was observed upon aggregation in methylcyclohexane (MCH) as studied spectroscopically (Van Herrikhuyzen et al. 2004). The aggregation for both the molecules in MCH was investigated by concentration-dependent UV–vis studies, from which aggregation constant (K_a) was found to be 1.5×10^7 M^{-1} for 3.13, whereas three orders of magnitude smaller K_a were noticed for 3.14. The authors explain this fact by assuming that the presence of methylene segment in 3.14 likely reduced the interacting van der Waals surfaces of the aromatic unit.

Another noteworthy work was the design of the dumbbell-shaped PDI derivative 3.15a that forms a supramolecular assembly in water (Zhang et al. 2007), which can further extend to supramolecular ribbons with increasing concentrations (X. Zhang et al. 2014). The presence of hydrophilic moiety triethylene glycol groups in 3.15a plays an important role in the formation of linear columnar stacks. The K_a for this stacking in water was found to be greater than 10^8 M^{-1}, which is higher than K_a of PDIs in MCH, and this is expected as the solvophobic interaction makes the π–π interaction stronger in water. When the hydrogens in the R_2 position (3.15) were replaced by a bulky aryloxy group, 3.15b leads to an anti-cooperative assembly by twisting the π-conjugated plane and thus reduces homopolymer formation. When both PDIs 3.15a and 3.15b (Figure 3.2) were assembled together, interestingly, very long bundled nanofibers were obtained as supramolecular copolymers. The composition of the obtained supramolecular block copolymer was proposed to be (PDI-3.15a$_m$PDI-3.15b$_2$)$_n$ based on NMR, optical and microscopic studies.

3.3.2.2   π–π Interaction with Hydrogen Bonding

The earlier section (only π–π stacking) described the formation of SPs by nonspecific π–π stacking that leads to irregular or "loose" stacking in the assembly. Herein we discuss columnar SPs obtained by the π–π stacking accompanied by more specific hydrogen bonding. To introduce columnar hydrogen-bonded architectures to π-conjugated molecules, C_3-symmetric molecules 1,3,5-benzene triamides (BTAs) (Figure 3.3a) have been utilized. The BTA molecule possesses a central benzene ring with three side chains that can be connected intermolecularly through amide functionalities (Yasuda et al. 1996; Hanabusa et al. 1997). X-ray diffraction studies (Lightfoot et al. 1999) and infrared spectroscopy have provided information that suggests that the BTA molecules undergo columnar stacking by virtue of a threefold intermolecular hydrogen bonding. The weak π–π interactions of the core benzene group are secondary to strong threefold hydrogen-bonding interactions. However, during the process of packing, helicity is introduced in the system as intermolecular hydrogen bonds rotate out of the plane (Figure 3.3b). The helicity of the SP is dependent on the substituent (R) attached to the nitrogens in the BTA molecule 3.16. The presence of no chiral center in R results in a mixture of both right- and left-handed helical columns in the SP, whereas one helicity is preferred over the other for homochiral alkyl chain (R), and this preference of helicity is believed to arise from the energy difference between the two helical columns.

In the previous section, we have elaborated the π–π interaction in PDI without having additional secondary interaction that can stack loosely in an organic solvent medium. In addition, we have observed how the hydrophobicity increases the strength of the assembly in water, to some extent. However, the addition of amide functionality enhances the formation of long SP in organic solvent, which can even form organogel. Such a compound is 3.17 (Figure 3.3c) where the π–π interacting

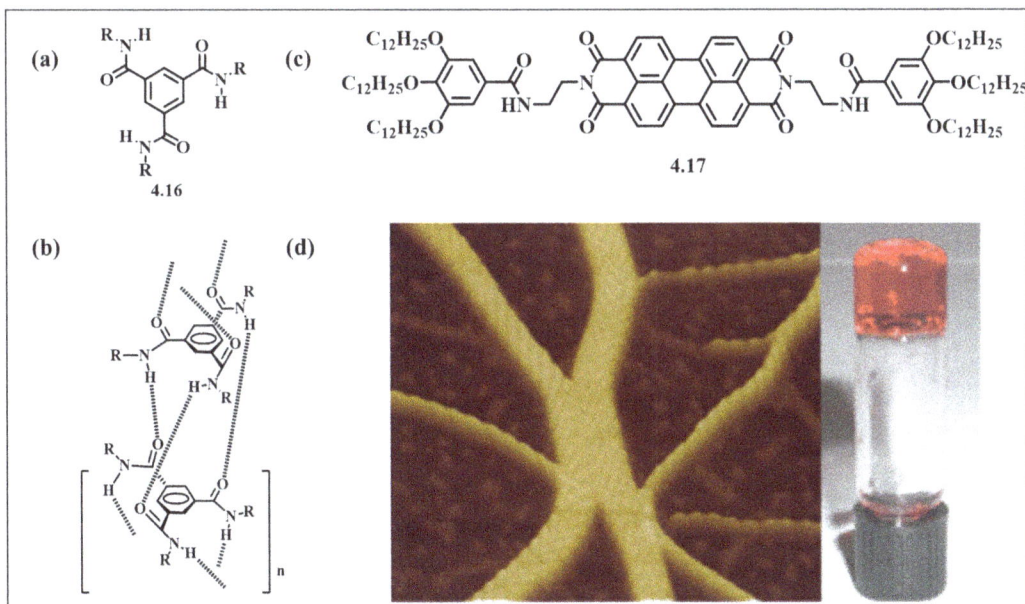

FIGURE 3.3 Molecules containing both π–π interaction and hydrogen-bonding motifs. (a) General representation of 1,3,5-benzene triamide (BTA) molecule 3.16, (b) the hydrogen-bonding pattern between BTA molecules that leads to helicity upon stacking, (c) Substituted PDI, 3.17, capable of forming long SPs and organogel, (d) AFM image for the diluted gel of 3.17 (left), organogel formed at 1.5 mM concentration (right). ((d) Reproduced with permission from Li et al. (2006) © 2006 The Royal Society of Chemistry.)

unit and hydrogen-bonding entity are introduced for suitable packing and self-assembly. As shown in studies (X. Q. Li et al. 2006; Ghosh et al. 2008), this molecule self-assembles into a one-dimensional helical aggregate that forms organogel with various organic solvents (Figure 3.3d).

3.3.3 Host–Guest Interactions

Host–guest interactions are the driving forces for the formation of host–guest complexes, which originally initiated the field of supramolecular chemistry by Lehn, Cram, and Pedersen (J.-M. Lehn 1988). As per nomenclature, two different molecules, namely host and guest are involved in the formation of a host–guest complex. The host comprises a cavity that can recognize the guest molecule provided that the size and shape of the guest fit with the cavity. Hence, they form a specific and strong supramolecular complex. It can be mentioned that host–guest interactions are very selective due to the multiple restrictions on the guest by the host in terms of size, shape, charge, and even polarity. The common host molecules utilized for SPs are cyclodextrins, cucurbit[8]urils, calixarenes, crown ethers, and pillar[n]arenes (Harada et al. 2009; Dong et al. 2014). The guest can be common organic molecules, which are accommodated in the cavity of host. The common hosts and their corresponding guests are listed in Table 3.1. For the construction of SPs, generally, two kinds of ditopic monomers featuring host (A) and guest (B) are designed. The first type is AB-type heteroditopic monomer and the second type is AA/BB-type homoditopic monomer. Various types of molecular designs containing these host and guest moieties can form SPs, which are summarized in Figure 3.4a (Yang et al. 2015). Both homo and heterotopic monomers experienced a real problem that is the competition between the formation of linear long polymers and short oligomers including intramolecular complexation, dimerization, and oligomeric cyclization. To troubleshoot the issue of the formation of short oligomers, higher concentration of the monomer should be used for polymerization that improves the yield of linear polymers, as such oligomers are preferably formed in a dilute solution.

TABLE 3.1

List of Common Host Molecules and Their Corresponding Guests to form Host–Guest Complexes and SPs

Common Host Molecules	Molecular Structure of Hosts	Corresponding Guests and Their Structures	
β - cyclodextrins		Viologen	Charged Alkanes
Cucurbit[8]uril	Adamantane　　Coumarin		
Calixaranes		Viologen	Charged Amines
Crown Ether	Methyl Viologen　　Charged Alkanes Charged Napthalenes　　Charged Anthracenes		
Pillaranes		Charged Imidazoles	DABCO

Crown ether-based linear SPs can be constructed based on the common principle of designing ditopic monomers containing a rigid linker. For instance, an AB-type heteroditopic monomer comprising of crown ether as host and dibenzylammonium salt as guest moieties was developed by the Stoddart group (Ashton et al. 1998; Cantrill et al. 2001). The host and guest moieties efficiently produced an intermolecular complex rather than an intramolecular one. However, as a downside, the dimer formation was favored over linear polymerization for this system.

An exciting example for the formation of linear SPs based on host–guest interaction was obtained by a rigid heteroditopic molecule 3.18 (Figure 3.4b), containing a crown-ether-macrocycle and rod-like paraquat moieties (Yamaguchi et al. 1998; Zheng et al. 2012). The DP increases with increasing monomer concentrations and at the monomer concentration of 2.0 M, a DP value of ~50 was found that corresponds to a molar mass of 51 kDa for a conventional polymer. Another set of homoditopic molecules 3.19 and 3.20 were designed for the construction of crown ether-based linear SPs (Figure 3.4c) (Huang and Gibson 2005; Huang et al. 2007; Zheng et al. 2012). These two monomers formed cyclic oligomers upon mixing in dilute solution and polymerization took place upon an increase in concentrations of monomers. The highest DP in the 60.3 mM solution was calculated to be 36.

3.3.4 AROMATIC DONOR–ACCEPTOR INTERACTIONS

Like the simple π–π interactions, aromatic donor–acceptor interactions alone are found to be too weak and less directional to initiate supramolecular polymerization, and thus directing groups are introduced on the periphery of an aromatic core to drive the formation of SPs (Das and Ghosh 2014). A π donor–acceptor host–guest complex can be formed by electron-deficient planar aromatic guests such as trinitrofluorenone, trinitrobenzene, and tetracyanobenzene, attached to a bisporphyrin tweezer as a host. Charge–transfer (CT) interaction between the host guest can then be exploited for initiating SP. A heteroditopic monomer, 3.21, was designed for this purpose, and it formed SPs by head–tail interaction (Figure 3.5a) (Haino et al. 2012). The 4,5,7-trinitrofluorenone-2-carboxylate (TNF) forms a head-to-tail complex with the bisporphyrin tweezer and thus leads to SPs. This complexation resulted in large upfield shifts in ^{1}H NMR for the TNF protons, confirming the TNF moiety was present within the porphyrins.

Another aromatic donor–acceptor system for supramolecular polymerization was constructed by Ghosh and co-workers using CT interaction between dialkoxynaphthalene (3.22, DAN) donor

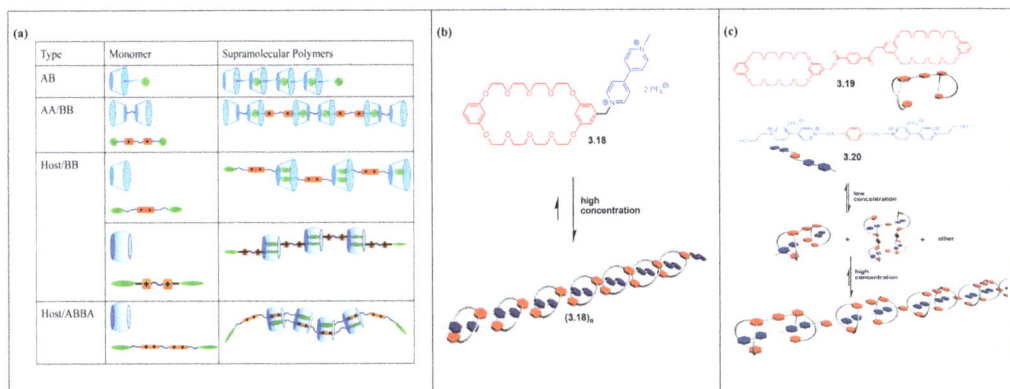

FIGURE 3.4 (a) Graphical representation of the homo (AA/BB) and heteroditopic (AB) monomers and their respective SPs (where A represents host moiety and B represents the guests). (b, c) Graphical illustration for linear crown ether-based SPs formation by an increase in the concentration of monomers; an AB-type (3.18) heteroditopic monomer (b) and AA/BB-type (3.19 & 3.20) homoditopic monomers (c). [(a) Reproduced with permission from Yang et al. (2015) © 2015 American Chemical Society. (b, c) Reproduced with permission from Zheng et al. (2012) © 2012 The Royal Society of Chemistry.]

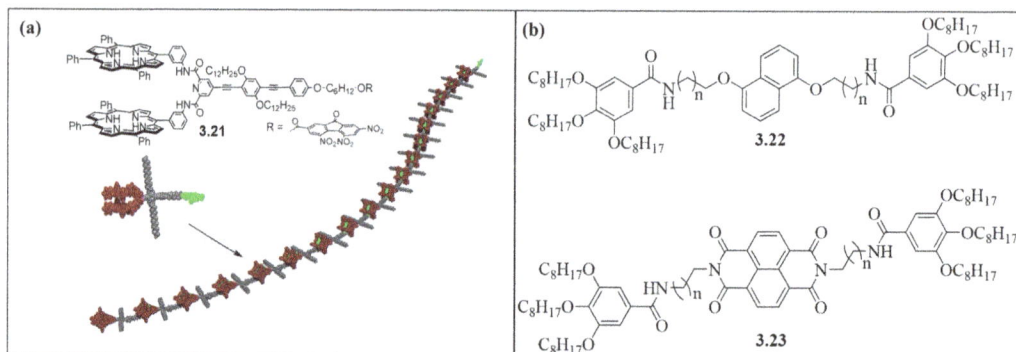

FIGURE 3.5 Supramolecular polymerization based on aromatic donor–acceptor interaction. (a) A complex (3.21) with bisporphyrin as π-donor and trinitrofluorenone carboxylate as π-acceptor that interacts in a head–tail manner. (b) Molecules exhibit charge–transfer interaction between a π-donor (dialkoxynaphthalene, 3.22) and π-acceptor (naphthalene diimide 3.23). ((a) Reproduced with permission from Haino et al. (2012) © 2012 Wiley-VCH Verlag GmbH & Co).)

and naphthalene diimide (3.23, NDI) acceptor (Figure 3.5b) (Das et al. 2011). The role of solvent is, however, crucial for the supramolecular assembly process. In moderately polar solvent tetrachloroethylene (TCE), a stable red gel is formed with a characteristic CT band, and the gel formation is believed to be contributed by both CT interaction and hydrogen bonding, which is less influential in the solvent. However, with increasing the polarity of the solvent, the CT interaction does not play a decisive role and H-bonding becomes the higher influential force for assembly. In the case of methylcyclohexane (MCH), a kinetically controlled red CT gel is formed, which then forms a thermodynamically stable gel within 5–6 h. Even this role of solvent can be further confirmed as 7% (v/v) of a protic solvent, MeOH results in a concomitant decrease of the CT band and the gel converts to the solution. The gel formed in TCE is thermally stable and shows a T_{gel} value of 72°C, and this stability is believed to be due to the intermolecular hydrogen bonding among the amide groups in the alternate donor–acceptor sequence.

3.4 MECHANISM

On the basis of mechanism of product formation, supramolecular polymerizations can be classified into two groups, isodesmic and cooperative polymerization. Cooperative mechanism can be subdivided into cooperative or anticooperative. These three mechanistic approaches can be understood by the monomer addition to the growing SP chain and are differentiated by the association constant. The association constant is further divided into nucleation constant (K_n) and elongation constant (K_e). The term degree of aggregation (α) is the numerical representation for the number of aggregated species at a time in the solution. An α value of 1 suggests complete aggregation while 0 represents no aggregation or a completely monomeric state (Hartlieb et al. 2020). Typically, a graph is plotted with α as a function of temperature or concentration, where the nature of the curve is helpful in differentiating the mechanistic processes (Smulders et al. 2010). The cooperativity factor (σ), defined as the ratio of nucleation constant to the elongation constant, $\sigma = K_n/K_e$ is also used to identify the involved mechanism in a supramolecular polymerization process (Von Krbek et al. 2017; Wehner and Würthner 2020).

The isodesmic mechanism is equivalent to the step-growth mechanism in covalent polymerization. Systems with the same Gibbs free energy for each step of polymerization process fall under the isodesmic mechanism. The association constant is found to be independent of the length of polymer. The entire process of polymerization is described by a single association constant (K_e) (Figure 3.6a), and hence, the value of σ is equaled to 1 considering $K_n = K_e$. The process of assembly is characterized

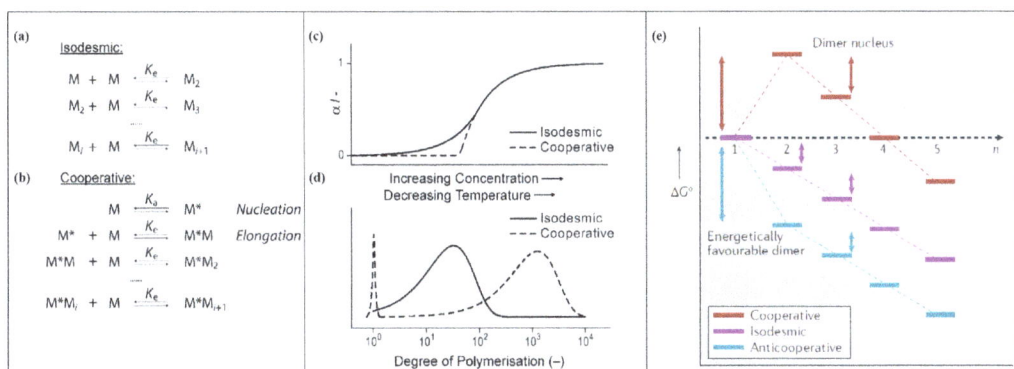

FIGURE 3.6   Representation for the mechanism of supramolecular polymerization. The steps involved in the supramolecular polymerization process: (a) isodesmic and (b) cooperative. (c) Graphical representation of the degree of association (α) versus concentration or temperature for the identification of isodesmic or cooperative mechanism. (d) Molecular weight distribution for an isodesmic and a cooperative self-assembly. (e) Graphical representation of the energy diagram involved in various supramolecular polymerization process: an isodesmic process (purple), cooperative process (red), and anti-cooperative process (blue). ((a–d) Reproduced with permission from (Smulders et al. 2010) © 2010 Wiley-VCH Verlag GmbH & Co). (e) Reproduced with permission from (Wehner and Würthner 2020) © 2020 Springer Nature.)

by the absence of both critical temperature and concentration and a plot between α and temperature or concentration gives a sigmoidal curve for the aggregated species, which is a characteristic of the isodesmic mechanism (Figure 3.6c) (De Greef et al. 2009; Smulders et al. 2010).

The chain growth mechanism in covalent polymerization is considered the equivalent of a cooperative mechanism in supramolecular polymerization. It is a nonlinear polymerization process that is often termed as nucleated polymerization and involves two important steps. The thermodynamically unfavorable nucleation step is followed by the spontaneous (energetically favored) elongation process. At higher temperature or at lower concentration, molecules remain in a monomeric state which starts to aggregate with decreasing temperature or with increasing concentration. The nuclei, composed of a few monomers, start to form when the system attains a critical concentration or temperature, and once nuclei are formed, the systems start to elongate rapidly to form larger assemblies or SP (Figure 3.6c). The cooperative mechanism is characterized by two different association constants: K_n and K_e (Figure 3.6b). The σ for this mechanism is less than 1 as $K_n < K_e$. Cooperative polymerization is characterized by critical temperature and concentration. Thus, plotting a graph between α as a function of temperature or concentration gives a curve similar to the isodesmic mechanism but with a sharp change near the nucleation step representing a critical temperature or concentration in the graph (De Greef et al. 2009; Smulders et al. 2010).

Anticooperative systems compared to cooperative mechanism can be differentiated by the equilibrium constant of nucleation and elongation steps. In the case of the anti-cooperative system, both nucleation and elongation steps are energetically favorable, where the nucleation is energetically more favorable than the elongation process (Figure 3.6e). Thus, a higher K_n is observed for the anti-cooperative mechanism. In this case, σ is found to be greater than 1 as $K_n > K_e$. The difference in σ is one of the characteristic differences between cooperative and anti-cooperative mechanisms (Wehner and Würthner 2020).

3.5   THERMODYNAMIC, KINETIC AND OUT OF EQUILIBRIUM SPs

Based on the energy landscape of the product formation, SPs are generally divided into three different types: thermodynamic, kinetic, and out of equilibrium (Figure 3.7) (Matern et al. 2019; Syamala and Würthner 2020). In supramolecular chemistry, the involved noncovalent interactions are relatively weak; hence, the SPs form and break down reversibly. As a result, many SPs are formed under thermodynamic control in which the chemistry is mainly focused on systems at equilibrium. Such an equilibrium system exists at the global minimum in the free energy landscape (Figure 3.7) and equilibrium supramolecular assemblies are still dynamic as that the supramolecular assemblies and the monomers remain in exchange with each other in the solution. In contrast, if strong noncovalent interactions or multiple noncovalent interactions are present in the assembly, the supramolecular polymerization process remains under the kinetic control and this may lead to kinetically trapped or metastable states (non-equilibrium states) (Figure 3.7), rather than the assembly corresponding to thermodynamic equilibrium. These assemblies reside at energetically higher local minima in the energy landscape but not in the global minima, and hence, they are placed under kinetic SP. In these non-equilibrium cases, the outcome of the assembly process strongly depends on the experimental protocols for the preparation of SP. There are many examples where the variation in the experimental procedures of preparation (such as solvent composition, temperature, pH, anionic strength, etc.) results in the formation of structurally very different aggregates from the same building blocks (Haedler et al. 2016). Within the kinetic regime, there is a possibility of a system that is out of equilibrium (also called far-from-equilibrium or dissipative self-assembly) where the system is not static but active in a cyclic manner and requires a continuous supply of energy to persist. In this case, a molecule lies at a lower energy level in the molecular state and then moves to the higher energy intermediate by the use of a fuel which forms an assembly (Figure 3.7). However, the assembly is transiently stable only as it degrades to the initial monomer and the entire process can be repeated multiple times under suitable conditions (Figure 3.7) (Sorrenti et al. 2017; Van Rossum et al. 2017; Matern et al. 2019).

In the case of thermodynamic assembly, the monomers are at slightly higher energy than the assembled state. The population density between assembled and monomeric states depends on the energy difference between the two states and the free energy available in the system. The assembly is in equilibrium as there is no net flow of matter and energy. The building blocks in the stacking can escape the assemblies, whereas the molecularly dissolved monomers can enter the assemblies, but this takes place at similar rates. However, the experimental protocols have no impact on the final equilibrium state (Smulders et al. 2010; Mattia and Otto 2015; Wehner and Würthner 2020). Still, the final equilibrium species may vary due to the alteration of conditions such as temperature, concentration, and solvent composition.

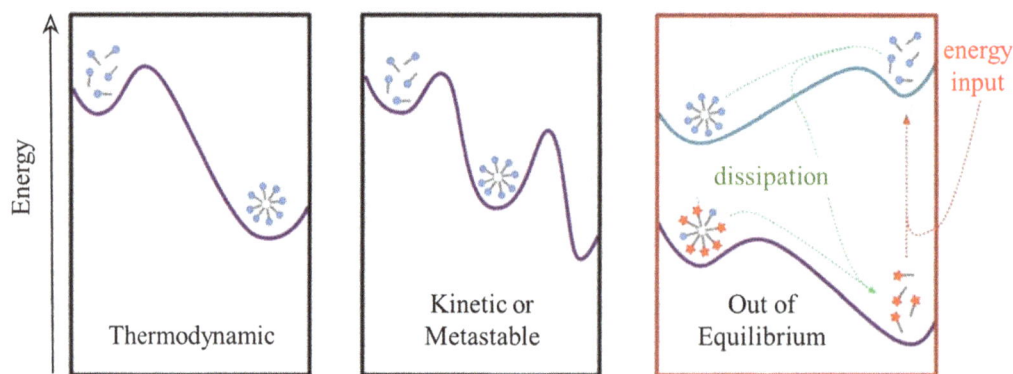

FIGURE 3.7 The energy diagram for various supramolecular polymerization processes. A thermodynamic system where the assembled molecule resides at the minima of energy (left), kinetic or metastable state where the molecule is trapped at other minima rather than the global minima of energy (middle) and out of equilibrium where assembly takes place only after some input of external energy (right). (Reproduced with permission from (Van Rossum et al. 2017) © 2017 The Royal Society of Chemistry.)

In the case of kinetic assembly, assemblies reside somewhat higher in the energy diagram. The kinetically assembled states are divided into two different categories: the metastable state and the kinetically trapped state (Sorrenti et al. 2017). When the energy required to pass the activation barrier to a thermodynamic state is high, the product is trapped in that state for a longer time, which is called the kinetically trapped kinetic product. A relatively lower energy barrier may lead to the escape of product sooner than expected and such a situation is termed as a metastable product. Products that are not thermodynamically stable are termed as kinetic non-equilibrium structure. The final outcome, in this case, is largely dependent on preparation methods, such as temperature, various kinds of solvent processing, external stimuli, and photochemical conversion. Kinetic, non-equilibrium structures are prone to go into the equilibrium state spontaneously over time or by the application of stimulus. The kinetic assembly is hence considered to be highly dependent on time. However, due to the faster formation of such assembly, characterization of these is relatively harder than thermodynamic assembly (Smulders et al. 2010; Wehner and Würthner 2020).

3.6 SP NETWORKS

SP networks are three-dimensional complex structures of crosslinked macromolecular chains linked by directional noncovalent bonds (Figure 3.8), and such SP networks are considered a fascinating class of soft materials. The major advantages of SP networks are stimuli-responsiveness, shape-memory, and self-healing abilities, and these emerging properties are originated from the versatile nature of synthetic polymer networks and flexible crosslinking (Seiffert and Sprakel 2012). A typical feature of SPs and their networks is that their dynamics is directed by two different timescales according to classical polymer physics (Yount et al. 2005). The first one is the timescale of

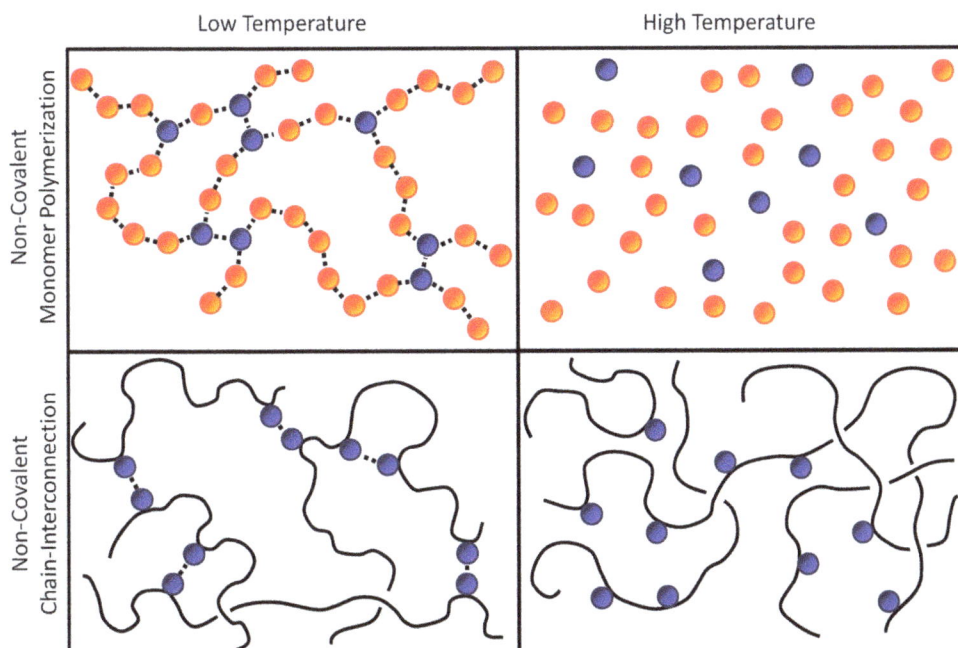

FIGURE 3.8 Schematic representation of two kinds of SP networks. Molecules form supramolecular polymeric chains by noncovalent interactions only (upper row) and another contains a network of covalently linked monomers that form the supramolecular polymeric chains by secondary interactions of side groups (lower row). In both cases, the systems are in a sol state at a higher temperature and forms gel at a low concentration. (Reproduced with permission from Seiffert and Sprakel (2012) © 2012 The Royal Society of Chemistry.)

formation and breakage of the SP network or interchain crosslinks. The second one is the timescale for the relaxation of polymer chains or chain segments. Similar to covalent counterparts, the most important feature of SP networks is their mechanical property, namely, viscoelastic nature (Seiffert and Sprakel 2012). The viscoelastic behavior is investigated by rheology and exhibits complexity because of the interplay of supramolecular chemistry and polymer physics.

3.7 SELF-HEALING SPs

One of the most attractive features of SP networks is their self-healing, the ability of the systems to spontaneously heal their damage, improving the lifetime of the materials (Campanella et al. 2018; Zhang et al. 2019; Sinawang et al. 2020). An ideal self-healing material should be able to repair itself quickly along with restoration of its mechanical properties (Zhang and Waymouth 2017). The key approaches employed to obtain self-healing supramolecular materials are physical, hierarchical, and chemical self-assembly. The physical approach mainly utilizes the incorporation of healing agents into the material that in turn facilitates the repairing process (Toohey et al. 2007). The hierarchical self-assembly approach exploits complex supramolecular structures and macromolecules at multiple levels (Elemans et al. 2003). This approach is inspired by nature, where complex molecules are formed by a combination of weak noncovalent interactions. Finally, the chemical approach utilizes the formation of multiple noncovalent bonds, the most reliable method for the preparation and healing process. Diels–Alder (DA) reactions, hydrogen bonds, π–π stacking, metal–ligand coordination, hydrophobic interactions, and host–guest interactions are among the forces that are exploited in a chemical method of self-healing material preparation.

The self-healing nature of the SP networks is originated from the reversibility and dynamic nature of the networks, which can repair damaged sites (Figure 3.9) (Campanella et al. 2018). "Stickers" as supramolecular functionalities generally exist in the supramolecular material (e.g. gel) to some extent and the "stickers" remain in equilibrium between open and closed states. The application of

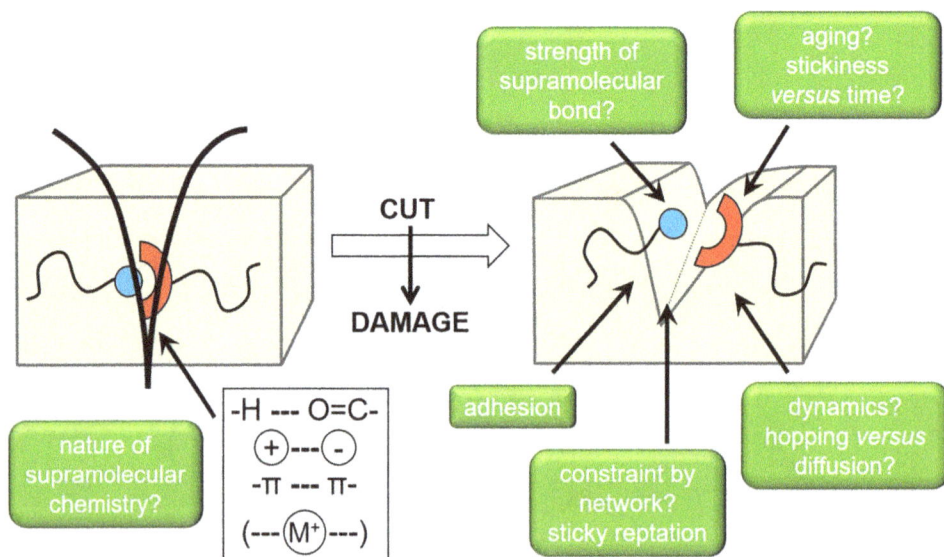

FIGURE 3.9 Self-healing supramolecular polymers. Representation of the loss of noncovalent interaction by mechanical stress and reformation of sticky groups, and list of the plausible effects like adhesion, aging, dynamicity in the formed interface of SP. (Reproduced with permission from(Campanella et al. (2018) ©2018 Wiley-VCH Verlag GmbH & Co.)

mechanical stress, for instance, physical rupture, can break the weak noncovalent bonds in the material, and this generates a number of sticky groups onto a freshly generated interface, which again can recombine to form initial "stickers" healing the material (Campanella et al. 2018). The early report of self-healing behavior of SP was made by the Leibler group (Cordier et al. 2008). In this report, aliphatic diacids and triacids were used to synthesize a number of oligomers featuring complementary hydrogen-bonding urea moieties. The self-assembly of these oligomers provides a glassy, plastic-like supramolecular material that displays elastomeric behavior upon heating. Upon breaking, this material was able to self-heal within a few minutes. Recovery of its initial mechanical properties on the repaired rubber was confirmed by rheology. It can be mentioned that the material shows limited healability for a short period, which is likely due to the formation of a new thermodynamically stable structure caused by the reorientation of the hydrogen-bonding groups over time. Another type of self-healing SP network is based on complementary π–π stacking interactions (Scott Lokey and Iverson 1995; Lokey et al. 1997; Greenland et al. 2008). This system is based on low-molecular-weight polyimides featuring multiple π-electron receptor sites that are capable of folding around π-electron-rich sites of a telechelic polysiloxane, thereby forming a complementary, π–π stacked SP network. The rapid healing of the networks was achieved at high temperatures as characterized by electron microscopy. The mechanism of healing likely happens due to partial dissociation of the complementary π–π interaction, followed by flow and re-association of the dissociated fragments.

3.8   CONCLUSIONS AND PERSPECTIVES

Conventional covalent polymers based on covalent bonds have excellent properties as materials. As a supplement to covalent polymers, SPs have been rapidly developing. The nature of weak noncovalent interactions confers reversibility and dynamic property to the SPs. SPs featuring such emerging properties can be utilized as stimulus-responsive, recyclable, and self-healing materials. Moreover, low viscosity melt for SP is ideal for its superior processing toward real application. In summary, we have discussed organic supramolecular polymers constructed from small organic molecules by various directional noncovalent interactions. SPs based on multiple-hydrogen-bonding, host–guest interactions, π–π stacking, and aromatic donor-acceptor along with their examples have been discussed. We have outlined preparation methods and characterization techniques of SPs along with the fundamentals of SP networks. We have highlighted the thermodynamic, kinetic, and out of equilibrium SPs and the accepted mechanisms of supramolecular polymerization such as isodesmic, cooperative, and anti-cooperative. Finally, the self-healing properties of SPs have been emphasized in this chapter. Unlike covalent counterparts, rational design and mechanistic understanding for SPs are not well understood yet. On the positive side, various functional moieties can be integrated through supramolecular copolymerization, leading to the engineering of functional materials and the development of such supramolecular copolymers which are needed to explore in the years to come.

REFERENCES

Adhikari, B., Aratsu, K., Davis, J., and Yagai, S. 2019. Photoresponsive circular supramolecular polymers: a topological trap and photoinduced ring-opening elongation. *Angew. Chem. Int. Ed.* 58(12):3764–3768.

Adhikari, B., Lin, X., Yamauchi, M., Ouchi, H., Aratsu, K., and Yagai, S. 2017. Hydrogen-bonded rosettes comprising π-conjugated systems as building blocks for functional one-dimensional assemblies. *Chem. Commun.* 53(70):9663–9683.

Adhikari, B., Yamada, Y., Yamauchi, M., Wakita, K., Lin, X., Aratsu, K., Ohba, T., Karatsu, T., Hollamby, M. J., and Shimizu, N., et al. 2017. Light-induced unfolding and refolding of supramolecular polymer nanofibres. *Nat. Commun.* 8(1):1–10.

Aida, T., Meijer, E. W., and Stupp, S. I. 2012. Functional supramolecular polymers. *Science.* 335(6070):813–817.

Armstrong, G., and Buggy, M. 2005 Hydrogen-bonded supramolecular polymers: a literature review. *J. Mater. Sci.* 40(3):547–559.

Ashton, P. R., Baxter, I., Cantrill, S. J., Fyfe, M. C. T., Glink, P. T., Stoddart, J. F., White, A. J. P., and Williams, D. J. 1998. Supramolecular daisy chains. *Angew. Chem. Int. Ed.* 37(9):1294–1297.

Babu, S. S., Praveen, V. K., and Ajayaghosh, A. 2014. Functional π-gelators and their applications. *Chem. Rev.* 114(4):1973–2129.

Berl, V., Schmutz, M., Krische, M. J., Khoury, R. G., and Lehn, J. M. 2002. Supramolecular polymers generated from heterocomplementary monomers linked through multiple hydrogen-bonding arrays - formation, characterization, and properties. *Chem. Eur. J.* 8(5):1227–1244.

Brunsveld, L., Folmer, B. J. B., Meijer, E. W., and Sijbesma, R. P. 2001. Supramolecular polymers. *Chem. Rev.* 101(12):4071–4097.

Campanella, A., Döhler, D., and Binder, W. H. 2018. Self-Healing in supramolecular polymers. *Macromolecules* 39(17):e1700739.

Cantrill, S. J., Youn, G. J., Stoddart, J. F., and Williams, D. J. 2001. Supramolecular daisy chains. *J. Org. Chem.* 66(21):6857–6872.

Chang, S. K., and Hamilton, A. D. 1988. Molecular recognition of biologically interesting Substrates: synthesis of an artificial receptor for barbiturates employing six hydrogen bonds. *J. Am. Chem. Soc.* 110(4):1318–1319.

Chen, S., Slattum, P., Wang, C., and Zang, L. 2015. Self-assembly of perylene imide molecules into 1d nanostructures: methods, morphologies, and applications. *Chem. Rev.* 115(21):11967–11998.

Chen, Z., Lohr, A., Saha-Möller, C. R., and Würthner, F. 2009. Self-assembled π-stacks of functional dyes in solution: structural and thermodynamic features. *Chem. Soc. Rev.* 38(2):564–584.

Chiper, M., Meier, M. A. R., Kranenburg, J. M., and Schubert, U. S. 2007. New insights into nickel(II), iron(II), and cobalt(II) bis-complex-based metallo-supramolecular polymers. *Macromol. Chem. Phys.* 208(7):679–689.

Corbin, P. S., and Zimmerman, S. C. 1998. Self-association without regard to prototropy. a heterocycle that forms extremely stable quadruply hydrogen-bonded dimers [16]. *J. Am. Chem. Soc.* 120(37):9710–9711.

Cordier, P., Tournilhac, F., Soulié-Ziakovic, C., and Leibler, L. 2008. Self-healing and thermoreversible rubber from supramolecular assembly. *Nature* 451(7181):977–980.

Das, A., and Ghosh, S. 2014. Supramolecular assemblies by charge-transfer interactions between donor and acceptor chromophores. *Angew. Chem. Int. Ed.* 53(8):2038–2054.

Das, A., Molla, M. R., Banerjee, A., Paul, A., and Ghosh, S. 2011. Hydrogen-bonding directed assembly and gelation of donor-acceptor chromophores: supramolecular reorganization from a charge-transfer state to a self-sorted state. *Chem. Eur. J.* 17(22):6061–6066.

De Greef, T. F. A., and Meijer, E. W. 2008. Materials science: supramolecular polymers. *Nature* 453:171–173.

De Greef, T. F. A., Smulders, M. M. J., Wolffs, M., Schenning, A. P. H. J., Sijbesma, R. P., and Meijer, E. W. 2009. Supramolecular polymerization. *Chem. Rev.* 109(11):5687–5754.

Dobrawa, R., Würthner, F. 2005. Metallosupramolecular approach toward functional coordination polymers. *J. Polym. Sci. Part A Polym. Chem.* 43(21):4981–4995.

Dong, S., Zheng, B., Wang, F., and Huang, F. 2014. Supramolecular polymers constructed from macrocycle-based host-guest molecular recognition motifs. *Acc. Chem. Res.* 47(7):1982–1994.

Elemans, J. A. A. W., Rowan, A. E., and Nolte, R. J. M. 2003. Mastering molecular matter. supramolecular architectures by hierarchical self-assembly. *J. Mater.Chem.* 13(11):2661–2670.

Folmer, B. J. B., Sijbesma, R. P., Versteegen, R. M., Van Der Rijt, J. A. J., and Meijer, E. W. 2000. Supramolecular polymer materials: chain extension of telechelic polymers using a reactive hydrogen-bonding synthon. *Adv. Mater.* 12(12):874–878.

Fouquey, C., Lehn, J.-M., and Levelut, A.-M. 1990. Molecular recognition directed self-assembly of supramolecular liquid crystalline polymers from complementary chiral components. *Adv. Mater.* 2(5):254–257.

Ghislaine, V., and Meijer, E. W. 2019. The construction of supramolecular systems. *Science* 363(6434):1396–1398.

Ghosh, S., Li, X. Q., Stepanenko, V., and Würthner, F. 2008. Control of H- and J-type π stacking by peripheral alkyl chains and self-sorting phenomena in perylene bisimide homo- and heteroaggregates. *Chem. Eur. J.* 14(36):11343–11357.

Görl, D., Zhang, X., and Würthner, F. 2012. Molecular assemblies of perylene bisimide dyes in water. *Angew. Chem. Int. Ed.* 51(26):6328–6348.

Greenland, B. W., Burattini, S., Hayes, W., and Colquhoun, H. M. 2008. Design, synthesis and computational modelling of aromatic tweezer-molecules as models for chain-folding polymer blends. *Tetrahedron* 64(36):8346–8354.

Haedler, A. T., Meskers, S. C. J., Zha, R. H., Kivala, M., Schmidt, H. W., and Meijer, E. W. 2016. Pathway complexity in the enantioselective self-assembly of functional carbonyl-bridged triarylamine trisamides. *J. Am. Chem. Soc.* 138(33):10539–10545.

Haino, T., Watanabe, A., Hirao, T., and Ikeda, T. 2012. Supramolecular polymerization triggered by molecular recognition between bisporphyrin and trinitrofluorenone. *Angew. Chem. Int. Ed.* 51(6):1473–1476.

Hanabusa, K., Koto, C., Kimura, M., Shirai, H., and Kakehi, A. 1997 Remarkable viscoelasticity of organic solvents containing trialkyl-1,3,5-benzenetricarboxamides and their intermolecular hydrogen bonding. *Chem. Lett.* 5:429–430.

Harada, A., Takashima, Y., and Yamaguchi, H. 2009. Cyclodextrin-based supramolecular polymers. *Chem. Soc. Rev.* 38(4):875–882.

Hartlieb, M., Mansfield, E. D. H., and Perrier, S. 2020. A guide to supramolecular polymerizations. *Polymer* 11(6): 1083–1110.

Hirschberg, J. H. K. K., Beijer, F. H., Van Aert, H. A., Magusin, P. C. M. M., Sijbesma, R. P., and Meijer, E. W. 1999. Supramolecular polymers from linear telechelic siloxanes with quadruple-hydrogen-bonded units. *Macromolecules* 32(8):2696–2705.

Hoeben, F. J. M., Jonkheijm, P., Meijer, E. W., and Schenning, A. P. H. J. 2005. About supramolecular assemblies of π-conjugated systems. *Chem. Rev.* 105(4):1491–1546.

Hollamby, M. J., Aratsu, K., Pauw, B. R., Rogers, S. E., Smith, A. J., Yamauchi, M., Lin, X., and Yagai, S. 2016. Simultaneous SAXS and SANS analysis for the detection of toroidal supramolecular polymers composed of noncovalent supermacrocycles in solution. *Angew. Chem. Int. Ed.* 55(34):9890–9893.

Huang, F., and Gibson, H. W. 2005. A supramolecular poly[3]pseudorotaxane by self-assembly of a homoditopic cylindrical bis(crown ether) host and a bisparaquat derivative. *Chem. Commun.* 13:1696–1698.

Huang, F., Nagvekar, D. S., Zhou, X., and Gibson, H. W. 2007. Formation of a linear supramolecular polymer by self-assembly of two homoditopic monomers based on the bis(m-phenylene)-32-crown-10/paraquat recognition motif. *Macromolecules* 40(10):3561–3567.

Hunter, C. A., and Sanders, J. K. M. 1990. The nature of π-π interactions. *J. Am. Chem. Soc.* 112(14):5525–5534.

Jorgensen, W. L., and Pranata, J. 1990. Importance of secondary interactions in triply hydrogen bonded complexes: guanine-cytosine vs uracil-2,6-diaminopyndine. *J. Am. Chem. Soc.* 112(5):2008–2010.

Kolomiets, E., Buhler, E., Candau, S. J., and Lehn, J. M. 2006. Structure and properties of supramolecular polymers generated from heterocomplementary monomers linked through sextuple hydrogen-bonding arrays. *Macromolecules* 39(3):1173–1181.

Krieg, E., Bastings, M. M. C., Besenius, P., and Rybtchinski, B. 2016. Supramolecular polymers in aqueous media. *Chem. Rev.* 116(4):2414–2477.

Krieg, E., Niazov-Elkan, A., Cohen, E., Tsarfati, Y., and Rybtchinski, B. 2019. Noncovalent aqua materials based on perylene diimides. *Acc. Chem. Res.* 52(9):2634–2646.

Langhals, H. 2005. Control of the interactions in multichromophores: novel concepts. perylene bis-imides as components for larger functional units. *Helv. Chim. Acta.* 88(6):1309–1343.

Lehn. J.-M. 1988. Supramolecular chemistry-scope and perspectives molecules, supermolecules, and molecular devices (nobel lecture). *Angew. Chem. Int. Ed.* 27(1):89–112.

Lewis, C. L., and Dell, E. M. 2016. A review of shape memory polymers bearing reversible binding groups. *J. Polym. Sci B Polym. Phys.* 54(14):1340–1364.

Li, M., Yamato, K., Ferguson, J. S., and Gong, B. 2006. Sequence-specific association in aqueous media by integrating hydrogen bonding and dynamic covalent interactions. *J. Am. Chem. Soc.* 128(39): 12628–12629.

Li, S. L., Xiao, T., Lin, C., and Wang, L. 2012. Advanced supramolecular polymers constructed by orthogonal self-assembly. *Chem. Soc. Rev.* 41(18):5950–5968.

Li, X. Q., Stepanenko, V., Chen, Z., Prins, P., Siebbeles, L. D. A., and Würthner, F. 2006. Functional organo-gels from highly efficient organogelator based on perylene bisimide semiconductor. *Chem. Commun.* 37:3871–3873.

Lightfoot, M. P., Mair, F. S., Pritchard, R. G., and Warren, J. E. 1999. New supramolecular packing motifs: π-stacked rods encased in triply-helical hydrogen bonded amide strands. *Chem. Commun.* 19:1945–1946.

Lokey, R. S., Kwok, Y., Guelev, V., Pursell, C. J., Hurley, L. H., and Iverson, B. L. 1997. A new class of poly-intercalating molecules. *J. Am. Chem. Soc.* 119(31):7202–7210.

Lutz, J. F., Lehn, J. M., Meijer, E. W., and Matyjaszewski, K. 2016. From precision polymers to complex materials and systems. *Nat. Rev. Mater.* 1(5):1–14.

Ma, X., and Tian, H. 2014. Stimuli-responsive supramolecular polymers in aqueous solution. *Acc. Chem. Res.* 47(7):1971–1981.

Martinez, C. R., and Iverson, B. L. 2012. Rethinking the term "Pi-stacking". *Chem. Sci.* 3(7):2191–2201.

Matern, J., Dorca, Y., Sánchez, L., and Fernández, G. 2019. Revising complex supramolecular polymerization under kinetic and thermodynamic control. *Angew. Chem. Int. Ed.* 58(47):16730–16740.

Mattia, E., and Otto, S. 2015. Supramolecular systems chemistry. *Nat. Nanotechnol.* 10(2):111–119.

Meier, M. A. R., Wouters, D., Ott, C., Guillet, P., Fustin, C. A., Gohy, J. F., and Schubert, U. S. 2006. Supramolecular ABA triblock copolymers via a polycondensation approach: synthesis, characterization, and micelle formation. *Macromolecules* 39(4):1569–1576.

Murray, T. J., and Zimmerman, S. C. 1992. New triply hydrogen bonded complexes with highly variable stabilities. *J. Am. Chem. Soc.* 114(10):4010–4011.

Noro, A., Matsushita, Y., and Lodge, T. P. 2008. Thermoreversible supramacromolecular ion gels via hydrogen bonding. *Macromolecules* 41(15):5839–5844.

Paulusse, J. M. J., Van Beek, D. J. M., and Sijbesma, P. P. 2007. Reversible switching of the sol-gel transition with ultrasound in rhodium(I) and iridium(I) coordination networks. *J. Am. Chem. Soc.* 129(8):2392–2397.

Petka, W. A., Harden, J. L., McGrath, K. P., Wirtz, D., and Tirrell, D. A. 1998. Reversible hydrogels from self-assembling artificial proteins. *Science* 281(5375):389–392.

Savyasachi, A. J., Kotova, O., Shanmugaraju, S., Bradberry, S. J., Ó'Máille, G. M., and Gunnlaugsson, T. 2017. Supramolecular chemistry: a toolkit for soft functional materials and organic particles. *Chem* 3(5): 764–811.

Schutte, W. J., Sluyters-Rehbach, M., and Sluyters, J. H. 1993. Aggregation of an octasubstituted phthalocyanine in dodecane solution. *J. Phys. Chem.* 97(22):6069–6073.

Scott Lokey, R., and Iverson, B. L. 1995. Synthetic molecules that fold into a pleated secondary structure in solution. *Nature* 375(6529):303–305.

Segarra-Maset, M. D., Nebot, V. J., Miravet, J. F., and Escuder, B. 2013. Control of molecular gelation by chemical stimuli. *Chem. Soc. Rev.* 42(17):7086–7098.

Seiffert, S., Sprakel, J. 2012. Physical chemistry of supramolecular polymer networks. *Chem. Soc. Rev.* 41(2):909–930.

Shankar, R., Jha, N., and Vasudevan, P. 1993. Synthesis of soluble phthalocyanines and study of their aggregation behavior in solution. *Indian J. Chem. A* 32(12):1029–1033.

Sijbesma, R. P., Beijer, F. H., Brunsveld, L., Folmer, B. J. B., Hirschberg, J. H. K. K., Lange, R. F. M., Lowe, J. K. L., and Meijer, E. W. 1997. Reversible polymers formed from self-complementary monomers using quadruple hydrogen bonding. *Science* 278(5343):1601–1604.

Sijbesma, R. P., and Meijer, E. W. 2003. Quadruple hydrogen bonded systems. *Chem. Comm.* 1:5–16.

Sinawang, G., Osaki, M., Takashima, Y., Yamaguchi, H., and Harada, A. 2020. Supramolecular self-healing materials from non-covalent cross-linking host-guest interactions. *Chem. Commun.* 56(32): 4381–4395.

Smulders, M. M. J., Nieuwenhuizen, M. M. L., De Greef, T. F. A., Van Der Schoot, P., Schenning, A. P. H. J., and Meijer, E. W. 2010. How to distinguish isodesmic from cooperative supramolecular polymerisation. *Chem. Eur. J.* 16(1):362–367.

Sorrenti, A., Leira-Iglesias, J., Markvoort, A. J., De Greef, T. F. A., and Hermans, T. M. 2017. Non-equilibrium supramolecular polymerization. *Chem. Soc. Rev.* 46(18):5476–5490.

Syamala, P. P. N., and Würthner, F. 2020. Modulation of the self-assembly of π-amphiphiles in water from enthalpy- to entropy-driven by enwrapping substituents. *Chem. Eur. J.* 26(38):8426–8434.

Ten Cate, A. T., and Sijbesma, R. P. 2002. Coils, rods and rings in hydrogen-bonded supramolecular polymers. *Macromol. Rapid Commun.* 23(18):1094–1112.

Thompson, C. B., and Korley, L. S. T. J. 2020. 100th anniversary of macromolecular science viewpoint: engineering supramolecular materials for responsive applications - design and functionality. *ACS Macro Lett.* 9(9):1198–1216.

Toohey, K. S., Sottos, N. R., Lewis, J. A., Moore, J. S., and White, S. R. 2007. Self-healing materials with microvascular networks. *Nat. Mater.* 6(8):581–585.

Van Herrikhuyzen, J., Syamakumari, A., Schenning, A. P. H. J., and Meijer, E. W. 2004. Synthesis of N-type perylene bisimide derivatives and their orthogonal self-assembly with p-type oligo(p-phenylene vinylene)S. *J. Am. Chem. Soc.* 126(32):10021–10027.

Van Nostrum, C. F., and Nolte, R. J. M. 1996. Functional supramolecular materials: self-assembly of phthalocyanines and porphyrazines. *Chem. Commun.* 21:2385–2392.

Van Rossum, S. A. P., Tena-Solsona, M., Van Esch, J. H., Eelkema, R., and Boekhoven, J. 2017. Dissipative out-of-equilibrium assembly of man-made supramolecular materials. *Chem. Soc. Rev.* 46(18):5519–5535.

Von Krbek, L. K. S., Schalley, C. A., and Thordarson, P. 2017. Assessing cooperativity in supramolecular systems. *Chem. Soc. Rev.* 46(9):2622–2637.

Wehner, M., and Würthner, F. 2020. Supramolecular polymerization through kinetic pathway control and living chain growth. *Nat. Rev. Chem.* 4(1): 38–53.

Wilson, A. J. 2007. Non-covalent polymer assembly using arrays of hydrogen-bonds. *Soft Matter.* 3(4):409–425.

Winter, A., and Schubert, U. S. 2016. Synthesis and characterization of metallo-supramolecular polymers. *Chem. Soc. Rev.* 45(19):5311–5357.

Würthner, F. 2004. Perylene bisimide dyes as versatile building blocks for functional supramolecular architectures. *Chem. Commun.* 4(14):1564–1579.

Würthner, F., Saha-Möller, C. R., Fimmel, B., Ogi, S., Leowanawat, P., and Schmidt, D. 2016. Perylene bisimide dye assemblies as archetype functional supramolecular materials. *Chem. Rev.* 116(3):962–1052.

Würthner, F., Thalacker, C., Diele, S., and Tschierske, C. 2001. Fluorescent J-type aggregates and thermotropic columnar mesophases of perylene bisimide dyes. *Chem. Eur. J.* 7(10):2245–2253.

Yagai, S., Kitamoto, Y., Datta, S., and Adhikari, B. 2019. Supramolecular polymers capable of controlling their topology. *Acc. Chem. Res.* 52(5):1325–1335.

Yamaguchi, N., Nagvekar, D. S., and Gibson, H. W. 1998. Self-organization of a heteroditopic molecule to linear polymolecular arrays in solution. *Angew. Chem. Int. Ed.* 37(17):2361–2364.

Yang, L., Tan, X., Wang, Z., and Zhang, X. 2015. Supramolecular polymers: historical development, preparation, characterization, and functions. *Chem. Rev.* 115(15):7196–7239.

Yasuda, Y., Iishi, E., Inada, H., and Shirota, Y. 1996. Novel low-molecular-weight organic gels : N, N′, N″-tristearyltrimesamide/organic solvent system. *Chem. Lett.* 7:575–576.

Yount, W. C., Loveless, D. M., and Craig, S. L. 2005. Strong means slow: dynamic contributions to the bulk mechanical properties of supramolecular networks. *Angew. Chem. Int. Ed.* 44(18):2746–2748.

Zeng, H., Miller, R. S., Flowers, R. A., and Gong, B. 2000. A highly stable, six-hydrogen-bonded molecular duplex. *J. Am. Chem. Soc.* 122(11):2635–2644.

Zhang, M., Yan, X., Huang, F., Niu, Z., and Gibson, H. W. 2014. Stimuli-responsive host-guest systems based on the recognition of cryptands by organic guests. *Acc. Chem. Res.* 47(7):1995–2005.

Zhang, Q., Tang, D., Zhang, J., Ni, R., Xu, L., He, T., Lin, X., Li, X., Qiu, H., and Yin, S., et al. 2019. Self-healing heterometallic supramolecular polymers constructed by hierarchical assembly of triply orthogonal interactions with tunable photophysical properties. *J. Am. Chem. Soc.* 141(44):17909–17917.

Zhang, X., Chen, Z., and Würthner, F. 2007. Morphology control of fluorescent nanoaggregates by co-self-assembly of wedge- and dumbbell-shaped amphophilic perylene bisimides. *J. Am. Chem. Soc.* 129(16):4886–4887.

Zhang, X., Gçrl, D., Stepanenko, V., and Wurthner, F. 2014. Hierarchical growth of fluorescent dye aggregates in water by fusion of segmented nanostructures. *Angew. Chem. Int. Ed.* 53(5):1270–1274.

Zhang, X., and Waymouth, R. M. 2017. 1,2-dithiolane-derived dynamic, covalent materials: cooperative self-assembly and reversible cross-linking. *J. Am. Chem. Soc.* 139(10):3822–3833.

Zheng, B., Wang, F., Dong, S., and Huang, F. 2012. Supramolecular polymers constructed by crown ether-based molecular recognition. *Chem. Soc. Rev.* 41(5):1621–1636.

4 Supramolecular Self-Healing Gels

Tooru Ooya
Kobe University

Ik Sung Cho
University of Illinois at Chicago

CONTENTS

4.1   INTRODUCTION

A gel is defined as "a polymer having a three-dimensional network structure insoluble in any solvent and its swelling body". So far, researchers have found that both organic and inorganic matters were capable of forming jellies. In the review summarized by Almdal et al., they said

> "Graham reported in 1864 on the unusual diffusion properties of jellies and studied the replacement of water with other liquids in jellies of silicic acid. He introduced the terms *hydrosol* and *hydrogel* for the liquid and gelatinous hydrates of silicic acid, respectively"

(Almdal et al. 1993; Graham 1864)

Based on this definition, hydrogels are classified as gels swollen in aqueous solvents. Hydrogels are hydrophilic polymer networks, and the amount of water is at least 20% in weight. The most significant nature of hydrogels is that they swell in water and shrink in the absence of water. To keep the network structure, the polymeric chain of hydrogels is physically or chemically crosslinked. The physically crosslinked hydrogels are the networks that are held together by the growth of physically connected aggregates (Figure 4.1). To prepare physically crosslinked hydrogels, various preparative methods have been explored by using noncovalent bonds, such as hydrophobic interactions, ionic interactions, and hydrogen bonding (Guenet 1992). The physical gel-related interactions are strong enough to form a semi-permanent junction in the polymeric network that maintains a large number of water molecules inside. The common drawback of physical crosslinking is that the gels formed are unstable and may quickly collapse at an unexpected rate. In particular, hydrogen bonding is one

DOI: 10.1201/9781003169130-5

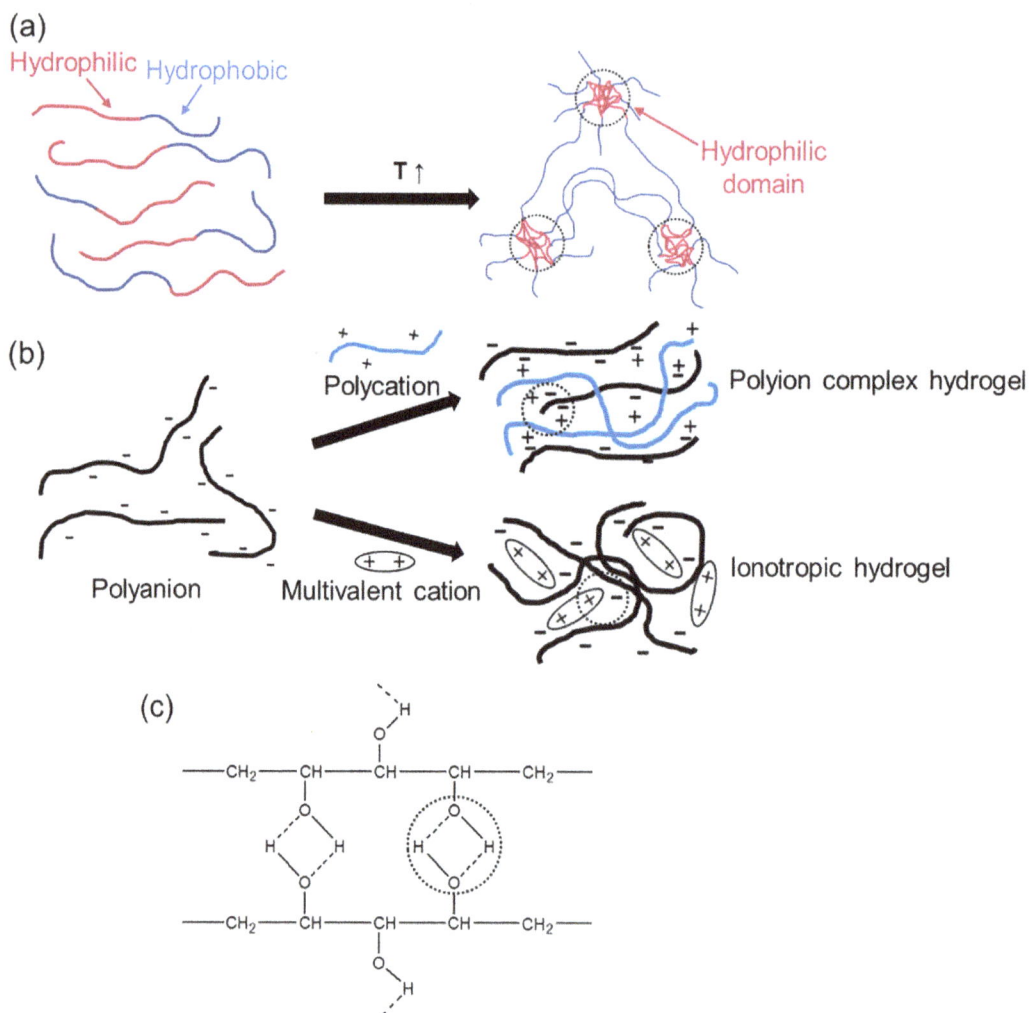

FIGURE 4.1 Schematic images of methods for the formation of physical hydrogels through (a) Hydrophobic interaction, (b) ionic interaction, (c) hydrogen bond.

of the important noncovalent bonds to make the network structure because we can think about the development of biological and/or biomimetic supramolecular architectures. Physically crosslinked hydrogels are prepared through hydrogen bonding by mixing two or more natural polymers, and the obtained hydrogels exhibit rheological change (Figure 4.1c). These blends give a gels-like texture compared with the individual polymers due to the extensive hydrogen bonding (Bajpai and Shrivastava 2005; Liu et al. 2005). Two kinds of natural polymers are mixed to form hydrogen bonding. For example, it is reported that gelatin-agar (Liu et al. 2005), hyaluronic acid-methylcellulose (Gupta et al. 2006), and starch-carboxymethyl cellulose (Ma et al. 2008) exhibit good hydrogel formation bearing excellent biocompatibility.

While physically crosslinked hydrogels have the general advantages of not requiring chemical modification or forming a gel without the addition of crosslinking agents, they also have disadvantages: It is difficult to control variables such as gelation time, network pore size, chemical functionalization, and degradation time (Lee and Tae 2007). Chemically crosslinked gels contain covalent bonds that are present between different polymer chains. Mechanical strength can be easily modulated according to the nature of the chemical bonds, and if the hydrogels have degradability, the degradation time is generally longer than physically crosslinked hydrogels (Vermonden et al. 2008).

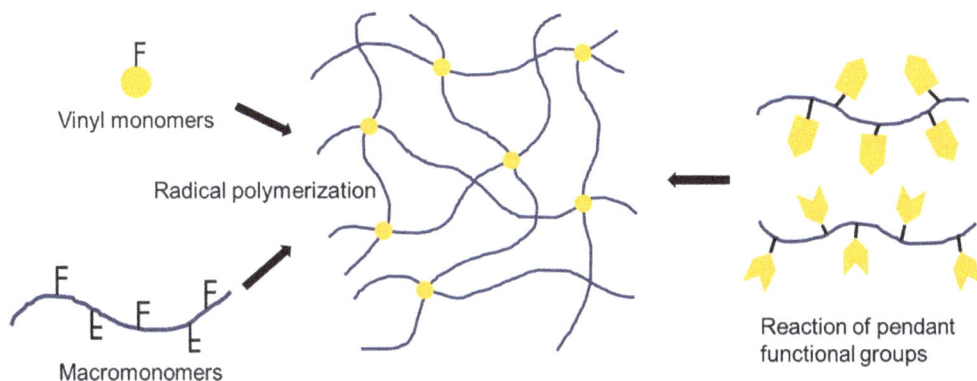

FIGURE 4.2 Schematic images of methods for the formation of chemical hydrogel.

Chemically crosslinked gels can be prepared by radical polymerization of low-molecular-weight monomers containing vinyl groups in the presence of a crosslinking agent (Figure 4.2). Redox polymerization is one of the most widely used methods for the synthesis of N-isopropylacrylamide (NIPAAm)-based hydrogels, using N, N, N, N-tertramethylethylenediamine (TEMED) as catalyst and ammonium persulfate (APS) as an initiator (Desai et al. 2012).

Although physically and chemically crosslinked hydrogels have been studied either in academic or in industrial research fields, those hydrogels are sometimes brittle and lack the ability to self-heal when the network is broken. Due to this disadvantage, the physically and chemically cross-linked hydrogels are greatly limited in their application in various biomedical fields. Alternatively, a novel class of crosslinked hydrogels, so-called *supramolecular hydrogels*, have been focused on in terms of unique functions, including self-healing, bioactivity, biodegradability, biostability, and biocompatibility (Dong et al. 2015). The supramolecular hydrogels are also applicable for drug delivery system (reza Saboktakin and Tabatabaei 2015) and tissue engineering (Saunders and Ma 2019). The main characteristics of supramolecular hydrogels are based on utilizing supramo-lecular binding motifs that rely on hydrogen bonding, electrostatic interactions, π–π interactions, host–guest interactions, hydrophobic interactions, or metal coordination (Yui 2002; Dong et al. 2015). Those supramolecular binding motifs can act as dynamic crosslinks between hydrophilic polymeric backbones to form hydrogels (Figure 4.3). Such dynamic nature is strongly dependent on the kinetics of the crosslinks, and the crosslinked density is strongly related to the equilibrium constants (Appel et al. 2012). In addition to the supramolecular binding motifs, recently, dynamic covalent chemistry focusing on Schiff's Base (imine) bonds, reversible hydrazine bonds, disulfide bonds, and reversible Diels–Alder (DA) reactions have been applied as supramolecular binding motifs to achieve tunable cleavage and self-healing properties of hydrogels for cellular engineering (Wang and Heilshorn 2015; Liu and Hsu 2018).

In this chapter, supramolecular self-healing hydrogels are summarized in terms of the mecha-nism of self-healing and hydrogel characteristics, including mechanical properties and biomedical applications.

4.2 SELF-HEALING SUPRAMOLECULAR HYDROGELS

4.2.1 TYPES OF SELF-HEALING SUPRAMOLECULAR HYDROGELS

Self-healing is defined as healing damages, restoring itself to normality by adding healing agents or by intrinsically. In the case of self-healing hydrogels, intrinsic healing is generally applied because the supramolecular binding motifs contribute to autonomous healing, which is defined as the prop-erty that enables a material to automatically heal and restore it to its original property (Figure 4.4). Autonomous self-healing usually occurs within a pre-prepared material network without an external

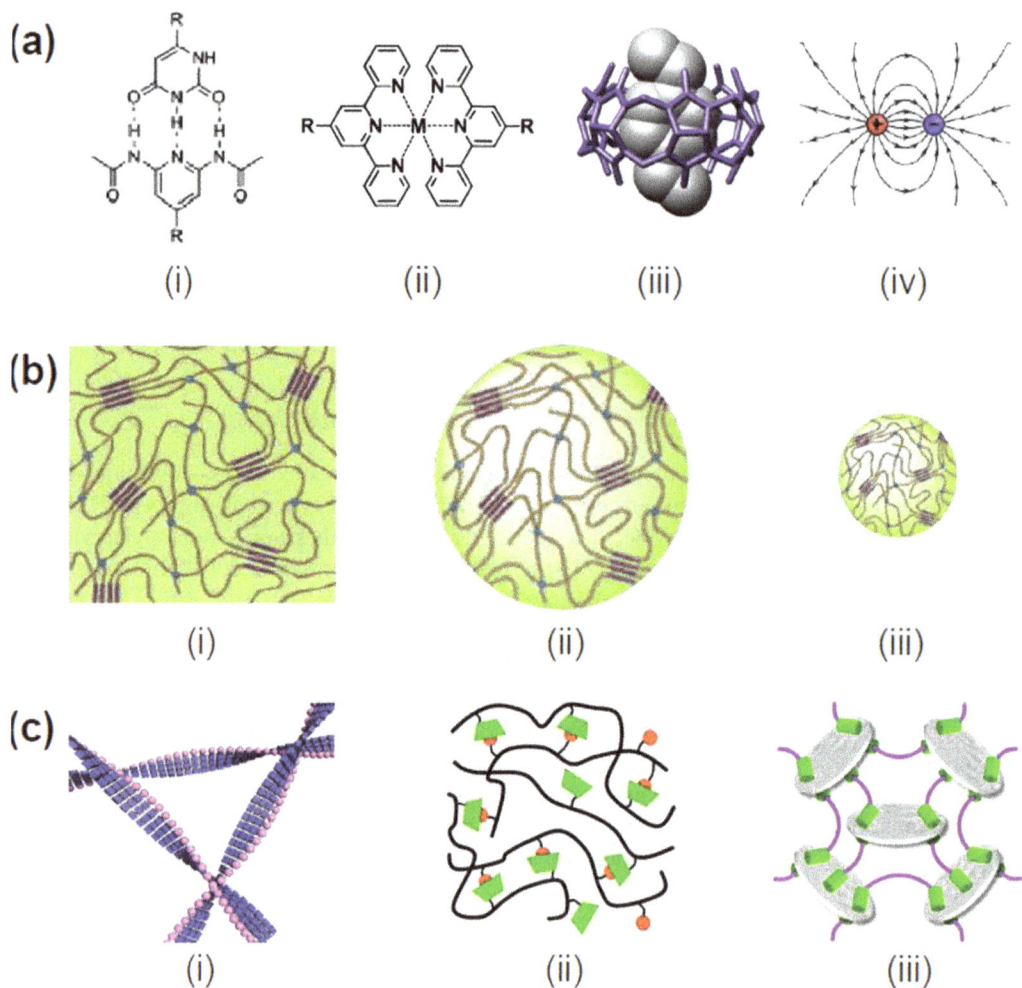

FIGURE 4.3 (a) Types of non-covalent interactions for the preparation of supramolecular hydrogels (i) hydrogen bonding, (ii) metal-ligand coordination, (iii) host–guest recognition and (iv) electrostatic interaction. (b) Classes of supramolecular hydrogels in view of the size: (i) macrohydrogel, (ii) microhydrogel, and (iii) nanohydrogel. (c) Classes of supramolecular hydrogels according to the types of the building blocks: (i) molecular hydrogel, (ii) supramolecular polymeric hydrogel, and (iii) supramolecular hybrid hydrogel. (Reproduced with permission from Dong (2015) © 2015 Royal Society of Chemistry.)

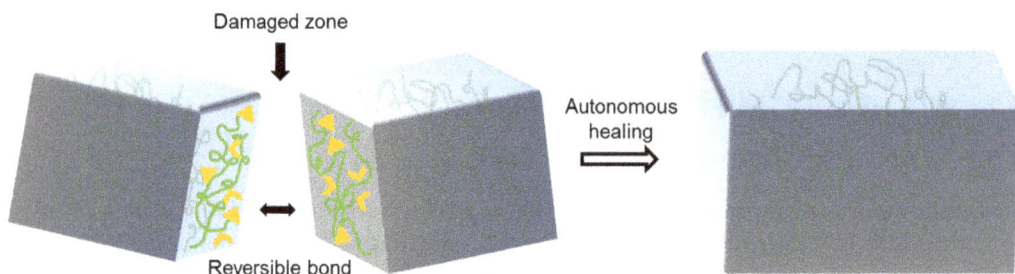

FIGURE 4.4 Schematic illustration of the autonomous self-healing process.

stimulus to promote the onset or progress of self-healing. Self-healing, self-repair, and self-recovery are used as synonyms, and represent ability of a material to recover mechanical properties after damage (Hager et al. 2010; Taylor and in het Panhuis 2016). Damage can be defined as undesired alterations at the molecular and/or macroscale, which can result in total or partial loss of an original functionality of the materials.

The purpose of the self-healing hydrogels concept is to extend the service lifetime of hydrogels and restore and/or maintain the original properties by implementing the concept of autonomous or induced repair, rather than improving their initial performance. This new trend in the development of hydrogels can increase the safety, durability, reliability of materials through damage management concept instead of damage prevention, assuming that damage is inevitable and that our materials have to be prepared for it. The design and development of self-healing mechanisms through synthetic pathways are often complex and difficult to control (Brochu et al. 2011). Unlike nature, synthetic hydrogels are usually not repairable after they have suffered catastrophic damage. This damage affecting the hydrogel structure leads to a decrease in the mechanical properties and functionality, resulting in a loss of time, material, and money to replace the damaged area. Self-healing mechanisms that are inspired or imitated by those encountered in natural systems are being applied to hydrogels (Blaiszik et al. 2010).

A traditional approach to long-term use of the structures and systems is to develop more tough, wear-resistant, and better materials. This means that the material has a higher performance than the material it replaces so that the material can be used for a longer time. Repairing materials after failure is usually done by adding fillers or adhesives that do not retain the original properties of the material. Often, self-healing systems require externally supplied agents or the supply of modest amounts of energy (Hillewaere and Du Prez 2015). However, breakage and cracking generally occur before significant damage is detected at the point of failure. Polymeric materials currently in use are not effective at preventing the progression of mechanical damage. Thus, the development of hydrogels that can quickly respond to damage and self-healing has emerged as a viable solution. For intrinsic self-healing in hydrogels, the self-healing capabilities are chemically or compositionally incorporated into the polymer structure through reversible bonding.

Ideally, the self-healing hydrogels should be produced with inexpensive and nontoxic/nonhazardous materials and techniques for use as biomaterials, and should not be prematurely degraded with respect to the application. In addition, autonomously self-healing hydrogel must: (i) respond spontaneously to damage, quickly and repeatedly without external stimuli, (ii) respond to damage from the micro- to macroscale, (iii) has mechanical, rheological, and biocompatibility suitable for the intended application, and (iv) maintain their original mechanical and rheological properties after self-healing. Studies on self-healing hydrogels include two main approaches; dynamic covalent bonds and non-covalent bonds (Table 4.1) (Taylor and in het Panhuis 2016). Self-healing properties

TABLE 4.1
Bond Types of Autonomous Self-Healing Supramolecular Hydrogels

Dynamic Covalent Bonds

- Schiff-base linkage
- Diels–Alder reaction
- Disulfide exchange chemistry
- Boronic esters and boronates

Non-Covalent Bonds

- Hydrogen bonding
- Electrostatic (ionic) interactions
- Host–guest interactions
- Hydrophobic interactions

of covalently crosslinked hydrogel often require a healing agent (monomer, oligomer, solvent, etc.) or the application of an external stimulus, such as pH, externally supplied agents, UV light.

4.2.1.1 Dynamic Covalent Bonding

Dynamic covalent bonding is a reversible bond that can break and re-form spontaneously or in response to stimuli. Depending on the condition, these bonds can be reversed like a noncovalent physical bond or permanent like original covalent bonds. Several research groups have exploited the reversibility of dynamic Schiff-based (imine) crosslinking to obtain self-healing properties. This linkage belongs to the family of dynamic covalent bonds and provides self-healing property by achieving the inherent dynamic equilibrium of bond association and dissociation in a hydrogel network.

Zang prepared multi-responsive, dynamic, and self-healing chitosan-based hydrogels (Zhang et al. 2011). These biocompatible hydrogels were synthesized by mixing a dibenzaldehyde-terminated telechelic poly(ethylene glycol) (PEG) and chitosan and were allowed to form Schiff-base linkages between the aldehyde groups in the PEG chain and the amino groups in chitosan. This hydrogel was found to be self-healable and stimuli-responsive (pH, amino acids, and vitamin derivatives) in drug delivery. Wei developed a polysaccharide-based self-healing injectable hydrogel system that mimics the biomechanical properties of brain tissue (Wei et al. 2016). A self-healing hydrogel was prepared in mild conditions using amino groups on *N*-carboxyethyl chitosan (CEC) and the aldehyde groups on oxidized sodium alginate (OSA), expressed as CEC-l-OSA hydrogel. This hydrogel exhibits cytocompatibility, water solubility, low-cost, and is abundant in nature. In particular, the imine linkage through the Schiff-base reaction provides a self-healing capability to the CEC-l-OSA hydrogels. Ding reported a self-healable polymeric hydrogel by Schiff-based crosslinking two natural polymers acrylamide-modified chitin (AMC) containing amino groups and oxidized alginate containing dialdehyde groups (Ding et al. 2015). The self-healing capability of this hydrogel depends on the polymer concentration and the surrounding pH environment. At pH 5, the imine linkages were not reversible, and no self-healing was observed. We have also prepared an injectable self-healing hydrogel by just mixing a glycol-chitosan (GC) and an oxidized dextran (ODEX) (Figure 4.5). The ODEX/GC hydrogels showed good self-healing ability under physiological conditions and kept the dynamic Schiff-base linkage at over 2 wt%. The release kinetics of a model protein was found to be controlled by varying the needle size during injection, due to modulation of the apparent size and shape of the fragmented hydrogels even in the self-healed state (Cho and Ooya 2018).

Many research studies have used Diels–Alder (DA) reaction in dynamic polymers. The DA reactions can be reacted in mild conditions but are reversible only at high temperatures through a dissociative retro-Diels–Alder (rDA) reaction. In addition, the DA reaction is classified as a "click" reaction due to its excellent selectivity and efficiency. One of the earliest studies of self-healing

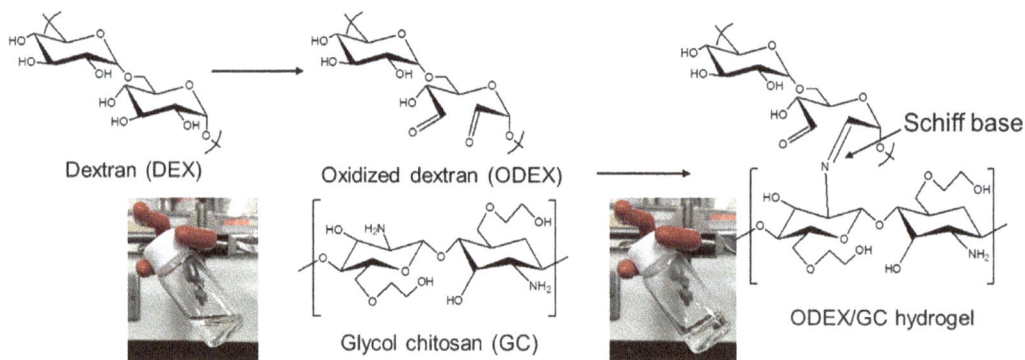

FIGURE 4.5 Synthetic scheme of dialdehyde modification of dextran and hydrogel formation mechanism through the Schiff-base linkage with glycol chitosan.

materials based on thermo-reversible DA was reported by Chen et al. (2002). The polymeric material was synthesized using a thermo-reversible DA reaction between multi-diene and multi-dienophile. The materials showed approximately 57% of the original fracture load after heating for 2 h at 150°C. An approach to improving self-healing materials is to combine DA reactions with hydrogen bonding. Recently, Schäfer and Kickelbick reported hybrid materials that contain two intrinsic self-healing forces: First, it is the reversible covalent bond formation of DA groups and the intrinsic hydrogen bonds that support DA groups (Schäfer and Kickelbick 2018).

Disulfide exchange is one of the most widely used chemistry for self-healing materials. Because of the relatively fast disulfide exchange reaction in the network, the self-healing phenomenon is observed under mild conditions or moderately low temperatures. Rekondo designed a novel poly(urea-urethane) thermoset elastomer with aromatic disulfide crosslinks, which is able to perform disulfide metathesis without a catalyst (Rekondo et al. 2014). The disulfide exchange reaction was also utilized for rapid sol–gel transition with self-healing properties: The disulfide exchange of 1,2-dithiolane-functionalized polymers can reform under neutral conditions (Yu et al. 2017; Zhang and Waymouth 2017). These materials do not require external intervention such as heat or light and exhibit remarkable self-healing efficiency at room temperature.

The dynamic covalent transesterification reactions of boronic acid with 1,2-diol or 1,3-diol cause the reversible formation of boronic ester under mild conditions. This reaction has been applied to self-healing or stimuli-responsive hydrogels with dynamic properties since rapid dynamic exchange without a catalyst is necessary in mild conditions and aqueous media. Smithmyer prepared hydrogels using boronic acid-functionalized polymers and showed their self-healing properties in neutral to acidic water (pH 4–7.4) (Smithmyer et al. 2018). Self-healing hydrogels were formed by mixing 2-acrylamidophenylboronic acid (2-APBA) monomer containing a copolymer with PVA or catechol-functionalized copolymers (Figure 4.6). Cash has also synthesized dynamic covalent boronic esters hydrogels through photoinduced radical thiol-ene click chemistry. Because of the reversibility of the boronic ester, the resulting materials exhibit self-healing in the absence of solvent at room temperature (Cash et al. 2015).

4.2.1.2 Non-Covalent Bonds

In other concepts of self-healing hydrogels, non-covalent bonds such as ionic bonding, hydrogen-bonding, host–guest interaction, hydrophobic bonding, and molecular diffusion and chain entanglement are generally used individually or in combination. However, hydrogels consisting of non-covalent bonds are inherently weaker compared to dynamic covalent network.

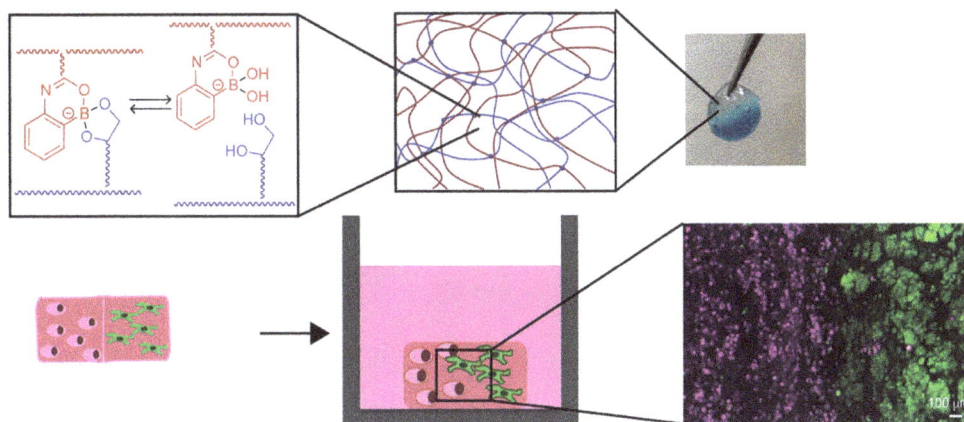

FIGURE 4.6 Synthetic image of self-healing hydrogels prepared by mixing 2-acrylamidophenylboronic acid (2-APBA) monomer containing a copolymer with PVA or catechol-functionalized copolymers. (Reproduced with permission from Smithmyer (2018) © 2018 American Chemical Society.)

Hydrogen bonding is a much weaker physical interaction than dynamic covalent crosslinks. The self-healing mechanism of hydrogen bonding is reversible crosslinking and is formed between the different polymer chains by electrostatic interactions that occur between the hydrogens and electronegative atoms of different residues. For example, Hou developed a new rapid self-integrating and shear-thinning hydrogel by combining a significant number of multiple-hydrogen-bond units on a DEX without any external interference (Hou et al. 2015). They chose a multiple-hydrogen-bond unit, which is a quadruple hydrogen-bond array named UPy (Ureidopyrimidinone) because the intermolecular bonding strength was much higher than that of a single hydrogen bond. The multifunctionalized DEX-UPy polymer can form a robust hydrogel and shows rapid self-healing properties.

Electrostatic (ionic) interactions have a relatively high strength of the interaction in aqueous media. The ionic self-healing mechanisms are caused by the reversible electrostatic interactions of counter-charged ions. This may include a charged ion or a charged polymer chain that is cross-linked with an oppositely charged polymer chain. Therefore, the physical nature of these electrostatic interactions can be applied excellently as a self-healing property. Bai reported a unique single-component zwitterionic hydrogel, driven by a new mechanism, that allows spontaneous healing under mild conditions without the need for additional healing reagents or the input of external energy (Bai et al. 2014).

Host–guest chemistry is composed of two or more molecules that cause unique molecular recognition and interactions by non-covalent binding. There is a unique structural relationship, in which one part is physically inserted and contained inside another moiety. Crosslinking of supramolecular assemblies by reversible non-covalent interactions can be used as a self-healing mechanism. For example, the Harada group fabricated self-healing cyclodextrin (CD)-guest gels crosslinked between polyacrylamide (PAAm) chains with inclusion complexes (Kakuta et al. 2013). The resulting CD-guest gels exhibit a self-assemble property without using any chemical crosslinkers, and when the gels are cut, cooperative host–guest complexation on the cut surfaces effectively rebuilds and restores the material strength (Figure 4.7). Recently, the Harada group summarized a review of supramolecular self-healing materials based on host–guest interaction (Sinawang et al. 2020).

Dense hydrophobic interactions occur when the aggregation of nonpolar material surfaces eliminates water molecules in an aqueous solution. The Micelle copolymerization technique has been studied to apply stable hydrophobic domains to hydrogels without affecting the hydrophilic chain. The rearrangement of the hydrophobic associations of crosslinked micelles within the hydrogel shows self-healing properties without the need for external stimuli. Tuncaboylu developed a micellar polymerization technique that uses reversible crosslinking, which is needed for autonomous self-healing (Tuncaboylu et al. 2012). Self-healing mechanisms are caused by the weakening of strong hydrophobic interactions due to the presence of surfactant molecules, and hydrophobic moieties can easily find partners on the other cut surface, resulting in self-healing of the damaged gel sample taking place in a short time.

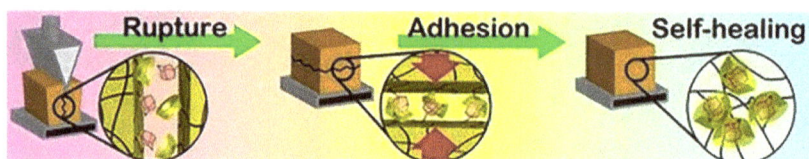

FIGURE 4.7   Image of self-healing cyclodextrin-guest gels. (Reproduced with permission from Kakuta (2013) © 2013 WILEY-VCH Verlag GmbH & Co. KGaA, Weinheim.)

4.3 TOUGH AND SELF-HEALING SUPRAMOLECULAR HYDROGELS

4.3.1 TOUGH HYDROGELS

As mentioned in the introduction part (Section 4.1), hydrogels are soft and wet polymer networks swollen with large amounts of water. However, most of hydrogels are mechanically weak or brittle, and they are not suitable for biological applications such as cartilage of the joints (Li et al. 2016). Many efforts have been made to develop highly tough hydrogels with new approaches and new gel systems, including double network hydrogels, slide-ring hydrogels, nanocomposite hydrogels, macromolecular microsphere composite hydrogels, and tetra-PEG hydrogels.

4.3.1.1 Double Network Hydrogels

Double network gels with good mechanical performance have been extensively developed (Gong et al. 2003). These gels consist of two interpenetrated polymer networks and are known as "double network (DN) gels". The DN gels, containing about 90 wt% water, possess both hardness and toughness comparable to cartilages and rubber (Na et al. 2004; Simha et al. 2004). The DN is generally synthesized through a two-step network formation: the first step is to form rigid, brittle, and tightly crosslinked gels and the second is to form soft, ductile, and loosely crosslinked networks with the first gels (Chen et al. 2015). In this system, the first network is commonly used as "sacrificial bonds" to protect the second network and increase crack propagation resistance. In this way, the gel network can effectively dissipate energy and improve mechanical properties (Chen et al. 2016). The networks of DN gels can be crosslinked by non-covalent or covalent bonds, which offer a rich phenomenon for gel networks and their relation to mechanical behaviors.

The classical methods have offered great generality and flexibility in preparing tough DN gels. However, there are some limitations to the multistep method: (i) the multistep are tedious and time-consuming; (ii) swelling and diffusion steps randomly distribute the molar ratio of the components and reduce repeatability; (iii) it is difficult to prepare hydrogels having different complex shapes because of the multistep. Only simple sheet or cylindrical shape DN gels have been reported so far in the literature; and (iv) the fracture of the fully chemically linked DN gel results in irreversible and permanent bond breaks, which makes it very difficult to be repaired and recovered from damage (Bu et al. 2017).

To overcome these limitations, many researchers developed a "one-pot" method to prepare hybrid physically and chemically crosslinked DN gels (Figure 4.8) (Chen et al. 2013; Sun et al. 2012). Since physical networks are often assembled by reversible interactions such as hydrogen bonding, ionic interaction, hydrophobic interactions, and metal-ligand coordination, π–π interactions, the physical network generally exhibits self-healing properties, and the associated mechanical properties can be recovered to some extent after damage or fatigue. Unlike the classical methods of preparing chemically linked DN gels, this one-pot method does not require swelling and diffusion processes; therefore, DN gels can be easily produced and optimized in a fast and controllable manner. This one-pot process only takes 1–2 h to complete DN gels, which is much faster than 1–2 days to complete the DN gel in a classical way.

4.3.1.2 Slide-Ring Hydrogels

Rotaxanes and polyrotaxanes have been extensively studied as typical mechanically interlocked molecular systems. They consist of a backbone string, rings, and two bulky end groups. The rings are threaded onto the axis string and are prevented from unthreading by the bulky end groups (Harada et al. 1992). When polyrotaxanes are crosslinked, the resulting supramolecular network is named as a slide-ring material. Chemically crosslinked gels are generally known to have a large inhomogeneous structure due to a gelation process that significantly reduces mechanical strength. Due to the fixed crosslinking of the chemical gel, polymer chains coexist with long and short chains.

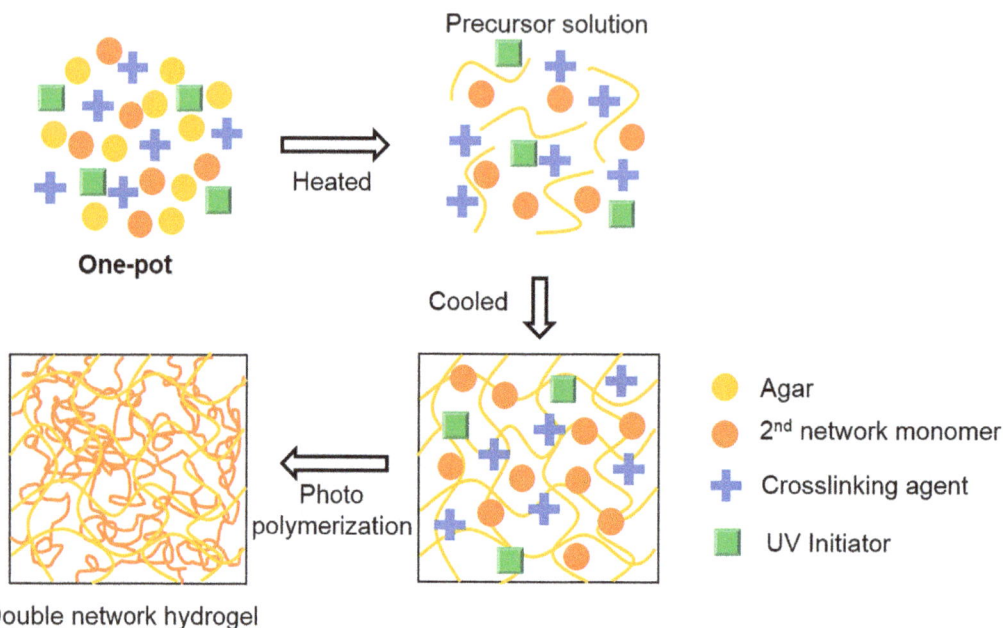

FIGURE 4.8   Schematic illustration of the one-pot method to prepare hybrid physically-chemically cross-linked DN gels.

As a result, the tensile stress concentrates on the shorter chains, which easily breaks the chemical gel. To solve this problem, several research groups have recently developed a slide-ring gel that can move freely on the supramolecular structures with topological characteristics. The polymer chains in topological gels are not covalently crosslinked like chemical gels and do not interact with each other like physical gels, but the backbone chains are mechanically or topologically inter-locked by figure-of-eight movable crosslinks (Imran et al. 2014; Ito 2010) (Figure 4.9). Thus, the figure-eight-shaped bridge can freely pass through the polymer chains to equalize the tension of the threaded polymer chains just like pulleys. This is called the *"pulley effect"*. The heterogeneous polymer lengths between the fixed crosslinks of the chemical gels result in breakage of the network by concentrating tensile strength on shorter network strands (Ito 2012). On the other hand, the cross-linking junctions in the slide-ring gel are not fixed and can move freely through the figure-of-eight crosslinks acting as pulleys. The movable crosslinking junctions in slide-ring hydrogels provide unique mechanical properties, quite different from the entanglement of polymer chains or the fixed junctions in chemical gels and rubbers (Liu et al. 2017). The first slide-ring hydrogel was synthe-sized by Okumura and Ito. By intermolecular crosslinking of the α-CDs contained in the polyro-taxanes, a transparent gel was obtained which has good tensile strength, low viscosity, and large swelling-ability (Okumura and Ito 2001).

Slide-ring hydrogels show extreme softness in the low elongation region and stretchability in the high elongation one. In addition, the moduli in both regions can be controlled by varying the num-ber of free rings and/or the crosslinking density in the slide-ring gel. Therefore, slide-ring materi-als are suitable for use as substitutes for various types of biomaterials. If biomaterials are made of slide-ring materials, they may fit better with native ones than fixed crosslinked materials. However, slide-ring hydrogels are not usable as biomaterials because the synthesis and crosslinking method of slide-ring hydrogels still involve free-radical polymerization, require high-pressure mercury lamps, long preparation time, and use toxic materials which are not biocompatible. Therefore, previously reported slide-ring hydrogels have only been used as reinforcing materials.

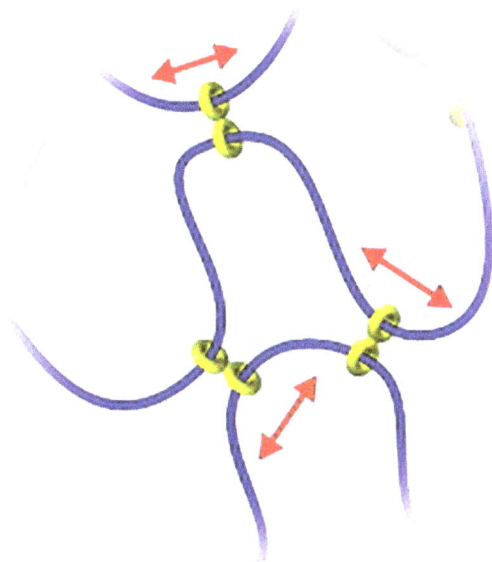

FIGURE 4.9 Schematic diagrams of slide-ring materials with freely movable figure-of-eight crosslinks acting like pulleys. (Reproduced with permission from (Ito 2010) © 2009 Elsevier Ltd.)

4.3.2 TOUGH AND SELF-HEALING SUPRAMOLECULAR HYDROGELS

Based on tough hydrogels such as DN and side-ring hydrogels, researchers have combined an idea to introduce self-healing properties into tough hydrogels. As described above, hydrogels with self-healing properties are generally inversely proportional to their mechanical strength because they are reversible crosslinking (dynamic covalent bonds and non-covalent bonds). Efficiencies and recoveries rate are improved as the lifetimes of the reversible crosslinks decreases. In contrast, excellent mechanical properties of hydrogels are based on the dissipation of mechanical energy by increasing the number of crosslinks, such as multiple crosslinks or molecular sliding as described above. High mechanical and self-healing properties are thought to be useful properties of functional materials; however, traditional wisdom has the limitation that these two different properties coexist within a single material.

In the case of DN hydrogels, Chen has fabricated a new type of agar/hydrophobically associated acrylamide (PAAm) DN gels that are fully physically crosslinked with a simple one-pot method (Chen et al. 2015). Interestingly, this DN gel shows not only ductile and tough properties but also rapid self-recovery at room temperature without any external stimuli. The self-healing behavior is attributed to the reconstruction of hydrophobic interactions between SDS micelles and alkyl groups. Liu and Li have prepared a κ-carrageenan/PAAm DN gels by combining an ionically crosslinked κ-carrageenan network with a covalently crosslinked PAAm network in a one-pot (Liu and Li 2016). Due to the thermo-reversible behavior of κ-carrageenan, the DN hydrogel also exhibited an excellent self-healing property for 20 min at 90°C. However, the self-healing DN gels studied so far still have limitations, such as the use of toxic agents during the synthesis process or the need for external stimuli to possess self-healing properties. We have designed a self-healable DN hydrogel with dynamic covalent bonds and hydrogen bonds, showing excellent mechanical properties and tuned cell attachment capacity (Cho and Ooya 2019). Agar formed the first network, while glycol chitosan and oxidized carboxylmethyl cellulose formed the second network in the resultant double network hydrogel (Figure 4.10). A simple one-pot fabrication allows the injection of hydrogels using a syringe. The mechanical properties of the double network hydrogel were improved compared to that of the parent single-network because they can effectively dissipate energy. Stress-strain curve showed that the hydrogel has good self-healing ability without heating or cut surface treatment. The incorporation of agar in the double network induced the enhanced protein adsorption, and the following cell attachments were governed by the adsorbed protein states.

In the case of slide-ring hydrogels, self-healing hydrogels have been developed using polyrotaxanes or pseudorotaxanes. The Harada group fabricated self-healing materials based on polymer gels crosslinked by boronate linkages, which is a dynamic covalent bond, between boronic acid of PAAm and diol compounds of polyrotaxnanes (Nakahata et al. 2016). These gels combine dynamic covalent bonds that allow the ring molecules to move freely along the axis of the polyrotaxane chain, and the sliding state of the crosslinks imparts rapid self-healing ability in both wet and semi-dry states. Li proposed a new strategy to construct self-supported supramolecular hydrogels through the hierarchically organic-inorganic hybridization of Laponite matrix with CD-based pseudopoly-rotaxanes (Li et al. 2017). The guanidinium cations bound to the CD are evenly distributed along the axial direction of pseudopolyrotaxanes and are more useful for realizing the multivalent interactions between the molecular glue and Laponite matrix. The obtained supramolecular hydrogels have self-healing capability because they are induced by non-covalent interactions. Alternatively, we fabricated a novel biocompatible self-healing slide-ring hydrogel consisting of glycol chitosan and hydroxypropylated polyrotaxane aldehyde (HP-PR-AH) (Cho and Ooya 2020). The movable cross-linking junctions in slide-ring hydrogels give unique mechanical behavior, such as fatigue-resistant, elastic, anticompressing due to the pully effect, which is quite different from the traditional hydrogels (Figure 4.11). Moreover, this hydrogel shows excellent self-healing performances because it is constructed through a dynamic covalent bond. *In vitro* cytotoxicity tests of HP-PR-AH and GC

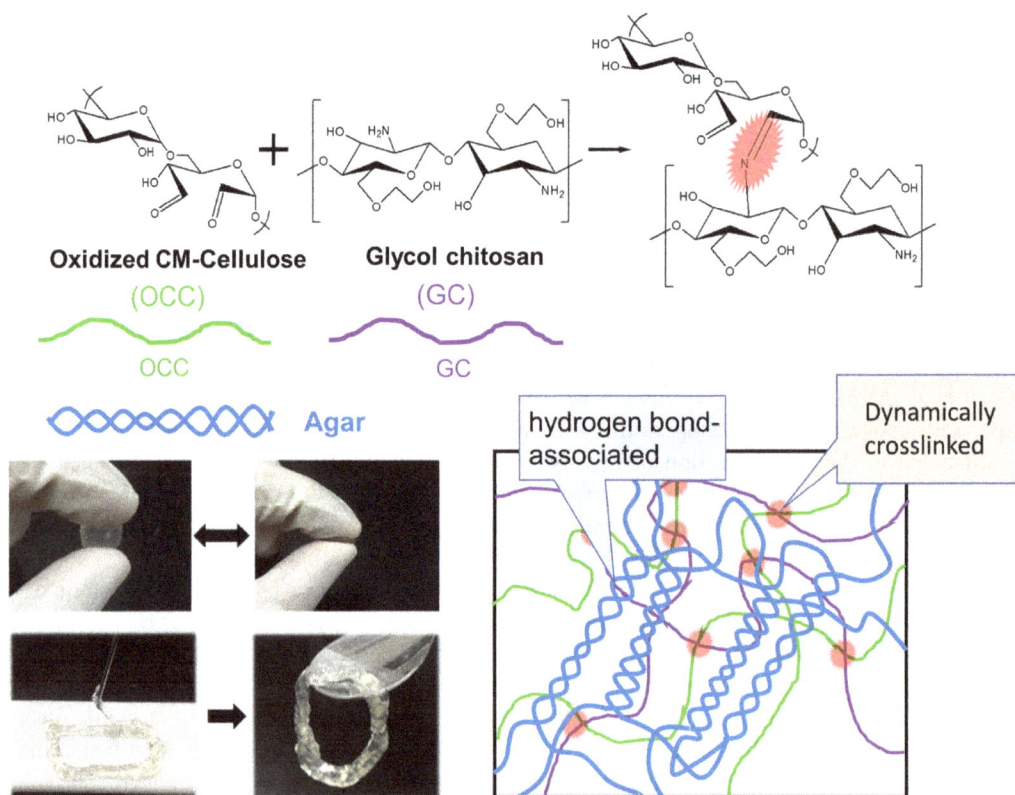

FIGURE 4.10 Illustration of the formation of the agar and the OCMC/GC double network hydrogel. (Reproduced with permission from Cho and Ooya (2019) © 2019 Elsevier B.V.)

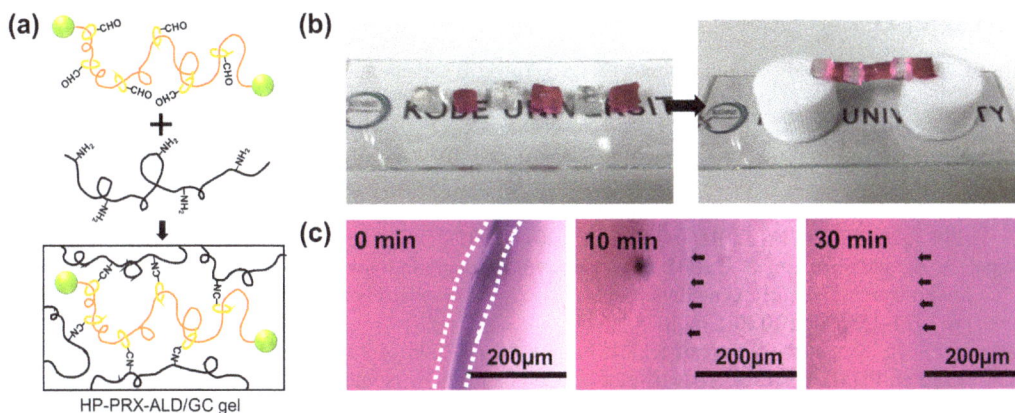

FIGURE 4.11 (a) Construction of the slide-ring hydrogel prepared by just mixing hydroxypropylated polyrotaxane aldehydes (HP-PRX-ALD) and glycol chitosan (GC) in an aqueous solution. (b) Freshly cut HP-PRX10-ALD/GC gels with the small columns and the self-healed cylinder bridging between two caps to support its own weight (left). Prepared HP-PRX10-ALD/GC gels were spliced and placed at room temperature (right). (c) Optical microscopy images of the HP-PRX$_{10}$-ALD/GC gel after being self-healed for 0, 10, and 30 min. (Reproduced with permission from Cho and Ooya (2020) © 2019 Wiley-VCH Verlag GmbH & Co. KGaA, Weinheim.)

were evaluated by the MTT assay and revealed that they showed at least 80% of cell viability after culture, which indicated that polymers are biologically safe and nontoxic. In addition, cell encapsulation test in the hydrogel matrix shows good cell proliferation, and can easily alter the hydrogel structure by self-healing properties, and it can be seen that the cells are connected and spread even at the self-healed interface. Based on their excellent mechanical properties and biocompatibility, the HP-PR-AH/GC gel is the first observation slide-ring hydrogel as a cell scaffold and can be considered a promising biomaterial.

4.4   CONCLUSIONS AND PERSPECTIVES

Self-healing supramolecular hydrogels maintain conventional hydrogels properties such as high water content, swelling, flexible nature, and biocompatibility. The incorporation of supramolecular binding motifs enables the hydrogels to show the dynamic nature of crosslinking, which exhibits the self-healing capabilities that are thought to be useful for cellular interaction. Self-healing supramolecular hydrogels have been explored for the 3D culture of various cell types as well as regeneration of dental and bone tissue, leading to soft tissue engineering such as dental pulp, cartilage, and cardiovascular tissues (Saunders and Ma 2019). Introducing a dynamic covalent bond into a physically crosslinked collagen hydrogel achieved tunable fast relaxation of the hydrogel matrix, which can preciously control tissue growth and development (Liu et al. 2020). Because of the robust nature of tough and self-healing supramolecular hydrogels, they are expected to be used as wearable electronics (Yang et al. 2020; Chen et al. 2019; Feng et al. 2020), 3D printing technology (Janarthanan et al. 2020; Sather et al. 2021), and biodegradable plastics (Li et al. 2021). Furthermore, self-healing supramolecular hydrogels can be applicable for biofouling by using zwitterionic copolymers (Ye et al. 2019) and by loading bioactive agents to achieve controlled release of them (Banerjee et al. 2020; Zhang et al. 2020). Recently, researchers are also developing the self-healing supramolecular coating technology for efficient corrosion protection (Zhao et al. 2020; Chen et al. 2020). The rational design of the tough and self-healing supramolecular hydrogels is expected for the development of both human-friendly and environmentally-friendly materials bearing smart functions.

REFERENCES

Almdal, K., Dyre, J., Hvidt, S., and Kramer, O. 1993. Towards a phenomenological definition of the term "gel". *Polymer Gels and Networks* 1 (1): 5–17. doi: 10.1016/0966–7822(93)90020–I.

Appel, E. A., del Barrio, J., Loh, X. J., and Scherman, O. A. 2012. Supramolecular polymeric hydrogels. *Chemical Society Reviews* 41 (18): 6195–6214. doi: 10.1039/c2cs35264h.

Bai, T., Liu, S., Sun, F., Sinclair, A., Zhang, L., Shao, Q., and Jiang, S. 2014. Zwitterionic fusion in hydrogels and spontaneous and time-independent self-healing under physiological conditions. *Biomaterials* 35 (13): 3926–3933. doi: 10.1016/j.biomaterials.2014.01.077.

Bajpai, A. K., and Shrivastava, J. 2005. In vitro enzymatic degradation kinetics of polymeric blends of cross-linked starch and carboxymethyl cellulose. *Polymer International* 54 (11). John Wiley & Sons, Ltd, pp. 1524–1536. doi: 10.1002/pi.1878.

Banerjee, S. L., Samanta, S., Sarkar, S., and Singha, N. K. 2020. A self-healable and antifouling hydrogel based on PDMS centered ABA tri-block copolymer polymersomes: a potential material for therapeutic contact lenses. *Journal of Materials Chemistry B* 8 (2). The Royal Society of Chemistry 226–243. doi: 10.1039/C9TB00949C.

Blaiszik, B. J., Kramer, S. L. B., Olugebefola, S. C., Moore, J. S., Sottos, N. R., and White, S. R., 2010. Self-healing polymers and composites. *Annual Review of Materials Research* 40 (1). Annual Reviews 179–211. doi: 10.1146/annurev-matsci-070909-104532.

Brochu, A. B., Craig, S. L., and Reichert, W. M. 2011. Self-healing biomaterials. *Journal of Biomedical Materials Research Part A* 96A (2). John Wiley & Sons, Ltd, 492–506. doi: https://doi.org/10.1002/jbm.a.32987.

Bu, Y., Shen, H., Yang, F., Yang, Y., Wang, X., and Wu, D. 2017. Construction of tough, in situ forming double-network hydrogels with good biocompatibility. *ACS Applied Materials & Interfaces* 9 (3): 2205–2212. doi: 10.1021/acsami.6b15364.

Cash, J. J., Kubo, T., Bapat, A. P., and Sumerlin, B. S. 2015. Room-temperature self-healing polymers based on dynamic-covalent boronic esters. *Macromolecules* 48 (7): 2098–2106. doi: 10.1021/acs.macromol.5b00210.

Chen, C., Xiao, G., He, Y., Zhong, F., Li, H., Wu, Y., and Chen, J. 2020. Bio-inspired superior barrier self-healing coating: self-assemble of graphene oxide and polydopamine-coated halloysite nanotubes for enhancing corrosion resistance of waterborne epoxy coating. *Progress in Organic Coatings* 139: 105402. doi: 10.1016/j.porgcoat.2019.105402.

Chen, J., Peng, Q., Thundat, T., and Zeng, H. 2019. Stretchable, injectable, and self-healing conductive hydrogel enabled by multiple hydrogen bonding toward wearable electronics. *Chemistry of Materials* 31 (12): 4553–4563. doi: 10.1021/acs.chemmater.9b01239.

Chen, Q., Chen, H., Zhu, L., and Zheng, J. 2015. Fundamentals of double network hydrogels. *Journal of Materials Chemistry B* 3 (18): 3654–3676. doi: 10.1039/C5TB00123D.

Chen, Q., Chen, H., Zhu, L., and Zheng, J. 2016. Engineering of tough double network hydrogels. *Macromolecular Chemistry and Physics* 217 (9): 1022–1036. doi: 10.1002/macp.201600038.

Chen, Q., Zhu, L., Chen, H., Yan, H., Huang, L., Yang, J., and Zheng, J. 2015. A novel design strategy for fully physically linked double network hydrogels with tough, fatigue resistant, and self-healing properties. *Advanced Functional Materials* 25 (10): 1598–1607. doi: 10.1002/adfm.201404357.

Chen, Q., Zhu, L., Zhao, C., Wang, Q., and Zheng, J. 2013. A robust, one-pot synthesis of highly mechanical and recoverable double network hydrogels using thermoreversible sol-gel polysaccharide. *Advanced Materials* 25 (30): 4171–4176. doi: 10.1002/adma.201300817.

Chen, X., Dam, M. A., Ono, K., Mal, A., Shen, H., Nutt, S. R., Sheran, K., and Wudl, F. 2002. A thermally re-mendable cross-linked polymeric material. *Science* 295 (5560): 1698–1702. doi: 10.1126/science.1065879.

Cho, I. S., and Ooya, T. 2018. An injectable and self-healing hydrogel for spatiotemporal protein release via fragmentation after passing through needles. *Journal of Biomaterials Science, Polymer Edition* 29 (2): 145–159. doi: 10.1080/09205063.2017.1405573.

Cho, I. S., and Ooya, T. 2019. Tuned cell attachments by double-network hydrogels consisting of glycol chitosan, carboxylmethyl cellulose and agar bearing robust and self-healing properties. *International Journal of Biological Macromolecules* 134: 262–268. doi: 10.1016/j.ijbiomac.2019.05.053.

Cho, I. S., and Ooya, T. 2020. Cell-encapsulating hydrogel puzzle: polyrotaxane-based self-healing hydrogels. *Chemistry - A European Journal* 26 (4): 913–920. doi: 10.1002/chem.201904446.

Desai, E. S., Tang, M. Y., Ross, A. E., and Gemeinhart, R. A. 2012. Critical factors affecting cell encapsulation in superporous hydrogels. *Biomedical Materials* 7 (2): 24108. doi: 10.1088/1748-6041/7/2/024108.

Ding, F., Wu, S., Wang, S., Xiong, Y., Li, Y., Li, B., Deng, H., Du, Y., Xiao, L., and Shi, X. 2015. A dynamic and self-crosslinked polysaccharide hydrogel with autonomous self-healing ability. *Soft Matter* 11 (20): 3971–3976. doi: 10.1039/C5SM00587F.

Dong, R., Pang, Y., Su, Y., and Zhu, X. 2015. Supramolecular hydrogels: synthesis, properties and their biomedical applications. *Biomater. Sci.* 3 (7): 937–954. doi: 10.1039/C4BM00448E.

Feng, E., Gao, W., Li, J., Wei, J., Yang, Q., Li, Z., Ma, X., Zhang, T., and Yang, Z. 2020. Stretchable, healable, adhesive, and redox-active multifunctional supramolecular hydrogel-based flexible supercapacitor. *ACS Sustainable Chemistry & Engineering* 8 (8): 3311–3320. doi: 10.1021/acssuschemeng.9b07153.

Gong, J. P., Katsuyama, Y., Kurokawa, T., and Osada, Y. 2003. Double-network hydrogels with extremely high mechanical strength. *Advanced Materials* 15 (14): 1155–1158. doi: 10.1002/adma.200304907.

Graham, T. 1864. XXXV.—On the properties of silicic acid and other analogous colloidal substances. *J. Chem. Soc.* 17: 318–327. doi: 10.1039/JS8641700318.

Guenet, J.-M. 1992. *Thermoreversible Gelation of Polymers and Biopolymers*. Academic Press.

Gupta, D., and Tator, C. H., and Shoichet, M. S. 2006. Fast-gelling injectable blend of hyaluronan and methylcellulose for intrathecal, localized delivery to the injured spinal cord. *Biomaterials* 27 (11): 2370–2379. doi: 10.1016/j.biomaterials.2005.11.015.

Hager, M. D., Greil, P., Leyens, C., Van Der Zwaag, S., and Schubert, U. S. 2010. Self-healing materials. *Advanced Materials* 22: 5424–5430. doi: 10.1002/adma.201003036.

Harada, A., Li, J., and Kamachi, M. 1992. The molecular necklace: a rotaxane containing many threaded α-cyclodextrins. *Nature* 356 (6367): 325–327. doi: 10.1038/356325a0.

Hillewaere, X. K. D., and Du Prez, F. E. 2015. Fifteen chemistries for autonomous external self-healing polymers and composites. *Progress in Polymer Science* 49–50: 121–153. doi: 10.1016/j.progpolymsci.2015.04.004.

Hou, S., Wang, X., Park, S., Jin, X., and Ma, P. X. 2015. Rapid self-integrating, injectable hydrogel for tissue complex regeneration. *Advanced Healthcare Materials* 4 (10): 1491–1495. doi: 10.1002/adhm.201500093.

Imran, A. B., Esaki, K., Gotoh, H., Seki, T., Ito, K., Sakai, Y., and Takeoka, Y. 2014. Extremely stretchable thermosensitive hydrogels by introducing slide-ring polyrotaxane cross-linkers and ionic groups into the polymer network. *Nature Communications* 5 (1): 5124. doi: 10.1038/ncomms6124.

Ito, K. 2010. Slide-ring materials using topological supramolecular architecture. *Current Opinion in Solid State and Materials Science* 14 (2): 28–34. doi: 10.1016/j.cossms.2009.08.005.

Ito, K. 2012. Novel entropic elasticity of polymeric materials: why is slide-ring gel so soft? *Polymer Journal* 44 (1): 38–41. doi: 10.1038/pj.2011.85.

Janarthanan, G., Shin, H. S., Kim, I.-G., Ji, P., Chung, E.-J., Lee, C., and Noh, I. 2020. Self-crosslinking hyaluronic acid–carboxymethylcellulose hydrogel enhances multilayered 3D-printed construct shape integrity and mechanical stability for soft tissue engineering. *Biofabrication* 12 (4): 45026. doi:10.1088/1758-5090/aba2f7.

Kakuta, T., Takashima, Y., Nakahata, M., Otsubo, M., Yamaguchi, H., and Harada, A. 2013. Preorganized hydrogel: self-healing properties of supramolecular hydrogels formed by polymerization of host–guest monomers that contain cyclodextrins and hydrophobic guest groups. *Advanced Materials* 25 (20): 2849–2853. doi: 10.1002/adma.201205321.

Lee, S.-Y., and Tae, G. 2007. Formulation and in vitro characterization of an in situ gelable, photo-polymerizable pluronic hydrogel suitable for injection. *Journal of Controlled Release* 119 (3): 313–319. doi: 10.1016/j.jconrel.2007.03.007.

Li, Y., Li, S., and Sun, J. 2021. Degradable poly(vinyl alcohol)-based supramolecular plastics with high mechanical strength in a watery environment. *Advanced Materials* 33 (13): 1–8. doi: 10.1002/adma.202007371.

Li, Z., Zhang, Y.-M., Wang, H.-Y., Li, H., and Liu, Y. 2017. Mechanical behaviors of highly swollen supramolecular hydrogels mediated by pseudorotaxanes. *Macromolecules* 50 (3): 1141–1146. doi: 10.1021/acs.macromol.6b02459.

Li, Z., Zheng, Z., Su, S., Yu, L., and Wang, X. 2016. Preparation of a high-strength hydrogel with slidable and tunable potential functionalization sites. *Macromolecules* 49 (1): 373–386. doi: 10.1021/acs.macromol.5b02359.

Liu, A., Wu, K., Chen, S., Wu, C., Gao, D., Chen, L., Wei, D., Luo, H., Sun, J., and Fan, H. 2020. Tunable fast relaxation in imine-based nanofibrillar hydrogels stimulates cell response through TRPV4 activation. *Biomacromolecules* 21 (9): 3745–3755. doi: 10.1021/acs.biomac.0c00850.

Liu, C., Kadono, H., Mayumi, K., Kato, K., Yokoyama, H., and Ito, K. 2017. Unusual fracture behavior of slide-ring gels with movable cross-links. *ACS Macro Letters* 6 (12): 1409–1413. doi: 10.1021/acsmacrolett.7b00729.

Liu, J., Lin, S., Li, L., and Liu, E. 2005. Release of theophylline from polymer blend hydrogels. *International Journal of Pharmaceutics* 298 (1): 117–125. doi: 10.1016/j.ijpharm.2005.04.006.

Liu, S., and Li, L. 2016. Recoverable and self-healing double network hydrogel based on κ-carrageenan. *ACS Applied Materials & Interfaces* 8 (43): 29749–29758. doi: 10.1021/acsami.6b11363.

Liu, Y., and Hsu, S. H. 2018. Synthesis and biomedical applications of self-healing hydrogels. *Frontiers in Chemistry* 6: 1–10. doi: 10.3389/fchem.2018.00449.

Ma, X., Chang, P. R., and Yu, J. 2008. Properties of biodegradable thermoplastic pea starch/carboxymethyl cellulose and pea starch/microcrystalline cellulose composites. *Carbohydrate Polymers* 72 (3): 369–375. doi: 10.1016/j.carbpol.2007.09.002.

Na, Y.-H., Kurokawa, T., Katsuyama, Y., Tsukeshiba, H., Gong, J. P., Osada, Y., Okabe, S., Karino, T., and Shibayama, M. 2004. Structural characteristics of double network gels with extremely high mechanical strength. *Macromolecules* 37 (14): 5370–5374. doi: 10.1021/ma049506i.

Nakahata, M., Mori, S., Takashima, Y., Yamaguchi, H., and Harada, A. 2016. Self-healing materials formed by cross-linked polyrotaxanes with reversible bonds. *Chem* 1 (5): 766–775. doi: 10.1016/j.chempr.2016.09.013.

Okumura, Y., and Ito, K. 2001. The polyrotaxane gel: a topological gel by figure-of-eight cross-links. *Advanced Materials* 13 (7): 485–487. doi: 10.1002/1521-4095(200104)13:7<485::AID-ADMA485>3.0.CO;2–T.

Rekondo, A., Martin, R., de Luzuriaga, A. R., Cabanero, G., Grande, H. J., and Odriozola, I. 2014. Catalyst-free room-temperature self-healing elastomers based on aromatic disulfide metathesis. *Materials Horizons* 1 (2): 237–240. doi: 10.1039/C3MH00061C.

reza Saboktakin, M., and Tabatabaei, R. M. 2015. Supramolecular hydrogels as drug delivery systems. *International Journal of Biological Macromolecules* 75: 426–436. doi: 10.1016/j.ijbiomac.2015.02.006.

Sather, N. A., Sai, H., Sasselli, I. R., Sato, K., Ji, W., Synatschke, C. V., and Zambrotta, R. T., 2021. 3D printing of supramolecular polymer hydrogels with hierarchical structure. *Small* 17 (5): 2005743. doi: 10.1002/smll.202005743.

Saunders, L., and Ma, P. X. 2019. Self-healing supramolecular hydrogels for tissue engineering applications. *Macromolecular Bioscience* 19 (1): 1800313. doi: 10.1002/mabi.201800313.

Schäfer, S., and Kickelbick, G. 2018. Double reversible networks: improvement of self-healing in hybrid materials via combination of diels–alder cross-linking and hydrogen bonds. *Macromolecules* 51 (15): 6099–6110. doi: 10.1021/acs.macromol.8b00601.

Simha, N. K., Carlson, C. S., and Lewis, J. L. 2004. Evaluation of fracture toughness of cartilage by micro-penetration. *Journal of Materials Science. Materials in Medicine* 15 (5): 631–639. doi: 10.1023/b:jmsm.0000026104.30607.c7.

Sinawang, G., Osaki, M., Takashima, Y., Yamaguchi, H., and Harada, A. 2020. Supramolecular self-healing materials from non-covalent cross-linking host–guest interactions. *Chemical Communications* 56 (32): 4381–4395. doi: 10.1039/D0CC00672F.

Smithmyer, M. E., and Deng, C. C., Cassel, S. E., LeValley, P. J., Sumerlin, B. S., and Kloxin, A. M. 2018. Self-healing boronic acid-based hydrogels for 3D co-cultures. *ACS Macro Letters* 7 (9): 1105–1110. doi: 10.1021/acsmacrolett.8b00462.

Sun, J.-Y., Zhao, X., Illeperuma, W. R. K., Chaudhuri, O., Oh, K. H., Mooney, D. J., Vlassak, J. J., and Suo, Z. 2012. Highly stretchable and tough hydrogels. *Nature* 489 (7414): 133–136. doi: 10.1038/nature11409.

Taylor, D. L., and in het Panhuis, M. 2016. Self-healing hydrogels. *Advanced Materials* 28 (41): 9060–9093. doi: 10.1002/adma.201601613.

Tuncaboylu, D. C., Sahin, M., Argun, A., Oppermann, W., and Okay, O. 2012. Dynamics and large strain behavior of self-healing hydrogels with and without surfactants. *Macromolecules* 45 (4): 1991–2000. doi: 10.1021/ma202672y.

Vermonden, T., Fedorovich, N. E., van Geemen, D., Alblas, J., van Nostrum, C. F., Dhert, W. J. A., and Hennink, W. E. 2008. Photopolymerized thermosensitive hydrogels: synthesis, degradation, and cyto-compatibility. *Biomacromolecules* 9 (3): 919–926. doi: 10.1021/bm7013075.

Wang, H., and Heilshorn, S. C. 2015. Adaptable hydrogel networks with reversible linkages for tissue engineering. *Advanced Materials* 27 (25): 3717–3736. doi: 10.1002/adma.201501558.

Wei, Z., Zhao, J., Chen, Y. M., Zhang, P., and Zhang, Q. 2016. Self-healing polysaccharide-based hydrogels as injectable carriers for neural stem cells. *Scientific Reports* 6 (1): 37841. doi: 10.1038/srep37841.

Yang, J., Yu, X., Sun, X., Kang, Q., Zhu, L., Qin, G., Zhou, A., Sun, G., and Chen, Q. 2020. Polyaniline-decorated supramolecular hydrogel with tough, fatigue-resistant, and self-healable performances for all-in-one flexible supercapacitors. *ACS Applied Materials & Interfaces* 12 (8): 9736–9745. doi:10.1021/acsami.9b20573.

Ye, Z., Zhang, P., Zhang, J., Deng, L., Zhang, J., Lin, C., Guo, R., and Dong, A. 2019. Novel dual-functional coating with underwater self-healing and anti-protein-fouling properties by combining two kinds of microcapsules and a zwitterionic copolymer. *Progress in Organic Coatings* 127: 211–221. doi: 10.1016/j.porgcoat.2018.11.021.

Yu, H., Wang, Y., Yang, H., Peng, K., and Zhang, X. 2017. Injectable self-healing hydrogels formed via thiol/disulfide exchange of thiol functionalized F127 and dithiolane modified PEG. *Journal of Materials Chemistry B* 5 (22): 4121–4127. doi: 10.1039/C7TB00746A.

Yui, N. 2002. *Supramolecular Design for Biological Applications.* CRC Press. https://books.google.co.jp/books?id=iUjRBQAAQBAJ.

Zhang, S. L., Li, T., Zhang, L., Azhar, U., Ma, J., Zhai, C., Zong, C., and Zhang, S. 2020. Cytocompatible and non-fouling zwitterionic hyaluronic acid-based hydrogels using thiol-ene "click" chemistry for cell encapsulation. *Carbohydrate Polymers* 236: 116021. doi: 10.1016/j.carbpol.2020.116021.

Zhang, X., and Waymouth, R. M. 2017. 1,2-dithiolane-derived dynamic, covalent materials: cooperative self-assembly and reversible cross-linking. *Journal of the American Chemical Society* 139 (10): 3822–3833. doi: 10.1021/jacs.7b00039.

Zhang, Y., Tao, L., Li, S., and Wei, Y. 2011. Synthesis of multiresponsive and dynamic chitosan-based hydrogels for controlled release of bioactive molecules. *Biomacromolecules* 12 (8): 2894–2901. doi: 10.1021/bm200423f.

Zhao, X., Wei, J., Li, B., Li, S., Tian, N., Jing, L., and Zhang, J. 2020. A self-healing superamphiphobic coating for efficient corrosion protection of magnesium alloy. *Journal of Colloid and Interface Science* 575: 140–149. doi: 10.1016/j.jcis.2020.04.097.

5 Coordination Polymers

Ganesan Prabusankar, Muneshwar Nandeshwar,
Suman Mandal, Sabari Veerapathiran,
and Kalaivanan Subramaniyam
Indian Institute of Technology Hyderabad

CONTENTS

5.1 INTRODUCTION

Historically, the unveiling of new molecules by a chemist has been a great contribution to human eudaimonia and prosperity. The crystal engineering of crystalline solids is a key technique to reveal the architecture of atoms in new molecules (Desiraju 2007; Braga and Grepioni 2005). Coordination metal complexes with unusual structural properties composed of both inorganic entity and organic linkers are known as coordination polymers (CPs) or metal–organic frameworks (MOFs) (Batten et al. 2012). CPs have established themselves as a key class of multifunctional materials in the last two decades. The structure of CPs can be controlled by selecting suitable metal ions or/and ligands (Batten 2012; Long and Yaghi 2009; Murray et al. 2009). The early inventions focused on simple ligands, which are generally preferred for reducing synthetic efforts and costs. Later, the use of structurally more refined ligands has been considered highly suitable, as they could lead to CP materials with novel topologies and properties (Zhang et al. 2015; Chen et al. 2014; Solovieva et al. 2017; Park et al. 2012). Thus, practically infinite combinations of CPs can be derived.

By comparing inorganic compounds with an organic polymer, the term "Coordination Polymer" was defined by Bailar et al. (1964). The CPs are hybrid materials, which consist of both inorganic and organic structures (Figure 5.1). The inorganic structures in CPs are metal ions, while the organic structures are donor ligands. The ligands are in general known as the "spacer" in CPs, as the organic ligand connects/separates the metal ions in the polymeric structure. The CPs represent extended structures such as infinite chain (Figure 5.1, II), sheet (Figure 5.1, III), or three-dimensional architecture (Figure 5.1, IV) or nano/micro particles (Figure 5.1, V) (Steed and Atwood 2009). Thus, a structure connected through coordination bonds in one direction along with supramolecular interaction can result in one-dimensional (1D) CPs. A structure connected through coordination bonds in two directions along with supramolecular interactions can result in two-dimensional (2D) CPs. A structure connected *via* coordination bonds in three directions with/without supramolecular interactions can produce three-dimensional (3D) CPs.

DOI: 10.1201/9781003169130-6

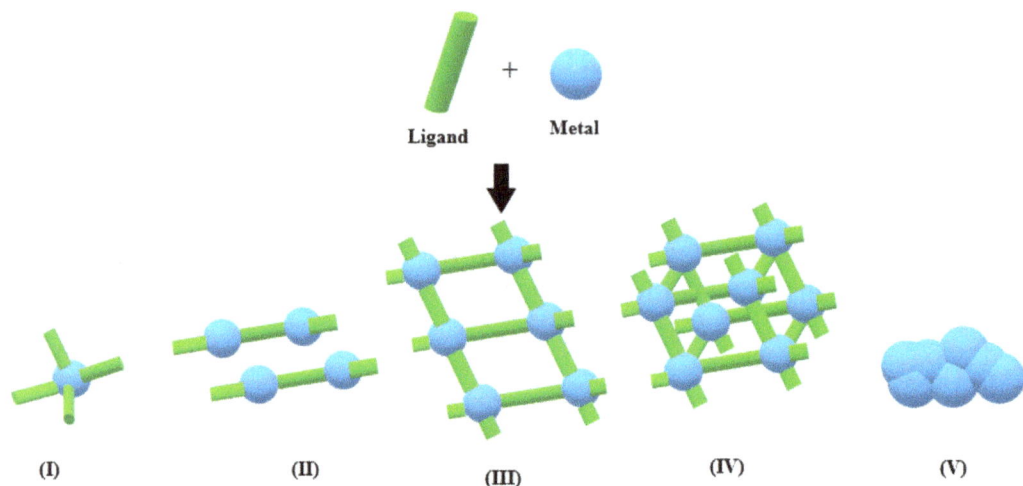

FIGURE 5.1 Schematic representation of the products obtained from the different bonding forces between metal ions and ligands. (I) Discrete molecule, (II) 1D coordination polymers, (III) 2D coordination sheet, (IV) 3D coordination materials, (V) Nano- and micro-particles.

5.2   CPs VS MOFs

However, the following two questions remain interesting. What are MOFs? How is it different from CPs? Three-dimensional metal-organic CPs may be called as MOFs. The classification of porous three-dimensional coordination polymer as a new term MOFs is very appropriate, while the extended one-dimensional or two-dimensional CPs are known as 1D CPs or 2D CPs (Biradha et al. 2009). The interactions between the host 3D framework and guest molecules cause a host-guest synergistic function in MOFs. MOFs with porous architectures can be prepared on a large scale from bottom-up synthesis.

The term MOF was introduced by Yaghi for the isolation of copper 4,4′-bipyridyl complex $Cu(4,4'-bpy)_{1.5}.NO_3(H_2O)_{1.25}$ with extended metal–organic frameworks (Yaghi and Li 1995). Later, several MOFs have been reported, in which MOF-5 depicts the zeolite-like framework in which inorganic core $[Zn_4O]^{6+}$ units are linked to an octahedral array of the 1,4-benzendicarboxylate organic linker to form a robust and highly porous cubic framework (Li et al. 1999) (Figure 5.2). In MOFs, the repeating unit of metal and ligand can be regarded as the secondary building units (SBUs). These SBUs are connected through the coordination bond to result in the final 3D structure or morphology (Rowsell and Yaghi 2004). More details about MOFs can be found in the following chapter.

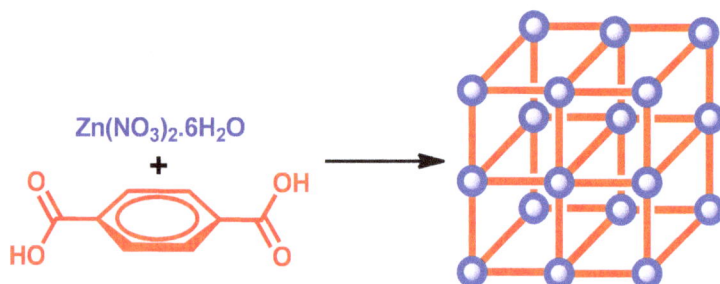

FIGURE 5.2 Synthesis of MOF-5.

5.3   APPLICATION OF CPs

The CPs consist of repeating metallic nodes connected by ligands (Figure 5.1). The structural networks can be easily modified for many applications using readily available yet highly tailorable ligand and metal precursors. Thus, the high surface areas along with chemically adjustable porosities (Ferey 2008) can provide the researcher with versatile applications, including in the field of heterogeneous catalysis (Ferey 2008; Liang et al. 2019; Babu et al. 2016), chemical sensing (Liu et al. 2020), energy conversion (Ferey 2008; Lustig and Li 2018; Li et al. 2018), gas storage, gas separation (Xue et al. 2019; Lippi and Cametti 2021), corrosion inhibitors (Saji 2019; Zhang et al. 2018), magnetism (Liu et al. 2016), drug delivery (Tu et al. 2019), bio-imaging (Zhu et al. 2019), photoluminescent (Suresh et al. 2015), and ion transport (Lee et al. 2020).

5.4   CLASSIFICATION OF CPs

Similar to organic polymers, the bulk properties of CPs are associated with the chemical nature of the building blocks, chain monodispersity, degree of polymerization, and intermolecular forces, which lead to extended 3D structures (Hoskins and Robson 1989). By carefully adjusting the confine conditions of self-assembly, it is possible to form suitable CP assemblies in solution. Thus, a general strategy for producing CPs requires the use of flexible or rigid or multi-donor ligands (Du et al. 2013) and suitable metal precursors to result in the assembly of 1D, 2D, or 3D structures (Figure 5.3) (Kurth and Higuchi 2006). To achieve this, the binding constants of metal ions with ligands have to be of optimal enduringness. Poor interactions can lead to discrete molecules, while too strong interactions can lead to nano- or micro-particle materials (Spokoyny et al. 2009). The depolymerization of CPs under suitable conditions can lead to the nano- or micro-particle materials. These nano- or micro-particle materials can be used in biomedical applications. Unlike the field of CPs, the synthesis and application of nano- or micro-particle materials derived from CPs are at a nascent stage. However, MOFs are not a suitable precursor to result in the nano- or micro-particle materials due to their static structures and bulk size. Similarly, the methods to control the particle size and shape are still rudimentary.

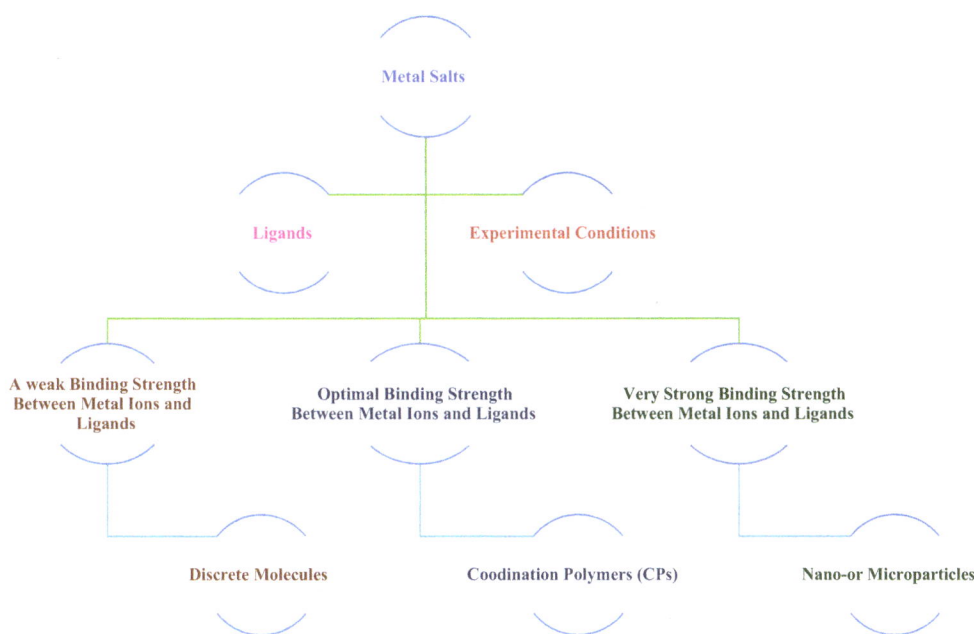

FIGURE 5.3   Methodology to isolate the coordination polymers.

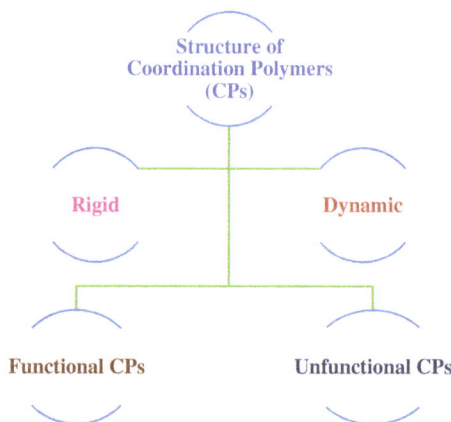

FIGURE 5.4   Types of coordination polymers.

Intermediate coordination strength between the metal ions and ligands can contribute to the CPs. The CPs formed by repeating metal-ligand coordinatively bound units can expand predominantly over one, two, or three dimensions. Through the choice of the metal ions, the design of the suitable ligands, the correct option of solvents, and the selection of experimental parameters, it is possible to determine a window of opportunity where CPs exist.

5.5   STRUCTURE OF CPs

The CPs can be rigid or much more dynamic due to the additional interactions (Friesea 2008) (Figure 5.4). In particular, the dynamic CPs can be isolated if the interaction between the metal ion and ligand is not too weak or too strong. As a result, the intermediate binding constants between metal ions and ligands imply a higher degree of freedom to the metal and ligand coordination. The large number of factors that are related to the chemical or physical properties, along with structural interactions, can also alter the whole entity. Hence, flexible porous CPs can be generated as a versatile host material, amplifying the effect. This is the significant advantage of dynamic CPs (Lee et al. 2020).

The chemical or physical functionalization of CP results in functional CPs (Kitagawa et al. 2004; Wang et al. 2017), while unfunctional CPs are structurally more attractive. The "functional" nature of CPs represents the potential applications of CPs. As listed in Section 5.2, the CPs and their functional properties have resulted in a significant impact on academic and industrial developments. Functional CPs can be rationally designed using metal ions or ligands or both, as these are the sources of functional properties.

According to the different compositions of ligands and metal ions, the CPs can be roughly classified into 1D, 2D, and 3D CPs (Figure 5.5) (Du et al. 2013; Loukopoulos and Kostakis 2019). The 1D CPs can exist in chain structures with a non-porous nature, while 2D CPs depict both porous and non-porous characterestics in layer structures. The 3D CPs are in the general porous type of materials. However, the rational design scheme to predict the structures of these CPs and synthetic strategies in defining the assemblies are not well-defined at this stage.

5.6   SUPRAMOLECULAR ASPECTS OF CPs

The structural packing of CPs can be determined by the secondary interactions, including π–π interactions or aromatic stacking (7.53-11.70 kJ/mol), CH-π interactions, hydrogen bonding (5–65 kJ/mol), van der Waals forces (<5 kJ/mol), dipole-dipole, ion-dipole, ion-ion (~250 kJ/mol), close packing, hydrophobic effect, and other chemical or physical factors such as ligand size, isomeric ligand effect, pH value, pressure, temperature, counter ion, solvent, and template, etc. (Figures 5.6

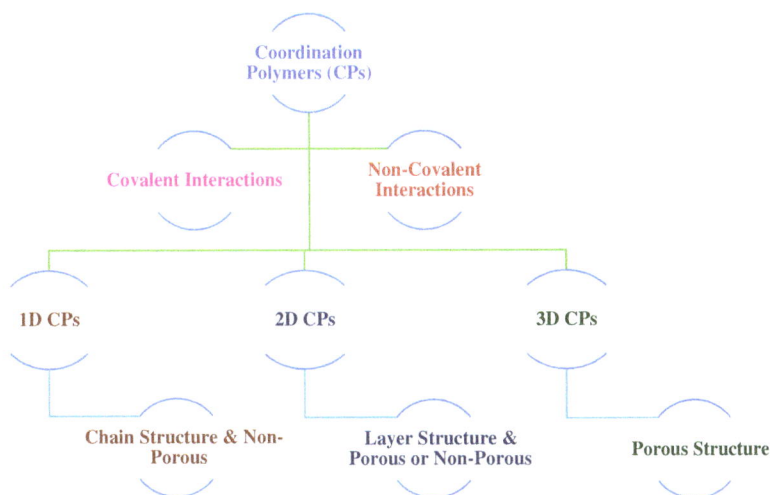

FIGURE 5.5   Structure of coordination polymers.

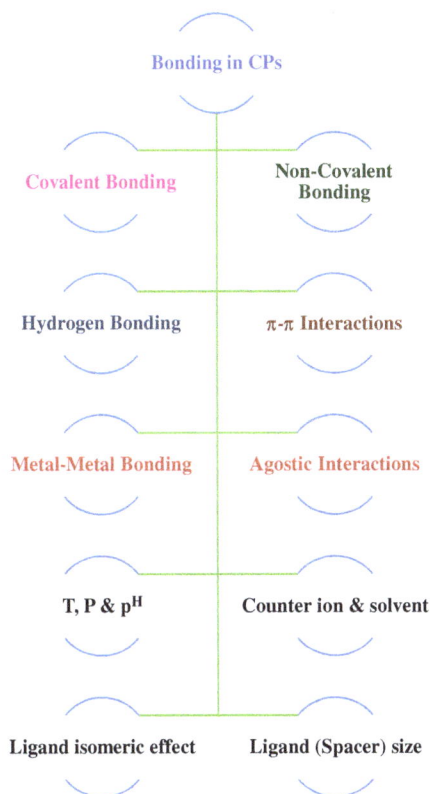

FIGURE 5.6   The selected influencing parameters in structural and functional properties of CPs. Both covalent and non-covalent interactions have a key role in determining the architecture of the CPs. The functional properties are influenced by the architecture of the CPs.

and 5.7) (Liang et al. 2008; Long 2010; Zheng et al. 2008; Du et al. 2009; Li et al. 2010; Li and Du 2011; Sun 2020; Lee 2020; Lippi and Cametti 2021). The primary (bond strength is about ~350 kJ/mol) and secondary interactions (bond strength is about <250 kJ/mol) are directed by the bridging ligands. Ligands with two or more oxygen, nitrogen, sulfur, phosphorous can depict different

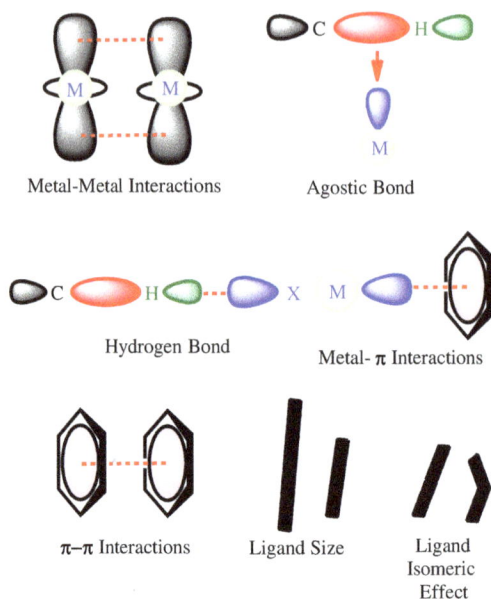

Metal-Metal Interactions   Agostic Bond

Hydrogen Bond   Metal- π Interactions

π–π Interactions   Ligand Size   Ligand
                                     Isomeric
                                     Effect

FIGURE 5.7 Some of the key structural and functional influencing factors in the supramolecular architecture of CPs.

binding abilities to metal ions for constructing diverse coordination networks. The rigid or flexible or both rigid and flexible ligands of mono-donor and hetero-donor ligands can be utilized with suitable metal ion precursors under different conditions to isolate the CPs with significant structural diversity and complexity.

The intermolecular or intramolecular interactions between C–H or C–C bond and metal can be regarded as the "agostic interactions" (Brookhart et al. 2007). The agostic bond is a 3-center-2-electron interaction between C–H and metal. The notion was introduced by Brammer et al. (1987). The bond distances of C–M and H–M are expected to be comparable in agostic interactions.

The metal-metal bond is an attractive interaction between dz^2 orbitals of metal centers, which is also known as the δ bond. The metal-metal bond can be supported by the bridging ligand or without the bridging ligand. The "metal-metal bonding interaction" was authenticated by Prof. F. Albert Cotton (Berry and Lu 2017; Krull 2018; Parkin 2010). In the representative example for the infinite quasi-1D zigzag coordination polymer PtCr(thiobenzoate)$_4$(NCS)]$_\infty$, the strong charge density along the Pr-Cr vector leads to strong Pt-Pt interactions in the solid-state (Figure 5.8) (Guillet et al. 2016).

The aromatic-aromatic interaction has been well accepted in supramolecular structures. Unfortunately, understanding the strength, directionality, control of aromatic-aromatic interactions are the key challenges. The interaction between aromatic π clouds in the crystal packing is known as the π–π interactions or π–π stacking. This can lead to the 1D, 2D, or 3D CPs. The term π–π interactions was rationalized by Martinez and Iverson in 2012 (Martinez and Iverson 2012). The π–π interactions or π–π stacking occurs due to the electrostatic effect of the polarized π systems. These interactions can be achieved when the aromatic groups are in T-shaped, offset-stacked, and face-centered stacking geometries. For example, as shown in 3D supramolecular coordination framework based on Cu(II) and zwitterionic 9,10-bis(imidazolylmethyl-N-acetate)anthracene bromate (BIMAA) as bridging anions, where the anthracene rings are aligned perfectly parallel to each other to enhance the π⋯π interactions (Figure 5.9) (Suresh et al. 2013).

C–H⋯π interaction is found between a soft acid and a soft base of the CPs. This attractive force occurs from the dispersion energy of aliphatic or aromatic CH groups, while coulombic energy is

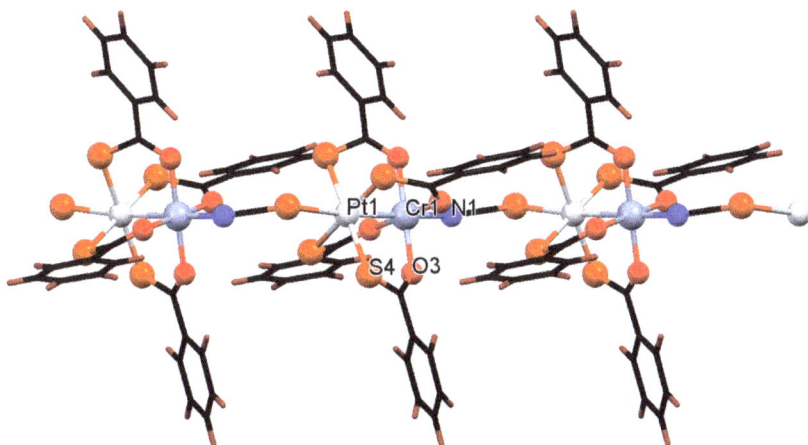

FIGURE 5.8 The infinite quasi-1D zigzag coordination polymer PtCr(thiobenzoate)$_4$(NCS)]$_\infty$ with metal-metal bonding.

the least important for this interaction (Nishio 2011; Nishio et al. 1998). Notably, the aromatic C–H⋯π interactions are stronger than the aliphatic C–H⋯π interactions. The directional property is important for the C–H⋯π interactions. The linear nature of C–H⋯π interactions can increase the strength of the hydrogen bond. The interaction between C–H and π cloud of aromatic fragment occurs cooperatively. The strength of the C–H⋯π interaction increases when the proton donating ability of the CH group is stronger.

One such classic example of 2D coordination polymer with C–H⋯π interactions is shown in Figure 5.10 (Ahmed et al. 2019). The two-dimensional coordination polymer, [Zn$_2$(fum)$_2$(4-phpy)$_4$(H$_2$O)$_2$]$_\infty$ (H$_2$fum = fumaric acid and 4-phpy = 4-phenyl pyridine), where the 14-membered metal-ligand ring with two Zn(II) centers is bridged by two fumarate dianions. This metal-ligand ring acts as a hydrogen acceptor with 4-phenyl pyridine to exhibit C–H⋯π interactions. Interestingly, the 14-membered metal-ligand ring on one layer interacts with the 4-phenyl pyridine ligand, coordinating with the metal center located at the adjacent layer in an edge-to-face C–H⋯π (metal-ligand ring) interaction with perfect T-shaped geometry.

Despite the vast applications of CPs, applications in corrosion science and technology are limited. The first comprehensive review with available literature was accounted in 2012 (Demadis et al. 2012) and later in 2019 (Saji 2019). The first authentic work on metal coordination polymer as a corrosion inhibitor at the metal surface was demonstrated by Demandis in 2005 (Demadis et al. 2005). The secondary interactions that exist in the CP seem to play a key role while grafting the CP on a metal surface (*vide supra*, Figure 5.6). For example, the reaction between zwitterionic HDTMP (Hexamethylenediamine-tetrakis(methylenephosphonate) and ZnCl$_2$ at p^H ~2.2 in a 1:1 ratio gave 3D CPs {Zn[(HO$_3$PCH$_2$)$_2$N(H)(CH$_2$)$_6$N(H)(CH$_2$PO$_3$H)$_2$]·H$_2$O}$_\infty$ (Zn-HDTMP) (Figure 5.11). The coordination environment of zinc(II) is fulfilled by the six phosphonate oxygen atoms in a distorted octahedral geometry. The polydentate mode of HDTMP in Zn-HDTMP results in the 3D CPs with a weak ZnO interactions. Zn-HDTMP coordination polymer was generated in situ on the surface of carbon steel. The Zn-HDTMP acts as a corrosion inhibitor by creating anticorrosive protective layers on the carbon steel surface and debited ~170% reduction in corrosion rate.

5.7 CONCLUSIONS AND PERSPECTIVES

This chapter implicates a broad overview of supramolecular aspects of CPs. A short classification for the CPs, as well as an overview of covalent and non-covalent interactions in the supramolecular architecture, are documented. Exceptionally, a large number of research articles have been published

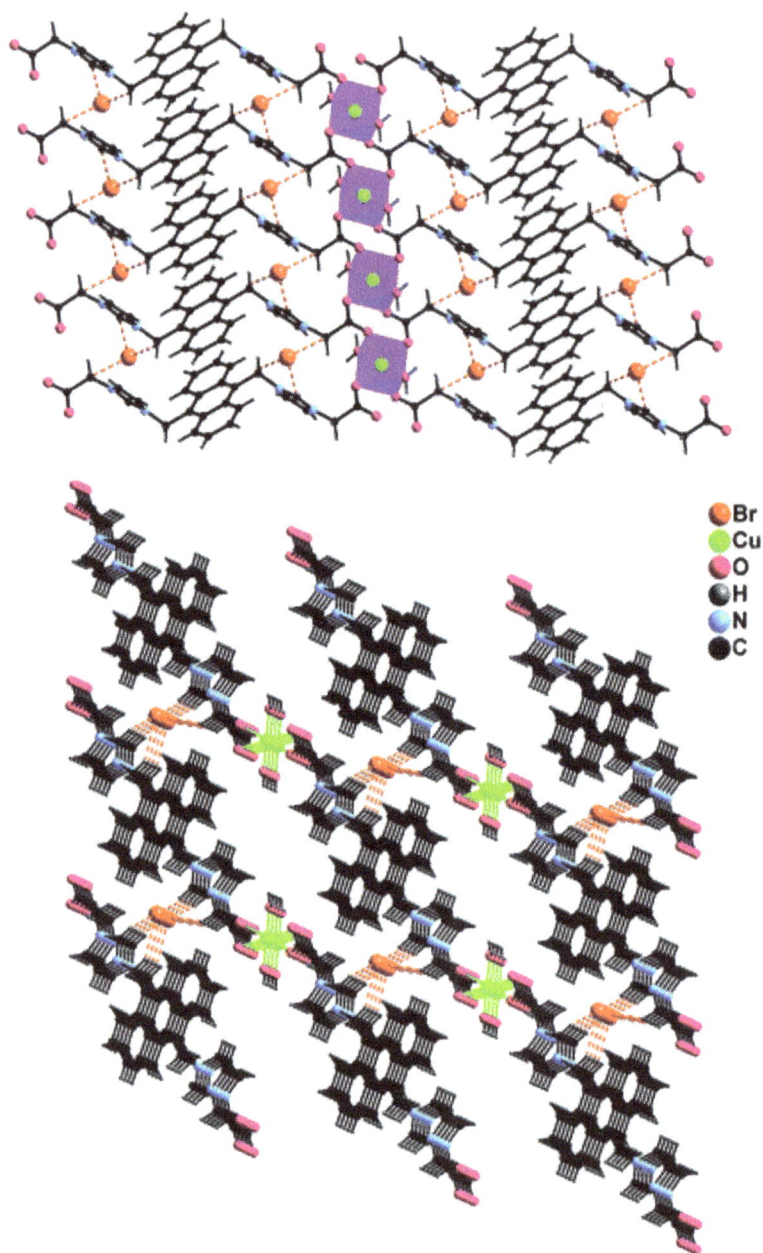

FIGURE 5.9 Top: A view of the 2D layer of $[\{(BIMAA)_2Cu(OH_2)_2\}_2Br_2]_\infty$ where polyhedra represent CuO_6; Bottom: A close view of the hydrogen bonding between the neighboring 2D layer in $[\{(BIMAA)_2Cu(OH_2)_2\}_2Br_2]_\infty$ viewed along the a-axis showing the 3D framework by π–π interactions or π–π stacking and hydrogen bonding between Br and C–H groups. (Reproduced with permission from Suresh (2013) © 2013 Royal Society of Chemistry.)

from 2011 to 2020. The number of papers published in the last 10 years has been accounted in Figure 5.12. Seventy thousand nine hundred fifty-eight research articles have been published on "coordination polymers." The nature of ligand and types of interactions in the resulting polymeric architectures are the key parameters in CPs design. However, there is still significant work to be done to understand the relationship between structural features and applications in corrosion.

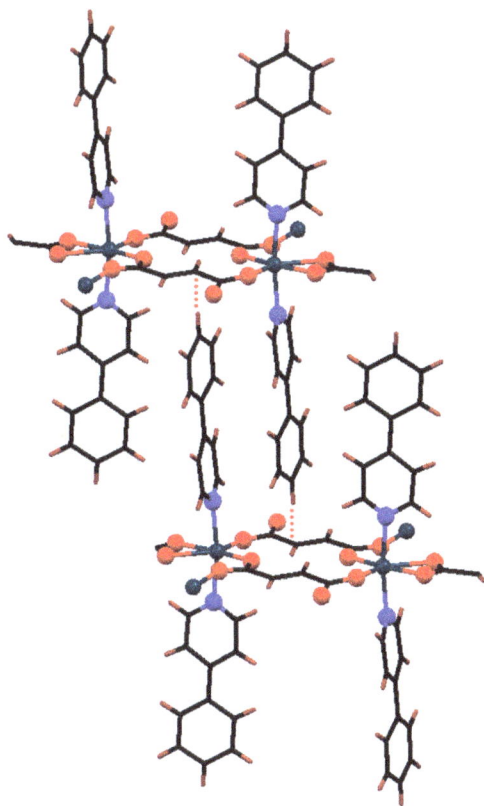

FIGURE 5.10 The C–H –π interactions induced 2D supramolecular coordination polymer based on Zn(II) 4-phenyl pyridine ligand and fumaric acid as bridging anions.

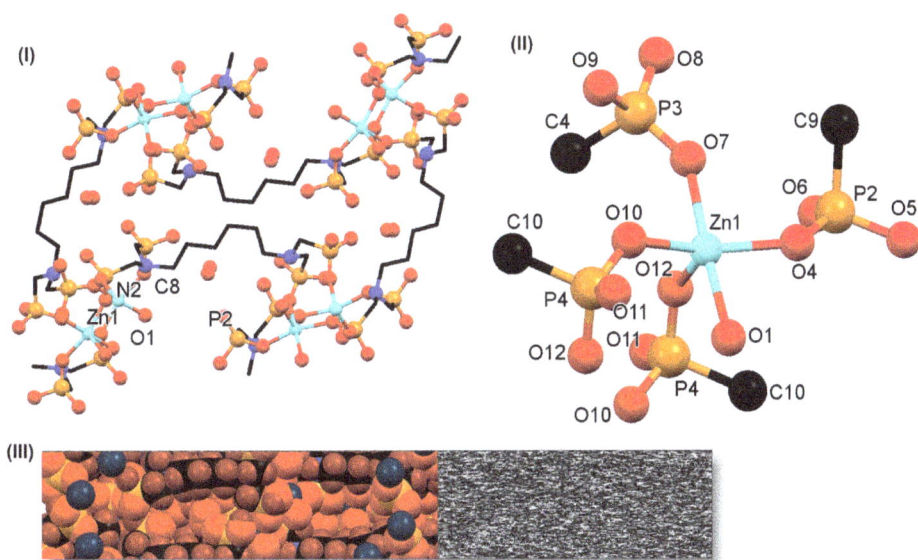

FIGURE 5.11 (I) 3D CPs of Zn-HDTMP. Hydrogen atoms have been omitted for clarity; (II) The coordination environment of zinc(II) in Zn-HDTMP; (III) Pictorial view of the anticorrosive protective film of Zn-HDTMP on the carbon steel surface.

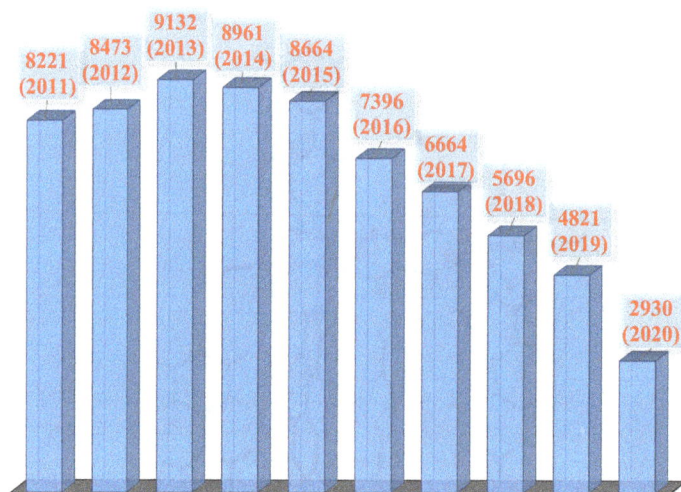

FIGURE 5.12 The number of papers published with coordination polymers from 2011 to 2020.

Although CPs are often reported to have excellent applications, these materials are ultimately useless if they are not stable on the metal surface. This could be due to the presence of liable coordination bonds that undergo hydrolysis to result in decomposition. Corrosion studies are generally conducted under harsh and non-ideal lab conditions. Therefore, the structural stability should be investigated using authenticated techniques. The disadvantages of these CPs mainly include the limited examples of corrosion inhibitors available up to date and the possible identification of the CPs in corrosion applications. New suitable CPs need to be developed for more complicated corrosion applications. The current strategy to use CPs in corrosion applications remains in the early stage. The corrosion performance of CP is not well related to the diverse structures of the available CPs. Those vast structures deserve future care for leading the rational designing of the corrosion systems, and we believe those exertions would be honoring.

REFERENCES

Ahmed, F., Ghosh, S. R., Halder, S., Guin, S., Alam, S. M., Ray, P. P., Jana, A. D., and Mir, M. H. 2019. Metal–ligand ring aromaticity in a 2D coordination polymer used as a photosensitive electronic device. *New J. Chem.* 43:2710–2717.

Arthur, D. E., Jonathan, A., Ameh, P. O., and Anya, C. 2013. A review on the assessment of polymeric materials used as corrosion inhibitor of metals and alloys. *Int. J. Indust. Chem.* 4(2):1–9.

Babu, C. N., Suresh, P., Srinivas, K., Sathyanarayana, A., Sampath, N., and Prabusankar, G. 2016. Catalytically active lead(II)-azolium coordination assemblies with diversified lead(II) coordination geometries. *Dalton Trans.* 45:8164–8173.

Bailar, J. C., Jr. 1964. Coordination polymer. *Prep. Inorg. React.* 1:1–57.

Batten, S. R., Champness, N. R., Chen, X.-M., Garcia-Martinez, J., Kitagawa, S., Öhrström, L., O'Keeffe, M., Suh, M. P., and Reedijk, J. 2012. Coordination polymers, metal–organic frameworks and the need for terminology guidelines. *Cryst. Eng. Comm.* 14:3001–3004.

Berry, J. F., and Lu, C. C. 2017. Metal–metal bonds: from fundamentals to applications. *Inorg. Chem.* 56(14):7577–7581.

Biradha, K., Ramanan, A., and Vittal, J. J. 2009, Coordination polymers versus metal–organic frameworks. *Cryst. Growth Des.* 9(7):2969–2970.

Braga D., and Grepioni, F. 2005. Making crystals from crystals: a green route to crystal engineering and polymorphism. *Chem. Commun.* 29:3635–3645.

Brammer, L., Charnock, J. M., and Goggin, P. L. 1987. Hydrogen bonding by cisplatin derivative: evidence for the formation of N–H ⋯ Cl and N–H ⋯ Pt bonds in [NPrn$_4$]$_2${[PtCl$_4$]·cis-[PtCl$_2$(NH$_2$Me)$_2$]}. *J. Chem. Soc. Chem. Commun.* 6:443–445.

Brookhart, M., Green, M. L. H., and Parkin, G. 2007. Agostic interactions in transition metal compounds. *PNAS* 104(17):6908–6914.

Chen, T. H., Popov, I., Kaveevivitchai, W., and Miljanic, O. S. 2014. Metal–organic frameworks: rise of the ligands. *Chem. Mater.* 26(15):4322–4325.

Demadis, K. D., Mantzaridis, C., Raptis, R. G., and Mezei, G. 2005. Metal–organotetraphosphonate inorganic–organic hybrids: crystal structure and anticorrosion effects of zinc hexamethylenediaminetetrak is(methylenephosphonate) on carbon steels. *Inorg. Chem.* 44:4469–4471.

Demadis, K. D., Papadaki, M., and Varouchas, D. 2012. Metal-phosphonate anticorrosion coatings. In Sharma, S. K., editor. *Green Corrosion Chemistry and Engineering: Opportunities and Challenges.* New Jersey: Wiley.

Desiraju, G. R. 2007. Crystal engineering: a holistic view. *Angew. Chem. Int. Ed.* 46:8342–8356.

Du, M., Li, C. P., and Guo, J. H. 2009. Unusual anion effect on the direction of three-dimensional (3-D) channel-like silver(I) coordination frameworks with isonicotinic acid N-oxide. *CrystEngComm.* 11:1536–1540.

Du, M., Li, C. P., Liu, C. S., and Fang, S. M. 2013. Design and construction of coordination polymers with mixed-ligand synthetic strategy. *Coord. Chem. Rev.* 257:1282–1305.

Ferey, G. 2008. Hybrid porous solids: past, present, future. *Chem. Soc. Rev.* 37:191–214.

Friesea, V. A., and Kurth, D. G. 2008. Soluble dynamic coordination polymers as a paradigm for materials science. *Coord. Chem. Rev.* 252(1–2):199–211.

Greer, H. F., Liu, Y., Greenaway, A., Wright, P. A., and Zhou, W. 2016. Synthesis and formation mechanism of textured MOF-5. *Cryst. Growth Des.* 16(4):2104–2111.

Guillet, J. L., Bhowmick, I, Shores, M. P., Daley, C. J. A., Gembicky, M. Golen, J. A., Rheingold, A. L., and Doerrer, L. H. 2016. Thiocyanate-ligated heterobimetallic {PtM} lantern complexes including a ferromagnetically coupled 1D coordination polymer. *Inorg. Chem.* 55(16):8099–8109.

Hoskins, B. F., and Robson, R. 1989. Infinite polymeric frameworks consisting of three dimensionally linked rod-like segments. *J. Am. Chem. Soc.* 111(50):5962–5964.

Kitagawa, S., Kitaura, R., and Noro, S. 2004. Functional porous coordination polymers. *Angew. Chem. Int. Ed.* 43:2334–2375.

Krull, C., Castelli, M., Hapala, P., Kumar, D., Tadich, A., Capsoni, M., Edmonds, M. T., Hellerstedt, J., Burke, S. A., Jelinek, P., and Schiffrin, A. 2018. Iron-based trinuclear metal-organic nanostructures on a surface with local charge accumulation. *Nature Chem.* 3211(9):1–7.

Kurth, D. G., and Higuchi, M. 2006. Transition metal ions: weak links for strong polymers. *Soft Matter.* 2:915–927.

Lee, J. S. M., Otake, K., and Kitagawa, S. 2020. Transport properties in porous coordination polymers. *Coord. Chem. Rev.* 421(213447):1–11.

Li, B., Fan, H. T., Zang, S. Q., Li, H. Y., and Wang, L. Y. 2018. Metal-containing crystalline luminescent thermochromic materials. *Coord. Chem. Rev.* 377:307–329.

Li, C. P., and Du, M. 2011. Role of solvents in coordination supramolecular systems. *Chem. Commun.* 47:5958–5972.

Li, C. P., Yu, Q., Chen, J., and Du, M. 2010. Supramolecular coordination complexes with 5-sulfoisophthalic acid and 2,5-bipyridyl-1,3,4-oxadiazole: specific sensitivity to acidity for Cd(II) species. *Cryst. Growth Des* 10(6):2650–2660.

Li, H., Eddaoudi, M., O'Keeffe, M., and Yaghi, O. M. 1999. Design and synthesis of an exceptionally stable and highly porous metal-organic framework. *Nature.* 402:276–279.

Liang, J., Huang, Y. B., and Cao, R. 2019. Metal–organic frameworks and porous organic polymers for sustainable fixation of carbon dioxide into cyclic carbonates. *Coord. Chem. Rev.* 378:32–65.

Lippi, M., and Cametti, M. 2021. Highly dynamic 1D coordination polymers for adsorption and separation applications, *Coord. Chem. Rev.* 430(213661):1–31.

Liu, J. Q., Luo, Z. D., Pan Y., Singh, A. K., Trivedi, M., and Kumar, A. 2020. Recent developments in luminescent coordination polymers: designing strategies, sensing application and theoretical evidences. *Coord. Chem. Rev.* 406(2131453):1–46.

Liu, K., Zhang, X., Meng, X., Shi, W., Cheng, P., and Powell, A. K. 2016. Constraining the coordination geometries of lanthanide centers and magnetic building blocks in frameworks: a new strategy for molecular nanomagnets. *Chem. Soc. Rev.* 45:2423–2439.

Long, J. R., and Yaghi, O. M. 2009. The pervasive chemistry of metal–organic frameworks. *Chem. Soc. Rev.* 38:1213–1214.

Long, L. S. 2010. pH effect on the assembly of metal–organic architectures. *CrystEngComm* 12:1354–1365.

Loukopoulos, E., and Kostakis, G. E. 2019. Recent advances in the coordination chemistry of benzotriazole-based ligands. *Coord. Chem. Rev.* 395:193–229.

Lustig, W. P., and Li, J. 2018. Luminescent metal–organic frameworks and coordination polymers as alternative phosphors for energy efficient lighting devices. *Coord. Chem. Rev.* 373:116–147.

Martinez, C. R., and Iverson, B. L. 2012. Rethinking the term "pi-stacking". *Chem. Sci.* 3:2191–2201.

Murray, L. J., Dinca, M., and Long J. R. 2009. Hydrogen storage in metal–organic frameworks. *Chem. Soc. Rev.* 38:1294–1314.

Nishio, M. 2011. The CH/π hydrogen bond in chemistry. Conformation, supramolecules, optical resolution and interactions involving carbohydrates. *Phys. Chem. Chem. Phys.* 13:13873–13900.

Nishio, M., Hirota, M., and Umezawa, Y. 1998. *The C–H/π Interaction: Evidence, Nature and Consequences,* Wiley-VCH, New York.

O'Keeffe M., and Yaghi, O. M. 2012. Deconstructing the crystal structures of metal–organic frameworks and related materials into their underlying nets. *Chem. Rev.* 112(2): 675–702.

Ovsyannikov, A., Solovieva, S., Antipin, I., and Ferlay, S. 2017. Coordination polymers based on calixarene derivatives: structures and properties. *Coord. Chem. Rev.* 352:151–186.

Park, S., Lee, S. Y., Park, K. M., and Lee, S. S. 2012. Supramolecular networking of macrocycles based on exo-coordination: from discrete to continuous frameworks. *Acc. Chem. Res.* 45(3):391–403.

Parkin, G. 2010. Structural and bonding: metal-metal bonding, In (Gonzalez-Gallardo, S., Prabusankar, G., Cadenbach, T., Gemel, C., Hopffgarten, M. V., Frenking, G., Fischer, R. F., Eds.), *Structure and Bonding of Metal-Rich Coordination Compounds Containing Low Valent Ga(I) and Zn(I) Ligands,* Springer, vol. 136, pp. 147–188.

Rosi, N. L., Eckert, J., Eddaoudi, M., Vodak, D. T., Kim, J., O'Keeffe, M., and Yagi, O. M. 2003. Hydrogen storage in microporous metal-organic frameworks. *Science* 300(5622):1127–1129.

Rowsell, J. L. C., and Yaghi, O. M. 2004. Metal–organic frameworks: a new class of porous materials. *Microp. Mesop. Mater.* 73:3–14.

Saji, V. S., 2019. Supramolecular concepts and approaches in corrosion and biofouling prevention. *Corros. Rev.* 37(3):187–230.

Spokoyny, A. M., Kim, D., Sumrein, A., and Mirkin, C. A. 2009. Infinite coordination polymer nano- and microparticle structures. *Chem. Soc. Rev.* 38: 1218–1227.

Steed, J. W., and Atwood, J. L. 2009. *Supramolecular Chemistry,* Wiley, Chichester, UK, 2nd Ed., pp. 538–589.

Sun, D., Ma, S., Simmons, J. M. Li, J. R., Yuan, D., and Zhou, H.-C. 2010. An unusual case of symmetry-preserving isomerism. *Chem. Commun.* 46:1329–1331.

Suresh, P., Babu, C. N., Sampath, N., and Prabusankar, G. 2015. Photoluminescent calcium azolium carboxylates with diversified calcium coordination geometry and thermal stability. *Dalton Trans.* 44:7338–7346.

Suresh, P., Radhakrishnan, S., Babu, C. N., Sathyanarayana, A., Sampath, N., and Prabusankar, G. 2013. Luminescent imidazolium carboxylate supported aggregate and infinite coordination networks of copper and zinc. *Dalton Trans.* 42:10838–10846.

Tu, D., Zheng, W., Huang, P., and Chen, X. 2019. Europium-activated luminescent nanoprobes: From fundamentals to bioapplications. *Coord. Chem. Rev.* 378:104–120.

Wang, H. Y., Cui, L., Xie, J. Z., Leong, C. F., D'Alessandro, D. M., and Zuo, J.-L. 2017. Functional coordination polymers based on redox-active tetrathiafulvalene and its derivatives. *Coord. Chem. Rev.* 345:342–361.

Xue, D. X., Wang, Q., and Bai, J. 2019. Amide-functionalized metal–organic frameworks: Syntheses, structures and improved gas storage and separation properties. *Coord. Chem. Rev.* 378:2–16.

Yaghi, O. M., Kalmutzki, M. J., and Diercks, C. S. 2019. Introduction to reticular chemistry: metal-organic frameworks and covalent organic frameworks. Chapter 1. In *Emergence of Metal-Organic Frameworks,* Wiley-VCH Verlag GmbH & Co. KGaA, pp. 1–28.

Yaghi, O. M., and Li, H. 1995. Hydrothermal synthesis of a metal-organic framework containing large rectangular channels. *J. Am. Chem. Soc.* 117:10401–10402.

Yu, Q., Zhang, X., Bian, H., Liang, H., Zhao, B., Yan, S., and Liao, D. 2008. pH-dependent Cu(II) coordination polymers with tetrazole-1-acetic acid: synthesis, crystal structures, EPR and magnetic properties. *Cryst. Growth Des.* 8:1140–1146.

Zhang, H., Zou, R., and Zhao, Y. 2015. Macrocycle-based metal-organic frameworks. *Coord. Chem. Rev.* 292:74–90.

Zhang, M., Ma, L., Wang, L., Sun, Y., and Liu, Y. 2018. Insights into the use of metal–organic framework as high-performance anticorrosion coatings. *ACS Appl. Mater. Interfaces* 10:2259–2263.

Zheng, B., Dong, H., Bai, J., Li, Y., Li, S., and Scheer, M. 2008. Temperature controlled reversible change of the coordination modes of the highly symmetrical multitopic ligand to construct coordination assemblies: experimental and theoretical studies. *J. Am. Chem. Soc.* 130:7778–7779.

Zhu, W., Zhao, J., Chen, Q., and Liu, Z. 2019. Nanoscale metal-organic frameworks and coordination polymers as theranostic platforms for cancer treatment. *Coord. Chem. Rev.* 398:113009.

6 Metal–Organic Frameworks and Their Host–Guest Chemistry

Suresh Babu Kalidindi
Andhra University

Marilyn Esclance DMello
Poornaprajna Institute of Scientific Research
Manipal Academy of Higher Education

Vasudeva Rao Bakuru
Andhra University

CONTENTS

6.1 INTRODUCTION

Metal–organic frameworks (MOFs) are porous crystalline materials with enormous surface areas and porosity (Zhou et al. 2012; Diercks et al. 2018). In MOFs, the metal ions/units (secondary building units, SBU) are connected with organic ligands through coordination covalent bonds. The directionality and strength of the coordination bonds play a vital role in the construction of these extended crystalline structures. MOFs can be envisioned as an extension of coordination complexes wherein the ligands are selected in such a way that extended framework solids are realized instead of molecular complexes. For instance, the copper acetate $\{Cu_2(CH_3CO_2)_4 \cdot 2H_2O\}$ has a paddle wheel structure (Figure 6.1a) in which one oxygen atom of each acetate is bound to one copper (Van Niekerk and Schoening 1953). The discrete units of copper acetate can be converted into an extended framework by replacing acetate with a 1, 3, 5-benzenetricarboxylic acid linker (Figure 6.1b and c). The resultant MOF is named as HKUST-1(Hong Kong University of Science and Technology) $\{Cu_3(BTC)_2(H_2O)_3\}$ and has an attractive Brunauer–Emmett–Teller (BET) surface area of ~2200 m^2/g (Chui et al. 1999). In this MOF, two water molecules are connected to each Cu-Paddle wheel to satisfy the coordination. These can be removed by heating at a high temperature. Similarly, MOF-5 $\{Zn_4(H_2BDC)_6\}$ can be envisioned as an extension of the tetranuclear zinc acetate molecular complex (Figure 6.1d). The extended structure of MOF-5 is obtained using a

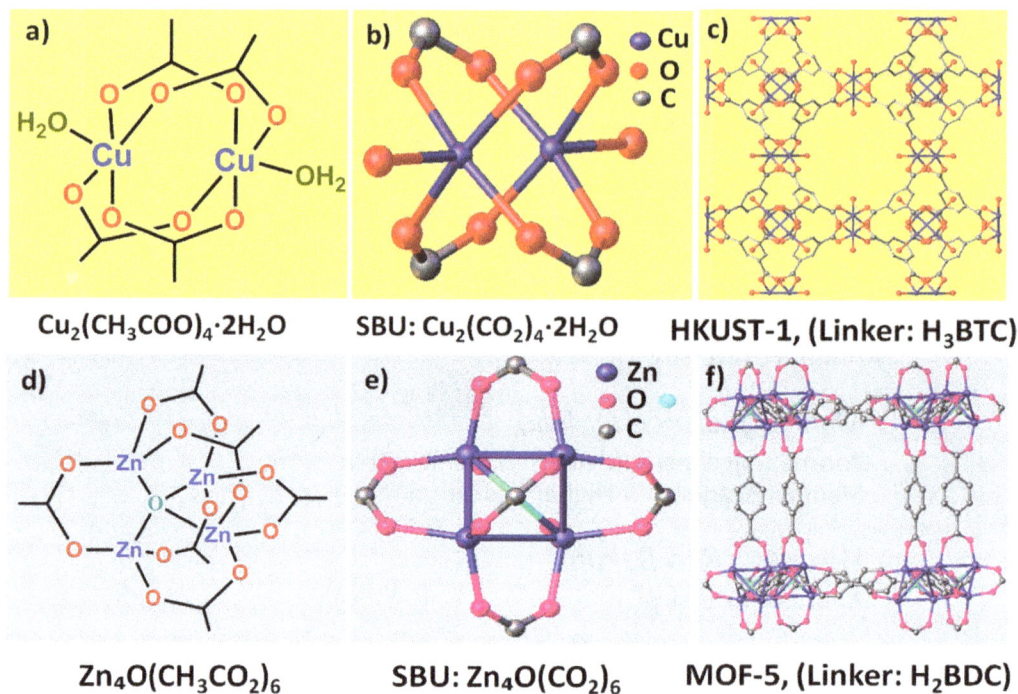

FIGURE 6.1   Structure of (a) dinuclear copper acetate hydrate, (b) and (c) secondary building unit (SBU) of HKUST-1 MOF, (d) tetranuclear zinc acetate, (e) and (f) SBU of MOF-5.

terephthalic acid linker instead of acetate (Figure 6.1e and f). The MOF-5 has a BET surface of around 3500 m^2/g (Rosi et al. 2003).

One of the attractive features of MOFs is their permanent porosity (Maurin et al. 2017). Most MOFs are synthesized under solvothermal conditions, and as-synthesized MOFs contain solvents in their pores. Most of the MOFs retain their structure and porosity when the solvent is removed from the pores (Lin et al. 2014), though sometimes it requires the use of softer methods such as supercritical CO$_2$ drying (Matsuyama 2018). The MOFs show attractive porosity and high surface areas up to 10,000 m^2/g (Table 6.1). These are several times higher when compared to traditional porous materials such as zeolites (Tao et al. 2006). Further, the porosity of MOFs is highly tunable (Wang et al. 2013). For example, by replacing terephthalic acid with 2-amino terephthalic acid, a Lewis basic functionality (-NH$_2$) can be introduced into the porosity of UiO-66 (Sun et al. 2013). The porosity of MOFs can be enhanced significantly by the use of longer linkers without altering the topology of the framework. This concept is known as isoreticular chemistry and is widely used for the synthesis of MOFs with enhanced porosity (Katz et al. 2013) (Figure 6.2).

The significant aspects concerning MOF design are (i) *Control over topology*: By apt selection of SBUs and organic linkers, it is possible to synthesize a targeted framework topology; (ii) *Control over porosity*: It is possible to modulate the porosity of frameworks by varying the size and nature of a structure without altering the underlying topology; (iii) *Pore functionalization*: Desired functionality can be included into the internal surface through organic linkers using direct/post-synthetic strategies. Thus, MOFs are customizable materials with well-defined porosity that can be manipulated at an atomic scale. MOFs, therefore, show a promising myriad of applications in several areas including gas storage/separation, biomedical imaging, optoelectronics, drug delivery, catalysis, and sensing (Furukawa et al. 2013).

TABLE 6.1
Different MOFs and Their Surface Area with Pore Volumes

S. No.	Name	Molecular Formula	Surface Area (m²/g)	Pore Volume (cm³/g)
1	UiO-66	$Zr_6O_4(OH)_4(H_2BDC)_6$	1391.0	0.52
2	NH$_2$-UiO-66	$Zr_6O_4(OH)_4(NH_2-H_2BDC)_6$	1070.0	0.45
3	UiO-67	$[Zr_6O_4(OH)_4 (H_2BPDC)_6$	2590.0	1.14
4	MIL-101	$Fe_3Cl(H_2O)_2O(H_2BDC)_3$	3192.0	1.86
5	MIL-53	$Al(OH)[H_2BDC]$	1410.0	0.54
6	MOF-5	$Zn_4(H_2BDC)_6$	3500.0	0.61
7	ZIF-8	$Zn(2-mIM)_2$	1813.0	0.65
8	HKUST-1	$Cu_3(H_3BTC)_2(H_2O)_3$	2200.0	0.19
9	MOF-74(Ni)	$Ni_2(H_2DOBDC)$	1418.0	0.84

H_2BDC, Benzene-1, dicarboxylic acid; NH_2-H_2BDC, 2-amino benzene-1, dicarboxylic acid; H_2BPDC, biphenyl-4, 4′-dicarboxylic acid; 2-mIM, 2-Methylimidazolate; H_3BTC, Benzene-1, 3, 5-tricarboxylic acid; H_2DOBDC, 2, 5-dihydroxyterephthalic acid.

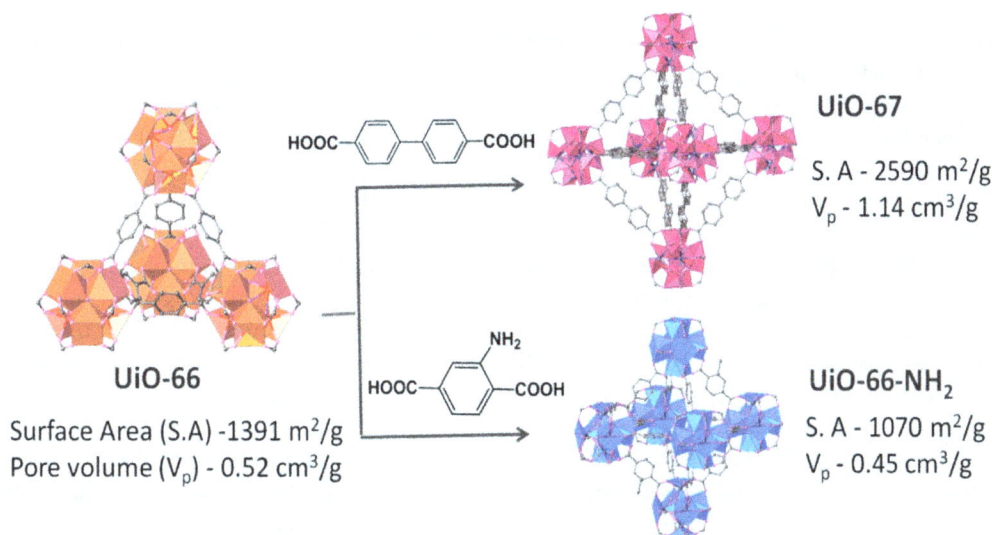

FIGURE 6.2 Iso-structural topology of UiO-66, UiO-67 and NH$_2$-UiO-66 frameworks and their surface areas with pore volume.

6.2 HOST–GUEST CHEMISTRY WITH MOFs

MOFs are versatile hosts for the incorporation of a variety of guest species due to the presence of tunable cavities and tailorable internal surfaces. Host–guest chemistry is one of the basic concepts of supramolecular chemistry. Host–guest chemistry, which was studied in solutions in the initial days, is now extended to porous solids such as MOFs (Bakuru et al. 2019; Cai et al. 2019). Guests whose size and shape match well with the pore window openings of the MOFs are immobilized within the cavities of the host framework generally by the weak and dynamic forces such as hydrogen bonding, van der Waals forces, π···π stacking, and electrostatic or dipole interactions. The stabilization of guest molecules (recognition process) inside the cavities of MOFs is driven by the type and number of preferential binding sites present in the framework. The weak supramolecular forces can be tuned through pore surface engineering both at SBU and organic linker. Thus, MOFs are interesting candidates for the study of host–guest chemistry (Cook et al. 2013).

6.3 HOST–GUEST INTERACTIONS WITHIN MOF CAVITIES

Various synthetic methodologies have been employed for the incorporation of guests into the cavities of MOFs, which include chemical vapor deposition, co-assembly of the guest during MOFs crystal growth, solution method, and cation exchange. The composites obtained through these methodologies find applications in catalysis, molecular separation, sensing, drug delivery, and magnetism.

In this book chapter, we have segregated the MOF-based host–guest chemistry based on the type of interaction between host–guest and discussed it with examples.

6.3.1 Hydrogen-Bonding Interactions

Hydrogen-bonding interactions are moderately weak electrostatic interactions. These types of interactions result from the dipole–dipole attraction of an electropositive hydrogen atom with a very electronegative atom, represented as C–H···X (where, X=N, O, Cl or F atom, etc.). Hydrogen-bonding interactions can be considered as one of the driving forces to assemble supramolecular functions in MOFs for various host–guest interactions. The directionality and reversibility of the hydrogen bonds are the major characteristics that deal with selective complexation of the host–guest type of interactions. Hence, a great variety of hydrogen donors/acceptors exist based on multipoint hydrogen-bonding concerned with host–guest complexes with MOFs.

The selective stabilization of xylene isomers through C–H···F contacts inside the cavities of ZU-61 MOF (where ZU=Zhejiang University and ZU-61=NbOFFIVE-bpy-Ni MOF) was effectively used in the separation of xylene isomers (Cui et al. 2020). The ZU-61 MOF was constructed from $NbOF_5^{2-}$ SBU and 4,4′-bipyridine (bpy) organic ligand. Industrial chemical separation of xylene isomers consisting of o-xylene (oX) m-xylene (mX), p-xylene (pX) and ethylbenzene (EB) is the most difficult mixture to separate. This is due to the identical molecular formula of these isomers as well as their negligible boiling points difference. The preferential adsorption sequence was in the order of oX>mX>pX, with a high uptake capacity of 3.2 mmol/g for oX and 3.4 mmol/g for mX at 333 K and 7.1 mbar.

The isomers were stabilized inside ZU-61 cavities with different degrees of hydrogen bonding as determined by single-crystal X-ray diffraction studies. In the case of oX adsorption in ZU-61, strong host–guest interaction through synergistic C–H···F contacts were responsible for oX binding (Figure 6.3a and b). As for pX adsorption, two types of adsorption sites were identified in the channels of ZU-61 as shown in Figure 6.3c–e. One of the adsorption sites for pX molecule is provided by the electronegative F atoms through synergistic C-H···F interactions with the hydrogen atoms in the xylene molecules. Whereas, two F atoms interacting with two methyl groups of pX molecule was considered as the second adsorption site. In the case of mX adsorption on ZU-61, a rotation of 6° of uncoordinated F sites with no structural deformation was detected. Herein, three F atoms interact with one mX molecule through synergistic C-H···F interactions (Figure 6.3f). Overall, electropositive $NbOF_5^{2-}$ anions are the main adsorption sites for binding xylene isomers through C-H···F hydrogen-bonding interactions.

Functional hydrogen-bonding donors such as amines, amides, urea, and thioureas present in the organic linker have been explored as charge neutral receptors for guest anions. For example, Kalidindi et al. investigated NH_2-UiO-66 for chemiresistive sensing of acidic gases like SO_2, NO_2, and CO_2 (UiO=University of Oslo). The detection of these acidic gases was studied based on the change in electrical resistance exhibited by the NH_2-UiO-66 MOF. The $-NH_2$ functionality of MOF can effectively form charge-transfer complexes through donor-acceptor hydrogen-bonding

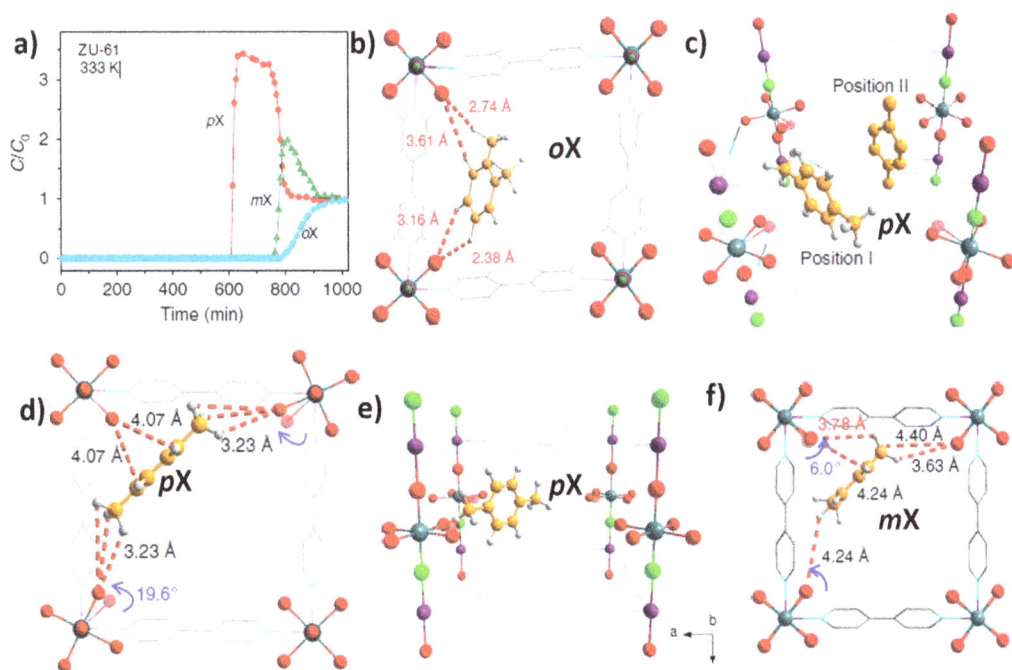

FIGURE 6.3 (a) Experimental column breakthrough results of ZU-61 for 1:1:1 pX/mX/oX separations with ZU-61 at 333 K. Single-crystal X-ray diffraction resolved structure of (b) ZU-61·oX. Crystal structure of ZU-61 after adsorption of o-xylene ZU-61·oX, (c–e) ZU-61·pX. Two adsorption positions of pX molecules in the unit cell of ZU-61 in different directions, and (f) mX in the pores of ZU-61. Color code: C, gray 50%; H, gray 10%; Nb, teal; Ni, violet; O, green; N, sky blue; F, red. (Reproduced from Cui et al. © 2020, Nature, under Creative Commons Attribution License.)

interaction with SO_2 gas compared to the other acidic gases as shown in Figure 6.4a and b (DMello et al. 2019). The NH_2-UiO-66 MOF was also explored for pyridine adsorption by S. H. Jhung and group. The basic adsorbate-pyridine interacted more favorably with increasing amino group content in UiO-66 MOF in both vapor and liquid phases as shown in Figure 6.4c and d. Herein, the hydrogen-bonding interactions were significant for increased adsorption of pyridine over NH_2-UiO-66 (Hasan et al. 2014).

6.3.2 π–π INTERACTIONS

Another class of important interaction in supramolecular host–guest chemistry of MOFs are π–π interactions. It is a type of a non-covalent interaction between aromatic π systems that involves a combination of dipole–dipole-induced dipole interaction. Nevertheless, these types of interactions have energies lower than 10 J/mol compared to that of hydrogen-bonding interaction energy of 25–40 kJ/mol. Although hydrogen-bonding interaction is the most common, sometimes it must cooperate with π–π interactions to stabilize these networks. The π–π interactions in MOFs can be utilized to fine-tune the orientation of the SBU, inducing the functionalities for a potential application.

Guest molecules containing π electron-rich donor and MOF host with π electron acceptor sites can effectively interact through weak π–π interactions. In this context, Genna et al. proposed a proof-of-concept study for pharmaceutical drug incorporation into MOFs and its removal from water through π–π stacking interaction. YCM-101 (YCM = Youngstown Crystalline Material) was made up of $InCl_3$ and 2,3,5,6-tetrafluoro-1,4-dicarboxylic acid as shown in Figure 6.5a and b. Initial studies were conducted on benzene and pyridine guest molecules. Two types of π–π stacking

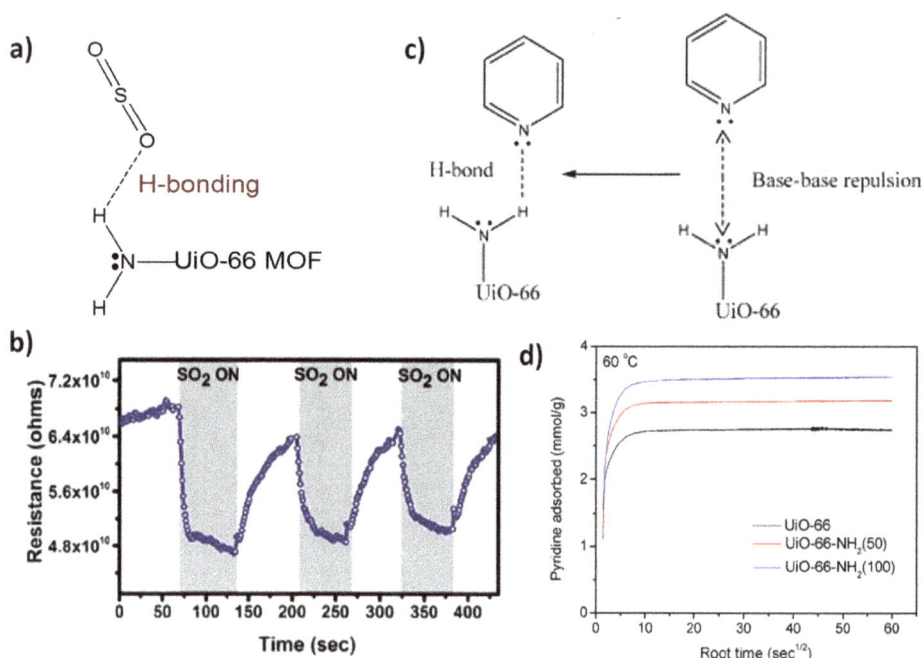

FIGURE 6.4 (a) Schematic representation of intermolecular hydrogen-bonding interaction between amine-functionalized UiO-66 MOF and SO_2 molecule, (b) Response-Recovery curves of chemiresistive gas sensing studies for SO_2 gas using NH_2-UiO-66, (c) Schematic representation of H-bond and base−base repulsion between pyridine and NH_2-UiO-66, and (d) Effect of temperature and amino group content of the UiO-66s on the kinetics of Py adsorptions at 60°C. ((b) Reproduced with permission from DMello et al. © 2019, Royal Society of Chemistry. (c) and (d) Reproduced with permission from Hasan et al. © 2014, American Chemical Society.)

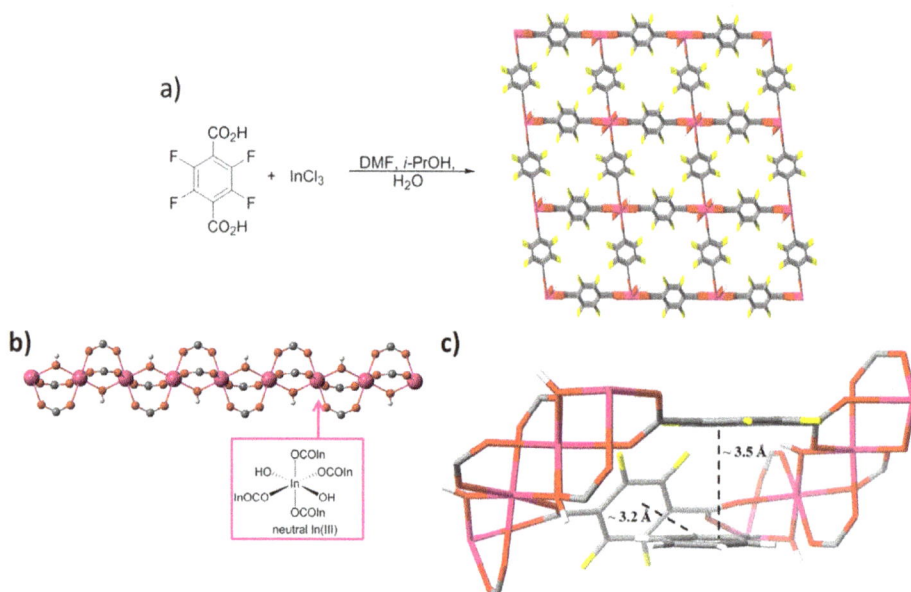

FIGURE 6.5 (a) Synthesis of YCM-101, (b) Cartoon and ChemDraw of SBU of YCM-101 and (c) π–π stacking interactions in YCM-101-benzene. (Reproduced with permission from DeFuria et al. (2016) © 2016, American Chemical Society.)

interactions were observed for benzene guest. One was a direct stacking interaction with two adjacent perfluorinated linkers and the other was a T-shaped one (Figure 6.5c). However, only direct stacking interaction was observed in the case of pyridine. The removal of pharmaceutical drug tetracycline hydrochloride by YCM-101 was conducted in an aqueous solution. Herein, the electron-deficient YCM-101 framework and electron-rich tetracycline guest interact with each other through π–π stacking. Approximately 800 ppm of tetracycline uptake by YCM-101 was observed from the initial 10,000 ppm concentration. However, the tetracycline doped YCM-101 was ineffective to recycle the pristine MOF. In the future, the development of perfluorinated MOF with more hydrophobic groups in structure can probably improve the inherent instability of these MOF structures (DeFuria 2016).

Significant progress has also been made in gas separation through a combination of hydrogen bonding and π–π stacking sites. NbOFFIVE-1-Ni MOF with a chemical composition of [Ni (NbOF$_5$) (C$_4$H$_4$N$_2$)$_2$ · 2H$_2$O]) was explored for propylene/propane separation (Antypov 2020). NbOFFIVE-1-Ni framework is made up of (NbOF$_5$)$^{2-}$ anions and pyrazine (C$_4$H$_4$N$_2$) linkers (Figure 6.6a and b). The (NbOF$_5$)$^{2-}$ anion interacts with pyrazine linkers to form an intraframework C-H···F contacts. The MOF (host) dynamics for guests (propylene/propane) are controlled by the guest entry and transport through the MOF channels. At 25°C and 1 bar pressure, NbOFFIVE-1-Ni adsorbs 58 mg/g of propylene and 3 mg/g of propane which increases as pressure and temperature are increased. From the structural experimental data and density functional theory, four close C-H···F

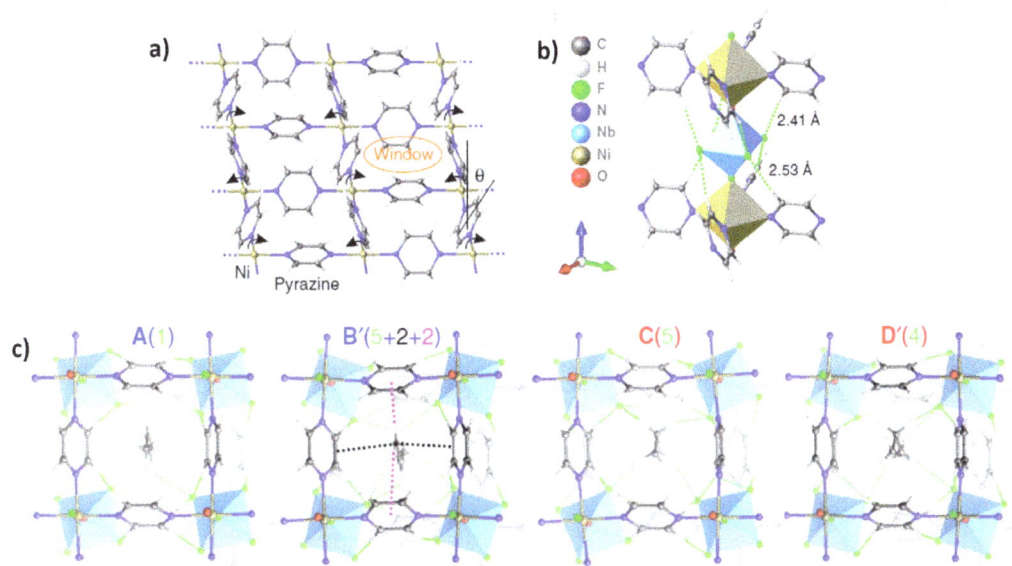

FIGURE 6.6 (a) The structure of the square-grid Ni-pyrazine layer with windows formed by four adjacent pyrazines. The arrows show the alternating tilts of pyrazine molecules within this layer, defined by θ, observed in all structures studied, (b) An octahedral (NbOF$_5$)$^{2-}$ anion forms two types of C-H···F contacts (green dotted lines, distances shown) with two adjacent Ni-pyrazine layers: shorter contacts with the top layer and longer contacts with the bottom layer. Each equatorial fluoride forms two contacts with pyrazines that are not equivalent, and (c) DFT calculations of guests near the window - The snapshots show the stable configurations for propylene (A and B′) and propane (C and D′) on each side of the window. The guests rotate by 90° as they pass through the window, owing to the shape of the channel. The green dotted lines show the C-H···F contacts shorter than 2.67 Å, the sum of van der Waals radii for H and F, with their total number indicated in brackets. Face-to-face (black dotted lines showing two carbon-to-double-bond distances of 3.18 Å each) and edge-to-face (magenta dotted line showing hydrogen-to-centroid distances of 2.75 and 3.35 Å) π–π interactions are shown for propylene in B′. (Reproduced with permission from Antypov et al. © 2020, Nature under Creative Commons Attribution License.)

contacts, π–π and C-H⋯ π interactions stabilized each propylene position near the pore window of the MOF. Whereas, depending on adjacent $(NbOF_5)^{2-}$ anion, each propane position is stabilized by up to five C-H⋯F contacts. Thus, the π–π interactions between the sp^2 carbon and the aromatic ring of pyrazine are overall responsible for propylene residing at the end of the cavity window (Figure 6.6c). In the case of propane, transport across the cavity is restricted due to the larger saturated propane guest.

6.3.3 Van der Waals Interactions

Van der Waals interactions are the weakest of all the intermolecular interactions that can occur between two or more atoms/molecules in close vicinity (0.3–0.6 nm). Three types of intermolecular interactions, namely (i) dipole–dipole interactions, (ii) dipole-induced dipole interactions, and (iii) spontaneous dipole-induced dipole interactions, contribute to the van der Waals forces. All these interactions can be responsible for the bond-free assembly of guest molecules inside the cavities of crystalline MOFs. This has been advantageously exploited for the storage/separation of small gas molecules such as H_2, CO_2, N_2, CH_4, and O_2 at very low temperatures inside MOF cavities. The van der Waal forces between the MOF internal surface and gas molecules play a vital role in determining the overall gas storage capacity/separation efficiencies.

Chun and Seo synthesized a Zn-based flexible MOF Zn(pydc)(DMA), (where, pydc = 3,5-pyridinedicarboxylate and DMA = N, N'-dimethylacetamide) under solvothermal conditions wherein, Zn^{2+} ion is coordinated to pydc linkers and DMA solvent molecules to yield large polyhedral crystals (Chun and Seo 2009). This gave rise to one- dimensional zig-zag channels with 8×5 Å^2 ellipsoidal cross section and the free passage is much smaller at 4.5×3 Å^2 (Figure 6.7a). N_2 (at 77 K) and Ar (at 87 K) gas adsorption isotherms showed unusual behavior for this MOF. There was no gas adsorption at low pressures. A breakthrough occurs at 17 and 26 kPa for N_2 and Ar respectively, and adsorption takes place inside the channels of MOF (Figure 6.7b and c). This was attributed to the slit-type gate opening phenomenon at breakthrough pressures. However, such a phenomenon is not observed in the case of kinetically smaller H_2 gas. The material was completely non-porous to H_2. In spite of small kinetic diameter, H_2 gas at 77 K failed to conduct the gate opening of the MOF channels due to very weak van der Waals interactions and high thermal energy. However, in

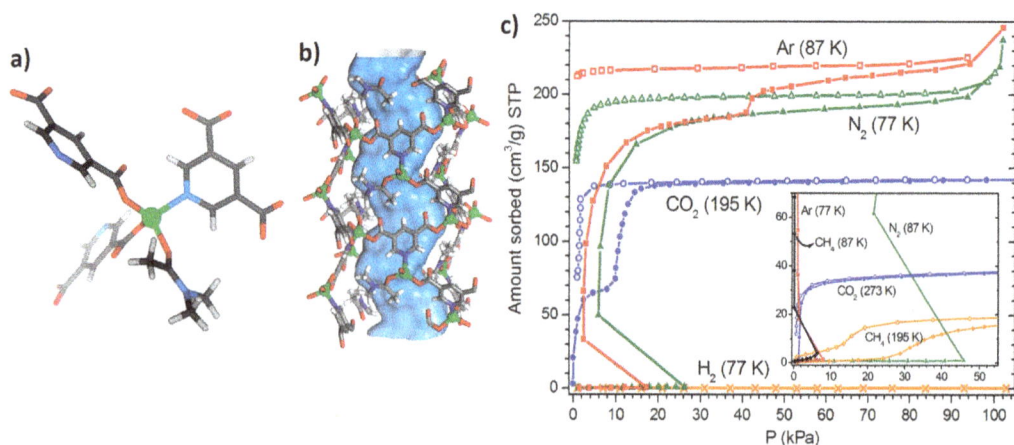

FIGURE 6.7 (a) Coordination environment of the Zn^{2+} ion, (The gray, white, red, green, and blue spheres represent carbon, hydrogen, oxygen, zinc, and nitrogen centers respectively). (b) Perspective side view of the 1D channel represented in blue color, and (c) Gas-sorption isotherms at various temperatures. Closed and open symbols denote adsorption and desorption, respectively. (Reproduced with permission from Chun and Seo © 2009 American Chemical Society.)

the case of CO_2 at 195 and 273 K, this breakthrough-like curve was absent and hence adsorbs a significant amount of CO_2 even at low pressures. The most favored adsorption of CO_2 in Zn(pydc)(DMA) MOF is attributed to the high quadrupole moment associated with the molecule. Thus, MOF sorption behavior is dependent on the strength of van der Waals forces between gas molecules and the framework rather than the size of the molecules.

6.4 CONCLUSIONS AND PERSPECTIVES

MOFs are customizable crystalline porous materials as it is possible to get control over topology, porosity, and pore chemistry using reticular chemistry principles. Herein, host–guest chemistry of MOFs is classified based on the type of interactions, namely hydrogen bonding, π–π stacking and van der Waals forces. ZU-61 MOF containing electropositive $NbOF_5^{2-}$ anions was effectively used in the separation of xylene isomers through C-H···F hydrogen-bonding interactions. In another example of functional hydrogen bond donors present on carboxylate-based linkers, NH_2-UiO-66 was used for SO_2 and pyridine adsorption. Further, describing the π–π stacking interactions, YCM-101 and NbOFFIVE-1-Ni MOF were studied for tetracycline uptake and propylene/propane separation respectively. Herein, both these MOFs displayed a combination of hydrogen bonding and π–π stacking sites. In the case of weak van der Waals interactions, Zn(pydc)(DMA) microporous $Mg(HCOO)_2$ and $Mn(HCOO)_2$ containing zig-zag channels were studied for small guest adsorption based on the pressure-temperature sensitive operation. Overall, MOFs with versatile porosity exhibited interesting shape/size-selective supramolecular host–guest chemistry through weak and dynamic interactions which find applications in separation/storage of industrially relevant molecules.

REFERENCES

Antypov, D., Shkurenko, A., Bhatt, P. M., Belmabkhout, Y., Adil, K., Cadiau, A., Suyetin, M., Eddaoudi, M., Rosseinsky, M. J., and Dyer, M. S. 2020. Differential guest location by host dynamics enhances propylene/propane separation in a metal-organic framework. *Nat. Commun.* 11(1):1–8.

Bakuru, V. R., Davis, D. and Kalidindi, S. B. 2019. Cooperative catalysis at the metal–MOF interface: hydrodeoxygenation of vanillin over Pd nanoparticles covered with a UiO-66 (Hf) MOF. *Dalton Trans.* 48(24):8573–8577.

Cai, H., Huang, Y. L., and Li, D. 2019. Biological metal–organic frameworks: structures, host–guest chemistry and bio-applications. *Coord. Chem. Rev.* 378:207–221.

Chui, S. S. Y., Lo, S. M. F., Charmant, J. P., Orpen, A. G., and Williams, I. D. 1999. A chemically functionalizable nanoporous material $[Cu_3(TMA)_2(H_2O)_3]_n$. *Science* 283(5405):1148–1150.

Chun, H. and Seo, J. 2009. Discrimination of small gas molecules through adsorption: reverse selectivity for hydrogen in a flexible metal– organic framework. *Inorg. Chem.* 48(21):9980–9982.

Cook, T. R., Zheng, Y. R., and Stang, P. J. 2013. Metal–organic frameworks and self-assembled supramolecular coordination complexes: comparing and contrasting the design, synthesis, and functionality of metal–organic materials. *Chem. Rev.* 113(1):734–777.

Cui, X., Niu, Z., Shan, C., Yang, L., Hu, J., Wang, Q., Lan, P. C., Li, Y., Wojtas, L., Ma, S., and Xing, H. 2020. Efficient separation of xylene isomers by a guest-responsive metal–organic framework with rotational anionic sites. *Nat. Commun.* 11(1):1–8.

DeFuria, M. D., Zeller, M. and Genna, D. T. 2016. Removal of pharmaceuticals from water via π–π stacking interactions in perfluorinated metal–organic frameworks. *Cryst. Growth Des.* 16(6):3530–3534.

Diercks, C. S., Kalmutzki, M. J., Diercks, N. J., and Yaghi, O. M. 2018. Conceptual advances from Werner complexes to metal–organic frameworks. *ACS Central Sci.* 4(11):1457–1464.

DMello, M. E., Sundaram, N. G., Singh, A., Singh, A. K., and Kalidindi, S. B. 2019. An amine functionalized zirconium metal–organic framework as an effective chemiresistive sensor for acidic gases. *Chem. Commun.* 55(3):349–352.

Furukawa, H., Cordova, K. E., O'Keeffe, M., and Yaghi, O. M. 2013. The chemistry and applications of metal-organic frameworks. *Science* 341(6149).

Hasan, Z., Tong, M., Jung, B. K., Ahmed, I., Zhong, C. and Jhung, S. H. 2014. Adsorption of pyridine over amino-functionalized metal–organic frameworks: attraction via hydrogen bonding versus base–base repulsion. *J. Phys. Chem. C* 118(36):21049–21056.

Katz, M. J., Brown, Z. J., Colón, Y. J., Siu, P. W., Scheidt, K. A., Snurr, R. Q., Hupp, J. T., and Farha, O. K. 2013. A facile synthesis of UiO-66, UiO-67 and their derivatives. *Chem. Commun.* 49(82):9449–9451.

Lin, Z. J., Lü, J., Hong, M., and Cao, R. 2014. Metal–organic frameworks based on flexible ligands (FL-MOFs): structures and applications. *Chem. Soc. Rev.* 43(16):5867–5895.

Matsuyama, K. 2018. Supercritical fluid processing for metal–organic frameworks, porous coordination polymers, and covalent organic frameworks. *J. Supercrit. Fluids* 134:197–203.

Maurin, G., Serre, C., Cooper, A., and Férey, G. 2017. The new age of MOFs and of their porous-related solids. *Chem. Soc. Rev.* 46(11):3104–3107.

Rosi, N. L., Eckert, J., Eddaoudi, M., Vodak, D. T., Kim, J., O'Keeffe, M., and Yaghi, O. M. 2003. Hydrogen storage in microporous metal-organic frameworks. *Science* 300(5622):1127–1129.

Sun, D., Fu, Y., Liu, W., Ye, L., Wang, D., Yang, L., Fu, X., and Li, Z. 2013. Studies on photocatalytic CO_2 reduction over NH_2-UiO-66 (Zr) and its derivatives: towards a better understanding of photocatalysis on metal–organic frameworks. *Chem. Eur. J.* 19(42):14279–14285.

Tao, Y., Kanoh, H., Abrams, L., and Kaneko, K. 2006. Mesopore-modified zeolites: preparation, characterization, and applications. *Chem. Rev.* 106(3):896–910.

Van Niekerk, J. N., and Schoening, F. R. L. 1953. X-ray evidence for metal-to-metal bonds in cupric and chromous acetate. *Nature* 171(4340):36–37.

Wang, C., Liu, D., and Lin, W. 2013. Metal–organic frameworks as a tunable platform for designing functional molecular materials. *J. Am. Chem. Soc.* 135(36):13222–13234.

Zhou, H. C., Long, J. R., and Yaghi, O. M. 2012. Introduction to metal–organic frameworks. *Chem. Rev.* 112(2):673–674.

7 Self-Assembled Metallo-Assemblies

Binduja Mohan and Sankarasekaran Shanmugaraju

Indian Institute of Technology Palakkad

CONTENTS

7.1 INTRODUCTION

The incredible dream of August Kekulé, *a snake seizing its own tail*, figured out the cyclic structure of benzene (Kekule 1865), then a variety of synthetic methods have been developed for the synthesis of previously unthinkable chemical structures, starting from simple organic molecules to large macromolecules like fullerene C_{60} (Boorum et al. 2001). In general, the conventional multi-step covalent synthesis of large macromolecules results in a low yield of the final product, and it is also difficult to obtain the expected product as it involves kinetically inert covalent bonds, which makes it impractical to attain finite molecular architecture (Stang and Olenyuk 1997). Over the past few decades, however, supramolecular self-assembly has become popular and proven to be a powerful and alternative synthetic method for the construction of targeted molecular architectures in high yield (Lehn 1995). The process "molecular self-assembly" is a unique phenomenon where two or more molecules or ions join together non-covalently to form complex structures (Steed et al. 2007). Nature's magic of designing complex natural systems such as the structure of DNA double helix, formation of micelles, and microtubules which are self-assembled from simple building blocks through non-covalent interactions boosts the relevance of supramolecular interactions in modern

DOI: 10.1201/9781003169130-8

science (Steed and Atwood 1995). The molecular self-assembly process has emerged as an inevitable component of supramolecular chemistry with interdisciplinary opportunities spanning the fields of chemistry, biology, and physics (Balzani and DeCola 1992). The inexorable march of supramolecular self-assembly unfolded the new door of design and synthesis of the aesthetic and intricate network of novel supramolecular architectures integrated with desired functional properties for various practical applications (Northrop et al. 2008). The following section explains the concept of the metal-ligand coordination-driven self-assembly strategy and its relevance in the construction of various self-assembled functional metallomacrocycles. This chapter is presented as a representative of supramolecular self-assembly and highlights the significant advances made in numerous applications of self-assembled supramolecular metallo-assemblies.

7.2 COORDINATION-DRIVEN SELF-ASSEMBLY

Among the various self-assembly processes, metal-ligand coordination-driven self-assembly has evolved to be the most efficient methodology for the successful construction of discrete supramolecular architectures of different topologies with fascinating functional properties (Leininger et al. 2000). The predictable directionality, high bond enthalpy (15–50 kcal/mol), and reversible nature of the kinetically labile metal-ligand coordination bond allow the easy access of self-corrected supramolecular architectures formation with stimuli-responsive properties (Stang et al. 1997). The important aspect of coordination-directed self-assembly relies on the judicious selection of pre-programmed complementary building units (a metal-based acceptor and an organic donor) that self-recognize themselves into ordered supramolecular structures (Fujita et al. 2005). Therefore, a desired supramolecular architecture can be achieved by the right combination of complementary building units in an appropriate stoichiometric ratio (Caulder and Raymond 1999). An acceptor is a chemical composite or moiety which can accommodate a lone pair of electrons offered by donor ligands. Acceptors are mostly transition metal-based coordination compounds or complexes that possess unoccupied or readily available coordination sites for incoming ligands. The unsaturated coordinating sites of the metal acceptors are usually occupied by labile substituents such as nitrate (NO_3^-), triflate ($CF_3SO_3^-$), perchlorate (ClO_4^-) anions, or solvent molecules, which then get displaced by the incoming donor ligands during the self-assembly process, resulting in a metal-ligand coordination bond. A variety of transition metal-based acceptors can be designed by the judicious selection of coordinatively unsaturated metal ions. As far as acceptor units are concerned, the square planar (Pt, Pd) and octahedral (Ru, Ir, Rh) metals are widely used for the construction of supramolecular architectures because of their stable coordination geometry and kinetic lability (Stang 2009; Therrien 2009) (see Figure 7.1 for examples). On the other hand, donors are in general, multidentate organic ligands capable of donating lone pair of electrons to the acceptors by keeping their chemical structure intact. Different organic ligands with two or more coordinating sites with bite angles ranging from 0° to 180° are successfully developed through the right functionalization of poly(hetero)cyclic aromatic hydrocarbons (Northrop et al. 2008). The electron-rich polypyridyl organic ligands are the favourite choices of donors because of their strong coordinating ability and controllable directionality (Tam et al. 2007) (see Figure 7.1 for examples).

Furthermore, the properties of supramolecular complexes depend on the functional groups present in them and coordination self-assembly provides a wealth of opportunities for the design of photo- and electrochemically active supramolecular complexes (Zheng et al. 2010). Using this simple coordination self-assembly strategy, numerous examples of aesthetically elegant functional supramolecular two-dimensional (2D) metallomacrocycles and three-dimensional (3D) metallocages with different topologies have been constructed in the past few decades (Chakrabarty et al. 2011). The collections of various supramolecular architectures developed to date employing different donors and acceptors are summarized in Figure 7.2. Referring to this molecular library, a plethora of desired 2D metallacycles and 3D metallocages can be built from their corresponding complementary building units (Chakrabarty et al. 2011). For instance, a molecular triangle can be

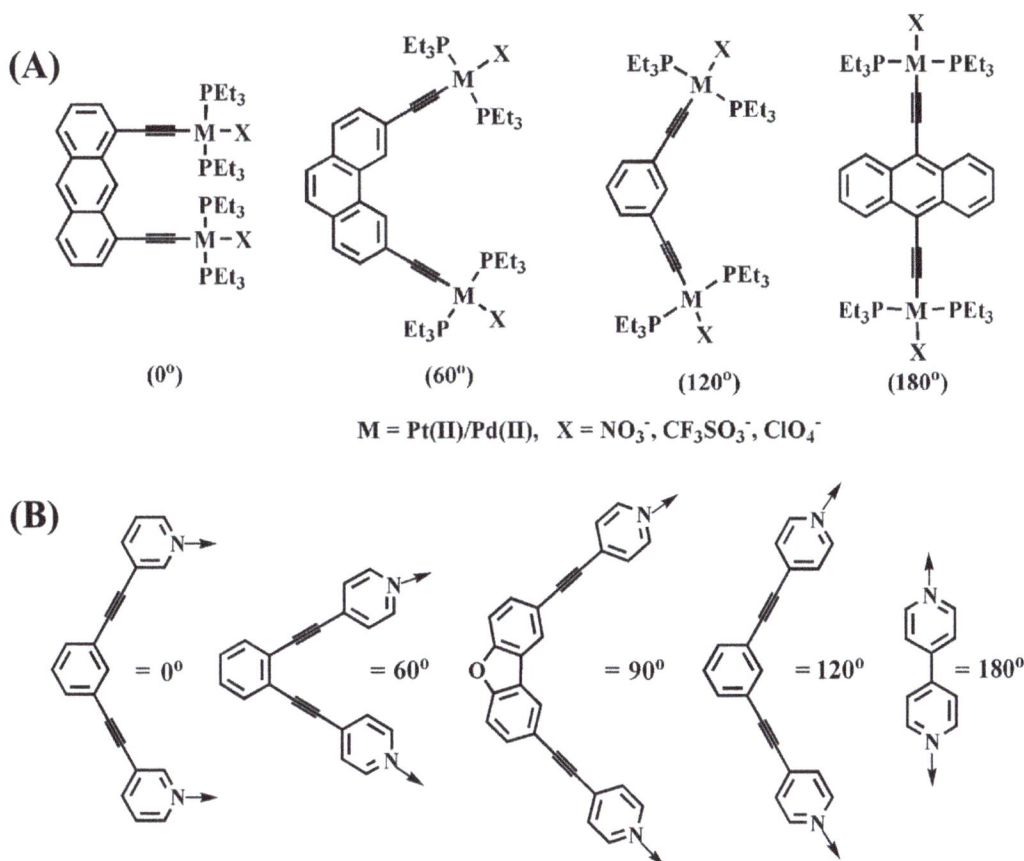

FIGURE 7.1 (a) Few representative examples of shape-selective acceptors and (b) shape-selective polypyridyl donors having different bite angles.

achieved by mixing three 60° donor units and complementary three 180° acceptors and *vice versa*. Likewise, 3D trigonal bipyramid can be obtained by reacting two tritopic building units possessing 60° bite angles with three ditopic angular 109° building units (Figure 7.2).

7.3   CONSTRUCTION OF TWO-DIMENSIONAL METALLOMACROCYCLES

In principle, the design of 2D metallomacrocycles is relatively simple and they can be synthesized by the linear combination of appropriate building blocks with definite bite angles. The fascinating world of 2D metallomacrocycles was explored by Fujita and co-workers in the early 1990s, they reported the first example of a metal-cornered molecular square (Fujita et al. 1990) which was then followed by the synthesis of numerous metallomacrocycles covering a range of sizes and geometry *i.e.*, from simple binuclear rhomboids to complex molecular polygons. In this section, we focus on a few representative examples of 2D metallomacrocycles built through a coordination-driven self-assembly process.

7.3.1   Self-Assembly of Molecular Rhomboids

The construction of binuclear molecular rhomboids is highly manageable and the easiest compared to the synthesis of other supramolecular complexes. The molecular rhomboids can be synthesized through [2+2] directional self-assembly of 120° ligand-donor and metal-acceptor units

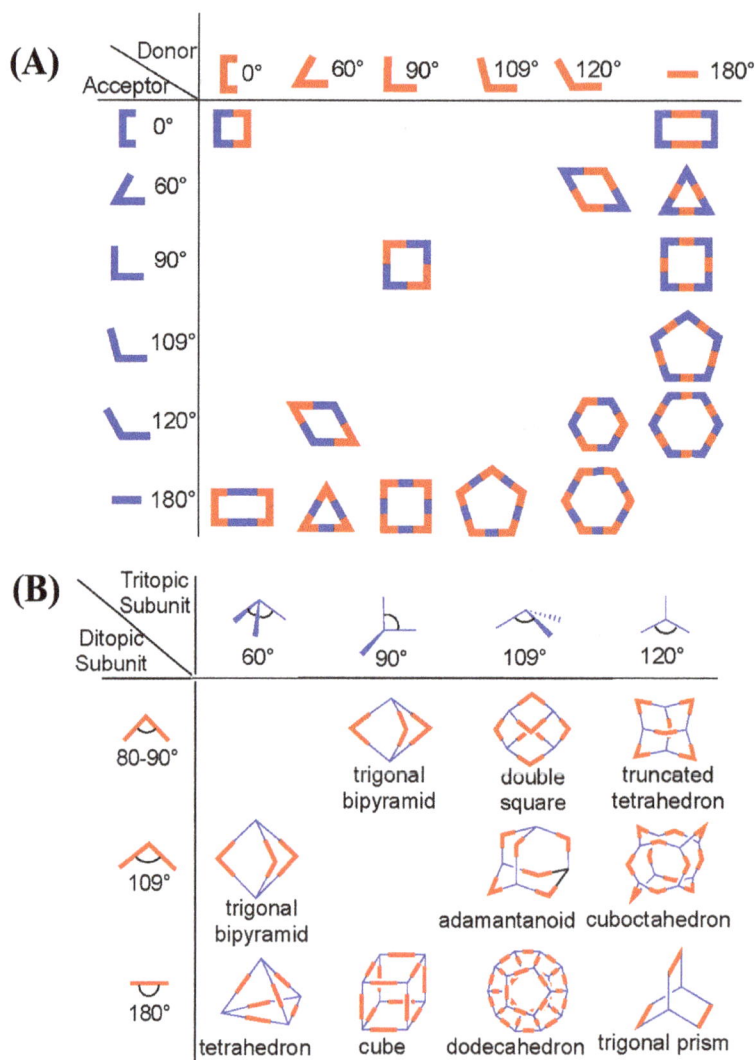

FIGURE 7.2 A molecular library for the rational design of (a) 2D metallomacrocycles and (b) 3D metallo-cages. (Reproduced with permission from Chakrabarty (2011) © 2011 American Chemical Society.)

(Chakrabarty et al. 2011). Maverick and co-workers reported the earliest example of [2+2] self-assembled molecular rhomboid by the self-assembly of the precursors, 1,7-bis(3-methylacetylac-etone)naphthalene donor and Cu(acetylacetonate)$_2$ acceptor (Maverick et. al. 1986). Other prime examples of water-soluble binuclear rhomboids were synthesized from the combination of either Pd(II) or Pt(II) based cis-blocked 90° acceptors and flexible donors (Fujita et al. 1996). Along with this, a series of cationic binuclear rhomboids have been prepared by the combination of Pd(II) or Pt(II) based cis-blocked 90° and silicon-based bipyridyl donor ligands. For instance, an equimo-lar mixture of dimethyl-bis(4-pyridyl)silane (D$_1$) and *cis*-[Pt(PEt$_3$)$_2$(OTf)$_2$] (A$_1$) resulted in silicon-containing molecular rhomboid (1) in high yield (Schmitz et al. 1999) (Figure 7.3a). The formation of self-assembled macrocyclic structures highly depends on the flexibility of the bridging linkers. In general, the use of flexible building units promotes the self-assembly of low nuclearity metal-lomacrocycles. It has been demonstrated that the introduction of a flexible spacer in between the coordination sites of bipyridyl linkers led to the equilibrium existence of a molecular rhomboid and interpenetrating[2]catenane structure (Lu et al. 2009). The self-assembly of a cis-blocked 90°

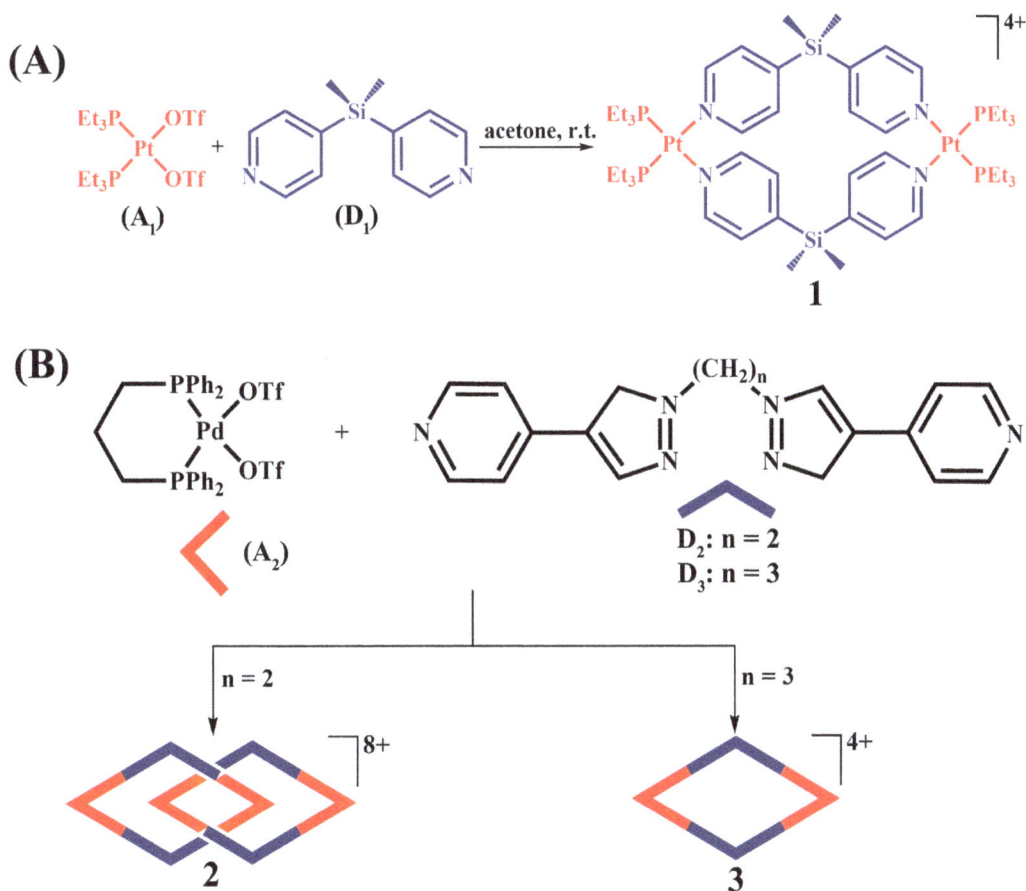

FIGURE 7.3 (a) Self-assembly formation of silicon-containing molecular rhomboid (1). (b) Schematic representation of the formation of interpenetrating [2]catenane (2) and rhomboid (3) self-assembled from flexible dipyridyl donors.

acceptor, [Pd(dppp)(OTf)$_2$] (A$_2$, dppp=diphenylphosphenopropane); with flexible dipyridyl donor, 1,2-bis(4-(4-pyridyl)pyrazolyl)ethane (D$_2$), resulted in [2]catenane (2) formation. Surprisingly, the same reaction with a more flexible dipyridyl donor, 1,3-bis(4-(4-pyridyl)pyrazole)propane (D$_3$), selectively yielded a cationic molecular rhomboid (3) (Figure 3.3b).

7.3.2 SELF-ASSEMBLY OF MOLECULAR TRIANGLES

Over the last few decades, there exists a rapid growth in the design and synthesis of numerous examples of molecular triangles because of their facile synthesis and fascinating applications (Chakrabarty et al. 2011). According to the direction bonding approach, the molecular triangle can be constructed in two different ways, (i) the self-assembly of three 180° metal acceptors with three 60° donor complementary donors and (ii) combination of three linear flexible donor ligands join with distorted metallo-corner acceptors to generate molecular triangles. The absence of metallo-corner acceptors with a 60° bite angle makes the second approach impractical. The self-assembly of a 60° dipyridyl donor, 4,7-phenanthroline (D$_4$), with a linear phenyl bridged dipalladium acceptor (A$_3$) produced a planar molecular triangle (4) (Hall et al. 1998) (Figure 7.4a). Another novel strategy for the synthesis of molecular triangles was developed by Newkome and his group in 2005. This method involves the self-assembly of octahedral metal acceptors such as Fe(II) or

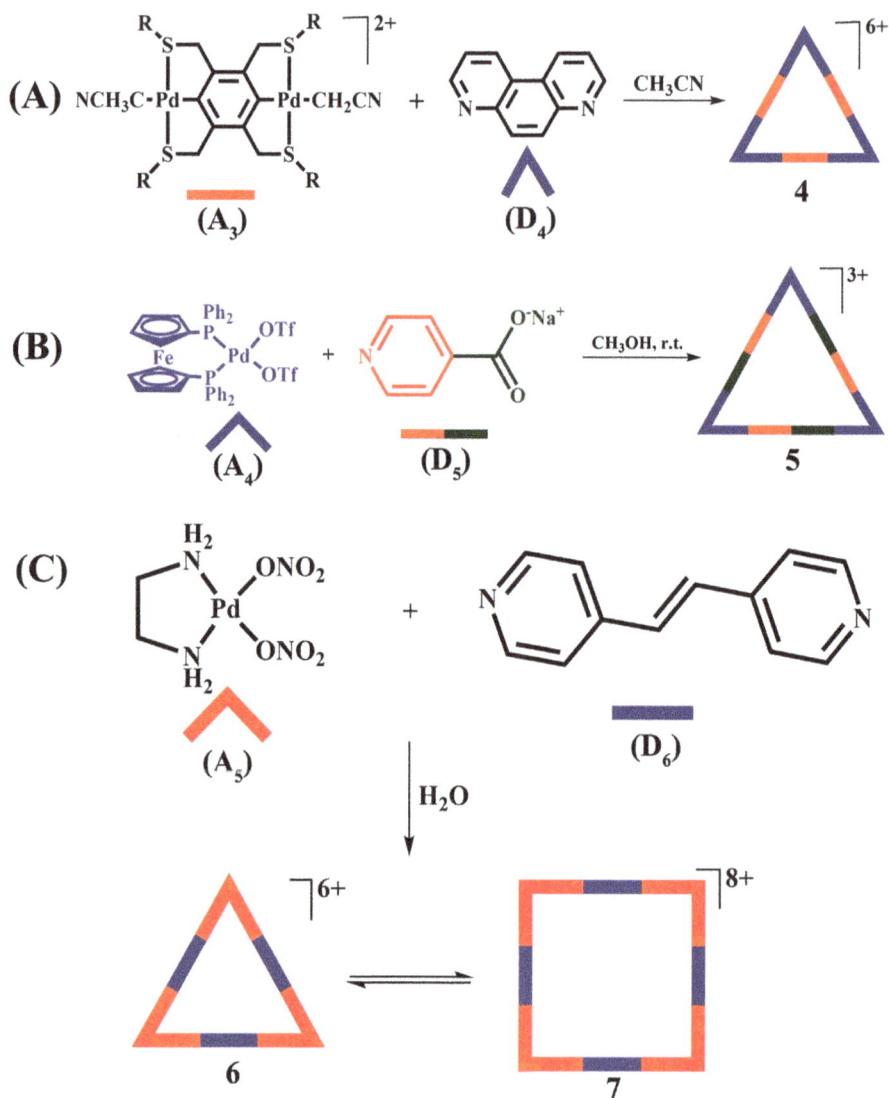

FIGURE 7.4 (a) Schematic representation of the formation of [3+3] self-assembled molecular triangle (4). (b) Selective formation of a head-tail type of single linkage isomeric molecular triangle (5) using ambidentate donor (D_5). (c) Equilibrium existence of [3+3] self-assembled triangle (6) and [4+4] self-assembled molecular square (7) in solution.

Ru(II) with terpyridine-based donor ligands with a bite angle of 60° between the two coordinating sites (Hwang et al. 2005). The use of a shape-selective 60° Pt(II) acceptor in combination with various bridging bipyridyl donors can generate unusual molecular triangles. Even though rigid bipyridyl ligands are commonly employed donor units, the use of flexible ambidentate donors in the self-assembly process is rare due to the possibility of forming various linkage isomeric products because of different bonding connectivity. However, Mukherjee and his group developed a facile route for the selective synthesis of single linkage isomeric molecular triangles by the combination of cis-blocked Pd(II)/Pt(II) metal-based 90° acceptors and several ambidentate donors. For example, the equimolar combination of *cis*-(dppf)Pd($H_2O)_2$(OTf)$_2$ [A_4; dppf=1,1′-bis(diphenylphospheno)ferrocene] and sodium nicotinate (D_5) lead to the selective isolation of single linkage isomeric molecular triangle (5) (Bar et al. 2009) (Figure 7.4b).

7.3.3 SELF-ASSEMBLY OF MOLECULAR SQUARES

Among 2D metallacycles, molecular squares are the most popular supramolecular architectures because of their facile synthesis, unique structural properties, and robust conformational stability (Chakrabarty et al. 2011). There are two pathways through which molecular squares can be prepared: (i) combining four 90° cis-blocked metal acceptors with four linear bridging donors, (ii) self-assembly of four 180° metal acceptors with four 90° donor ligands. Fujita and co-workers reported the first-ever example of Pd(II) molecular [4+4] square through coordination self-assembly reaction between 90° cis-protected palladium(II) acceptor, cis-[Pd(en)(NO$_3$)$_2$] (en=ethylene-diamine) and 4,4′-bipyridine, and this square was found to recognized 1,3,5-trimethoxybenzene organic guest molecule in an aqueous medium (Fujita et al. 1991). Adopting this facile method, various functional molecular squares based on functional groups such as porphyrins, calixarene, crown ethers were reported (Stang 1998; Chakrabarty et al. 2011). Stang et al. designed numerous examples of molecular squares employing cis-protected Pd(II) or Pt(II) metal-based 90° acceptors (Olenyuk et al. 1998). Hupp and co-workers have pioneered the rational design of neutral chiral molecular squares from various organic linkers combined with Re(I)-containing acceptor units (Slone et al. 1996). Also, several electrochemically active nanoscopic molecular squares based on perylene bisimide linkers were reported by the Würthner group (Würthner and Sautter 2000). In addition to the exclusive self-assembly formation of a molecular square, in some cases, the reaction of 90° acceptors with linear linker led to the formation of an equilibrium mixture of both square and triangle macrocycles due to the lack of thermodynamic preference during the self-assembly processes. It was Fujita et al. who first experimentally observed the formation of an equilibrium mixture of a molecular triangle (6) and square (7) from self-assembly of diamine-protected 90° metal acceptor (cis-[Pd(en)(NO$_3$)$_2$] (A$_5$) with flexible dipyridyl donor ligand, $trans$-1,2-bis(4-pyridyl)ethylene (D$_6$) (Fujita et al. 1996) (Figure 7.4c). Among these macrocycles, triangle (6) was found to be a major product due to the entropy preferences and upon increasing the concentration of reaction mixture, the equilibrium shifted towards molecular square (7) as a major product as the enthalpy gain compensated for the loss of entropy. The equilibrium existence of triangle and square was further demonstrated by Schalley and Stang groups employing various amide-based flexible donors and different ancillary linkers. It was observed that the bulky ancillary ligands favour the triangle as a major product, but the use of less bulky ancillary ligands prefers square as a major product (Weilandt et al. 2008; Schweiger et al. 2002).

7.3.4 SELF-ASSEMBLY OF MOLECULAR RECTANGLE

Molecular rectangles are rarely reported 2D metallacycle because of the difficulty in synthesizing them from cis-blocked Pd(II)/Pt(II)-based 90° corner acceptors. Generally, three-component [4+2+2] molecular self-assembly of cis-protected 90° acceptor and two distinct rigid linear donors are expected to produce a molecular rectangle, but usually, this reaction results in the generation of two different molecular squares of varied sizes due to lack of enthalpy preference and absence of selective recognition between two different reactive linker units (Chakrabarty et al. 2011). To overcome this problem and to selectively construct a molecular rectangle, Suss-Fink and co-workers ascertained a novel two-component design principle through the self-assembly of 0° clip-type acceptor with a linear rigid donor. For instance, a cationic molecular rectangle (8) was prepared by a 1:1 molar combination of oxalato-bridged binuclear p-cymene-Ru(II) clip-type acceptor (A$_6$; [Ru$_2$(μ-η^4-C$_2$O$_4$)-(MeOH)$_2$(η^6-p-iPr-C$_6$H$_4$Me)$_2$][CF$_3$SO$_3$]$_2$) and linear donor, 4,4′-bipyridine (D$_7$) (Yan 1997) (Figure 7.5a). A series of molecular rectangles were reported following this simple two-component self-assembly method and these rectangular structures were further exploited for selective host-guest complex formations (Chakrabarty et al. 2011; Fujita et al. 2005). As alluded to above, if ambidentate donors are used instead of rigid symmetrical dipyridyl donors, then self-assembly reaction would lead to two kinds of linkage isomeric rectangles. This is because of the

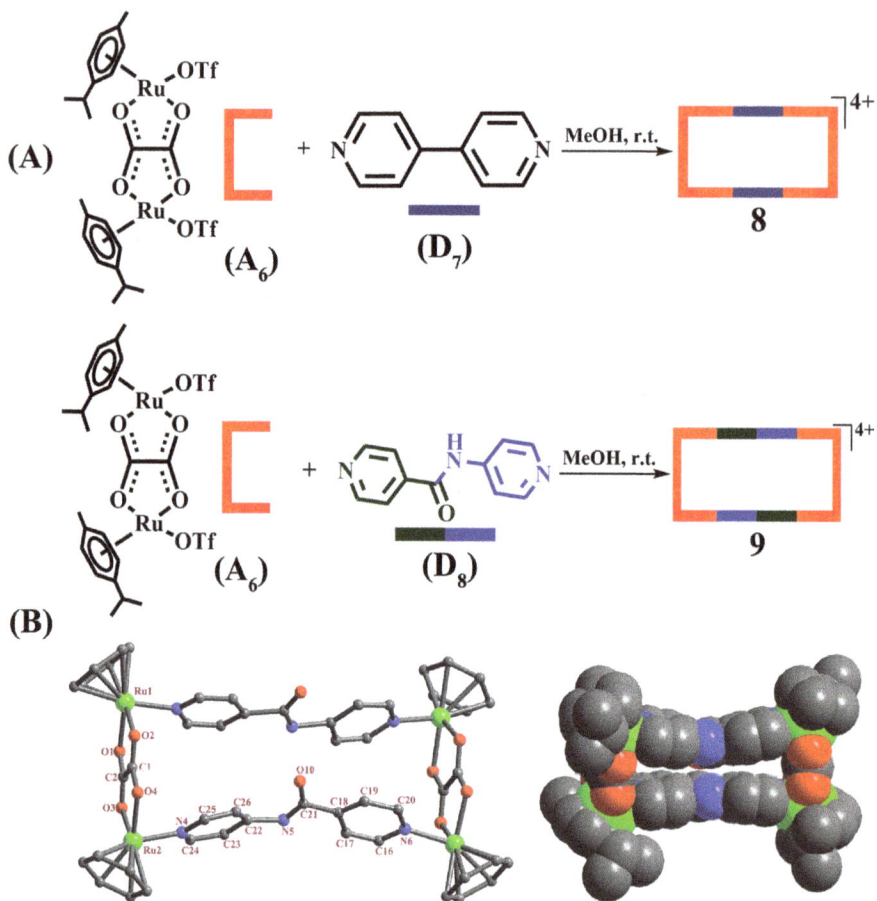

FIGURE 7.5 (a) Schematic representation of the formation of [2+2] self-assembled Ru(II) molecular rectangle (8) from symmetrical donor ligand (D_7). (b) Selective self-assembly formation of single linkage isomeric molecular rectangle (9) from ambidentate donor (D_8) and its X-ray crystal structure. (Reproduced with permission from Shanmugaraju (2011) © 2011 American Chemical Society.)

different bond connectivity of ambidentate donors, and thus, isolation of the desired product is challenging. However, the reaction of acceptor A_6 and an amide-based linear ambidentate ligand (D_8, N-(4-Pyridyl)isonicotinamide) produced a single linkage isomeric rectangle (9) in quantitative yield (Figure 7.5b) (Shanmugaraju et al. 2011). The uniform electronic charge distribution throughout the skeleton facilitated the formation of a single linkage isomeric rectangle among the mixture rectangle formation.

7.3.5 SELF-ASSEMBLY OF HIGHER-ORDER POLYGONS

Along with numerous small polygons highlighted above, several higher-order 2D supramolecular ensembles were also reported (Chakrabarty et al. 2011; Stang 1998). Molecular pentagons and hexagonal supramolecular assemblies are very common macrocycles among various higher-order ensembles. The combination of five 180° building blocks and five angular matching units generates pentagonal architectures, whereas hexagonal assemblies can be prepared by joining six 120° units with either six complementary linear units or three complementary 120° linker units (Chakrabarty et al. 2011). The interaction of divalent octahedral transition metal cation Fe(II) with 9-hexyl-3,6-bis(terpyridine)carbazole (D_9) yielded a molecular pentagon (10) (Hwang et al. 2005) (Figure 7.6a).

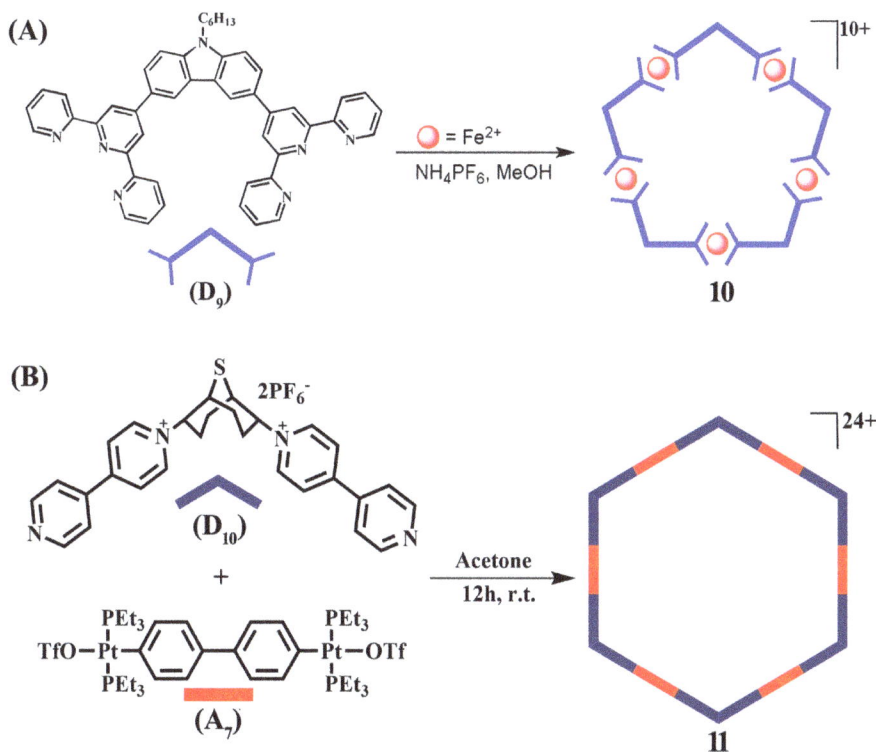

FIGURE 7.6   (a) Schematic representation of the formation of [5+5] self-assembled molecular pentagon (10). (b) [6+6] self-assembly formation of molecular hexagon (11).

Likewise, the self-assembly of bimetallic Pt(II) linear acceptor (A_7) with 120° dipyridyl linker unit (D_{10}) produced molecular hexagonal architecture (11) (Figure 7.6b). The complementary pathway i.e., the combination of D_{10} with bimetallic Pt(II) acceptor, 4,4′-Bis(*trans*-Pt(PEt$_3$)$_2$(OTf))benzophenone, with 120° bite angle generates [3+3] self-assembled molecular hexagon (Yang et al. 2006).

7.4   CONSTRUCTION OF THREE-DIMENSIONAL METALLOCAGES

Unlike 2D metallomacrocycles, the synthesis of finite 3D metallocages is difficult and challenging because of their inherent structural complexity, and thus it requires a more careful design principle. The main criteria for the construction of a 3D metallocage structure are that one of the reacting building blocks must have more than two coordinating sites. For instance, the self-assembly of three ditopic 0° clip-type donors with two equivalent tritopic planar acceptors having 120° bite angles can produce a molecular trigonal prismatic cage (Chakrabarty et al. 2011). As it is listed in Figure 7.2, vast numbers of 3D supramolecular architectures can be made from their corresponding building blocks. The following section provides a detailed discussion on coordination self-assembly formation of selected examples of 3D supramolecular architectures.

7.4.1   SELF-ASSEMBLY OF TRUNCATED TETRAHEDRA AND OCTAHEDRA

The molecular truncated tetrahedra can be synthesized through face-directed self-assembly of four 120° tritopic donor units with six ditopic units with a 90° bite angle (Chakrabarty et al. 2011). Raymond and co-workers used the symmetry interaction approach to construct various supramolecular tetrahedral cage-like architectures from different tritopic donor units and tri/tetravalent metal cations such as In(III), Al(III), Ti(IV), Ga(III) (Saalfrank et al. 2002). The self-assembly of tetravalent Ti(IV) cation and a catecholamide trigonal tris-bidentate donor (D_{11}) produced an anionic tetrahedral architecture (12) (Figure 7.7a). Likewise, Stang and co-workers adopted a face-directed self-assembly procedure for the creation of numerous examples of truncated tetrahedral cages from cis-blocked 90° acceptor as corner units (Fan et al. 2000). Molecular octahedrons are less common in the literature, due to the complexity in the design of suitable shape-selective building blocks. Shionoya and co-workers managed to design a set of octahedron metallocages through a face-directed self-assembly approach using several divalent d^5-d^{10} transition metal-based

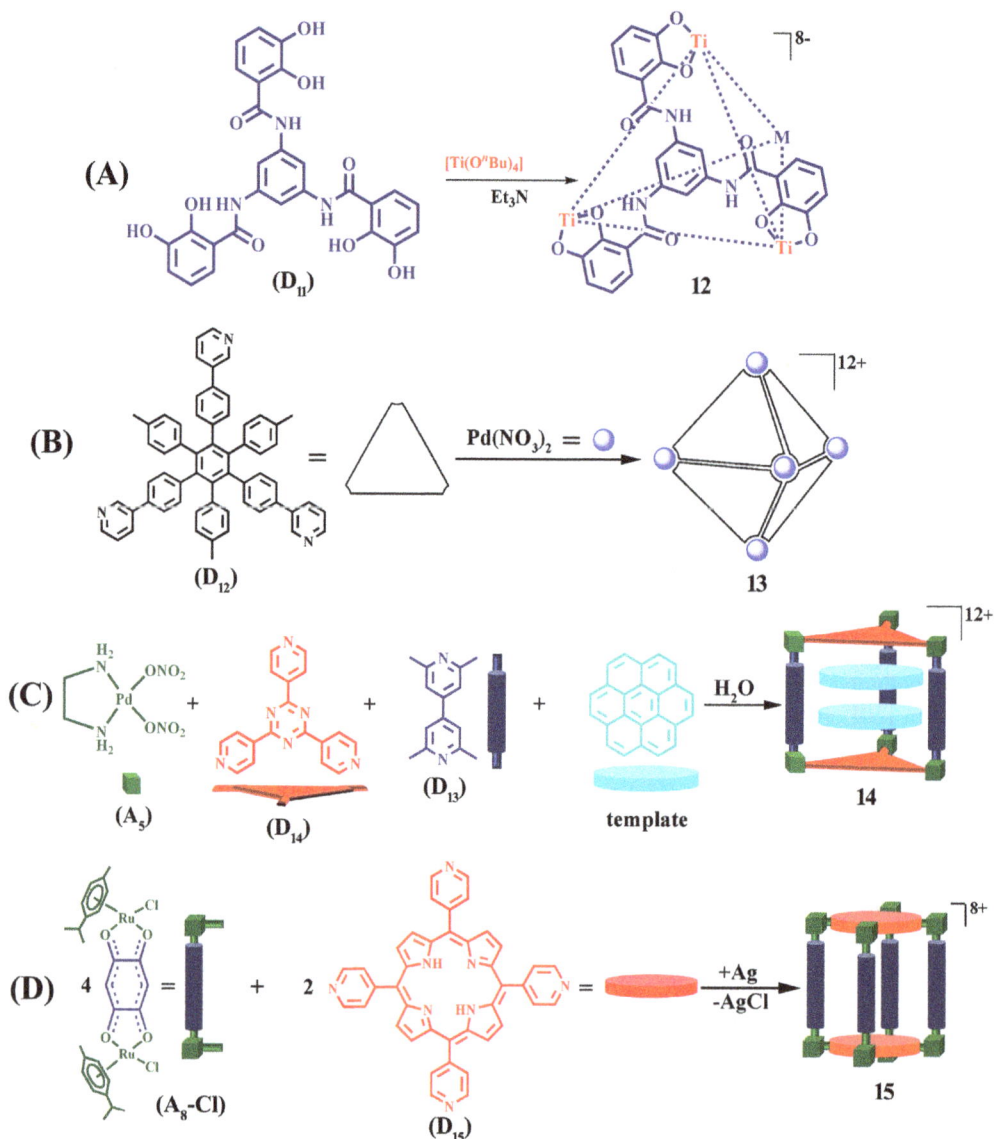

FIGURE 7.7 (a) Schematic representation of the formation of face-directed [4+4] self-assembled molecular tetrahedron anionic cage (12). (b) Two-component face-directed [6+6] self-assembly of molecular octahedron (13) from 3-pyridyl tritopic donor (D_{12}) reacting with Pd(NO$_3$)$_2$. (c) Construction of trigonal prism (14) through three-component template-assisted self-assembly. (d) The formation of a porphyrin containing tetragonal prism (15) through two-component self-assembly.

acceptors and pseudo-C_3-symmetric tritopic donor units. For example, an octahedron cage (13) can be synthesized from an 8:6 molar mixture of 3-pyridyl-based tritopic donor (D_{12}) reacting with $Pd(NO_3)_2$ (Hiraoka et al. 2006) (Figure 7.7b).

7.4.2 SELF-ASSEMBLY OF TRIGONAL AND TETRAGONAL PRISMS

The power and versatility of the coordination self-assembly strategy have further been extended and successfully used for the construction of complex supramolecular architectures. According to the general design strategy, a trigonal prismatic cage can rationally be synthesized through three-component directional self-assembly of a 6:3:2 molar mixture of 90° *cis*-protected metal acceptor, a bidentate linear donor ligand, and a tritopic planar unit (Chakrabarty et al. 2011). Following this three-component self-assembly procedure, several cationic trigonal prisms were developed by Fujita and co-workers. The formation of trigonal prism (14) from six equivalents of 90° acceptor A_5, three equivalents of tetramethyl-4,4′-bipyridine (D_{13}), and two equivalents of 2,4,6-tripyridyl-1,3,5-triazine (D_{14}) in the presence of coronene as a template is the best example to highlight the design trigonal prism formation (Yoshizawa et al. 2005) (Figure 7.7c). Another facile approach for the design of trigonal prism is two-component coordination self-assembly, and this approach is preferred over multicomponent due to the less entropy loss and thus the designed prismatic structure can be isolated in quantitative yield. In light of this Mukherjee, Stang, and several other research groups have reported several trigonal prisms self-assembled from clip-shaped precursors and complementary C_3-symmetric tritopic linkers (Zheng et al. 2010; Ghosh and Mukherjee 2008; Barry et al. 2008). Tetragonal prisms are rarely observed in Archimedean solid because of the complexity in the synthesis and selection of proper building blocks. Adopting the two-component self-assembly of four clip-shaped building blocks with 0° bite angle with two C_4-symmetric tetratopic donor units is expected to form a tetragonal prism. Therrien and co-workers have reported a cationic tetragonal prism (15) obtained by reacting a porphyrin-based tetratopic donor, 5,10,15,20-tetra(4-pyridyl)porphyrin (D_{15}), with quinonato bridged *p*-cymene-Ru(II) based binuclear "clip" type acceptor (A_8, $[Ru_2(dobq)\text{-}(MeOH)_2(\eta^6\text{-}p\text{-}iPr\text{-}C_6H_4Me)_2][CF_3SO_3]_2$ this was in-situ prepared from its chloro-derivative $[Ru_2(dobq)\text{-}(MeOH)_2(\eta^6\text{-}p\text{-} iPr\text{-}C_6H_4Me)_2]Cl_2$) ($A_8$-Cl) reacting with AgOTf to replace Cl. Tetragonal prism (15) was found to strongly binds with human telomeric quadruplex DNA as well as duplex DNA (Barry et al. 2009) (Figure 7.7d). A similar kind of tetragonal prism was designed by the Lu group employing π-electron-rich tetrapyridyl donor, 1,2,4,5-tetraethynyl(4-pyridyl)benzene, reacted with one equivalent of $Re_2(CO)_{10}$ in high boiling alcoholic media (Manimaran et al. 2003).

7.5 APPLICATIONS OF SELF-ASSEMBLED SUPRAMOLECULAR COORDINATION COMPLEXES

The past few decades have seen tremendous advancement in the numerous applications of self-assembled supramolecular coordination complexes in various fields. The directional coordination-self-assembly is a convenient synthetic method for the easy construction of thermodynamically stable and rigid metallo-assemblies integrated with fascinating functional properties (Chakrabarty et al. 2011; Cook et al. 2015). In addition, various functional metallo-assemblies can be constructed either by selecting the appropriate pre-functionalized building units or through the post-synthetic functionalization process. Compared to 2D metallacycles, 3D metallocages are more beneficial because they provide a confined nanoscopic cavity that can be used to selectively encapsulate guest analytes of matching sizes and shapes through non-covalent interactions (Cook and Stang 2015). In this section, we exemplified various applications of self-assembled metallacycles and metal-locages in detail.

7.5.1 METALLO-ASSEMBLIES FOR ANION SENSING

Anions are omnipresent and play important roles in various chemical, biological and environmental processes (Martínez-Máñez and Sancenón 2003). In the recent past, supramolecular anion chemistry received special attention and substantial efforts have been devoted to the development of molecular receptors for anion sensing (Beer and Gale 2001). Among the various reported sensor systems for anions, amide-based sensors are commonly used for anion sensing because the amide N-H group can form strong hydrogen bonding with anions (Sessler et al. 2003). In general, the deprotonation of amide N-H upon binding with basic anions acts as a signalling output and thus indicates the presence and binding strength of anions (Sessler et al. 2003). In this context, in 2010, Shanmugaraju et al. reported a carbazole-amide-based bifunctional Pt(II) molecular square (16) and employed it as a potential chemosensor for selective sensing of pyrophosphate ($P_2O_7^{4-}$) anion in solution (Shanmugaraju et al. 2010). A very first example of carbazole-based shape-selective diplatinum organometallic 90° acceptor, 3,6-bis[*trans*-platinum(PEt$_3$)$_2$(NO$_3^-$)(ethynyl)]carbazole (A_9), was synthesized adopting the established Sonagashira-coupling reaction. The equimolar self-assembly of A_9 with an amide-containing flexible dipyridyl donor, 1,3-bis(4-pyridyl)isophthalamide (D_{16}), afforded a [2+2] self-assembled square 16 in good yield (Figure 7.8a). The X-ray diffraction analysis revealed that, due to the strong torsional angle strain, square 16 assumes a sigmoidal-like shape instead of an expected ideal planar square type of structure in the solid-state (Figure 7.8a). Molecular square 16 is composed of an amide-based receptor that recognizes anions through hydrogen-bonding interactions and carbazole-based fluorophore acts as signalling units

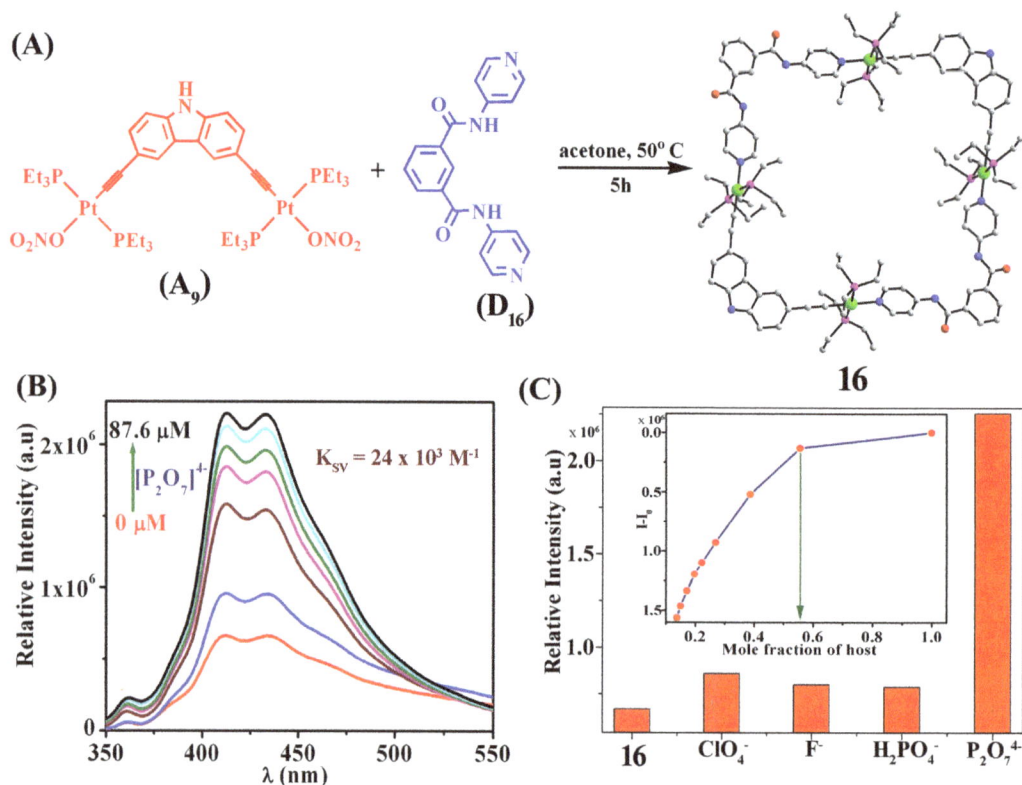

FIGURE 7.8   (a) Self-assembly of [2+2] molecular squares 16 from a novel organometallic 90° acceptor (A_9) and flexible ditopic donor (D_{16}). (b) Changes in the emission intensity of 16 in DMF solution upon the mixing of $P_2O_7^{4-}$ ion and (c) relative change in emission intensity measure for various anions (insert: Job's plot). (Reproduced with permission from Shanmugaraju (2010) © 2010 American Chemical Society.)

that communicate the anion recognition. The DMF solution of 16 was found to be weakly emissive due to the photo-induced electron transfer (PET) process from amide groups to carbazole moiety. But, with the gradual addition of $P_2O_7^{4-}$ anion the fluorescence emission intensity was drastically enhanced (Figure 7.8b). Notable, the addition of other anions such as F^-, ClO_4^- and $H_2PO_4^-$ showed no moderate influence on the fluorescence intensity (Figure 7.8c). The high selectivity for $P_2O_7^{4-}$ anion was attributed to its large size, compared to other anions, that engenders it to fit within the cleft-shaped amide receptor site of 16. The observed emission enhancement after binding with $P_2O_7^{4-}$ anion is due to the blocking of the PET process. From the titration studies, the Stern-Volmer binding constant was calculated to be 24×10^3 M^{-1} and the Job's plot analysis indicated a 1:1 mode of binding (Figure 7.8c). All these studies confirm that molecular square 16 can be used as a turn-on fluorescence sensor for selective sensing and quantification of $P_2O_7^{4-}$ anion, an important species that play a vital role in various biological processes such as energy transduction, DNA polymerization, and involves in various enzymatic reactions.

Following this simple two-component self-assembly method, Chi and co-workers synthesized an arene-ruthenium-based molecular rectangle and used it as a selective fluorescence sensing system for biologically relevant multi-carboxylate anions such as oxalate, tartrate, and citrate (Vajpayee et al. 2011). The coordination self-assembly reaction of the arene-ruthenium naphtha-cenedione acceptor with the amide-based dipyridyl donor, N, N'-4-Dipyridyl oxalamide, yielded a [2+2] molecular rectangle in high yield. The solid-state structural analysis confirmed the formation of rectangular geometry, and it was found that the large internal cavity was filled with solvent diethyl ether. The distance between the two amidic groups is 7.28 Å and this cavity size is sufficient to form the stable host-guest complex. The absorption spectrum of the rectangle in methanol showed several bands at 377, 567, and 611 nm corresponding to π–π* and MLCT transition corresponds to an arene-ruthenium moiety. Upon photo-excitation at 377 nm, the rectangle displayed a weak fluorescence emission due to the PET process from acceptor fluorophore to the amide-donor. Considering the presence of amide functional groups and fluorescence behaviour in the visible region, the rectangle was used as a fluorescence chemosensor for multi-carboxylate anions. Fluorescence titration studies with several monoanions such as F^-, Br^-, Cl^-, and CH_3COO^- exhibited a week binding affinity and caused almost no perturbation in emission intensity. In contrast, the incremental mixing of multi-carboxylate anions like oxalate, citrate, and tartrate elicited several-fold emission enhancement due to their strong hydrogen-bonding interactions that inhibit the PET process. Notably, no binding was observed for flexible dicarboxylate anions such as malonate and succinate, which implies that perfect geometrical complementarity between anions and rectangle is responsible for the increased binding affinity for dicarboxylate anions. It was found from Job's plot analysis that rectangle forms 1:1 host-guest complexes with anions and the Stern-Volmer binding constant estimated to be in the range of 10^3–10^4 M^{-1}. To further probe the binding interactions, ^1H NMR titration experiments were carried out and the results showed a sharp single at $\delta = 11.47$ ppm which disappeared in the presence of 2.0 equivalents of oxalate anions. This further supported that hydrogen-bonding interactions between amide NH protons and oxalate anion were responsible for its strong binding affinity with the rectangle.

7.5.2 NITROAROMATIC EXPLOSIVES SENSING

In the past several years, we have seen an increased research interest in the design of reliable fluorescence sensors for trace detection of jeopardizing chemical explosives. The nitro (NO_2) containing aromatic compounds such as picric acid (PA), and 2,4,6-trinitrotoluene (TNT) are well-known secondary chemical explosives and are the primary constituent of various commercially available chemical explosives (Shanmugaraju et al. 2015). Besides their explosive nature, nitroaromatic compounds (NACs) are known for serious health menace and safety concerns. The prolonged exposure to the saturated vapours of NACs causes several health issues such as severe headaches, abnormal liver function, anaemia, neurological disorders, and cancer (Shanmugaraju et al. 2017). Therefore, there is an urgent need to develop practically feasible sensor systems for the trace analysis and quantification of NACs in soil and groundwater. Due to the presence of electron-withdrawing NO_2 groups, in general, NACs are π-electron-deficient, and thus, they can form an efficient donor-acceptor type of charge-transfer (CT) complex with electron-rich sensors. In light of this, Shanmugaraju et al. developed an anthracene-ethynyl conjugated electron-rich Pt(II) molecular rectangle for fluorescence sensing of NACs, in particular for selective sensing of PA (Shanmugaraju et al. 2011). The finite molecular rectangle was synthesized by a two-component self-assembly reaction of a binuclear organometallic acceptor, 4,4′-bis[*trans*-Pt(PEt$_3$)$_2$(O$_3$SCF$_3$)(ethynyl)]biphenyl and an anthracene based clip-type dipyridyl donor, 1,8-bis(4-pyridyl)ethynylanthracene. Owing to the extended π-conjugation and electron-richness, the rectangle showed strong emission intensity at 433 and 458 nm, and solution-state quantum yield of 0.12. The strong fluorescence emission intensity of the rectangle was quenched completely in the presence of PA in the DMF solution. The Stern-Volmer binding constant was calculated to be $K_{SV}=5.0\times10^6$ M^{-1} and the strong binding affinity was also reflected by a sharp-visual colour change of rectangle from green to colourless after binding to PA. The observed strong fluorescence quenching was ascribed to the ground-state CT complex formation, and it follows a static-quenching mechanism. The UV–vis absorption and ^{1}H NMR titration studies were further confirmed the ground-state static-quenching mechanism. All these studies suggested that the rectangle is a potential fluorescence sensor for selective sensing of PA in solution. In continuation, the same group reported a 3D tetragonal prismatic cage for fluorescence sensing NACs both in the solution and vapour phase (Shanmugaraju et al. 2012). The self-assembly of a novel star-shaped pyrene tetra-platinum acceptor, 1,3,6,8-tetrakis[*trans*-Pt(PEt$_3$)$_2$(I)(ethynyl)] pyrene (A$_{10}$) and amide-functionalized dipyridyl clip-type donor, 1,3-bis(3-pyridyl)isophthalamide (D$_{17}$) produced a [2+4] tetragonal prism 17 in pure form (Figure 7.9a). A pulsed-gradient spin-echo NMR experiment determined the internal diameter of 2.02 nm of 17 and the energy-minimized structure showed that the distance between above and below pyrene fluorophore is 1.41 nm, which is too large to accommodate electron-deficient NACs through non-covalent interactions. The initial emission intensity of 17 was rapidly attenuated upon the mixing of TNT in chloroform solution and the binding constant was found to be $K_{SV}=9.7\times10^5$ M^{-1}, which is much higher than the several reported fluorescence sensors for TNT (Figure 7.9b). The non-fluorescent ground-state CT complex formation was the reason for the rapid depletion of emission intensity. Interestingly, similar sensing studies with other interfering analytes did not show any such changes in the emission intensity and thus confirmed the high selectivity of both 17 and acceptor A$_{10}$ for TNT (Figure 7.9c). To meet the practical demand, the vapour phase sensing studies were also carried out using the freshly made thin film from 17. The film was emissive and upon exposure to the saturated vapours of 2,4-dinitrotoluene (DNT) quenched the emission intensity and the vapour phase sensing was highly reversible. Therefore, the tetragonal prim 17 can be viewed as a potential fluorescence sensor for selective, sensitive, and reversible sensing of TNT both in the solution and vapour phase.

7.5.3 SUPRAMOLECULAR CATALYSIS

The unique and intricate characteristics of the natural enzyme catalysis drive researchers to design an abiotic alternative with all the imperative features of the enzymatic system. Some of the essential traits of enzyme catalysis are robust binding affinity towards the substrate but weaker interaction with the product, the presence of a hydrophobic cavity to perform host-guest interaction, and stabilization of intermediates or transition state (Kaphan et al. 2015). Over the past few years, chemists have focused on the development of supramolecular catalysts as their aesthetic architecture mimics natural enzymatic systems. Fujita et al. designed a cationic Pd(II) coordination cage 18 through the self-assembly of Pd(II) 90° acceptor A$_5$ and symmetric tripyridyl donor ligand D$_{14}$ (Murase et al. 2012). Cage 18 possesses a large internal cavity, and it was used as an efficient catalyst for Knoevenagel condensation of aromatic aldehydes in water under neutral conditions. The hydrophobic cationic cavity facilitates the stabilization of the anionic intermediate and promotes

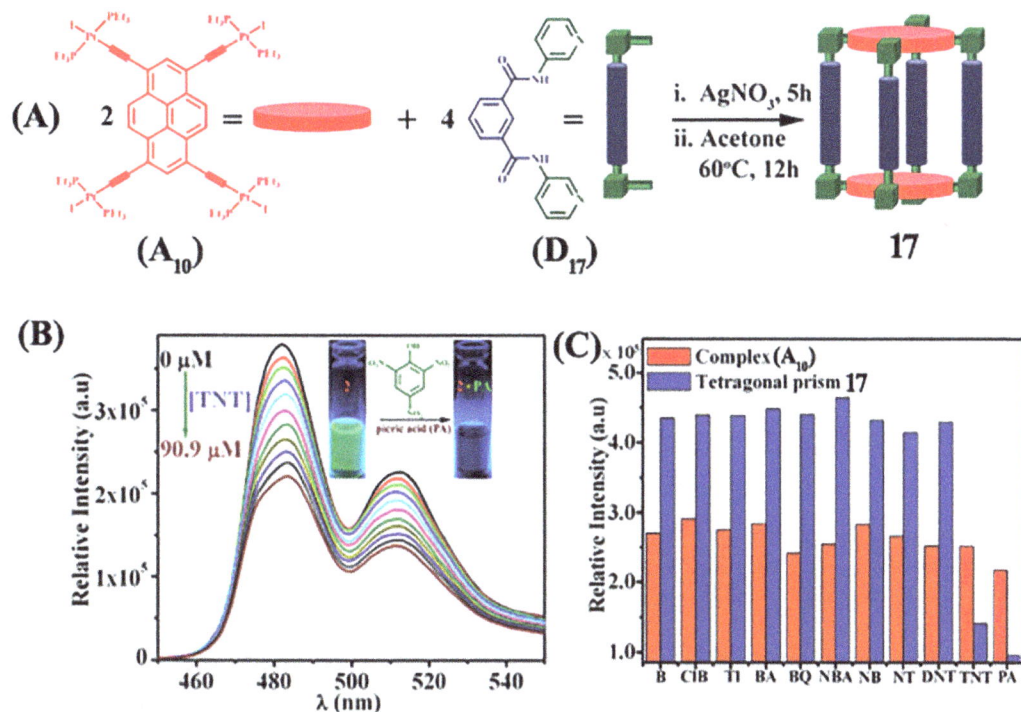

FIGURE 7.9 (a) [2+4] Self-assembly of a tetragonal prism (17) from pyrene-based star-shaped Pt(II) acceptor in combination with an amide-based dipyridyl donor. (b) Decrease in emission intensity of prism 17 upon the addition of TNT (inset: visual colour change of 17 upon the addition of TNT under UV light) and (c) Relative changes in the emission intensity of 17 and acceptor A_{10} upon the addition of different competing analytes. (Reproduced with permission from Shanmugaraju (2012) © 2012 American Chemical Society.)

condensation reaction. The cage spontaneously evacuates the condensed product as it is too huge to fully fit inside the cage. It was noticed that even 1 mol% of the cage can effectively catalyse condensation reaction to produce around 96% product yield. Normally, Knoevenagel condensation of 9-anthracene aldehyde is extremely complicated due to the steric factor, but the same reaction within the cage cavity happens smoothly with around 63% yield. Since the cage is a highly positively charged species, it favours electron-rich aldehydes such as methoxy- and amino-substituted naphthaldehydes as the guest species (Figure 7.10a).

In 2016, Mukherjee and co-workers synthesized template-free multicomponent self-assembly of a 3D trigonal prism 19 from $Pd(NO_3)_2$, a clip-type donor (3,3′-(^{1}H-1,2,4-triazole-3,5-diyl)dipyridine, D_{19}) and ditopic urea "strut" (di(4-pyridylureido)benzene, D_{18}) in 1:1:1 molar ratio. The prism has the potency to act as a hydrogen-bond-donor catalyst for Michael and Diels-Alder reactions in the aqueous medium (Howlader et al. 2016) (Figure 7.10b). The formation of prism 19 was confirmed using NMR spectroscopy and the molecular structural features were verified using single-crystal XRD analysis, which showed the shortest distance between two adjacent centroids of the triangles as 7.056 Å confirming the existence of hydrogen-bonding between the neighbouring urea units. Upon binding of water-soluble nitro-olefins, e.g., 1-(2-nitro vinyl)naphthalene], the characteristics of N-H vibrations were shifted from 3061–3330 to 3363 cm^{-1} confirms the strong hydrogen bond formation between urea and the nitro group. This was further supported by the UV–vis absorption study and the guest encapsulation was also reflected by sharp colour changes. It was found that even the addition of 1 mol% of the prism 19 can efficiently catalyse the reaction to produce around 80% of the final product at room temperature, but in the absence of prism, the yield was very less. Despite the mild reaction conditions, highly sensitive N-Boc protected nitro-olefins and even bulky 1/2-naphthyl

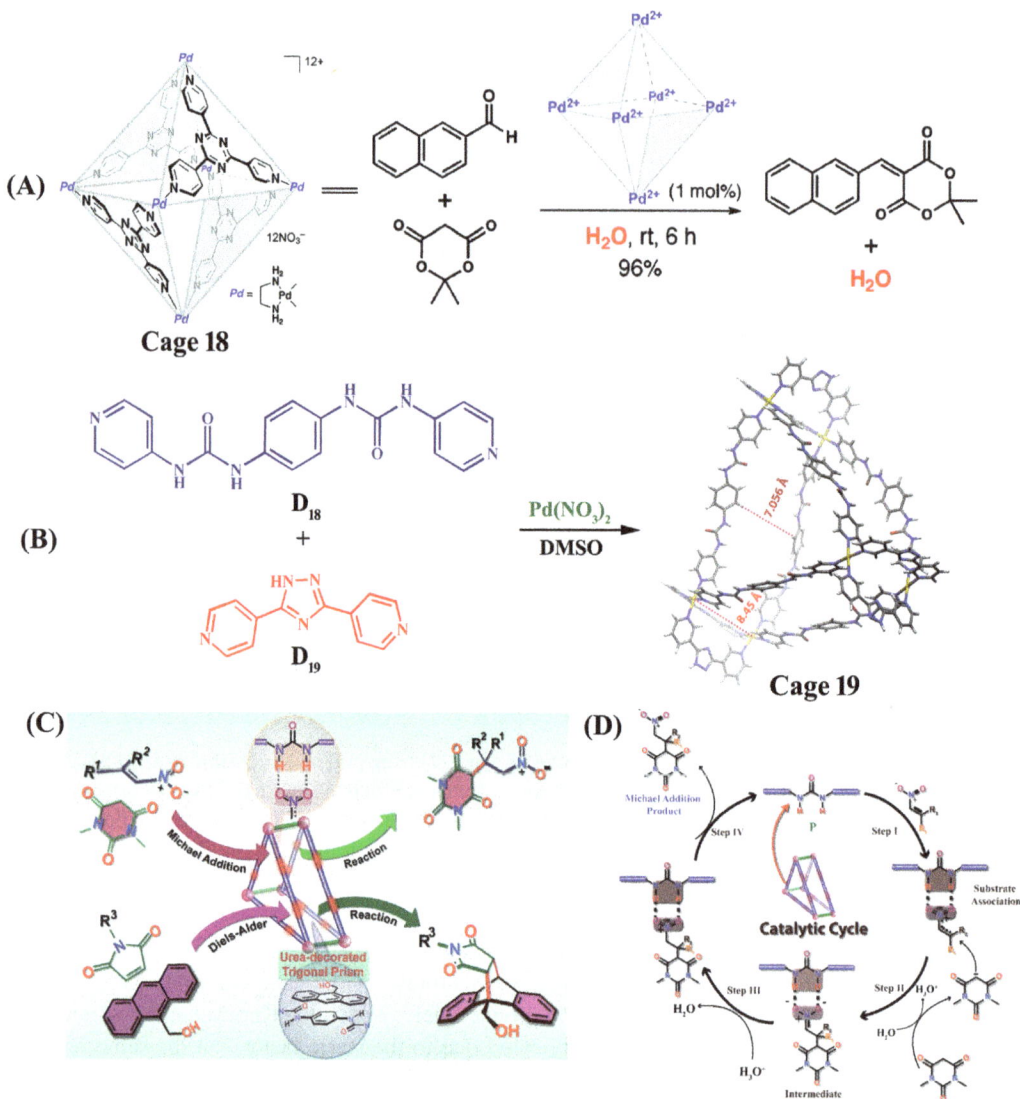

FIGURE 7.10 (a) Octahedral Pd(II) cage (18) catalysed Knoevenagel condensation reaction of the aromatic aldehyde with Meldrum's in H_2O. (b) Schematic representation of the three-component self-assembly formation of Pd(II) 3D prism (19) and its molecular structure. (c) Graphical representation of the use of cage 19 as a supramolecular catalyst for Michael addition and Diels-alder reactions. (d) The proposed catalytic cycle for the Michael addition reaction within the cage 19. (Reproduced with permission from Murase (2012), Howlader (2016) © 2011 and 2016 American Chemical Society.)

nitro-olefins underwent excellent Michael reaction with promising yield. Notably, Michael's addition on bulkier pyrene-substituted maleimide, the reaction yields were tremendously decreased implying the importance of the cavity size for the catalysis (Figure 7.10c and d). Besides, the use of prism 19 for Michael's reaction, it was also acting as a catalyst for the Diels-alder reaction. About 5 mol% of the prism can generate 84% of Diels-alder adduct at room temperature with a reaction time of around 60 h (Figure 7.10c). A chiral metallosalen-based octahedral coordination cage was synthesized by heating a 1:2:1 mixture of Mn(H$_2$L)Cl (H$_2$L = 1,2-cyclohexane diamine-N, N′-bis-(3-tert-butyl-5-(carboxyl)salicylide), H$_4$TBSC (*p*-tert-butylsulfonylcalix[4]arene) and Zn(OAc)$_2$.2H$_2$O

(Tan et al. 2018). The chiral cage structure was confirmed by single-crystal X-ray diffraction, and it was clear that the $[Zn_4(\mu^4\text{-}H_2O)(TBSC)_2]$ tetramers occupy the space between adjacent layers, and these tetramers were involved in various hydrophobic interactions with the adjoining cages through tert-butyl groups. This chiral cage was employed as an adequate supramolecular catalyst for the generation of enantiomerically pure secondary alcohols from racemic secondary alcohol and for the epoxidation of olefins to generate attractive stereoselective products. The rigid cage structure could formulate an interior hydrophobic cavity for the reactants as well as enhance the catalytic property, and resist deactivation by stabilizing the metallosalen moieties.

7.5.4 METALLO-ASSEMBLIES FOR BIOSENSING

There exists a great interest in the design and synthesis of functional supramolecular coordination complexes that can respond to bio-related external stimuli because of their potential practical applications. In this context, Xu and co-workers synthesized a supramolecular hexagonal metallacycle adorned with naphthalimide fluorophore and the hexagon was further used as a fluorescence sensor for sensing and quantification of H^+ ion (He et al. 2014). Probing the concentration of H^+ ion is important since it plays a significant role in various chemical and biological processes and several diseases such as peptic ulcers, gastroesophageal reflux, and cancer are directly linked to the concentration of H^+ ion. The author had selected naphthalimide as a fluorescence signalling unit because of its good photostability, interesting photophysical characteristics and high quantum yield, and feasible biocompatibility. Molecular hexagon was readily obtained in excellent yield by mixing the naphthalimide containing 120° dipyridyl donor with an equimolar concentration of complementary 120° diplatinum acceptor in dichloromethane (for further information see the cited reference). The naphthalimide fluorophore was attached to the dipyridyl donor ligand by non-conjugation which prevents the direct conjugation effect of self-quenching of intrinsic fluorescence intensity. The hexagon architecture was nearly non-fluorescence in basic pH because of the efficient intramolecular PET process from the electron-donor 'N' atom on N-methyl piperazine moiety to the electron-deficient 1,8-naphthalimide fluorophore. When the pH of the sensing medium changed into acidic from 7.5 to 3.5, the emission intensity of hexagon was enhanced gradually and increased up to 75-fold enhancement. The turn-on emission was attributed to the protonation of the 'N' atom on N-methyl piperazine that inhibits the PET process. Moreover, a perfect linear relation was observed for the enhancement of emission intensity monitored at 514 nm for the increasing concentration of H^+ ion ($0 \rightarrow 60\,\mu M$). Thus, the hexagon can be a feasible fluorescence probe for both sensing and quantification of H^+ ion solution.

Heparin, a sulfate glycosaminoglycan polymer, is an anticoagulant drug that is commonly used to treat blood clotting. It has been demonstrated that monitoring the concentration of heparin is extremely important during anticoagulation surgery since the excess concentration of heparin could cause several catastrophic complications like haemorrhages and thrombocytopenia (Lever and Page 2002). The heparin polymer chain is negatively charged because of the presence of many sulfates and carboxylate groups. So, it can form effective aggregates with positively charged metallacycles through multiple electrostatic interactions. Given this fact, Yang and co-workers developed an AIE active tetraphenyl ethylene-based cationic organoplatinum(II) hexagonal metallacycle and it was successfully employed as a fluorescence sensor for selective detection of heparin in aqueous media (Chen et al. 2015). The TPE-based hexagon 20 was obtained through a two-component [3+3] self-assembly reaction between a TPE containing 120° donor (D_{20}) and a 120° diplatinum acceptor (A_{11}, 4,4'-Bis(trans-Pt(PEt$_3$)$_2$(OTf))benzophenone) in a 1:1 molar ratio in dichloromethane. Hexagon 20 was found to be weekly emissive in the heparin-free stage and the presence of negatively charged heparin resulted in the higher-order aggregate formation, typical behaviour for a TPE-based system (Figure 7.11a). The aggregate formation through multiple electrostatic interactions facilitated an efficient emission enhancement at 486 nm, due to the restriction of TPE free rotation in the aggregate

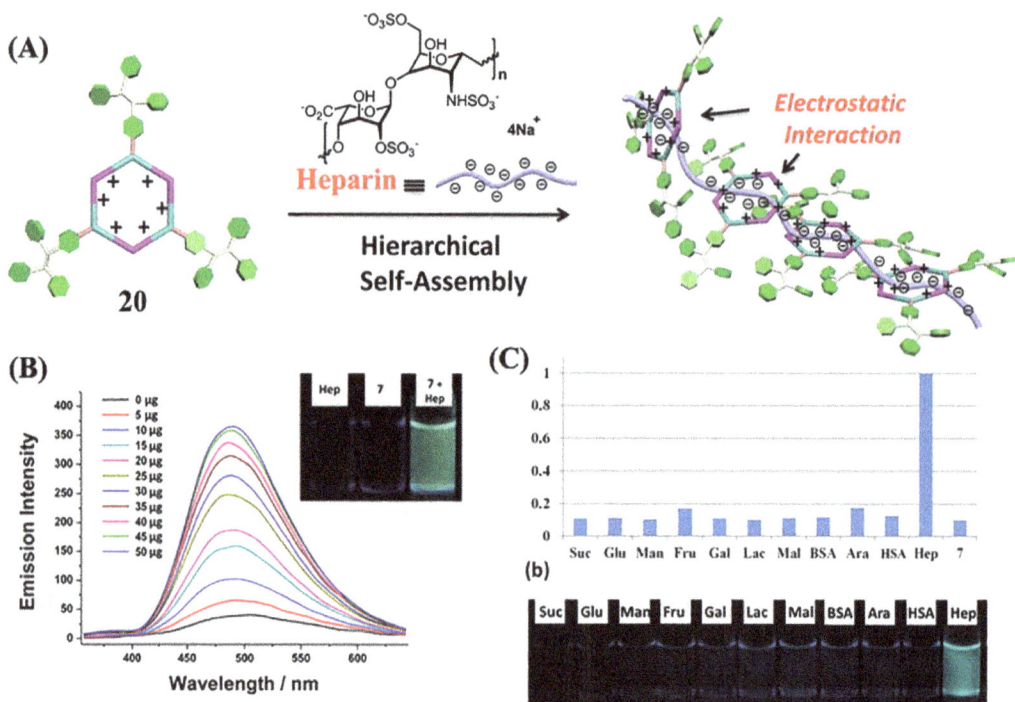

FIGURE 7.11 (a) Schematic representation of the electrostatic binding and aggregation of TPE-based molecular hexagon 20 with heparin. (b) Changes in emission spectra of 20 upon increasing the concentration of heparin (insert: visual colour changes for 20 before and after the addition of heparin). (c) Selectivity plot with various potential interfering biomolecules and the corresponding visual colour changes under UV light. (Reproduced with permission from Chen (2015) © 2015 American Chemical Society.)

form. Moreover, the emission enhancement of 20 was linear to the concentration of heparin indicating that it can be used for the highly selective quantification of heparin at low mM concentration and hence this method could find clinical application in anticoagulant therapy (Figure 7.11b and c).

7.5.5 ANTICANCER-ACTIVE METALLO-ASSEMBLIES

Cancer is uncontrolled cell growth and fast-growing threat that hits billions of people every year. Therefore, there is an urgent requirement for sophisticated drugs which can specifically target cancer cells and biologically compatible with normal cells (Galanski et al. 2005). After the discovery of cis-platin by Barnett Rosenberg et al., several metal complexes have been examined and used as anticancer agents, and a few metal complexes have already entered into the clinical trials (Rosenberg et al. 1969). However, low efficacy, poor water solubility, hazardous side effects, and reduced pharmacological activities are some of the issues faced by these complexes when using them as potential anticancer drugs. The development of supramolecular coordination complexes built from anticancer-active appropriate building blocks could easily overcome all these drawbacks. In this context, Chi et al. developed a series of arene-ruthenium metallacycles with potential application in anticancer activity and cellular pharmacology (Dubey et al. 2015). The [2+2] self-assembly of Ru(II) acceptors having O∩O chelating ligands, $[Ru_2(\mu\text{-}\eta^4\text{-}C_2O_4)\text{-}(MeOH)_2(\eta^6\text{-}p\text{-}iPr\text{-}C_6H_4Me)_2][CF_3SO_3]_2$ (A_6) or $[Ru_2(dobq)\text{-}(MeOH)_2(\eta^6\text{-}p\text{-}iPr\text{-}C_6H_4Me)_2][CF_3SO_3]_2$ (A_8) or $[Ru_2(donq)(H_2O)_2(\eta^6\text{-}p\text{-}iPr\text{-}C_6H_4Me)_2][CF_3SO_3]_2]$ (A_{12}) separately with two distinct pyridyl organic ligands, bis(pyridin-4-yl)-1,2,4,5-tetrazine (D_{21}) or 2,5-bis(pyridin-4-ylethynyl)furan (D_{22}), produced a series of arene-ruthenium metallacycles (21–26). All metallacycles were fully characterized by using various spectroscopic methods and their anticancer efficacy was examined using both in-vitro and in-vivo studies. The growth inhibitory activity of all the metallacycles and their corresponding building blocks were measured against AGS and HCT-15 human cancer cell lines by estimating the cell viability through exposing the cell lines to the increasing concentration of the metallacycles for about 24 h. The metallacycles 21 and 24 showed no effect upon exposure to both cancer cells and but metallacycles 22 and 25 were weakly active against AGS and HCT-15 cancer cells with an IC_{50} value in the range of 105 and 180 and >200 μM, respectively. However, the metallacycles 23 and 26 displayed strong anticancer activity against AGS cancer cells with IC_{50} value equal to 30 and 23 μM, respectively, which is quite lower than that of clinically used cisplatin (IC_{50} = 107 μM) and moderate activity against HCT-15 cells (IC_{50} 29 and 27 μM, respectively). The in-vivo pharmacodynamic studies were also performed using a hollow-fibre (HF) assay. The HCT-15 cells loaded in semipermeable hollow-fibres were instilled into the subcutaneous (SC) and intraperitoneal (IP) parts of host mice and were dosed with 100 μg per day of supramolecular complexes 23 and 26 and examined for a week. The HCT-15 cancer cell growth (IP) was restrained to about 14.2% and 21.8% by the injected complexes 23 and 26. Metallacycle 26 showed higher efficacy compared to metallacycle 23 but lower than the efficiency of cisplatin (37.9%). Autophagy (programmed cell death) activity of complex 23 and 26 was examined using the monodansylcadaverine (MDC) staining method. Autophagic vacuoles accumulated during the mature autophagy process can be detected by this staining method. Upon the treatment with complexes 23 and 26, the MDC-labelled autophagic vacuoles tremendously increased. Also, at a lower concentration (0–5 μM), as there is an increase in the concentration of the complexes, there is a corresponding increase in the autophagy activity. In particular, complex 26 exhibited more than a 2-fold increment in autophagy activity.

The first-ever Ru(II) and Ir(III) based metalla-rectangles 27–30 employing an anticancer-active boron-dipyrromethene (BODIPY)-based linker, 2,6-di-(4-pyridyl)-1,3,5,7-tetramethyl-8-phenyl-4,4-difluoroboradiazaindacene (D_{23}) was reported by Gupta and co-workers in 2016 (Gupta et al. 2016) (Figure 7.12a). Rectangles 28 and 30 were fully characterized by using X-ray diffraction analysis (Figure 7.12b). The fluorescence nature of the BODIPY core and the AIE-based emission associated with it promote easy visualization of metallacycles within cancer cells. The assembly showed a strong emission peak around 537 nm, which corresponds to the BODIPY moiety. Dose-dependent response studies were performed to understand the anticancer properties of metallo-rectangles towards various cancer cell lines such as lung (A549), breast (MCF-7), cervical (HeLa), and brain (U87) cancer cells. Rectangle 27 inhibited the proliferation of MCF-7, HeLa, and U87 cancer cells with a higher rate of cell death than the normal cell (WI-38 cells), whereas rectangle 29 selectively inhibited the growth of MCF-7 cancer cell (Figure 7.12c). The rectangles 28 and 30 showed notable action against all the cancer cell lines, but the mortality rate of cancer cells compared to normal cells (WI-38) was higher only in brain cells. The presence of extensively conjugated naphthoquinone ligands and increased nuclearity were the reason for the increased anticancer potency of 28 and 30. The BODIPY ligand D_{23} alone does not show any notable anticancer activity, which ruled out its role in the anticancer therapeutic nature of the rectangles. The confocal fluorescence microscope was used as a tracking device to locate the rectangles in the cells and the result showed a medium to bright green fluorescence emission upon excitation at 488 nm. These studies confirmed that BODIPY containing metallo-assemblies can find applications as theragnostic agents for both diagnostic and therapeutic effects in cancer treatment.

A unique Ru^{II}_8 molecular cage 31–34 with anticancer ability was developed through coordination-driven self-assembly of Ru(II)-based acceptors having various O∩O bridging ligands [$Ru_2(\mu$-η^4-$C_2O_4)$-$(MeOH)_2(\eta^6$-p-iPr-$C_6H_4Me)_2][CF_3SO_3]_2$, [$Ru_2(dobq)$-$(MeOH)_2(\eta^6$-p- iPr-$C_6H_4Me)_2][CF_3SO_3]_2$, [$Ru_2(donq)(H_2O)_2(\eta^6$-$p$-iPr-$C_6H_4Me)_2][CF_3SO_3]_2$ and [$Ru_2(dhtq)(H_2O)_2(\eta^6$-$p$-iPr-$C_6H_4Me)_2][CF_3SO_3]_2$ separately reacting with tetrapyridyl donor, N, N, N′, N′-tetra(pyridin-4-yl)benzene-1,4-diamine (D_{24}) (Adeyemo et al. 2017). The MTT assay showed excellent anticancer activity for metallo-assemblies against different cancer cells. Among the four cages, 33 and 34 exhibited significant anticancer activity against both HeLa and A549 cancer cell lines with IC_{50} value much lower than the commonly

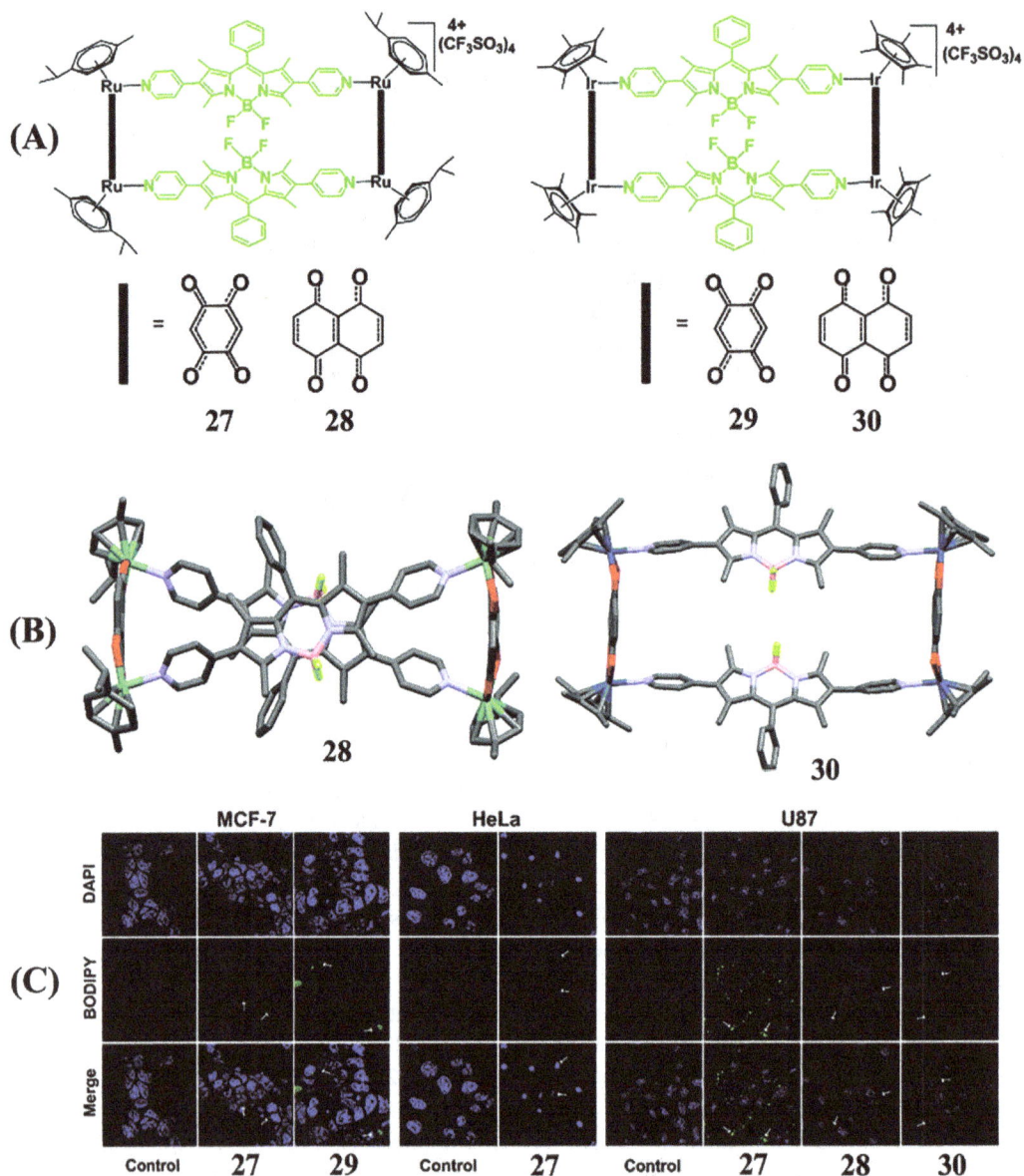

FIGURE 7.12 (a) Two-component self-assembly formation of BODIPY based molecular rectangles 27–30. (b) X-ray crystal structure of rectangle 28 and 30. (c) Confocal fluorescence images of rectangles 27–30 localized within different cell lines. (Reproduced with permission from Gupta (2016) © 2016 The Royal Society of Chemistry.)

used cancer drug cisplatin. In comparison to cisplatin, 31 and 32 showed a moderate anticancer activity with poor IC_{50} value against A549 and HeLa cancer cell lines, which highlights the role of the aromatic moieties of acceptor units on anticancer activity.

7.5.6   Drug Delivery using Metallo-Assemblies

The targeted drug delivery is the most feasible method for selective transportation of pharmaceutically active ingredients to the diseased sites (Chakrabarty et al. 2011). Usually, the free drugs are encapsulated in a specific delivery vehicle to enhance the pharmacological properties and therapeutic efficacy. The primary requirement for a drug delivery system is a suitable internal cavity to properly encapsulate the drug molecules. Over the past few years, various metallosupramolecular architectures have been reported as effective drug delivery systems because they provide a confined nanoscopic cavity that can encapsulate numerous guest molecules. Crowley and his group developed a stimuli-responsive Pd(II) metallocage which can encapsulate as well as deliver anticancer agent cisplatin (Lewis et al. 2012). The discrete metallocage was synthesized through the self-assembly of tripyridyl angular donor ligand, 2,6-bis(pyridin-3-ylethynyl)pyridine (L) with Pd(II) based acceptor unit $[Pd(CH_3CN)_4](X)_2$ [where $X = BF_4^-$ and SbF_6^-]. The reversible stimuli-responsive disassembly/ reassembly is a requisite for an efficient drug delivery system as the delivery vehicle should be able to show targeted release of an encapsulated drug from the supramolecular architecture to the particular site where the drug is required. Upon adding and eliminating proper competing ligands such as 4-dimethyl aminopyridine (DMAP) and $Bu_4N^+Cl^-$, $[Pd_2(L)_4](X)_4$ cages showed stimuli-responsive behaviour and can reversibly be disassembled and reassembled which makes these Pd(II) cages an adequate drug delivery vehicle. The central cavities of the cage are aligned with pyridine units creating a lantern-shape architecture and these pyridine units exhibited hydrogen-bonding interactions between the cisplatin amine units and the cages, which resulted in strong host-guest complex formation. Upon the addition of minute quantities of D_2O to the cisplatin bounded cage in CD_3CN, the expulsion of the two cisplatin moieties from the cage was observed. On replacing the central pyridine donor unit with the central benzene unit (1,3-bis(pyridin-3-ylethynyl)benzene), the corresponding metallocage showed no interaction with cisplatin drug molecule which indicates the relevance of hydrogen-bonding interaction for the host-guest chemistry. Treating the cisplatin encapsulated cage with competing ligands like DMAP or $Bu_4N^+Cl^-$ led to the dissociation of the cage and release of the drug molecule to the targeted site. Therefore, Pd(II) metallocage acts as an excellent drug delivery vehicle.

Mukherjee and co-workers developed a tetrafacial water-soluble Pd(II) self-assembled molecular barrel, from tetrapyridyl donor (D_{24}) and cis-blocked 90° acceptor A_5, that can act as an aqueous carrier for hydrophobic curcumin (an active ingredient of turmeric) (Bhat et al. 2017). The formation of a barrel-shaped architecture was confirmed using diffraction analysis. This kind of architecture has enormous application in biological systems. For example, the β-barrel proteins support the easy diffusion of smaller ions and fragments through the cell membranes. In addition, this barrel has an internal hydrophobic cavity that facilitates the easy encapsulation of hydrophobic curcumin in an aqueous medium. Hence, this barrel helps to solubilize curcumin in water, subsequently enhancing its bioavailability. In addition to the solubility, it also shields the photosensitive curcumin from sunlight/UV radiations. A large portion of incident photons was absorbed by the aromatic wall of the cage having high absorption cross-sectional area restricting them from the photodegradation of encapsulated curcumin. There occurs notable progress in the cellular uptake as well as the solubility of curcumin in an aqueous medium when encapsulated in the molecular barrel. In an aqueous medium, free curcumin showed negligible activity against cancer cells at room temperature due to the lack of solubility, whereas curcumin encapsulated in a barrel exhibited an IC_{50} value of approximately 14 μM. From the above points, it is very clear that the water-soluble self-assembled Pd-based molecular barrier can efficiently act as a drug delivery vehicle for curcumin. Lippard and his group synthesized Pt(II) cages for the encapsulation of Pt(IV) based prodrugs for delivery of clinically tested anticancer drug cisplatin (Zheng et al. 2015). The octahedral cage 35 was obtained through self-assembly of the 90° acceptor, cis-$[Pt(en)(NO_3)_2]$ (A_{13}) and triazine-based tripyridyl donor (D_{14}) in a 6:4 stochiometric ratio. The host-guest complex was produced by the mixing of 4:1 equivalences of adamantyl Pt(IV) prodrug with cage 35, respectively. The adamantyl unit of Pt(IV) prodrug interacts with cage moiety resulting in a strong host-guest complex formation (Figure 7.13a). The synthesized host-guest complex was successfully characterized using 1D and 2D NMR spectroscopy. Since the synthesized supramolecular cage architecture is positively charged, it could easily enter the cancer cell. Ascorbic acid and other biological reductants help in the reduction of this prodrug and release of cisplatin to the cell which arrests the

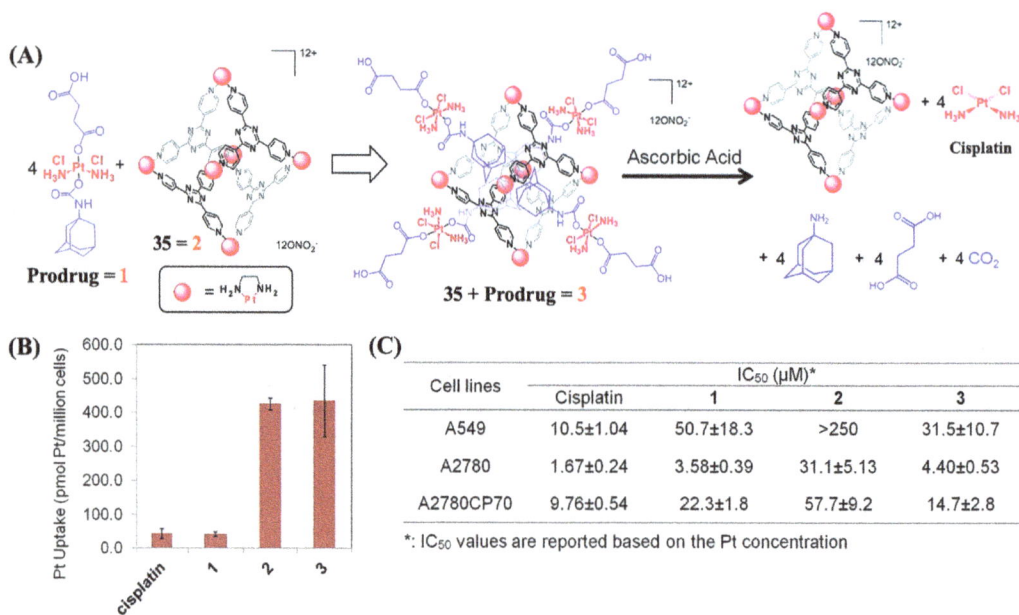

FIGURE 7.13 (a) Schematic representation of host-guest complexation between Pt(II) octahedral cage 35 with adamantly based prodrug guest molecule and the subsequent delivery of Cisplatin through ascorbic acid reduction. (b) Percentage of cell uptake of cisplatin, cage 35, and its precursor ligand measured using A2780CP70 cell line. (c) Cytotoxicity profiles of cisplatin and host-guest complexes including the precursors against different cancer cell lines. (Reproduced with permission from Zheng (2015) Under Creative Commons Licence © 2015 The Royal Society of Chemistry.)

cell cycle and facilitates apoptosis (Figure 7.13b). Compared to cisplatin, the self-assembled complex with guest Pt(IV) prodrug shows relatively better cytotoxicity against various human cancer cell lines such as A549 (lung carcinoma), A2780 (ovarian carcinoma), and A2780CP70 (ovarian carcinoma resistant to cisplatin) cells, however, the cytotoxicity of the prodrug alone or the cage alone is lower than cisplatin (Figure 7.13c). This proves the importance and advantages of using self-assembled metallo-cages in drug delivery applications.

7.6 CONCLUSIONS AND PERSPECTIVES

In this chapter, we have systematically highlighted the design, synthesis, and a few selected applications of various self-assembled 2D metallacycles and 3D metallocages synthesized through coordination-driven self-assembly processes. The modular coordination-driven self-assembly strategy is a flourishing area of research in the current era of supramolecular chemistry. By selecting the appropriate complementary building blocks and controlling the stoichiometric ratio of them, a variety of self-assembled coordination assemblies can be generated employing simple 90° acceptors to large shape-selective organometallic building units. A vast library of 2D and 3D supramolecular architectures of different topologies integrated with desired functional properties are developed, and several of them have been successfully realized for various practical applications. Even though there is a substantial advancement made in this area, the characterizations of self-assembled supramolecular structures are still challenging and complicated as it is hard to obtain diffraction quality suitable single crystals of large supramolecular structures for X-ray diffraction analysis. Nowadays, synchrotron-based analysis has been used for the successful elucidation of the geometry of self-assembled structures, but it is an expensive technique. To date, only a few selected metal cations, due to their stable coordination geometry, are used to design transition metal-based acceptors, and efforts must be devoted to developing a more general synthetic strategy for the design of metallo-assemblies from all kinds of metal ions, which would allow us to construct assorted supramolecular assemblies with unique functional properties. The multicomponent self-assembly process has been widely studied to generate a variety of supramolecular complexes, but the selective self-assembly of the formation of desired metallo-assemblies employing this strategy is challenging as it can yield a mixture of products, and thus, more research efforts must be focused on this direction. Lastly, the practical applications of self-assembled metallocycles in biomedicine, material sciences, and many other areas remain relatively less explored, and in the future, more research attention should be focused on exploring various real-world applications of self-assembled metallocycles and aiming to mimic various biological metabolic processes and enzymatic functions. Even though we have not highlighted the importance of molecular self-assembly in relevance to corrosion and biofouling prevention (see some of the later chapters), the information presented on supramolecular self-assembly concepts and design approaches could be helpful to researchers and be beneficial for the development of new technologies in this area.

7.7 ACKNOWLEDGMENT

We gratefully acknowledge the Science and Engineering Research Board (EMEQ Award EEQ/2018/000799 to SS), India, and the Indian Institute of Technology Palakkad (IITPKD) for financial support.

REFERENCES

Adeyemo, A. A., Shettar, A., Bhat, I. A., Kondaiah, P., and Mukherjee, P. S. 2017. Self-assembly of discrete Ru$^{II}_8$ molecular cages and their in vitro anticancer activity. *Inorg. Chem.* 56:608–617.

Balzani, V., and DeCola, L. (Eds.) 1992. *Supramolecular Chemistry.* Kluwer Academic: The Netherlands.

Bar, A. K., Chakrabarty, R., Chi, K.-W., Batten, S. R., and Mukherjee, P. S. 2009. Synthesis and characterisation of heterometallic molecular triangles using ambidentate linker: self-selection of a single linkage isomer. *Dalton Trans.* 17:3222–3229.

Barry, N. P. E., Abd Karim, N. H., Vilar, R., and Therrien, B. 2009. Interactions of Ruthenium coordination cubes with DNA. *Dalton Trans.* 48:10717–10719.

Barry, N. P. E., Govindaswamy, P., Furrer, J., Süss-Fink, G., and Therrien, B. 2008. Organometallic boxes built from 5,10,15,20-tetra(4-pyridyl)porphyrin panels and hydroxyquinonato-bridged diruthenium clips. *Inorg. Chem. Commun.* 11:1300–1303.

Beer, P. D., and Gale, P. A. 2001. Anion recognition and sensing: the state of the art and future perspectives. *Angew. Chem. Int. Ed.* 40:486–516.

Bhat, I. A., Jain, R., Siddiqui, M. M., Saini, D. K., and Mukherjee P. S., 2017. Water-soluble Pd$_8$L$_4$ self-assembled molecular barrel as an aqueous carrier for hydrophobic curcumin. *Inorg. Chem.* 56:5352–5360.

Boorum, M. M., Vasilev, Y. V., Drewello, T., and Scott, L. T. 2001. Groundwork for a rational synthesis of C$_{60}$: cyclodehydrogenation of a C$_{60}$H$_{30}$ polyarene. *Science* 294:828–831.

Caulder, D. L., and Raymond, K. 1999. Supermolecules by design. *Acc. Chem. Res.* 32:975–982.

Chakrabarty, R., Mukherjee, P. S., and Stang, P. J. 2011. Supramolecular coordination: self-assembly of finite two- and three-dimensional ensembles. *Chem. Rev.* 111:6810–6918.

Chen, L.-J., Ren, Y.-Y., Wu, N.-W., Sun, B., Ma, J.-Q., Zhang, L., Tan, H., Liu, M., Li, X., and Yang, H.-B. 2015. Hierarchical self-assembly of discrete organoplatinum(II) metallacycles with polysaccharide via electrostatic interactions and their application for heparin detection. *J. Am. Chem. Soc.* 137:11725–11735.

Cook, T. R., and Stang, P. J. 2015. Recent developments in the preparation and chemistry of metallacycles and metallacages via coordination. *Chem. Rev.* 115:7001–7045.

Dubey, A., Jeong, Y. J., Jo, J. H., Woo, S., Kim, D. H., Kim, H., Kang, S. C., Stang, P. J., and Chi, K. W. 2015. Anticancer activity and autophagy involvement of self-assembled arene–ruthenium metallacycles. *Organometallics* 34:4507–4514.

Fan, J., Schmitz, M., and Stang, P. J. 2000. Archimedean solids: transition metal mediated rational self-assembly of supramolecular-truncated tetrahedra. *Proc. Natl. Acad. Sci.* 97:1380–1384.

Fujita, M., Aoyagi, M., and Ogura, K. 1996. Macrocyclic dinuclear complexes self-assembled from (en) Pd(NO₃)₂ and pyridine-based bridging ligands. *Inorg. Chim. Acta* 246:53–57.

Fujita, M., Sasaki, O., Mitsuhashi, T., Fujita, T., Yazaki, J., Yamaguchi, K., and Ogura, K. 1996. On the structure of transition-metal-linked molecular squares. *Chem. Commun.* 1996:1535–1536.

Fujita, M., Tominaga, M., Hori, A., and Therrien, B. 2005. Coordination assemblies from a Pd(II)-cornered square complex. *Acc. Chem. Res.* 38:369–378.

Fujita, M., Yazaki, J., and Ogura, K. 1990. Preparation of a macrocyclic polynuclear complex, [(en)Pd(4,4′-bpy)]₄(NO₃)₈ (en= ethylenediamine, bpy= bipyridine), which recognizes an organic molecule in aqueous media. *J. Am. Chem. Soc.* 112:5645–5647.

Fujita, M., Yazaki, J., and Ogura, K. 1991. Spectroscopic observation of self-assembly of a macrocyclic tetranuclear complex composed of Pt²⁺ and 4,4′-bipyridine. *Chem. Lett.* 20:1031–1032.

Galanski, M., Jakupec, M. A., and Keppler, B. K. 2005. Update of the preclinical situation of anticancer platinum complexes: novel design strategies and innovative analytical approaches. *Curr. Med. Chem.* 12:2075–2094.

Ghosh, S., and Mukherjee, P. S. 2008. Self-assembly of a nanoscopic prism via a new organometallic Pt₃ acceptor and its fluorescent detection of nitroaromatics. *Organometallics* 27:316–319.

Gupta, G., Das, A., Ghate, N. B., Kim, T. H., Ryu, J. Y., Lee, J. Mandal, N., and Lee, C. Y. 2016. Novel BODIPY-based Ru (II) and Ir (III) metalla-rectangles: cellular localization of compounds and their antiproliferative activities. *Chem. Commun.* 52:4274–4277.

Hall, J., Loeb, S. J., Shimizu, G. K. H., and Yap, G. P. A. 1998. Supramolecular arrays of 4,7-phenanthroline complexes: self-assembly of molecular Pd₆ hexagons. *Angew. Chem. Int. Ed.* 37:121–123.

`He, M.-L., Wu, S., He, J., Abliz, Z., and Xu, L. 2014. Construction of a naphthalimide-containing hexagonal metallocycle via coordination-driven self-assembly and its fluorescence detection of protons. *RSC Adv.* 4:2605–2608.

Hiraoka, S., Harano, K., Shiro, M., Ozawa, Y., Yasuda, N., Toriumi, K., and Shionoya, M. 2006. Isostructural coordination capsules for a series of 10 different d⁵–d¹⁰ transition-metal ions. *Angew. Chem. Int. Ed.* 45:6488–6491.

Howlader, P., Das, P., Zangrando, E., and Mukherjee, P. S. 2016. Urea-functionalized self-assembled molecular prism for heterogeneous catalysis in water. *J. Am. Chem. Soc.* 138:1668–1676.

Hwang, S.-H., Moorefield, C. N., Fronczek, F. R., Lukoyanova, O., Echegoyen, L., and Newkome, G. R. 2005. Construction of triangular metallomacrocycles: [M₃(1,2- bis(2,2′:6′, 2″-terpyridin-4-yl-ethynyl)benzene)₃] [M=Ru(II), Fe(II), 2Ru(II)Fe(II)]. *Chem. Commun.* 2005:713–715.

Kaphan, D. M., Levin, M. D., Bergman, R. G., Raymond, K. N., and Toste, F. D. 2015. A supramolecular microenvironment strategy for transition metal catalysis. *Science* 350:1235–1238.

Kekule, A. 1865. Sur la constitution des substances aromatiques. *Bull. Soc. Chim. Fr.* 3:98–110.

Lehn, J.-M. 1995. *Supramolecular Chemistry: Concepts and Perspectives*, VCH: Weinheim.

Leininger, S., Olenyuk, B., and Stang, P. J. 2000. Self-assembly of discrete cyclic nanostructures mediated by transition metals. *Chem. Rev.* 100:853.

Lever, R., and Page, C. P. 2002. Novel drug development opportunities for heparin. *Nat. Rev. Drug. Discov.* 1:140–148.

Lewis, J. E. M., Gavey, E. L., Cameron, S. A., and Crowley, J. D. 2012. Stimuli-responsive Pd₂L₄ metallosupramolecular cages: towards targeted cisplatin drug delivery. *Chem. Sci.* 3:778–7841.

Lu, J., Turner, D. R., Harding, L. P., Byrne, L. T., Baker, M. V., and Batten, S. R. 2009. Octapi interactions: self-assembly of a Pd-based [2]catenane driven by eightfold π interactions. *J. Am. Chem. Soc.* 131:10372–10373.

Manimaran, B., Thanasekaran, P., Rajendran, T., Liao, R.-T., Liu, Y.-H., Lee, G.-H., Peng, S.-M., Rajagopal, S., and Lu, K.-L. 2003. Self-assembly of octarhenium-based neutral luminescent rectangular prisms. *Inorg. Chem.* 42:4795–4797.

Martínez-Máñez, R., and Sancenón, F. 2003. Fluorogenic and chromogenic chemosensors and reagents for anions. *Chem. Rev.* 103:4419–4476.

Maverick, A. W., Buckingham, S. C., Yao, Q., Bradbury, J. R., and Stanley, G. G. 1986. Intramolecular coordination of bidentate Lewis bases to a cofacial binuclear Copper(II) complex. *J. Am. Chem. Soc.* 108:7430–7431.

Murase, T., Nishijima, Y., and Fujita, M. 2012. Cage-catalyzed knoevenagel condensation under neutral conditions in water. *J. Am. Chem. Soc.* 134:162–164.

Northrop, B. H., Chercka, D., and Stang, P. J. 2008. Carbon-rich supramolecular metallacycles and metallacages. *Tetrahedron* 64:11495–11503.

Northrop, B. N., Yang, H.-B., and Stang, P. J. 2008. Coordination-driven self-assembly of functionalized supramolecular metallacycles. *Chem. Commun.* 2008:5896–5908.

Olenyuk, B., Fechtenkötter, A., and Stang, P. J. 1998. Molecular architecture of cyclic nanostructures: use of co-ordination chemistry in the building of supermolecules with predefined geometric shapes. *J. Chem. Soc. Dalton Trans.* 1998:1707–1728.

Rosenberg, B., Vancamp, L., Trosko, J., and Mansour, V. H. 1969. Platinum compounds: a new class of potent antitumour agents. *Nature* 222:385–386.

Saalfrank, R. W., Glaser, H., Demleitner, B., Hampel, F., Chowdhry, M. M., Schünemann, V., Trautwein, A. X., Vaughan, G. B. M., Yeh, R., Davis, A. V., and Raymond, K. N. 2002. Self-assembly of tetrahedral and trigonal antiprismatic clusters $[Fe_4(L^4)_4]$ and $[Fe_6(L^5)_6]$ on the basis of trigonal tris-bidentate chelators. *Chem. Eur. J.* 8:493–497.

Schmitz, M., Leininger, S., Fan, J., Arif, A. M., and Stang, P. J. 1999. Preparation and solid-state properties of self-assembled dinuclear Platinum(II) and Palladium(II) rhomboids from carbon and silicon tectons. *Organometallics* 18:4817–4824.

Schweiger, M., Seidel, S.R., Arif, A.M., and Stang, P.J. 2002. Solution and solid-state studies of a triangle-square equilibrium: anion-induced selective crystallization in supramolecular self-assembly. *Inorg. Chem.* 41:2556–2559.

Sessler, J. L., Camilo, S., and Gale, P. A. 2003. Pyrrolic and polypyrrolic anion binding agents. *Coord. Chem. Rev.* 240:17–55.

Shanmugaraju, S., Bar, A. K., Joshi, S. A., Patil, Y. P., and Mukherjee, S. 2011. Constructions of 2d-metallamacrocycles using half-sandwich Ru^{II}_2 precursors: synthesis, molecular structures, and self-selection for a single linkage isomer. *Organometallics* 30:1951–1960.

Shanmugaraju, S., Dabadie, C., Byrne, K., Savyasachi, A. J., Umadevi, D., Schmitt, W., Kitchen, J. A., and Gunnlaugsson, T. 2017. A supramolecular Tröger's base derived coordination zinc polymer for fluorescent sensing of phenolic-nitroaromatic explosives in water. *Chem. Sci.* 8:1535–1546.

Shanmugaraju, S., Harshal, J., Patil, Y. P., and Mukherjee, P. S. 2012. Self-assembly of a Pt^{II}_8 tetragonal prism from a new Pt^{II}_4 organometallic star-shaped acceptor and its nitroaromatic explosives sensing. *Inorg. Chem.* 51:13072–13074.

Shanmugaraju, S., Joshi, S. A., and Mukherjee, P. S. 2011. Self-assembly of metallamacrocycles using a dinuclear organometallic acceptor: synthesis, characterization, and sensing study. *Inorg. Chem.* 50:11736–11745.

Shanmugaraju, S., and Mukherjee, P. S. 2015. Self-assembled molecular sensors for nitroaromatics. *Chem. Eur. J.* 21:6656–6666.

Slone, R. V., Hupp, J. T., Stern, C. L., and Albrecht-Schmitt, T. E. 1996. Self-assembly of luminescent molecular squares featuring octahedral Rhenium corners. *Inorg. Chem.* 35:4096–4097.

Würthner, F., and Sautter, A. 2000. Highly fluorescent and electroactive molecular squares containing perylene bisimide ligands. *Chem. Commun.* 445–446. https://doi.org/10.1039/A909892E

Stang, P. J. 1998. Molecular architecture: coordination as the motif in the rational design and assembly of discrete supramolecular species-self-assembly of metallacyclic polygons and polyhedra. *Chem. Eur. J.* 4:19–27.

Stang, P. J. 2009. From solvolysis to self-assembly. *J. Org. Chem.* 74:2–20.

Stang, P. J., and Olenyuk, B. 1997. Self-assembly, symmetry, and molecular architecture: coordination as the motif in the rational design of supramolecular metallacyclic polygons and polyhedra. *Acc. Chem. Res.* 30:502.

Steed, J. W., and Atwood, J. L. 1995. *Supramolecular Chemistry*, 1st Ed., Academic Press: New York.

Steed, J. W., Turner, D. R., and Wallace, K. J. 2007. *Concepts in Supramolecular Chemistry and Nanochemistry*, John Wiley & Sons: England.

Tam, A. Y.-Y., Wong, K. M.-C., Wang, G.-X., and Yam, V. W.-W. 2007. Luminescent metallogels of Platinum(II) terpyridyl complexes: interplay of metal···metal, π–π and hydrophobic–hydrophobic interactions on gel formation. *Chem. Commun.* 2007:2028–2030.

Tan, C., Jiao, J., Li, Z., Liu, Y., Han, X., and Cui, Y. 2018. Design and assembly of a chiral metallosalen-based octahedral coordination cage for supramolecular asymmetric catalysis. *Angew. Chem. Int. Ed.* 57:2085–2090.

Therrien, B. 2009. Arene ruthenium cages: boxes full of surprises. *Eur. J. Inorg. Chem.* 2009:2445–2453.

Vajpayee, V., Song, Y. H., Lee, M. H., Kim, H., Wang, M., Stang, P. J., and Chi, K. W. 2011. Self-assembled arene-ruthenium-based rectangles for the selective sensing of multi-carboxylate anions. *Chem. Eur. J.* 17:7837–7844.

Weilandt, T., Troff, R. W., Saxell, H., Rissanen, K., and Schalley, C. A. 2008. Metallo-supramolecular self-assembly: the case of triangle-square equilibria. *Inorg. Chem.* 47:7588–7598.

Würthner, F., and Sautter, A. 2000. Highly fluorescent and electroactive molecular squares containing perylene bisimide ligands. *Chem. Commun.* 2000:445–446.

Yan, H., Süss-Fink, G., Neels, A., and Stoeckli-Evans, H. 1997. Mono-, di- and tetra-nuclear *p*-cymenerutheium complexes containing oxalato ligands. *J. Chem. Soc. Dalton Trans.* 1997:4345–4350.

Yang, H.-B., Das, N., Huang, F., Hawkridge, A. M., Díaz, D. D., Arif, A. M., Finn, M. G., Muddiman, D. C., and Stang, P. J. 2006. Incorporation of 2,6-Di(4,4′-dipyridyl)-9-thiabicyclo[3.3.1]nonane into discrete 2D supramolecules via coordination-driven self-assembly. *J. Org. Chem.* 71: 6644–6647.

Yoshizawa, M., Nakagawa, J., Kumazawa, K., Nagao, M., Kawano, M., Ozeki, T., and Fujita, M. 2005. Discrete stacking of large aromatic molecules within organic-pillared coordination cages. *Angew. Chem. Int. Ed.* 44: 1810–1813.

Zheng, Y.-R., Suntharalingam, K., Johnstone, T. C., and Lippard, S. J. 2015. Encapsulation of Pt(IV) prodrugs within a Pt(II) cage for drug delivery. *Chem. Sci.* 6:1189–1193.

Zheng, Y.-R., Zhao, Z., Wang, M., Ghosh, K., Pollock, J. B., Cook, T. R., and Stang, P. J. 2010. A facile approach toward multicomponent supramolecular structures: selective self-assembly via charge separation. *J. Am. Chem. Soc.* 132:16873–16882.

Section II

Applications in Corrosion Protection

A

Protective Coatings (Chapters 8 to 14)

B

Corrosion Inhibitors (Chapters 15 to 19)

8 Self-Healing Mechanisms in Smart Protective Coatings

Alicja Stankiewicz and Katarzyna Zielińska
coat-it sp. z o. o.

CONTENTS

8.1   INTRODUCTION

Compared to conventional materials, self-healing materials demonstrate the ability to restore their properties after mechanical and functional damage. Given the lability of interactions inside or between supramolecular entities, these species are of particular attention to scientists designing self-healing materials as they can easily exchange their constituents. Self-healing mechanisms are generally broadly classified into two categories: intrinsic and extrinsic ones (Sanka et al. 2019). Intrinsic self-healing mechanisms rely on dynamic bonds and reversible interactions embedded in a coating material that can be broken and renewed. This is realised mostly in two ways through physical interactions and chemical interactions, as shown in Figure 8.1. The materials healing in an extrinsic way are based on the presence of a healing agent and a catalyst/hardener, which are delivered in the form of capsules, microvascular systems or hollow fibres. Even though the former mechanism is mostly associated with supramolecular chemistry, the latter draws from it as well. It is worth mentioning that one of the first successfully applied self-healing materials was utilising capsules and was based on the ring-opening metathesis polymerisation (ROMP) of dicyclopentadiene.

DOI: 10.1201/9781003169130-10

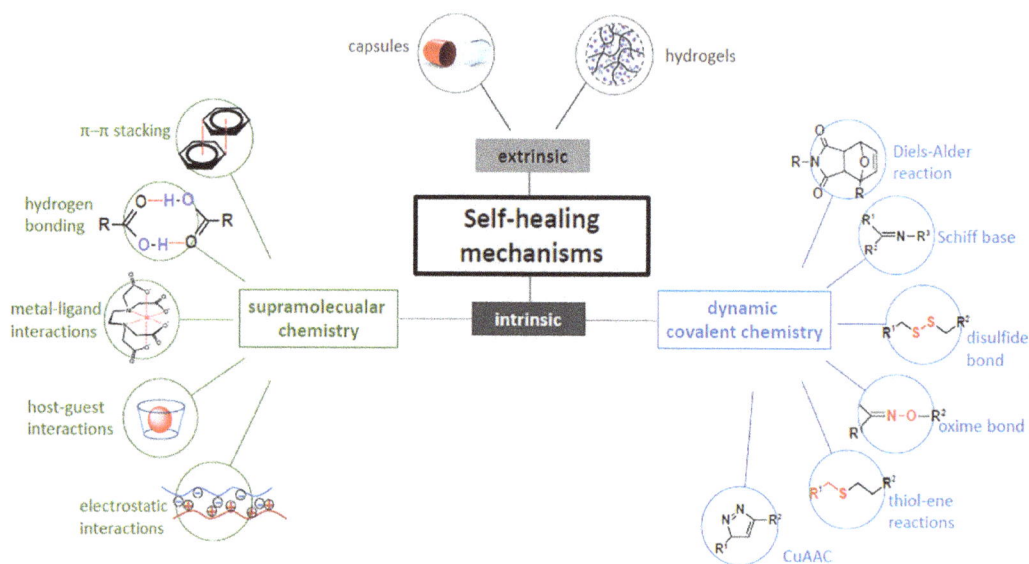

FIGURE 8.1 Schematic representation of mechanisms used in synthesis of self-healing materials.

The healing agent released from the capsules to the crack plane, in combination with The Grubbs' catalyst, yielded static fracture recovery of an epoxy matrix (White et al. 2001).

The molecular chemistry is also furnished with dynamic attributes, as some covalent bonds may break and develop reversibly. It serves as a rich source of components of self-healing materials. Thus, considering that dynamic chemistry encompasses as reactional and motional dynamics, some examples of self-healing materials based on dynamic covalent bonds (chemical interactions in Figure 8.1) are also discussed here.

In terms of corrosion protection, substances respond to pH changes, electrons flow or UV light can all be used for a coating material programmed to heal itself. Importantly, an excellent candidate for reversible systems requires spontaneous formation, high efficiency, no or only inoffensive by-products (like water molecules) and no toxic catalysts.

Taking into consideration the extensive literature, this chapter presents only selected examples from the field to demonstrate the concepts and applications of supramolecular chemistry in self-healing coating materials.

8.2 DYNAMIC INTERACTIONS

Supramolecular components held together by noncovalent interactions, such as $\pi-\pi$ stacking, hydrogen bonding or metal ion coordination, form systems dynamic in nature – so-called dynamic polymer networks (DPNs). The weak nature of supramolecular interactions allows dissociation of dynamic bonds upon the stress or mechanical damage. Nonetheless, these bonds can easily re-associate without the intervention of an external trigger. Such a feature is highly favourable for material selection in self-healing applications. Some examples of self-healing coatings based on dynamic interaction mechanisms are presented below.

8.2.1 $\pi-\pi$ STACKING

$\pi-\pi$ stacking refers to noncovalent interactions between aromatic rings containing π-bonds. These interactions occur when the plane aromatic rings are stacked parallel to one another. Common structures of aromatic compounds are presented in Figure 8.2. The strength of the interaction depends on many factors, including the size and geometry of the aromatic core, the nature of the substituents

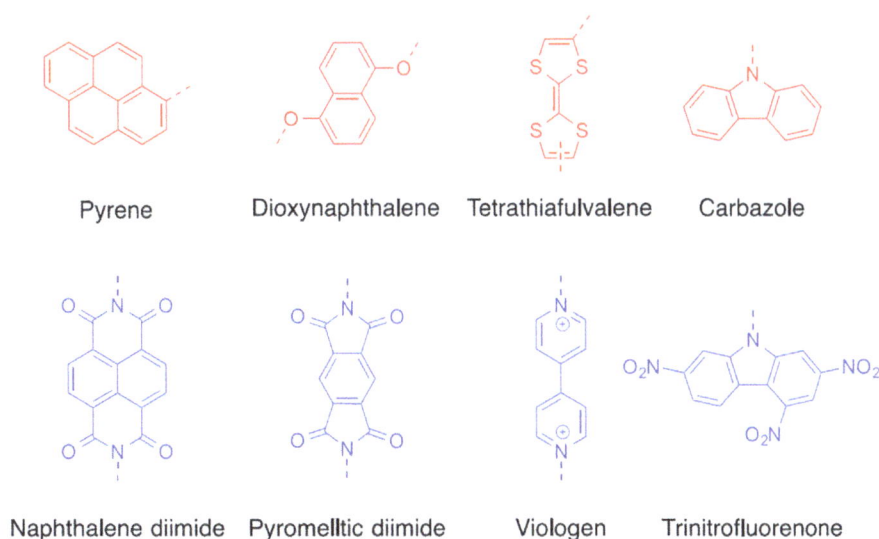

FIGURE 8.2 Exemplary structures of donor (red) and acceptor (blue) compounds. (Reproduced with permission from Hayes (2015) © 2015, Springer International Publishing Switzerland.)

and the solvent. There are numerous examples of supramolecular materials that rely on π–π interactions, e.g., linear or crosslinked polymers, gels or vesicles.

In general, they are designed to contain π-electron-poor receptor sites along with π-electron-rich groups. As it was shown in chain-folding, healable materials. The blend containing N, N′-bis(2-hydroxy-ethyl)naphthalene-1,4,5,8-tetracarboxylic-diimide (NDI) residues separated by an ethylenedioxy-bisethylamine unit and a commercially available amine-terminated poly(dimethyl siloxane) (PDMS) was proven to contain approximately five chain-folding residues in the backbone. These complementary π-electron-rich and π-electron-poor receptors exhibited healable characteristics in the solid state in response to temperature (Burattini et al. 2009a; Hart et al. 2015). The benefit of the PMDS (and similar polymeric compounds) employment is the simplicity of improving the mechanical performance of the materials without influencing the ability to form a stable cross-linked network supported by π–π stacking. This could be realised by introducing amine-terminated co-oligomers (Burattini et al. 2009b) or another class of supramolecular interactions like hydrogen bonding (Burattini et al. 2010). A mechanism for thermo-reversible healing behaviour involves disruption of the intermolecular π–π stacking cross-links at elevated temperatures. This allows a flow of low glass transition temperature (T_g) component and the restoration of blend properties. When the temperature is decreased, π–π stacks are re-established, ensuring good damage recovery.

π–π stacking can serve as a driving force in the modification of carbon nanotubes (CNT) in nanocomposite materials with conjugated polymers as matrices. An example of such a material is the CNT/polyimide nanocomposite. Compared to the poor polyimide-based matrix, the composite material shows greater mechanical strength, Young's modulus, electrical conductivity and thermal stability (Jiang et al. 2020). Another carbonaceous compound – graphene – was used directly in anticorrosion application to protect steel substrate. Waterborne epoxy coatings containing perylenebisimide (PBI)/graphene (GR) presented enhanced corrosion protection; however, no self-healing effect was reported (Lao et al. 2021).

8.2.2 Hydrogen Bonding

Hydrogen bonding is a special type of dipole–dipole attraction between a hydrogen atom covalently bonded to a very electronegative atom such as an N, O or F and another very electronegative atom. Typical bonding energy for H-bonding interaction is in the range of 5–30 kJ/mol, which is almost ten times lower compared to covalent bonds (approx. 345 kJ/mol for C–C bond). Despite low bonding energy, this type of interaction has a great influence on polymer viscoelasticity, degree of crystallinity and phase separation.

Polyurethanes (PU) are polymers most often used in designing protective coatings with self-healing properties based on hydrogen-bonds restoration mechanism (Deflorian et al. 2013; Montano et al. 2021; Nardeli et al. 2020; Yan 2018). The self-healing action could be moisture- or temperature-driven (Gadwal 2021). In general, the PU structure consists of two types of segments – rigid and flexible ones – which have a high tendency to interact through hydrogen bonding. To increase the number of hydrogen bonds, the covalent bonds in PU can be partially substituted with hydrogen bonds by the introduction of ureidopyrimidinone (UPy) groups (Yan 2018) or maleic anhydride (Bao et al. 2006). UPy motifs dimerize strongly through quadruple hydrogen bonds. The coating containing UPy can heal itself upon UV radiation.

PU coatings can be used to protect various metallic substrates. One example is bio-based PU coatings derived from vegetable oils employed to produce self-healing protective layers for Al alloy substrate. In this case, the coating with a larger number of flexible segments (polyesters) exhibited better anticorrosion protection and self-healing properties. The flexible segments contributed to a highly developed crosslinked structure due to the formation of hydrogen bonds in greater quantity (Nardeli et al. 2020). Interestingly, softer PUs are unable to sufficiently restore barrier properties. This feature is attributed to PUs with a higher urea/urethane ratio (Montano et al. 2021). Thus, when designing the protective material with the use of Pus, careful consideration of the final properties should be carried out.

Polymers reach in free hydroxyl groups like poly(vinyl alcohol) (PVA) and poly(acrylic acid) (PAA) can form supramolecular complexes, suitable for highly transparent and scratch-healing coating. The efficient intrinsic scratch-healing is attributed to the considerable number of free hydroxyl groups at the scratched interfaces (Jucius et al. 2019).

The introduction of monomers like acrylic acid and 2-acrylamido-2-methyl-1-propanesulfonic acid (AMPS) into resins such as phenolic novolac type epoxy can furnish them with self-repair properties. The polar hydrogen bond donors significantly enhanced the adhesion strength, surface wetting behaviour and the three-dimensional structure of the coatings. This influenced its transition temperature and improved mechanical properties (Shen et al. 2016).

Hydrogen bonds are responsible for the restoration of protective coatings consisting of natural polyphenols like tannic acid (TA) and polyethylene glycol (PEG) (Du et al. 2016; Zhang X. et al. 2016). After drying, TA and PEG coatings exhibited robustness and good adhesion to various substrates, while in the presence of water, they were soft and healable, capable of sealing micrometer-sized cracks (Du et al. 2016).

8.2.3 INTERACTIONS INVOLVING METALS

Having preferential coordination geometries and being thermodynamically stable, metal ions can be successfully used in the steric control of many self-assembling processes. So-called metallo-polymers combine some characteristics of metals (e.g., bioactivity, increased conductivity, molecular magnetism) and polymers (e.g., film formation) (Winter and Schubert 2016). Metal complexes exhibit bi-stability – two or more different states – thus can be switched by an external input to produce an output signal at the molecular level (Bazzicalupi et al. 2014). Such materials can show phenomenal reversible responses to a variety of stimuli, including thermal, mechanical, chemical and light. Another feature that makes them ideal constituents of self-healing materials. Among metals, transition ones and lanthanides have been efficiently employed so far. Histidine, as a ligand, is the most versatile in terms of possible inter- and intramolecular interactions, especially when coordinating divalent transition metal ions (Zechel et al. 2019).

Properties of the final material can be tailored by tuning the polymer composition and by choosing different types of ligands and metal ions, and thus influencing the strength of interactions between them. This gives rise to different types of metallopolymers, as shown in Figure 8.3. In type-I polymers, the metal ions/complexes are bound to the polymer at the side chain or the end group by (i) electrostatic interaction, (ii) coordination bond or (iii) covalent bonding. The type-II polymers have metal centres as part of the main chain, and they are incorporated either by (i) covalent linkages or by (ii) metal-to-ligand coordination. The type-III polymers consist of networks with metal ions embedded into a matrix through physical interactions. Weaker and more dynamic metal-polymer complexes – so-called metallosupramolecular polymers – are chosen for self-healing materials (Bode et al. 2013).

The first example is a biomimetic supramolecular polymer equipped with a ligand macromolecule carrying multiple tridentate ligand 2,6-bis(1,2,3-triazol-4-yl)pyridine (BTP) units synthesised through CuAAC in the polymer backbone together with zinc and/or europium salts (Yuan et al. 2013). Kinetically labile metal–ligand interactions are sensitive to mechanical loading. Upon stretching, the complexes are pulled out of the hard phases, acting as sacrificial bonds, and realising the local strain energy. Such material can form films, so it could be suitable for the development of self-healing coatings with tunable mechanical performance.

A self-healing multiphase polymer was obtained by implementing a pervasive network of dynamic Zn–imidazole interactions in the soft matrix of a hard/soft two-phase brush acrylate copolymer system (Mozhdehi et al. 2014). The mechanical and dynamic properties of this material can be programmed through the backbone composition and degree of polymerisation, the brush density, the ligand density, and the ligand/metal ratio. Possessing the dynamic Zn^{2+}-imidazole complexes residing in the dynamic soft phase of a two-phase polymer system, the material showed excellent self-healing properties in the solid state.

Optimisation of the copolymer structure and the number of cross-links allows tuning the properties of metallopolymers. An example of such a methodology is the incorporation of iron *bis*-terpyridine complexes into a polymer network based on methacrylates (Bode et al. 2013). The number of crosslinking units strongly influenced the thermal and mechanical properties of the resulting polymer. At the same time, polymers with more mobile and flexible backbone were more prone to self-repair action. The schematic representation of polymer, metal–ligand structure and self-healing action is presented in Figure 8.4.

It is also possible to merge different types of interactions to form self-healing polymers. For example, metal–metal (Pt···Pt) and π–π interactions by incorporating a cyclometalated platinum(II) complex, Pt(6-phenyl-2,2′-bipyridyl)Cl (acting as crosslinking agent) into a PDMS backbone. The obtained polymer was used to produce an elastic film (Mei et al. 2016). The molecular-level damage of the metal complex network directly results in the activation of the autonomous healing process. The outstanding stretchability and self-healing features of the PDMS-Pt coating were ascribed to

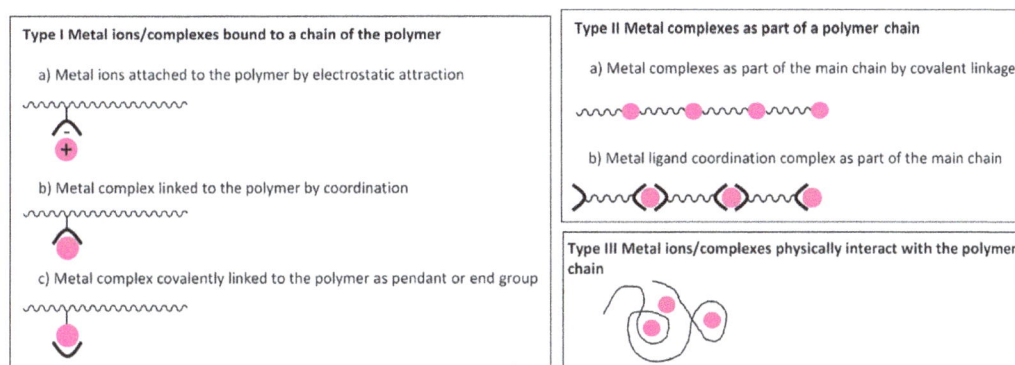

FIGURE 8.3 Schematic representation of the general architectures of metal-containing polymers. (Reproduced with permission from Wild (2011) © 2011 The Royal Society of Chemistry.)

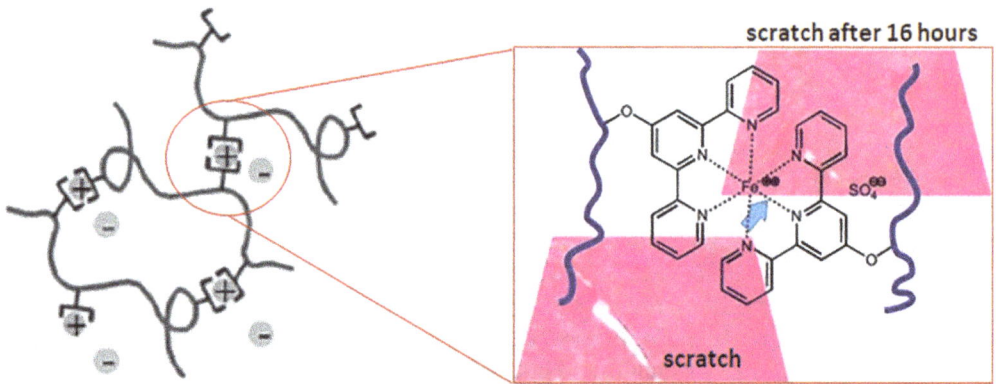

FIGURE 8.4 Schematic representation of iron-bisterpyridine polymer and their ability for self-healing. (Reproduced with permission from Bode (2013) © 2013 WILEY-VCH Verlag GmbH & Co. KGaA, Weinheim.)

the supportive activity of introduced π–π interactions. In addition, the polymer's properties are insensitive to surface ageing, thanks to the lack of the affinity of Pt(II) complexes for water.

8.2.4 HOST–GUEST INTERACTIONS

The host–guest interaction relies on macrocyclic molecules. Among them, cyclodextrin (CD) has been paid particular attention, due to its easy availability and inexpensive cost. CDs can be acquired from their starch precursors, like potato, rice, and corn. CD is an oligosaccharide consisting of six, seven or eight glucose units (α, β, or γ -CD, respectively). As a cone-shaped cavity, in Figure 8.5, CDs can accommodate a great variety of functional 'guest' molecules by taking advantage of its geometric compatibility. The interplay between the CDs as a host and the guest molecules is rooted in hydrophobic interactions. The inner hydrophobic environment of the CDs can hold small hydrophobic molecules or small portions of the polymeric structure. The exterior portion of the cavity made from the hydroxyl group is responsible for aqueous solubility and allows the hydroxyl group to reciprocate with the hydrophilic components (Wankar et al. 2020). The formation and dissociation

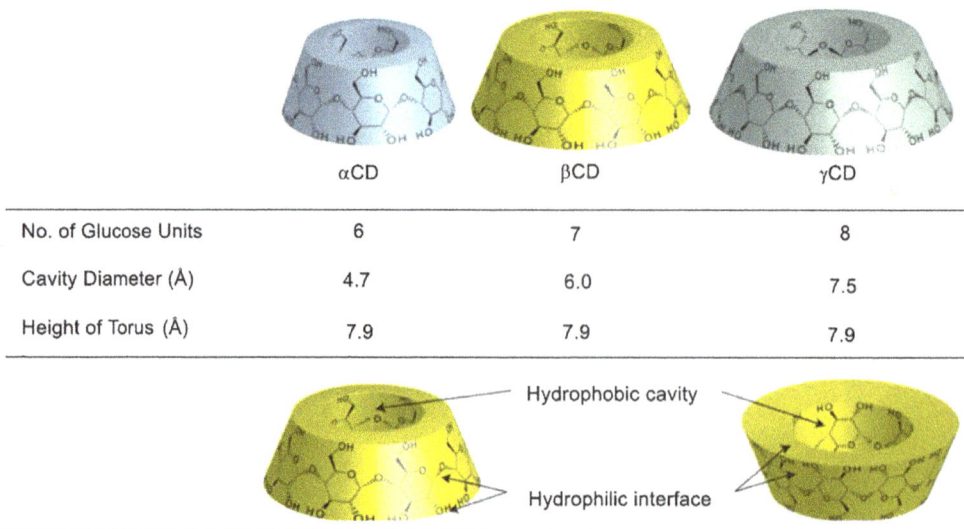

	αCD	βCD	γCD
No. of Glucose Units	6	7	8
Cavity Diameter (Å)	4.7	6.0	7.5
Height of Torus (Å)	7.9	7.9	7.9

FIGURE 8.5 Schematic representation of cyclodextrin. (Reproduced with permission from Mohamadhoseini (2020) © 2020 Elsevier B.V.)

of the host–guest complexes are closely related to environmental conditions, like pH and temperature. It makes the CD-based host–guest system responsive to the corrosion action, as pH changes are involved in this process.

CDs systems can be used to produce protective coatings, e.g., in a layer-by-layer (LbL) process, where the CD acts as a coupling agent, Figure 8.6 (Xu et al. 2016). Polyethylenimine-β-cyclodextrin (PEI-β-CD) and ferrocene-modified chitosan (Fc-CHT) were used to obtain low-fouling, antimicrobial, and biocorrosion inhibition multilayers coatings on stainless steel. PEI-β-CD has an antifouling property because of the presence of hydrophilic PEI, while Fc-modified chitosan is responsible for antimicrobial and fouling resistance. This successful production of such a multilayer barrier could be the first step to design enhanced protective coatings with additional features like self-healing properties.

Another example where the β-CD was used – a UV-shielding coating with a self-healing effect. This coating material was polymerised with the inclusion of the host–guest complex – β-cyclodextrin-titanium dioxide nanoparticles/2-hydroxyethyl-methacrylate-co-butyl acrylate (β-CD-TiO$_2$/P(HEMA-co-BA)). CD grafted with TiO$_2$ nanoparticles (ultraviolet shielding agent) was used as the host molecule, while adamantane grafted with a reactive monomer (HEMA) was used as the guest. Upon mechanical damage, the interaction between host–guest molecules is interrupted, resulting in many free host and guest molecules as depicted in Figure 8.7. The damaged side

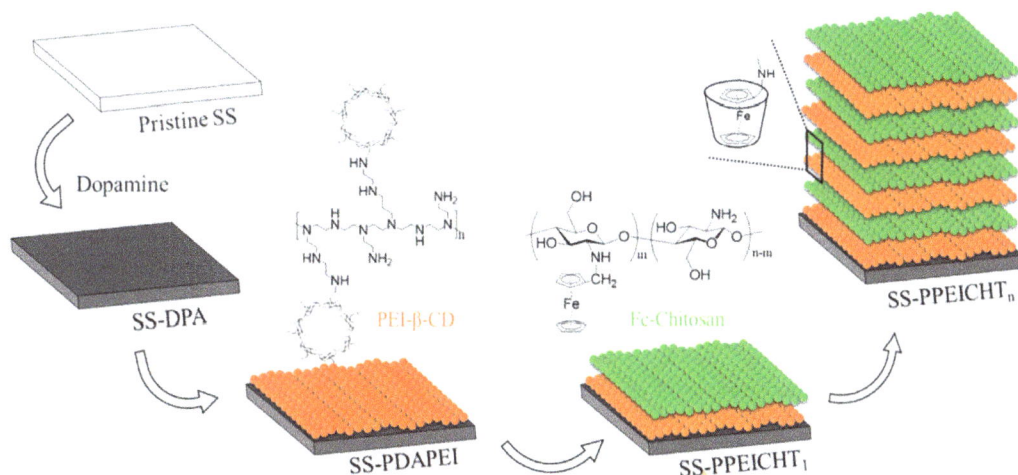

FIGURE 8.6 Schematic representation of the deposition of multi-layered barrier coating by a layer-by-layer method. (Reproduced with permission from Xu (2016) © 2016 American Chemical Society.)

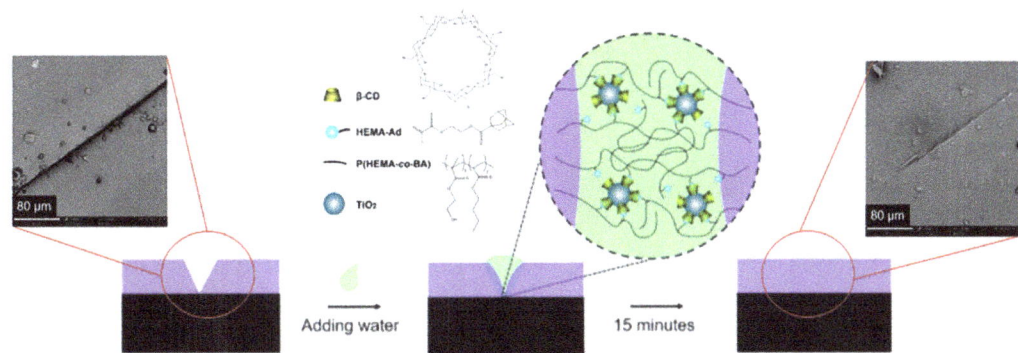

FIGURE 8.7 Self-healing mechanism of the β-CD-TiO$_2$/P(HEMA-co-BA) coating. (Reproduced with permission from Peng (2019) © MDPI under Creative Commons Attribution License.)

planes expand after contact with moisture and eventually touch each other. This causes entanglement of the polymer molecular chains, and the free host and guest molecules on the cross-section are re-integrated (Peng et al. 2019). The overall mechanical properties of the coating were improved compared to the unmodified one as the host and guest groups acted as a crosslinking agent during the polymerisation process. The coating exhibited good hydrophobic property (water contact angle >90° at normal humidity) and UV absorption rate >90% in the range of 200–350 nm. Multiple self-healing properties were confirmed.

The host–guest interactions between β-CD and azobenzene were employed to fabricate the self-healing system in the epoxy matrix (Hu et al. 2018). Acrylamido azobenzene (AAAB) moieties (as a guest component) were introduced into the epoxy acrylate network through copolymerisation. The CD moieties can incorporate azobenzene moieties in epoxy and act as noncovalent macrolinkers. The reciprocal actions between β-CD and azobenzene unlock or lock the connection between the epoxy network and graphene nanosheets (GN), Figure 8.8. In this way, the intrinsic self-healing ability and excellent mechanical property were introduced to the resin. Near infra-red (NIR) light was used as a trigger, as the graphene can in situ transform NIR light into thermal energy to start the host–guest interactions.

8.2.5   ELECTROSTATIC INTERACTIONS

The LbL self-assembly of charged entities has recently attracted attention for the preparation of thin films and coatings. The LbL process is based on the stepwise deposition of oppositely charged species (e.g., polyelectrolytes, nanoparticles, inhibitors, dyes, proteins, in coating application two-three species are used) on a substrate, as presented in Figure 8.9. The properties of obtained coating depend on the type of charged species and adsorption conditions, and not too much extent on the substrate (Glinel et al. 2002). Properly chosen polyelectrolytes exhibit very good adhesion to the surface. They not only provide an effective protective barrier but can seal surface defects.

The LbL coatings held together thanks to coulombic (electrostatic) attractions exhibit rapid self-healing behaviour. By changing environmental conditions, such as pH, it is possible to cause deprotonation of the charged ions and thus tune the mechanical properties of a material. An example of LbL anticorrosion coating, prepared by deposition of polyelectrolyte and inhibitor layers on an Al alloy surface contained positively charged poly(ethyleneimine), negatively charged poly(styrene sulfonate) and inhibitor (8-hydroxyquinoline). It is assumed that such a multilayer system provides three ways of corrosion protection. It passivates metal by controlled release of an inhibitor. The polyelectrolyte layer acts as a buffer to pH changes at the corrosive area and coating defects are self-cured due to the mobility of the polyelectrolyte chains (Andreeva et al. 2008).

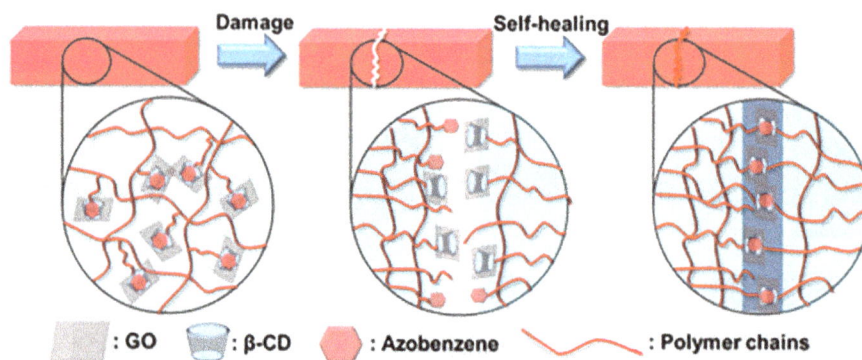

FIGURE 8.8   Schematic representation of the self-healing mechanism of epoxy composite films. (Reproduced with permission from Hu (2018) © 2018 American Chemical Society.)

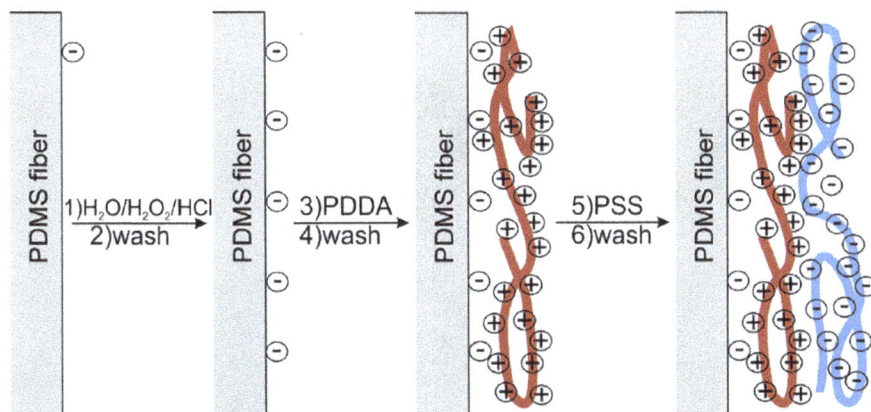

FIGURE 8.9   Schematic representation of layer-by-layer self-assembly. (Reproduced with permission from Zielińska (2014) © 2014 Elsevier B.V.)

8.3   MICRO/NANOPARTICLES

Another method of introduction of self-healing properties into the coating is the incorporation of micro- and nanoparticles into the coating matrix (particles are distributed in the continuous phase). This approach is seen mostly in polymeric coatings (Liu et al. 2012; Sauvant-Moynot et al. 2008), less often into a sol–gel (Borisova et al. 2011; Borisova et al. 2013), and sporadically in a metallic coating, e.g., electrochemical nickel (Moustafa et al. 2013). The micro/nanoparticles can be divided into two types: capsules and hydrogels. There are a plethora of systems where nanoparticles were introduced into the coating matrix. Here, we focus on those based on supramolecular chemistry.

8.3.1   CAPSULES

The encapsulation of healing agents (Brown et al. 2003; Dry and Sottos 1995; White et al. 2001) and corrosion inhibitors (Shchukin and Möhwald 2007; Skorb et al. 2009) was one of the earliest approaches to obtain self-healing materials (Montemor 2014). The concept of self-healing mechanism based on micro/nanocapsules consists of a few stages (Samadzadeh et al. 2010):

1. Damage is inflicted on the material resulting in rupturing the container.
2. A generated 'mobile phase' releases a healing agent.
3. Damage is removed by local mending reactions through physical interaction or chemical bonding.
4. When the healing process is completed, the previously 'mobile phase' is immobilised.
5. The initial properties of the material are restored.

There are four principal methods of obtaining capsules: emulsification and precipitation, LBL assembly, coacervation and internal phase separation. So far, five approaches of the encapsulation of the self-healing agent have been described as efficient, Figure 8.10 (Kosarli et al. 2020). The single capsule solution is presented in Figure 8.10a. Such a system consists of at least one restoring agent from reactive chemicals, solvents, or low melting point metals. In the second approach (Figure 8.10b) along with the capsules, catalyst is dispersed within the matrix. The third solution is based on a healing reaction between two healing components in a form of fluid. One is released from the capsule, while the other experiences phase separation, Figure 8.10c. Two reactive liquid healing agents could be incorporated also as double-system capsules, Figure 8.10d. In Figure 8.10e,

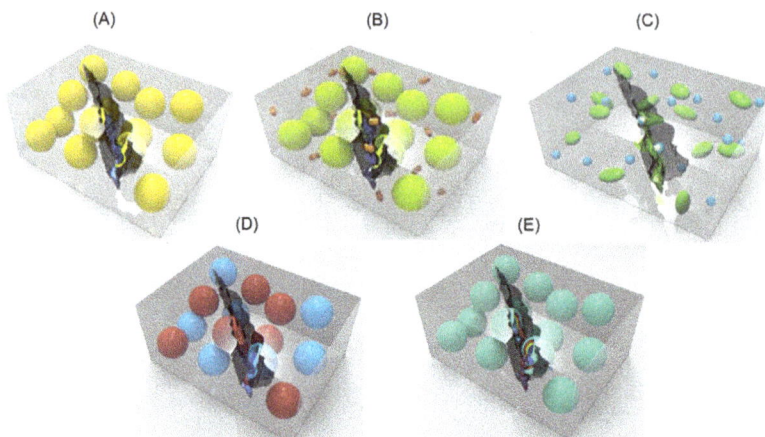

FIGURE 8.10 The encapsulation approaches: (a) single capsules, (b) capsule (green)/dispersed catalyst (orange), (c) phase-separated droplet/capsules (green), (d) double-capsule and (e) all-together microcapsules (multiple shell walls depicted with different colours). (Reproduced with permission from Zhu (2015) © 2015 Elsevier Ltd.)

the monomer and the catalyst are included in one multilayer capsule. The substances are separated by the inner wall – an all-together attitude.

Supramolecular containers can be based on electrostatic interactions between hollow mesoporous zirconia nanospheres (HMZSs) as nanocontainers and L-Carnosine (L-CAR) as a green corrosion inhibitor. L-CAR-loaded HMZSs were incorporated into the waterborne epoxy coating (Wang et al. 2015). L-CAR-loaded HMZSs exhibited an acid/alkaline responsive controlled release property. It allowed to respond to pH variation occurring on micro-anodic and micro-cathodic zones of the electrochemical corrosion process and subsequently release mixed-typed corrosion inhibitors. Another example of increasing the lifespan of the epoxy relies on the inclusion of nanocontainers built with multi-walled carbon nanotubes (MWCNTs) and β-CD loaded with benzimidazole (BZ) (He et al. 2016). CD having a special molecular structure with a hydrophobic inner cavity and a hydrophilic exterior as described in Section 8.2.4 is also often used to encapsulate the widest range of guest molecules based on host–guest inclusion interaction. The host–guest interaction triggered by substrate corrosion initiated the release of the corrosion inhibitor. The prolongation of the coating material lifetime was confirmed by the scarification test. Another example of inhibitor encapsulation based on CD is the benzotriazole (BTA)-loaded nanocontainers (Liu et al. 2018). The combination of the CD with the impermeable graphene nanosheets greatly impeded the electrolyte penetration and corrosion extension of the carbon steel substrate. The release of BTA from containers caused the formation of adsorption layers on an exposed metal surface. In these two latter examples, CD-based nanocontainers were delivered to the protective coating with the support of carbonaceous materials.

Despite their clear advantages, coatings containing capsules have one major drawback: they are sacrificial, and self-healing is possible only in the case of the first damage to the coating. In addition, in the case of polymeric coatings, this healing mechanism to perform properly and to arrest the progression of damage requires that the polymerisation of the encapsulated monomers is sufficiently fast.

8.3.2 HYDROGELS

Hydrogels microparticles (microgels) and nanoparticles (nanogels), polymeric three-dimensional networks have also got attention as carriers of active substances for corrosion prevention (Latnikova et al. 2012; Stankiewicz et al. 2015, 2018). They consist of a three-dimensional polymeric network crosslinked by physical and/or chemical methods. In the case of the physical crosslinking, being exposed to appropriate external stimuli, the hydrogel changes reversibly into the corresponding precursor polymer chains (Fennell and Huyghe 2019). Hydrogels exhibit superior features compared to the other conventional polymers. The most important one is that they possess a high degree of flexibility similar to natural tissue due to large water content (Ahmed and del Campo 2015). They can change their volume as a response to different stimuli, Figure 8.11. This attribute makes them a very attractive substrate for self-healing materials design.

The ability of hydrogels to repair the coating damages can be based on a few mechanisms:

- dissociation/association of dynamic bonds within the gel polymeric network,
- gels may serve as containers of healing agents and/or corrosion inhibitors,
- volumetric expansion of the gel at the crack gap,
- self-assembly of gel micro/nanoparticles on the surface.

In terms of corrosion prevention hydrogels, responsiveness to pH changes is a very desirable feature. During the corrosion process, local changes of pH in the anodic and cathodic sides of electrochemical processes are observed. An active response to these changes can slow down the corrosion progress. Among hydrogels, those crosslinked by hydrogen bonding, metal coordination, host–guest interactions, dynamic covalent or 'click' chemistry deserve particular attention. Promising candidates are listed below:

- crosslinked by hydrogen bonding;
  - copolymers containing ureidopyrimidinone (UPy) which could assemble into dimers by four-fold hydrogen bonds – e.g., DMAEMA (2-(dimethylamino)ethyl methacrylate) copolymerized with 2-(3-(6-methyl-4-oxo-1,4dihydropyrimidin-2-yl)ureido)ethyl

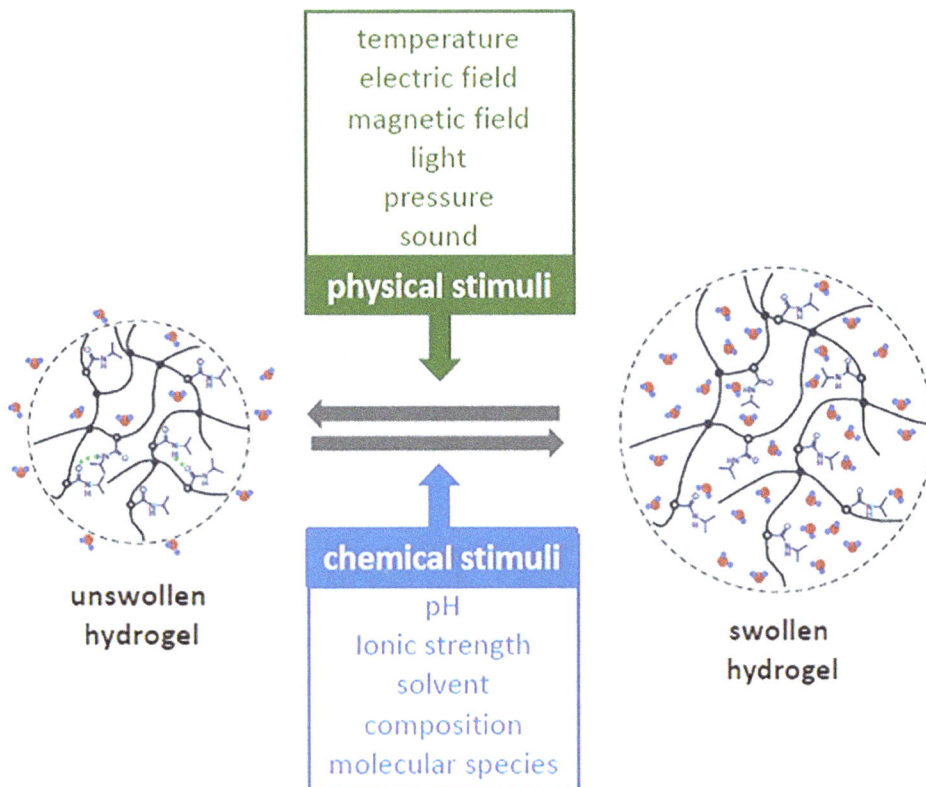

FIGURE 8.11   Types of stimuli for volume transition of hydrogels.

methacrylate (SCMHBMA) (Cui 2012) or multiblock copolymer based on poly(ethylene glycol) (PEG) matrix (Guo et al. 2014)

- crosslinked by metal coordination;
  - complexes of ferric ions with linear and branched PEG, which is end-functionalised with one to four 3,4-dihydroxyphenylalanine (dopa) groups – the metal–ligand interaction forms mono-complexes at pH below 5, bis-complexes about pH 8 and tris-complexes above pH 8 (Holten-Andersen et al. 2011; Lee et al. 2002)
  - poly(acrylamide-coacrylic acid) (P(AAm-co-AAc)) copolymer crosslinked by carboxyl-Fe^{3+} coordination bonds – the coordination bond is controlled by varying pH (Zheng et al. 2016)
- crosslinked by host–guest interactions;
  - redox-responsive ferrocene moieties as guest functionalities and β-CD moieties as host functionalities in acrylamide (AAm) crosslinked by N, N′-methylenbisacrylamide (BIS) – an increase of the pH to above 6 triggers the swelling of the responsive hydrogel (ferrocene moieties) (Ma et al. 2014)
  - covalently crosslinked polyacrylamide (pAAm)-based hydrogels possessing either β-CD or dansil (5-(dimethylamino)naphthalene-1-sulfonyl) (Dns) moieties – below pH 3 a protonation of the Dns group increasing its hydrophilicity initiates the disaggregation of the macroscopic assembly of hydrogels (Zheng Y. et al. 2013)
  - poly(acrylamide) gels carrying phenylboronic acid (PB) and catechol moieties (CAT) – the assembly and disassembly of PB and CAT is reversibly switched by varying the pH (Nakahata et al. 2014)
- crosslinked by dynamic covalent chemistry;
  - boronate esters prepared from boronic acids and 1,2- or 1,3-diols – the reversibility of a boronate ester is largely dependent on the pH of the solution and the pK_a of the boronic acid component (Deng et al. 2015)
  - 2-methacryloyloxyethyl phosphorylcholine (MPC) polymeric matrix containing boronic ester – the hydrogel can be dissociated by lowering the pH (Chen et al. 2018)
- crosslinked by 'click' chemistry;
  - hydrogel crosslinking by Schiff base formation, e.g., amino and aldehyde groups generate an imine linkage – the uncoupling and recoupling of this pseudo-covalent bond resulting in self-healing capability of the hydrogel are dependent on the surrounding pH (Zhang et al. 2011).

To the best of the authors' knowledge, none of the hydrogels mentioned above has been used as a carrier of active substances in protective coatings yet. However, pH-responsive supramolecular containers beyond hydrogels have been successfully utilised in self-healing protective coatings (He et al. 2016; Wang et al. 2015).

Besides, hydrogels with supramolecular interactions embedded within polymeric matrices, there are a variety of examples of self-healing hydrogel systems taking advantage of mechanisms rooted in the noncovalent interaction between the hydrogel molecules (He et al. 2019). It is well known that hydrogel particles can arrange themselves on surfaces (Zielińska et al. 2016), which can give rise to smart, self-healing coatings. The viability of such an approach was proven in silica material (Kolmakov et al. 2009). It was shown that it is possible to introduce to the material the self-replenish properties by hardening of labile bonds between nanoscopic gel particles that are assembled into a macroscopic network. Under mechanical stress, bonds between individual gel particles can break and readily re-form with reactive groups on a neighbouring particle. In addition, they proved that the presence of labile bonds can significantly improve the tensile strength of the coating.

For example, zwitterionic microgels are constructed in a way employing the combination of covalent crosslinking inside each microgel and supramolecular interactions between microgel particles. The self-healing mechanism in this material is a consequence of the zwitterionic fusion mechanism, which integrates strong hydration, zwitterion pair attraction, and H-bonding (Sinclair et al. 2018). Rapid self-healing action has been reported in bioinspired hydrogels built from poly(acrylic acid)–poly-(acrylamide)–poly(dopamine) (PAAc–PAM–PDA) matrix containing poly(N-isopropylacrylamide) (PNIPAM) microgels (He et al. 2019). This material exhibits exceptional adhesiveness to various substrates including hydrophilic, hydrophobic, or metallic surfaces. So far, they have been applied mostly in biomedical applications; however, they have the potential to be used as a protective coating.

It is crucial to equip hydrogel-based materials with good mechanical properties. So far, a few approaches to achieve that have been described, e.g., hydrogel obtained by combining both covalent crosslinking and ion-pairing of Fe^{3+}- acrylic acid (Lin et al. 2015); polyelectrolyte hydrogels containing ionic bonds composed of polyampholytes (Sun et al. 2013); or flexible poly(ionic liquid) hydrogel electrolytes built with 3-(1-vinyl-3-imidazolio)propanesulfonate/2-acrylamido-2-methyl-propane sulphonic acid (ZIW/AMPS) (Zhou et al. 2016). The latter one is incredibly interesting because it exhibits extremely high conductivities over 1 S/m at room temperature and low activation energy. This feature is beneficial when tackling corrosion problems, as a corrosion mechanism based on electrochemical reactions, so being conductive such a hydrogel can sense and react on electrons flow.

8.4 'CLICK' CHEMISTRY

The concept for conducting organic reactions that do not yield side products has found particular attention in creating self-healing materials. It is due to the high efficiency of such reactions and the fact that can be carried out under a variety of mild conditions. There are several efficient reactions used to produce functional synthetic molecules and organic materials, which are grouped under the name 'click reactions', e.g., Diels–Alder (DA), Schiff base, disulphide bond reshuffling reaction, oxime, and thiol-ene reactions (thiol-Michael addition and free radical thiol-ene addition), as well as Cu(I) catalysed alkyne-azide cycloaddition. The main characteristics of these processes embrace solvent-less (or aqueous) reaction conditions and susceptibility to a wide range of available substrate compounds, as well as reaction's orthogonality. However, the self-healing process using these dynamic bonds requires some external stimulus, such as heat, UV light, pH, or force, to trigger these reversible reactions.

8.4.1 DIELS–ALDER

A DA cycloaddition reaction not only can form carbon–carbon bonds but also can form hetero-atom–heteroatom bonds. It is the thermal reversible reaction that makes it well suited for self-healing materials development. Thermally reversible crosslinked polymers and hydrogels based on a DA mechanism are used as encapsulants or coatings. Direct DA cycloaddition reactions involving multifunctional monomers, DA cycloaddition of linear polymers bearing pendant furan and/or maleimide, or a cross-linker containing a DA linkage can be used to synthesise polymers network (Tasdelen 2011).

Copolymers consisting of biobased diethylitaconate and furfuryl methacrylate moieties were copolymerised through the DA reaction with a bismaleimide (Turkenburg et al. 2017). The material was successfully applied on a steel substrate. Self-healing of damaged coatings was achieved through a temporal de-crosslink of the polymer network upon heat treatment. Better results were obtained for lower crosslink density. Another example of DA reaction used in the production of anticorrosive protective coatings was the synthesis of DA adduct and its crosslinking in an epoxy resin. 1,1′methylenedi-4,1-phenylene-bismaleimmide, 4,4′-diaminodiphenylmethane, O, O′-bis(2-aminopropyl) polypropylene glycol-block-polyethylene glycol-block-polypropylene glycol (Jeff 500) and diglycidyl ether of bisphenol A (DGEBA) were used. A self-healing cycle included the first heating at 120°C to achieve uniform morphological healing, then annealing at 90°C for full restoration of mechanical properties. The re-establishment of coating protective properties against saline solution corrosion was achieved (Dello Iacono et al. 1990). The unhealed coating exhibited signs of substrate oxidation, while the healed one preserved it.

The healing efficiency of DA coating materials based on diglycidyl ether of bisphenol A can be amended by the application of solvent (Pratama et al. 2012). In this case, the coating swelled the polymer network, helping the contact between crack surfaces and lowering the T_g. The polymer can achieve a rubbery state at room temperature. While the polymer is expanded, the nodules on the crack surfaces are mechanically interlocked. The solvent evaporation returned the polymer to a glassy state leaving the crack surfaces interconnected.

8.4.2 Schiff Base Approach

Schiff base approach refers to the reaction between the substances containing carbonyls with amino groups. The reactants are simply mixed and stirred at room temperature resulting in complete polymerisation. The final product contains C=N double bonds. These C=N bonds break when the pH value goes down below 3 (Xin and Yuan 2012). The sensitivity of the imine linkage to pH value makes this system capable of the development of anticorrosive protective materials. Furthermore, the Schiff bases have been widely employed as corrosion inhibitors for various metallic substrates including Al alloys, Cu or steel (Heydari et al. 2018; Ma et al. 2002; Morad and Sarhan 2008; Nazir et al. 2020). The high inhibition efficiency of the Schiff bases comes from electron density present around the hetero atom, which causes the formation of coordinate-covalent bond and interaction of π-orbital with metal (Sethi et al. 2007). Having these features, the Schiff bases are capable to inhibit corrosion processes while implemented into the coating material.

2,4,6-Triazine hydrazide-modified biobased cardanol was subjected to condensation polymerisation with terephthaldehyde to produce Schiff base polyamines. This cardanol Schiff base polymer was combined with the epoxy matrix to produce a self-healing protective coating for steel substrate (Atta et al. 2020). The chelation of the triazine hydrazine groups with the steel substrate during the curing of the epoxy film eliminates the coating porosity, thus preventing the penetration of chloride ions from the corrosive medium (tested in the salt spray chamber). The self-healing effect of the cardanol Schiff base polymer relies on the exchange reaction between imine, amine and aldehyde end groups. These reactions include hydrolysis, imine metathesis and transamination. The presented material exhibits also superhydrophobic properties. In addition, the appropriate composition of the Schiff base material can lead to tune flammability properties and fire hazard-related features (Waśkiewicz et al. 2013). These features include the heat release rate (HRR), smoke extinction area (SEA) or total heat release (THR). The lower the feature values the safer the material is. A significant decrease in HRR and SEA numbers was achieved for the epoxy coating containing N, N′-bis(2-hydroxybenzelidene)-4,4′-oxydianiline compared with the commercially available epoxy resin. Further modification of the Schiff base-epoxy coating with nanoparticles and metal complexes (Ni, Cu, Cr or Co) can improve the mechanical properties of the final product, which is an important feature when creating more durable materials (Fadl et al. 2019). The Schiff base ligand, p-phenylamine-N(4-chlorosalicylaldenemine) (HL), and its metal (Ni(II) and Cr(III)) complexes composed with silica nanoparticles as pigment modifiers for steel epoxy coating gave increased protective properties. The mechanism of the corrosion of the steel substrate prevention is presented in Figure 8.12.

The inhibition effect relies on: (i) the protective performance distinguished by the impact of π- electrons adsorption of aromatic rings (present in epoxy resin); (ii) the oxygen atoms sole pair and the azomethine group presented in HL and its metal complexes; (iii) polar terminals inducing robust electrostatic physical bonding between the steel substrate and the epoxy film modified with HL and its metal complexes; (iv) impermeable shield of the well-dispersed amorphous metal complex pigment molecules creating interfering platelets, thus impeding the aggressive atmosphere penetration; (v) the fully dispersed silica nanoparticles reinforcing the inter-coating adhesion and supporting the

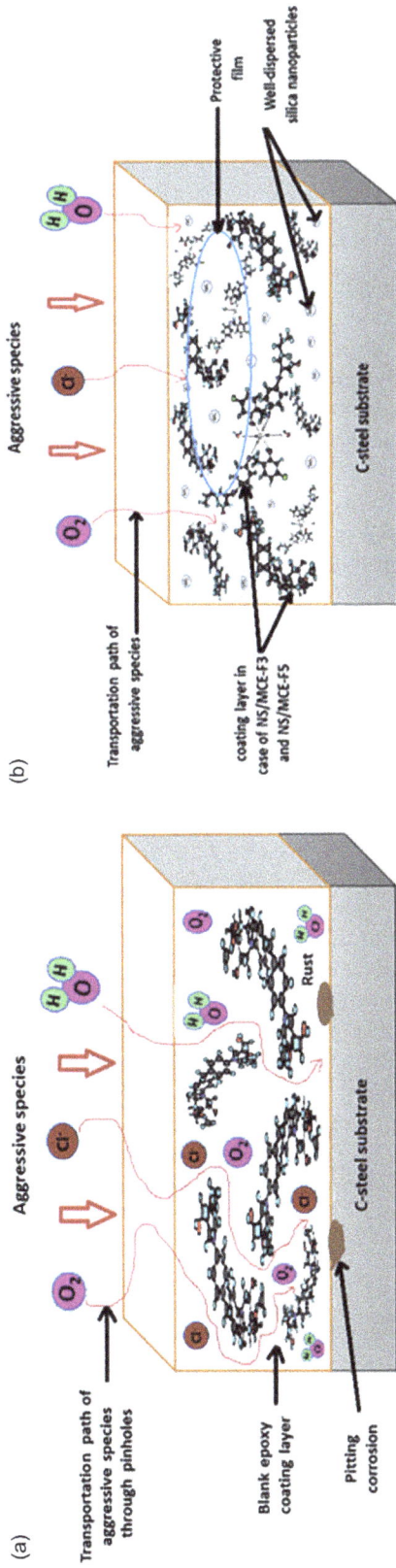

FIGURE 8.12 The protective mechanism of (a) epoxy coating and (b) epoxy coating containing silica particles and Schiff base nickel (NS/MCE-F3) and chromium (NS/MCE-F5) complexes. (Reproduced with permission from Fadl (2019) © 2019 Elsevier B.V.)

surface protection. A similar approach was utilised for the Al alloy substrate. AA2024-T3 alloy was coated with Schiff base orthosilicate (Bonetti et al. 2019). First, the functionalisation of silica surface extracted from rice husk ash with 1,2-dicloroetane was carried out. Then, the Schiff base was anchored to the modified silica surface. The Schiff base orthosilicate coating shows excellent possibilities for complexation with Cu cations originated from the corrosion of the intermetallic Al_2CuMg phase, thus preventing further dissolution of the Al alloy surface.

8.4.3 DISULPHIDE BOND RESHUFFLING

Self-healing action based on disulphide bonds can be carried out at low-temperature conditions without any external stimuli. The thiol radicals, generated during mechanical breakage of disulphide bonds, can rapidly exchange with other disulphide bonds, assuring fast self-healing and preventing fracture of the material. Dynamic disulphide exchange reaction allowed the introduction of self-reparation mechanism into multifunctional – antifouling and antibacterial hydrogel coatings. Coatings were synthesised through the highly efficient thiol-ene photopolymerisation on the stainless steel surface in a 'grafting-through' approach using (poly(ethylene glycol)methyl ether methacrylate (PEGMA), *N*-hydroxyethyl acrylamide (HEAA) and 2-(methacryloyloxy)ethyl trimethylammonium chloride (META)) monomers and bis(2-methacryloyl)oxyethyl disulphide – a disulphide-containing cross-linker (Yang et al. 2015).

8.4.4 OXIME BOND FORMATION

The oxime bond formation can be described as highly efficient and chemoselective, taking place in aqueous solvents and producing water as a by-product (Collins et al. 2016). It is the reaction between aldehydes or ketones and hydroxylamines. So far, oxime chemistry has been mainly employed for bioconjugation (Ulrich et al. 2014). However, there is a huge potential for the usage of the oxime compounds to functionalise polymers or hydrogels with self-healing properties, as the oxime bond is prone to dynamic interactions upon pH changes.

The reversibility of oxime-linked macromolecules could be achieved also by competitive exchange of the hydrolytically stable ketoxime in the presence of small molecule alkoxyamine or carbonyl compounds or by temperature changes (Liu et al. 2017; Mukherjee et al. 2014).

Polyoximes and hexamethylene diisocyanate were utilised for the catalyst-free preparation of poly(oxime-urethanes) at an ambient temperature. Thermally reversible covalent polymers were obtained in this way (Liu et al. 2017). The new polymer exhibits the repairable performance for crosslinked networks due to the oxime-promoted transcarbamoylation reaction, Figure 8.13.

Specifically for corrosion prevention, oxime compounds were employed for the surface modification of Zn. The protective film of (2E)-2-(hydroxylamino)-1, 2-diphenylethanol was created on the Zn substrate by immersion method (Achary and Naik 2013). Having electroactive >C=N– and >N–OH groups, the oxime compound forms a complex with Zn that avoids the contact of aggressive species with the metal surface.

8.4.5 THIOL-ENE REACTIONS

The Michael addition reaction can be characterised generally as the reaction of an enolate-type nucleophile in the presence of a catalyst to an α, β-unsaturated carbonyl. The reaction is straightforward, vigorous and can result in C–C bond formation under relatively facile reaction conditions. The thiol-Michael addition 'click' reaction was applied to modify graphene oxide (GO) with commercial acrylate phosphorous monomer for obtaining corrosion protective waterborne epoxy coatings (Huang et al. 2020). γ-Mercaptopropyl triethoxysilane acted as a bridge to obtain thiol-capped GO. The increase of the waterborne epoxy polymer lifespan was achieved through the formation of

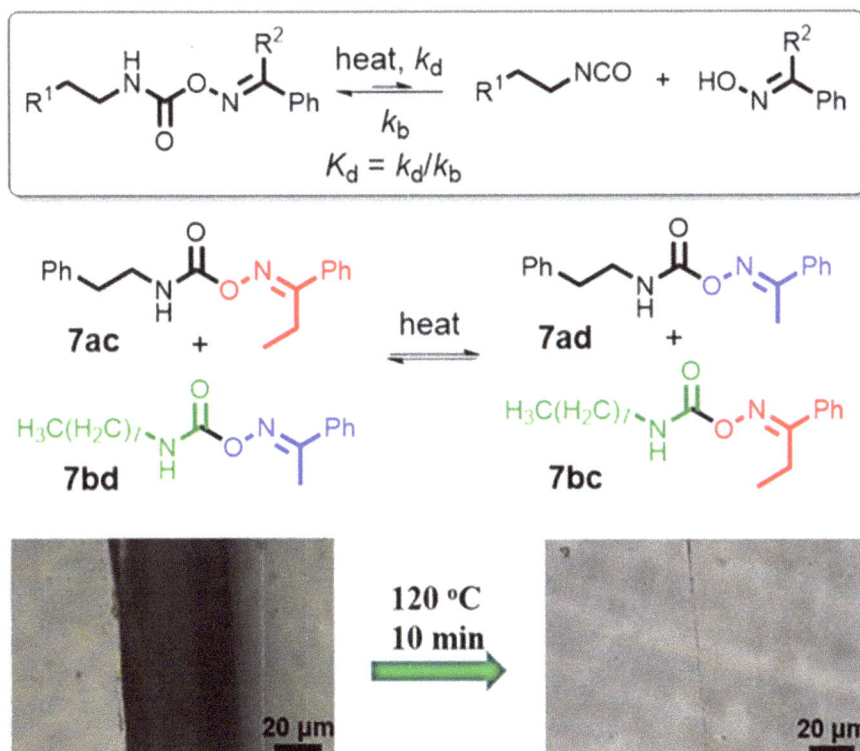

FIGURE 8.13 The oxime-enabled transcarbamoylation reaction through a dissociative approach and microscopic images of two pieces of polymeric film before (left) and after (right) thermal self-healing. (Reproduced with permission from Liu (2017) © 2017 American Chemical Society.)

a labyrinth effect inside the coating causing the elongation of the corrosive medium's penetration path, thus delaying metal corrosion. No self-healing effect has been reported.

The free radical thiol-ene addition based on the creation of a thiol to an electron-rich/electron-poor double bond has been gaining increased interest for the preparation of antifouling UV-cured coatings (Chen et al. 2010; Resetco et al. 2017). Examples include thiol-ene 'click' networks from amphiphilic fluoropolymers (Imbesi et al. 2012) or poly(ethylene glycol)-based thiol-ene hydrogel coatings (Lundberg et al. 2010). The anticorrosion attribute has been confirmed for hybrid silica sol–gel coating obtained by in situ thiol-ene 'click' reaction for Cu (Peng et al. 2014) and mild steel protection (Taghaviksih et al. 2016). Self-healing feature in thiol-ene type coating materials has been reported for the thiol-ene fluorinated siloxane (T-FAS), PDMS elastomer and hydrophobic fumed silica nanoparticles (Zhang H. et al. 2016). This coating was applied to cotton fabric. The self-healing mechanism of the coating immersed in aggressive media derives from polymer chain interdiffusion.

The visible-light photocatalytic thiol-ene 'click' reaction could be employed as a healing action in polymeric-based coating materials exposed to radiation. Photo-initiated polymerisation is related to the presence of unsaturated groups, mostly (meth)acrylates. In the process of thiol-acrylate photopolymerisation, the Michael-type addition reaction between the thiol and acrylates occurs simultaneously, resulting in a mixed-mode polymerisation (Cramer and Bowman 2001). This feature can be used to tune hydrogel properties by adjusting the ratio of thiol and acrylate.

8.4.6 COPPER (I) CATALYSED ALKYNE-AZIDE CYCLOADDITION

The last examples, in this section, embrace coatings produced by Cu(I)-catalysed azide/alkyne cycloaddition (CuAAC). Fast crosslinking and extensive molecular and chemical diversity, combined

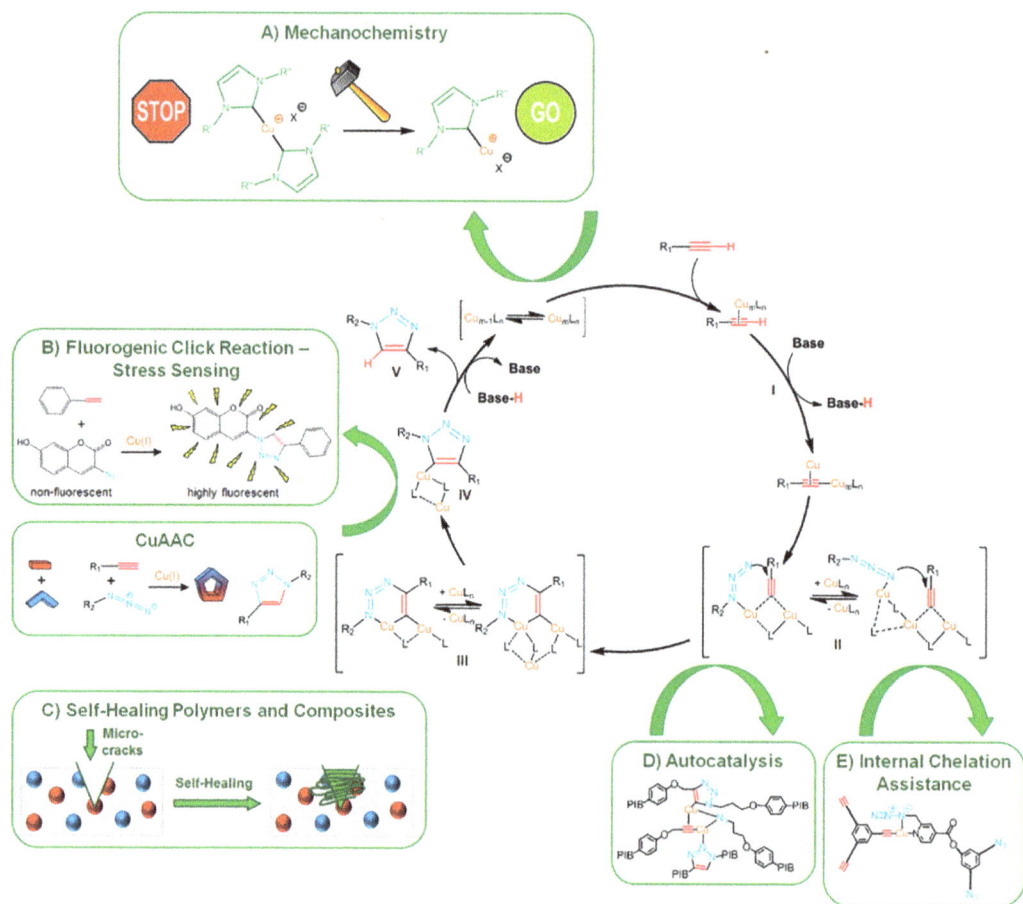

FIGURE 8.14 Mechanisms of CuAAC activities: (a) mechanochemistry by catalysts; (b) fluorogenic 'click' reactions; (c) thermally responsive and stress sensing materials; (d) autocatalytic and (e) internal chelation systems. (Reproduced with permission from Döhler (2017) © 2017 American Chemical Society.)

with an excellent catalyst design and capability to fluoresce, make the CuAAC, a particularly attractive concept for self-healing materials (Döhler et al. 2017). The reactivity of CuAAC compounds is easily adjustable by careful selection of the ligands and can be activated in various ways – through 'click' reactions or mechanochemistry, autocatalytically or with the use of catalysts. Schematic representation of these processes is illustrated in Figure 8.14.

The application of the CuAAC in self-healing materials is uncomplicated due to the fast, efficient and substrate-independent forming of chemical bonds. It is crucial when materials need repair after being damaged. Therefore, the mechanical force strategy for restoration can be used for multivalent liquid azides and alkynes in rapid crosslinking chemistries. Crosslinking chemistries based on the CuAAC reaction could be applied to a capsule-based self-healing approach.

8.5   SELF-HEALING MATERIALS IN PRACTICAL USAGE

The global self-healing materials market is one of the fastest-growing ones (CAGR of 46.1% from 2019 to 2025). Its size in 2018 was estimated at USD 291.4 million (Grand View Research). With the biggest demand in building and construction, energy and automotive sectors, the coating segment is estimated to reach 28.3% market share by 2025. It is a consequence of the rising demand for the product from the automotive and aerospace industries. Major factors driving this growth are as follows (Grand View Research; Markets and Markets):

- an increase in the operation lifetime of paints and coatings;
- the reduction in the maintenance cost;
- government policy focus on legislations/regulations mandating longer service guarantees/ warranties;
- knowledge transfer and closer collaboration at early-stage development.

However, it is hard to estimate how big a share of this market has self-healing coatings based on supramolecular chemistry.

The latest patent publications (Espacenet Patent Search) reveal a variety of recipes for self-healing coating materials and confirm increasing interest in supramolecular chemistry-based solutions. Some of the examples are presented underneath:

- Self-healing supramolecular nonlinear polymers containing ureido groups, crosslinked by hydrogen bonds; KR102052583B1.
- Supramolecular copolymer self-repairing coating material (containing dithioester derivative chain transfer agent, organic free radical initiator and Fe(II) salt as crosslinking agent); CN103951838A.
- Supramolecular polymer based on a three-dimensional conductive network (containing carbon nanostructure or a metal nanowire) having self-healing function; KR101803782B1.
- A self-repairing oleophobic coating (consisting of supramolecular polymer, PDMS-Cat-M, where the ligand compound PDMS-Cat is obtained by copolymerization of dopamine hydrochloride (DOPA), isophorone diisocyanate (IPDI) and amino-terminated polydimethylsiloxane (PDMS), and M is a metal ion); CN112210064A.
- Self-healing clearcoat formulation based on polyimide and capable of forming charge transfer complexes; KR20190057767A.

So far, there are a few commercially available solutions for protective coatings based on self-healing materials. The prominent players and their solutions operating in the global self-healing coating materials market are listed below:

- AkzoNobel N.V. (the Netherlands) with the two-component clearcoat Sikkens Autoclear LV Exclusive. AkzoNobel also supports SAS Nanotechnologies LLC (US) and their technology of smart microcapsules production.
- Arkema SA (France) with their revolutionary self-healing rubber called Reverlink™ based on the concept of supramolecular chemistry.
- Autonomic Materials Inc. (US) provides (i) microcapsules that contain healing agents – a blend of resins, corrosion inhibitors and adhesion promoters; (ii) AMP-UP™ 100- a single component self-healing waterborne epoxy primer.
- Bayer (Germany) designed crosslinked PU coatings, which heal themselves thanks to hydrogen bonds reposition.
- NEI Corporation offers NANOMYTE® MEND – self-healing coating materials based on PUs.
- Covestro AG (Germany) supplies Bayhydrol® U PU dispersions with excellent physical properties, scratch resistance and self-healing effects.

8.6   MODELLING SELF-HEALING MECHANISMS

Modelling and computational simulation is a complementary tool that can help in the optimisation of materials properties and the cost reduction of the experimental development of materials (Barbero et al. 2005; Bluhm et al. 2014; Javierre 2019). The computational approach has been successfully applied in the case of self-healing coatings. The modelling strategy depends strongly on the type of material the coating is made of and healing mechanisms involved.

Simulation of material response to mechanical damage on a macroscopic level can be achieved by models based on continuum damage-healing mechanics or by applying cohesive models of crack growth and healing. In the first approach, the presence of cracks is homogenized through the computational domain. Damage and healing variables/states are introduced to account for the loss and recovery of mechanical integrity of the material (Barbero et al. 2005). The definition of variables is a key factor determining the output of the model (Voyiadjis and Kattan 2009; Voyiadjis et al. 2011). Cohesive models of crack growth and healing combine fracture mechanics and continuum damage mechanics. It helps to understand the properties of both original material and material after a healing event (which are typically different) (Maiti and Geubelle 2006; Ponnusami et al. 2018; Yang et al. 2001). In the case of polymeric self-healing materials exhibiting DPNs behaviour, the rearrangement of dandling polymeric chains can be simulated on a microscopic level using i.e. coarse-grained approach (Lyakhova et al. 2014) or dissipative particle dynamics method (Esteves et al. 2013; Esteves et al. 2014). The computations allowed, i.e. to get more insight into the behaviour of polymer chains at the interface (Diffusion-reaction theory), to define the minimum thickness of the polymer layer necessary for providing optimal self-healing ability and to influence the surface on coating repair process (Wang et al. 2017).

For systems based on encapsulated healing agents, computational analyses are capable of determining material parameters which facilitate self-repair such as the ratio of capsule wall thickness to capsule size (Gilabert et al. 2017) or the relationship of material and capsule stiffness (Ponnusami et al. 2015). There have been attempts as well to model the leaching of healing agents and their subsequent transport to the crack site (Batchelor 1990; Prosek and Thierry 2004; Wang et al. 2004).

The interactions, formation and re-formation of labile bonds between gel particles forming self-healing materials is another example of the processes being modelled. Individual particles can be simulated through lattice spring model while particles network can be modelled e.g. through a hierarchical Bell model (Duki et al. 2011; Iyer et al. 2013; Kolmakov et al. 2009; Salib et al. 2011; Iyer et al. 2013).

8.7   CONCLUSIONS AND PERSPECTIVES

Supramolecular chemistry and the application of dynamic bonds give a lot of options to design and synthesise self-healing coatings. They significantly reduce the loss that results from corrosion of the materials and prolongs service life of the coating and coated substrate. It is believed that materials of this type are the future of coating technology in automotive, aviation, marine, construction, and oil and gas industries.

Unfortunately, the dynamic nature of the coating material is often connected with compromised mechanical properties. It can be overcome, to a large extent, by incorporation into coating matrix nanomaterials, by increasing the number of host and guest sites or by application of dual (physical and chemical) crosslinking methods.

Another challenge connected with all coatings is the minimisation of their environmental impact, both during the production and their lifespan. It is particularly hard because a lot of green alternatives fail to meet or exceed the performance requirements of current materials (Hughes et al. 2010). One of the obvious trends for greener coatings is to avoid the use of solvents and volatile organic components. Scaling up production of self-healing coatings from pilot to industrial scale is still a dominant challenge. Another restrain in using self-repair materials is their cost – they are currently more expensive than conventional materials. Materials whose repair mechanisms are rooted in supramolecular chemistry have one more disadvantage, not resolved yet, their reversibility and repeatability decrease over time.

Hopefully, the intensive research in the area of self-healing coatings will help to overcome those obstacles. The first examples of solutions available on the market prove that it is possible to obtain sustainable, commercially viable self-repair coating.

REFERENCES

Achary, G., and Naik, Y. A. 2013. Surface modification of zinc with an oxime for corrosion protection in chloride medium, *J. Chem.* 2013:239747.

Ahmed, E. M. 2015. Hydrogel: preparation, characterization, and applications: a review, *J. Adv. Res.* 6:105–121.

Andreeva, D. V., Fix, D., Mohwald, H., and Shchukin, D. G. 2008. Self-healing anticorrosion coatings based on pH-sensitive polyelectrolyte/inhibitor sandwichlike nanostructures, *Adv. Mater.* 20:2789–2794.

Atta, A. M., Ahmed, M. A., Al-Lohedan, H. A., and El-Faham, A. 2020. Functionalization of silica with triazine hydrazide to improve corrosion protection and interfacial adhesion properties of epoxy coating and steel substrate, *Coatings* 10: 327–346.

Bao, L., Lan, Y., and Zhang, S. 2006. Synthesis and properties of waterborne polyurethane dispersions with ions in the soft segments, *J. Polym. Res.* 13:507–514.

Barbero, E. J., Greco, F., and Lonetti, P. 2005. Continuum damage-healing mechanics with application to self-healing composites, *Int. J. Damage Mech.* 14:51–81.

Batchelor, B. 1990. Leach models: theory and application, *J. Hazard. Mater.* 24:255–266.

Bazzicalupi, C., Bianchi, A., García-España, E., and Delgaro-Pinar, E. 2014. Metals in supramolecular chemistry, *Inorg. Chim. Acta* 417:3–26.

Bluhm, J., Specht, S., and Schröder, J. 2014. Modeling of self-healing effects in polymeric composites, *Arch. Appl. Mech.* 85:1469–1481.

Bode, S., Zedler, L., Schacher, F. H., Dietzek, B., Schmitt, M., Popp, J., Hager, M. D., and Schubert, U. S. 2013. Self-healing polymer coatings based on crosslinked metallosupramolecular copolymers, *Adv. Mater.* 25:1634–1638.

Bonetti, S., Spengler, R., Petersen, A., Aleixo, L. S., Merlo, A. A., and Tamborim, A. M. 2019. Surface-decorated silica with Schiff base as an anticorrosive coating for aluminium alloy 2024-T3, *Appl. Surf. Sci.* 475:684–694.

Borisova, D., Möhwald, H., and Shchukin, D. G. 2011. Mesoporous silica nanoparticles for active corrosion protection, *ACS Nano* 5:1939–1946.

Borisova, D., Möhwald, H., and Shchukin, D. G. 2013. Influence of embedded nanocontainers on the efficiency of active anticorrosive coatings for aluminum alloys part ii: influence of nanocontainer position, *ACS Appl. Mater. Interface* 5:80–87.

Brown, E. N., Kessler, M. R., Sottos, N. R, and White, S. R. 2003. In situ poly(urea-formaldehyde) microencapsulation of dicyclopentadiene, *J. Microencapsul.* 20:719–730.

Burattini, S., Colquhoun, H. M., Fox, J. D., Friedmann, D., Greenland, B. W., Harris, P. J. F., Hayes, W., Mackay, M. E., and Rowan, S. J. 2009b. A self-repairing, supramolecular polymer system: healability as a consequence of donor-acceptor pi-pi stacking interactions, *Chem. Commun.* 44:6717–6719.

Burattini, S., Colquhoun, H. M., Greenland, B. W., and Hayes, W. 2009a. A novel self-healing supramolecular polymer system, *Faraday Discuss.* 143:251–264.

Burattini, S., Greenland, B. W., Merino, D. H., Weng, W., Seppala, J., Colquhoun, H. M., Hayes, W., Mackay, M. E., Hamley, I. W., and Rowan, S. J. 2010. A healable supramolecular polymer blend based on aromatic pi-pi stacking and hydrogen-bonding interactions, *J. Am. Chem. Soc.* 132:12051–12058.

Chen, Y., Diaz-Dussan, D., Wu, D., Wang, W., Peng, Y. Y., Benozir Asha, A., Hall, D. G., Ishihara, K., and Narain, R. 2018. Bioinspired self-healing hydrogel based on benzoxaborole-catechol dynamic covalent chemistry for 3D cell encapsulation, *ACS Macro Lett.* 7:904–908.

Chen, Z., Chisholm, B. J., Patani, R., Wu, J. F., Fernando, S., Jogodzinski, K., and Webster, D. C. 2010. Soy-based UV-curable thiol-ene coatings, *J. Coat. Technol. Res.* 7:603–613.

Collins, J., Xiao, Z., Müllner, M., and Connal, L. A. 2016. The emergence of oxime click chemistry and its utility in polymer science, *Polym. Chem.* 7:3812–3826.

Cramer, N. B., and Bowman, C. N. 2001. Kinetics of thiol-ene and thiol-acrylate photopolymerizations with real-time fourier transform infrared, *J. Polym. Sci. Part A: Polym. Chem.* 39:3311–3319.

Cui, J., and del Campo, A. 2012. Multivalent H-bonds for self-healing hydrogels, *Chem. Commun.* 48:9302–9304.

Deflorian, F., Rossi, S., and Scrinzi, E. 2013. Self-healing supramolecular polyurethane coatings: preliminary study of the corrosion protective properties, *Corros. Eng. Sci. Technol.* 48:147–154.

Dello Iacono, S., Martone, A., and Amendola, E. 2018. Corrosion-resistant self-healing coatings, *AIP Conf. Proc.* 1990:020010.

Deng, C. C., Brooks, W. L. A., Abboud, K. A., and Sumerlin, B. S. 2015. Bioinspired self-healing hydrogel based on benzoxaborole-catechol dynamic covalent chemistry for 3D cell encapsulation, *ACS Macro Lett.* 4:220–224.

Döhler, D., Michael, P., and Binder, W. H. 2017. CuAAC-based click chemistry in self-healing polymers, *Acc. Chem. Res.* 50:2610–2620.

Dry, C. M., and Sottos, N. 1996. Passive smart self-repair in polymer matrix composite materials, *Smart Mater. Proc. SPIE* 1993:438–444.

Du, Y., Qiu, W.-Z., Wu, Z. L., Ren, P.-F., Zheng, Q., and Xu, Z.-K. 2016. Water-triggered self-healing coatings of hydrogen-bonded complexes for high binding affinity and antioxidative property, *Adv. Mater. Interfaces* 3:1600167.

Duki, S. F., Kolmakov, G. V., Yashin, V. V., Kowalewski, T., Matyjaszewski, K., and Balazs, A. C. 2011. Modeling the nanoscratching of self-healing materials, *J. Chem. Phys.* 134:084901.

Esteves, A. C. C., Lyakhova, K., van der Ven, L. G. J., van Benthem, R. A. T. M., and de With, G. 2013. Surface segregation of low surface energy polymeric dangling chains in a cross-linked polymer network investigated by a combined experimental–simulation approach, *Macromolecules* 46:1993–2002.

Esteves, A. C. C., Lyakhova, K., van Riel, J. M., van der Ven, L. G. J., van Benthem, R. A. T. M., and de With G. 2014. Self-replenishing ability of cross-linked low surface energy polymer films investigated by a complementary experimental-simulation approach, *J. Chem. Phys.* 140:124902.

Fadl, A. M., Abdou, M. I., Laila, D., and Sadeek, S. A. 2019. Application insights of Schiff base metal complex/SiO$_2$ hybrid epoxy nanocomposite for steel surface coating: correlation the protective behavior and mechanical properties with material loading, *Prog. Org. Coat.* 136:105226.

Fennell, E., and Huyghe, J. M. 2019. Chemically responsive hydrogel deformation mechanics: a review, *Molecules* 24:3521.

Gadwal, I. 2021. A brief overview on preparation of self-healing polymers and coatings via hydrogen bonding interactions, *Macromol* 1:18–36.

Gilabert, F. A., Garoz, D., and van Paepegem, W. 2017. Macro- and micro-modeling of crack propagation in encapsulation-based self-healing materials: application of XFEM and cohesive surface techniques, *Mater. Des.* 130:459–478.

Glinel, K., Moussa, A., Jonas, A. M., and Laschewsky, A. 2002. Influence of polyelectrolyte charge density on the formation of multilayers of strong polyelectrolytes at low ionic strength. *Langmuir* 18:1408–1412.

Grand View Research, https://www.grandviewresearch.com/industry-analysis/self-healing-materials, accessed on 15th March 2021.

Guo, M., Pitet, L. M., Wyss, H. M., Vos, M., Dankers, P. Y., and Meijer, E. W. 2014. Tough stimuli-responsive supramolecular hydrogels with hydrogen-bonding network junctions, *J. Am. Chem. Soc.* 136:6969–6977.

Hart, L. R., Nguyen, N. A., Harries, J. L., Mackay, M. E., Colquhoun, H. M., and Hayes W. 2015. Perylene as an electron-rich moiety in healable, complementary π–π stacked, supramolecular polymer systems, *Polymer* 69:293–300.

Hayes, W., and Greenland, B. W. 2015. Donor–acceptor π–π stacking interactions: from small molecule complexes to healable supramolecular polymer networks, *Adv. Polym. Sci.* 268:143–166. doi: 10.1007/978-3-319-15404-6_4.

He, X., Liu, L., Han, H., Shi, W., Yang, W., and Lu, X. 2019. Bioinspired and microgel-tackified adhesive hydrogel with rapid self-healing and high stretchability, *Macromolecules* 52:72–80.

He, Y., Zhang, C., Wu, F., and Xu, Z. 2016. Fabrication study of a new anticorrosion coating based on supramolecular nanocontainer, *Synth. Met.* 212:186–194.

Heydari, H., Talebian, M., Salarvand, Z., Raeissi, K., Bagheri, M., and Golozar, M. A. 2018. Comparison of two Schiff bases containing *O*-methyl and nitro substitutes for corrosion inhibiting of mild steel in 1 M HCl solution, *J. Mol. Liq.* 254:177–187.

Holten-Andersen, N., Harrington, M. J., Birkedal, H., Lee, B. P., Messersmith, P. B., Lee, K. Y., and Waite, J. H. 2011. pH-induced metal-ligand cross-links inspired by mussel yield self-healing polymer networks with near-covalent elastic moduli, *Proc. Natl. Acad. Sci.* 108:2651–2655.

Hu, Z., Zhang, D., Lu, F., Yuan, W., Xu, X., Zhang, Q., Liu, H., Shao, Q., Guo, Z., and Huang, Y. 2018. Multistimuli-responsive intrinsic self-healing epoxy resin constructed by host–guest interactions, *Macromolecules* 51:5294–5303.

Huang, H., Tian, Y., Xie, Y., Mo, R., Hu, J., Li, M., Sheng, X., Jianga, X., and Zhang, X. 2020. Modification of graphene oxide with acrylate phosphorus monomer via thiol-Michael addition click reaction to enhance the anti-corrosive performance of waterborne epoxy coatings, *Prog. Org. Coat.* 146:105724.

Hughes, A. H., Cole, I. S., Muster, T. M., and Varley, R. J. 2010. Designing green, self-healing coatings for metal protection, *NPG Asia Mater.* 2:143–151.

Imbesi, P. M., Raymond, J. E., Tucker, B. S., and Wooley, K. L. 2012. Thiol-ene "click" networks from amphiphilic fluoropolymers: full synthesis and characterization of a benchmark anti-biofouling surface, *J. Mater. Chem.* 22:19462–19473.

Iyer, B. V. S., Salib, I. G., Yashin, V. V., Kowalewski, T., Matyjaszewski, K., and Balazs, A. C. 2013. Modeling the response of dual cross-linked nanoparticle networks to mechanical deformation, *Soft Matter* 9:109–121.

Javierre, E. 2019. Modeling self-healing mechanisms in coatings: approaches and perspectives, *Coatings* 9:122–141.

Jiang, Q., Zhang, Q., Wu, X., Wu, L., and Lin, J. H. 2020. Exploring the interfacial phase and π–π stacking in aligned carbon nanotube/polyimide nanocomposites, *Nanomaterials* 10:1158–1170.

Jucius, D., Lazauskas, A., and Gudaitis R. 2019. Multiple hydrogen-bonding assisted scratch–healing of transparent coatings, *Coatings* 9:796–806.

Kolmakov, G. V., Matyjaszewski, K., and Balazs, A. C. 2009. Harnessing labile bonds between nanogel particles to create self-healing materials, *ACS Nano* 3:885–892.

Kosarli, M., Bekas, D., Tsirka, K., and Paipetis, A. S. 2020. Capsule-based self-healing polymers and composites, In *Self-Healing Polymer-Based Systems*, pp. 259–278, Elsevier, Amsterdam. doi: 10.1016/B978-0-12-818450-9.00010-6.

Lao, L., Liu, K., Ren, L., Yu, J., Cheng, J., Li, Y., and Lu, S. 2021. Improving corrosion protection and friction resistance of Q235 steel by combining noncovalent action and rotating coating method, *ACS Omega* 6:7434–7443.

Latnikova, A., Grigoriev, D., Schenderlein, M., Möhwald, H., and Shchukin, D. 2012. A new approach towards "active" self-healing coatings: exploitation of microgels, *Soft Matter* 8:10837–10844.

Lee, B. P., Dalsin, J. L., and Messersmith, P. B. 2002. Synthesis and gelation of DOPA-modified poly(ethylene glycol) hydrogels, *Biomacromolecules* 3:1038–1047.

Lin, P., Ma, S., Wang, X., and Zhou, F. 2015. Molecularly engineered dual-crosslinked hydrogel with ultrahigh mechanical strength, toughness, and good self-recovery, *Adv. Mater.* 27:2054–2059.

Liu, C., Zhao, H., Hou, P., Qian, B., Wang, X., Guo, C., and Wang, L. 2018. Efficient graphene/cyclodextrin-based nanocontainer, *ACS Appl. Mater. Interfaces* 10:36229–36239.

Liu, W. X., Zhang, C., Zhang, H., Zhao, N., Yu, Z. X., and Xu, J. 2017. Oxime-based and catalyst-free dynamic covalent polyurethanes, *J. Am. Chem. Soc.* 139:8678–8684.

Liu, X., Zhang, H., Wang, J., Wang, Z., and Wang, S. 2012. Preparation of epoxy microcapsule based self-healing coatings and their behavior, *Surf. Coat. Technol.* 206:4976–4980.

Lundberg, P., Bruin, A., Klijnstra, J. W., Nystrom, A. M., Johansson, M., Malkoch, M., and Hult A. 2010. Poly(ethylene glycol)-based thiol-ene hydrogel coatings-curing chemistry, aqueous stability, and potential marine antifouling applications, *ACS Appl. Mater. Interfaces* 2:903–912.

Lyakhova, K., Esteves, A. C. C., van de Put, M. W. P., van der Ven, L. G. J., van Benthem, R. A. T. M., and de With, G. 2014. Simulation-experimental approach to investigate the role of interfaces in self-replenishing composite coatings, *Adv. Mater. Interfaces* 1:1400053–1400063.

Ma, C., Li, T., Zhao, Q., Yang, X., Wu, J., Luo, Y., and Xie, T. 2014. Supramolecular Lego assembly towards three-dimensional multi-responsive hydrogels, *Adv. Mater.* 26:5665–5669.

Ma, H., Chen, S., Niu, L., Zhao, S., Li, S., and Li, D. 2002. Inhibition of copper corrosion by several Schiff bases in aerated halide solutions, *J. Appl. Electrochem.* 32:65–72.

Maiti, S., and Geubelle, P. H. 2006. Cohesive modeling of fatigue crack retardation in polymers: crack closure effect, *Eng. Fract. Mech.* 73:22–41.

Markets and Markets, https://www.marketsandmarkets.com/Market-Reports/self-healing-material-market-46412119.html, accessed on 15th March 2021.

Mei, J. F., Jia, X. Y., Lai, J. C., Sun, Y., Li, C. H., Wu, J. H., Cao Y., You, X. Z., and Bao, Z. 2016. A highly stretchable and autonomous self-healing polymer based on combination of Pt···Pt and π-π interactions, *Macromol. Rapid Commun.* 37:1667–1675.

Mittal, V. 2014. Self-healing anti-corrosion coatings for applications in structural and petrochemical engineering, In *Handbook of Smart Coatings for Materials Protection*, pp. 183–197, Woodhead Publishing, Cambridge. https://doi.org/10.1533/9780857096883.2.183.

Mohamadhoseini, M., and Mohamadnia, Z. 2020. Supramolecular self-healing materials via host-guest strategy between cyclodextrin and specific types of guest molecules, *Coord. Chem. Rev.* 432:213711.

Montano, V., Vogel, W., Smits, A., van der Zwaag, S., and Garcia, S. J. 2021. From scratch closure to electrolyte barrier restoration in self healing polyurethane coatings, *ACS Appl. Polym. Mater.* 3:2802–2812.

Montemor, M. F. 2014. Functional and smart coatings for corrosion protection: a review of recent advances, *Surf. Coat. Technol.* 258:17–37.

Morad, M. S., and Sarhan, A. A. O. 2008. Application of some ferrocene derivatives in the field of corrosion inhibition, *Corros. Sci.* 50:744–753.

Moustafa, E. M., Dietz, A., and Hochsattel, T. 2013. Manufacturing of nickel/nanocontainer composite coatings, *Surf. Coat. Technol.* 216:93–99.

Mozhdehi, D., Ayala, S., Cromwell, O. R., and Guan, Z. 2014. Self-healing multiphase polymers via dynamic metal-ligand interactions, *J. Am. Chem. Soc.* 136:16128–16131.

Mukherjee, S., Bapat, A. P., Hill, M. R., and Sumerlin, B. S. 2014. Oximes as reversible links in polymer chemistry: dynamic macromolecular stars, *Polym. Chem.* 5:6923–6931.

Nakahata, M., Mori, S., Takashima, Y., Hashidzume, A., Yamaguchi, H., and Harada, A. 2014. pH- and sugar-responsive gel assemblies based on boronate–catechol interactions, *ACS Macro Lett.* 3:337–340.

Nardeli, J. V., Fujiwara, C. S., Taryba, M., Montemor, M. F., and Benedetti, A. V. 2020. Self-healing ability based on hydrogen bonds in organic coatings for corrosion protection of AA1200, *Corros. Sci.* 177:108984.

Nazir, U., Akhter, N., Janjua, N. K., Adeel Asghar, M., Kanwal, S., Maryum Butt, T., Sani, A., Liaqat, F., Hussain, R., and Ullah Shah F. 2020. Biferrocenyl Schiff bases as efficient corrosion inhibitors for an aluminium alloy in HCl solution: a combined experimental and theoretical study, *RSC Adv.* 10:7585–7599.

Peng, L., Lin, M., Zhang, S., Li, L., Fu, Q., and Hou, J. 2019. A self-healing coating with UV-shielding property, *Coatings* 9:421–434.

Peng, S., Zeng, Z., Zhao, W., Li, H., Chen, J., Han, J., and Wu, X. 2014. Novel functional hybrid silica sol–gel coating for copper protection *via in situ* thiol–ene click reaction, *RSC Adv.* 4:15776–15781.

Ponnusami, S. A., Krishnasamy, J., Turteltaub, S., and van der Zwaag, S. 2018. A cohesive-zone crack healing model for self-healing materials, *Int. J. Solids Struct.* 134:249–263.

Ponnusami, S. A., Turteltaub, S., and van der Zwaag, S. 2015. Cohesive-zone modelling of crack nucleation and propagation in particulate composites, *Eng. Fract. Mech.* 149:170–190.

Pratama, P. A., Peterson, M. A., and Palmese, G. R. 2012. Diffusion and reaction phenomena in solution-based healing of polymer coatings using the Diels–Alder reaction, *Macromol. Chem. Phys.* 213:173–181.

Prosek, T., and Thierry, D. 2004. A model for the release of chromate from organic coatings, *Prog. Org. Coat.* 49:209–217.

Resetco, C., Hendriks, B., Badia, N., and Du Prez, F. 2017. Thiol–ene chemistry for polymer coatings and surface modification – building in sustainability and performance, *Mater. Horizons* 4:1041–1053.

Salib, I. G., Kolmakov, G. V., Gnegy, C. N., Matyjaszewski, K., and Balazs, A. C. 2011. Role of parallel reformable bonds in the self-healing of cross-linked nanogel particles, *Langmuir* 27:3991–4003.

Samadzadeh, M., Boura, S. H., Peikari, M., Kasiriha, S. M., and Ashrafi, A. 2010. A review on self-healing coatings based on micro/nanocapsules, *Prog. Org. Coat.* 68:159–164.

Sanka, R. V. S. P., Krishnakumar, B., Leterrier, Y., Pandey, S., Rana, S., and Michaud, V. 2019. Soft self-healing nanocomposites, *Front. Mater.* 6:137.

Sauvant-Moynot, V., Gonzalez, S., and Kittel, J. 2008. Self-healing coatings: an alternative route for anticorrosion protection, *Prog. Org. Coat.* 63:307–315.

Sethi, T., Chaturvedi, A., Upadhyay, R. K., and Mathur, S. P. 2007. Corrosion inhibitory effects of some Schiff's bases on mild steel in acid media, *J. Chil. Chem. Soc.* 52:1206–1213.

Shchukin, D. G., and Möhwald, H. 2007. Surface-engineered nanocontainers for entrapment of corrosion inhibitors, *Adv. Funct. Mater.* 17:1451–1458.

Shen, L., Zheng, J., Wang, Y., Lu, M., and Wu, K. J. 2016. Influence of hydrogen bonding interactions on the properties of ultraviolet-curable coatings, *Appl. Polym. Sci.* 133:43113–43122.

Sinclair, A., O'Kelly, M. B., Bai, T., Hung, H. C., Jain, P., and Jiang, S. 2018. Self-healing zwitterionic microgels as a versatile platform for malleable cell constructs and injectable therapies, *Adv. Mater.* 1803087:1–8.

Skorb, V., Shchukin, D. G., Möhwald, H., and Sviridov, D. V. 2009. Photocatalytically-active and photocontrollable coatings based on titania-loaded hybrid sol–gel films, *J. Mater. Chem.* 19:4931–4937.

Stankiewicz, A., Jagoda, Z., Zielińska, K., and Szczygieł, I. 2015. Gelatin microgels as a potential corrosion inhibitor carriers for self-healing coatings: preparation and codeposition, *Mater. Corros.* 66:1391–1396.

Stankiewicz, A., Kefallinou, Z., Jagoda, Z., Mordarski, G., and Spencer, B. F. 2018. Surface functionalisation by the introduction of self-healing properties into electroless Ni-P coatings, *Electrochim. Acta* 297:427–434.

Sun, T. L., Kurokawa, T., Kuroda, S., Ihsan, A. B., Akasaki, T., Sato, T., Haque, M. A., Nakajima, T., and Gong, J. P. 2013. Physical hydrogels composed of polyampholytes demonstrate high toughness and viscoelasticity, *Nat. Mater.* 12:932–937.

Taghaviksih, M., Subianto, S., Kumar Dutta, N., and Choudhury, N. R. 2016. Novel thiol-ene hybrid coating for metal protection, *Coatings* 6:17.

Tasdelen, M. A. 2011. Diels–Alder "click" reactions: recent applications in polymer and material science, *Polym. Chem.* 2:2133–2145.

Turkenburg, D. H., Durant, Y., and Fischer, H. R. 2017. Bio-based self-healing coatings based on thermoreversible Diels-Alder reaction, *Prog. Org. Coat.* 111:38–46.

Ulrich S., Boturyn D., Marra A., Renaudet O., and Dumy P. 2014. Oxime ligation: a chemoselective click-type reaction for accessing multifunctional biomolecular constructs, *Chem. Eur. J.*, 20:34–41.

Voyiadjis, G., Shojaei, A., Li G., and Kattan P. 2011. Continuum damage-healing mechanics with introduction to new healing variables, *Int. J. Damage Mech.* 21:391–414.

Voyiadjis, G. Z., and Kattan, P. I. 2009. A comparative study of damage variables in continuum damage mechanics, *Int. J. Damage Mech.* 18:315–340.

Wang, H., Presuel, F., and Kelly, R. G. 2004. Computational modeling of inhibitor release and transport from multifunctional organic coatings, *Electrochim. Acta* 49:239–255.

Wang, M. D., Liu, M. Y., and Fu J. J. 2015. An intelligent anticorrosion coating based on pH-responsive smart nanocontainers fabricated *via* a facile method for protection of carbon steel, *J. Mater. Chem.* 3:6423–643.

Wang, Q., Gao, Z., and Yu, K. 2017. Interfacial self-healing of nanocomposite hydrogels: theory and experiment, *J. Mech. Phys. Solids* 109:288–306.

Wankar, J., Kotla, N. G., Gera, S., Rasala, S., Pandit, A., and Rochev, Y. A. 2020. Recent advances in host–guest self-assembled cyclodextrin carriers: implications for responsive drug delivery and biomedical engineering, *Adv. Funct. Mater.* 30:1909049.

Waśkiewicz, S., Zenkner, K., Langer, E., Lenartowicz, M., and Gajlewicz, I. 2013. Organic coatings based on new Schiff base epoxy resins, *Prog. Org. Coat.* 76:1040–1045.

White, S. R., Sottos, N. R., Geubelle, P. H., Moore, J. S., Kessler, M. R., Sriram, S. R., Brown, E. N., and Viswanathan, S. 2001. Autonomic self-healing of polymer composites, *Nature* 409:794–797.

Wild, A., Winter, A., Schlütter, F., and Schubert, U. S. 2011. Advances in the field of π-conjugated 2,2′:6′,2″-terpyridines, *Chem. Soc. Rev.* 40:1459–1511.

Winter, A., and Schubert, U.S. 2016. Synthesis and characterization of metallo-supramolecular polymers, *Chem. Soc. Rev.* 45:5311–5357.

Xin, Y., and Yuan, J. 2012. Schiff's base as a stimuli-responsive linker in polymer chemistry, *Polym. Chem.* 3:3045–3055.

Xu, G., Pranantyo, D., Xu, L., Neoh, K. G., Kang, E. T., and Teo, S. L. M. 2016. Antifouling coatings via tethering of hyperbranched polyglycerols on biomimetic anchors, *Ind. Eng. Chem. Res.* 55:10906–10915.

Yan, X. et al. 2018. Quadruple H-bonding cross-linked supramolecular polymeric materials as substrates for stretchable, antitearing, and self-healable thin film electrodes, *J. Am. Chem. Soc.* 140:5280–5289.

Yang, B., Mall, S., and Ravi-Chandar, K. 2001. A cohesive zone model for fatigue crack growth in quasibrittle materials, *Int. J. Solids Struct.* 38:3927–3944.

Yang, W., Tao, X., Zhao, T., Weng, L., Kang, E. T., and Wang, L. 2015. Antifouling and antibacterial hydrogel coatings with self-healing properties based on dynamic disulfide exchange reaction, *Polym. Chem.* 6:7027–7035.

Yuan, J., Zhang, H., Hong, G., Chen, Y., and Chen, G. 2013. Using metal–ligand interactions to access biomimetic supramolecular polymers with adaptive and superb mechanical properties, *J. Mater. Chem. B* 1:4809–4818.

Zechel, S., Hager, M. D., Priemel, T., and Harrington, M. J. 2019. Healing through histidine: bioinspired pathways to self-healing polymers via imidazole–metal coordination, *Biomimetics* 4:20–41.

Zhang, H., Ma, Y., Tan, J., Fan, X., Liu, Y., Gu, J., Zhang, B., Zhang, H., and Zhang, Q. 2016. Robust, self-healing, superhydrophobic coatings highlighted by a novel branched thiol-ene fluorinated siloxane nanocomposites, *Compos. Sci. Technol.* 137:78–86.

Zhang, X., Ren, P.-F., Yang, H.-C., Wan, L.-S., and Xu, Z.-K. 2016. Co-deposition of tannic acid and diethylenetriamine for surface hydrophilization of hydrophobic polymer membranes, *Appl. Surf. Sci.* 360:291–297.

Zhang, Y., Tao, L., Li, S., and Wei, Y. 2011. Synthesis of multiresponsive and dynamic chitosan-based hydrogels for controlled release of bioactive molecules, *Biomacromolecules* 12:2894–2901.

Zheng, S. Y., Ding, H., Qian, J., Yin, J., Wu, Z. L., Song, Y., and Zheng, Q. 2016. Metal-coordination complexes mediated physical hydrogels with high toughness, stick–slip tearing behavior, and good processability, *Macromolecules* 49:9637–9646.

Zheng, Y., Hashidzume, A., and Harada A. 2013. pH-responsive self-assembly by molecular recognition on a macroscopic scale, *Macromol. Rapid Commun.* 34:1062–1066.

Zhou, T., Gao, X., Dong, B., Sun, N., and Zheng, L. 2016. Poly(ionic liquid) hydrogels exhibiting superior mechanical and electrochemical properties as flexible electrolytes, *J. Mater. Chem. A* 4:1112–1118.

Zhu, D. Y., Rong, M. Z., and Zhang, M. Q. 2015. Self-healing polymeric materials based on microencapsulated healing agents: from design to preparation, *Prog. Polym. Sci.* 4950:175–220.

Zielińska, K., Sun, H., Campbell, R. A., Zarbakhsh, A., and Resmini, M. 2016. Smart nanogels at the air/water interface: structural studies by neutron reflectivity, *Nanoscale* 8:4951–4960.

Zielińska, K., and van Leeuwen, H. P. 2014. Polyelectrolyte coatings prevent interferences from charged nanoparticles in SPME speciation analysis, *Anal. Chim. Acta* 844:44–47.

9 Polymeric Capsules as Nanocontainers in Surface Coatings

Sarah B. Ulaeto
Rhema University

T. P. D. Rajan
CSIR-National Institute for Interdisciplinary
Science and Technology (CSIR-NIIST)

CONTENTS

9.1 INTRODUCTION

Polymeric capsules are constructs containing a core surrounded by a polymeric shell network. They are essential for storing various active substances and are usually externally triggered to release their contents (Cui et al. 2014; Zarket and Raghavan 2017; Deng et al. 2020). These stimuli-responsive polymeric constructs can be synthesized either by template-assisted techniques or by template-free techniques (Cui et al. 2014) and can be obtained in the nanoscale. These polymer-based constructs serve as encapsulating shells at the nanometer scale and entrapping active substances within their core. They have been classed differently from the nanospheres. While the polymer nanocapsules' hollow core may contain an oily and/or aqueous active substance, the shell is the polymer network that controls the release of the encapsulated material(s). On the other hand, the polymer nanospheres entrap the active substances within their polymer network and adsorb the same onto their surface. Therefore, it is a continuous polymer network-based shell with no fluid-like core (Khalil et al. 2017; Deng et al. 2020; Zielińska et al. 2020). Morphology differences between polymer nanocapsules and nanospheres are presented in Figure 9.1a.

The hollow core may contain triglycerides as part of the structural components in the core meant to either dissolve or disperse the active substances encapsulated within the core. The oily component of the core can be a liquid or solid lipid, and if the oily core is the main active substance, then its release is controlled by the action of the polymeric shell wall. The polymeric wall can be either

DOI: 10.1201/9781003169130-11

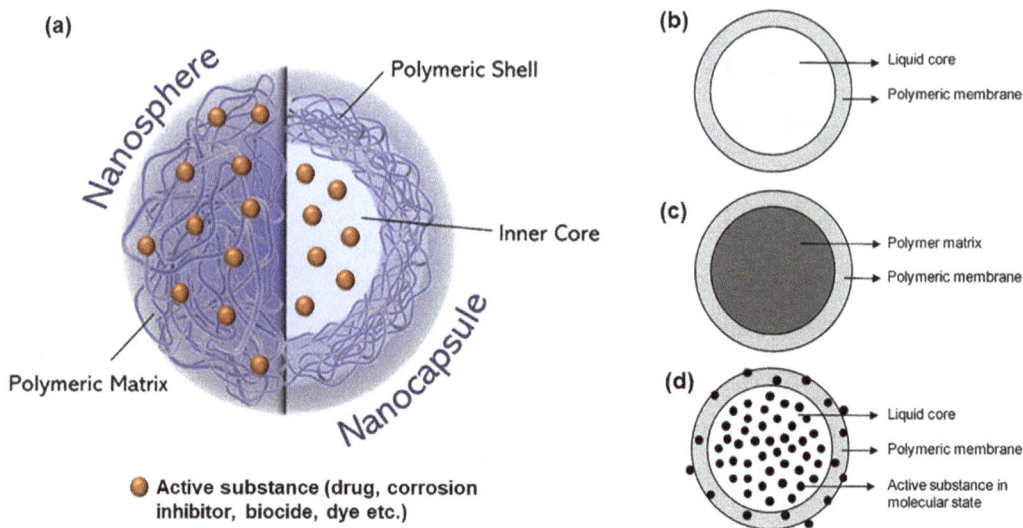

FIGURE 9.1   (a) Schematic of morphology differences between polymer nanocapsules and nanospheres. (b) Nanocapsular structural differences. ((a) Reproduced with permission from Gagliardi et al. (2021) © 2021 Authors, under Creative Commons Attribution License. (b) Reproduced with permission from Mora-Huertas et al. (2010) © 2009 Elsevier B.V.)

crystalline, semi-crystalline, or amorphous polymer. It protects the active substance(s) against external factors and is also the diffusion barrier (Poletto et al. 2011). Furthermore, the polymeric nanocapsules can be either ionic or non-ionic, which depends on the chemical nature of the polymer and surfactant involved in its preparation. The charges could be negative (anionic) or positive (cationic) on the ionic polymeric nanocapsule surfaces. At the same time, non-ionic surfaces can be obtained when neutral hydrophilic or hydrophobic polymers are used (Poletto et al. 2011). The oily core serves in lipophilic molecule encapsulation, and the aqueous core promotes the delivery of molecules with hydrophilic nature (Deng et al. 2020). The nanocapsule structure is further illustrated in Figure 9.1b, distinguishing between the liquid core, the polymer matrix within the core area, and when active substances are dispersed in the liquid core. The structures realized are meant to preserve the contents of the core for its tailored function.

Several polymers were employed in the synthesis of functional nano-reservoir systems, including polycaprolactone (PCL), polylactide-co-glycolide (PLGA), polyethylene glycol (PEG), PCL-PEG-PCL, PEG-polypropylene glycol (PPG)-PEG, PLGA-PEG, and Euragrit® (EPO, RS100, L100–55) (Deng et al. 2020; Zielińska et al. 2020). Poly(glycolic acid) (PGA), poly(lactic acid) (PLA) and poly gamma glutamic acid (γ-PGA) have been used (Khalil et al. 2017). Poly(4-styrene sulfonate) (PSS)/poly(allylamine hydrochloric acid) (PAH), polymethacrylic acid (PMA)/poly(vinyl pyrrolidone) (PVPON), poly(4-styrene sulfonate) (PSS)/protamine(Pro), dextran(Dex)/protamine(Pro), poly(4-styrene sulfonate) (PSS)/poly(diallyldimethylammonium) chloride (PDADMAC), poly(4-styrene sulfonate) (PSS)/diazoresin (Dar), poly(diethylaminoethylmethacrylate) (PDEAEM)-co-pyridyldisulfideethylmethacrylate (PPDSM), and polydopamine (Larrañaga et al. 2017), poly(acrylic acid), poly((ethylene glycol dimethylacrylate)-co-(hydroxyethyl methacrylate)), polypyrrole, and so on (Bentz and Savin 2018) were also reported. Furthermore, polysaccharides (alginate, polycyclodextrin, heparin, etc.), chitosan, protein-based polymers (e.g., albumin), dextran sulfate, etc. (Deng et al. 2020) have been used to mention a few. With diverse synthesis protocols, it is possible to obtain various polymer capsules that differ in size, composition, morphology, and properties. Some of the sort after properties include high surface area, high colloidal stability, high loading and storage capacity, stimuli-responsive reversible swelling and contraction, and highly tuneable size. Applications of polymer capsules are vast; some areas are catalysis and sensing, controlled drug and vaccine delivery, cosmetics and antigen carriers, energy storage, fabrication of micro-and nano-reactors, separation membranes, and surface coatings. The wide application of polymer nanocontainers is connected to its wide variety of useable polymers and synthesis protocols.

9.2 SYNTHESIS OF POLYMERIC NANOCAPSULES

Polymeric capsules can be synthesized in several ways, which include: layer-by-layer (LbL) assembly, emulsion-based polymerization, surface-initiated polymerization (SIP) (Huang and Voit 2013; Khalil et al. 2017), self-assembly, single-step polymer adsorption, bioinspired assembly, ultrasound assembly (Cui et al. 2014; Khalil et al. 2017), phase separation and suspension polymerization (Liu et al. 2018). The solvent displacement method is also a key process for preparing polymeric nanocapsules and is referred to as nanoprecipitation (Zielińska et al. 2020). Amongst these techniques, LbL, single-step polymer adsorption, bioinspired assembly, SIP, ultrasound assembly, phase separation method, and suspension polymerization are classified as template-assisted methods. However, self-assembly is an essential template-free method with varying approaches. Unlike the template-assisted route, the template-free method is a convenient route and generates less waste during the process. Selected template-assisted methods are discussed below.

9.2.1 TEMPLATE-ASSISTED METHODS

a. **Layer-by-layer assembly (LbL)**: this technique requires a sacrificial template which is usually removed in the process of obtaining the desired capsule (Cui et al. 2014). LbL involves using either covalent or electrostatic interactions, single polymer assembly, or hydrogen bonding to deposit polymer layers on a sacrificial core, followed by dissolution and purification steps. The sacrificial template controls the size, shape, and encapsulation method; e.g. includes melamine formaldehyde (MF), poly(methyl methacrylate) (PMMA), and polystyrene (PS). The processing step of LbL polymer capsules is illustrated in Figure 9.2 (Huang and Voit 2013).

b. **Emulsion polymerization**: this is one of the oldest techniques used, and it is very versatile. The emulsions are used as soft templates, the reactions are mild, and the particle morphologies are highly spherical. In the process, the emulsifier is partitioned; this reduces the active stabilizer and affects the particle size, which becomes mostly larger than 100 nm irrespective of increasing surfactant concentration (Bentz and Savin 2018). The nano-emulsion template method can be categorized into emulsion-diffusion/evaporation, emulsion-coacervation, and double emulsion (Deng et al. 2020). The emulsion-based method for the nanocapsule synthesis is contained in Figure 9.2.

c. **SIP**: the polymeric nanocapsule is prepared by growing the initiating polymers on the nano-sized template; thereafter, core removal is initiated, creating the hollow core surrounded by surface-attached polymer brushes (Huang and Voit 2013; Cui et al. 2014). Solid templates such as gold and silica nanoparticles, tin dioxide, gold@silica nanoparticles (Au@SiNPs), and so on (Bentz and Savin 2018) are used but face the challenge of removal under harsh chemical conditions. Harsh conditions like hydrofluoric acid for SiO_2 core and hydrochloric acid among other acids for metal nanocores. This inorganic nanoparticle templating process limits the use of biomaterials. Soft templates like vesicles and emulsion droplets have been used to overcome the abovementioned challenges and can be removed under mild conditions. Polymerization techniques applied so far to achieve SIP polymer nanocapsules are (i) surface-initiated nitroxide mediated radical polymerization (NMRP), (ii) surface-initiated atom transfer radical polymerization (ATRP), and (iii) surface-initiated reversible addition-fragmentation chain-transfer (RAFT) (Huang and Voit 2013). Polymeric materials such as polyacetylene, poly(caprolactone-b-ethylene glycol) (PCL-b-PEG), poly(ethyleneglycol dimethacrylate-co-methacrylic acid) (P(EGDMA-co-MAA), ethylene glycol dimethacrylate and methacrylate acid, polydopamine copolymer, resol, and Pluronic F127, block polymers of polyferrocenyldimethylsilane)-b-poly(2-vinylpyridine) (PFS-b-P2VP) and homopolymers of PFS, etc. have been grafted onto the inorganic cores to obtain the polymer nanocapsule shells (Bentz and Savin 2018). An illustration of the polymeric capsules prepared through this route is presented in Figure 9.2.

d. **Single-step polymer adsorption**: These are amongst the template-directed synthesis of polymer nanocapsules. This route minimizes the steps involved in the synthesis. In this approach, mesoporous silica templates, bromoisobutyramide templates, and polyrotaxane templates have been used, to mention a few (Cui et al. 2014).

e. **Nanoprecipitation**: In this method, two categories of solvents are involved; an organic solvent and an aqueous solvent. The polymer is dissolved in organic solvents such as ethanol, dioxane, acetone, and so on and is added dropwise with controlled stirring into an aqueous phase. The polymer nanocapsules precipitate as the solvent diffuses out of the nanodroplets. Surfactants are employed to ensure the stability of the colloidal suspension. Loading of the active substance can be achieved during the process when the active substance is first dissolved in oil prior to the emulsification step in the polymer solution (Mora-Huertas et al. 2010; Zielińska et al. 2020). This method is otherwise known as interfacial deposition or solvent displacement method, and different opinions surrounding the mechanism of nanocapsule formation have existed in the past (Mora-Huertas et al. 2010).

9.2.2 Template-Free Method

Self-assembly: this approach can be achieved in various ways, including γ-ray irradiation assisted self-assembly for which ^{60}Co has been used, inside-out Ostwald ripening process (where spheres can develop into capsules with time), plasma-enhanced chemical vapor deposition process

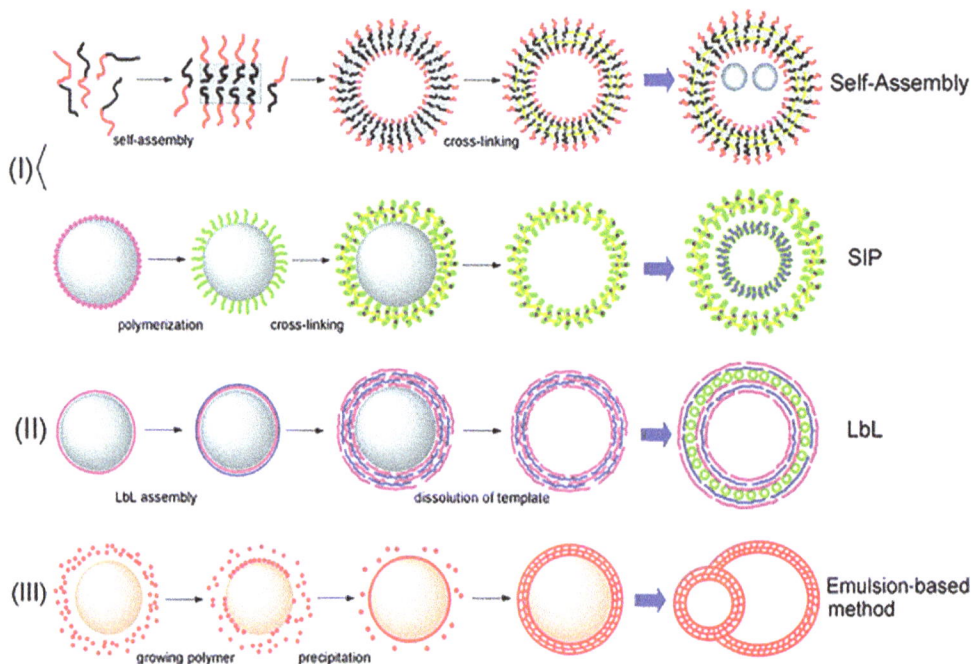

FIGURE 9.2 Schematic of selected preparation methods of polymeric capsules; (I) Self-Assembly and SIP, (II) LbL, and (III) Emulsion-based method. (Reproduced with permission from (Huang & Voit 2013) © 2013 The Royal Society of Chemistry.)

(where capsule sizes depend on the method of plasma activation), and so on. Furthermore, the plasma activation method could be either pulsed or continually applied (Bentz and Savin 2018). The self-assembly method of polymer capsule synthesis is also presented in Figure 9.2.

9.3 CHARACTERIZATIONS OF POLYMERIC NANOCAPSULES

The characterizations of polymer nanocapsules involve determining the size, shape, surface charge, crystallinity, particle trafficking, and hydrophobic/hydrophilic nature. In the determination of shape, microscopy techniques, such as transmission electron microscopy (TEM), scanning electron microscopy (SEM), and atomic force microscopy (AFM), are used. TEM, SEM, AFM, and dynamic light scattering techniques (DLS) are employed in size determination. In determining the zeta potential, which reveals the surface charge, Zetasizer is required, and for crystallinity, X-ray diffraction technique (XRD) is used. For particle trafficking, fluorescent labeling is needed, and in determining the hydrophobic or hydrophilic character, contact angle measurements are checked. The molecular weight of the polymer is decisive in the nanoparticle size to be obtained, the efficiency of encapsulation, and the rate of polymer degradation (Kothamasu et al. 2012; Khalil et al. 2017). The size exclusion chromatography (SEC) enables the determination of the polymer molar mass distribution. SEC also involves determining the influence of the components used in the formulation of the nanoparticles on the polymerization process. Information on polymer degradation and chemical reactions between the active substances and the shell can also be determined. The surface chemistry of the nanoparticles can be determined using X-ray photoelectron spectroscopy (Zielińska et al. 2020).

In the morphology determination using TEM, sample dehydration may occur due to the high-energy electron beam used. The resultant effect is shrinking and possible bursting of the nanocapsule(s). The preservation of the hydrated structure of the polymer nanocapsule(s) and its content can be achieved with the Cryo-TEM. The contact mode of the AFM tip is also capable of deforming the soft polymeric nanocapsules; therefore, tapping or intermittent contact is preferred. The determination of the internal radial structure and multicomponent of polymer nanoparticles can be achieved using the small-angle X-ray scattering (SAXS) and small-angle neutron scattering (SANS) techniques. The glass transition temperature, T_g, and melting temperature of the polymer capsule(s) as well as encapsulated active substance(s) are examined using the differential scanning calorimetry (DSC), while the thermal degradation profile of the samples is examined with thermogravimetric analysis (TGA).

Furthermore, the chemical composition of the polymer nanocapsule and its active substance is determined with the attenuated total reflectance-Fourier-transform infrared (ATR-FTIR) spectroscopy. Raman spectroscopy is an advantage in reducing the interference from liquid water. High-performance liquid chromatography (HPLC) or batch UV-visible spectrophotometry are relevant to quantify the released active substance(s) from the nanocapsules (Shakiba et al. 2020).

9.4 POLYMERIC CAPSULE LOADING AND RELEASE

The active substances within the polymer capsules can be encapsulated either during the capsule synthesis process or after the capsules are prepared as a post-grafting step. During the synthesis process encapsulation strategy, the active substances remain enclosed after the template is removed. In contrast, the post-synthesis loading of active substances requires a blend of the already synthesized nanoparticles and the active substance(s) through ultrasonication or by application of vacuum pressure. The loading processes earlier mentioned physically entrap the active substances; however, it is possible to covalently bond the active substances to the polymer chain during the loading process. The release of the active substance after a bond cleavage can also be triggered by a pH change (Pitakchatwong et al. 2018).

The release of the encapsulated active substances occurs through different mechanisms. This includes (i) burst release, (ii) diffusion, and (iii) degradation of the polymer shell of the capsule. Burst release occurs when the concentration of encapsulated active constituent is very high such that the active substance is in excess even at the surface of the nanoparticles. Diffusion usually occurs from higher concentration regions to lower concentration regions and is affected by dilution. Increasing the size of the nanoparticles, increasing cross-linking, and reducing porosity contributes to controlling the diffusion of the active substances, especially where a delay is required. Also, swelling and relaxation of the polymer nanocapsule occur during the diffusion process (Shakiba et al. 2020). Diffusion processes can be evaluated using the UV/Visible spectroscopy technique and/or the SERS technique under the influence of pH. The techniques enable the determination of the specific conditions for inhibitor release and action. For example, Tavandashti et al. evaluated the release of mercaptobenzothiazole (MBT) from polyaniline (PANI) capsules using the SERS technique and reported that basic pH was the most effective releasing medium (Figure 9.3). In conducting polymers such as PANI, the pH of the solution strongly influences ion transport, and protonation or deprotonation of PANI is possible in acidic or basic media (Pirhady Tavandashti et al. 2016).

Degradation may be photoinitiated, pH stimulated, hydrolytic degradation, and enzyme triggered (Shakiba et al. 2020), as is illustrated in Figure 9.4. pH-assisted degradation of the polymeric capsules could be acidic or basic triggered, releasing the encapsulated active substance(s). At a neutral pH, physically entrapped compounds can be released depending on the nature of the active compound, but the possibility of releasing a covalently linked compound can be challenging due to poor solubility and bond cleavage (Pitakchatwong et al. 2018).

Furthermore, some polymer capsules release their contents due to their thermo-responsive property. In this scenario, heating the polymer above its specific lower critical solution temperature shrinks the polymer, making the polymeric chains hydrophobic, and preventing the release of its core. However, when the temperature decreases, the reverse occurs, and the encapsulated core content can be released (Zarket and Raghavan 2017).

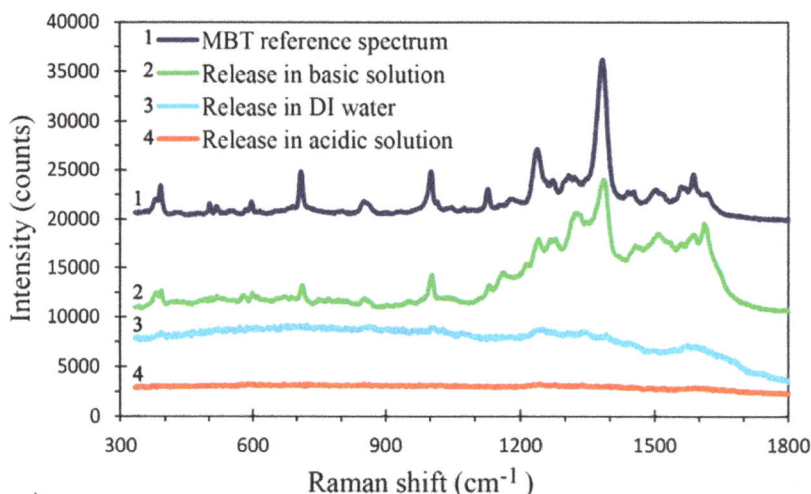

FIGURE 9.3 SERS spectra showing pH-influenced release of MBT from PANI capsules. (Reproduced with permission from Pirhady Tavandashti et al. (2016) © 2016 Elsevier Ltd.)

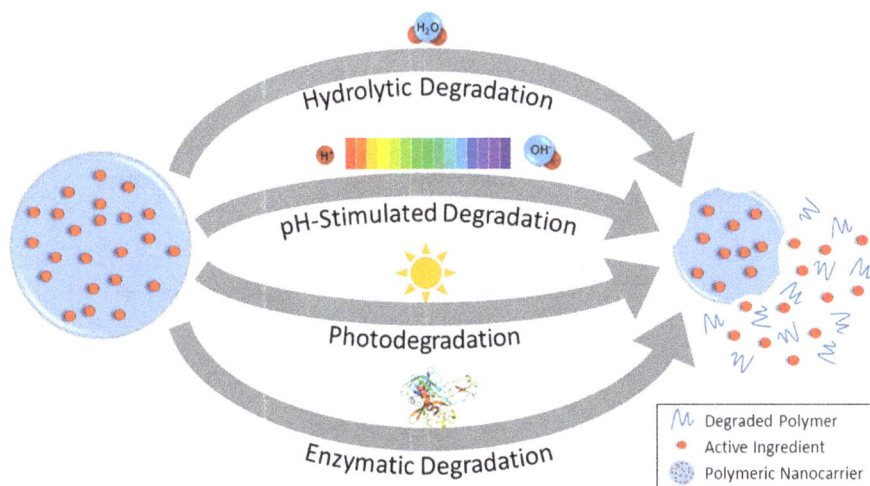

FIGURE 9.4 Illustration of active substance release from the nanocapsules through polymer matrix degradation. (Reproduced with permission from Shakiba et al. (2020) © 2019 The Royal Society of Chemistry.)

9.5 POLYMERIC NANOCAPSULES IN SURFACE COATINGS

Nanoparticles in coatings contribute to filling voids, deflecting crack paths, encapsulating and adsorbing active substances, and bridging between crack planes, which impact the performance of the coating applied to the surface it is meant to sustain. Irrespective of the synthesis route employed, the nanocapsules need to be diffusion-ready to enable the release of the contents entrapped in their core. The polymeric nanocapsules encase the active substances prepared by any synthesis protocols to yield nanocomposites that can be dispersed in surface coatings for specific functions. For example, Bahmani et al. prepared a PEG nanocomposite coating for near-infrared (NIR) imaging and phototherapeutic applications. The poly(allylamine) hydrochloride nanocapsule dispersed in the coating was prepared by a self-assembly method and contained indocyanine green in its core for the aforementioned application (Bahmani et al. 2011). Kasper et al. prepared a Zein nanocapsule biofilm for reducing oral infection/disease by delaying the recolonization of the tooth. The Zein nanocapsule of 110–235 nm prepared by the self-assembly method was loaded with S-phenyl-L-cysteine sulfoxide as the active substance (H. Kasper et al. 2016). Grillo et al. designed a chitosan surface coating on PCL nanocapsules loaded with atrazine. The chitosan coating interfered with the release of the atrazine herbicide from the nanocapsules. The 200–500 nm poly-e-caprolactone nanocapsules were prepared by interfacial deposition of preformed polymer (Grillo et al. 2014).

For corrosion protection, the nanocapsules contain corrosion inhibitors (illustrated in Figure 9.5) and function through a trigger-release mechanism in the defective zones of any nanocomposite-coated metallic substrate. The corrosion inhibitors diffuse out to suppress corrosion reaction, deactivate the corrosive species and provide a passive film within the defective zone. The new barrier film formed heals the corrosion site and terminates the electrochemical reactions responsible for corrosion as it hinders the ingress of corrosive species. Anticorrosive surface coatings requiring active substances loaded in polymer nanocapsules focus on self-healing and form a passive protective film within the damaged zones of the coating. The coating material in which the corrosion inhibitor-loaded nanocapsules are dispersed may be organic, inorganic, or ceramic-based. However, compatibility between the loaded contents and the coating type is essential; otherwise, the protective performance of the composite coating will be compromised.

Furthermore, other coating requirements for tailored functionalities aside from corrosion protection can be encased in polymeric nanocapsules. These include biocides, pigments, dyes, other relevant coating additives, and so on. Electrochemical impedance spectroscopy (EIS), scanning vibration

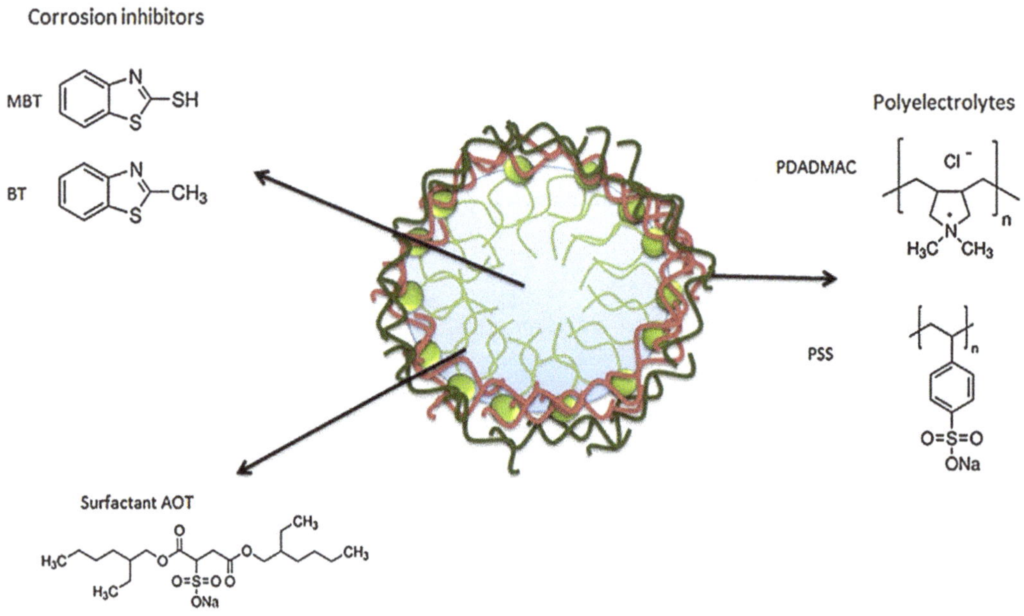

Corrosion inhibitors

MBT

BT

Surfactant AOT

Polyelectrolytes

PDADMAC

PSS

FIGURE 9.5 Mercaptobenzothiazole (MBT) and 2-methylbenzothiazole (BT) as examples of corrosion inhibitors encapsulated in polyelectrolyte nanocapsules in the presence of a surfactant. (Reproduced with permission from Kopeć et al. (2015) © 2015 Elsevier.)

electrode technique (SVET), and salt spray tests are the most widely used techniques to determine the efficiency of the passive films formed and monitor the corrosion and inhibition processes.

The release of the corrosion inhibitor containing heteroatoms and its adsorption onto the metal surface is a donor–acceptor reaction between the inhibitor heteroatoms or pie electrons and the vacant d-orbital of the metal surface. The barrier film formed prevents access of the corrosive species to the reaction site preventing corrosion (Li et al. 2014; Kopeć et al. 2015; Qian et al. 2017). EIS analysis provides both the time and frequency domain behavior of investigated electrochemical systems. EIS is widely used to evaluate the inhibitor release and self-healing of defective coatings on metal substrates. The corrosion inhibitor concentration released in the corrosive medium is dependent on its solubility in the medium (Li et al. 2014; Kopeć et al. 2015). The impedance values for inhibiting or corrosive systems are usually analyzed over a wide frequency range. Quantitative data can be extracted from the fitted circuit models either from the software installed in the electrochemical workstation or from secondary software such as ZSimpWin, ZView, EC Lab, and so on. (Qian et al. 2017; Ulaeto et al. 2019; Kim et al. 2020).

Anodic currents over a scribed area during the immersion time in the corrosive medium can be monitored by SVET analysis. When the corrosion inhibitors are released into such defective zones, the currents are expected to be lowered and remain negligible due to the passive film formation from the release of sufficient inhibiting substance(s). The SVET analysis shows that the electrochemical reactions encouraging corrosion are controlled (Li et al. 2014; Pirhady Tavandashti et al. 2016; Grigoriev et al. 2017). If the inhibitors are depleted or insufficient or are not available, an increase in the anodic current will be observed as the corrosive ions can approach the metal surface unhindered. This approach results in an attack on the metal, giving rise to any form of corrosion suitable for the specific aggressive environment.

Compatibility and high loading capacity are also advantaging when polymeric carriers are involved (Grigoriev et al. 2017). However, inhibitor-loaded polymeric nanocontainers dispersed in water-based polymer resins are not triggered for premature release of the core contents as organic solvents are not involved in the coating preparation, which can dissolve the nanocontainers. Therefore, the choice of resin (water-borne or solvent-borne) is crucial to blend with the loaded polymeric carriers to obtain smooth coatings and control the premature release of the active core or degradation of the capsule.

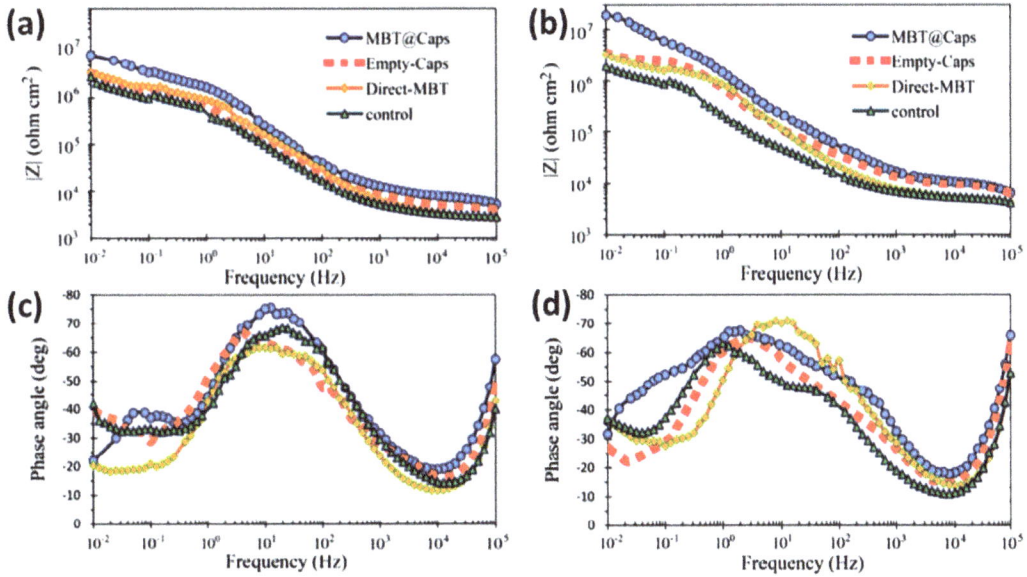

FIGURE 9.6 Bode impedance and phase angle plots of the scratched samples after (a, c) 1-day immersion, and (b, d) 7 days immersion in 0.3 wt% NaCl solution. (Reproduced with permission from Pirhady Tavandashti et al. (2016) © 2016 Elsevier Ltd.)

Figure 9.6 represents bode and phase angle plots of the different samples investigated. The unloaded epoxy ester coating as the control, empty PANI capsule loaded coating, MBT dispersed coating, and MBT loaded PANI in epoxy ester coating immersed in the corrosive salt solution (0.3 wt% NaCl solution) for up to 7 days. The impedance values at the lower frequency domain were an order of magnitude more significant for the encapsulated inhibitor-loaded coatings compared to the control coating after 7 days. The phase angle plots show the corresponding time constants of the sample coatings in response to the electrochemical reactions. Oxide layers were formed, but the highest resistance was due to the inhibitor-capsule-based coating, whose oxide resistance increased with immersion time. As a result, the defect was covered with a passive layer that prevented further penetration of the corrosive species.

The salt spray test is widely used for industrial corrosion analysis, especially for long hours of testing in very harsh salt fog conditions up to 5 wt% NaCl. Immersion times from 2 h extending up to 24, 150, 1000, 3700 h, etc., (Guin et al. 2014; Kopeć et al. 2015; Grigoriev et al. 2017) have been employed to test the efficiency of the anticorrosive coatings.

Table 9.1 presents selected examples of polymeric capsules as nanocontainers in anticorrosive surface coatings.

9.6 POLYMERIC MICROCAPSULES IN SURFACE COATINGS

There are several works available on polymeric capsules at microscales that are not discussed above. Polymeric microcapsules are mostly spherical particles not more than 1 mm in diameter, irrespective of the synthesis route. Templated and template-free methods are also used to prepare the microcapsules giving rise to a modifiable particle consisting of an inner core and an outer shell. As usual, the core template is etched out for templated synthesis, allowing the hollow for active substance(s) to be encapsulated for diverse applications. Template-free methods include modified self-emulsification (Yu et al. 2011), self-assembly (Discher et al. 1999), and so on. The factors that influence the formation and morphology of microcapsules include the ratio of the core–shell material, the agitation rate during the synthesis, the temperature of the suspended solution, and the chemical properties of the shell (Koh and Park 2017).

Microcapsules have encapsulated corrosion inhibitors and other relevant additives for corrosion control of metals and alloys exposed to aggressive environments. The microcapsules can also be

TABLE 9.1

Selected Examples of Polymeric Capsules as Nanocontainers in Anticorrosive Surface Coatings

Active Substance(s) in the Core	Polymeric Shell	Nanocapsule Synthesis Route	Capsule Size (nm)	Type of Coating	Ref.
Methyl diphenyl diisocyanate	Urea formaldehyde	In situ oil-in-water emulsion polymerization	100–800	Sol-gel	Guin et al. (2014)
Mercaptobenzo-thiazole (MBT)	Polyaniline (PANI)	Soft template method	-	Epoxy ester	Pirhady Tavandashti et al. (2016)
Ethanolamine, Diethanolamine, Triethanolamine, Propylamine, Dipropylamine, 5-Amino-1-pentanol	Poly(methyl methacrylate -co- methacrylic acid -co- butyl acrylate) (inner shell) and polystyrene (outer shell)	Multi-stage emulsion polymerization	100	Polyurethane	Choi et al. (2013)
Neem oil	Urea formaldehyde	In situ oil-in-water emulsion polymerization	65	Epoxy polyamide	Bagale et al. (2018)
2-Methylbenzothiazole (BT) and 2-Mercaptobenzothiazole (MBT)	Polyelectrolyte -Poly(diallyldimethyl ammonium chloride) and poly(sodium 4-styrenesulfonate) (polycation–polyanion pair)	-	100	Water-based epoxy	Kopeć et al. (2015)
Quaternized alkyl pyridine	Polypyrrole/sodium dodecyl benzene sulfonate/ quaternized alkyl pyridine	In situ chemical polymerization	-	Epoxy	Javadian et al. (2020)
Benzotriazole (BTA)	Polystyrene capsule and additional polyethylenimine adsorbed layer to control spontaneous leakage	-	74	Epoxy and Sol-gel	Li et al. (2014)

conveniently tuned to have functional properties such as photochromic (Hu et al. 2016), thermo-chromic (Zhu et al. 2018; Pedaballi et al. 2019), and pH sensitivity (Kanellopoulos et al. 2017), to mention a few. Mohammadloo et al. investigated the anticorrosion performance of linseed oil, 8-hydroxyquinolein (8-HQ), and cerium acetate as active cores encapsulated in urea formaldehyde microcapsules and blended into epoxy coatings. Mixed inhibitors loaded in the microcapsules (8-HQ+Ce) and dispersed in the epoxy coating revealed enhanced anticorrosion performance for mild steel in 3.5 wt% NaCl after 168 h of immersion. Specifically, no remarkable decrease in imped-ance at 0.01 Hz was observed. The impedance value after 1 h was 8.62×10^9 ohm. cm^2, and after 168 h, it was 1.98×10^9 ohm. cm^2 at 0.01 Hz (Eivaz Mohammadloo et al. 2019).

Koh & Park examined the coating performance of triazole- and oleate-derivative-based corro-sion inhibitors-loaded microcapsules dispersed in commercial, automotive coating under salt spray test for a maximum of 35 days. The loaded commercial-grade polyurethane microcapsules (15 wt%) spread in a commercial primer had an automotive surface coating applied over the primer. Defects were induced in the coatings under examination, and the degree of rust was less in the inhibitor-loaded coatings than in the unloaded coatings (Koh and Park 2017).

Zhang et al. evaluated the inhibiting role of a room-temperature curable bio-based epoxy ester. The microcapsule was a poly(urea formaldehyde)-epoxy ester, and the epoxy-amine coating was used. The resistive behavior at 0.01 Hz of the samples was evaluated for 14 days in 3.5 wt% NaCl. The neat defect-free coating and the microcapsule-loaded scratched coating were evaluated with EIS and salt fog test. The impedance value at 0.01 Hz for the loaded coating with an induced defect was 2.97×10^{11} Ohm cm^2, and after 7 days reduced to 3.81×10^7 Ohm cm^2, while the control coating was two orders of magnitude less. After 14 days, the new film formed in the defective region pro-duced a second semicircle attributed to possible weak interactions and the inability to protect the underlying substrate completely. Although the protective performance of the modified coating was better than the unmodified coating (Zhang et al. 2018).

Another study on urea formaldehyde microcapsule was reported by Neto et al. The active core was tung oil encapsulated in the microcapsule for self-healing of damaged commercial alkyd coat-ing (MCOF) for the protection of carbon steel. The induced defect caused the microcapsule to rupture, releasing the tung oil after 168 h of immersion in 3.5 wt% NaCl. The effect of the tung oil released from the modified commercial alkyd coating was compared with commercial alkyd coating modified with zinc phosphate (TCF) and also compared with the unmodified commercial alkyd coating (TSF). 1.95 wt% each of tung oil and zinc phosphate were used for modifying the alkyd coating. MCOF provided the highest impedance amongst the examined coatings implying electrolyte penetration difficulty due to the new protective tung oil-based film within the defect. The prepared microcapsules were stable for up to 60 days in the alkyd diluent, suggesting industrial viability (Cordeiro Neto et al. 2020).

9.7   CONCLUSIONS AND PERSPECTIVES

Research on polymer capsules as nanocontainers is a very attractive field and is still undergoing a lot of innovative development. The application in surface coatings is also vast as surface coatings are of various kinds and for diverse functions. In this chapter, emphasis is on the application of polymeric nanocapsules in anticorrosive coatings but anticorrosive coatings are also a constantly developing research area as emerging coatings have displayed multiple functions. Therefore, a quick mention of a few other applications of polymeric nanocapsules could not be overlooked. Similarly, a brief discussion on microcapsules is not neglected either. Polymer nanocapsules in anticorrosive surface coatings are essential for carrying and preserving the active inhibitive sub-stances and dyes which have major functions during coating damage and repair. This ensures that defects in coatings do not propagate as ordinarily expected. Furthermore, the encapsulated active substances can be tailored to provide effective antimicrobial and antifouling properties in the surface coatings.

ACKNOWLEDGMENT

The first author would like to appreciate every encouragement from Dr. Onwubiko Ngozi Dike toward the completion of the work.

REFERENCES

Bagale UD, Sonawane SH, Bhanvase BA, Kulkarni RD, Gogate PR. 2018. Green synthesis of nanocapsules for self-healing anticorrosion coating using ultrasound-assisted approach. *Green Process Synth.* 7(2):147–159. https://www.degruyter.com/document/doi/10.1515/gps-2016-0160/html.

Bahmani B, Gupta S, Upadhyayula S, Vullev VI, Anvari B. 2011. Effect of polyethylene glycol coatings on uptake of indocyanine green loaded nanocapsules by human spleen macrophages in vitro. *J. Biomed. Opt.* 16(5):051303. http://biomedicaloptics.spiedigitallibrary.org/article.aspx?doi=10.1117/1.3574761.

Bentz KC, Savin DA. 2018. Hollow polymer nanocapsules: synthesis, properties, and applications. *Polym. Chem.* 9(16):2059–2081.

Choi H, Kim KY, Park JM. 2013. Encapsulation of aliphatic amines into nanoparticles for self-healing corrosion protection of steel sheets. *Prog. Org. Coat.* 76(10):1316–1324. https://www.sciencedirect.com/science/article/pii/S0300944013000982.

Cordeiro Neto AG, Pellanda AC, de Carvalho Jorge AR, Floriano JB, Coelho Berton MA. 2020. Preparation and evaluation of corrosion resistance of a self-healing alkyd coating based on microcapsules containing Tung oil. *Prog. Org. Coat.* 147:105874. https://doi.org/10.1016/j.porgcoat.2020.105874.

Cui J, van Koeverden MP, Müllner M, Kempe K, Caruso F. 2014. Emerging methods for the fabrication of polymer capsules. *Adv. Colloid Interface Sci.* 207(1):14–31. http://dx.doi.org/10.1016/j.cis.2013.10.012.

Deng S, Gigliobianco MR, Censi R, Di Martino P. 2020. Polymeric nanocapsules as nanotechnological alternative for drug delivery system: current status, challenges and opportunities. *Nanomaterials* 10(5):847. https://www.mdpi.com/2079-4991/10/5/847.

Discher BM, Won YY, Ege DS, Lee JCM, Bates FS, Discher DE, Hammer DA. 1999. Polymersomes: tough vesicles made from diblock copolymers. *Science* 284(5417):1143–1146. https://science.sciencemag.org/content/284/5417/1143.

Eivaz Mohammadloo H, Mirabedini SM, Pezeshk-Fallah H. 2019. Microencapsulation of quinoline and cerium based inhibitors for smart coating application: anti-corrosion, morphology and adhesion study. *Prog. Org. Coat.* 137:105339. https://doi.org/10.1016/j.porgcoat.2019.105339.

Gagliardi A, Giuliano E, Venkateswararao E, Fresta M, Bulotta S, Awasthi V, Cosco D. 2021. Biodegradable polymeric nanoparticles for drug delivery to solid tumors. *Front. Pharmacol.* 12:1–24. https://www.frontiersin.org/articles/10.3389/fphar.2021.601626/full.

Grigoriev D, Shchukina E, Shchukin DG. 2017. Nanocontainers for self-healing coatings. *Adv. Mater. Interfaces* 4(1):1600318. http://doi.wiley.com/10.1002/admi.201600318.

Grillo R, Rosa AH, Fraceto LF. 2014. Poly(ε-caprolactone) nanocapsules carrying the herbicide atrazine: effect of chitosan-coating agent on physico-chemical stability and herbicide release profile. *Int. J. Environ. Sci. Technol.* 11(6):1691–1700. http://link.springer.com/10.1007/s13762-013-0358-1.

Guin AK, Nayak S, Bhadu MK, Singh V, Rout TK. 2014. Development and performance evaluation of corrosion resistance self-healing coating. *ISRN Corros.* 2014:1–7 https://www.hindawi.com/archive/2014/979323/.

Hu L, Lyu S, Fu F, Huang J. 2016. Development of photochromic wood material by microcapsules. *BioResources* 11(4):9547–9559. http://ojs.cnr.ncsu.edu/index.php/BioRes/article/view/8390.

Huang X, Voit B. 2013. Progress on multi-compartment polymeric capsules. *Polym. Chem.* 4(3):435–443. http://xlink.rsc.org/?DOI=C2PY20636F.

Javadian S, Ahmadpour Z, Yousefi A. 2020. Polypyrrole nanocapsules bearing quaternized alkyl pyridine in a green self-healing coating for corrosion protection of zinc. *Prog. Org. Coat.* 147:105678. https://linkinghub.elsevier.com/retrieve/pii/S0300944019316947.

Kanellopoulos A, Giannaros P, Palmer D, Kerr A, Al-Tabbaa A. 2017. Polymeric microcapsules with switchable mechanical properties for self-healing concrete: synthesis, characterisation and proof of concept. *Smart Mater. Struct.* 26(4):045025. https://iopscience.iop.org/article/10.1088/1361-665X/aa516c

Kasper SH, Hart R, Bergkvist MA, Musah RC, Cady N. 2016. Zein nanocapsules as a tool for surface passivation, drug delivery and biofilm prevention. *AIMS Microbiol.* 2(4):422–433. http://www.aimspress.com/article/10.3934/microbiol.2016.4.422.

Khalil I, Burns A, Radecka I, Kowalczuk M, Khalaf T, Adamus G, Johnston B, Khechara M. 2017. Bacterial-derived polymer poly-y-glutamic Acid (y-PGA)-based micro/nanoparticles as a delivery system for antimicrobials and other biomedical applications. *Int. J. Mol. Sci.* 18(2):313. http://www.mdpi.com/1422-0067/18/2/313.

Kim C, Karayan AI, Milla J, Hassan M, Castaneda H. 2020. Smart coating embedded with pH-responsive nanocapsules containing a corrosion inhibiting agent. *ACS Appl. Mater. Interfaces* 12(5):6451–6459. https://pubs.acs.org/doi/10.1021/acsami.9b20238.

Koh E, Park S. 2017. Self-anticorrosion performance efficiency of renewable dimer-acid-based polyol microcapsules containing corrosion inhibitors with two triazole groups. *Prog. Org. Coat.* 109:61–69. http://dx.doi.org/10.1016/j.porgcoat.2017.04.021.

Kopeć M, Szczepanowicz K, Mordarski G, Podgórna K, Socha RP, Nowak P, Warszyński P, Hack T. 2015. Self-healing epoxy coatings loaded with inhibitor-containing polyelectrolyte nanocapsules. *Prog. Org. Coat.* 84:97–106. https://linkinghub.elsevier.com/retrieve/pii/S0300944015000594.

Kothamasu P, Kanumur H, Ravur N, Maddu C, Parasuramrajam R, Thangavel S. 2012. Nanocapsules: the weapons for novel drug delivery systems. *BioImpacts* 2(2):71–81. /pmc/articles/PMC3648923/.

Larrañaga A, Lomora M, Sarasua JR, Palivan CG, Pandit A. 2017. Polymer capsules as micro-/nanoreactors for therapeutic applications: current strategies to control membrane permeability. *Prog. Mater. Sci.* 90:325–357 https://linkinghub.elsevier.com/retrieve/pii/S0079642517301019.

Li GL, Schenderlein M, Men Y, Möhwald H, Shchukin DG. 2014. Monodisperse polymeric core-shell nanocontainers for organic self-healing anticorrosion coatings. *Adv. Mater. Interfaces* 1(1):1300019. http://doi.wiley.com/10.1002/admi.201300019.

Liu J, Fan X, Xue Y, Liu Y, Song L, Wang R, Zhang H, Zhang Q. 2018. Fabrication of polymer capsules by an original multifunctional, active, amphiphilic macromolecule, and its application in preparing PCM microcapsules. *New J. Chem.* 42(8):6457–6463. http://xlink.rsc.org/?DOI=C8NJ00546J.

Mora-Huertas CE, Fessi H, Elaissari A. 2010. Polymer-based nanocapsules for drug delivery. *Int. J. Pharm.* 385(1–2):113–142. https://linkinghub.elsevier.com/retrieve/pii/S0378517309007273.

Pedaballi S, Li C-C, Song Y-J. 2019. Dispersion of microcapsules for the improved thermochromic performance of smart coatings. *RSC Adv.* 9(42):24175–24183. http://xlink.rsc.org/?DOI=C9RA04740A.

Pirhady Tavandashti N, Ghorbani M, Shojaei A, Mol JMC, Terryn H, Baert K, Gonzalez-Garcia Y. 2016. Inhibitor-loaded conducting polymer capsules for active corrosion protection of coating defects. *Corros. Sci.* 112:138–149. http://dx.doi.org/10.1016/j.corsci.2016.07.003.

Pitakchatwong C, Schlegel I, Landfester K, Crespy D, Chirachanchai S. 2018. Chitosan nanocapsules for pH-triggered dual release based on corrosion inhibitors as model study. *Part Part Syst. Charact.* 35(7):1800086. http://doi.wiley.com/10.1002/ppsc.201800086.

Poletto FS, Beck RCR, Guterres SS, Pohlmann AR. 2011. Polymeric nanocapsules: concepts and applications. In *Nanocosmetics and Nanomedicines*, Springer Berlin Heidelberg, p. 49–68. http://link.springer.com/10.1007/978-3-642-19792-5_3.

Qian B, Song Z, Hao L, Wang W, Kong D. 2017. Self-healing epoxy coatings based on nanocontainers for corrosion protection of mild steel. *J. Electrochem. Soc.* 164(2):C54–C60. https://iopscience.iop.org/article/10.1149/2.1251702jes.

Shakiba S, Astete CE, Paudel S, Sabliov CM, Rodrigues DF, Louie SM. 2020. Emerging investigator series: polymeric nanocarriers for agricultural applications: synthesis, characterization, and environmental and biological interactions. *Environ. Sci. Nano* 7(1):37–67. http://xlink.rsc.org/?DOI=C9EN01127G.

Ulaeto SB, Nair AV, Pancrecious JK, Karun AS, Mathew GM, Rajan TPD, Pai BC. 2019. Smart nanocontainer-based anticorrosive bio-coatings: evaluation of quercetin for corrosion protection of aluminium alloys. *Prog. Org. Coat.* 136:105276. https://linkinghub.elsevier.com/retrieve/pii/S0300944019301523.

Yu X, Zhao Z, Nie W, Deng R, Liu S, Liang R, Zhu J, Ji X. 2011. Biodegradable polymer microcapsules fabrication through a template-free approach. *Langmuir* 27(16):10265–10273. https://pubs.acs.org/doi/abs/10.1021/la201944s.

Zarket BC, Raghavan SR. 2017. Onion-like multilayered polymer capsules synthesized by a bioinspired inside-out technique. *Nat. Commun.* 8(1):193. http://dx.doi.org/10.1038/s41467-017-00077-7.

Zhang C, Wang H, Zhou Q. 2018. Preparation and characterization of microcapsules based self-healing coatings containing epoxy ester as healing agent. *Prog. Org. Coat.* 125:403–410. https://doi.org/10.1016/j.porgcoat.2018.09.028.

Zhu X, Liu Y, Li Z, Wang W. 2018. Thermochromic microcapsules with highly transparent shells obtained through in-situ polymerization of urea formaldehyde around thermochromic cores for smart wood coatings. *Sci. Rep.* 8(1):4015. http://dx.doi.org/10.1038/s41598-018-22445-z.

Zielińska A, Carreiró F, Oliveira AM, Neves A, Pires B, Venkatesh DN, Durazzo A, Lucarini M, Eder P, Silva AM, et al. 2020. Polymeric nanoparticles: production, characterization, toxicology and ecotoxicology. *Molecules* 25(16):3731. http://www.mdpi.com/1422-0067/18/2/313.

10 Cyclodextrins in Surface Coating Applications

Puspalata Rajesh and T.V. Krishna Mohan
BARC Facilities

CONTENTS

10.1 INTRODUCTION

Supramolecular compounds can be classified into (i) molecular self-assemblies, (ii) host–guest complexes, and (iii) molecules built into specific shapes (Saji 2019). Cyclodextrin (CD), a supramolecule, was first discovered by Villiers in 1891 as "cellulosine" (Szejtli 1988). They are natural molecules obtained from the enzymatic degradation of starch. The nature and type of resulting CDs depend on the degradation reaction conditions, the nature of starch and the transferase enzyme used. The typically used enzyme is cyclodextrin glucosyltransferase (Jin 2013). CDs are crystalline, water-soluble, non-reducing cyclic oligosaccharides (D-glucopyranose or α–1,4-glycosidic linkages) connected edge to edge. Depending on the number of glucose units, CDs are denoted as α, β, and γ (α-6, β-7, and γ-8), and they are the three naturally occurring forms identified and isolated by Schardinger (Figure 10.1). Therefore, CDs are also known as "Schardinger dextrins" (Jin 2013; Crini 2014).

CDs, having fewer than six glucose residues, do not exist, probably because of steric hindrance (Sundararajan and Rao 1970). CDs with more than eight units are denoted as large CDs (Ueda 2002). The most attractive feature of CD is the formation of a self-assembled ring or toroid-shaped hydrophobic cavity due to the ring of glucosidic oxygen and C–H groups. The smaller and the larger openings of the toroid are exposed to the solvent by primary and secondary hydroxyl groups, respectively. The exterior of the torus structure is hydrophilic. The cavity mimics nature and is the basis for creating biomimetic functional architectures by the formation of inclusion complexes with various compounds (guests) in the truncated cone-shaped CD (host). The guest compounds range from polar substances such as acids, amines, and halogens to highly apolar aliphatic and aromatic hydrocarbons, and even rare gases (Crini 2014). The driving forces for this host–guest complex formation are the release of enthalpy-rich water molecules from the cavity and the interactions are mainly van der Waals forces, dipole–dipole interactions, charge–transfer interactions, hydrogen bonding, and solvent effects. The bond strength of these intermolecular bonds is weaker than that of a normal covalent bond (Schubert et al. 2003). The stability of the complexes depends on the shape and size of the guests and how well they fit inside the CD cavity.

Due to weak interactions, these inclusion complex formations are reversible equilibrium reactions and play a critical role in their application. The schematic for different stoichiometric complex formations is depicted in Figure 10.2. Usually, the host to guest ratio in encapsulated complexes is

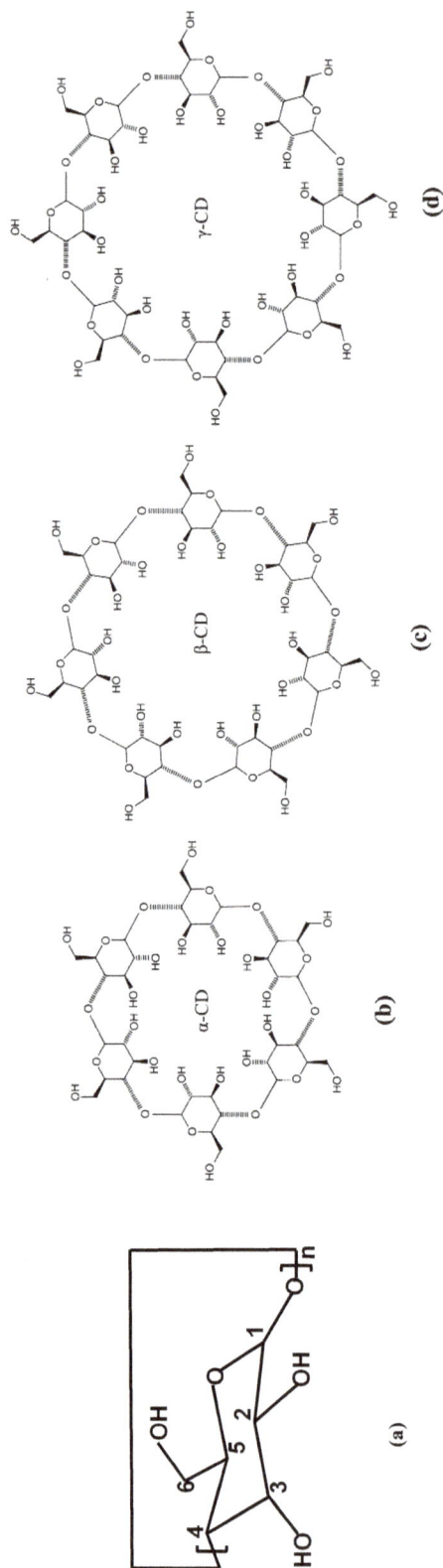

FIGURE 10.1 Chemical structure of the three main types of naturally occurring CDs; (a) general formula (b) α-CD ($n=6$), (c) β-CD ($n=7$) and (d) γ-CD ($n=8$). (Reproduced from Cyclodextrin.svg - Wikimedia Commons. © Stanisław Skowron, under Creative Commons Licence.)

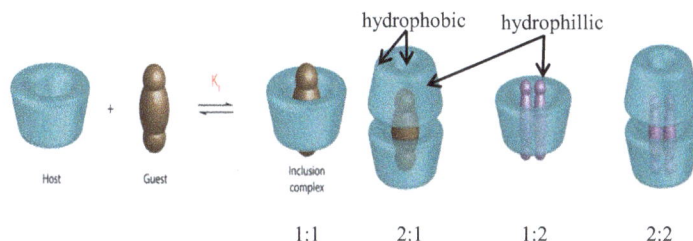

hydrophobic   hydrophillic

Host   Guest   inclusion complex

1:1   2:1   1:2   2:2

FIGURE 10.2 Schematic illustrations for the host (CD)–guest concept utilised in supramolecular chemistry of CDs. (Reproduced with permission from Crini et al. (2018) © Springer International Publishing AG, part of Springer Nature 2018.)

1:1 in dilute solutions. However, nonstoichiometric 2:1, 1:2, 2:2 or even higher-order equilibrium exists at higher concentrations or with specific guests or in the solid-state (Szejtli 1988). K_f is the stability constant for inclusion in complex formation. The higher the K_f value, the more stable the inclusion complex is, and the less dissociation occurs. The properties of α-, β-, and γ-CDs are given in Table 10.1.

The most widely investigated CD is the β-CD due to its low cost and favourable properties (Szejtli 1998; Uekama et al. 1998; Chernykh and Brichkin 2010). Since β-CD is sparingly soluble, it can be easily separated and purified by crystallisation, whereas the more soluble α- and γ- CDs require expensive and time-consuming chromatographic techniques for their separation. CD polymers are cheap in cost and recognised as safe by Food and Drug Administration (FDA) (no daily intake limit for α- and γ-). In pure aqueous solutions (concentrations >1% w/v), CD molecules self-assemble to form nanoparticles (diameter ~20–200 nm) and the aggregation increases with increasing concentration (Loftsson et al. 2004; He et al. 2008). These aggregates have structures similar to micelles and can dissolve water-insoluble lipophilic molecules. The complex aggregates of CDs with drugs or corrosion inhibitors are at nano/microscale and are held by weak hydrogen bonds and hydrophobic forces. These aggregates dissociate readily upon dilution or under suitably controlled conditions.

The supramolecular chemistry of CDs has opened up a vast area of multidisciplinary research and applications. These remarkable encapsulation properties can modify and/or improve the physical, chemical, and/or biological characteristics of the guest molecule (Crini et al. 2018; Szejtli 1988).

TABLE 10.1

Properties of α-, β-, and γ CDs

Property	α-CD	β-CD	γ-CD
Number of glucopyranose units	6	7	8
Molecular weight (g/mol)	972	1135	1297
Solubility in water at 25°C (w/v%)	14.5	1.85	23.2
Outer diameter (Å)	14.7	15.3	17.5
Cavity diameter (Å)	5.1	6.2	8.1
Height of torus (Å)	7.8	7.8	7.8
Cavity volume (Å^3)	174	262	427
Melting range (°C)	255–260	255–265	240–245
Water of crystallisation	10.2	13–15	8–18
Water molecules in cavity	6	11	17
Price (US$/g pharma-grade)	1.0	0.025	0.8

Source: Szejtli (1988), Amiri and Rahimi (2016).

Complexation with dextrins enhances the solubility of poorly soluble substances in water, protects air-sensitive substances against atmospheric oxidation, masks unwanted physiological effects and reduces the loss of highly volatile substances. Applications of CDs/modified CDs are reported in the field of catalysis of reactions (Bender and Komiyama 1978; Szejtli 1988), and soil remediation to remove organic pollutants and dyeing fabrics (Hedges 1998). Due to their excellent biocompatibility, they are used in biomedical applications like solubilising non-polar hydrophobic drugs as nano reservoirs, thereby increasing bioavailability and preventing aggregation (stability) in aqueous mixtures (Szejtli 1988; Hedges 1998; Uekama et al. 1998). They are applied as implantable drug delivery depots, based on the unique ability of CD subunits to reversibly bind and release drug compounds through non-covalent interactions. Non-inclusion complexes formed by the interaction of the outer hydroxyl groups are also used in pharmaceutical applications (Bilensoy 2011; Kurkov and Loftsson 2013).

Extensive work has been carried out since the 1970s exploring the encapsulation properties and their applications (Duchêne 1987; Szejtli 1988; Dodziuk 2006; Schneider 2012; Amiri and Amiri 2017; Fourmentin et al. 2018). The design and synthesis of these "molecular machines" formed the basis for two Noble Prizes in chemistry for the years 1987 and 2016. These recognitions have generated a renewed interest in host–guest chemistry. Investigations on CDs have broken the entire boundary between different disciplines, such as chemistry, biochemistry, biology, health science, and agriculture. CDs in both food and cosmetics are for the molecular encapsulation of flavours and fragrances. It has forced scientists to work together to discover its new theoretical and practical concepts, impacts, and applications in many sectors. A simulated approach of controlled drug release (Uekama et al. 1998) in drug delivery systems is applied to organic molecules having inhibition efficiency by entrapping or complexing them in CD. They can be effectively used for the immobilisation of various species for their potential application as biofouling and anti-corrosion coatings. In this chapter, the surface coating applications of inclusion complexes of CDs are discussed.

10.2 CDs IN SURFACE COATINGS

The surface of a material exhibits a critical role in deciding its performance and success in its application in medical and non-medical fields. Medical devices including catheters, prosthesis, vascular grafts, and bone implants benefit from surface grafting of CDs. In the non-medical sector, the surface of products can impact their commercial value from their appearance, and operational durability (Learn et al. 2020a). Various reported applications of CD-based surface coatings include sustained delivery of antibiotics (Thatiparti et al. 2010; Cyphert et al. 2017), drugs (Vermet et al. 2014; Haley et al. 2019; Chai et al. 2019), pesticides (Romi et al. 2005), and fragrances (Abdelkader et al. 2019) and efficient uptake and retention of pollutants (Martel et al. 2002; Wang et al. 2020) or dyes (Ghoul et al. 2010).

Reports on short-chain dextrins and CDs as corrosion inhibitors have been available since the 1970s on titanium and aluminium (Shibad and Balachandra 1976; Talati and Modi 1976). Inhibitors with high solubility cannot provide long-term active corrosion protection of the coated alloy because of the very fast release of the inhibitor from the coating. Moreover, the osmotic pressure can cause the formation of blisters and delaminations of the coating during immersion, which can significantly affect the coating integrity and corrosion protection. Therefore, inhibitive compounds with low solubility might be preferable candidates for incorporation into coatings. CDs are important in this direction.

10.2.1 CDs IN ANTI-CORROSION COATINGS

Cost-effective, uniform, and adherent coatings are being utilised to control corrosion problems. Protective surface coatings prevent corrosion of the underlying substrate by passive protection (purely physical barrier protection), active protection (corrosion inhibitors are incorporated into the coating), or by both active and passive methods (Saji and Thomas 2007; Montemor 2015). Encapsulated corrosion inhibitors in the form of their inclusion complexes with CDs are expected to be more easily trapped within the cross-linked nano-porous coating material, making the inhibitor more difficult to leach out and thus prolonging the inhibition effect. Such smart coatings combine both active coatings and passive protection by releasing inhibitors on demand (Plawecka et al. 2014; Maia et al. 2016; Latnikova et al. 2011). Typically, organic corrosion inhibitors (aromatic and heterocyclic compounds) are suitable candidates for inclusion in complex formation with CDs (Rekharsky and Inoue 1998).

Khramov et al. introduced corrosion inhibitors, mercapto benzo thiazole (MBT) and 2-mercapto benzimidazole (MBI) to the sol–gel formulations as inclusion host/guest complexes with β-CD $(2.6 \times 10^{-3}$ M) for corrosion protection of AA2024 Al alloy (Khramov et al. 2004). The formation constants (log K) of the inclusion complexes of β-CD with MBT and MBI determined through a spectrophotometric method were 2.10 ± 0.07 and 2.17 ± 0.06, respectively. Their defect healing studies by scanning vibrating electrode technique (SVET) showed that the films doped with MBT and MBI provided superior corrosion protection when compared with the undoped ones. Moreover, the coatings doped with entrapped inhibitors outperformed those made by simple addition, due to slow, prolonged release of the inhibitor at the corrosion sites followed by self-healing as observed from the current density map. The defect healing ability was further improved when the hybrid coating was followed by an organic topcoat. Their study also revealed that ionisable inhibitors, such as mercapto benzimidazole sulfonate and thio-salicylic acid, were too strongly attached to the sol–gel matrix and were not released during corrosion and hence displayed far weaker inhibition effects than the non-ionisable inhibitors (MBT and MBI) (Khramov et al. 2005). These studies inspired extensive systematic screening of several CDs with different inhibitors.

Fan et al. investigated the corrosion inhibition properties of supramolecular host–guest complexes of β-CD/hydroxypropyl-β-CD (higher soluble derivative) and octadecylamine (ODA) on mild steel in boiler condensate water (Fan et al. 2014a, b). The complex changed from 1:1 to 2:1 at high β-CD concentrations (>4 mM). XRD and SEM results showed the formation of a monomolecular

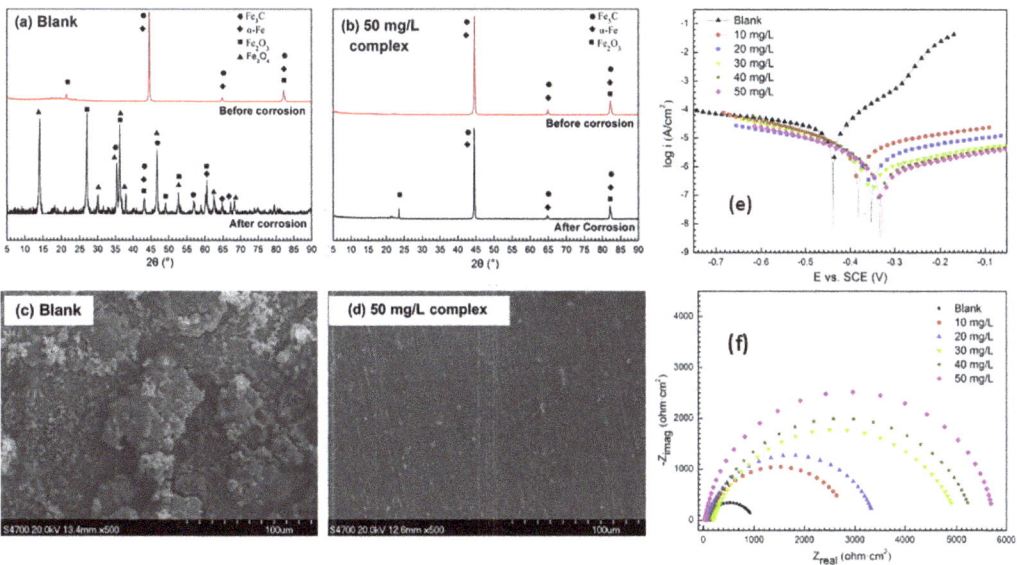

FIGURE 10.3 XRD and SEM studies of carbon steel without (a, c) and with (b, d) inhibitor addition in condensate water. (e) Polarisation curves and (f) EIS plots recorded as a function of inhibitor concentration $(40 \pm 5°C$, pH 5.68 ± 5). (Reproduced with permission from Fan et al. (2014a) © 2014 Elsevier Ltd.)

hydrophobic surface layer (1.8 nm thickness) by releasing ODA molecules and providing a protective effect (Figure 10.3).

When the bare sample revealed severe uniform corrosion (Figure 10.3c) with the appearance of rust (Fe_3O_4 and γ-Fe_2O_3 phases) in the XRD (Figure 10.3a), the sample immersed in the 50 mg/L ODA solution showed a smooth surface (Figure 10.3d). The inhibitor polarised both the anodic and the cathodic curves, indicating a mixed protection mechanism. The polarisation (Figure 10.3e) and Nyquist (Figure 10.3f) plots showed an increase in inhibition efficiency with an increase in inhibitor concentration. The capacitive behaviour in electrochemical impedance spectroscopy (EIS) and the mixed inhibition by polarisation suggest that the inhibitor works by surface adsorption. Similar slow release and enhanced inhibition efficiency of the inclusion complexes of β-CD has been reported in various applications, such as the oil and gas sectors (Zou et al. 2011; He et al. 2014). The corrosion rate of carbon steels in HCl was significantly reduced using a chitosan modified-β-CD coating (Liu et al. 2015).

In all the above cases, complexation with CDs conferred only a slow, steady release of the guest molecule but not a "delivery on-demand" or response to any external stimuli. However, Amiri and Rahimi (Amiri and Rahimi 2016, 2014; Rahimi and Amiri 2016, 2015) used α, β, and γ forms of CD-based inclusion complex containing MBT/MBI, to create on-demand release anti-corrosion coatings where the hydrophobic cavity of the CD has encapsulated the benzyl part of the MBT/MBI. XRD studies indicated the formation of head-to-head channel-type and cage-type crystal structures for α-CD/MBT and α-CD/MBI inclusion complexes, respectively. The results are summarised in Figure 10.4. The SEM and TEM images of the coating containing corrosion inhibitor and γ-CD nanocontainer via inclusion complex formation are shown in Figure 10.4a and b. In γ-CD as the cavity size was bigger, the inhibitor solubility was more and uniformly dispersed smaller nanocapsules (~40–80 nm range) were observed. Corrosion initiation of the coated 2024 Al substrate occurred after 6 days of salt spray on the bare sample and increased with an increase in exposure time. However, the samples with the nano-container-based coatings containing corrosion inhibitors did not rust even after 1000 h. Figure 10.4c shows the typical result of the salt spray test, which was supported by their electrochemical measurements of 5% NaCl. The amount of drug released (Figure 10.4d) was measured from the absorbance at 237 and 239 nm for MBT and MBI, respectively. It showed the controlled release of inhibitors from the nanocontainers (formed under

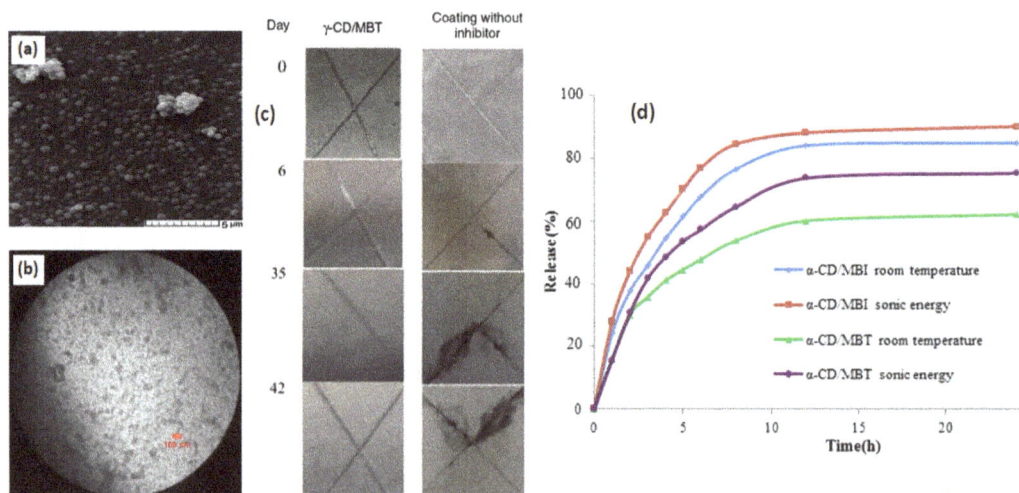

FIGURE 10.4 SEM (a) and TEM (b) image of γ-CD/MBT. Results of the 1000-h salt spray test (c). Release profile from coated Al alloy (d). ((a–c) Reproduced with permission from Amiri and Rahimi (2016) © 2016, American Coatings Association. (d) Reproduced with permission from Rahimi and Amiri (2015) © 2014, Springer Science Business Media Dordrecht.)

various conditions) over time. This was due to the reversibility of the hydrogen bonds associated with the formation of inclusion complexes between α-CDs and the inhibitors (Rahimi and Amiri 2015; Amiri and Rahimi 2016).

Studies on polymer coatings of β-CD[MBT] in polyvinyl butyral (PVB) showed passivation of Zn in KCl (Altin et al. 2017, 2018, 2019). The β-CD[MBT] containing PVB was applied on top of metallic Zn and the progress of corrosion was monitored. Initially, cathodic delamination slowed down with time, and the potential of the defect as measured by a scanning Kelvin probe (SKP) moved away from the potential of corroding Zn towards the potential of passive Zn. Consequently, a defect that was corroding initially was healed due to the inhibitor release. The presence of β-CD increased the solubility of the corrosion inhibitor and the coating facilitates healing by increasing accessibility of the inhibitor to the defective, corroding surface. If the film can be held in equilibrium with monomeric MBT in solution, i.e. if MBT remains present in solution, the film will reform in case of mechanical damage inflicted on the surface (Chen and Erbe 2018). If an external effect drives the electrode potential up, the film itself will be oxidised to 2,2′-dibenzothiazole disulphide (DBTA) which is the oxidative dimer of MBT, rather than oxidising the metal. DBTA as a dimer is still able to participate in the formation of the complex coordination polymer type inhibiting film. In this situation, the MBT film acts as a "sacrificial inhibitor," as it is oxidised instead of metal. The self-healing function for polymer-based materials can be achieved in two main ways: (i) reversible cross-linking of the broken chemical bonds and (ii) repair of the defect by the healing agent released from micro-nano-reservoirs.

However, a proper trigger can affect the inhibitor/sol–gel interactions and therefore provide the release of the inhibitor. In several cases, the release of organic molecular species from the hybrid sol–gel matrix can be described by the pH-dependent triggered release mechanism, which often implies protonation or de-protonation of the nitrogen atom in functional groups (Vreugdenhil and Woods 2005). When the ζ-potential of the sol–gel matrix becomes comparable with the charge of the organic molecule, the inhibitor is 'pushed' out of the coating. Such triggering of the desorption processes can provide an 'intelligent' release of the corrosion inhibitor only in places of local pH changes originating from localised corrosion processes (Khramov et al. 2005). In another case, a functional nano-reservoir based on multi-walled carbon nanotubes (MWCNTs) surface mounted with β-CDs showed anti-corrosion performance and self-healing capabilities in NaCl solution after loading with benzimidazole inhibitor (He et al. 2016).

10.2.2 CDs in Anti-Biofouling Coatings

CDs are widely reported as a host for biocide encapsulation, improving physicochemical properties like aqueous solubility, decreased vapour pressure and stability (Véronique and Leclercq 2014). Controlled release of biocides by encapsulation enhances their shelf life and effectiveness for long-term biocidal activity. The increased solubility of slightly soluble biocides as inclusion complex increases the bioavailability and makes them environmentally friendly. Véronique and Leclercq have reviewed CDs in biocidal applications, highlighting several examples of biocide-CD inclusion complexes of organic, amphiphilic, and heavy-metal biocides (Véronique and Leclercq 2014). There is a clear relationship between the binding constants and the antimicrobial activity of the preservatives on different strains. If the binding constant of the inclusion complex is higher, a higher concentration of biocide is required to have the same minimal inhibition effect as a free biocide. Highly water-soluble substances showed low inactivation whereas lipophilic substances presented strong inactivation in the presence of 2-hydroxypropyl (HP)-β-CD due to the sequestration of hydrophobic biocides in CDs. Therefore, an appropriate biocide and CD must be selected to maintain an acceptable biocidal activity (Lehner et al. 1994). Though CDs show some effectiveness towards pathogenic activity, usually a higher concentration (>5 mM) is required to be maintained for efficient biocidal activity. But the virucidal activity is reported to be boosted by a synergistic effect when CDs are combined with di-n-decyl dimethyl ammonium chloride, reducing the virucide requirement (Leclercq et al. 2012, 2016).

Wen et al. have reported outstanding antimicrobial activity from a novel antimicrobial packaging material against Gram-positive and Gram-negative bacteria fabricated by the co-precipitation method by integrating cinnamon essential oil (CEO)/β-CD inclusion complex into polylactic acid nanofibre film (Wen et al. 2016). The minimum inhibitory concentration of the nanofilm against *E. coli* and *S. aureus* was approximately 1 mg/mL as compared to the equivalent CEO concentration of 11.35 µg/mL. Cai et al. developed a host–guest chemistry-based biomimetic strategy for the fabrication of antifouling Ti-oxide surfaces modified with immobilised catecholic derivatives of dopamine which act as guests. The surfaces were further functionalised with zwitterionic and hydrophilic β-CD. The zwitterionic and hydrophilic functionalised surfaces showed reduced adsorption of bovine plasma fibrinogen protein and bacterial adhesion (Cai et al. 2016).

To maximise coating uniformity and adherence of CD-based polyurethane coatings on unreceptive polypropylene substrates (which are extensively used in biomedical and non-medical commercial products), non-thermal plasma activation of polypropylene substrates was utilised (Learn et al. 2020b). The coating was first evaluated using contact angle measurement and uniformity was assessed through SEM. Adherence was evaluated as per the ASTM method of lap-shear testing and XPS was used to establish evidence of interfacial covalent bonds. An overview of this work is presented in Figure 10.5.

A decrease in contact angles from >120° to <60° after 20 min of plasma exposure suggested increased wettability of plasma on polypropylene (PP) substrates with the increase in plasma treatment duration. XPS results also revealed enhancement in the atomic % of oxygen, suggesting the presence of surface hydroxyl groups, increasing the polarity and degree of interfacial covalent bonding with the polymeric CD coating. Their study revealed that low-pressure non-thermal plasma treatment of PP surfaces for 10 min improved coating adherence regardless of the hexamethylene diisocyanate cross-linker amount.

Furthermore, owing to hydrophilic properties imparted by the intrinsically polar exterior of CD subunits, the coatings have shown potential for mitigating events of biofouling, through resistance to non-specific protein adsorption, mammalian cell adhesion, and bacterial attachment (Learn et al.

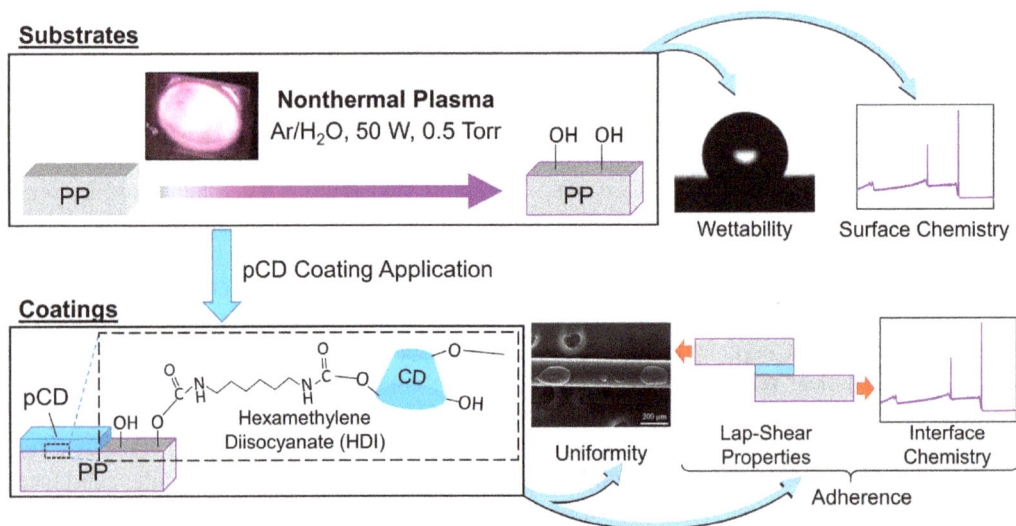

FIGURE 10.5 Effects of non-thermal plasma on polypropylene (PP) substrate were investigated in terms of wettability and surface chemistry and effects on polymerised CD (pCD) coatings were explored in terms of uniformity and adherence. (Reproduced with permission from Learn et al. (2020b) © MDPI under Creative Commons Attribution Licence.)

2020a). Polymerised CD coatings have been reported to be applied to prevent infection on polyester surgical fabrics, prosthetic meshes, and metallic orthopaedic screws (Grafmiller et al. 2016; Thatiparti et al. 2010). The affinity properties of CD polymers make them very well-suited for active anti-biofouling materials as they can release biocides over longer durations (Chen et al. 2019). As compared to purely diffusion-based active anti-biofouling materials, these encapsulated biocides can be refilled more easily once the reservoir is depleted.

CD-based coating materials may be appropriate for industrial or medical applications for which biofouling-resistant and/or drug-delivering surfaces are required. β-CD-salicylate was synthesised by trans-esterification followed by electrodeposition on the stainless steel surface by anodic electro-oxidation polymerisation of the salicylate. The corrosion resistance of β-CD-covered stainless steel demonstrated antibacterial properties for *E. coli* with antibiotic chloramphenicol drug encapsulated in the CD cavity. The drug loading was evaluated using an electrochemical quartz crystal microbalance (Bin et al. 2016).

A recent report on nano-composite coatings of β-CD grafted functionalised magnetite nanoparticles has shown better biocompatibility for in-vivo application (Agotegaray et al. 2020). An optimum system in terms of size and surface charge was obtained depending on the ratios of magnetite nanoparticles and β-CD with improved localisation of drug delivery and increased specificity in *C. elegans* worms. Figure 10.6 shows the schematic of improvement in biocompatibility of magnetic nano-composite coating with β-CD for biomedical applications.

10.2.3 CDs Coatings in Other Applications

The CD-based coatings have found applications not only in the biomedical sector, pharmacy, pharmacotherapy, biology, and biotechnology but also in the textile industry by providing clothes for transdermal delivery. The release of fragrance from laundry dryer sheets can be controlled by complexing the fragrance with CDs (Toan 1996; Schofield and Badyal 2011). The fragrance in general

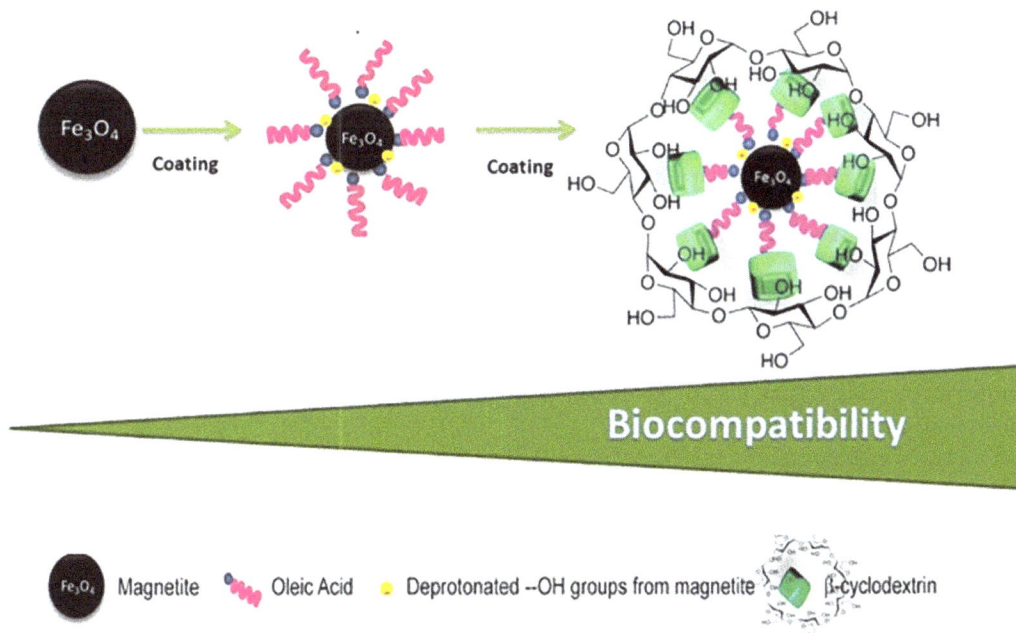

FIGURE 10.6 Biocompatibility of magnetic nano-composite coating with β-CD for biomedical applications. (Reproduced with permission from Agotegaray et al. (2020) © 2020, Springer Science Business Media, LLC, part of Springer Nature.)

is composed of many components; some are more volatile than others. Some of the highly volatile components are easily lost, but once they are complexed with CD, they are retained so that the character of the fragrance is not lost. Because of the moisture present during the drying process, some of the fragrance was released, but most of it was transferred to the fabric. When the fabric is remoistened by perspiration or other moisture, the fragrance is released from the complex to give an impression of freshness.

The release of the flavour was observed to be extended when a flavour complex was incorporated into chewing gum (Sato et al. 1992). The gum with the CD complexes containing 3-levo-menthoxy-propane-1,2-diol had a higher flavour impact and a greater sensation of coolness and a desirably firmer texture than the gum with uncomplexed methoxy-propanediol (Patel and Hvizdos 1992). The off-taste from coatings of cans was reduced or eliminated by using CDs coating extracted from methylene chloride (Bobo 1993).

Some applications of these solid inclusion complexes in the food industry can be found in the literature. For example, the incorporation of β-CD/trans-cinnamaldehyde inclusion complex into a based chitosan edible coating improved the shelf life of fresh-cut melon (Pereira et al. 2014) and papaya (Brasil et al. 2012). They can also ensure an optimal flavour and nutritional quality of fruits and vegetables until consumption. Sugammadex, the first CD derivative approved as an active pharmaceutical ingredient, is a neuromuscular reversal agent with practically no side effects (Bom et al. 2002). It has revolutionised recovery from the anaesthesia effect. Once the steroidal drugs enter the cavity of the Sugammadex molecule, their action is prevented and almost an instantaneous recovery of neuromuscular function occurs.

Recent advances in the development of innovative functionalised materials such as CD-silica/magnetite have shown an impact on analytical and environmental chemistry, catalysis, and so on. with improved binding affinities for chemical substances such as metal ions, dyes, pesticides, and drugs (Chen et al. 2014; Wang et al. 2015; Belyakova and Lyashenko 2014). While Chen et al. (2014) synthesised CD-modified cellulose nanocrystals supported by silica coating for the removal of two model drugs (procaine hydrochloride and imipramine hydrochloride), Wang et al. proposed core-shell superparamagnetic Fe_3O_4@CD composites for host–guest adsorption of polychlorinated biphenyls (Wang et al. 2015). Belyakova's group used these organo-silicates for adsorption and removal of toxic metals such as Hg(II), Cd(II) and Zn(II) and pollutants such as p-nitrophenol utilising nano-sized pores (Belyakova and Lyashenko 2014).

10.3 CONCLUSIONS AND PERSPECTIVES

A brief account has been given on the chemistry aspects behind the mechanism of employing CD-based inclusion complexes and tuning their properties suitable for specific applications, namely, corrosion and biofouling, which are of technological importance. The major strength of CD is its peculiar nature to interact non-covalently with guest molecules, while the weakness is that only molecules with suitable size, geometry, and inherent solubility can form inclusion complexes. Furthermore, scale-up of CD derivatives with high purity, yield, and cost are the limiting factors in their clinical applications.

CD-based coatings represent a useful class of emerging materials explored for wide novel applications by modifying the surfaces of the materials to achieve desired functions without compromising the bulk properties. Applications of these CD-based coatings include the unique property of sustained, controlled release, and targeted delivery of inhibitors, drugs, pesticides, or fragrances that can be stimulated.

Furthermore, owing to hydrophilic properties imparted by the intrinsically polar exterior of CD, specific formulations of polymerised CD applied as coatings have shown immense potential for mitigating biofouling through resistance to protein, mammalian cells, and bacterial adhesion. In addition, due to the weak nature of their non-covalent molecular interactions, the activation processes of these coatings are reversible so that the coating gets healed or regenerated and can be reused multiple times, thus imparting extended functionality. It is essential to have insight into different factors that can affect complex formation to formulate widely investigated CD complexes with unique and desirable properties to achieve controlled, sustained, or targeted release of the guest molecule. The prospects of CDs and their derivatives are encouraging for possible applications in new unexplored fields. Much effort is desirable for the successful translation of many of these laboratory innovations to practical field applications.

REFERENCES

Abdelkader, M. B., Azizi, N., Baffoun, A., Chevalier, Y., and Majdoub, M. 2019. "Fragrant microcapsules based on β-cyclodextrin for cosmetotextile application." *J. Renewable Mater.* 7:1347–1362. https://doi.org/10.32604/jrm.2019.07926.

Agotegaray, M., Blanco, M. G., Campelo, A., García, E., Zysler, R., Massheimer, V., Rosa, M. J. D., and Lassalle, V. 2020. "β-cyclodextrin coating: improving biocompatibility of magnetic nanocomposites for biomedical applications." *J. Mater. Sci.: Mater. Med.* 31(2):1–11. https://doi.org/10.1007/s10856-020-6361-4.

Altin, A., Krzywiecki, M., Sarfraz, A., Toparli, C., Laska, C., Kerger, P., Zeradjanin, A., Mayrhofer, K. J. J., Rohwerder, M., and Erbe. A. 2018. "Cyclodextrin inhibits zinc corrosion by destabilizing point defect formation in the oxide layer." *Beilstein J. Nanotechnol.* 9:936–944. https://doi.org/10.3762/bjnano.9.86.

Altin, A., Rohwerder, M., and Erbe. A. 2017. "Cyclodextrins as carriers for organic corrosion inhibitors in organic coatings." *J. Electrochem. Soc.* 164(4):C128–C134. https://doi.org/10.1149/2.0481704jes.

Altin, A., Vimalanandan, A., Sarfraz, A., Rohwerder, M., and Erbe. A. 2019. "Pretreatment with a β-cyclodextrin-corrosion inhibitor complex stops an initiated corrosion process on zinc." *Langmuir* 35(1):70–77. https://doi.org/10.1021/acs.langmuir.8b03441.

Amiri, S., and Amiri. S. 2017. *Cyclodextrins Properties and Industrial Applications.* Wiley. ISBN: 978-1-119-24752-4.

Amiri, S., and Rahimi, A. 2014. "Preparation of supramolecular corrosion-inhibiting nanocontainers for self-protective hybrid nanocomposite coatings." *J. Polym. Res.* 21(10):566. https://doi.org/10.1007/s10965-014-0566-5.

Amiri, S., and Rahimi, A. 2016. "Anticorrosion behavior of cyclodextrins/inhibitor nanocapsule-based self-healing coatings." *J. Coat. Technol. Res.* 13(6):1095–1102. https://doi.org/10.1007/s11998-016-9824-2.

Belyakova, L. A., and Lyashenko, D. Y. 2014. "Nanoporous functional organosilicas for sorption of toxic ions." *Russ. J. Phys. Chem. A* 88(3):489–493. https://doi.org/10.1134/S0036024414030030.

Bender, M. L., and Komiyama, M. 1978. *Cyclodextrin Chemistry.* Springer-Verlag Berlin Heidelberg. https://doi.org/10.1142/8630.

Bilensoy, E. 2011. Cyclodextrins in Pharmaceutics, Cosmetics and Biomedicine: Current and Future Industrial Applications. John Wiley, Hoboken. https://doi.org/10.1002/9780470926819.

Bin, C., Jie C., Liming, Y., Guochen, Z., and Guowei, D. 2016. "Functional β-cyclodextrin coating by electrodeposition on stainless steel for drug loading and release." *Chem. Res. Chin. Univ.* 32(2):278–283. https://doi.org/10.1007/s40242-016-5328-y.

Bobo, W. S. 1993. Interior can coating compositions containing cyclodextrins. US Patent, 5,177,129. https://patents.google.com/patent/US5177129A/en.

Bom, A., Bradley, M., Cameron, K., Clark, J. K., Egmond, J. V., Feilden, H., and MacLean, E. J., et al. 2002. "A novel concept of reversing neuromuscular block: chemical encapsulation of rocuronium bromide by a cyclodextrin-based synthetic host." *Angew. Chem. Int. Ed.* 41(2):266–270. https://doi.org/10.1002/1521-3773(20020118)41:2<265::aid-anie265>3.0.co;2-q.

Brasil, I. M., Gomes, C., Puerta-Gomez, A., Castell-Perez, M. E., and Moreira, R. G. 2012. "Polysaccharide-based multilayered antimicrobial edible coating enhances quality of fresh-cut papaya." *LWT - Food Sci. Technol.* 47(1):39–45. https://doi.org/10.1016/j.lwt.2012.01.005.

Cai, X.Y., Li, N.N., Chen, J.C., Kang, E-T., and Xu, L.Q. 2016. "Biomimetic anchors applied to the host-guest antifouling functionalization of titanium substrates." *J. Colloid Interf. Sci.* 475:8–16. https://doi.org/https://doi.org/10.1016/j.jcis.2016.04.034.

Callow, J. A., and Callow, M. E. 2011. "Trends in the development of environmentally friendly fouling-resistant marine coatings." *Nat. Commun.* 2(1):1–10. https://doi.org/10.1038/ncomms1251.

Chai, F., Maton, M., Degoutin, S., Vermet, G., Simon, N., Rousseaux, C., Martel, B., and Blanchemain, N. 2019. "In vivo evaluation of post-operative pain reduction on rat model after implantation of intraperitoneal pet meshes functionalised with cyclodextrins and loaded with ropivacaine." *Biomaterials* 192:260–270. https://doi.org/10.1016/j.biomaterials.2018.07.032.

Chen, H., Li, L., Ma, Y., Mcdonald, T. P., and Wang, Y. 2019. "Development of active packaging film containing bioactive components encapsulated in β-cyclodextrin and its application." *Food Hydrocoll.* 90:360–366. https://doi.org/10.1016/j.foodhyd.2018.12.043.

Chen, L., Berry, R. M., and Tam, K. C. 2014. "Synthesis of β-cyclodextrin-modified cellulose nanocrystals $(CNCs)@Fe_3O_4@SiO_2$ superparamagnetic nanorods." *ACS Sustain. Chem. Eng.* 2(4):951–958. https://doi.org/10.1021/sc400540f.

Chen, Y-H., and Erbe, A. 2018. "The multiple roles of an organic corrosion inhibitor on copper investigated by a combination of electrochemistry-coupled optical in situ spectroscopies." *Corros. Sci.* 145:232–238. https://doi.org/10.1016/j.corsci.2018.09.018.

Chernykh, E. V., and Brichkin, S. B. 2010. "Supramolecular complexes based on cyclodextrins." *High Energy Chem.* 44:83–100. https://doi.org/10.1134/S0018143910020013.

Crini, G. 2014. "Review: a history of cyclodextrins." *Chem. Rev.* 114(21):10940–10975. https://doi.org/10.1021/cr500081p.

Crini, G., Fourmentin, S., Fenyvesi, É., Torri, G., Fourmentin, M., and Morin-Crini, N. (2018) Fundamentals and applications of cyclodextrins. In Fourmentin S., Crini G., and Lichtfouse E. (eds), *Cyclodextrin Fundamentals, Reactivity and Analysis. Environmental Chemistry for a Sustainable World.* Springer. https://doi.org/10.1007/978-3-319-76159-6.

Cyphert, E. L., Zuckerman, S. T., Korley, J. N., and von Recum, H. A. 2017. "Affinity interactions drive post-implantation drug filling, even in the presence of bacterial biofilm." *Acta Biomater.* 57:95–102. https://doi.org/10.1016/j.actbio.2017.04.015.

Dodziuk, H. 2006. *Cyclodextrin and Their Complexes: Chemistry, Analytical Methods, and Applications.* Wiley-VCH Verlag GmbH & Co. KGaA. https://doi.org/10.1002/3527608982.

Duchêne, D., ed. 1987. *Cyclodextrins and Their Industrial Uses.* France. ISBN.2-86411-019-9.

Fan, B., Wei, G., Zhang, Z., and Qiao, N. 2014a. "Characterization of a supramolecular complex based on octadecylamine and β-cyclodextrin and its corrosion inhibition properties in condensate water." *Corros. Sci.* 83:75–85. https://doi.org/10.1016/j.corsci.2014.01.043.

Fan, B., Wei, G., Zhang, Z., and Qiao, N. 2014b. "Preparation of supramolecular corrosion inhibitor based on hydroxypropyl-β-cyclodextrin/octadecylamine and its anti-corrosion properties in the simulated condensate water." *Anti-Corros. Method M* 61(2):104–111. https://doi.org/10.1108/ACMM-04-2013-1256.

"File:Cyclodextrin.Svg - Wikimedia Commons." n.d. Accessed April 27, 2021. https://commons.wikimedia.org/wiki/File:Cyclodextrin.svg.

Fourmentin, S., Crini, G., and Lichtfouse, E. (eds) 2018. Cyclodextrin applications in medicine, food, environment and liquid crystals. In *Environmental Chemistry for a Sustainable World.* Springer. https://doi.org/10.1007/978-3-319-76162-6.

Ghoul, Y. E., Martel, B., Achari, A. E., Campagne, C., Razafimahefa, L., and Vroman, I. 2010. "Improved dyeability of polypropylene fabrics finished with β-cyclodextrin–citric acid polymer." *Polym. J.* 42: 804–811. https://doi.org/10.1038/pj.2010.80.

Grafmiller, K. T., Zuckerman, S. T., Petro, C., Liu, L., von Recum, H. A., Rosen, M. J., and Korley, J. N. 2016. "Antibiotic-releasing microspheres prevent mesh infection in vivo." *J. Surg. Res.* 206(1):41–47. https://doi.org/10.1016/j.jss.2016.06.099.

Haley, R. M., Qian, V. R., Learn, G. D., and Von Recum, H. A. 2019. "Use of affinity allows anti-inflammatory and anti-microbial dual release that matches suture wound resolution." *J. Biomed. Mater. Res.* 107A:1434–1442. https://doi.org/10.1002/jbm.a.36658.

He, Y., Fu, P., Shen, X., and Gao, H. 2008. "Cyclodextrin-based aggregates and characterization by microscopy." *Micron* 39(5):495–516. https://doi.org/10.1016/j.micron.2007.06.017.

He, Y., Yang, Q., and Xu, Z. 2014. "A supramolecular polymer containing β-cyclodextrin as corrosion inhibitor for carbon steel in acidic medium." *Russ. J. Appl. Chem.* 87(12):1936–1942. https://doi.org/10.1134/S1070427214120234.

He, Y., Zhang, C., Wu, F., and Xu, Z. 2016. "Fabrication study of a new anticorrosion coating based on supramolecular nanocontainer." *Synth. Met.* 212:186–94. https://doi.org/10.1016/j.synthmet.2015.10.022.

Hedges, A. R. 1998. "Industrial applications of cyclodextrins." *Chem. Rev.* 98(5):2035–2044. https://doi.org/10.1021/cr970014w.

Jin, Z. Y. 2013. *Cyclodextrin Chemistry: Preparation and Application.* World Scientific Publishing Co. Pte. Ltd.

Khramov, A. N., Voevodin, N. N., Balbyshev, V. N., and Donley, M. S. 2004. "Hybrid organo-ceramic corrosion protection coatings with encapsulated organic corrosion inhibitors." *Thin Solid Films* 447–448:549–557. https://doi.org/10.1016/j.tsf.2003.07.016.

Khramov, A. N., Voevodin, N. N., Balbyshev, V. N., and Mantz, R. A. 2005. "Sol-gel-derived corrosion-protective coatings with controllable release of incorporated organic corrosion inhibitors." *Thin Solid Films* 483 (1–2):191–196. https://doi.org/10.1016/j.tsf.2004.12.021.

Kurkov, S. V., Loftsson, T. 2013. "Cyclodextrins." *Int. J. Pharm.* 453(1):167–180. https://doi.org/10.1016/j.ijpharm.2012.06.055.

Latnikova, A., Grigoriev, D. O., Hartmann, J., Möhwald, H., and Shchukin, D. G. 2011. "Polyfunctional active coatings with damage-triggered water-repelling effect." *Soft Matter* 7(2):369–372. https://doi.org/10.1039/C0SM00842G.

Learn, G. D., Lai, E. J., and von Recum, H. A. 2020a. "Cyclodextrin polymer coatings resist protein fouling, mammalian cell adhesion, and bacterial attachment." *BioRxiv* 1–23. https://doi.org/10.1101/2020.01.16.909564.

Learn, G. D., Lai, E. J., and von Recum, H. A. 2020b. "Nonthermal plasma treatment improves uniformity and adherence of cyclodextrin-based coatings on hydrophobic polymer substrates." *Coatings* 10(11):1–18. https://doi.org/10.3390/coatings10111056.

Leclercq, L., Dewilde, A., Aubry, J. M., and Véronique, N. R. 2016. "Supramolecular assistance between cyclodextrins and didecyldimethylammonium chloride against enveloped viruses: toward eco-biocidal formulations." *Int. J. Pharm.* 512(1):273–281. https://doi.org/10.1016/j.ijpharm.2016.08.057.

Leclercq, L., Lubart, Q., Dewilde, A., Aubry, J. M., and Véronique, N. R. 2012. "Supramolecular effects on the antifungal activity of cyclodextrin/di-n-decyldimethylammonium chloride mixtures." *Eur. J. Pharm. Sci.* 46(5): 336–345. https://doi.org/10.1016/j.ejps.2012.02.017.

Lehner, S. J., Müller, B. W., and Seydel, J. K. 1994. "Effect of hydroxypropyl-β-cyclodextrin on the antimicrobial action of preservatives." *J. Pharm. Pharmacol.* 46(3):186–191. https://doi.org/10.1111/j.2042-7158.1994.tb03775.x.

Liu, Y., Zou, C., Yan, X., Xiao, R., Wang, T., Li. M. 2015. "β-cyclodextrin modified natural chitosan as a green inhibitor for carbon steel in acid solutions." *Ind. Eng. Chem. Res.* 54(21):5664–5672. https://doi.org/10.1021/acs.iecr.5b00930.

Loftsson, T., Másson, M., and Brewster, M. E. 2004. "Self-association of cyclodextrins and cyclodextrin complexes." *J. Pharm. Sci.* 93(5):1091–1099. https://doi.org/https://doi.org/10.1002/jps.20047.

Maia, F, Yasakau, K. A., Carneiro, J., Kallip, S., Tedim, J., Henriques, T., Cabral, A., Venâncio, J., Zheludkevich, M. L., and Ferreira, M. G. S. 2016. "Corrosion protection of AA2024 by sol–gel coatings modified with MBT-loaded polyurea microcapsules." *Chem. Eng. J.* 283:1108–1117. https://doi.org/https://doi.org/10.1016/j.cej.2015.07.087.

Martel, B., Thuaut, P. L., Bertini, S., Crini, G., Bacquet, M., Torri, G., and Morcellet, M. 2002. "Grafting of cyclodextrins onto polypropylene nonwoven fabrics for the manufacture of reactive filters. III. Study of the sorption properties." *J. Appl. Polym. Sci.* 85(8):1771–1778. https://doi.org/https://doi.org/10.1002/app.10682.

Montemor, M. F. 2015. *Smart Composite Coatings and Membranes: Transport, Structural, Environmental and Energy Applications.* U.K.: Woodhead Publishing Series in Composites Science and Engineering vol. 64. https://doi.org/10.1016/C2013-0-16518-X.

Patel, M. H., and Hvizdos, S. A. 1992. Cooling agent/cyclodextrin complex for improved flavor release. US Patent 5,165,943.

Pereira, M. S., Carvalho, W. M., Alexandrino, A. C., Bezerra, H. C. P., Rodrigues, M. C. P., Figueiredo, R. W., Maia, G. A., Figueiredo, M. A. T., and Brasil, I. M. 2014. "Freshness retention of minimally processed melon using different packages and multilayered edible coating containing microencapsulated essential oil." *Int. J. Food Sci. Technol.* 49: 2192–2203. https://doi.org/10.1111/ijfs.12535.

Plawecka, M., Snihirova, D., Martins, B., Szczepanowicz, K., Warszynski, P., and Montemor, M. F. 2014. "Self healing ability of inhibitor-containing nanocapsules loaded in epoxy coatings applied on aluminium 5083 and galvanneal substrates." *Electrochim. Acta* 140:282–293. https://doi.org/https://doi.org/10.1016/j.electacta.2014.04.035.

Rahimi, A., and Amiri, S. 2015. "Self-healing hybrid nanocomposite coatings with encapsulated organic corrosion inhibitors." *J. Polym. Res.* 22(1):624. https://doi.org/10.1007/s10965-014-0624-z.

Rahimi, A., and Amiri, S. 2016. "Self-healing anticorrosion coating containing 2-mercaptobenzothiazole and 2-mercaptobenzimidazole nanocapsules." *J. Polym. Res.* 23(4):83. https://doi.org/10.1007/s10965-016-0973-x.

Rekharsky, M. V., and Inoue, Y. 1998. "Complexation thermodynamics of cyclodextrins." *Chem. Rev.* 98, 1875–1917. https://pubs.acs.org/sharingguidelines.

Romi, R., Nostro, P. L., Bocci, E., Ridi, F., and Baglioni, P. 2005. "Bioengineering of a cellulosic fabric for insecticide delivery via grafted cyclodextrin." *Biotechnol. Progr.* 21(6):1724–1730. https://doi.org/https://doi.org/10.1021/bp050276g.

Rosenhahn, A., Schilp, S., Kreuzer, H. J., and Grunze, M. 2010. "The role of 'inert' surface chemistry in marine biofouling prevention." *Phys. Chem. Chem. Phys.* 12(17):4275–4286. https://doi.org/10.1039/C001968M.

Saji, V. S. 2019. "Supramolecular concepts and approaches in corrosion and biofouling prevention." *Corros. Rev.* 37(3):187–230. https://doi.org/10.1515/corrrev-2018-0105.

Saji, V. S., and Thomas, J. 2007. "Nanomaterials for Corrosion Control." *Curr. Sci.* 92(1):51–55. http://www.jstor.org/stable/24096821.

Sato, Y, Suzuki, Y., Ito, K., and Shingawa, T. 1992. Flavor and taste composition for a chewing gum. US Patent 5,156,866. https://patents.google.com/patent/US5156866A/en.

Schneider, H. J. 2012. *Applications of Supramolecular Chemistry*. CRC Press, Boca Raton, FL.

Schofield, W. C. E., and Badyal, J. P. S. 2011. "Controlled fragrant molecule release from surface-tethered cyclodextrin host–guest inclusion complexes." *ACS Appl. Mater. Interfaces* 3:2051–2056. https://doi.org/10.1021/am200281x.

Schubert, U. S., Hochwimmer, G., Schmatloch, S., and Hofmeier, H. 2003. "A novel class of smart materials." *Eur. Coat. J.* 6:28–34.

Schultz, M. P., Bendick, J. A., Holm, E. R., and Hertel, W. M. 2011. "Economic impact of biofouling on a naval surface ship." *Biofouling* 27(1):87–98. https://doi.org/10.1080/08927014.2010.542809.

Shibad, P. R., and Balachandra, J. 1976. "Behaviour of titanium and its alloy with hydrofluoric acid in HCl and H2SO4 solutions with addition agents." *Anti-Corros. Method M* 23(12):16–28. https://doi.org/10.1108/eb007024.

Sundararajan, P. R., and Rao, V. S. R. 1970. "Conformational studies on cycloamyloses." *Carbohydr. Res.* 13(3):351–358. https://doi.org/10.1016/S0008-6215(00)80592-3.

Szejtli, J. 1988. *Topics in Inclusion Science: Cyclodextrin Technology*. Springer Science+Business Media, B.V. U.K. 10.1007/978-94-015-7797-7.

Szejtli, J. 1998. "Introduction and general overview of cyclodextrin chemistry." *Chem. Rev.* 98(5):1743–1753. https://doi.org/10.1021/cr970022c.

Talati, J. D., and Modi, R. M. 1976. "Colloids as corrosion inhibitors for aluminium-copper alloy in sodium hydroxide." *Anti-Corros. Method. Mater.* 23(1):6–9. https://doi.org/10.1108/eb006996.

Thatiparti, T. R., Shoffstall, A. J., and Recum, H. A. 2010. "Cyclodextrin-based device coatings for affinity-based release of antibiotics." *Biomaterials* 31(8):2335–2347. https://doi.org/https://doi.org/10.1016/j.biomaterials.2009.11.087.

Toan, T. 1996. *Proceedings of the Eighth International Symposium on Cyclodextrins* (J. Szejtli and L. Szente), Springer. doi: 10.1007/978-94-011-5448-2.

Ueda, H. 2002. "Physicochemical properties and complex formation abilities of large-ring cyclodextrins." *J. Incl. Phenom. Macrocycl. Chem.* 44 (1): 53-56. https://doi.org/10.1023/A:1023055516398.

Uekama, K., Hirayama, F., and Irie, T. 1998. "Cyclodextrin Drug Carrier Systems." *Chem. Rev.* 98:2045–2076. https://pubs.acs.org/sharingguidelines.

Vermet, G, Degoutin, S., Chai, F., Maton, M., Bria, M., Danel, C., Hildebrand, H. F., Blanchemain, N., and Martel, B. 2014. "Visceral mesh modified with cyclodextrin for the local sustained delivery of ropivacaine." *Int. J. Pharm.* 476(1):149–159. https://doi.org/https://doi.org/10.1016/j.ijpharm.2014.09.042.

Véronique, N. R., and Leclercq, L. 2014. "Encapsulation of biocides by cyclodextrins: toward synergistic effects against pathogens." *Beilstein J. Org. Chem.* 10:2603–2622. https://doi.org/10.3762/bjoc.10.273.

Vreugdenhil, A. J., and Woods, M. E. 2005. "Triggered release of molecular additives from epoxy-amine sol–gel coatings." *Prog. Org. Coat.* 53(2):119–125. https://doi.org/https://doi.org/10.1016/j.porgcoat.2005.02.004.

Wang, M., Liu, P., Wang, Y., Zhou, D., Ma, C., Zhang, D., and Zhan, J. 2015. "Core–shell superparamagnetic Fe_3O_4@β-CD composites for host–guest adsorption of polychlorinated biphenyls (PCBs)." *J. Colloid Interface Sci.* 447:1–7. https://doi.org/https://doi.org/10.1016/j.jcis.2015.01.061.

Wang, Z., Guo, S., Zhang, B., Fang, J., and Zhu, L. 2020. "Interfacially crosslinked β-cyclodextrin polymer composite porous membranes for fast removal of organic micropollutants from water by flow-through adsorption." *J. Hazard. Mater.* 384:121187. https://doi.org/https://doi.org/10.1016/j.jhazmat.2019.121187.

Wen, P., Zhu, D. H., Feng, K., Liu, F. J., Lou, W. Y., Li, N., Zong, M. H., and Wu, H. 2016. "Fabrication of electrospun polylactic acid nanofilm incorporating cinnamon essential oil/β-cyclodextrin inclusion complex for antimicrobial packaging." *Food Chem.* 196:996–1004. https://doi.org/https://doi.org/10.1016/j.foodchem.2015.10.043.

Zou, C. J., Liao, W. J., Zhang, L., and Chen, H. M. 2011. "Study on acidizing effect of β-cyclodextrin-PBTCA inclusion compound with sandstone." *J. Petrol. Sc. Eng.* 77(2):219–225. https://doi.org/https://doi.org/10.1016/j.petrol.2011.03.010.

11 Mesoporous Silica Nanoparticles in Surface Coating Applications

Idalina Vieira Aoki and Brunela Pereira da Silva
Polytechnic School of the University of São Paulo

Jesus Marino Falcón-Roque
Universidad San Ignacio de Loyola

CONTENTS

11.1 INTRODUCTION

Porous materials are classified into three groups based on their pore sizes following the International Union of Pure and Applied Chemistry (IUPAC): microporous, mesoporous, and macroporous materials with pore sizes of <2, 2–50, and >50 nm, respectively (Ariga et al. 2012; Soler-Illia and

DOI: 10.1201/9781003169130-13

Azzaroni 2011; Liu et al. 2011). Mesoporous silica nanoparticles (MSNs) present individual hexagonal ordered pores of narrow size distribution, allowing incorporation of larger molecules, such as corrosion inhibitors into the mesopores. These long-range non-connected porous channels are favorable attributes for inhibitor loading and can tune their release kinetics.

Various types of silica nanoparticles (NPs) are used for loading corrosion inhibitors that include solid and dense, mesoporous, hollow with a mesoporous shell, and highly ordered mesoporous. The main methods of synthesis are hard and soft template and one-pot synthesis. Hard and soft template methods associated with the hydrothermal process are mainly used to obtain hollow and ordered mesoporous silica particles like SBA-15 (Santa Barbara Amorphous) and MCM-41 (Mobil Composition of Matter no. 41). Recently, a new mesoporous silica resembling a crumpled paper-like ball was synthesized using a single-step sol–gel hydrothermal method with a relatively large surface area of $602\,m^2/g$ and a high pore volume of $0.97\,cm^3/g$ (Soltani et al. 2020).

Among the most popular and commercially available mesoporous silica, it is worth citing Santa Barbara Amorphous-15 (SBA-15). It is a highly stable mesoporous silica developed by researchers at the University of California at Santa Barbara. This mesoporous silica is characterized by cylindrical and uniform pores, ordered in a hexagonal structure, with pore diameter varying from 4.6 to 30 nm. SBA silica has a high internal surface area, which makes it adequate for various applications such as environmental adsorption and separation, catalysis, and encapsulation of drugs or corrosion inhibitors (Zhao et al. 1998a, b; Kruk et al. 2000; Kumaran et al. 2006; Janus et al. 2020).

Another commercially available mesoporous silica is the MCM-41 (Mobil Composition of Matter no. 41), the first from the family of mesoporous materials synthesized by researchers at the Mobil Oil Company in 1992 (Beck et al. 1992). The constituents are silicate and aluminosilicate. It presents a great diversity of synthesis procedures and the possibility of introducing several metals into its structural network, either by isomorphic substitution or post-synthesis impregnation. In addition, these materials also have a large surface area, which can reach up to more than $1000\,m^2/g$, high thermal stability, possibility of controlling pore size, and hydrophobicity. Such characteristics make MCM-41 a promising material for catalysis and are used in adsorption processes of various molecules such as corrosion inhibitors and drugs and ion exchange (Dyer et al. 2009; Alipour and Nasirpouri 2017; Vallet-Regí et al. 2004; ACS Chem-a). Another one is FDU-12, a three-dimensional mesoporous material with a superior 3D channel, which is ideal for mass transfer and guest molecule diffusion. It is large-pore type silica, a highly structured and face-centric cubic mesoporous material, called cage-like porous material (ACS Chem-b).

11.2   SYNTHESIS OF MESOPOROUS SILICA NANOPARTICLES

The Stöber method is the first and most commonly used method for synthesizing mesoporous solid silica particles using a sol–gel technique associated with the soft template and hydrothermal processes. The Stöber method (Stöber et al. 1968) involves hydrolysis of silane molecules (e.g., tetraethyl orthosilicate (TEOS), sodium silicate, or tetramethyl orthosilicate (TMOS) in an alcohol–water solution in the presence of a catalyst (e.g., ammonium hydroxide). The hydrolyzed molecules start polymerizing among themselves and make larger molecules (oligomers), which deposit solid and dense silica nanoparticles. The organic components are then removed by calcination, thus creating MSNs.

11.2.1   Soft Templating for Dense MSNs

Spherical dense MSNs were obtained by using TEOS as an Si source and an aqueous solution of surfactant (CTAB) that permits the formation of oil/water (o/w) emulsion with micelles as templates (Omid et al. 2020). A hexagonal arrangement of micelles was obtained using Si source (TEOS) and structure-directing surfactants (triblock polymer, pluronicP123, and CTAB). After condensation of silanols to SiO_2 in an acidic or alkaline medium, the particles are calcinated to destroy the organic surfactants, leading to the highly ordered mesoporous silica, as illustrated in Figure 11.1 (Seljak et al. 2020).

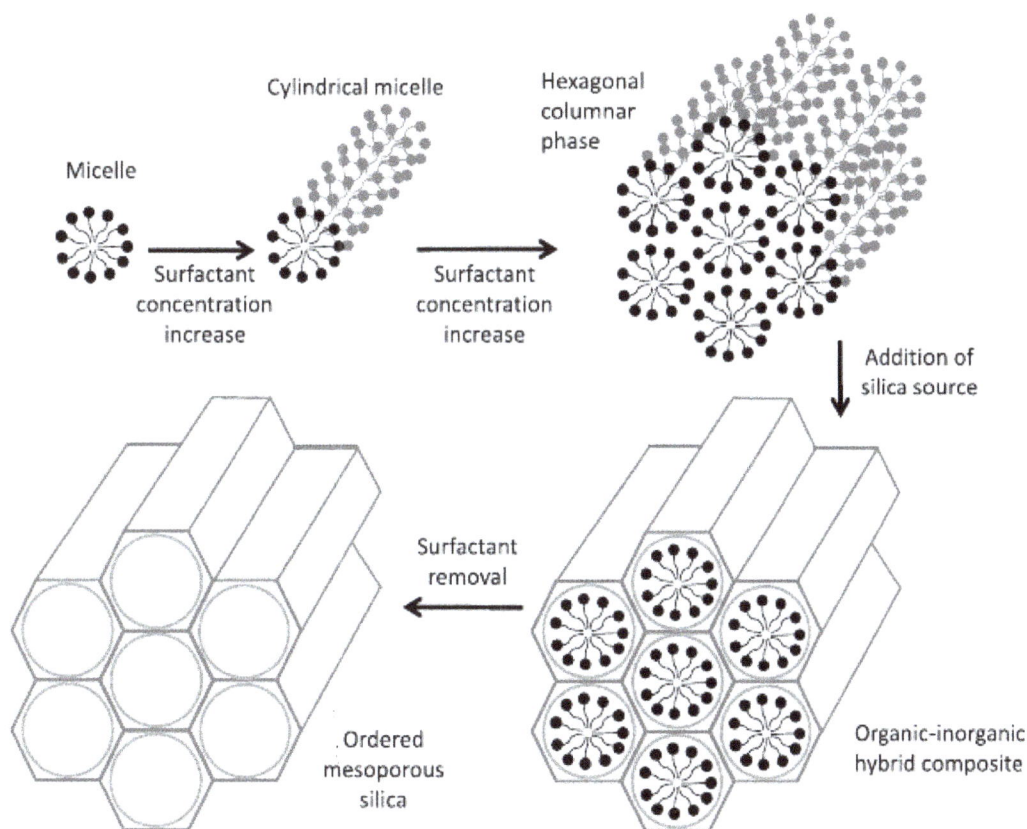

FIGURE 11.1 Schematic of synthesis of MSNs. (Reproduced with permission from Seljak et al. (2020) © 2020 Elsevier B.V.)

Silica being a high strength material, removing organic templates does not cause any damage to the structure and the particle morphology, but creates a well-organized porous structure (Ariga et al. 2012; Falcón et al. 2016; Ijaz et al. 2020). SBA-15 and MCM-41 mesoporous silica were also obtained by the sol–gel soft templating method (Janus et al. 2020; Vanichvattanadecha et al. 2020).

11.2.2 HARD TEMPLATING FOR HOLLOW MSNs

Among different hollow particles such as ZnO, TiO_2, SiO_2, and ZrO_2 (Yao et al. 2021; Chenan et al. 2014a, b), hollow MSNs (HMSNs) are highly preferred because of their low cost, facile synthesis process, well-known chemistry, and wide-ranging applications. It must be noted that the inherent nature of HMSNs synthesis results in micro- or mesoporous shells (silica shells having micro-or mesopores). The porosity of the HMSNs refers to the pores in the shell and not to the size of the central hollow cavity. Figure 11.2 shows different synthesis strategies to obtain the HMSNs. They can be obtained via soft or hard template methods (Figure 11.2). The first one in the figure represents a soft template and the others are hard templates.

A simple strategy to prepare HMSNs using a hard template involves the use of polystyrene beads, which are available in a very small size. TEOS and surfactant were used to assemble and form the silica shell on the polystyrene beads. They can be calcinated to destroy the spherical template or be easily removed by solvent extraction. The release of the template creates pores in the hard silica shell. In 2006, periodic mesoporous organosilica (PMO) NPs were obtained by a sol–gel process from organo-functional bi-alkoxysilanes in the presence of structure-directing (templating) agents

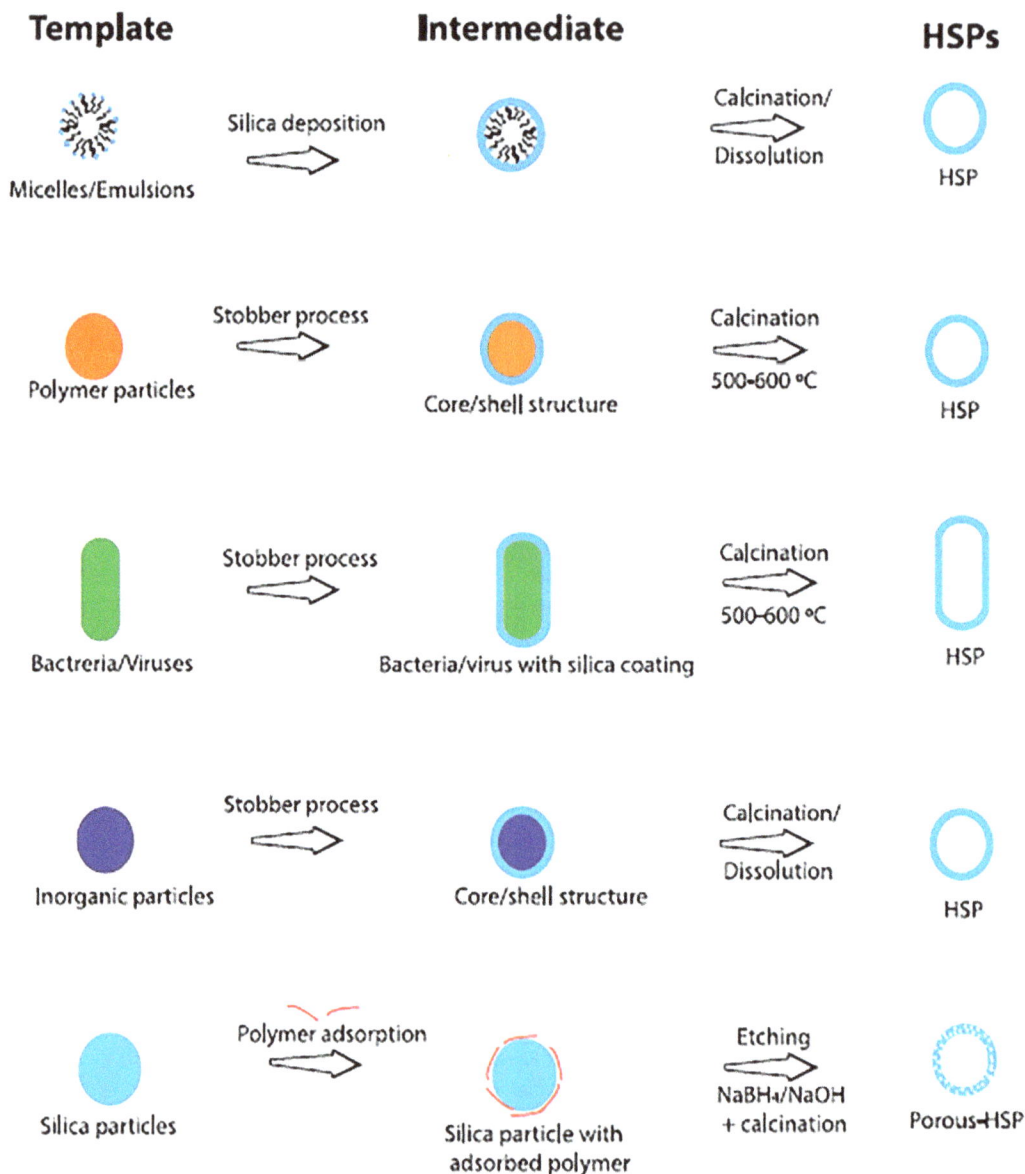

FIGURE 11.2 Schematic illustration of common strategies used for the synthesis of hollow silica nanoparticles (HMSNs). (Reproduced with permission from Sharma and Polizos (2020) © 2020, by the authors, licensee MDPI, under Creative Commons Attribution License.)

as surfactants. PMO particles normally have a size in the broad range of 100–1000 nm in diameter. Hollow PMO particles were also obtained using two structure-directing or templating surfactants (CTAB and FC-4) (Croissant et al. 2015; Li et al. 2020).

The hard template method involves the formation of a silica layer (shell) on a hard-core particle such as silica or magnetite (Fe_3O_4) or other metal oxides or sulfides such as CuO, CuS, or AgO. This strategy permits to obtain magnetic particles to be used in the delivery of drugs to specific organs in the human body (Pooresmaeil et al. 2020; Nemec and Kralj 2021; Teng et al. 2020) or particles with antimicrobial or antifouling properties that can be used as pigments in antifouling coatings (Ruggiero et al. 2019; Sun et al. 2020; Liu et al. 2019). To make core–shell and hollow $TiO_2@SiO_2$ mesoporous particles, the template needs to be dissolved, whereas, for dense $SiO_2@$

TiO$_2$, this is not needed (Li et al. 2021). Comparing soft and hard templating strategies, generally, one can conclude that soft templating methods are simpler and faster since they do not require the preparation and removal of a hard template. Hard templating strategies are attractive as they easily enable the incorporation of metal and metal oxide NPs into MSNs or PMOs when designing magnetic or catalytic MSNs/PMOs. Currently, only soft templating approaches produce MSNs with long-range molecular periodicity of functionalized organic fragments, required for specific properties and applications (Croissant et al. 2015).

11.2.3 ONE-POT SYNTHESIS METHOD

The one-pot synthesis approach (the sol–gel Stöber method) is very simple and does not involve the calcination step at the end of MSNs preparation. The molecules that one intends to encapsulate in the silica NPs are added during the sol–gel process and the precipitated silica entraps them. It is a synthesis method and a method to encapsulate, and it is discussed further in this chapter (see Section 11.3).

11.2.4 DENDRIMERS-LIKE MESOPOROUS SILICA

Dendrimers are highly branched, globular, nanoscopic macromolecules composed of two or more tree-like dendrons emanating from a central core, which can be either a single atom or an atomic group (Sarkar et al. 2010). They are built up of three-dimensional "branch cells" that are organized into concentric layers (usually referred to as "generations") around the core, and these branch cells can be considered dendritic analogues of traditional linear polymer repeat units. They can be divided into two main types: interior branch cells, which form the intramolecular dendrimer armature, and exterior (or surface) branch cells, which carry either reactive or inert end-groups (Sarkar et al. 2010). Fabrication of dendritic mesoporous silica nanospheres (DMSNs) was performed for application in drug-delivery carriers, catalysis, and nanodevices owing to their unique open three-dimensional dendritic superstructures with large pores and highly accessible internal surface areas (Hao et al. 2020; Siva et al. 2021).

Reports on other organized MSNs were also available (Suteewong et al. 2013) including cage-like arrangements (Eggenhuisen et al. 2012; Suteewong et al. 2013; Du et al. 2019a).

11.2.5 MESOPOROUS SILICA STRUCTURE RESEMBLING CRUMPLED PAPER BALLS

Soltani et al. developed a single-step sol–gel, soft templating, and hydrothermal route to synthesize novel mesoporous crumpled paper-like silica balls with a relatively large surface area and unique morphology. The scheme of the synthesis process of this special mesoporous silica and its aspect by FE-SEM are shown in Figure 11.3. EM micrographs of the sample indicated that this mesoporous silica comprises hollow and crumpled paper-like balls resulting from the crumpling of thin sheets of mesoporous silica (Figure 11.3) (Soltani et al. 2020).

11.3 USE OF MSNS FOR ENCAPSULATING CORROSION INHIBITORS

Porous nanomaterials having large surface areas, widely open and interconnected pore structures, and thermal and mechanical stability are desirable candidates for encapsulation (Walcarius 2013). Among the different types of inorganic nanocontainers, the most important are MSNs (Saji 2019; Wang et al. 2012; Zea et al. 2018a), mesoporous TiO$_2$ (Abu-Thabit and Hamdy 2016), mesoporous ZrO$_2$ (Chenan et al. 2014a), clay minerals such as halloysite (Shchukina et al. 2017; Shchukina et al. 2018; Abdullayev et al. 2009; Falcón et al. 2015) and those formed based on multilayers of polyelectrolytes (Möller et al. 2007; Shchukin and Grigoriev 2012; Shchukin 2013; Zahidah et al. 2017). Silica and clay are highly preferred as they are abundant and cheap. Layered double hydroxides are also used to encapsulate corrosion inhibitors (Bendinelli et al. 2019; Zheludkevich et al. 2010).

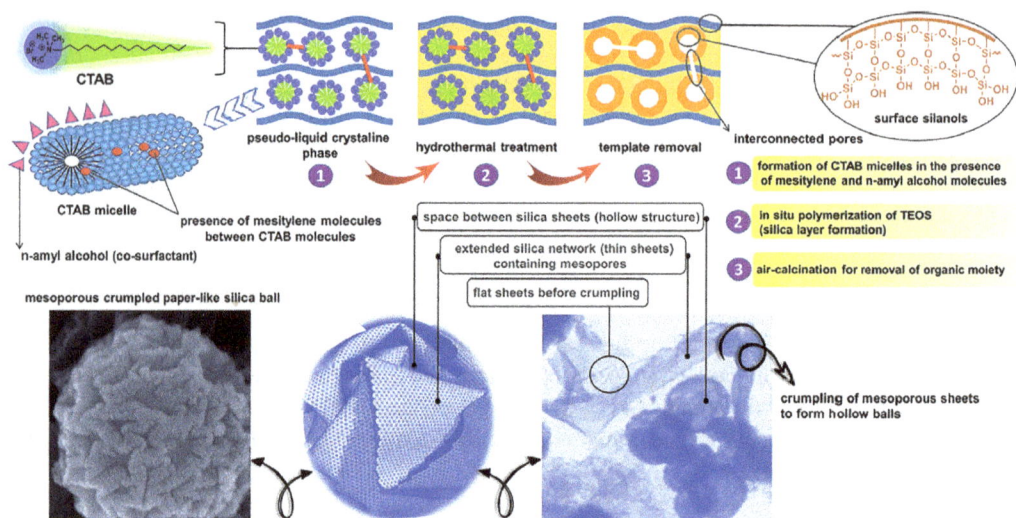

FIGURE 11.3 Schematic representation of the synthesis of mesoporous crumpled paper-like silica balls. (Reproduced with permission from Soltani et al. (2020) © 2020 Elsevier B.V.)

Currently, interest in the use of nanoscale containers has grown significantly, leading to the development of new technologies for producing these materials (Möller et al. 2007; Zahidah et al. 2017; Wang et al. 2012; Haddadi et al. 2019). The use of larger and microscale containers can damage the integrity of the coatings due to the creation of pores and defects, reducing their barrier properties (Shchukin and Möhwald 2007a). In addition, smaller diameters improve the dispersion of nanocontainers in the polymeric matrix (Zahidah et al. 2017) and facilitate easy assimilation in thin coatings (10–20 µm) (Shchukin 2013). Shchukin and Möhwald, in the pioneer paper on the topic, considered that nanocontainers should have diameters ideally smaller than 300–400 nm (Shchukin and Möhwald 2007a).

11.3.1 LOADING METHODS

A suitable loading procedure is expected to afford a large quantity of loaded inhibitor and controlled release kinetics over time. Among the different loading methods, the most used are adsorption, Layer-by-Layer (LbL), vacuum, and simultaneous synthesis and loading in the one-pot method.

11.3.1.1 Adsorption

A typical adsorption method involves soaking the MSNs in the inhibitor solution till saturation of adsorption sites (Speybroeck et al. 2009; Rajput et al. 2020) with subsequent solvent evaporation. It is simple and the adsorption can be enhanced by the suitable functionalization of the silica particles.

11.3.1.2 Vacuum

Another process to load molecules and compounds in MSNs is the vacuum method. In this process, the MSNs are put together with the corrosion inhibitor solution in water or an organic solvent in a closed vessel with subsequent application of reduced pressure or vacuum. As air is pumped from the pores, the inhibitor solution occupies the air sites inside the MSN. The loading is followed by centrifugation of the nanocontainers suspended in water at high rpm values (5000 rpm). The process is typically repeated 3–4 times for effective encapsulation of inhibitors inside the MSN pores and subsequently dried in air or in an oven at 50°C–60°C to evaporate the residual solvent (Price et al. 2001; Shchukina et al. 2018; Shi et al. 2017; Shchukina et al. 2017; Zheludkevich et al. 2007a, b; Falcón et al. 2016).

11.3.1.3   Layer-by-Layer

The LbL electrostatic deposition technique of electrically charged molecules has attracted significant research attention, where thin and uniform films with a wide range of properties can be prepared on various substrates or particles (Ghosh 2006). Dense silica particles without porosity can
be used as a base to encapsulate corrosion inhibitors to form layers of polyelectrolytes of different
charges, allowing the inhibitor to be deposited, trapped between the layers of adsorbed polyelectrolytes, forming layers on top of the silica particles.

The films reach thicknesses ranging from 5 to 500 nm on an inert particle. The forces of attraction
between the layers formed by this technique are primarily electrostatic, but they can also involve
hydrogen bonding or other types of non-covalent interactions (Shchukin et al. 2006; Andreeva et al.
2008). The "walls" made of polyelectrolytes that form different layers have semi-permeability and
sensitivity to a variety of chemical and physical environmental conditions into which the NPs are
inserted. Usually, these walls are permeable to macromolecules at pH <3 in such a way that, at
neutral pH, they are closed. A possible explanation for these permeability properties is related to
the interactions between the polyelectrolyte molecules within the layer (wall) (Shchukin et al. 2006;
Andreeva et al. 2008).

The possibility of maneuvering the capsule walls between an open and closed state provides
an efficient tool for loading and releasing active materials in a controlled manner (Andreeva et al.
2010). For example, the capsules can be loaded at low pH and then increasing pH to capture the
material inside the capsule (Andreeva et al. 2008; Grigoriev et al. 2017). The application methods
of different polyelectrolytes layers range from dip coating, spin coating, and spraying (Abu-Thabit
and Hamdy 2016). Figure 11.4 illustrates the LbL technique for encapsulating the corrosion inhibitor within the layers of two different polyelectrolytes (polyethyleneimine and polystyrene sulfonate)
deposited on silica nanoparticles.

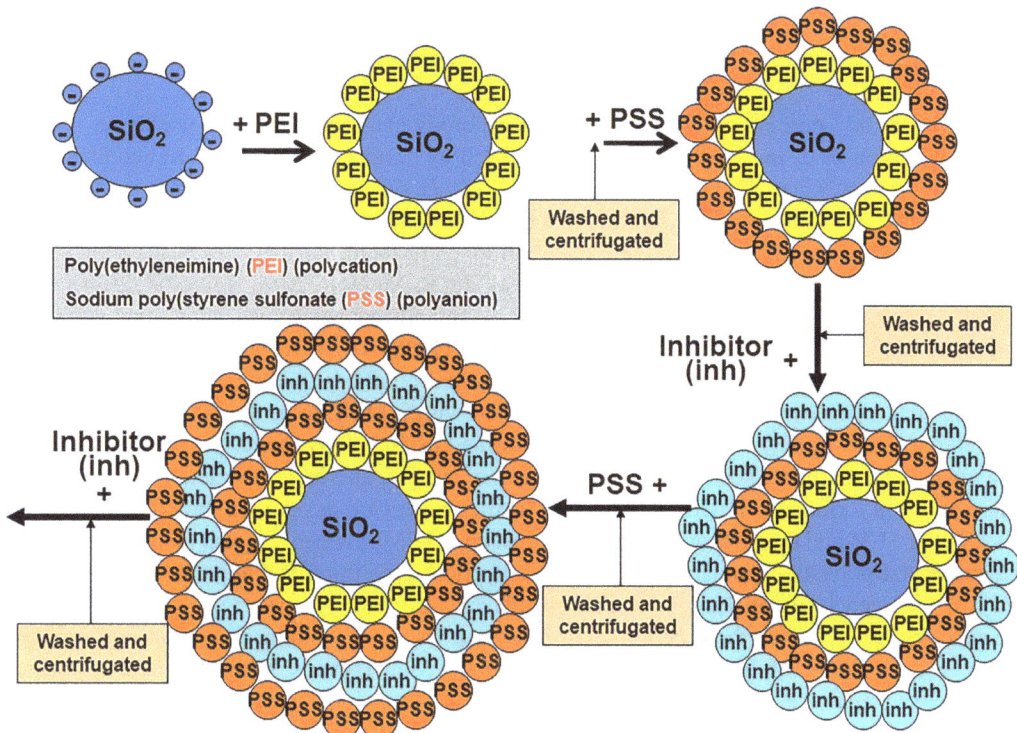

FIGURE 11.4   Scheme showing the different steps for the encapsulation of an active agent (corrosion inhibitor) by the Layer-by-Layer technique.

In several works, the desired molecules were encapsulated in MSNs, and thereafter, additional layers of polyelectrolytes were used to close (act like *gatekeepers*) to impart controlled release of the encapsulated molecules (Abu-Thabit and Hamdy 2016; Carneiro et al. 2015; Saji 2019; Zea et al. 2015, 2018a).

11.3.1.4   Supercritical and Liquid CO_2

CO_2 in a supercritical state (supercritical carbon dioxide, $SCCO_2$) is a non-toxic, non-flammable, and inexpensive solvent. CO_2 is a non-polar solvent, and it does not compete with the adsorption sites of MSNs. This loading technique could reduce the time for loading when compared to other methods. No contamination occurs with this clean solvent, which is frequently used to encapsulate drugs. Moreover, it can be easily removed from the MSNs.

Compared to the adsorption method, the $SCCO_2$ technique improved the loading of ibuprofen onto MCM-41 and fenofibrate loading onto ordered mesoporous silica. Additionally, the drug penetrated deeper into the mesopores, resulting in a longer sustained release time (Bouledjouidja et al. 2016).

11.3.1.5   One-Pot Synthesis

The one-step or one-pot approach to the synthesis of silica nanocontainers has attracted significant recent research attention. The advantages are the easy synthesis route and the better performance of the resulting reservoirs (Chen et al. 2008; Kou et al. 2014; Liu et al. 2017; Maia et al. 2012; Xu et al. 2018; Zhao et al. 2016).

During the conventional synthesis of MSNs, the soft template is removed using a calcination stage at higher temperatures (up to 600°C), as the pores must be free for further incorporation of molecules. In one-pot synthesis and loading, the template is an inhibitor molecule-filled micelle (soft template) and the silica co-precipitates with the desired molecule, encapsulating it; thus, the removal step is not needed. On the other hand, it should be noted that this type of synthesis can negatively affect the ordered mesoporous structure (Wan et al. 2016). Deng et al. presented a facile and controllable one-pot approach for the preparation of amino-functionalized hollow MSNs (Deng et al. 2020).

Usually, in this process, the synthesis of silica porous nanospheres makes use of an innovative combination of dynamic self-assembly controlled by surfactant with stabilized condensation of silica precursor (Chen et al. 2008; Maia et al. 2012). Here, typically, a surfactant is added to a catalytic medium (acidic or basic). Because the hydrophobic corrosion inhibitors have high solubility in hydrophobic micelles, the inhibitors get entrapped in them. Further, TEOS is added slowly to the precursor solution under stirring (Chen et al. 2008; Xu et al. 2018; Kou et al. 2014). Chen et al. synthesized silica nanospheres using CTAB as an anthropogenic surfactant, ethyl ether as a co-solvent, ammonia solution (25%–28%) as a catalyst, TEOS as a silica precursor and pyrene as an encapsulated drug, entrapping this substance in the silica containers in a single-step.

Maia et al. synthesized silica nanocapsules loaded with 2-MBT, to be applied as a water-based coating on Al 2024 alloy. The synthesis was performed using CTAB as the surfactant, ethyl ether as a co-solvent, ammonia solution (25%–28%) as the catalyst, and balance deionized water. After microemulsion formation, TEOS was added as a silica precursor, developing the porous nanocapsules. Their SEM and transmission electron microscopy (TEM) images revealed the desired spherical shape and porosity. Through the nitrogen adsorption–desorption technique, it was concluded that the obtained isotherms are typical of multilayer adsorption in mesoporous materials, with the surface areas (BET method) being $151 \, m^2/g$ for the system of containers without inhibitor and $168 \, m^2/g$ for the MSN-inhibitor system (Maia et al. 2012). More recently, Xu et al. synthesized silica NPs in a one-step process based on the high solubility of organic inhibitors in the hydrophobic cores of template micelles. A novel one-step method was employed in developing the containers free of organic solvents, in which SiO_2-CTAB- benzotriazole (CTAB-BTA) was produced, as shown in Figure 11.5 (Xu et al. 2018). A schematic of the traditional multi-step method is also shown.

Traditional Multi-step Synthesis

Self- and
Co-assembly

Template
Removal

Inhibitor
Loading

Inhibitor-loaded
Nanocontainers

Present One-step Synthesis

Silica Source: TEOS   Template: CTAB   Inhibitor: BTA

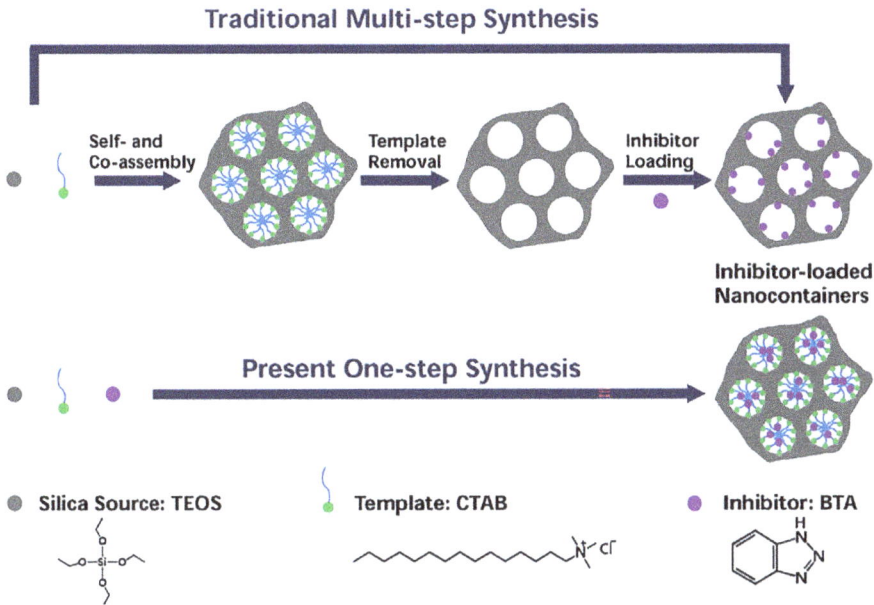

FIGURE 11.5 Schematics showing the synthesis of inhibitor-loaded mesoporous SiO_2 nanocontainers by the traditional multi-step method and the one-step route discussed in this section. (Reproduced with permission from Xu et al. (2018) © 2018 2018 Elsevier Ltd.)

Table 11.1 presents a list of major published works on MSNs with loaded corrosion inhibitors/drugs. As the table shows, the most frequently used loading is a low pressure or vacuum method, and then comes the adsorption method. This is followed by the LbL and the one-pot synthesis. The $SCCO_2$ method is used mainly for medical purposes. The efficiency of encapsulation in this report varies from 10% to 37.7%. The highest loading efficiency was obtained for $SCCO_2$ (37.7%). After loading, the surface area was diminished, as the adsorbed molecules occupied the most active sites. The most commonly encapsulated inhibitors were BTA, 2-MBT, and 8-HQ. Molybdate, phosphomolybdate, dodecylamine, fluoride, caffeine, and sulfamethazine were also used. In most of the papers, NaCl solution was used to study the release and self-healing.

11.4 CHARACTERIZATION OF MSNs

There are many analytical techniques to characterize MSNs. The most frequently employed are briefly mentioned here with selected references as examples.

11.4.1 LOADING EFFICIENCY AND THERMAL STABILITY BY THERMOGRAVIMETRIC ANALYSIS

The loading efficiency is commonly determined by thermogravimetric analysis (TGA). Several works used the TGA technique with MSNs (Shchukina et al. 2018; Yao et al. 2021; Lamaka et al. 2008; Abdullayev et al. 2009; Fix et al. 2009; Skorb et al. 2009). The TGA and differential thermogravimetric (DTG) curves of polyurethane (PU) coating loaded with MSNs revealed improved thermal stability of the PU coating in the presence of MSNs (Pergal et al. 2021). Differential scanning calorimetry (DSC) has been used to assess the melting transition of the encapsulated substances in MSNs. Thermoporometry/DSC of confined water was used to quantify the entrance sizes in a series of mesoporous silica materials with a cage-like pore structure, like SBA-16 and FDU-12. With the DSC, entrance sizes of materials with cages of up to 15 nm were determined, which could not be assessed by nitrogen physisorption (Eggenhuisen et al. 2012; Iza et al. 2000).

TABLE 11.1

Types and Methods of Loading of Corrosion Inhibitors or Drugs onto Mesoporous Silica Nanoparticles

Loading Method	Corrosion Inhibitor/Drug	% of Loading	MSN Type	Surface Area (m^2/g)	Ref.
LbL	BTA	N/A	Dense silica nanoparticles	N/A	Shchukin et al. (2006)
LbL	BTA	N/A	Dense silica nanoparticles	N/A	Shchukin and Möhwald (2007b)
LbL	Dodecylamine	N/A	Dense silica nanoparticles	N/A	Falcón et al. (2014)
LbL	BTA	N/A	Dense silica nanoparticles	N/A	Feng and Cheng (2017)
Vacuum	BSS	N/A	MSNs	N/A	Skorb et al. (2009)
Vacuum	BTA	N/A	MSNs	780	Hollamby et al. (2011)
Vacuum	BTA	N/A	HMSNs	1255	Chen and Fu (2012)
Vacuum	Caffeine	20.2	HMSNs	1141.2	Fu et al. (2013)
Vacuum	2-MBT	N/A	MSNs	873.4	Qiao et al. (2015)
Vacuum	Dodecylamine	N/A	MSNs	N/A	Falcón et al. (2016)
Vacuum	8-HQ	N/A	SBA-15	550–600	Shi et al. (2017)
Vacuum	Sulfamethazine	N/A	MSNs	883.2	Yeganeh et al. (2019a)
Vacuum	BTA	~12	MSNs	641	Xiong et al. (2019)
Vacuum	2-MBT	N/A	MSNs	925	Ouyang et al. (2020)
Vacuum	BTA	12.58	HMSNs	589.5	Zhou et al. (2020)
Adsorption	2-MBT	20	MSNs and HMSNs	1000	Borisova et al. (2013a)
Adsorption	2-MBT	N/A	HMSNs	940.57	Chenan et al. (2014b)
Adsorption	Fluoride	N/A	MSNs	776.2	Yeganeh and Saremi (2015)
Adsorption	Sodium phosphomolybdate	N/A	MSNs	926.53	Zea et al. (2018a)
Adsorption	Sodium molybdate	N/A	MSNs	776.2	Yeganeh et al. (2019b)
Adsorption	2-MBT	N/A	MSNs, SBA-15	609	Amini et al. (2020)
Adsorption	8-HQ	N/A	Dendrimer- like mesoporous silica	1322.4	Siva et al. (2021)
One-pot synthesis and loading	2-MBT	10.0	MSNs	151	Maia et al. (2012)
One-pot synthesis and loading	BTA	22.31	HMSNs	263.19	Zhao et al. (2017)
One-pot synthesis and loading	BTA	N/A	MSNs	1222	Xu et al. (2018)
One-pot synthesis and loading	BTA	16.5	MSNs	619.6	Ma et al. (2021)
Supercritical CO_2	Dexamethasone	N/A	MCM-41 and SBA-15	-	Matos et al. (2013)
Supercritical CO_2	Fenofibrate	N/A	MSNs	450–600	Bouledjouidja et al. (2016)
Supercritical CO_2	Mangiferin	N/A	MSNs	342	García-Casas et al. (2018)
Supercritical CO_2	Zedoary oil	37.73	MSNs	675.53	Jia et al. (2019)
Supercritical CO_2	Benzoic acid	N/A	Mesoporous silica aerogel	580	Singh et al. (2021)

BTA, Benzotriazole; BSS, 2-(Benzothiazol-2-ylsulfanyl)-succinic acid; LbL, Layer-by-Layer; 2-MBT, 2-mercaptobenzo-thiazole; 8-HQ, 8-hydroxy quinoline.

11.4.2 Morphology by Electron Microscopy

The morphology of MSNs is typically assessed by scanning electron microscopy (FE-SEM or SEM). The elemental composition can be assessed by associated energy dispersive spectroscopy (EDS). TEM provides a fine porous structure and geometry. The TEM images of mesoporous silica, SBA-15 are presented in Figure 11.6 (Falcón et al. 2016). It is possible to observe the formation of well-defined parallel channels on the synthesized sample, where the cylindrical mesopores are uniformly and radially aligned (Figure 11.6a). The TEM images show clearly the well-ordered hexagonal arrangement of mesopores with an internal diameter of about 6–7 nm (Figure 11.6b). Figure 11.6c is a front view of the pore entrance in the mesoporous silica fiber walls.

Multiple tiny Ag (MT-Ag) NPs were loaded onto hollow mesoporous organosilica spheres (MT-Ag@HMOS). TEM images of MT-Ag@HMOS revealed numerous Ag NPs, distributed inside the cavity as well as on the PMO organosilica shells (Figure 11.7a and b). Ag NPs without aggregation were observed, exhibiting a uniform size of nearly 3 nm in diameter (Figure 11.7c). EDS analysis has shown the distributions of C, O, Si, and S elements in the shell (Figure 11.7d–g), with the core and shell rich in Ag (Figure 11.7h), demonstrating that particles located inside and on the outside of organosilica spheres are Ag NPs (Xu et al. 2020).

11.4.3 Size of Nanoparticles by Dynamic Light Scattering

The hydrodynamic diameters of relevant NPs were measured by dynamic light scattering (DLS) (Cui et al. 2019). The hydrodynamic diameter of MSN@OH was 164 nm on average. With the decoration of MSNs with poly(g-benzyl-L-glutamate) (PBLG), the hydrodynamic diameter significantly increased (ca. 250 nm), which was larger than those calculated from TEM analysis. Two factors explained this discrepancy: (i) the formation of a hydrate layer around the NPs in an aqueous solution; (ii) the occurrence of aggregation in the test dispersant system. The Zeta Potential of MSN@PBLG was more negative than MSN@OH due to anionic carboxy groups in PBLG (Cui et al. 2019).

11.4.4 X-Ray Diffraction

The crystalline structure or level of crystallinity of the samples is usually identified by X-ray powder diffraction (XRD). The XRD pattern of calcined mesoporous silica showed three main peaks, the most intense one at 2θ of 1.01°, which was assigned to the (100) plane, and the other two peaks at 1.68° and 1.9° were respectively assigned to the (110) and (200) planes. Here, the silica sample had no typical crystalline structure, and only three peaks appeared at very small 2θ angles. XRD

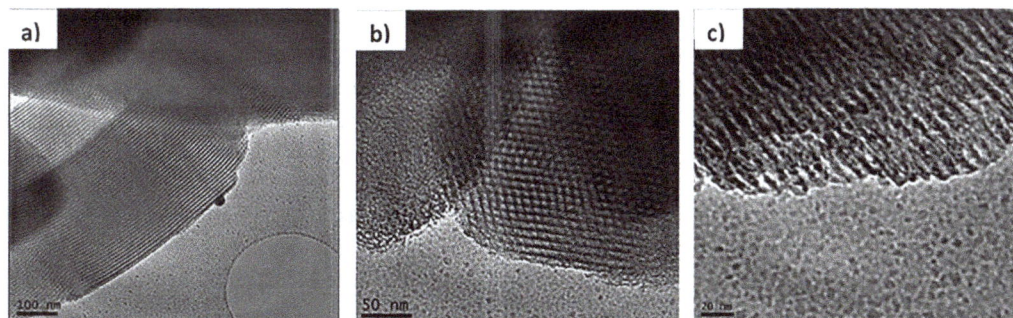

FIGURE 11.6 TEM images of calcined mesoporous silica (a) aligned pore channels and (b, c) entrance of hexagonal pores into the silica fiber walls. (Reproduced with permission from Falcón et al. (2016) © 2015 Elsevier B.V.)

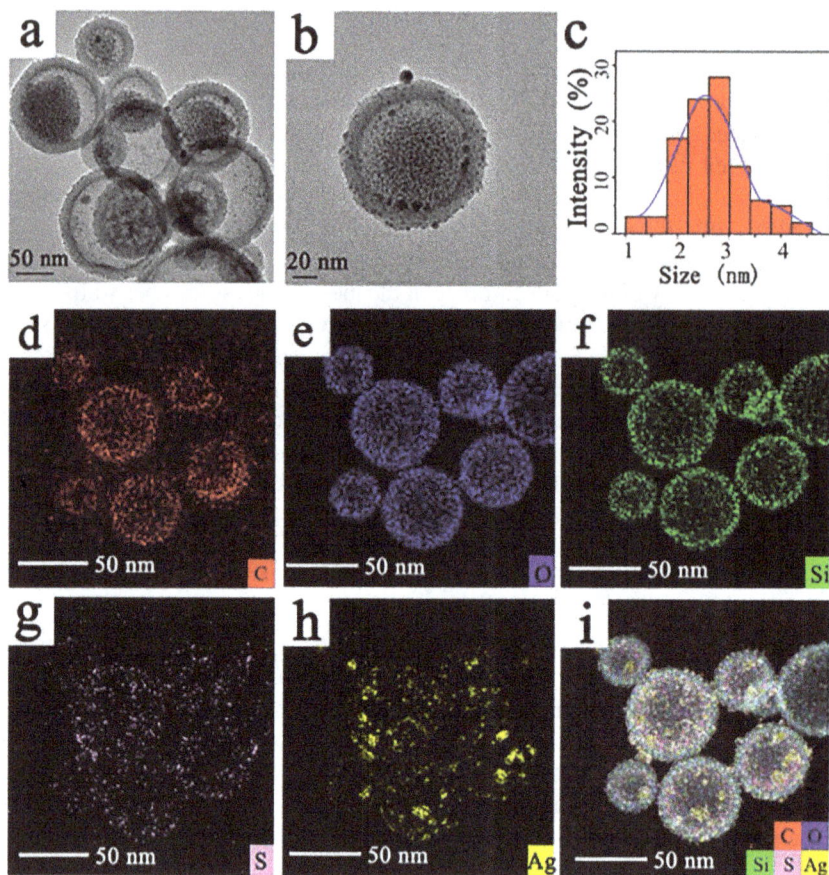

FIGURE 11.7 TEM images of MT-Ag@HMOS (a, b), particle size distribution of Ag NPs (c), EDS elemental mapping images of C, O, Si, S, and Ag (d–h) and their merged image (i). (Reproduced with permission from Xu et al. (2020) © 2020 Taiwan Institute of Chemical Engineers. Published by Elsevier B.V.)

at higher 2θ values showed the absence of peaks, but a single wide band with a maximum at ~25°, confirming the typical amorphous structure (Falcón et al. 2016).

11.4.5 N$_2$ Adsorption–Desorption Isotherms

The Brunauer–Emmett–Teller (BET) surface area of the prepared materials could be determined from the liquid N$_2$ adsorption–desorption isotherms. The pore size can be determined by the Barrett-Joyner-Halenda (BJH) method (Falcón et al. 2016). Typically, the pore size, pore volume, and surface area of MSNs decrease on inhibitor loading (He et al. 2017). MSNs usually show isotherm Type IV according to IUPAC classification, showing a hysteresis loop at about p/p_0 around 0.4, typical of condensation of N$_2$ in uniform mesoporous materials. The second hysteresis at a high relative pressure ($p/p_0 > 0.8$) corresponds to the interparticle voids for capillary condensation of N$_2$, which is considered the textural porosity of large pores (Nguyen et al. 2021). Furthermore, from N$_2$ adsorption–desorption isotherms, three well-distinguished regions were evident: (i) monolayer–multilayer adsorption (at low p/p_0 ratio), (ii) capillary condensation of N$_2$ at large pores (at around p/p_0 of 0.75), and (iii) multilayer adsorption on the outer particle surfaces (at high p/p_0) (Zhao et al. 1998a; Zhao et al. 1998b).

Nitrogen adsorption–desorption isotherm plots and pore size distribution curve for calcined mesoporous silica, SBA-15 was obtained by Falcón et al. This mesoporous silica owns typical Type IV isotherm and a clear type-H1 hysteresis loop, which is representative of an adsorbent material with a narrow distribution of relatively uniform pores. The pore size distribution indicated that the silica sample had well-defined uniform pore dimensions with an average pore radius of 2–3 nm. From the BET and BJH models, the specific surface area and total pore volume of MSNs were calculated to be 725 and 0.63 cm^3/g, respectively (Falcón et al. 2016).

11.4.6 FOURIER-TRANSFORM INFRARED SPECTROSCOPY

The functional chemical groups in the samples could be readily identified by Fourier-transform infrared spectroscopy (FTIR). This technique is used to establish the silica structure, its functionalization, and encapsulation. After amine grafting, the number of -OH groups on SBA-15 decreased significantly (Nguyen et al. 2021). The presence of N-H peaks identified the attachment of aminosilanes onto MSNs. The FTIR spectra of functionalized MSNs (with aminosilane and Eriochrome Black T) revealed the presence of Si-OH, Si-O-Si and Si-OH related to MSNs and N-H and azo groups corresponding to aminosilane and Eriochrome Black T (Ashrafi-Shahri et al. 2019).

11.5 ANALYTICAL METHODS FOR MONITORING THE RELEASE OF ENCAPSULATED COMPOUNDS

The release of encapsulated corrosion inhibitors or other substances can be detected or analyzed by many analytical techniques. Here, we provide a concise description of the selected instrumentations.

The corrosion inhibitor (molybdate) release from MSNs in a wider range of pH values was measured by ICP-OES. The release of molybdate inhibitor from loaded MSNs at various pHs was determined during 3 days of immersion. The amount of molybdate release in the alkaline media was much higher than in the acidic media, and that was attributed to the powder's surface potential in the solution. For pH up to 8, the rate of molybdate release increased sharply with time and then reached a constant value. The diffusion of molybdate in primary stages was easy and fast and obeyed the logarithmic law. With longer immersion time, the rate of release is reduced (Keyvani et al. 2017).

BTA-loaded MSNs were dispersed in water (pH 7), and HCl solutions (at pH 1.5 and 4.5) at a concentration of 0.030 mg/mL, and the BTA release was monitored by UV spectrophotometry. Calibration curves were used to determine the BTA concentration from the absorbance of the band between 220 and 300 nm. An increase in optical density was observed at low wavelengths, mainly due to the rise in the concentration of released BTA (Castaldo et al. 2020).

The release profiles of 2-MBT have been monitored by High-Performance Liquid Chromatography (HPLC) coupled with a diode-array ultraviolet (UV-DAD) detector. An isocratic mixture of water and acetonitrile has been used. The correlation coefficients of the calibration curves obtained were higher than 0.999. Samples of 5 mg of each nanocontainer (Si-NC, Si-MNP, MNP-impregnated) were dispersed in 100 mL of solvent (H$_2$O or ethanol) and monitored at time intervals between 0 and 120 days (Ruggiero et al. 2019). Different release kinetics have been obtained for different nanocontainers.

Electrochemical techniques such as electrochemical impedance spectroscopy (EIS) can be used as an indirect method to monitor the release of encapsulated corrosion inhibitors (Falcón et al. 2014, 2016, 2015). Figure 11.8 shows the impedance diagrams for coated carbon steel containing 1 wt% of nanocontainers loaded with dodecylamine, after different immersion times in 0.1 mol/L NaCl solution at pH 2. At the initial immersion times (1–3 h), there was only a slight increase in the capacitive arc's diameter as shown in Nyquist diagrams whose values oscillate around 750 Ω cm^2. The diameter increased significantly for longer immersion (16 h), reaching the impedance component up to 2500 Ω cm^2. These results demonstrate the efficient release of the inhibitor with time, the enhanced corrosion resistance and the permeability properties of polyelectrolyte walls which are open at pH values <3 (Falcón et al. 2014).

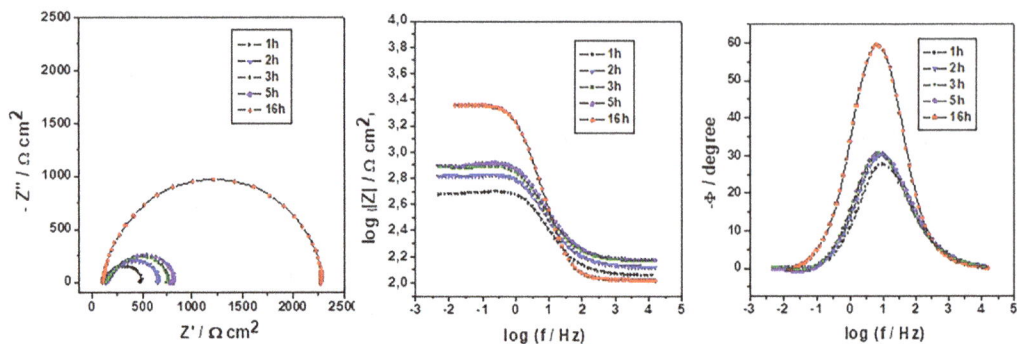

FIGURE 11.8 Impedance diagrams: Nyquist (a) and bode (b, c) plots of coated (containing 1 wt% of nano-containers with encapsulated dodecylamine) carbon steel after different immersion times in 0.1 mol/L NaCl at pH 2. (Reproduced with permission from Falcón et al. (2014) © 2013 Elsevier Ltd.)

Linear kinetics for release was determined. At pH 2, the ratio between the impedance modulus for the samples with and without nanocontainers increased quickly for longer immersion times (slope of 0.0039) compared to that of shorter times. However, at pH 6.2 and pH 9, this increase was more gradual (slope of 0.0010 for both), the release was slower due to the resistance of nano-container walls which are almost closed at these pH conditions. From the slopes of the kinetic curves, it was possible to affirm that at pH 2, the inhibitor release rate was ~3.9 times higher compared to pH 6.2 and pH 9 cases (Falcón et al.2014).

Falcón et al. have first employed Infrared Spectroscopy for *in situ* monitoring of the release. The intense peak at $1596 cm^{-1}$ was monitored during the experiment to prove the release of the dodecylamine from the nanocontainers. A 3D graph of IR spectra showed three peaks; the first peak was located at $1100 cm^{-1}$ and the second one at $1640 cm^{-1}$ and that was attributed to the overlaid peaks of the polystyrene sulfonate and polyethyleneimine. The third peak corresponded to dodecylamine. The study showed that besides the release of dodecylamine, there was also a release of polyelectrolytes from the nanocontainer walls. This system is sensitive to the pH shift in the acidic region, releasing the inhibitor on demand when the anodic process starts (Falcón et al. 2014).

11.6   MSNs-LOADED PROTECTIVE COATINGS - TECHNIQUES TO EVALUATE SELF-HEALING

Currently, an innovative theme in the area of protective coatings is the so-called self-repairing coatings. This type of coating with self-healing power contains additives that allow the polymeric film to be remade at the defect site or include corrosion inhibitors that will act in the defect area of the coating. Combining functionality and self-repair characteristics in coatings gives rise to a more comprehensive class, the so-called smart coatings. One of the strategies to obtain smart coatings is the use of synthetic (mesoporous silica) or natural nano/microcapsules (clays such as halloy-site, hydrotalcite, montmorillonite, and bentonite). The nano/microcapsules are incorporated in the coating during its formulation or application. They will allow the controlled output of the active agents according to the external stimulus for their releases, such as pH variation, temperature varia-tion, mechanical damage, radiation, and others. The direct incorporation of corrosion inhibitors into the polymeric matrix of coatings can lead to undesired reactions between the matrix and the inhibitors, which is the main reason to avoid the direct addition of anti-corrosion compounds in the coatings, which contributes vastly to the development of technology for the synthesis of nano reservoirs containing self-repairing agents (Wei et al. 2015; Zheludkevich et al. 2007a, b; Shchukin and Grigoriev 2012; Shchukin 2013; Thakur and Kessler 2015). The MSNs constitute the *extrinsic* self-healing concept, having the curing mechanism carried out by an agent not part of the material

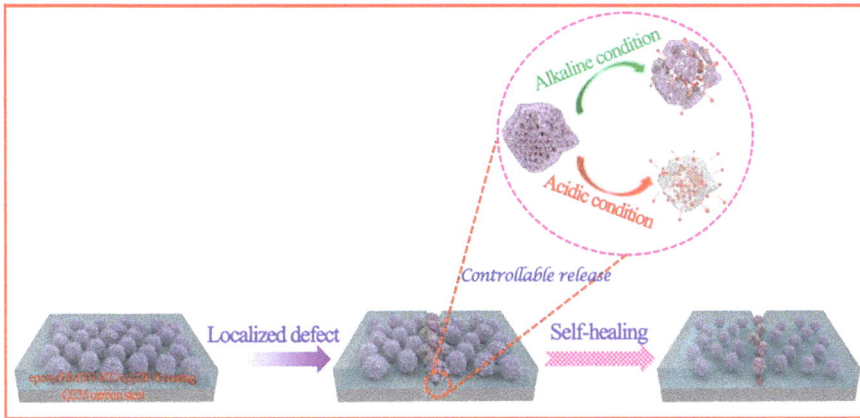

FIGURE 11.9 Schematic protection mechanism for the smart anticorrosion coating incorporated with BTA loaded HMSNs with pH-responsive gatekeepers. (Reproduced with permission from Zhou et al. (2020) © 2019 Elsevier B.V.)

itself but a compound of interest added to the coating (see Chapter 8). Figure 11.9 presents a scheme showing the self-healing mechanism in a doped coating with BTA-loaded HMSNs.

As evident from Table 11.2, a significant extent of works has been dedicated to coatings incorporated with MSNs. The data in the table shows that the most commonly used type of silica nanocontainers is a sol–gel soft template made and calcined MSNs. HMSNs follow this. Only a few papers used dense silica particles on which external surface inhibitors have been loaded by the LbL method.

The most frequently used substrate is carbons steel, followed by Al and Mg alloys. Galvanized steel and zinc were used in a few works. Only one or two works appeared on tin and Cu. The most used type of coatings with MSNs is hybrid sol–gel coatings, followed by epoxy coatings, alkyd primers, acrylic and polyester coatings, in this order. Two papers reported the embedding of MSN loaded with corrosion inhibitors in metallic coatings, one in Ni and another in Zn electrodeposited metallic coatings. BTA was the most frequently used inhibitor for loading, followed by molybdate, phosphomolybdate, 2-MBT, and 8-HQ (see Table 11.2).

The following section briefs the major techniques used to evaluate the self-healing properties of the coatings. The most used technique is EIS, followed by scanning vibrating electrode technique (SVET), and salt spray testing. EIS is performed in coated samples with and without a provoked defect. The high impedance at low frequencies corresponds to better inhibition due to self-healing. From Figure 11.10, the self-healing effect is clear for the coating loaded with inhibitor-containing MSNs when compared to the coating loaded with MSNs only (without inhibitor, empty). The key factor to obtain good EIS results as a technique to evaluate self-healing properties is to perform reproducible defects in the coating to expose the same area to the electrolyte. Using a microdrill tool of 200 μm diameter, and checking the defect site with an optical microscope, it is possible to achieve reliable results.

SVET, a local electrochemical technique, is a powerful method to study self-healing properties in a region of defect. The method is based on the measurement of tiny potential variations created by ionic current flows in two vertical positions of the probe vibration caused by the reactions occurring in the active surface of the material. The ionic current maps allowed the quantification and extension of the corrosion processes on the studied material surface within the exposed region. In the ionic current maps, the red color represents the ionic currents resulting from anodic reactions (oxidation process), and the blue represents ionic currents resulting from cathodic reactions (reduction process) (Karavai et al. 2010; Bastos et al. 2017). Figure 11.11 shows the results of Falcón et al. on an alkyd primer loaded with MSNs with encapsulated dodecylamine. For the doped coating

TABLE 11.2

Self-Healing Properties of Coatings Incorporated with MSNs Loaded with Corrosion Inhibitors

Techniques Used to Evaluate Self-Healing	Type of MSNs	Substrate	Type of Coating	Corrosion Inhibitor Used	Ref.
EIS and SVET	Dense silica and LbL inhibitor loading	AA 2024	Sol–gel hybrid coating	BTA	Shchukin et al. (2006)
EIS and SVET	Dense silica and LbL inhibitor loading	AA 2024	Sol–gel	BTA	Zheludkevich et al. (2007b)
EIS and SVET	N/A	AA–2024	Sol–gel, SiO$_2$/ZrO$_2$	8-HQ	Lamaka et al. (2008)
SVET	MSNs	AA 2024	Sol–gel	2-benzothiazol-2-sulfamyl succine	Skorb et al. (2009)
SVET	MSNs	AA 2024	Sol–gel	BTA	Grigoriev et al. (2009)
SVET	MSNs	AA 2024	Sol–gel	BTA	Borisova (2011)
SVET and EIS	MSNs	Galvanized steel	Polyester primer	BTA	Hollamby et al. (2011)
EIS	MSNs, one-pot	AA 2024	Water based epoxy	2-MBT	Maia et al. (2012)
EIS and salt spray	Na-MMT and MSN (MCM–41)	Mild steel	Epoxy	No inhibitor	Wang et al. (2012)
SVET	MSNs	Mild steel	Polyester coating	BTA and a biocide	Zheng et al. (2013)
SVET	MSNs	AA–2024	Water-borne epoxy	2-MBT	Borisova et al. (2013a)
EIS machu test					
SVET	MSNs	AA–2024	Sol–gel, SiO2/ZrO2	2-MBT	Borisova et al. (2013b)
EIS					
SVET	HMSNs – PS template + capping polyelectrolytes	AA–2024	GPTMS + TEOS (SNAP coating)	Caffeine	Fu et al. (2013)
EIS	MSN functionalized with aminosilane	Mild steel	Electrodeposited polypirrole	Molybdate	Yeganeh et al. (2014)
EIS	HMSNs	Ferritic steel (9Cr 1Mo)	Sol–gel, SiO$_2$/ZrO$_2$	2-MBT (72% loading)	Chenan et al. (2014b)
EIS, SVET, salt spray	Dense silica	Mild steel	Alkyd	Dodecylamine	Falcón et al. (2014)
EIS	MSNs	Mg alloy	Epoxy	2-MBT	Qiao et al. (2015)
EIS, polarization	MSNs	Zn	Sol–gel	Rhodamine 6G dye	Albert et al. (2015)
Polarization	MSNs capped with poly (diallyldimethylammonium chloride)	Mild steel	MSNs with molybdate	Phosphomolybdate	Zea et al. (2015)
EIS	MSNs + aminosilane	Mg	Alkyd	Fluoride ions	Yeganeh and Saremi (2015)
EIS	MSNs	Mild steel	Epoxy	Molybdate	Yeganeh and Keyvani (2016)
XPS					
EIS, polarization	MSNs modified with polyaniline	Mild steel	Epoxy	BTA	Wang et al. (2016)

(Continued)

TABLE 11.2 (Continued)

Self-Healing Properties of Coatings Incorporated with MSNs Loaded with Corrosion Inhibitors

Techniques Used to Evaluate Self-Healing	Type of MSNs	Substrate	Type of Coating	Corrosion Inhibitor Used	Ref.
Polarization, electrochemical noise	MSNs (MCM-41) and ash	Mild steel	Polythiophene	Fe(III)	Gutiérrez-Díaz et al. (2016)
SVET, EIS	MSNs+HMAP (apatite)+fluorsilane (hydrophobic)	Mg alloy	Self-assembled nanophase particle (SNAP) barrier coating	2-hydroxy H methoxy acetophenone	Ding et al. (2016)
Polarization	Silica xerogel	Galvanized steel	Sol-gel	BTA+fluorsilane	Cotolan et al. (2016)
SVET, SECM	Mesoporous silica film	AA 2024	Silica sol-gel+acrylic top coat	BTA	Recloux et al. (2016)
SVET, EIS salt spray	MSNs, SBA-15	Mild steel	Alkyd primer	dodecylamine	Falcón et al. (2016)
EIS	MSNs+β-cyclodextrin monofunctionalized with a ferrocene moiety	AA2024	Ce(IV)-doped ZrO_2-SiO_2 sol-gel	CA	Wang et al. (2017)
SVET, EIS, salt spray	MSNs, SBA-15	AA2024-T3	Epoxy	8-HQ	Shi et al. (2017)
Salt spray	MSNs+halloysite	Mild steel	Polyester powder	8-HQ	Shchukina et al. (2017)
EIS, polarization	MSNs MCM-41	Zn	Electrodeposition, 3-mercaptopropyltrimethoxysilane	Molybdate	Alipour and Nasirpouri (2017)
EIS	HMSNs	Al	Sol-gel, SiO_2/ZrO_2	BTA	Zhao et al. (2017)
EIS, polarization	MSNs, MCM-41	Mg alloy	Ni, electrodeposited	NaF	Xie et al. (2017)
LEIS, SKP	HMSNs	Mild steel	Sol-gel, SiO_2/ZrO_2 coating	Phosphomolybdate	Zea et al. (2017a)
SKP	HMSNs	Mild steel	Sol-gel	Phosphomolybdate	Zea et al. (2017b)
Salt spray	MSNs siloyd C803 and halloysite	Mild steel	Epoxy powder coating	8-HQ	Shchukina et al. (2018)
EIS, salt spray	MSNs film	AA2024-T3	Mesoporous sol-gel film+epoxy	No inhibitor	Thai et al. (2018)
SKP	MSNs+aminosilane	Mild steel	No coating	Phosphomolybdate+zinc chromate	Zea et al. (2018a)
EIS	MSNs+aminosilane	Mild steel	Epoxy	Molybdate	Yeganeh et al. (2018)
SVET	MSNs, one-pot	Cu	Alkyd primer	BTA	Xu et al. (2018)
Polarization	MSNs, PDDA	Mild steel	No coating	molybdate	Zea et al. (2018b)
Chemical and abrasion resistance	MSNs, SBA-15	Tin plate, carbon steel	Epoxy resin+benzoxazine	No inhibitor	Li et al. (2019)
EIS, polarization	MSNs+tannic acid+Fe(III) (gatekeepers)	Mild steel	Water-borne alkyd coating	BTA	Qian et al. (2019a)

(Continued)

TABLE 11.2 (Continued)

Self-Healing Properties of Coatings Incorporated with MSNs Loaded with Corrosion Inhibitors

Techniques Used to Evaluate Self-Healing	Type of MSNs	Substrate	Type of Coating	Corrosion Inhibitor Used	Ref.
EIS, noise, polarization	MSNs, one-pot	Mild steel	Epoxy	Molybdate	Yeganeh et al. (2019b)
EIS, polarization	MSNs, MCM-41 + sulfur silane	Mild steel	Zn	No inhibitor	Alipour and Nasirpouri (2019)
SVET, EIS, salt spray	MSNs + supramolecular nanovalves.	AA 2024	Sol–gel, SiO_2/ZrO_2	BTA	Wang et al. (2019)
EIS	MSNs + aminosilane	Mild steel	Sol–gel, TiO_2/SiO_2	Erio–chrome black T + Fe (III)	Ashrafi-Shahri et al. (2019)
EIS	MSNs, one-pot	Zn	Mesoporous silica coating	BTA	Szabó et al. (2019)
EIS, salt spray	MSNs + graphene oxide sheets + aminosilanes	AA 2024	Sol–gel, ZrO_2/SiO_2	BTA	Xiong et al. (2019)
EIS salt spray	MSNs, MCM-48 + polydopamine as gatekeeper	mild steel	Water-borne alkyd	BTA	Qian et al. (2019b)
EIS	MSNs (1%)	Mild steel	Epoxy coating	sulfamethazine	Yeganeh et al. (2019a)
SVET, EIS	MSNs + graphene oxide sheets	Mild steel	Sol–gel + PVB top coating	BTA	Du et al. (2019b)
EIS, salt spray	MSNs, succinic acid + aminosilane functionalized + capped with PEI	Mild steel	Epoxy	BTA	Wen et al. (2020)
EIS, polarization	MSNs + perfluorooctyltriethoxysilane	Tin plate, carbon steel	Epoxy	No inhibitor	Tang et al. (2020)
EIS	MSNs, SBA-15 + piperazine	Mild steel	Epoxy silicone	2-MBT	Amini et al. (2020)
UV accelerated test, HCl vapor test	MSNs, vacuum loading	Cu	Acrylic	BTA	Castaldo et al. (2020)
EIS, polarization	MSNs silanized + LDH gate keepers	Mg AZ31	SNAP, sol–gel	2-MBT	Ouyang et al. (2020)
EIS, SKP salt spray	HMSNs + ZIF gatekeeper	Mild steel	Epoxy composite	BTA	Zhou et al. (2020)
SVET, EIS, salt spray	MSNs loaded with BTA and capped with polyelectrolytes	AA 2024	Water-borne epoxy	BTA	Udoh et al. (2020)
EIS, polarization	Hydrophobic MSNs + hexamethyl disilazane	Mild steel	Stearic acid grafted chitosan	No inhibitor	Shamsheera et al. (2020)
SVET, EIS	Dendrimer-like MSN capped with PEI	Mild steel	Acrylic	8-HQ	Siva et al. (2021)
EIS, salt spray	MSNs, sol–gel	Mild steel	Water-borne epoxy	8-HQ, BTA	Židov et al. (2021)

SVET, scanning vibrating electrode technique; SECM, scanning electrochemical microscopy; LEIS, local electrochemical impedance spectroscopy; SKP, scanning Kelvin probe; CA, p-coumaric acid.

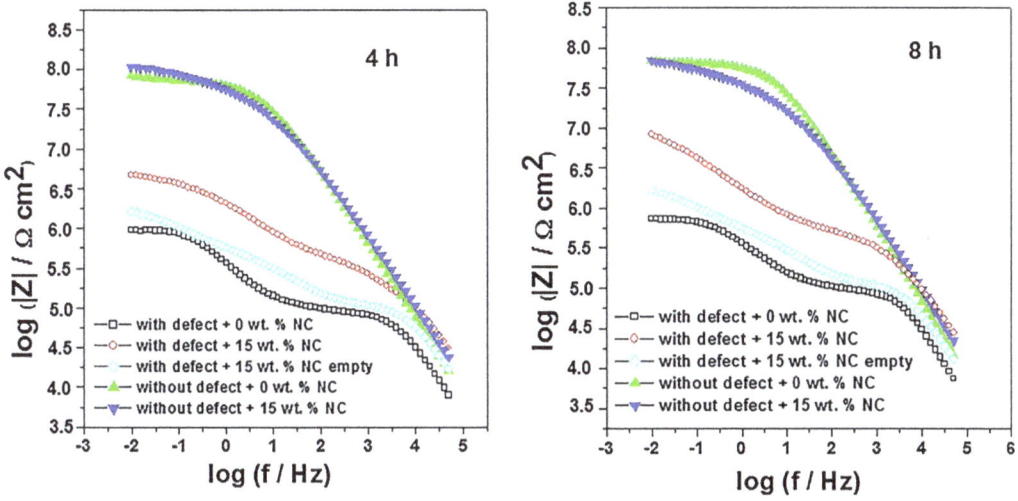

FIGURE 11.10 Bode plots for carbon steel coated with alkyd primer doped with 0 or 15 wt% of MSNs loaded/ not loaded with dodecylamine. The plots were recorded after 4 and 8 h of immersion in 0.01 mol/L NaCl with/ without a provoked defect. (Reproduced with permission from Falcón et al. (2016) © 2015 Elsevier B.V.)

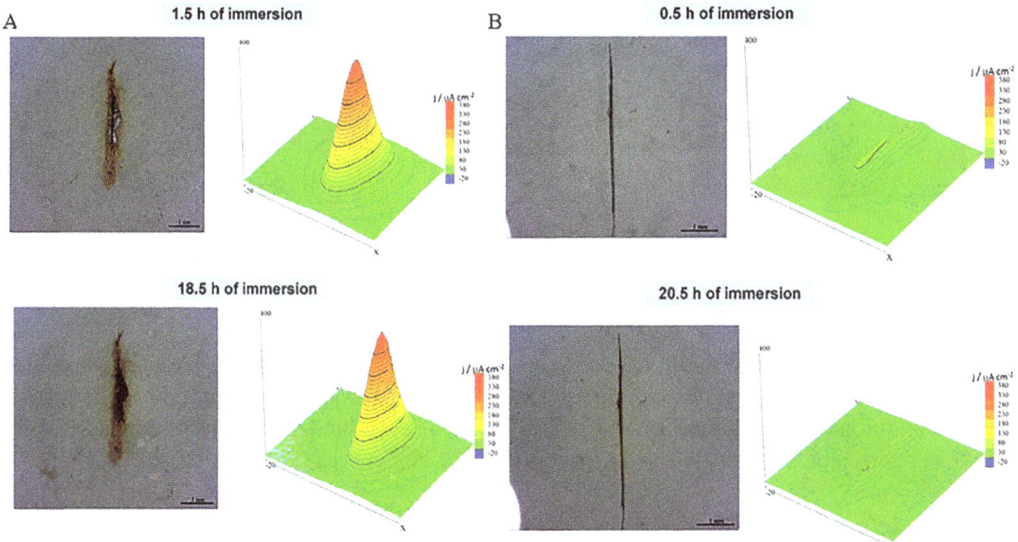

FIGURE 11.11 SVET maps of ionic currents measured above the surface of carbon steel coated with alkyd primer: (a) without and (b) with 10 wt% of dodecylamine loaded MSNs, after different immersion times in 0.01 mol/L NaCl. Reproduced with permission from Falcón et al. (2014) © 2013 Elsevier Ltd.)

(Figure 11.11b), the ionic currents above the defect site are very low, near-zero, proving the release of the corrosion inhibitor, which has been adsorbed on the carbon steel surface exposed in the scribe. For the bare, high ionic current densities appear on the defect site during the whole immersion time (Figure 11.11a) (Falcón et al. 2014).

The accelerated corrosion test (salt spray test) can be used to prove self-healing properties. Figure 11.12 presents the SSC test results obtained by Falcón et al. for a double layer alkyd coating with the primer layer doped with MSNs with encapsulated dodecylamine inhibitor. It is possible to see that after 720 h, the coated samples without MSNs presented severe corrosion attack, especially around the provoked defect area, besides intense blistering under the coating, which indicates a

Without inhibitor loaded silica

15 wt.% of inhibitor loaded silica

0 h

Without inhibitor loaded silica

15 wt.% of inhibitor loaded silica

720 h

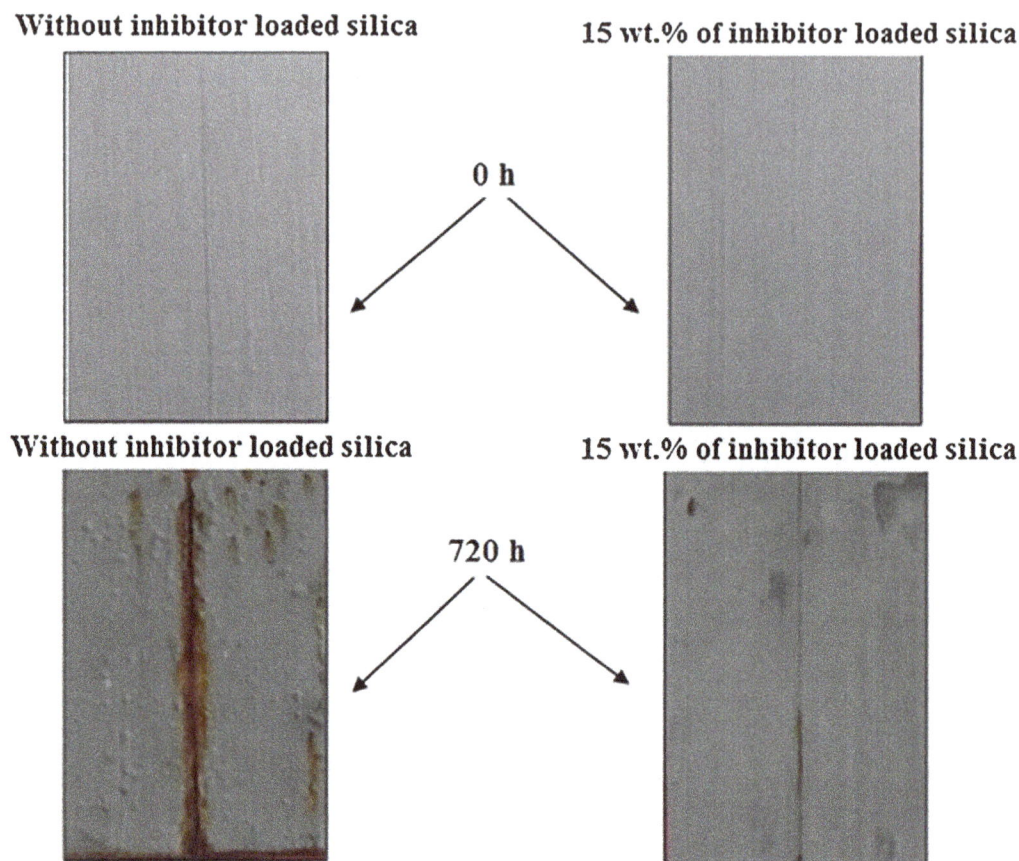

FIGURE 11.12 Coated samples with two layers of alkyd paint with a primer containing 0 or 15 wt% of MSNs loaded with inhibitors, before and after 720 h of exposure in the salt spray chamber. (Reproduced with permission from Falcón et al. (2016) © 2015 Elsevier B.V.)

permeation of the aggressive solution toward the metal surface through the sharp tool-made defect. It is possible to conclude that the addition of 15 wt% of mesoporous silica into the primer coating provided additional protection against corrosion of carbon steel, and after 720 h of exposure, there were fewer corrosion products and blistering around the defect area (Falcón et al. 2016).

11.7 CONCLUSIONS AND PERSPECTIVES

This chapter is intended to present the most significant information on the synthesis methods, characterization techniques, and loading processes of MSNs. The kinetics of the release of encapsulated compounds was also looked at. Special emphasis was given to the loading of MSNs in coatings and how the active protection and self-healing effects can be assessed.

MSNs have gained immense attention and have been a preferred choice among various nanocarriers to achieve corrosion inhibitor/drug encapsulation. The relatively easy synthesis is the main favorable point. Adequate functionalization of MSNs helps to better interact with the encapsulated compounds, resulting in higher encapsulation efficiency and controlled release. Toward this objective, the concept and approaches in supramolecular chemistry are a valuable tool to find the best way to optimize encapsulation and the release of adsorbed molecules on MSNs (Wang et al. 2017; Wang et al. 2019; Ding et al. 2017; Qian et al. 2019a; Saji 2019; Ouyang et al. 2020). The most recent papers have just used these concepts to obtain novel MSNs for anticorrosive coatings (Wen et al. 2020). Another significant trend is the development of engineered MSNs that resemble polymeric molecules called dendrimers (Siva et al. 2021), which enhances the loading capacity of regular MSNs. A bright future seems to come with novelties in MSN to encapsulate corrosion inhibitors or drugs.

ACKNOWLEDGMENTS

The authors are thankful to CNPq - Conselho Nacional de Desenvolvimento Científico e Tecnológico – Brasil for financial support and scholarships to researchers (processes 310504/2020-1 and 140187/2017-0).

REFERENCES

Abdullayev, E., R. Price, D. Shchukin, and Y. Lvov. 2009. "Halloysite tubes as nanocontainers for anticorrosion coating with benzotriazole." *ACS Applied Materials & Interfaces* 1 (7): 1437–1443. doi:10.1021/am9002028.

Abu-Thabit, N. Y., and A. S. Hamdy. 2016. "Stimuli-responsive polyelectrolyte multilayers for fabrication of self-healing coatings – a review." *Surface and Coatings Technology 303 (Part B)* Elsevier B.V.: 406–424. doi:10.1016/j.surfcoat.2015.11.020.

"ACS Chem-a, ACS Material Advanced Chemicals Supplier." a https://www.acsmaterial.com/mcm-41.html.

"ACS Chem-b, ACS Material Advanced Chemicals Supplier." b https://www.acsmaterial.com/fdu-12.html.

Albert, E., N. Cotolan, N. Nagy, G. Sáfrán, G. Szabó, L. Maria Mureşan, and Z. Hórvölgyi. 2015. "Mesoporous silica coatings with improved corrosion protection properties." *Microporous and Mesoporous Materials* 206 (C): 102–113. doi:10.1016/j.micromeso.2014.12.021.

Alipour, K., and F. Nasirpouri. 2017. "Smart anti-corrosion self-healing zinc metal-based molybdate functionalized-mesoporous-silica (MCM-41) nanocomposite coatings." *RSC Advances* 7 (82): 51879–51887. doi:10.1039/c7ra06923e.

Alipour, K. and F. Nasirpouri. 2019. "Effect of morphology and surface modification of silica nanoparticles on the electrodeposition and corrosion behavior of zinc-based nanocomposite coatings." *Journal of The Electrochemical Society* 166 (2): D1–D9. doi:10.1149/2.0191902jes.

Amini, M., R. Naderi, M. Mahdavian, and A. Badiei. 2020. "Effect of piperazine functionalization of mesoporous silica type SBA-15 on the loading efficiency of 2-mercaptobenzothiazole corrosion inhibitor." *Industrial and Engineering Chemistry Research* 59 (8): 3394–3404. doi:10.1021/acs.iecr.9b05261.

Andreeva, D. V., D. Fix, H. Möhwald, and D. G. Shchukin. 2008. "Self-healing anticorrosion coatings based on PH-sensitive polyelectrolyte/inhibitor sandwichlike nanostructures." *Advanced Materials* 20 (14): 2789–2794. doi:10.1002/adma.200800705.

Andreeva, D. V., E. V. Skorb, and D. Shchukin. 2010. "Layer-by-layer polyelectrolyte/inhibitor nanostructures for metal corrosion protection." *Applied Materials & Interfaces* 2: 1954–1962. doi: 10.1021/am1002712.

Ariga, K., Q. Ji, G. J. Richards, and J. P. Hill. 2012. "Soft capsules, hard capsules, and hybrid capsules." *Soft Materials* 10 (4): 387–412. doi:10.1080/1539445x.2010.523751.

Ashrafi-Shahri, S. M., F. Ravari, and D. Seifzadeh. 2019. "Smart organic/inorganic sol-gel nanocomposite containing functionalized mesoporous silica for corrosion protection." *Progress in Organic Coatings* 133: 44–54. doi:10.1016/j.porgcoat.2019.04.038.

Bastos, A. C., M. C. Quevedo, O. V. Karavai, and M. G. S. Ferreira. 2017. "Review—on the application of the scanning vibrating electrode technique (svet) to corrosion research." *Journal of The Electrochemical Society* 164 (14): C973–C990. doi:10.1149/2.0431714jes.

Beck, J. S., J. C. Vartuli, W. J. Roth, M. E. Leonowicz, C. T. Kresge, K. D. Schmitt, C. T. W. Chu, et al. 1992 "A new family of mesoporous molecular sieves prepared with liquid crystal templates." *Journal of the American Chemical Society* 114 (27): 10834–10843. doi:10.1021/ja00053a020.

Bendinelli, E. V., I. V. Aoki, O. Barcia, and I. C. P. Margarit-Mattos. 2019. "Kinetic aspects of Mg-Al layered double hydroxides influencing smart corrosion protective behavior." *Materials Chemistry and Physics* 238: 121883. doi:10.1016/J.MATCHEMPHYS.2019.121883.

Borisova, D., D. Akçakayiran, M. Schenderlein, H. Möhwald, and D. G. Shchukin. 2013a. "Nanocontainer-based anticorrosive coatings: effect of the container size on the self-healing performance." *Advanced Functional Materials* 23 (30): 3799–3812. doi:10.1002/adfm.201203715.

Borisova, D., H. Mohwald, and D. G. Shchukin. 2011. "Mesoporous silica nanoparticles for active corrosion protection." *ACS Nano* 5 (3): 1939–1946. doi:10.1021/Nn102871v.

Borisova, D., H. Möhwald, and D. G. Shchukin. 2013b. "Influence of embedded nanocontainers on the efficiency of active anticorrosive coatings for aluminum alloys part II: influence of nanocontainer position." *ACS Applied Materials and Interfaces* 5 (1): 80–87. doi:10.1021/am302141y.

Bouledjouidja, A., Y. Masmoudi, M. V. Speybroeck, L. Schueller, and E. Badens. 2016. "Impregnation of fenofibrate on mesoporous silica using supercritical carbon dioxide." *International Journal of Pharmaceutics* 499 (1–2): 1–9. doi:10.1016/j.ijpharm.2015.12.049.

Carneiro, J., A. F. Caetano, A. Kuznetsova, F. Maia, A. N. Salak, J. Tedim, N. Scharnagl, M. L. Zheludkevich, and M. G. S. Ferreira. 2015. "Polyelectrolyte-modified layered double hydroxide nanocontainers as vehicles for combined inhibitors." *RSC Advances* 5 (50): 39916–39929. doi:10.1039/c5ra03741g.

Castaldo, R., M. Salzano de Luna, C. Siviello, G. Gentile, M. Lavorgna, E. Amendola, and M. Cocca. 2020. "On the acid-responsive release of benzotriazole from engineered mesoporous silica nanoparticles for corrosion protection of metal surfaces." *Journal of Cultural Heritage* 44: 317–324. doi:10.1016/j.culher.2020.01.016.

Chen, H., J. He, H. Tang, C. Yan, H. Chen, J. He, H. Tang, and C. Yan. 2008. "Porous silica nanocapsules and nanospheres : dynamic self-assembly synthesis and application in controlled release porous silica nanocapsules and nanospheres : dynamic self-assembly synthesis and application in controlled release." *Chemistry of Materials* 20 (18): 5894–5900. doi:10.1021/cm801411y.

Chen, T., and J. Fu. 2012. "PH-responsive nanovalves based on hollow mesoporous silica spheres for controlled release of corrosion inhibitor." *Nanotechnology* 23: 235605. doi:10.1088/0957-4484/23/23/235605.

Chenan, A., S. Ramya, R. P. George, and U. Kamachi Mudali. 2014a. "Hollow mesoporous zirconia nanocontainers for storing and controlled releasing of corrosion inhibitors." *Ceramics International* 40: 10457–10463. doi:10.1016/j.ceramint.2014.03.016.

Chenan, A., S. Ramya, R. P. George, and U. Kamachi Mudali. 2014b. "2-mercaptobenzothiazole-loaded hollow mesoporous silica-based hybrid coatings for corrosion protection of modified 9Cr-1Mo ferritic steel." *Corrosion* 70: 496–511. doi:10.5006/1090.

Cotolan, N., S. Varvara, E. Albert, G. Szabó, Z. Hórvölgyi, and L. M. Mureşan. 2016. "Evaluation of corrosion inhibition performance of silica sol–gel layers deposited on galvanised steel." *Corrosion Engineering Science and Technology* 51 (5): 373–382. doi:10.1080/1478422X.2015.1120404.

Croissant, J., G. X. Cattoën, M. W. Chi Man, J. O. Durand, and N. M. Khashab. 2015. "Syntheses and applications of periodic mesoporous organosilica nanoparticles." *Nanoscale* 7 (48): 20318–20334. doi:10.1039/C5NR05649G.

Cui, Y., R. Deng, X. Li, X. Wang, Q. Jia, E. Bertrand, K. Meguellati, and Y. W. Yang. 2019. "Temperature-sensitive polypeptide brushes-coated mesoporous silica nanoparticles for dual-responsive drug release." *Chinese Chemical Letters* 30 (12): 2291–2294. doi:10.1016/j.cclet.2019.08.017.

De Matos, M. B. C., A. P. Piedade, C. Alvarez-Lorenzo, A. Concheiro, M. E. M. Braga, and H. C. de Sousa. 2013. "Dexamethasone-loaded poly(E-caprolactone)/silica nanoparticles composites prepared by supercritical CO2 foaming/mixing and deposition." *International Journal of Pharma* 456: 269–281. doi:10.1016/j.ijpharm.2013.08.042.

Deng, S., C. X. Cui, L. Liu, L. Duan, J. Wang, Y. Zhang, and L. Qu. 2020. "A facile and controllable one-pot synthesis approach to amino-functionalized hollow silica nanoparticles with accessible ordered mesoporous shells." *Chinese Chemical Letters* 32: 1177–1180 doi:10.1016/j.cclet.2020.09.002.

Ding, C., Y. Liu, M. Wang, T. Wang, and J. Fu. 2016. "Self-healing, superhydrophobic coating based on mechanized silica nanoparticles for reliable protection of magnesium alloys." *Journal of Materials Chemistry A* 4 (21): 8041–8052. doi:10.1039/c6ta02575g.

Ding, C. D., L. Tong, and J. J. Fu. 2017. "Quadruple stimuli-responsive mechanized silica nanoparticles: a promising multifunctional nanomaterial for diverse applications." *Chemistry - A European Journal* 23 (60): 15041–15045. doi:10.1002/chem.201704245.

Du, G., Y. Song, N. Li, X. Lijian, C. Tong, Y. Feng, T. Chen, and J. Xu. 2019a. "Cage-like hierarchically mesoporous hollow silica microspheres templated by mesomorphous polyelectrolyte-surfactant complexes for noble metal nanoparticles immobilization." *Colloids and Surfaces A: Physicochemical and Engineering Aspects* 575: 129–139. doi:10.1016/j.colsurfa.2019.04.088.

Du, P., J. Wang, H. Zhao, G. Liu, and L. Wang. 2019b. "Graphene oxide encapsulated by mesoporous silica for intelligent anticorrosive coating: studies on release models and self-healing ability." *Dalton Transactions* 48 (34): 13064–13073. doi:10.1039/c9dt02454a.

Dyer, A., J. Newton, and M. Pillinger. 2009. "Synthesis and characterisation of mesoporous silica phases containing heteroatoms, and their cation exchange properties. part 1. Synthesis of Si, Al, B, Zn substituted MCM-41 materials and their characterisation." *Microporous and Mesoporous Materials* 126 (1–2): 192–200. doi:10.1016/j.micromeso.2009.06.002.

Eggenhuisen, T. M., G. Prieto, H. Talsma, K. P. De Jong, and P. E. De Jongh. 2012. "Entrance size analysis of silica materials with cagelike pore structure by thermoporometry." *The Journal of Physical Chemistry C* 116 (44): 23383–23393. doi:10.1021/jp3070213.

Falcón, J. M., F. F. Batista, and I. V. Aoki. 2014. "Encapsulation of dodecylamine corrosion inhibitor on silica nanoparticles." *Electrochimica Acta* 124: 109–118. doi:10.1016/j.electacta.2013.06.114.

Falcón, J. M., L. M. Otubo, and I. V. Aoki. 2016. "Highly ordered mesoporous silica loaded with dodecylamine for smart anticorrosion coatings." *Surface and Coatings Technology* 303: 319–329. doi:10.1016/j.surfcoat.2015.11.029.

Falcón, J. M., T. Sawczen, and I. V. Aoki. 2015. "Dodecylamine-loaded halloysite nanocontainers for active anticorrosion coatings." *Frontiers in Materials* 2: 1–13 doi:10.3389/fmats.2015.00069.

Feng, Y., and Y. F. Cheng. 2017. "An intelligent coating doped with inhibitor-encapsulated nanocontainers for corrosion protection of pipeline steel." *Chemical Engineering Journal* 315: 537–551. doi:10.1016/j.cej.2017.01.064.

Fix, D., D. V. Andreeva, Y. M. Lvov, D. G. Shchukin, and H. Möhwald. 2009. "Application of inhibitor-loaded halloysite nanotubes in active anti-corrosive coatings." *Advanced Functional Materials* 19 (11): 1720–1727. doi:10.1002/adfm.200800946.

Fu, J., T. Chen, M. Wang, N. Yang, S. Li, Y. Wang, and X. Liu. 2013. "Acid and alkaline dual stimuli-responsive mechanized hollow mesoporous silica nanoparticles as smart nanocontainers for intelligent anticorrosion coatings." *ACS Nano* 7 (12): 11397–11408. doi:10.1021/nn4053233.

García-Casas, I., A. Montes, D. Valor, C. Pereyra, and E. J. Martínez de la Ossa. 2018. "Impregnation of mesoporous silica with mangiferin using supercritical CO_2." *The Journal of Supercritical Fluids* 140: 129–136. doi:10.1016/j.supflu.2018.06.013.

Ghosh, S. K. 2006. *Functional Coatings: By Polymer Microencapsulation.* ISBN: 978-3-527-31296-2.

Grigoriev, D., E. Shchukina, and D. G. Shchukin. 2017. "Nanocontainers for self-healing coatings." *Advanced Materials Interfaces* 4 (1): 1600318. doi:10.1002/admi.201600318.

Grigoriev, D. O., K. Köhler, E. Skorb, D. G. Shchukin, and H. Möhwald. 2009. "Polyelectrolyte complexes as a 'smart' depot for self-healing anticorrosion coatings." *Soft Matter* 5 (7): 1426. doi:10.1039/b815147d.

Gutiérrez-Díaz, J. L., J. Uruchurtu-Chavarín, M. Güizado-Rodríguez, and V. Barba. 2016. "Steel protection of two composite coatings: polythiophene with Ash or MCM-41 particles containing iron(III) nitrate as inhibitor in chloride media." *Progress in Organic Coatings* 95: 127–135. doi:10.1016/j.porgcoat.2016.03.005.

Haddadi, S. A., T. B. Kohlan, S. Momeni, S. A. Ramazani, and M. Mahdavian. 2019. "Synthesis and application of mesoporous carbon nanospheres containing walnut extract for fabrication of active protective epoxy coatings." *Progress in Organic Coatings* 133: 206–219. doi:10.1016/j.porgcoat.2019.04.046.

Hao, P., B. Peng, B. Q. Shan, T. Q. Yang, and K. Zhang. 2020. "Comprehensive understanding of the synthesis and formation mechanism of dendritic mesoporous silica nanospheres," *Nanoscale Advances* 2 (5): 1792–1810. doi:10.1039/d0na00219d.

He, Y., S. Liang, M. Long, and H. Xu. 2017. "Mesoporous silica nanoparticles as potential carriers for enhanced drug solubility of paclitaxel." *Materials Science and Engineering C* 78: 12–17. doi:10.1016/j.msec.2017.04.049.

Hollamby, M. J., D. Fix, I. Dönch, D. Borisova, H. Möhwald, and D. Shchukin. 2011. "Hybrid polyester coating incorporating functionalized mesoporous carriers for the holistic protection of steel surfaces." *Advanced Materials* 23 (11): 1361–1365. doi:10.1002/adma.201003035.

Ijaz, A., M. B. Yagci, C. W. Ow-Yang, A. L. Demirel, and A. Mikó. 2020. "Formation of mesoporous silica particles with hierarchical morphology." *Microporous and Mesoporous Materials* 303: 110240. doi:10.1016/j.micromeso.2020.110240.

Iza, M., S. Woerly, C. Danumah, S. Kaliaguine, and M. Bousmina. 2000. "Determination of pore size distribution for mesoporous materials and polymeric gels by means of DSC measurements: thermoporometry." *Polymer* 41 (15): 5885–5893. doi:10.1016/S0032-3861(99)00776-4.

Janus, R., M. Wądrzyk, M. Lewandowski, P. Natkański, P. Łątka, and P. Kuśtrowski. 2020. "Understanding porous structure of SBA-15 upon pseudomorphic transformation into MCM-41: non-direct investigation by carbon replication." *Journal of Industrial and Engineering Chemistry* 92: 131–144. doi:10.1016/j.jiec.2020.08.032.

Jia, J., X. Liu, K. Wu, X. Zhou, and F. Ge. 2019. "Loading zedoary oil into PH-sensitive chitosan grafted mesoporous silica nanoparticles via gate-penetration by supercritical CO2 (GPS)." *Journal of CO2 Utilization* 33: 12–20. doi:10.1016/j.jcou.2019.05.010.

Karavai, O. V., A. C. Bastos, M. L. Zheludkevich, M. G. Taryba, S. V. Lamaka, and M. G. S. Ferreira. 2010. "Localized electrochemical study of corrosion inhibition in microdefects on coated AZ31 magnesium alloy." *Electrochimica Acta* 55 (19): 5401–5406. doi:10.1016/j.electacta.2010.04.064.

Keyvani, A., M. Yeganeh, and H. Rezaeyan. 2017. "Application of mesoporous silica nanocontainers as an intelligent host of molybdate corrosion inhibitor embedded in the epoxy coated steel." *Progress in Natural Science: Materials International* 27 (2): 261–267. doi:10.1016/j.pnsc.2017.02.005.

Kou, L., Y. Wang, Y. Dong, D. Han, B. Ni, and Z. Yan. 2014. "One-step synthesis of three-dimensionally ordered macro–mesoporous silica–alumina composites." *Materials Letters* 121: 212–214. doi:10.1016/j.matlet.2014.01.144.

Kruk, M., M. Jaroniec, C. Ko, R. Ryoo. 2000. "Characterization of the porous structure of SBA-15." *Chemistry of Materials* 12: 1961–1968. doi:10.1021/cm000164e.

Kumaran, G. M., S. Garg, K. Soni, M. Kumar, L. D. Sharma, G. M. Dhar, and K. S. R. Rao. 2006. "Effect of Al-SBA-15 support on catalytic functionalities of hydrotreating catalysis I. Effect of variation of Si/Al ratio on catalytic functionalities." *Applied Catalysis A* 305 (2): 123–129. doi: 10.1016/j.apcata.2006.02.057.

Lamaka, S. V., D.G. Shchukin, D.V. Andreeva, M. L. Zheludkevich, H. Mohwald, and M. G. S. Ferreira. 2008. "Sol-gel/polyelectrolyte active corrosion protection system." *Advanced Functional Materials* 18: 3137–3147. doi:10.1002/adfm.200800630.

Li, H., L. Raehm, C. Chamay, J.-O. Durand, and R. Pleixats. 2020. "Preparation and characterization of novel mixed periodic mesoporous organosilica nanoparticles." *Materials* 13: 1–16. doi:10.3390/ma13071569.

Li, L., X. Chen, X. Xiong, X. Wu, Z. Xie, and Z. Liu. 2021. "Synthesis of hollow TiO2@SiO2 spheres via a recycling template method for solar heat protection coating." *Ceramics International* 47 (2): 2678–2685. doi:10.1016/j.ceramint.2020.09.117.

Li, X., S. Zhao, W. Hu, X. Zhang, L. Pei, and Z. Wang. 2019. "Robust superhydrophobic surface with excellent adhesive properties based on benzoxazine/epoxy/mesoporous SiO 2." *Applied Surface Science* 481: 374–378. doi:10.1016/j.apsusc.2019.03.114.

Liu, J., S. Li, Y. Fang, and Z. Zhu. 2019. "Boosting antibacterial activity with mesoporous silica nanoparticles supported silver nanoclusters." *Journal of Colloid and Interface Science* 555: 470–479. doi:10.1016/j.jcis.2019.08.009.

Liu, J.; S. Z. Qiao, Q. H. Hu, G. Q. Lu, 2011. "Magnetic nanocomposites with mesoporous structures: synthesis and applications." *Small* 7 (4): 425–443. doi:10.1002/smll.201001402.

Liu, Y. H., J. B. Xu, J. T. Zhang, and J. M. Hu. 2017. "Electrodeposited silica film interlayer for active corrosion protection." *Corrosion Science* 120: 61–74. doi:10.1016/j.corsci.2017.01.017.

Ma, L., J. Wang, D. Zhang, Y. Huang, L. Huang, P. Wang, H. Qian, X. Li, H. Terryn, and J. Mol. 2021. "Dual-action self-healing protective coatings with photothermal responsive corrosion inhibitor nanocontainers." *Chemical Engineering Journal* 404: 127118. doi:0.1016/j.cej.2020.127118.

Maia, F., J. Tedim, A. D. Lisenkov, A. N. Salak, M. L. Zheludkevich, and M. G. S. Ferreira. 2012. "Silica nanocontainers for active corrosion protection." *Nanoscale* 4(4) 1287–1298. doi:10.1039/c2nr11536k.

Möller, K., J. Kobler, and T. Bein. 2007. "Colloidal suspensions of nanometer-sized mesoporous silica." *Advanced Functional Materials* 17 (4): 605–612. doi:10.1002/adfm.200600578.

Nemec, S., and S. Kralj. 2021. "A versatile interfacial coassembly method for fabrication of tunable silica shells with radially aligned dual mesopores on diverse magnetic core nanoparticles." *ACS Applied Materials and Interfaces* 13 (1): 1883–1894. doi:10.1021/acsami.0c17863.

Nguyen, C. H., C. C. Fu, Z. H. Chen, T. T. V. Tran, S. H. Liu, and R. S. Juang. 2021. "Enhanced and selective adsorption of urea and creatinine on amine-functionalized mesoporous silica SBA-15 via hydrogen bonding." *Microporous and Mesoporous Materials* 311: 110733. doi:10.1016/j.micromeso.2020.110733.

Omid, S., R. Sheykholeslami, M. Etminanfar, and J. Khalil-allafi. 2020. "Toward a facile synthesis of spherical sub-micron mesoporous silica : effect of surfactant concentration" *Journal of Ultrafine Grained and Nanostructured Materials* 53 (1): 31–38. doi:10.22059/jufgnsm.2020.01.05.

Ouyang, Y., L. X. Li, Z. H. Xie, L. Tang, F. Wang, and C. J. Zhong. 2020. "A self-healing coating based on facile PH-responsive nanocontainers for corrosion protection of magnesium alloy." *Journal of Magnesium and Alloys* doi:10.1016/j.jma.2020.11.007.

Pergal, M. V., J. Brkljačić, G. Tovilović-Kovačević, M. Špírková, I. D. Kodranov, D. D. Manojlović, S. Ostojić, and N. Knežević. 2021. "Effect of mesoporous silica nanoparticles on the properties of polyurethane network composites." *Progress in Organic Coatings* 151: 1–13. doi:10.1016/j.porgcoat.2020.106049.

Pooresmaeil, M., S. Javanbakht, S. B. Nia, and H. Namazi. 2020. "Carboxymethyl cellulose/mesoporous magnetic graphene oxide as a safe and sustained ibuprofen delivery bio-system: synthesis, characterization, and study of drug release kinetic." *Colloids and Surfaces A: Physicochemical and Engineering Aspects* 594: 124662. doi:10.1016/j.colsurfa.2020.124662.

Price, R. R., B. P. Gaber, and Y. Lvov. 2001. "In-vitro release characteristics of tetracycline HC1, khellin and nicotinamide adenine dineculeotide from halloysite; a cylindrical mineral." *Journal of Microencapsulation* 18 (6): 713–722. doi:10.1080/02652040010019532.

Qian, B., M. Michailidis, M. Bilton, T. Hobson, Z. Zheng, and D. Shchukin. 2019a. "Tannic complexes coated nanocontainers for controlled release of corrosion inhibitors in self-healing coatings." *Electrochimica Acta* 297: 1035–1041. doi:10.1016/j.electacta.2018.12.062.

Qian, B., Z. Zheng, M. Michailids, N. Fleck, M. Bilton, Y. Song, G. Li, and D. Shchukin. 2019b. "Mussel-inspired self-healing coatings based on polydopamine-coated nanocontainers for corrosion protection." *ACS Applied Materials and Interfaces* 11 (10): 10283–10291. doi:10.1021/acsami.8b21197.

Qiao, Y., W. Li, G. Wang, X. Zhang, and N. Cao. 2015. "Application of ordered mesoporous silica nanocontainers in an anticorrosive epoxy coating on a magnesium alloy surface." *RSC Advances* 5 (59): 47778–47787. doi:10.1039/c5ra05266a.

Rajput, S. M., M. Kuddushi, A. Shah, D. Ray, V. K. Aswal, S. K. Kailasa, and N. I. Malek. 2020. "Functionalized surfactant based catanionic vesicles as the soft template for the synthesis of hollow silica nanospheres as new age drug carrier." *Surfaces and Interfaces* 20: 100596. doi:10.1016/j.surfin.2020.100596.

Recloux, I., Y. Gonzalez-Garcia, M. E. Druart, F. Khelifa, Ph Dubois, J. M.C. Mol, and M. G. Olivier. 2016. "Active and passive protection of AA2024-T3 by a hybrid inhibitor doped mesoporous sol–gel and top coating system." *Surface and Coatings Technology* 303: 352–361. doi:10.1016/j.surfcoat.2015.11.002.

Ruggiero, L., E. Di Bartolomeo, T. Gasperi, I. Luisetto, A. Talone, F. Zurlo, D. Peddis, M. A. Ricci, and A. Sodo. 2019. "Silica nanosystems for active antifouling protection: nanocapsules and mesoporous nanoparticles in controlled release applications." *Journal of Alloys and Compounds* 798: 144–148. doi:10.1016/j.jallcom.2019.05.215.

Saji, V. S. 2019. "Supramolecular concepts and approaches in corrosion and biofouling prevention." *Corrosion Reviews* 37 (3): 187–230. doi:10.1515/corrrev-2018-0105.

Sarkar, A., P. I. Carver, T. Zhang, A. Merrington, K. J. Bruza, J. L. Rousseau, S.E. Keinath, and P. R. Dvornic. 2010. "Dendrimer-based coatings for surface modification of polyamide reverse osmosis membranes." *Journal of Membrane Science* 349 (1–2): 421–428. doi:10.1016/j.memsci.2009.12.005.

Seljak, K. B., P. Kocbek, and M. Gašperlin. 2020. "Mesoporous silica nanoparticles as delivery carriers: an overview of drug loading techniques." *Journal of Drug Delivery Science and Technology* 59: 101906. doi:10.1016/j.jddst.2020.101906.

Shamsheera, K. O., A. R. Prasad, and A. Joseph. 2020. "Extended protection of mild steel in saline and acidic environment using stearic acid grafted chitosan preloaded with mesoporous-hydrophobic silica (MhSiO2)." *Surface and Coatings Technology* 402: 126350. doi:10.1016/j.surfcoat.2020.126350.

Sharma, J., and G. Polizos. 2020. "Hollow silica particles: recent progress and future perspectives." *Nanomaterials* 10 (8): 1–22. doi:10.3390/nano10081599.

Shchukin, D. G. 2013. "Container-based multifunctional self-healing polymer coatings." *Polymer Chemistry* 4 (18): 4871–4877. doi:10.1039/c3py00082f.

Shchukin, D. G., and D. O. Grigoriev. 2012. "The use of nanoreservoirs in corrosion protection coatings." In *Corrosion Protection and Control Using Nanomaterials*, pp. 264–282. ISBN: 9781845699499.

Shchukin, D. G., and H. Möhwald. 2007a. "Self-repairing coatings containing active nanoreservoirs." *Small* 3 (6): 926–943. doi:10.1002/smll.200700064.

Shchukin, D. G., and H. Möhwald. 2007b. "Surface-engineered nanocontainers for entrapment of corrosion inhibitors" *Functional Materials* 17: 1451–1458. doi:10.1002/adfm.200601226.

Shchukin, D. G., M. Zheludkevich, K. Yasakau, S. Lamaka, M. G. S. Ferreira, and H. Möhwald. 2006. "Layer-by-layer assembled nanocontainers for self-healing corrosion protection." *Advanced Materials* 18 (13): 1672–1678. doi:10.1002/adma.200502053.

Shchukina, E., D. Shchukin, and D. Grigoriev. 2017. "Effect of inhibitor-loaded halloysites and mesoporous silica nanocontainers on corrosion protection of powder coatings." *Progress in Organic Coatings* 102: 60–65. doi:10.1016/j.porgcoat.2016.04.031.

Shchukina, E., D. Shchukin, and D. Grigoriev. 2018. "Halloysites and mesoporous silica as inhibitor nanocontainers for feedback active powder coatings." *Progress in Organic Coatings* 123: 384–389. doi:10.1016/j.porgcoat.2015.12.013.

Shi, H., L. Wu, J. Wang, F. Liu, and E. H. Han. 2017. "Sub-micrometer mesoporous silica containers for active protective coatings on AA 2024-T3." *Corrosion Science* 127: 230–239. doi:10.1016/j.corsci.2017.08.030.

Singh, N., M. Vinjamur, and M. Mukhopadhyay. 2021. "In vitro release kinetics of drugs from silica aerogles loaded by different models and conditions using supercritical CO2." *The Journal of Supercritical Fluids* 170: 105142. doi:10.1016/j.supflu.2020.105142.

Siva, T., S. S. Sreeja Kumari, and S. Sathiyanarayanan. 2021. "Dendrimer like mesoporous silica nano container (DMSN) based smart self healing coating for corrosion protection performance." *Progress in Organic Coatings* 154: 106201. doi:10.1016/j.porgcoat.2021.106201.

Skorb, E. V., D. Fix, D. V. Andreeva, H. Mohwald, and D. G. Shchukin. 2009. "Surface-modified mesoporous SiO2 containers for corrosion protection." *Advanced Functional Materials* 19: 2373–2379. doi:10.1002/adfm.200801804.

Soler-Illia, G. J. A. A., and O. Azzaroni 2011. "Multifunctional hybrids by combining ordered mesoporous materials and macromolecular building blocks." *Chemical Society Reviews* 40 (2): 1107–1150. doi:10.1039/C0CS00208A.

Soltani, R., A. Marjani, and S. Shirazian 2020. "Novel mesoporous crumpled paper-like silica balls." *Materials Letters* 281: 128230. doi:10.1016/j.matlet.2020.128230.

Speybroeck, M. V., V. Barillaro, T. D. Thi, R. Mellaerts, J. Martens, J. V. Humbeeck, J. Vermant, P. Annaert, G. V. Den Mooter, and P. Augustijns. 2009. "Ordered mesoporous silica material SBA-15: a broad-spectrum formulation platform for poorly soluble drugs." *Journal of Pharmaceutical Sciences* 98 (8): 2648–2658. doi:10.1002/jps.21638.

Stöber, W., A. Fink, and E. Bohn. 1968. "Controlled growth of monodisperse silica spheres in the micron size range." *Journal of Colloid And Interface Science* 26 (1): 62–69. doi:10.1016/0021-9797(68)90272-5.

Sun, J., Y. Fan, P. Zhang, X. Zhang, Q. Zhou, J. Zhao, and L. Ren. 2020. "Self-enriched mesoporous silica nanoparticle composite membrane with remarkable photodynamic antimicrobial performances." *Journal of Colloid and Interface Science* 559: 197–205. doi:10.1016/j.jcis.2019.10.021.

Suteewong, T., H. Sai, R. Hovden, D. Muller, M. S. Bradbury, S. M. Gruner, and U. Wiesner. 2013. "Multicompartment mesoporous silica nanoparticles with branched shapes: an epitaxial growth mechanism." *Science* 340 (6130): 337–341. doi:10.1126/science.1231391.

Szabó, G., E. Albert, J. Both, L. Kócs, G. Sáfrán, A. Szöke, Z. Hórvölgyi, and L. Maria Mureşan. 2019. "Influence of embedded inhibitors on the corrosion resistance of zinc coated with mesoporous silica layers." *Surfaces and Interfaces* 15: 216–223. doi:10.1016/j.surfin.2019.03.007.

Tang, G., T. Ren, Z. Yan, L. Ma, X. Hou, and X. Huang. 2020. "Preparation and anticorrosion resistance of a self-curing epoxy nanocomposite coating based on mesoporous silica nanoparticles loaded with perfluorooctyl triethoxysilane." *Journal of Applied Polymer Science* 137 (36): 1–11. doi:10.1002/app.49072.

Teng, Y., Y. Du, J. Shi, and P. W. T. Pong. 2020. "Magnetic iron oxide nanoparticle-hollow mesoporous silica spheres:fabrication and potential application in drug delivery." *Current Applied Physics* 20 (2): 320–325. doi:10.1016/j.cap.2019.11.012.

Thai, T. T., M. E. Druart, Y. Paint, A. T. Trinh, and M. G. Olivier. 2018. "Influence of the sol-gel mesoporosity on the corrosion protection given by an epoxy primer applied on aluminum alloy 2024 –T3." *Progress in Organic Coatings* 121: 53–63. doi:10.1016/j.porgcoat.2018.04.013.

Thakur, V. K., and M. R. Kessler. 2015. "Self-healing polymer nanocomposite materials: a review." *Polymer* 69: 369–383. doi:10.1016/j.polymer.2015.04.086.

Udoh, I. I., H. Shi, F. Liu, and E. H. Han. 2020. "Microcontainer-based waterborne epoxy coatings for AA2024-T3: effect of nature and number of polyelectrolyte multilayers on active protection performance." *Materials Chemistry and Physics* 241: 1–15 doi:10.1016/j.matchemphys.2019.122404.

Vallet-regí, M., J. C. Doadrio, R. P. Doadrio, I. Izquierdo-barba, A. Ramila, R. P. Delreal, and J. Pérez-pariente. 2004. "A new property of MCM-41: hexagonal ordered mesoporous materials as a matrix for the amoxicillin." *Solid State Ionics* 172: 435–439. doi:10.1016/j.ssi.2004.04.036.

Vanichvattanadecha, C., W. Singhapong, and A. Jaroenworaluck. 2020. "Different sources of silicon precursors influencing on surface characteristics and pore morphologies of mesoporous silica nanoparticles." *Applied Surface Science* 513: 145568. doi:10.1016/j.apsusc.2020.145568.

Walcarius, A. 2013. "Mesoporous materials and electrochemistry." *Chemical Society Reviews* 42 (9): 4098–4140. doi:10.1039/c2cs35322a.

Wan, M. M., Y.Y. Li, T. Yang, T. Zhang, X. D. Sun, and J. H. Zhu. 2016. "In situ loading of drugs into mesoporous silica SBA-15." *Chemistry - A European Journal.* 22 (18): 6294–6301. doi:10.1002/chem.201504532.

Wang, H., M. Gan, L. Ma, T. Zhou, H. Wang, S. Wang, W. Dai, and X. Sun. 2016. "Synthesis of polyaniline-modified mesoporous-silica containers for anticorrosion coatings via in-situ polymerization and surface-protected etching." *Polymers for Advanced Technologies* 27 (7): 929–937. doi:10.1002/pat.3750.

Wang, N., K. Cheng, H. Wu, C. Wang, Q. Wang, and F. Wang. 2012. "Effect of nano-sized mesoporous silica MCM-41 and MMT on corrosion properties of epoxy coating." *Progress in Organic Coatings* 75 (4): 386–391. doi:10.1016/j.porgcoat.2012.07.009.

Wang, T., J. Du, S. Ye, L. Tan, and J. Fu. 2019. "Triple-stimuli-responsive smart nanocontainers enhanced self-healing anticorrosion coatings for protection of aluminum alloy." *ACS Applied Materials and Interfaces* 11 (4): 4425–4438. doi:10.1021/acsami.8b19950.

Wang, T., L. Tan, C. Ding, M. Wang, J. Xu, and J. Fu. 2017. "Redox-triggered controlled release systems-based Bi-layered nanocomposite coating with synergistic self-healing property." *Journal of Materials Chemistry A.* 5 (4): 1756–1768. doi:10.1039/c6ta08547d.

Wei, H., Y. Wang, J. Guo, N. Z. Shen, D. Jiang, X. Zhang, and X. Yan, et al. 2015. "Advanced micro/nanocapsules for self-healing smart anticorrosion coatings." *Journal of Materials Chemistry A* 3: 469–480. doi:10.1039/c4ta04791e.

Wen, J., J. Lei, J. Chen, L. Liu, X. Zhang, and L. Li. 2020. "Polyethylenimine wrapped mesoporous silica loaded benzotriazole with high PH-sensitivity for assembling self-healing anti-corrosive coatings." *Materials Chemistry and Physics* 253: 123425. doi:10.1016/j.matchemphys.2020.123425.

Xie, Z. H., D. Li, Z. Skeete, A. Sharma, and C. Jian Zhong. 2017. "Nanocontainer-enhanced self-healing for corrosion-resistant Ni coating on Mg alloy." *ACS Applied Materials and Interfaces* 9 (41): 36247–36260. doi:10.1021/acsami.7b12036.

Xiong, L., J. Liu, Y. Li, S. Li, and M. Yu. 2019. "Enhancing corrosion protection properties of sol-gel coating by PH-responsive amino-silane functionalized graphene oxide-mesoporous silica nanosheets." *Progress in Organic Coatings* 135: 228–239. doi:10.1016/j.porgcoat.2019.06.007.

Xu, J. B., Y. Q. Cao, L. Fang, and J. M. Hu. 2018. "A one-step preparation of inhibitor-loaded silica nanocontainers for self-healing coatings." *Corrosion Science* 140: 349–362. doi:10.1016/j.corsci.2018.05.030.

Xu, P., Z. Wu, W. Dai, Y. Wang, M. Zheng, X. Su, and Z. Teng. 2020. "Hollow mesoporous organosilica nanospheres with multiple tiny silver nanoparticles: a polymer mediated growth approach and catalytic application." *Journal of the Taiwan Institute of Chemical Engineers* 117: 287–293. doi:10.1016/j.jtice.2020.12.015.

Yao, P., A. Zou, Z. Tian, W. Meng, X. Fang, T. Wu, and J. Cheng. 2021. "Construction and characterization of a temperature-responsive nanocarrier for imidacloprid based on mesoporous silica nanoparticles." *Colloids and Surfaces B: Biointerfaces* 198 (2020): 111464. doi:10.1016/j.colsurfb.2020.111464.

Yeganeh, M., and A. Keyvani. 2016. "The effect of mesoporous silica nanocontainers incorporation on the corrosion behavior of scratched polymer coatings." *Progress in Organic Coatings* 90: 296–303. doi:10.1016/j.porgcoat.2015.11.006.

Yeganeh, M., and M. Saremi. 2015. "Corrosion inhibition of magnesium using biocompatible alkyd coatings incorporated by mesoporous silica nanocontainers." *Progress in Organic Coatings* 79: 25–30. doi:10.1016/j.porgcoat.2014.10.015.

Yeganeh, M., M. Saremi, and H. Rezaeyan. 2014. "Corrosion inhibition of steel using mesoporous silica nanocontainers incorporated in the polypyrrole." *Progress in Organic Coatings* 77 (9): 1428–1435. doi:10.1016/j.porgcoat.2014.05.007.

Yeganeh, M., M. Omidi, and T. Rabizadeh. 2019b. "Anti-corrosion behavior of epoxy composite coatings containing molybdate-loaded mesoporous silica." *Progress in Organic Coatings* 126: 18–27. doi:10.1016/j.porgcoat.2018.10.016.

Yeganeh, M., N. Asadi, M. Omidi, and M. Mahdavian. 2019a. "an investigation on the corrosion behavior of the epoxy coating embedded with mesoporous silica nanocontainer loaded by sulfamethazine inhibitor." *Progress in Organic Coatings* 128: 75–81. doi:10.1016/j.porgcoat.2018.12.022.

Yeganeh, M., S. M. Marashi, and N. Mohammadi. 2018. "Smart corrosion inhibition of mild steel using mesoporous silica nanocontainers loaded with molybdate." *International Journal of Nanoscience and Nanotechnology* 14 (2): 143–151.

Zahidah, K. A., S. Kakooei, M. C. Ismail, and P. B. Raja. 2017. "Halloysite nanotubes as nanocontainer for smart coating application: a review." *Progress in Organic Coatings* 111: 175–185. doi:10.1016/j.porgcoat.2017.05.018.

Zea, C., J. Alcántara, R. Barranco-García, J. Simancas, M. Morcillo, and D. de la Fuente. 2017a. "Anticorrosive behavior study by localized electrochemical techniques of sol–gel coatings loaded with smart nanocontainers." *Journal of Coatings Technology and Research* 14 (4): 841–850. doi:10.1007/s11998-017-9936-3.

Zea, C., J. Alcántara, R. Barranco-García, M. Morcillo, and D. De La Fuente. 2018b. "Synthesis and characterization of hollow mesoporous silica nanoparticles for smart corrosion protection." *Nanomaterials* 8 (7): 478. doi:10.3390/nano8070478.

Zea, C., R. Barranco-García, B. Chico, I. Díaz, M. Morcillo, and D. De La Fuente. 2015. "Smart mesoporous silica nanocapsules as environmentally friendly anticorrosive pigments." *International Journal of Corrosion* 2015: 1–8. doi:10.1155/2015/426397.

Zea, C., R. Barranco-García, J. Alcántara, B. Chico, M. Morcillo, and D. de la Fuente. 2017b. "Hollow mesoporous silica nanoparticles loaded with phosphomolybdate as smart anticorrosive pigment." *Journal of Coatings Technology and Research* 14 (4): 869–878. doi:10.1007/s11998-017-9924-7.

Zea, C., R. Barranco-García, J. Alcántara, J. Simancas, M. Morcillo, and D. de la Fuente. 2018a. "PH-dependent release of environmentally friendly corrosion inhibitor from mesoporous silica nanoreservoirs." *Microporous and Mesoporous Materials* 255: 166–173. doi:10.1016/j.micromeso.2017.07.035.

Zhao, D., D. Liu, and Z. Hu. 2017. "A smart anticorrosion coating based on hollow silica nanocapsules with inorganic salt in shells." *Journal of Coatings Technology and Research* 14 (1): 85–94. doi:10.1007/s11998-016-9830-4.

Zhao, D., J. Feng, Q. Huo, N. Melosh, G. H. Fredrickson, B. F. Chmelka, and G. D. Stucky. 1998a. "Triblock copolymer syntheses of mesoporous silica with periodic 50 to 300 angstrom pores." *Science* 279 (5350): 548–552. doi:10.1126/science.279.5350.548.

Zhao, D., Q. Huo, J. Feng, B. F. Chmelka, and G. D. Stucky 1998b. "Nonionic triblock and star diblock copolymer and oligomeric sufactant syntheses of highly ordered, hydrothermally stable, mesoporous silica structures." *Journal of the American Chemical Society* 120: 6024–6036. doi:10.1021/ja974025i.

Zhao, S., Y. Zhang, Y. Zhou, X. Sheng, C. Zhang, M. Zhang, and J. Fang. 2016. "One-step synthesis of core-shell structured mesoporous silica spheres templated by protic ionic liquid and CTAB." *Materials Letters* 178: 35–38. doi:10.1016/j.matlet.2016.04.182.

Zheludkevich, M. L., D. G. Shchukin, K. A. Yasakau, and M. G. S. Ferreira. 2007b. "Anticorrosion coatings with self-healing effect based on nanocontainers impregnated with corrosion inhibitor." *Chemistry of Materials* 100: 402–411. doi: 10.1021/cm062066k.

Zheludkevich, M. L., K. A. Yasakau, A. C. Bastos, O. V. Karavai, and M. G. S. Ferreira. 2007a. "On the application of electrochemical impedance spectroscopy to study the self-healing properties of protective coatings." *Electrochemistry Communications* 9 (10): 2622–2628. doi:10.1016/j.elecom.2007.08.012.

Zheludkevich, M.L., S. K. Poznyak, L.M. Rodrigues, D. Raps, T. Hack, L.F. Dick, T. Nunes, and M.G.S. Ferreira. 2010. "Active protection coatings with layered double hydroxide nanocontainers of corrosion inhibitor." *Corrosion Science* 52 (2): 602–611. doi:10.1016/j.corsci.2009.10.020.

Zheng, Z., X. Huang, M. Schenderlein, D. Borisova, R. Cao, H. Möhwald, and D. Shchukin. 2013. "Self-healing and antifouling multifunctional coatings based on PH and sulfide ion sensitive nanocontainers." *Advanced Functional Materials* 23 (26): 3307–3314. doi:10.1002/adfm.201203180.

Zhou, C., Z. Li, J. Li, T. Yuan, B. Chen, X. Ma, D. Jiang, X. Luo, D. Chen, and Y. Liu. 2020. "Epoxy composite coating with excellent anticorrosion and self-healing performances based on multifunctional zeolitic imidazolate framework derived nanocontainers." *Chemical Engineering Journal* 385: 1–17. doi:10.1016/j.cej.2019.123835.

Židov, B., Z. Lin, I. Stojanović, and L. Xu. 2021. "Impact of inhibitor loaded mesoporous silica nanoparticles on waterborne coating performance in various corrosive environments." *Journal of Applied Polymer Science* 138 (1): 1–19. doi:10.1002/app.49614.

12 MOFs-Based Surface Coatings

A. Madhan Kumar
King Fahd University of Petroleum & Minerals

CONTENTS

12.1 INTRODUCTION

The development of supramolecular chemistry has been witnessed by the discovery of various coordination structures with tunable forms and dimensions ranging from 1 to 100 nm through impulsive secondary interactions, including hydrogen bonding, charge transfer, van der Waals, π–π stacking, and dipole-dipole interactions. Encouraged by monotonous needs, the progression in research accomplishments has been facilitated to emphasize structural supramolecular strategies. In this way, metal-organic frameworks (MOFs), a kind of coordination polymer with ultra-high porosity, have garnered massive attention in the past few years owing to their unprecedented features, including designable porosity with controllable pore size, various configurations, high surface areas and precise adsorption interactions. The features that are exclusively different from the conventional porous materials such as zeolites, activated carbon and porous silica have fascinated imperative research interests in a wide range of applications, including CO_2 capture, energy storage, catalysis, gas separation, drug delivery, biomedical, anticorrosion, self-healing, and self-cleaning coatings (Melgar et al. 2015; Gangu et al. 2016).

In general, MOFs are constructed by combining inorganic nodes (metal ions/clusters) with organic linkers (OLs) connected through coordination bonds in which the metallic ions act as the core. In contrast, the OLs act as a bridge between them, creating orderly-packed 3D networks. Ligands and nodes are designated as secondary building units (SBUs). The OLs are usually comprised of carboxylates, sulfonates, phosphonates, and heterocyclic compounds, whereas the metal ions mainly involve alkali metals, alkaline earth metals, transition metals, actinides and lanthanides. MOFs are synthesized into various forms, such as linear, planar, triangular and square. Based on the constituent organic and inorganic SBUs and their nature of bonding, the geometry, topology, and characteristics of MOFs can considerably differ. The assembly of various metallic ions and OLs provides the resultant MOFs with wide-ranging framework topologies (Meng et al. 2020; Qiu et al. 2021).

More than 20,000 diverse MOFs have been publicized during the past few decades (Furukawa et al. 2013). Thus, it has accomplished a substantial expansion, permitting MOFs to be categorized as one of the utmost state-of-the-art research focuses in materials science. The capacity to prompt post-synthetic functionalization in numerous MOF forms has yet to untie a new way for attempting challenging applications using MOF chemistry (Begum et al. 2019). In recent years, efficient investigations have been conducted to improve the characteristics and performance of MOFs through the addition of novel functional materials, including polymers, carbon nanostructures, biomolecules

DOI: 10.1201/9781003169130-14

and metallic nanocrystals (Meng et al. 2020). The developed composites succeed in improving the MOF characteristics and provide further benefits from the interactions between them. Thus, it is considered the most efficient strategy to design and build distinct MOFs-based composites to attain the synergistic features of MOFs and functional materials in different applications.

Protective coatings have been widely employed in the industrial sectors for the corrosion protection of metallic structures. Though protective coatings have been subjected to vast improvement over the past few decades, there are still several shortcomings, such as less UV resistance, inadequate abrasion resistance, and insufficient multifunctional characteristics such as superhydrophobicity and self-healing. To fulfill the inadequacy of protective coatings, fabricating multifunctional coatings encapsulated in porous structures, such as nanocontainers, has been a popular research topic. The representative nanomaterials, including carbon nanostructures (graphene oxide, carbon nanotubes, etc.), nano-SiO_2, montmorillonites (MMT) and halloysite, able to be encapsulated with desired constituents, have been broadly inspected for protective coatings (Kumar et al. 2015a, b; Kalaivasan et al. 2017). However, researchers are in search of novel hybrid nanomaterials with inorganic and organic functional groups containing abundant ability to interact with the polymeric chain to attain the desired characteristics.

In recent years, the integration of MOFs into protective coatings has been renowned as an excellent strategy by many researchers to achieve the desired functional characteristics (Zhou et al. 2020; Zheng et al. 2019; Falcaro et al. 2011). For instance, the homogeneous distribution of nanomaterials in polymeric resins is continually a challenging task in the fabrication of polymer nanocomposite coatings. As a blend of inorganic and organic constituents, MOFs have the benefit of functional groups to have a homogeneous distribution in the polymeric coatings. In addition, effective interfacial interaction between the polymeric chain and MOFs is probably attributed to the large surface areas and reactive functional groups of MOFs (Peng et al. 2013). In this book chapter, the comprehensive research analysis on MOF-based coatings is emphasized to enlighten its impact on the corrosion protection of metallic components. The significant advantage of these kinds of MOF-based coatings lies in their capability to be multifunctional coatings. Those characteristics can be altered by controlling their porosity, dimension, and chemical configuration. The anticorrosion, superhydrophobic, self-healing and biomedical applications of MOFs-based coatings have been discussed in this regard.

12.2 FABRICATION APPROACHES

Several methods have been established to prepare MOFs as coatings that consist of the molten linker method, seeding method, self-assembled monolayers, electroless metal oxidation, liquid phase epitaxy, UV lithography, microwave-assisted synthesis, and colloidal chemical solution deposition. Most of the mentioned techniques have been efficient in attaining MOF coatings. However, they typically require prolonged duration and procedures, high temperatures, and specified instruments (Stassen et al. 2013; Mao et al. 2013; Zhu et al. 2012). In contrast, electrochemical methods permit the fabrication of MOF coatings with less processing time at room temperature with simple instrument facilities. Both cathodic and anodic electrochemical methods have been utilized for attaining MOF coatings. In the cathodic process, the high pH due to water reduction results in anionic and deprotonated OLs that can interact with the metallic cations added to the solution to form MOF films. In contrast, the oxidation occurring at the metallic anode delivers the required metallic ions to form the MOF coating during the anodic process (Worrall et al. 2016).

In addition to designing and altering polymeric coatings, researchers can improve the performance and targeted features of polymers through the production of composites. Such alterations are intended to improve obtainable characteristics or import novel, non-inherent features to the polymer matrix by reinforcing this distinctive kind of framework. MOF-based polymer composites have attained remarkable consideration in recent years, revealing that MOFs can be utilized to enhance polymer performance in desired applications (Lu et al. 2014; Yang et al. 2021). MOF-based polymer composites have been synthesized using different strategies, including in-situ polymeriza-tion on MOF nanopores, consumption of polymer-tethered ligands, polymerization of ligands and incorporation of polymeric molecules into MOF nanopores, as shown in Figure 12.1 (Yang et al. 2021; Kalaj et al. 2020).

12.3   MOFs-BASED ANTICORROSION COATINGS

Owing to the attractive crystalline structure, the MOF surfaces have further been functionalized to make them compatible and responsive for constructing MOFs-based polymer composites. The existence of OLs provides them with the opportunity to be more compatible with the polymeric matrix. The reinforcement of highly porous MOFs with preferred topography into thermoset poly-meric chains, including epoxy (EP) resins, could facilitate the curing process. Jouyandeh et al. revealed that the reinforcement of MIL-101 (0.1 wt% loading) into EP resin considerably improved (by 63%) the heat release during the curing reaction. This substantial increase in the cure enthalpy is regarded as beneficial based on the Cure Index because of the diffusion of EP within the MOFs 3D nanoporous structure, consequently prompting the cross-linking reactions (Figure 12.2) (Jouyandeh et al. 2020). The reinforcement of MOFs into EP coatings influences their final features, including mechanical, anticorrosion, anti-wear, and other desired characteristics (Seidi et al. 2020).

It is expected that the high interaction ability of MOFs with organic and inorganic constituents permits the creation of high cross-linked EP/MOF coatings that can further enhance the coat-ing's barrier protection. Furthermore, the capacity to utilize the diverse metallic ions and OLs permits the formation of MOFs with pH, magnetic, pressure molecular, and thermo sensitivities. Utilizing the stated features, stimuli-responsive MOFs can smartly provide inhibitor release as a response to the environmental variations caused by the EP coatings encapsulated in anticorrosive MOF nanocontainers. Thus, polymeric coatings reinforced with such stimuli-responsive MOF nanocontainers deliver strong barrier performance to the coatings against aggressive species and

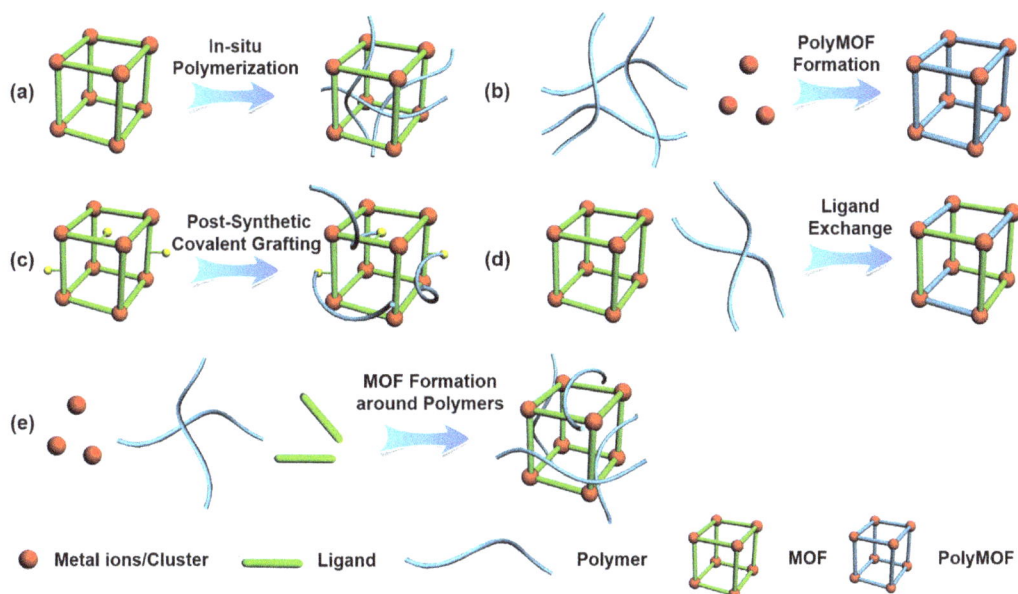

FIGURE 12.1   Representative synthesis methods of MOF composites: (a) In-situ polymerization, (b) Formation of MOF from polymer ligands, (c) Post-synthetic covalent grafting, (d) Exchange of ligands and (e) Formation of MOF around pores. (Reproduced with permission from Yang et al. (2021) © 2020 The Authors. Published by Elsevier B.V.)

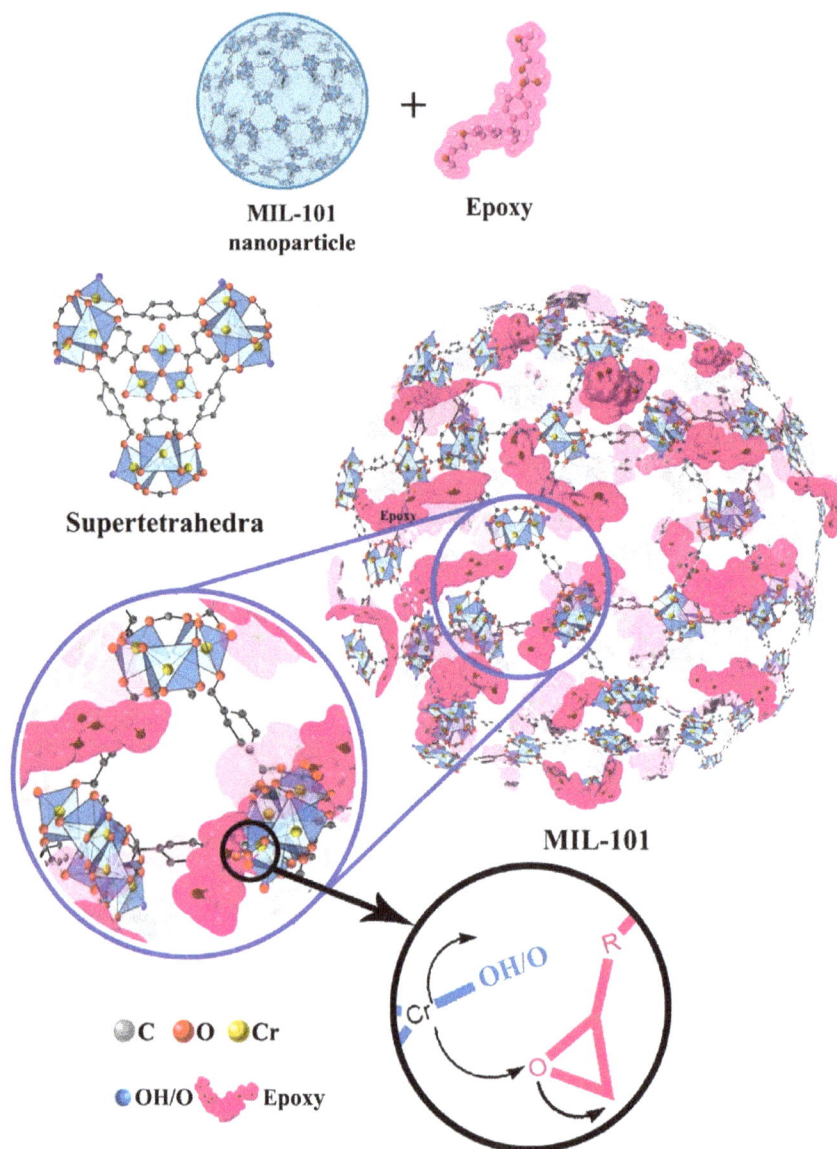

FIGURE 12.2 Schematic representation of the distribution of epoxy resin into the pores of MIL-100 and its possible reaction in between. (Reproduced with permission from Jouyandeh et al. (2020) © 2020 Elsevier B.V.)

inhibit metal corrosion by the corrosion-sensing and self-healing approach through controlled release of encapsulated corrosion inhibitors (Zhang et al. 2018a) (see next section).

Kumaraguru et al. synthesized three different acid-based MOFs (Co-MOF, Ni-MOF and Cu-MOF). They incorporated the prepared MOFs separately into acrylic coatings to protect mild steel from corrosion. The results confirmed that Ni-MOF exhibited better surface protection than Co-MOF and Cu-MOF (Kumaraguru et al. 2017). Ren et al. encapsulated an eco-friendly corrosion inhibitor, zinc gluconate (ZnG) inside the porous ZIF-8 and the prepared ZIF-8/ZnG was subsequently incorporated into EP coatings on Mg alloy. The results validated the improved corrosion-resistant performance and interfacial adhesion to the Mg surface (Ren et al. 2020). Wei et al. synthesized composites based on graphene oxide (GO) with Cu-MOFs and Mn-MOFs, and integrated them into acrylic coatings on steel (Wei et al. 2020). Their findings validated that the surface protective behavior of MOFs/GO/acrylic coating significantly improved due to the synergetic physical barrier effect of GO with the chemical adsorption effect of MOFs.

Ren et al. prepared a polyaniline/MIL-101 composite by incorporating polyaniline into the pores of MIL-101 through in-situ polymerization, and the resultant composite was then incorporated into an EP coating on an Mg alloy substrate (Ren et al. 2020). Their results confirmed that incorporating PANI inside the MIL-101 pores increased the long-term anticorrosion of the EP coating by delaying the permeation rate of corrosive ions, lengthening the penetration pathways, and offering strong barrier protection. Mohammadpour and Zare encapsulated 2-aminobenzothiazol (2-ABT) into MOF of Cu-1,3,5-tricarboxylate (Cu-BTC) and reinforced the prepared 2ABT/Cu-BTC composites into a sol–gel processed silica coating on Ni alloy (Mohammadpour and Zare 2021). The controlled release of the 2ABT from Cu-BTC was assessed using electrochemical studies. The results revealed that the release of 2ABT was ~40%, depending on the weak electrostatic interaction of 2ABT outside of the MOF. The improved anticorrosive performance of 2ABT/Cu-BTC composite coating was mainly attributed to the reaction between the π electrons of nitrogen and sulfur in the 2ABT moieties and the metallic surface. Similarly, Yin et al. also prepared Cu-based MOFs encapsulated with benzotriazole (BTA) molecules and reinforced the resultant MOFs into EP coating (Yin et al. 2019). Their results revealed that adding inhibitor-loaded Cu/MOFs improved the water contact angle (CA) and enhanced its anticorrosion performance. Wang et al. grafted dopamine (DA) onto the surface of the MOF-5 and the resultant DA/MOF-5 was incorporated into the EP coating. A 0.5 wt% loading was found to be optimum for improving the anticorrosion behavior. The addition of DA was helpful in increasing the adhesion strength between the coating and the base substrate (Wang et al. 2017).

Tarzanagh et al. prepared MIL-53 from Al(III) and terephthalic acid using the hydrothermal treatment and incorporated these MOFs into a hybrid sol–gel coating (Tarzanagh et al. 2019). Qiu et al. prepared ultrathin MOF nanosheets based on central Cu^{2+} ions and porphyrin ligands through a bottom-up approach and then incorporated them into the EP coating. The Cu-MOFs in the coating were found at the sites of artificial scratches and they inhibited the transitory permeation of aggressive species (Qiu et al. 2021). Zhang and Liu prepared a ZIF-8-based coating with varied surface morphologies using simple hydrothermal growth of ZnAl–NO_3 layers followed by solvothermal treatment with ZIF-8 precursors. The improved corrosion performance of the 2-methylimidazole/ZIF-8–ZnAl–NO_3 coating was attributed to the reduced grain boundary defect density (Zhang and Liu 2020). Liu et al. explored an adaptable strategy to prepare different MOFs over various metallic substrates using polydopamine (PDA) as an effective adhesive polymer (Liu et al. 2020). The enhanced corrosion-resistant behavior of the PDA/MOF coating was attributed to the compact structure, well-grown crystals, surface hydrophobicity, efficient obstruction of aggressive species, and improved adhesion strength.

Wang et al. recently utilized ZIF-8 in EP coating along with the selective "turn-on" chemosensor extracted from rhodamine B for Cu(II) to detect the early stages of corrosion on Cu alloys (Wang et al. 2020). They found that the early stages of Cu corrosion can be detected with high specificity and sensitivity of chemosensor against Cu(II) with no essential damage to the coating. Motamedi et al. synthesized novel MOFs based on the nanoceria (NC) and Ce(III)-imidazole network (CIN) and reinforced them with an EP coating to enhance their thermo-mechanical and anticorrosion characteristics. Their results revealed that the EP with NC/CIN exhibited excellent thermal stability up to 800°C with remarkable mechanical performance in tensile and viscoelastic tests. The EP/NC/CIN-coating exhibited vital barrier features together with self-healing performance (Motamedi et al. 2020). The below section discusses more work on self-healing coatings.

12.4   SELF-HEALING STRATEGIES

Numerous reports have explored the utilization of MOFs as nanocontainer materials in enhancing the corrosion protection performance of polymeric coatings through the self-healing approach. Cao et al. prepared BTA encapsulated cerium MOF, and covered it with a silane coupling agent/tetra ethyl orthosilicate (TEOS) to construct a self-healing EP coating (Cao et al.

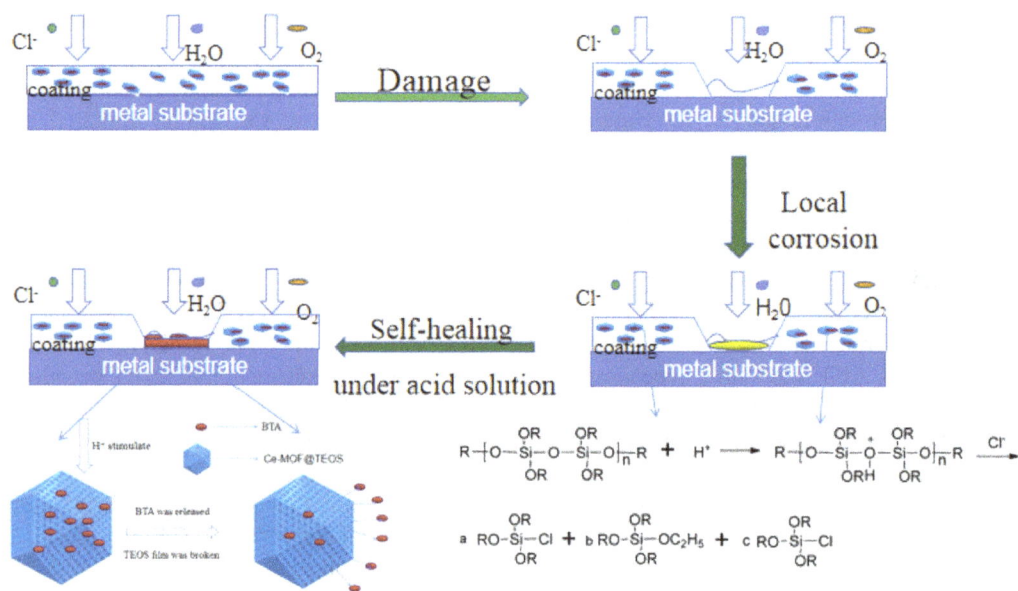

FIGURE 12.3 Schematic illustration of the self-healing mechanism behind the BTA/CeMOF/TEOS coating. (Reproduced with permission from Cao et al. (2019) © 2019 Elsevier B.V.)

2019). The BTA/Ce-MOF/TEOS coating exhibited improved homogeneous dispersion and presented efficient self-healing activity as shown in Figure 12.3. At the anodic sites, hydrogen ions were formed due to the hydrolysis of metallic ions and interacted with TEOS. Consequently, the chloride ions from the medium also reacted with TEOS, causing damage to the TEOS film and the release of BTA from Ce-MOF.

Tian et al. prepared a triazole derivative and incorporated it into ZIF-8. The triazole-filled ZIF-8 coating exhibited excellent anticorrosion performance through the controlled release of triazole (Tian et al. 2018). In another work, researchers have explored the utilization of zinc phosphate (ZP) as an encapsulation inhibitor into Mn-MOFs, as pH-responsive self-healing inhibitive pigments in the acrylic coating (Chen et al. 2021). The release of ZP from ZP/Mn MOFs occurred while the coating was damaged and/or local change of pH in the coating interface.

Zhao et al. prepared the GO/MOFs nanocontainers encapsulated with BTA and reinforced them with a polyvinyl butyral (PVB) coating, and assessed their corrosion protection on Cu and bronze substrates. The passive (barrier) and active (self-healing) corrosion-resistant behaviors were provided by the strong barrier characteristics of GO and the pH-sensitive controlled discharge of encapsulated BTA from the MOFs/GO nanocontainers (Zhao et al. 2020). Similarly, Cao et al. also prepared MOF nanocontainers and encapsulated them with BTA with the wrapping of TEOS and then added GO nanosheets. The prepared nanocomposites were incorporated into the EP coating and assessed for their anticorrosive behavior on steel in NaCl solution. The EP nanocomposite coating exhibited improved active and passive corrosion protection due to the synergetic effect of barrier features of GO and the inhibition effect of BTA from MOFs with pH-sensitive film (Cao et al. 2020).

Ramezanzadeh et al. explored the utilization of Zr-based ZIF-8 in the EP coating to target both active and passive corrosion protection (Ramezanzadeh et al. 2021a, b). The authors prepared Zr-MOF using OLs such as terephthalic acid (UIO-66: UIO) and 2-aminoterephthalic acid (NH$_2$-UIO: UIN). Further, to attain homogeneous dispersion of MOFs in EP resins, they have also modified the surface of MOFs using glycidyl methacrylate (G/UIN). The water solubility measurements confirmed the controlled release of the encapsulated inhibitor from UIN, UIO, and G/UIN

(Ramezanzadeh et al. 2021b). The EP film chemistry changed in the presence of Zr-based MOF and that was decisive in the improved coating performance (Ramezanzadeh et al. 2021a, b).

Xiong et al. utilized ZIF-8 formed on GO sheets to incorporate salicylaldehyde to enhance the anticorrosion performance of PVB coating on an Al alloy (Xiong et al. 2019). Alipanah et al. encapsulated MIL-88A in an EP/polyamide coating. The theoretical simulations confirmed the interfacial attachment of complexes based on Fe(III) and fumarate on the metal/coating interface (Alipanah et al. 2021). Lashgari et al. prepared ZIF-67 on GO nanosheets and modified them with a silane coupling agent to improve the active sites for the interfacial interaction, with subsequent incorporation into the EP matrix. Due to the changes in pH at anodic sites, the damage to ZIF-67 occurred and released the inhibiting constituents (Co^{2+} and 2-methyl imidazole, 2MIA) from the GO surface. The released 2MIA remained in a protonated form at anodic sites and interacted with Fe^{+2}, reducing the corrosion rate. Further, 2MIA can also form chelates and complexes of Co-2MIA-Fe on the steel surface. Conversely, the released Co ions also react with OH-, forming cobalt hydroxide at the cathodic sites. Thus, adsorption of the released 2MIA and Co ions simultaneously reduced the anodic and cathodic reactions at the metal/coating interface (Lashgari et al. 2021).

In summary, the addition of MOFs to the polymeric coatings primarily covers the pores and micro-cracks, thereby increasing the barrier performance of the coatings. In contrast, polymeric coatings comprising inhibitor-incorporated MOFs release the inhibiting constituents when the aggressive ions promptly infuse from the micro defects and lead to a change in the local pH. The released constituents produce a film at the locality of the cracked regions to effectively avoid corrosion reaction and further damage in a self-repairing route.

12.5 SUPERHYDROPHOBIC FEATURES

In recent years, a range of synthetic methodologies have been established to fabricate super-hydrophobic (SHPC) MOFs materials. The surface wettability of MOFs is directly governed by the nature of the ligands. In addition, the porosity of the MOFs can also be easily altered by chemical modification to achieve the target surface roughness to attain perfect surface wettability. Some excellent reviews about hydrophobic MOFs have been recently published (Xie et al. 2020; Jayaramulu et al. 2019).

Zhang et al. fabricated SHPC ZIF-7 coating with a micro/nano flower-like architecture on Cu using in situ ligand-solvothermal transformation, without any activation or modification. ZnO sheets were initially prepared on a Cu specimen and then reacted with an imidazole ligand in solution, obtaining an ultrathin flake-like ZnO array hierarchical structure with nanometric protrusions by the solvothermal method. The ligands acted as an etching reagent to seize the metal ions from the ZnO template, and the ZnO arrays were transformed in-situ to ZIF-7. The CA of the ZIF-7 coating was 151.31° (Zhang et al. 2017).

Chen et al. prepared a self-cleaning SHPC coating based on ZIF-8 nanoparticles, low surface energy perfluorooctyl-triethoxy silane (POTS) and EP resin (Chen et al. 2021). The resultant EP/POTS/ZIF-8 coating exhibited high CA of ~168.2° with different substrates as shown in Figure 12.4d–f, with excellent self-cleaning properties (Figure 12.4(1–6)). The surface also showed superior chemical and mechanical durability (Figure 12.4(7–9)). The CA retained high values even after weathering in the air for 300 days or exposure to NaCl solution for 60 days (Chen et al. 2021). Wu et al. fabricated core/shell structured SiO_2/ZIF-8 SHPC coatings on Mg alloy. The coating exhibited a CA of 153°, attributed to the rhombic dodecahedral ZIF surface with a micro/nano hierarchical structure. The SiO_2/ZIF-8 SHPC coating showed good mechanical durability during friction and wear tests and exhibited improved corrosion-resistant behavior in a 3.5% NaCl solution (Wu et al. 2017). Several works not mentioned here explored highly hydrophobic MOF coatings (Zhang et al. 2018a, b; Yi et al. 2021).

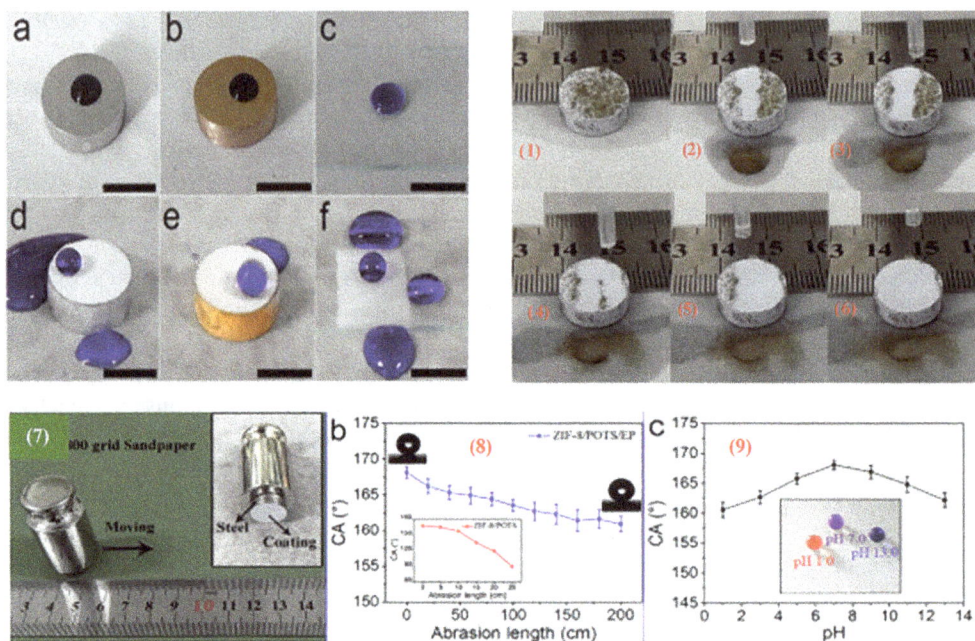

FIGURE 12.4 The photographs of water droplets on the surfaces of uncoated (a) steel, (b) copper, (c) glass, and ZIF-8/POTS/EP-coated (d) steel, (e) copper, and (f) glass. (1–6) Photographs showing self-cleaning property of the coated surface. (7) Schematic of abrasion test, and (8) effect of abrasion length on the CA. (9) Effect of pH values of water droplets on the CA of ZIF-8/POTS/EP coating. (Reproduced with permission from Chen et al. (2021) © 2020 Elsevier B.V.)

12.6   CONCLUSIONS AND PERSPECTIVES

MOFs-based coatings could be the next generation of multifunctional protective self-healing coatings. The versatile and tunable features make them the most favorable materials for protective coatings. Despite the flourishing progress in recent years, there are still a few shortcomings and concerns about MOFs that should be reasonably deliberated for the future utilization of more effective and durable MOF-based coatings. The significant demerits restraining the further applications of MOFs are the cost of the materials, heterogeneous distribution due to accumulation, and the negative effects of interacting with other constituents of coatings. The incorporation of macromolecular organic constituents into MOFs is quite a challenging task that could result in the aggregation of the nanomaterials. Considerable efforts should be made to overcome the above drawbacks and improve the desired performance of the MOF materials. Adapted MOFs can be produced by designing proper metallic cations and organic ligands for extending their multifunctional applications. Alternatively, simple production routes and the variety of abundant raw materials permit the opportunity of manufacturing industrial-scale MOFs-based coatings. With the suitable selection of OLs and the proper alteration of the reactive functional constituents concerning the target polymeric matrix and consuming MOFs as nanocontainers for inhibitive reagents, their prospects beyond these kinds of multifunctional coatings appear to be optimistic.

REFERENCES

Alipanah, N., Yaria, H., Mahdavian, M., Ramezanzadeh, B., and Bahlakeh, G. 2021. MIL-88A (Fe) filler with duplicate corrosion inhibitive/barrier effect for epoxy coatings: electrochemical, molecular simulation, and cathodic delamination studies. *J. Ind. Eng. Chem.* 97: 200–215.

Begum, S., Hashem, T., Tsotsalas, M., Woll, C., and Alkordi, M. H. 2019. Electrolytic conversion of sacrificial metal–organic framework thin films into an electrocatalytically active monolithic oxide coating for the oxygen-evolution reaction. *Energy Technol.* 7: 1900967.

Cao, K., Yu, Z. D., and Yin, D. 2019. Preparation of Ce-MOF@ TEOS to enhance the anticorrosion properties of epoxy coatings. *Prog. Org. Coat.* 135: 613–621.

Cao, K., Yua, Z., Cao, K., Yua, Yin, Z. D., Chen, L., Jiang, Y., and Zhu, L. 2020. Fabrication of BTA-MOF-TEOS-GO nanocomposite to endow coating systems with active inhibition and durable anticorrosion performances. *Prog. Org. Coat.* 143: 105629.

Chen, H., Wang, F., Fan, H., Hong, R., and Li. W. 2021. Construction of MOF-based superhydrophobic composite coating with excellent abrasion resistance and durability for self-cleaning, corrosion resistance, anti-icing, and loading-increasing research. *Chem. Eng. J.* 408: 127343.

Chen, Z., He, K., Wei, R., Lv, Y., Liu Z., and Han, G. C. 2021. Synthesis of Mn-MOFs loaded zinc phosphate composite for water-based acrylic coatings with durable anticorrosion performance on mild steel. *RSC Adv.* 11: 3371.

Falcaro, P., Hill, A. J., Nairn, K. M., Jasieniak, J., Mardel, J. I., Bastow, T. J., Mayo, S. C., Gimona, M., Gomez, D., Whitfield, H. J., Riccò, R., Patelli, A., Marmiroli, B., Amenitsch, H., Colson, T., Villanova, L., and Buso, D. 2011. A new method to position and functionalize metal-organic framework crystals. *Nat. Commun.* 2: 237.

Furukawa, H., Cordova, K. E., O'Keeffe, M., and Yaghi, O. M. 2013. The chemistry and applications of metal-organic frameworks. *Science* 341: 1230444–1230445.

Gangu, K. K., Maddila, S., Mukkamala, S. B., and Jonnalagadda, S. B. 2016. A review on contemporary metal-organic framework materials. *Inorg. Chim. Acta.* 446: 61–74.

Jayaramulu, K., Geyer, F., Schneemann, A., Kment, Š., Otyepka, M., Zboril, R., Vollmer, D., and Fischer, R. A. 2019. Covalent graphene-MOF hybrids for high-performance asymmetric supercapacitors. *Adv. Mater.* 33: 1900820.

Jouyandeh, M., Tikhani, F., Shabanian, M., Movahedi, F., Moghari, S., Akbari, V., Gabrion, X., Laheurte, P., Vahabi, H., and Saeb, M. R. 2020. Synthesis, characterization, and high potential of 3D metal–organic framework (MOF) nanoparticles for curing with epoxy. *J. Alloys Compd.* 829: 1154547.

Kalaivasan, N., and Syed Shafi, S. 2017. Enhancement of corrosion protection effect in mechano chemically synthesized polyaniline/MMT clay nanocomposites. *Arab. J. Chem.* 10: S127–S133.

Kalaj, M., Bentz, K. C., Ayala, S. J., Palomba, J. M., Barcus, K. S., Katayama, Y., and Cohen, S. M. 2020. MOF-polymer hybrid materials: from simple composites to tailored architectures. *Chem. Rev.* 120: 8267–8302.

Kumar, A. M., Latthe, S. L., Sudhagar, P., Obot, I. B., and Gasem, Z. M. 2015. In-situ synthesis of hydrophobic SiO2-PMMA composite for surface protective coatings: experimental and quantum chemical analysis. *Polymer* 77: 79–86.

Kumar, A. M., Rahman, M. M., and Gasem, Z. M. 2015. A promising nanocomposite from CNTs and nanoceria: nanostructured Nanostructured fillers in polyurethane coatings for surface protection. *RSC Adv.* 5: 63537–63544.

Kumaraguru, S., Pavulraj, R., and Mohan, S. 2017. Influence of cobalt, nickel and copper-based metal-organic frameworks on the corrosion protection of mild steel. *Trans. Inst. Met. Finish* 95: 131–136.

Lashgari, S. M., Yari, H., Mahdavian, M., Ramezanzadeh, B., Bahlakeh, G., and Ramezanzadeh, M. 2021. Synthesis of graphene oxide nanosheets decorated by nanoporous zeolite-imidazole (ZIF-67) based metal-organic framework with controlled-release corrosion inhibitor performance: experimental and detailed DFT-D theoretical explorations. *J. Hazard. Mater.* 404: 124068.

Liu, X. B., Yue, T., Qi, K., Xia, B. Y., Chen, Z., Qiu, Y., and Guo, X. 2020. Probe into metal-organic framework membranes fabricated via versatile polydopamine-assisted approach onto metal surfaces as anti-corrosion coatings. *Corros. Sci.* 177: 108949.

Lu, C., Ben, T., Xu, S., and Qiu, S. 2014. Electrochemical synthesis of a microporous conductive polymer based on a metal–organic framework thin film, *Angew. Chem. Int. Ed.* 53: 6454–6458.

Mao, Y., Shi, L., Huang, H., Cao, W., Li, J., Sun, L., Jin, X., and Peng, X. 2013. Room temperature synthesis of free-standing HKUST-1 membranes from copper hydroxide nanostrands for gas separation, *Chem. Commun.* 49: 5666–5668.

Melgar, V. M. A., Kim, J., and Othman, M. R. 2015. Zeolitic imidazolate framework membranes for gas separation: a review of synthesis methods and gas separation performance. *J. Ind. Eng. Chem.* 28: 1–15.

Meng, J., Liu, X., Niu, C., Pang, Q., Li, J., Liu, F., Liu, Z., and Mai, L. 2020. Advances in metal–organic framework coatings: versatile synthesis and broad applications. *Chem. Soc. Rev.* 49: 3142–3186.

Mohammadpour, Z., and Zare, H. R. 2021. The role of embedded 2-ABT@Cu-BTC MOF on the anti-corrosion performance of electro-assisted deposited silica sol-gel composite film. *Mater. Chem. Phys.* 267: 124590.

Motamedi, M., Ramezanzadeh, M., Ramezanzadeh, B., and Mahdavian, M. 2020. One-pot synthesis and construction of a high performance metal-organic structured nano pigment based on nanoceria decorated cerium (III)-imidazole network (NC/CIN) for effective epoxy composite coating anti-corrosion and thermo-mechanical properties improvement. *Chem. Eng. J.* 38215: 122820.

Peng, S., Bie, B., Sun, Y., Liu, M., Cong, H., Zhou, W., Xia, Y., Tang, H., Deng, H., and Zhou, X. 2018. Metal-organic frameworks for precise inclusion of single-stranded DNA and transfection in immune cells. *Nat. Commun.* 9: 1293.

Qiu, S., Su, Y., Zhao, H., Wang, L., and Xue, Q. 2021. Ultrathin metal-organic framework nanosheets prepared via surfactant-assisted method and exhibition of enhanced anticorrosion for composite coatings. *Corros. Sci.* 178: 109090.

Ramezanzadeh, M., Tati, A., Bahlakeh, G., and Ramezanzadeh, B. 2021a. Construction of an epoxy composite coating with exceptional thermo-mechanical properties using Zr-based NH2-UiO-66 metal-organic framework (MOF): experimental and DFT-D theoretical explorations. *Chem. Eng. J.* 408: 127366.

Ramezanzadeh, M., Tati, A., Bahlakeh, G., and Ramezanzadeh, B. 2021b. Development of an active/barrier bi-functional anti-corrosion system based on the epoxy nanocomposite loaded with highly-coordinated functionalized zirconium-based nanoporous metal-organic framework (Zr-MOF). *Chem. Eng. J.* 408: 127361.

Ren, B., Chen, Y., Li, Y., Li, W., Gao, S., Li, H., Cao, R., Li, Y., and Li, H. F. 2020. Rational design of metallic anti-corrosion coatings based on zinc gluconate@ZIF-8. *Chem. Eng. J.* 384: 123389.

Ren, B., Li, Y., Meng, D., Li, J., Gao, S., and Cao, R. 2020. Encapsulating polyaniline within porous MIL-101 for high-performance corrosion protection. *J. Colloid Interface Sci.* 579: 842–852.

Seidi, F., Jouyandeh, M., Taghizadeh, M., Taghizadeh, A., Vahabi, H., Habibzadeh, S., Formela, K., and Sae, M. R. 2020. Metal-organic framework (MOF)/epoxy coatings: a review. *Materials* 13: 2881. doi: 10.3390/ma13122881.

Stassen, I., Campagnol, N., Fransaer, J., Vereecken, P., De Vos, D., and Ameloot, R. 2013. Solvent-free synthesis of supported ZIF-8 films and patterns through transformation of deposited zinc oxide precursors. *Cryst. Eng. Commun.* 15: 9308–9311.

Tarzanagh, J. Y., Seifzadeh, D., Rajabalizadeh, Z., Habibi-Yangjeh, A., Khodayari, A., and Sohrabnezh S. 2019. Sol-gel/MOF nanocomposite for effective protection of 2024 aluminum alloy against corrosion. *Surf. Coat. Technol.* 380: 125038.

Tian, H., Li, W., Liu, A., Gao, X., Han, P., Ding, R., Yang, C., and Wang, D. 2018. Controlled delivery of multi-substituted triazole by metal-organic framework for efficient inhibition of mild steel corrosion in neutral chloride solution. *Corros. Sci.* 131: 1–16.

Wang, H., Fan, Y., Tian, L., Zhao, J., and Ren, L. 2020. Colorimetric/fluorescent dual channel sensitive coating for early detection of copper alloy corrosion. *Mater. Lett.* 265: 127419.

Wang, N., Zhang, Y., Chen, J., Zhang, J., Fang, Q. 2017. Dopamine modified metal-organic frameworks on anti-corrosion properties of waterborne epoxy coatings. *Prog. Org. Coat.* 109: 126–134.

Wei, W., Liu, Z., Wei, R., Han G. C., and Liang, C. 2020. Synthesis of MOFs/GO composite for corrosion resistance application on carbon steel. *RSC Adv.* 10: 29923–29934.

Worrall, S. D., Mann, H., Rogers, A., Bissett, M. A., Attfield, M. P., and Dryfe, R. A. W. 2016. Electrochemical deposition of zeolitic imidazolate framework electrode coatings for supercapacitor electrodes. *Electrochim. Acta* 197: 228–240.

Wu, C., Liu, Q., Chen, R., Liu, J., Zhang, H., Li, R., Takahashi, K., Liu, P., and Wang, J. 2017. Fabrication of ZIF-8@SiO2 micro/nano hierarchical superhydrophobic surface on AZ31 magnesium alloy with impressive corrosion resistance and abrasion resistance. *ACS Appl. Mater. Interfaces* 9: 11106–11115.

Xie, L. H., Xu, M. M., Liu, X. M., Zhao, M. J., and Li, J. R. 2020. Hydrophobic metal–organic frameworks: assessment, construction, and diverse applications. *Adv. Sci.* 7: 1901758.

Xiong, L., Liu, J., Yu, M., and Li, S. 2019. Improving the corrosion protection properties of PVB coating by using salicylaldehyde@ ZIF-8/graphene oxide two dimensional nanocomposites. *Corros. Sci.* 146: 70.

Yang, S., Karve, V. V., Justin, A., Kochetygov, I., Espín, J., Asgari, M., Trukhina, O., Sun, T. D., Peng, L., and Queen, W. L. 2021. Enhancing MOF performance through the introduction of polymer guests. *Coord. Chem. Rev.* 427: 213525.

Yi, B., Wong, Y. L., Hou, C., Zhang, J., Xu, Z., and Yao, X. 2021. Coordination-driven assembly of metal–organic framework coating for catalytically active super hydrophobic Surface. *Adv. Mater. Interface* 8: 2001202.

Yin, D., Yu, Z., Chen, L., and Cao, K. 2019. Enhancement of the anti-corrosion performance of composite epoxy coatings in presence of BTA-loaded copper-based metal-organic frameworks. *Int. J. Electrochem. Sci.* 14: 4240–4253.

Zhang, G., Zhang, J., Su, P., Xu, Z., Li, W., Shen C., and Meng, Q. 2017. Non-activation MOF arrays as a coating layer to fabricate a stable superhydrophobic micro/nano flower-like architecture. *Chem. Commun.* 53: 8340.

Zhang, M., and Liu, Y. 2020. Enhancing the anti-corrosion performance of ZIF-8-based coatings via microstructural optimization. *New J. Chem.* 44: 2941.

Zhang, M., Ma, L., Wang, L. L., Sun, Y., and Liu, Y. 2018a. Insights into the use of metal–organic framework as high-performance anticorrosion coatings. *ACS Appl. Mater. Interfaces* 110: 2259–2263.

Zhang, X., Zhang, Y., Wang, Y., Gao, D., and Zhao, H. 2018b. Preparation and corrosion resistance of hydrophobic zeolitic imidazolate framework (ZIF-90) film @Zn-Al alloy in NaCl solution. *Prog. Org. Coat.* 115: 94.

Zhao, Y., Jiang, F., Chen, Y. Q., and Hu, J. M. 2020. Coatings embedded with GO/MOFs nanocontainers having both active and passive protecting properties. *Corros. Sci.* 168: 108563.

Zheng, Q., Li, J., Yuan, W., Liu, X., Tan, L., Zheng, Y., Yeung, K. W. K., and Wu, S. 2019. Metal organic frameworks incorporated polycaprolactone film for enhanced corrosion resistance and biocompatibility of Mg alloy. *ACS Sustainable Chem. Eng.* 7: 118114–18124.

Zhou, C., Li, Z., Li, J., Yuan, T., Chen, B., Ma, X., Jiang, D., Luo, X., Chen, D., and Liu, Y. 2020. Epoxy composite coating with excellent anticorrosion and self-healing performances based on multifunctional Zeolitic imidazolate framework derived nanocontainers. *Chem. Eng. J.* 385: 123835.

Zhu, M., Jasinski, J. B., and Carreon, M. A. 2012. Growth of Zeolitic imidazolate framework-8 crystals from the solid-liquid interface. *J. Mater. Chem.* 22: 7684–7686.

13 Graphene-Based Anti-Corrosion Coatings: Performance and Mechanisms

Md Julker Nine and Dusan Losic
The University of Adelaide

CONTENTS

13.1 INTRODUCTION

Corrosion of metals is one of the unresolved problems in modern civilization as it affects every metallic structure in major industries, including marine, petroleum, nuclear, defence, and transportation industries. The corrosion cost is so high that an old US survey in 2002 showed that corrosion costs six cents for every dollar of gross domestic product (GDP) in the USA (Koch et al. 2002). The cost in China alone is 3.34% (~310 billion USD) of the GDP in 2015 (Hou et al. 2017). The overall worldwide cost associated with corrosion problems and their maintenance (estimated by NACE International) contributes between 3.5% and 5.2% (~average of 4.35%) of world GDP. The NACE research postulates that corrosion control practices may save between 15% and 35% of the corrosion cost, which could be between $375 and $875 billion USD per annum. However, effective corrosion control materials, such as hexavalent chromium (Cr(VI)), cadmium (Cd), lead (Pb), copper (Cu), and cobalt (Co), which were in use for a long time, have recently been banned or allowed for restricted use only due to their known negative environmental impact and carcinogenic effect on humans (Bal and Kasprzak 2002; Templeton and Liu 2010; L'Azou et al. 2014, EC Commission Decision 2002/525). Among them, the use of Cd-plating for electronic and electrical devices has been restricted since July 1, 2006, due to its hazardous effects on the environment (Directive 2002/95/EC). The use of Co is also envisioned to be restricted soon by European Union (EU) regulations where manufacturers must classify and label cobalt-containing coatings (Greimel et al. 2013; Park et al. 2017). The known hazard of lead paint urges scientists and coating industries to find alternatives of lead, which is still, to some extent, in use in making paints today (Tanquerel des Planches and Dana 1848).

In this circumstance, graphene (G), as the thinnest and structurally flexible coating with its chemical inertness (Liao et al. 2014), and impermeable properties (Berry 2013), appeared as a light in the protective coating industry (Nine et al. 2015; Nine and Losic 2021). In general, G is a monolayer crystalline carbon-based material discovered in 2004 (Allen et al. 2010), that inspired

DOI: 10.1201/9781003169130-15

exploring a new set of materials called "2D materials" (Hassan et al. 2020). This pioneering 2D material is an atomically thin hexagonal crystal structure made of carbon atoms, wherein the atoms are sp^2-hybridized (Figure 13.1a). The crystal structure with hexagonally packed carbon atoms in single- and double-layer G was captured by a high-resolution transmission scanning microscope (HRTEM), as shown in Figure 13.1b. Beyond the application of energy, corrosion research is another intensive area, where G was studied most among many other coating applications of this wonder material (Nine et al. 2015). The outcomes of industrial translation of G technology have started to make an impact, which can be realized by the number of patents (~53,644) filed, particularly between the period from 2004 to 2017 (Yang et al. 2018). This is the right time to realize the practical anti-corrosion application of G in different industries, particularly for use in the industrial heat-exchanger, petroleum and marine industries.

13.2 GRAPHENE CHEMISTRY AND SUPRAMOLECULAR FUNCTIONAL APPROACHES

|||Surface functionalization and structural tailoring of G were mostly inherited and adopted from the knowledge achieved from the surface functionalization of carbon nanotubes (CNTs) (Balasubramanian and Burghard 2005; Banerjee et al. 2005). Despite the availability of high-quality and large-area chemical vapour deposition (CVD) grown G, (Kobayashi et al. 2013), the multiple functionalization approaches were extensively studied on exfoliated G materials derived from natural graphite (Georgakilas et al. 2012). The mechanical and chemical exfoliation of G leave defects in the structure, such as vacancy of atoms, folds, nonuniform layers, and inconsistent lateral sizes, which were later recognized to be useful for various energy and environmental applications, including coatings (Bracamonte et al. 2014). The primary functionalization could be the top-down synthesis of G through chemical exfoliation using modified or improved Hummer's method leading to the formation of oxidized G, but G has to sacrifice its conductive properties (Hummers and Offeman 1958; Chen et al. 2013). The synthesized graphene oxide (GO) possesses numerous oxygen functional groups such as hydroxyl, carboxyl, ketone, and epoxy; hence, GO can function as the precursor or starting materials for further tailoring of the G structure (Compton et al. 2010). The second-degree chemical modification and functionalization of GO enable us to produce useful derivatives (e.g. N, P doped G), which could be an excellent candidate for energy, sensing and coating applications (Sahoo et al. 2012; Zhu et al. 2014; Nine et al. 2015; Yap et al. 2021). However, the reduction of GO can partially restore electrical conductivity, making it functional for electronic applications, while leaving a defecting holy structure in reduced graphene oxide (rGO)

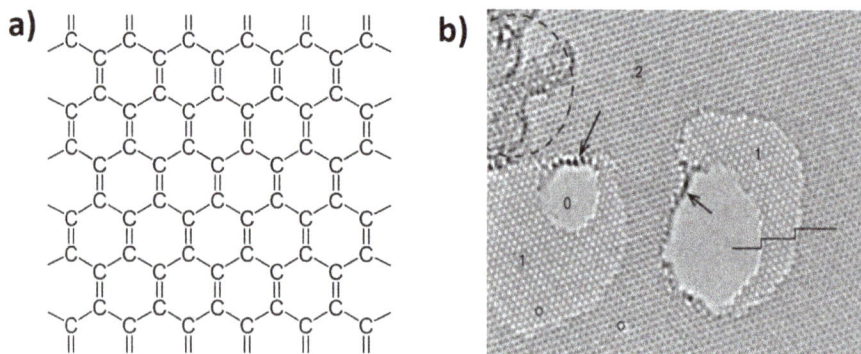

FIGURE 13.1 (a) Schematic of G skeleton (Nine and Losic 2021), (b) HRTEM image of G, showing a hole (vacuum), monolayer and bilayer areas, contamination, and edges (Meyer 2014). ((a) Reproduced with permission from Nine and Losic (2021) © 2021 Elsevier B.V. (b) Reproduced with permission from Meyer (2014) © 2014 Woodhead Publishing Limited.)

(Eigler et al. 2012). The advanced functionalization of G enables it to bond with other materials by creating different covalent and non-valent bonds. In general, these functionalization methods can be classified into several categories as shown in Figure 13.2, which are namely substitutional doping, nanoparticle immobilization, covalent and non-covalent functionalization (Georgakilas et al. 2012; Rodríguez-Pérez et al. 2013). Substitutional elemental doping can be done by mechanical and chemical modification of G which can allow embedding other atoms (e.g. nitrogen (N), Bornon (B), Phosphorus (P), oxygen (O), and sulphur (S)) in the G structure by replacing carbon (Poh et al. 2013; Ullah et al. 2017). Organic molecules such as chromophores, diazonium compounds and polymers can be attached to the G structure through covalent functionalization (Ariga et al. 2012). Beyond these modification routes, non-covalent interactions of G principally involves π-system with other anionic, cationic molecules and hydrogen bond to enhance structural and chemical properties (Georgakilas et al. 2012). This improved functionalization of G derivatives enables immobilizing inorganic nanostructures (nanoparticles, nanocrystals, and quantum dots) in functionalized graphene (FG). Such chemistry and flexibility of FG-derivatives enable these materials to self-assemble and function as building blocks of different 1D, 2D and 3D superstructures such as fibre, thin-film, nano-sphere, aerogel, and so on (Nine et al. 2017). These superstructures opened a new door for versatile application of G and G derivatives.

13.3 ANTI-CORROSION MECHANISMS OF GRAPHENE DERIVATIVES

Chemical inertness (Liao et al. 2014), thermal stability (Campos-Delgado et al. 2009), structural flexibility (Bunch et al. 2008), and most importantly, impermeability to gas (Berry 2013), and ions (Su et al. 2014; Nine et al. 2015), can be considered as primary properties of G to be the best material for anti-corrosion applications. In this section, the corrosion barrier mechanism of G will be discussed.

In general, the locations of weak bonds, highly localized orbitals, and dangling bonds influence chemical reactivity, which is absent in the G structure. The chemical inertness of G emanates from

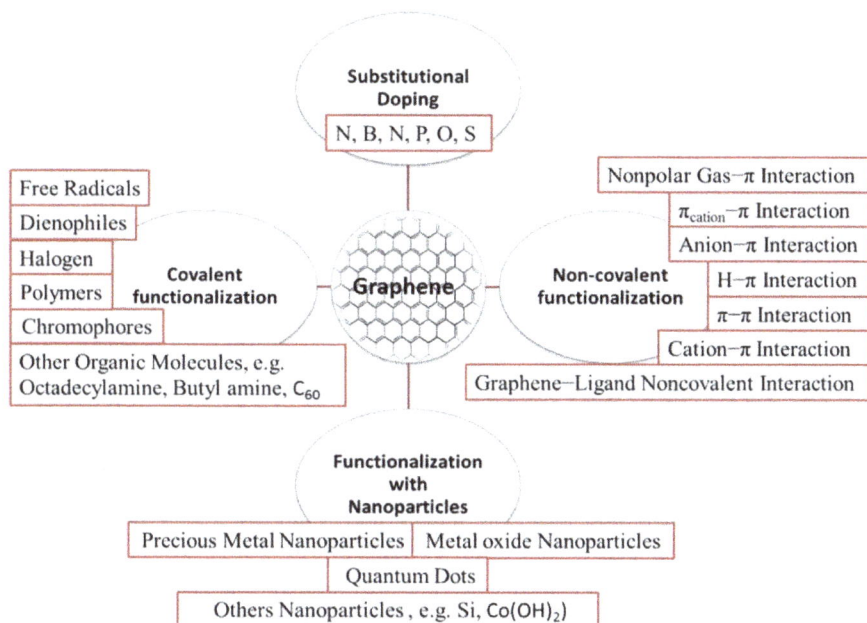

FIGURE 13.2 Functional and supramolecular approaches of G. Doping, Covalent, and non-covalent functionalization routes of G.

its intrinsic structure and the type of carbon–carbon (C-C) bond in the hexagon. The sp^2 hybridization results in the planar equilibrium geometry for aromatic carbons. Covalent chemistry on the G basal plane is hindered by a high kinetic barrier associated with rearrangements of the carbon lattice. The p_z atomic orbitals in G are strongly coupled and stabilized in this delocalized π-bonding system that keeps the G reluctant to form a covalent bond with other materials and provides a self-passivating property (Yan et al. 2012). The disruption of such a stable chemical structure is thermodynamically unfavourable and requires high energy radicals localized on adjacent carbons, which is highly challenging to perform in ambient conditions (Johns and Hersam 2013).

The impermeable nature of G emerged from its small geometric pore in the structure as shown in Figure 13.3a (Berry 2013). The pore diameter in the hexagon is as small as 0.246 nm, including the nuclei of carbon atoms, while the C-C bond length is 0.14 nm. If the van der Waals radii (0.11 nm) of

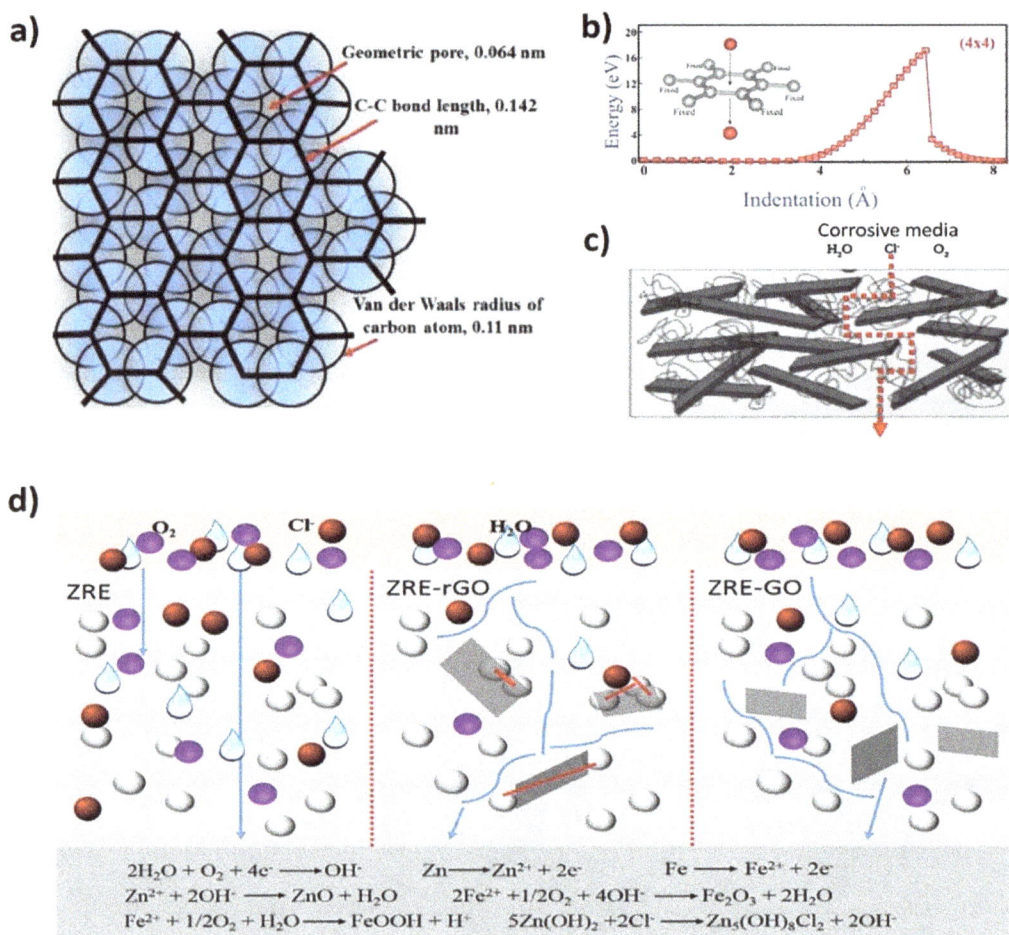

FIGURE 13.3 Anti-corrosion mechanism of G and G derivatives. (a) Close packed atomic structure of the single-layer G sheet (Nine et al. 2015). (b) Energy barrier of singlet (nonmagnetic) O_2, which is forced to pass from the top to the bottom side of G (Topsakal et al. 2012). (c) Tortuous path effect mechanism of G in a polymer matrix (Compton et al. 2010). (d) Illustration of the corrosion protective mechanism for different coatings with cathodic protection (Zhou et al. 2019). ((a) Reproduced with permission from Nine et al. (2015) © 2015 The Royal Society of Chemistry. (b) Reproduced with permission from Topsakal et al. (2012) © 2012 American Physical Society. (c) Reproduced with permission from Compton et al. (2010) © 2010 WILEY-VCH Verlag GmbH & Co. KGaA, Weinheim. (d) Reproduced with permission from Zhou et al. (2019) © 2019 Elsevier Ltd.)

the carbon atoms are taken into account, the geometric pore reduces to 0.064 nm. This small opening into the closely-packed hexagon in the G lattice is even smaller than the helium atom (Bunch et al. 2008). In addition, the dense and delocalized electron cloud of the π–π conjugated carbon network blocks the gap within its hexagonal aromatic rings. It poses a repelling field to reactive atoms or molecules, providing a physical separation between the refined metal surface and environmental reactants (Liao et al. 2014). This energy barrier for a single-layer coating of G is high enough to block oxygen diffusion to the underlying metal interface. The magnitude of this energy barrier varies with the path of the atomic and molecular permeation through the GR lattice, as shown in Figure 13.3b, suggesting that GR can be the thinnest ever known corrosion barrier (Topsakal et al. 2012). Therefore, both inertness and small geometric pores in the G lattice enable this material to display an excellent barrier to reactive gases, liquids, salts, and acids (Su et al. 2014; Nine et al. 2015).

Posing a tortuous path effect or longer travel distance in the matrix (e.g. polymer, ceramic, metal) is another barrier mechanism when G is used as a composite filler. Based on the alignment and orientation of G in the matrix, although two models were discussed (modified-Nielsen and Cussler models) in the literature, here Nielsen model was shown in Figure 13.3c (Compton et al. 2010). The G-embedded polymer matrix creates a nano-barrier wall effect and does not allow the reactive molecules to get diffused easily through the matrix. This tortuosity is one of the main factors that can enhance the barrier properties of the composite matrix by involving stacking and homogeneous dispersion of G flakes into the matrix. For example, if the nanofillers of G are stacked randomly or horizontally, it enhances tortuosity resulting in good barrier properties. To improve the tortuosity, the stacking of nanosheets should be vertically aligned to the direction from which diffusion occurs (Cui et al. 2016). Factors such as crystallinity of polymer matrix, G aspect ratio, G sheets orientation, polymer matrix with the dispersion of fillers and also the exfoliation of filler can affect the tortuosity of corrosive ions diffusion path.

Apart from the inertness, barrier effect, and tortuosity of G, it can play a significant role in increasing the cathodic protection of zinc (Zn)-rich coatings by providing a long-range electrical connection as shown in Figure 13.3d. Most of the Zn-rich epoxy coatings suffer from a lack of interconnection and fail to provide electrical conductivity that results in unused Zn-particles left in the matrix (Cubides and Castaneda 2016). The use of conductive G in Zn-rich coatings can enable to reach percolation threshold with a lower Zn-concentration compared with the traditional Zn-rich epoxy (Zhou et al. 2019). G, as an electrically conductive 2D material with its large surface area, is capable of providing a large-area electrical connection to maximize the use of Zn- particles when resting apart from each other in the matrix (the details of this mechanism will be discussed in Section 13.4.2).

13.4 ANTI-CORROSION GRAPHENE COATINGS AND THEIR EVOLUTION

13.4.1 GRAPHENE COATING BY BOTTOM-UP APPROACHES

In the beginning, the corrosion study was mainly associated with the G grown by CVD methods (Chen et al. 2011). Some pioneering research reported CVD-grown G as an excellent anti-corrosion coating to protect refined metals (Chen et al. 2011; Kirkland et al. 2012; Prasai et al. 2012; Singh Raman et al. 2012). Chen et al. (2011) claimed for the first time to reveal the ability of CVD-grown G to protect the Cu and Cu/Ni alloy surface from air oxidation. They displayed that G prevents the formation of an oxide on the protected metal surfaces only one atom away from reactive environments, as depicted in Figure 13.4a. They further tested the corrosion-resistant ability of G -coated Cu-coin under 30% H_2O_2 treatment for 2 min as well as at elevated temperature. The unprotected part of the Cu penny turned a dark shade of brown, as shown in Figure 13.4b, while the protected coin maintained its original glossy appearance. Similar outcomes were reported by Kirkland et al. (2012), and Singh Raman et al. (2012), wherein G was grown on Cu and tested by potentiodynamic polarization experiments in NaCl solution. The electrochemical corrosion test in an aggressive chloride

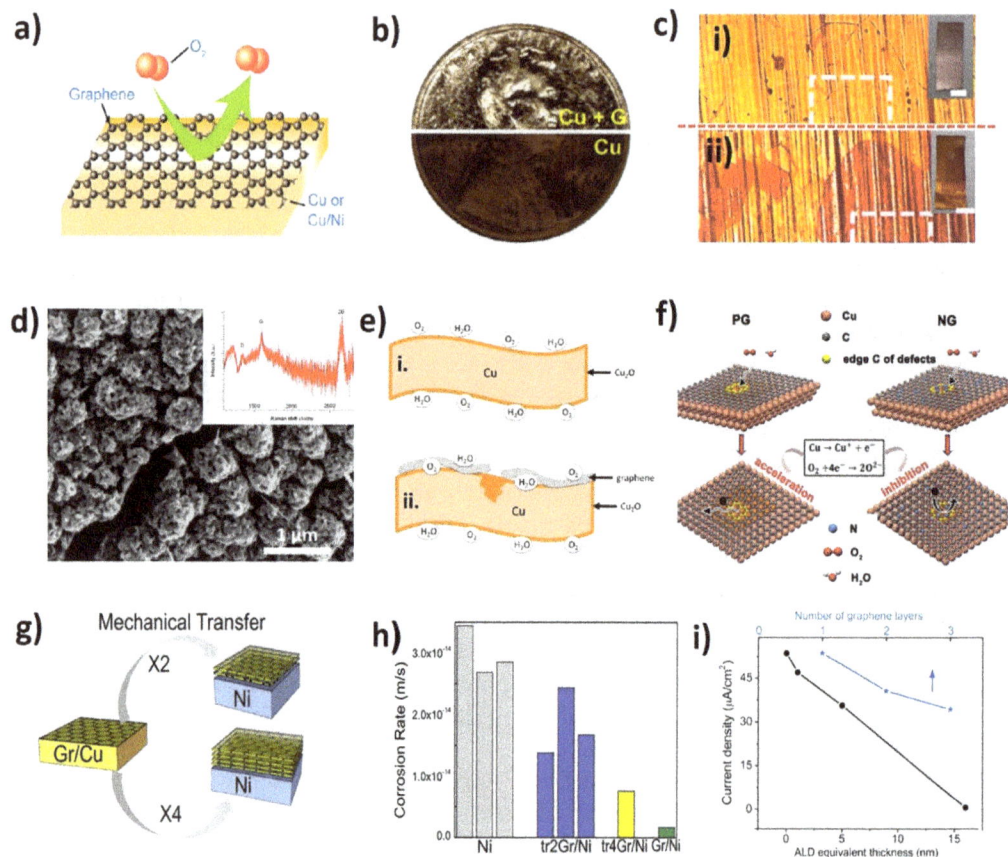

FIGURE 13.4 Progress in CVD-grown G-coatings for improved corrosion protection, (a) Illustration depicting a G sheet as a chemically inert diffusion barrier, (b) Photograph showing a G-coated (upper) and uncoated (lower) penny after H_2O_2 treatment (30%, 2 min) (Chen et al. 2011). (c) Optical micrographs of (i) an as-prepared G/Cu sample, and (ii) a 6-month-aged G/Cu sample (Zhou et al. 2013). (d) Complete surface oxidation (at 250°C) of Cu foil covered with monolayer CVD G. (e) Schematic of native Cu oxide formation as compared with G-assisted Cu corrosion. (i) Bare Cu forms a passivating native oxide, (ii) A G coating on Cu slows diffusion of O_2/H_2O but assists long-term Cu corrosion in a number of ways (Schriver et al. 2013). (f) Schematic diagrams of the corrosion mechanisms for the pristine-G (PG) and N-doped G (NG)-coated on Cu foils (Ren et al. 2018). (g) Mechanically transferred multilayer CVD-grown G onto the Ni surface, (h) Corrosion rates of bare Ni samples and the samples where G was transferred onto Ni substrate (Prasai et al. 2012). (i) Peak corrosion current density for Cu/G/ALD samples with different ALD thickness compared to peak current for Cu/G samples with 1–3 layers (Hsieh et al. 2014). ((a, b) Reproduced with permission from Chen et al. (2011) © 2011 American Chemical Society. (c) Reproduced with permission from Zhou et al. (2013) (doi:10.1021/nn402150t) © 2013 American Chemical Society. Further permission related to the material excerpted should be directed to the ACS. (d, e) Reproduced with permission from Schriver et al. (2013) © 2013 American Chemical Society. (f) Reproduced with permission from Ren et al. (2018) © 2018 The Royal Society of Chemistry. (g, h) Reproduced with permission from Prasai et al. (2012) © 2012 American Chemical Society. (i) Reproduced with permission from Hsieh et al. (2014) © 2014 American Chemical Society.)

environment revealed that the impedance of Cu increases dramatically. The anodic and cathodic current densities of coated Cu become nearly 1–2 orders of magnitude smaller than the control. The observations were found to be counterintuitive as graphite in contact with metals increases metallic corrosion. However, it did not take much time for others to report CVD-grown G as worse than nothing for the long-term oxidation barrier (Schriver et al. 2013; Zhou et al. 2013). Zhou et al. (2013) reported an enhanced corrosion effect of Cu foil in the presence of G coating at room temperature for 6 months of exposure to the air, as shown in Figure 13.4c. The outcome reveals that the simple layer G completely fails to protect the Cu substrate at temperatures below ~300°C. The G -coated Cu surface turns a uniform matte black when exposed to an atmosphere at 250°C for 17 h. The elemental peak of the oxidized surface indicates the formation of CuO with the morphology of small flower-like balls after the treatment at 250°C as shown in Figure 13.4d. However, the G coating was found to slow down the diffusion of reactive molecules (e.g. O_2, H_2O), but it assists in long-term corrosion by maintaining a conductive pathway across the metal surface and to selected regions of the substrate, as shown in Figure 13.4e. Defected G may cause non-uniform oxidation due to having strong and weak points on the plane and promotes stress and cracking of the oxide layer.

The conductive G on and through the cracks further promotes corrosion as graphitic materials are cathodic to many corrosion-prone metals (Schriver et al. 2013). To address the issue, N doing of CVD-grown G was proposed by Ren et al. (2018) towards long-term corrosion protection of Cu as shown in Figure 13.4f. The N-doped G can significantly reduce the electrical conductivity of G, hence preventing electron transport between Cu and N-doped G interfaces. The outcome shows both short- and long-term corrosion are inhibited for 3 months at ambient conditions. The mechanical transfer of multilayer CVD-grown G was introduced by Prasai et al. (2012) to improve the corrosion-resistant efficiency of single layer CVD-grown G (Figure 13.4g). The transfer of four layers of CVD-grown G on the Ni substrate improves corrosion resistance by 20 folds measured from Tafel analysis as displayed in Figure 13.4h. Later, a similar mechanical transfer of CVD-grown G was introduced wherein two G layers were sandwiched by three layers of polyvinyl butyral on aluminium (Al) alloys showed long-term corrosion protection for 3 months when exposed to simulated seawater (Yu et al. 2018). Another effective engineering for improved corrosion resistance involves atomic layer deposition (ALD) to grow an ultrathin Al_2O_3 layer on CVD-grown G to provide an additional passivation layer on it. The top ALD-Al_2O_3 layer was deposited to make up for the existing faults or surface defects of G. It is reported that a ~5 nm equivalent thickness of Al_2O_3 layer on CVD-grown G renders the surface as inert as obtained by three layers of G. The complete passivation effect was realized by ~16 nm equivalent film thickness, which resulted in an inhibition efficiency of >99%.

13.4.2 GRAPHENE-BASED COMPOSITE COATINGS

The polymer matrix is extensively studied to form a G-based composite (Mu et al. 2021), while the use of G as filler in ceramic (Porwal et al. 2013), and the metal matrix are investigated in parallel (Jin et al. 2019; Güler and Bağcı 2020). In polymer/G composite, the engineering of coatings was designed to exploit properties of different G derivatives (e.g. G, GO, rGO) at their best. The barrier effect in terms of the tortuous path effect in the matrix showed significant improvement in the anti-corrosion performance. G has shown great promise in corrosion control, but GO is the most studied derivative due to its structural advantage of having numerous oxygen functional groups. These groups in the GO basal plane and edges can change the Van der Waals forces acting upon them, resulting in easy access to the polymeric matrix.

Electrophoretic deposition (EPD) of negatively charged GO is able to form uniform coatings on a metal substrate (Figure 13.5a), which decreases corrosion current density and provides a positive shift in corrosion potential (He et al. 2013). In another study, Singh et al. (2013) showed that in the presence of polymeric isocyanate crosslinked with hydroxy-functional acrylic (PIHA), GO could acquire a high positive charge, enabling EPD of the composite. The EPD approach deposited a thin composite coating (~40 nm) of PIHA/GO on the Cu substrate, exhibiting excellent oxidation resistance characteristics under a severe chlorine ion environment. A hydrophobic treatment on the EPD coating significantly enhances the water repellency, leading to enhanced corrosion resistance. However, the superior barrier effect of the GO composite was achieved against ions, gases, and acids, when GO was combined with polyvinyl alcohol (PVA) (Su et al. 2014), which later inspired us to develop a simple air-brush spray based on PVA/GO composite coatings on

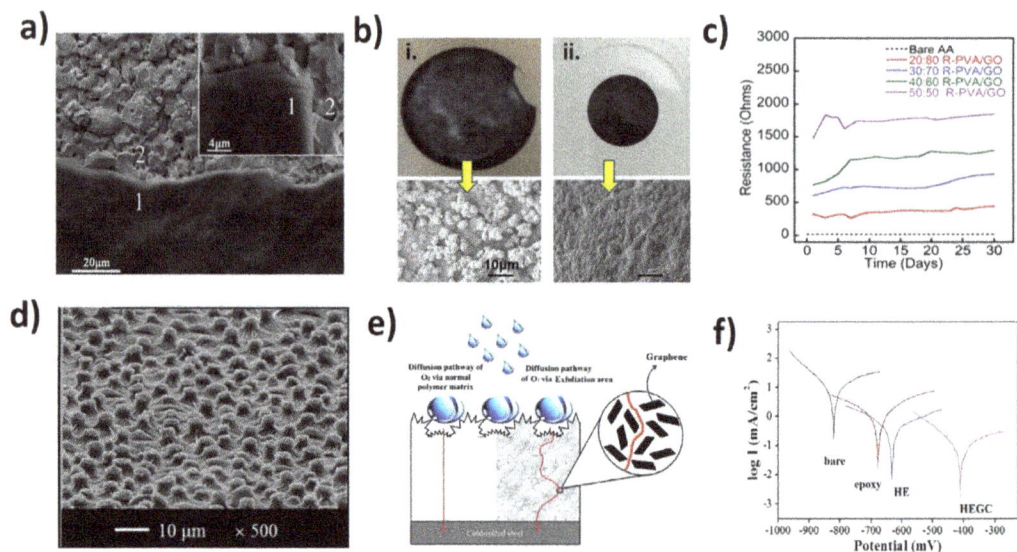

FIGURE 13.5 Engineering of polymer-based G composite. (a) Scanning electron microscope (SEM) images of the GO coating on the NdFeB surface, Area 1 represents EPD-GO coated NdFeB while Area 2 represents NdFeB (He et al. 2013). (b) Digital images of (i) bare Al alloy and (ii) PVA/rGO (30:70) samples after 4 days of electrochemical testing. (c) Time dependence of Rp for the bare Al alloy at the initial stage and coated Al alloy over 30 days of exposure (De and Lutkenhaus 2018). (d) the "Xanthosoma sagittifolium" leaf-like hydrophobic surface of the composite, (e) Schematic representation of hydrophobic surface and oxygen following a tortuous path through epoxy and EGC materials, (f) Tafel plots for bare, epoxy-coated, HE-coated, and HEGC-coated CRS electrodes measured at 25°C ± 0.5°C (Chang et al. 2014). ((a) Reproduced with permission from He et al. (2013) © 2013 Elsevier B.V. (b, c) Reproduced with permission from De and Lutkenhaus (2018) © 2018 The Royal Society of Chemistry. (d–f) Reproduced with permission from Chang et al. (2014) © 2013 Elsevier Ltd.)

Al alloy. An adhesion-promoting layer of branched polyethyleneimine (B-PEI) between the substrate and coatings provided excellent corrosion resistance followed by the reduction process of coatings (PVA/rGO). The PVA/rGO coatings were evaluated in a 0.2 M NaCl solution for 30 days. The coating with 60–70 wt% of rGO exhibited the best performance with little to no corrosion in comparison to the bare aluminium alloy, as shown in Figure 13.5b and c.

Epoxy resin is one of the most frequently used polymers used to combine with G derivatives. The alteration of the weight percentage of GO in the epoxy matrix was found to affect the corrosion efficiency as the epoxy hardener has a significant effect on the formation of bonds between the polymer matrix and GO (Chen et al. 2017). When GO is added as a nanofiller with an epoxy coating, it enhances the protection against corrosion and does not allow corrosive agents to get diffused on the surface of the metal. It also contributes to enhancing anodic corrosion as GO is electrically insulative. The modification of the surface chemistry to create hydrophobicity of G-based composite coatings was found to improve the anti-corrosion efficiency as elevated water contact (CA) angle repel reactive moisture and provides an air gap between reactants and coatings (Singh et al. 2013; Chang et al. 2014; Nine et al. 2015). Singh et al. (2013) showed that a simple hydrophobic treatment with silicone fluids (KF-99) on PIHA/GO composite coatings significantly enhanced the CA, hence enhancing corrosion resistance. Later, a coating of epoxy/G composites (EGCs) as corrosion inhibitors with hydrophobic surfaces (HEGC) was developed as shown in Figure 13.5d (Chang et al. 2014).

The mechanism of hydrophobicity and gas barrier effect in HEGC was envisaged to result in low-wettability, while the dispersed G in the epoxy matrix can increase the tortuosity of the oxygen diffusion pathway, as shown in Figure 13.5e. It was clear that the HEGC coating outperforms all other coatings when compared to the corrosion potential and the corrosion current in the Tafel analysis shown in Figure 13.5f. The finding claims threefold protection for metals from corrosion in this study. First, the cured epoxy coating acted as an excellent barrier to limit the accessibility of moisture and oxygen to the metal surface. Second, the hydrophobicity repelled the moisture and further reduced the water/corrosive media adsorption on the epoxy surface, preventing the underlying metal from corrosion. Finally, the dispersed G in the epoxy matrix increased the tortuosity of the oxygen diffusion pathway (Chang et al. 2014). Recently, Nine et al. (2015) used rGO to enhance the ionic barrier in polydimethylsiloxane (PDMS) treated porous superhydrophobic composite coatings of DE/TiO$_2$. The coating justified the barrier properties by showing excellent corrosion resistance (96.78%) in the potentiodynamic polarization study. Furthermore, HCl$_{(aq)}$ was used on the surface to investigate further insight into coating barrier properties, wherein the presence of rGO in superhydrophobic coatings slows down ionic diffusion rate by half.

The metal-based G composite is an interesting inclusion in the anti-corrosion study of G (Jin et al. 2019). Inspired by CVD-grown G-coated Cu, this work presented G encapsulated Cu nano-flakes which were used as a building block for assembling bulk G/Cu composites, as shown in Figure 13.6a. The bulk composite prepared in this way was expected to exhibit improved anti-corrosion properties. The SEM image in Figure 13.6b shows that the presence of superficial G-coatings on the Cu grain is highly visible. The as-prepared bulk from G-coated individual Cu-flake significantly reduced the corrosion rate, which was measured at one-third of the uncoated sample (Figure 13.6c). Although the concept of the study is interesting, there are a couple of fundamental issues with the work. The degree of mechanical strength among G encapsulated Cu flakes as the metallic bond

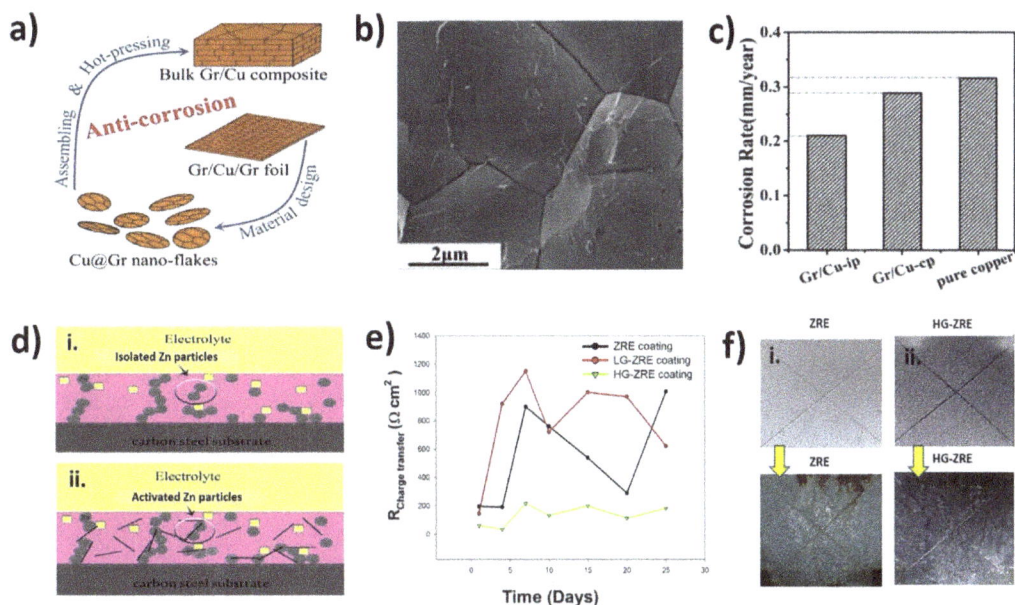

FIGURE 13.6 Engineering of metal-based G composite (a) Schematic illustration of material design based on a building block strategy. Inspired by CVD-grown G anti-corrosion role as shown in G/Cu/G foil, G encapsulated Cu(Cu@G) nano-flakes are designed and then used as the building blocks for assembling bulk G/Cu composites, (b) the Cu@G flakes, (c) corrosion rate of the Cu, G/Cu-ip and G/Cu-cp samples (Jin et al. 2019). (d) Schematic representation of the percolating structure of (i) Zn-rich epoxy (ZRE), and (ii) high G content-ZRE (HG-ZRE) coatings before immersion in NaCl solution, (e) Evolution of charge transfer resistance as a function of immersion time in 3.5% NaCl solution for various coating systems, (f) Photographs of ZRE and HG-ZRE samples before after 1000h of salt spray test, (i) ZRE sample (upper-before test, bottom- after test) and (ii) HG-ZRE sample (upper-before test, bottom- after test) (Hayatdavoudi and Rahsepar 2017). ((a–c) Reproduced with permission from Jin et al. (2019) © 2018 Elsevier Ltd. (d–f) Reproduced with permission from Hayatdavoudi and Rahsepar (2017) © 2017 Elsevier B.V.)

will be abolished, and the presence of galvanic coupling between G and Cu may promote long-term oxidation in Cu flakes, as discussed in Section 13.4.1. However, the use of electrically conductive G came into effect when the approach of transforming G into an anode was implemented. It was done by covering it with a more active anode substance such as Zn. The use of G as a conductive filler in Zn-rich coatings was reported to increase the lifetime of coatings and assist in improving barrier properties and percolation of the coating (Gergely et al. 2015; Punith Kumar et al. 2015; Hayatdavoudi and Rahsepar 2017; Ramezanzadeh et al. 2017; Shen et al. 2019). G can slow down the self-corrosion of Zn-particles due to the presence of oxygen and water, leading to rapid loss of the electrical connection between the adjacent Zn-particles. G as a nanofiller improves the electrical continuity between the metal substrate and Zn-particles, enhancing the process of percolation that significantly decreases the quantity of pigmentation of Zn (Gergely, Pászti et al. 2015). The degree of cathodic polarization of high content GR (0.4 wt%) in Zn-rich epoxy (HG-ZRE) in comparison to low content (0.1 wt%) was revealed better (Hayatdavoudi and Rahsepar 2017). The 0.4 wt% is considered satisfactory loading of GR, which could be attributed to the more uniform activation of Zn-particles as shown in Figure 13.6d-(ii). The final value of the mixed interface potential depends on the relative active surface area of the Zn-particles and steel substrate. Higher Zn to steel active surface area results in a more cathodic polarization of the substrate and causes a negative shift of mixed interface potential. On the other hand, for a Zn-particle to be considered as an active sacrificial species, it must be electrically connected to the steel substrate while immersed in the electrolyte. The measured charge transfer resistance value for HG-ZRE was significantly lower than for ZRE and LG-ZRE coatings. The presence of G-activated sacrificial Zn-particles with a low percolation resistance of coating indicates that the HG-ZRE coating provides superior cathodic protection performance as compared with those of ZRE and LG-ZRE coatings. Furthermore, Figure 13.6f-(i) and (ii) exhibits the outcome (photographs) of coating specimens before and after 1000 h of salt spray test on ZRE and HG-ZRE substrates. As seen, while the specimen with the ZRE coating displayed red rust formation on the scribed areas, white rust was formed on the surface of the HG-ZRE coatings, which revealed the active sacrificial action of Zn-particles over long immersion times in the presence of a high amount of graphene nanosheets.

Apart from the CVD-grown G and solid composite coatings, G derivatives were investigated as corrosion-inhibiting agents mixed with oil and other acidic fluids (Mayavan et al. 2013; Gupta et al. 2017; Baig et al. 2019; Haruna et al. 2019; Hegde et al. 2020; Haruna et al. 2021). G derivatives, particularly GO, were functionalized with other corrosion-inhibiting agents or polymeric compounds, such as diethylenetriamine (Baig et al. 2019), diazo pyridine (Gupta et al. 2017), Garcinia gummigutta (Hegde et al. 2020), and Cyclodextrin (Haruna et al. 2019), to develop advanced corrosion inhibitors. The as-prepared functionalized derivatives were dispersed in 1 M HCl or in other suitable organic solvents for use as corrosion inhibitors for mild steel or other metal substrates. All these relevant results involving corrosion inhibitors showed significant improvement in corrosion resistance compared to their control substrates.

13.5   CONCLUSION AND PERSPECTIVES

Graphene, with its inertness and barrier properties, as well as conductive nature, draws great attention from researchers to exploit this material in different ways for anti-corrosion applications. The tailoring flexibility of graphene with elemental and molecular doping, covalent and non-covalent bonds further allows the material to be used in various forms to introduce G-enhanced properties in the coating matrix. Engineering of coatings includes physical and chemical modification, doping, as well as multilayer graphene transfer, and the passivation concept of CVD-grown graphene. The composites are also in gradual development through various strategies of using different graphene derivatives, including the use of pristine graphene as barrier filler, crosslinking GO with polymers, and graphene as a conductive filler in Zn-rich coatings. Finally, graphene derivatives were also investigated as corrosion inhibitor agents to protect metals. To date, the progress in the development of graphene-based anti-corrosion coatings seems to be accomplished from a conceptual point of view. This is the right time to make an industrial translation of graphene-based coating technologies to fill the gap of previously used effective coating constituents.

REFERENCES

Allen, M. J., V. C. Tung, and R. B. Kaner (2010). "Honeycomb carbon: a review of graphene." *Chemical Reviews* 110(1): 132–145.

Ariga, K., Q. Ji, M. J. McShane, Y. M. Lvov, A. Vinu, and J. P. Hill (2012). "Inorganic nanoarchitectonics for biological applications." *Chemistry of Materials* 24(5): 728–737.

Baig, N., D. S. Chauhan, T. A. Saleh and M. A. Quraishi (2019). "Diethylenetriamine functionalized graphene oxide as a novel corrosion inhibitor for mild steel in hydrochloric acid solutions." *New Journal of Chemistry* 43(5): 2328–2337.

Bal, W. and K. S. Kasprzak (2002). "Induction of oxidative DNA damage by carcinogenic metals." *Toxicology Letters* 127(1–3): 55–62.

Balasubramanian, K. and M. Burghard (2005). "Chemically functionalized carbon nanotubes." *Small* 1(2): 180–192.

Banerjee, S., T. Hemraj-Benny and S. S. Wong (2005). "Covalent surface chemistry of single-walled carbon nanotubes." *Advanced Materials* 17(1): 17–29.

Berry, V. (2013). "Impermeability of graphene and its applications." *Carbon* 62: 1–10.

Bracamonte, M. V., G. I. Lacconi, S. E. Urreta and L. E. F. Foa Torres (2014). "On the nature of defects in liquid-phase exfoliated graphene." *The Journal of Physical Chemistry C* 118(28): 15455–15459.

Bunch, J. S., S. S. Verbridge, J. S. Alden, A. M. van der Zande, J. M. Parpia, H. G. Craighead and P. L. McEuen (2008). "Impermeable atomic membranes from graphene sheets." *Nano Letters* 8(8): 2458–2462.

Campos-Delgado, J., Y. Kim, T. Hayashi, A. Morelos-Gómez, M. Hofmann, H. Muramatsu, M. Endo, H. Terrones, R. Shull and M. Dresselhaus (2009). "Thermal stability studies of CVD-grown graphene nanoribbons: defect annealing and loop formation." *Chemical Physics Letters* 469(1–3): 177–182.

Chang, K.-C., M.-H. Hsu, H.-I. Lu, M.-C. Lai, P.-J. Liu, C.-H. Hsu, W.-F. Ji, T.-L. Chuang, Y. Wei, J.-M. Yeh and W.-R. Liu (2014). "Room-temperature cured hydrophobic epoxy/graphene composites as corrosion inhibitor for cold-rolled steel." *Carbon* 66: 144–153.

Chen, C., S. Qiu, M. Cui, S. Qin, G. Yan, H. Zhao, L. Wang and Q. Xue (2017). "Achieving high performance corrosion and wear resistant epoxy coatings via incorporation of noncovalent functionalized graphene." *Carbon* 114: 356–366.

Chen, J., B. Yao, C. Li and G. Shi (2013). "An improved Hummers method for eco-friendly synthesis of graphene oxide." *Carbon* 64: 225–229.

Chen, S., L. Brown, M. Levendorf, W. Cai, S.-Y. Ju, J. Edgeworth, X. Li, C. W. Magnuson, A. Velamakanni, R. D. Piner, J. Kang, J. Park and R. S. Ruoff (2011). "Oxidation resistance of graphene-coated Cu and Cu/Ni alloy." *ACS Nano* 5(2): 1321–1327.

Compton, O. C. and S. T. Nguyen (2010). "Graphene oxide, highly reduced graphene oxide, and graphene: versatile building blocks for carbon-based materials." *Small* 6(6): 711–723.

Compton, O. C., S. Kim, C. Pierre, J. M. Torkelson and S. T. Nguyen (2010). "Crumpled graphene nanosheets as highly effective barrier property enhancers." *Advanced Materials* 22(42): 4759–4763.

Cubides, Y. and H. Castaneda (2016). "Corrosion protection mechanisms of carbon nanotube and zinc-rich epoxy primers on carbon steel in simulated concrete pore solutions in the presence of chloride ions." *Corrosion Science* 109: 145–161.

Cui, Y., S. I. Kundalwal and S. Kumar (2016). "Gas barrier performance of graphene/polymer nanocomposites." *Carbon* 98: 313–333.

De, S. and J. L. Lutkenhaus (2018). "Corrosion behaviour of eco-friendly airbrushed reduced graphene oxide-poly(vinyl alcohol) coatings." *Green Chemistry* 20(2): 506–514.

"Directive 2002/95/EC of The European Parliament and of The Council of 27 January 2003 on the restriction of the use of certain hazardous substances in electrical and electronic equipment." *Official Journal of the European Communities* L37: 19–37.

EC Commission Decision -2002/525/EC: Commission decision of 27 June 2002 amending Annex II of Directive 2000/53/EC of the European Parliament and of the council on end-of-life vehicles, (C(2002) 2238). https://op.europa.eu/en/publication-detail/-/publication/e1bfc49e-c555-4dca-9a44-4596c66d084c

Eigler, S., C. Dotzer and A. Hirsch (2012). "Visualization of defect densities in reduced graphene oxide." *Carbon* 50(10): 3666–3673.

Georgakilas, V., M. Otyepka, A. B. Bourlinos, V. Chandra, N. Kim, K. C. Kemp, P. Hobza, R. Zboril and K. S. Kim (2012). "Functionalization of graphene: covalent and non-covalent approaches, derivatives and applications." *Chemical Reviews* 112(11): 6156–6214.

Gergely, A., Z. Pászti, J. Mihály, E. Drotár and T. Török (2015). "Galvanic function of zinc-rich coatings facilitated by percolating structure of the carbon nanotubes. Part I: Characterization of the nano-size particles." *Progress in Organic Coatings* 78: 437–445.

Greimel, K. J., V. Perz, K. Koren, R. Feola, A. Temel, C. Sohar, E. Herrero Acero, I. Klimant and G. M. Guebitz (2013). "Banning toxic heavy-metal catalysts from paints: enzymatic cross-linking of alkyd resins." *Green Chemistry* 15(2): 381–388.

Güler, Ö. and N. Bağcı (2020). "A short review on mechanical properties of graphene reinforced metal matrix composites." *Journal of Materials Research and Technology* 9(3): 6808–6833.

Gupta, R. K., M. Malviya, C. Verma, N. K. Gupta and M. A. Quraishi (2017). "Pyridine-based functionalized graphene oxides as a new class of corrosion inhibitors for mild steel: an experimental and DFT approach." *RSC Advances* 7(62): 39063–39074.

Haruna, K., L. M. Alhems and T. A. Saleh (2021). "Graphene oxide grafted with dopamine as an efficient corrosion inhibitor for oil well acidizing environments." *Surfaces and Interfaces* 24: 101046.

Hassan, K., M. J. Nine, T. T. Tung, N. Stanley, P. L. Yap, H. Rastin, L. Yu and D. Losic (2020). "Functional inks and extrusion-based 3D printing of 2D materials: a review of current research and applications." *Nanoscale* 12(37): 19007–19042.

Haruna, K., T. A. Saleh, I. B. Obot and S. A. Umoren (2019). "Cyclodextrin-based functionalized graphene oxide as an effective corrosion inhibitor for carbon steel in acidic environment." *Progress in Organic Coatings* 128: 157–167.

Hayatdavoudi, H. and M. Rahsepar (2017). "A mechanistic study of the enhanced cathodic protection performance of graphene-reinforced zinc rich nanocomposite coating for corrosion protection of carbon steel substrate." *Journal of Alloys and Compounds* 727: 1148–1156.

He, W., L. Zhu, H. Chen, H. Nan, W. Li, H. Liu and Y. Wang (2013). "Electrophoretic deposition of graphene oxide as a corrosion inhibitor for sintered NdFeB." *Applied Surface Science* 279: 416–423.

Hegde, M. B., S. R. Nayak, K. N. S. Mohana and N. K. Swamy (2020). "Garcinia gummigutta vegetable oil–graphene oxide nano-composite: an efficient and eco-friendly material for corrosion prevention of mild steel in saline medium." *Journal of Polymers and the Environment* 28(2): 483–499.

Hou, B., X. Li, X. Ma, C. Du, D. Zhang, M. Zheng, W. Xu, D. Lu and F. Ma (2017). "The cost of corrosion in China." *npj Materials Degradation* 1(1): 4.

Hsieh, Y.-P., M. Hofmann, K.-W. Chang, J. G. Jhu, Y.-Y. Li, K. Y. Chen, C. C. Yang, W.-S. Chang and L.-C. Chen (2014). "Complete corrosion inhibition through graphene defect passivation." *ACS Nano* 8(1): 443–448.

Hummers, W. S. and R. E. Offeman (1958). "Preparation of graphitic oxide." *Journal of the American Chemical Society* 80(6): 1339–1339.

Jin, B., D.-B. Xiong, Z. Tan, G. Fan, Q. Guo, Y. Su, Z. Li and D. Zhang (2019). "Enhanced corrosion resistance in metal matrix composites assembled from graphene encapsulated copper nanoflakes." *Carbon* 142: 482–490.

Johns, J. E. and M. C. Hersam (2013). "Atomic covalent functionalization of graphene." *Accounts of Chemical Research* 46(1): 77–86.

Kirkland, N. T., T. Schiller, N. Medhekar and N. Birbilis (2012). "Exploring graphene as a corrosion protection barrier." *Corrosion Science* 56: 1–4.

Kobayashi, T., M. Bando, N. Kimura, K. Shimizu, K. Kadono, N. Umezu, K. Miyahara, S. Hayazaki, S. Nagai, Y. Mizuguchi, Y. Murakami and D. Hobara (2013). "Production of a 100-m-long high-quality graphene transparent conductive film by roll-to-roll chemical vapor deposition and transfer process." *Applied Physics Letters* 102(2): 023112.

Koch, G. H., M. P. Brongers, N. G. Thompson, Y. P. Virmani and J. H. Payer (2002). Corrosion cost and preventive strategies in the United States (No. FHWA-RD-01–156).

L'Azou, B., I. Passagne, S. Mounicou, M. Treguer-Delapierre, I. Puljalte, J. Szpunar, R. Lobinski and C. Ohayon-Courtes (2014). "Comparative cytotoxicity of cadmium forms (CdCl2, CdO, CdS micro- and nanoparticles) in renal cells." *Toxicology Research* 3(1): 32–41.

Liao, L., H. Peng and Z. Liu (2014). "Chemistry makes graphene beyond graphene." *Journal of the American Chemical Society* 136(35): 12194–12200.

Mayavan, S., T. Siva and S. Sathiyanarayanan (2013). "Graphene ink as a corrosion inhibiting blanket for iron in an aggressive chloride environment." *RSC Advances* 3(47): 24868–24871.

Meyer, J. C. (2014). 5-Transmission Electron Microscopy (TEM) of Graphene. In *Graphene* (V. Skákalová and A. B. Kaiser), Woodhead Publishing, pp. 101–123.

Mu, J., F. Gao, G. Cui, S. Wang, S. Tang and Z. Li (2021). "A comprehensive review of anticorrosive graphene-composite coatings." *Progress in Organic Coatings* 157: 106321.

Nine, M. J. and D. Losic (2021). 13-Application of graphene in protective coating industry: prospects and current progress. In *Handbook of Modern Coating Technologies* (M. Aliofkhazraei, N. Ali, M. Chipara, N. Bensaada Laidani and J. T. M. De Hosson), Amsterdam, Elsevier, pp. 453–492.

Nine, M. J., M. A. Cole, D. N. H. Tran and D. Losic (2015). "Graphene: a multipurpose material for protective coatings." *Journal of Materials Chemistry A* 3(24): 12580–12602.

Nine, M. J., M. A. Cole, L. Johnson, D. N. H. Tran and D. Losic (2015). "Robust superhydrophobic graphene-based composite coatings with self-cleaning and corrosion barrier properties." *ACS Applied Materials & Interfaces* 7(51): 28482–28493.

Nine, M. J., T. T. Tung and D. Losic (2017). 9.04-self-assembly of graphene derivatives: methods, structures, and applications. In *Comprehensive Supramolecular Chemistry II* (J. L. Atwood), Oxford, Elsevier, pp. 47–74.

Park, H., D. S. Kim, S. Y. Hong, C. Kim, J. Y. Yun, S. Y. Oh, S. W. Jin, Y. R. Jeong, G. T. Kim and J. S. Ha (2017). "A skin-integrated transparent and stretchable strain sensor with interactive color-changing electrochromic displays." *Nanoscale* 9(22): 7631–7640.

Poh, H. L., P. Šimek, Z. Sofer and M. Pumera (2013). "Halogenation of graphene with chlorine, bromine, or iodine by exfoliation in a halogen atmosphere." *Chemistry – A European Journal* 19(8): 2655–2662.

Porwal, H., S. Grasso and M. J. Reece (2013). "Review of graphene–ceramic matrix composites." *Advances in Applied Ceramics* 112(8): 443–454.

Prasai, D., J. C. Tuberquia, R. R. Harl, G. K. Jennings and K. I. Bolotin (2012). "Graphene: corrosion-inhibiting coating." *ACS Nano* 6(2): 1102–1108.

Punith Kumar, M. K., M. P. Singh and C. Srivastava (2015). "Electrochemical behavior of Zn–graphene composite coatings." *RSC Advances* 5(32): 25603–25608.

Ramezanzadeh, B., M. H. Mohamadzadeh Moghadam, N. Shohani and M. Mahdavian (2017). "Effects of highly crystalline and conductive polyaniline/graphene oxide composites on the corrosion protection performance of a zinc-rich epoxy coating." *Chemical Engineering Journal* 320: 363–375.

Ren, S., M. Cui, W. Li, J. Pu, Q. Xue and L. Wang (2018). "N-doping of graphene: toward long-term corrosion protection of Cu." *Journal of Materials Chemistry A* 6(47): 24136–24148.

Rodríguez-Pérez, L., M. Á. Herranz and N. Martín (2013). "The chemistry of pristine graphene." *Chemical Communications* 49(36): 3721–3735.

Sahoo, N. G., Y. Pan, L. Li and S. H. Chan (2012). "Graphene-based materials for energy conversion." *Advanced Materials* 24(30): 4203–4210.

Schriver, M., W. Regan, W. J. Gannett, A. M. Zaniewski, M. F. Crommie and A. Zettl (2013). "Graphene as a long-term metal oxidation barrier: worse than nothing." *ACS Nano* 7(7): 5763–5768.

Shen, L., Y. Li, W. Zhao, L. Miao, W. Xie, H. Lu and K. Wang (2019). "Corrosion protection of graphene-modified zinc-rich epoxy coatings in dilute NaCl solution." *ACS Applied Nano Materials* 2(1): 180–190.

Singh, B. P., B. K. Jena, S. Bhattacharjee and L. Besra (2013). "Development of oxidation and corrosion resistance hydrophobic graphene oxide-polymer composite coating on copper." *Surface and Coatings Technology* 232: 475–481.

Singh Raman, R. K., P. Chakraborty Banerjee, D. E. Lobo, H. Gullapalli, M. Sumandasa, A. Kumar, L. Choudhary, R. Tkacz, P. M. Ajayan and M. Majumder (2012). "Protecting copper from electrochemical degradation by graphene coating." *Carbon* 50(11): 4040–4045.

Su, Y., V. G. Kravets, S. L. Wong, J. Waters, A. K. Geim and R. R. Nair (2014). "Impermeable barrier films and protective coatings based on reduced graphene oxide." *Nature Communications* 5(1): 4843.

Tanquerel des Planches L., and S. L. Dana (1848). *A Treatise from the French of L. Tanquerel Des Planches: With Notes and Additions on the Use of Lead Pipe and Its Substitutes.* Lowell, MA, D. Bixby and Company, p. 441.

Templeton, D. M. and Y. Liu (2010). "Multiple roles of cadmium in cell death and survival." *Chemico-Biological Interactions* 188(2): 267–275.

Topsakal, M., H. Şahin and S. Ciraci (2012). "Graphene coatings: an efficient protection from oxidation." *Physical Review B* 85(15): 155445.

Ullah, S., P. A. Denis and F. Sato (2017). "Triple-doped monolayer graphene with boron, nitrogen, aluminum, silicon, phosphorus, and sulfur." *ChemPhysChem* 18(14): 1864–1873.

Yan, L., Y. B. Zheng, F. Zhao, S. Li, X. Gao, B. Xu, P. S. Weiss and Y. Zhao (2012). "Chemistry and physics of a single atomic layer: strategies and challenges for functionalization of graphene and graphene-based materials." *Chemical Society Reviews* 41(1): 97–114.

Yang, X., X. Yu and X. Liu (2018). "Obtaining a sustainable competitive advantage from patent information: a patent analysis of the graphene industry." *Sustainability* 10(12): 4800.

Yap, P. L., M. J. Nine, K. Hassan, T. T. Tung, D. N. H. Tran and D. Losic (2021). "Graphene-based sorbents for multipollutants removal in water: a review of recent progress." *Advanced Functional Materials* 31(9): 2007356.

Yu, F., L. Camilli, T. Wang, D. M. A. Mackenzie, M. Curioni, R. Akid and P. Bøggild (2018). "Complete long-term corrosion protection with chemical vapor deposited graphene." *Carbon* 132: 78–84.

Zhou, F., Z. Li, G. J. Shenoy, L. Li and H. Liu (2013). "Enhanced room-temperature corrosion of copper in the presence of graphene." *ACS Nano* 7(8): 6939–6947.

Zhou, S., Y. Wu, W. Zhao, J. Yu, F. Jiang, Y. Wu and L. Ma (2019). "Designing reduced graphene oxide/zinc rich epoxy composite coatings for improving the anticorrosion performance of carbon steel substrate." *Materials & Design* 169: 107694.

Zhu, J., D. Yang, Z. Yin, Q. Yan and H. Zhang (2014). "Graphene and graphene-based materials for energy storage applications." *Small* 10(17): 3480–3498.

14 Supramolecular Intrinsic Self-Healing in Polyurethane Coatings

Mohammad Mizanur Rahman, Bashirul Haq, and Viswanathan S. Saji
King Fahd University of Petroleum and Minerals

CONTENTS

14.1 Introduction 239
14.2 Polyurethanes...240
14.3 Intrinsic Self-Healing 240
14.4 Conclusions and Perspectives 244
References..244

14.1 INTRODUCTION

Polymer coatings are at the forefront of the coatings market, where the dominant are those based on epoxy, acrylate, and polyurethanes (PUs) (Chattopadhyay et al. 2007; Chattopadhyay and Webster 2009; Habib et al. 2021). Due to the unique qualities of PUs such as barrier resistance, adhesive strength, mechanical strength, and UV-degradation resistance, they are mainly used as the top layer in industrial multi-layer coatings (Chattopadhyay and Webster 2009; Rahman et al. 2013). Even though PU-only coatings are well known for their outstanding protective properties at the initial periods of service life, the protection typically gets compromised with time (Xie et al. 2009; Rahman 2020). Significant efforts have been dedicated to enhancing the mechanical strength, thermal stability, and barrier resistance of PU coatings. Different methodologies, such as anchoring multi-functional groups and adding nano-fillers and cross-linkers, were explored. Unfortunately, most of these coatings failed to utilize the self-healing potential of PUs and hence were unable to restore the protective properties once corrosion was initiated.

The early-stage micro-damages in a coating in service life can be easily rectified, provided the coating possesses self-healing property. A coating can be made self-healable in both intrinsic and extrinsic ways (see Chapter 8). An extrinsic healable coating involves adding a healing agent (e.g. capsule carrying a corrosion inhibitor) to the pristine coating. On the other hand, an intrinsic healable coating relies on the healing ability through existing base functional groups and bonds. Most of the reported works on self-healing coatings make use of extrinsic microencapsulation-based techniques. Despite the better self-healing effect, the capsule approaches rely closely on the type and the nature of the trigger, such as temperature, pH, and the capsules' parameters. The intrinsic self-healing is advantageous in this aspect. If optimized precisely, they could heal and restore the properties without adding a trigger-controlled external component (Deflorian et al. 2013; Saji 2019; Nardeli et al. 2020; Montano et al. 2021).

The intrinsic self-healing in PUs could be attributed to both dynamic covalent chemistry and supramolecular chemistry (Willocq et al. 2020; Saji 2019). Willocq et al. described self-healing through dynamic covalent chemistry (reversible covalent chemistries) under three classifications,

DOI: 10.1201/9781003169130-16

239

namely, (i) thermo-responsive, (ii) photo-responsive, and (iii) chemo-responsive, as triggers for chemical changes in dynamic covalent interactions. These processes rely on several reversible reactions, such as Diels–Alder. On the other hand, the self-healing by supramolecular chemistry, which is the subject of this chapter, was explained under four heads, namely, (i) hydrogen bonding, (ii) metal–ligand coordination, (iii) π–stacking, and (iv) ionic interaction (Willocq et al. 2020). These are the four major non-covalent interactions that could facilitate supramolecular intrinsic self-healing in PUs. Conceivably, the most important one among them is hydrogen bonding.

Only a few works have been dedicated to evaluating the role of non-covalent interactions in the anti-corrosion properties of PUs-based barrier coatings. This is because the passive barrier effect (attributed to covalent bonding) is the primary corrosion protection mechanism in PUs. This chapter discusses only the hydrogen bonding interactions in PU coatings and their role in self-healing, which is the most reported in this line.

14.2 POLYURETHANES

An organic coating that contains urethane groups (-NHCOO-) is usually termed as PU coating, though it is ubiquitous to have urea (-NHCONH-) groups along with urethanes in the base PU chain (Figure 14.1). PU can also contain ethers, esters, and several aromatic groups. The nature of monomers used for the synthesis determines their functionalization possibilities and desired properties. In the last two decades, several novel monomers and polymerization methods were invented, leading to various PUs (Chattopadhyay and Webster 2009; Rahman 2020).

Fossil fuel-based monomers mainly dominated the PU coating market. Many of them are highly toxic. In the last two decades, scientists have invented novel eco-friendly synthesis routes using renewable resources such as vegetable oils and lignin. The typical organic solvent-based PU coatings are not attractive in terms of environmental legislation. The recent trend is eco-friendly waterborne PU coatings (Chattopadhyay and Webster 2009; Aguirresarobe et al. 2021).

PUs typically consist of soft- and hard- segments. The soft segments are usually composed of polyols, whereas diisocyanate and ethylenediamine/diol constitute hard segments. Different molecular weight polyether polyol, polyester polyol, acrylate, and hydroxy-terminated siloxane are the most common soft segments, whereas 4,4′-methylenebis(phenyl isocyanate), toluene 2,4-diisocyanate, isophorone diisocyanate, 4,4′-methylenebis(cyclohexyl isocyanate), and hexamethylene diisocyanate, are the typical hard segments. The soft segments control the elastomeric properties, whereas the hard segments are responsible for the hardness and mechanical strength. The coating properties are highly dependent on the ratio of the soft and hard segments (Chattopadhyay and Webster 2009). The functional groups, which come from the soft and hard segments, make non-covalent interactions and ultimately determine the supramolecular structure.

14.3 INTRINSIC SELF-HEALING

Several works are available on the intrinsic self-healing properties of PUs (Willocq et al. 2020). However, only a few works have explored their applicability in self-healing anti-corrosion coatings. Hydrogen-bonds restoration (reversible crosslinking) is the most important supramolecular self-healing mechanism in PUs (Deflorian et al. 2013; Yan et al. 2018; Nardeli et al. 2020; Montano et al. 2021).

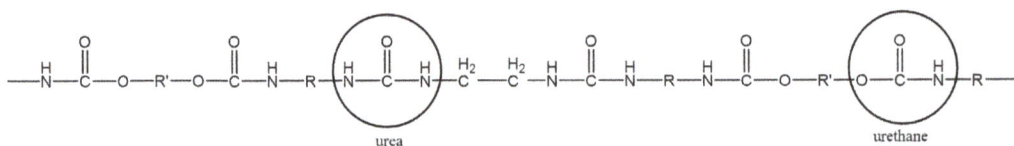

FIGURE 14.1 Two major groups (urethane and urea) in the typical PU structure.

The reassociation of hydrogen bonds after damage initiation is responsible for the healing effect. Such reassociation can be triggered by other stimuli such as temperature, light, and electrostatic interaction. The polar groups in a PU structure could effortlessly make hydrogen bonds.

Though hydrogen bonds are expected to enhance self-healing, a significantly greater hydrogen bonding could sacrifice the coating barrier properties. Deflorian et al. investigated the effect of hydrogen bonding in PUs on barrier protection and self-healing. The authors replaced covalent bonds in the coating structure with hydrogen bonds to yield enhanced self-healing. The coating was self-healable up to 25% of the replacement of covalent bonds with hydrogen bonds. They observed that such a replacement does not affect the coating barrier property, and hence the coating displays better corrosion protection. The authors used different ureidopyrimidinone (UPy) contents during the coating preparation. The available electronegative atoms of UPy create hydrogen bonds which ultimately form a compact cross-linkable structure with different chains. This was reflected in the glass transition temperature (T_g), which increased with an increase in UPy content in the coating (Figure 14.2a). Anti-corrosion properties were evaluated by electrochemical impedance spectroscopy (EIS) (Figure 14.2b and c). The plots correspond to two types of coatings: PU coatings with and without 25% UPy. A clear variation in the Nyquist plot is visible after 24 h of immersion, though initially, the data were almost similar. The 25% UPy PU coating displayed better impedance behaviour attributed to the improved corrosion protection, conceivably associated with the modification. The coating capacitance can determine the water diffusion trends. The coating capacitance evaluated for the two coatings (Figure 14.2d) also supports the above observation. The lower capacitance values confirmed better water diffusion resistance for the 25% UPy coating. The differences in the two coatings' capacitance values were low at early periods, increasing with the increase in immersion time (Deflorian et al. 2013).

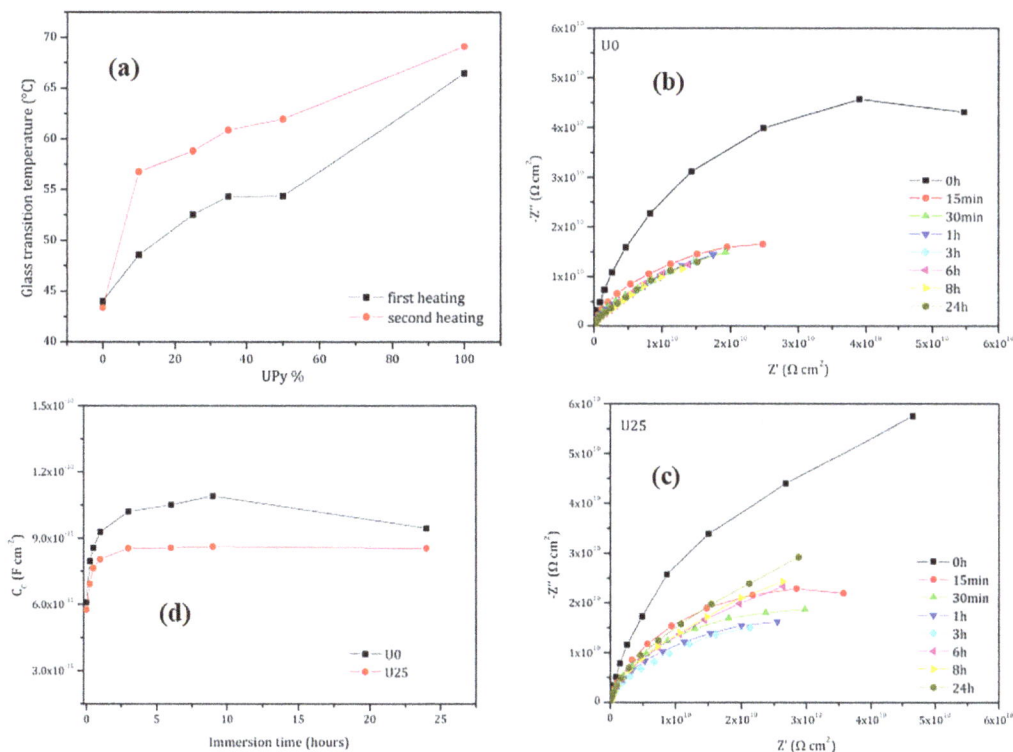

FIGURE 14.2 (a) Glass transition temperature as a function of % UPy, (b, c) Nyquist plots of the coated samples with immersion time for (b) bare PU (U0) and (c) 25% UPy PU coating (U25), (d) Coating capacitance vs immersion time. (Reproduced with permission from Deflorian (2013) © 2013 Taylor & Francis.)

Nardeli et al. showed that PU coating having higher flexible segments (polyesters) exhibited better anti-corrosion protection and self-healing properties. The authors prepared two different formulated coatings, namely, Coat-I and Coat-II. The formulations were different based on the polyester (flexible segment) and the prepolymer (rigid segment) ratio. This ratio was 1.5 (higher flexible segment) for Coat-I and 1 (lower flexible segment) for Coat-II. Hydrogen bonds were more in Coat-I than in Coat-II. They have employed localized EIS to measure the self-healing of scratched coatings. The ratio (%) of the maximum admittance values (A/A_0) was used (Figure 14.3). A lower value of 100% signifies self-healing, whereas values above 100% indicate the progressed corrosion. The value remained 100% for Coat-I, whereas for Coat-II, higher values were recorded after 16 h of immersion. These results confirm the better self-healing capacity of Coat-I (Nardeli 2020).

They also schematically explained the reason for self-healing for Coat-II. The higher polyester content contributes to a better cross-linked structure due to the greater level of hydrogen bonds, which ultimately improves self-healing and corrosion resistance (Figure 14.4).

More recently, Montano et al. have shown that the coating's scratch healing and barrier restoration efficiency are closely dependent on the soft/hard block ratio (χ_{SF}), which was affected by the extent of polyols, chain extenders, and isocyanates. They proved that the PU design is critical in the coating's protective properties as the χ_{SF} was crucial for restoring the barrier resistance. A typical soft/hard block PU coating is shown in Figure 14.5. The soft block was made from long-chain polyol and isocyanate, whereas the hard block was made from diisocyanate and short-chain polyol (Montano 2021).

Softer PU (when soft block content was high) showed faster self-healing, though the barrier restoration was not up to the mark. At the same time, harder PU (when soft block content was low), which was comparatively slower in self-healing, had a high barrier restoring capacity. PUs with lower soft phase segments led to a higher urea/urethane ratio, which influenced the healing efficiency. The authors measured the urea fraction and showed its effect on corrosion resistance (Figure 14.6). Harder PU coating, which had a high urea fraction, slowed down the healing but displayed excellent barrier protection. Even though the softer PU coating healed the scratch effectively, restoring barrier property was not achieved ultimately. The work suggests that careful optimization of the soft/hard block and urea/urethane content is a requisite for formulating PU coatings with balanced self-healing and barrier protection (Montano 2021).

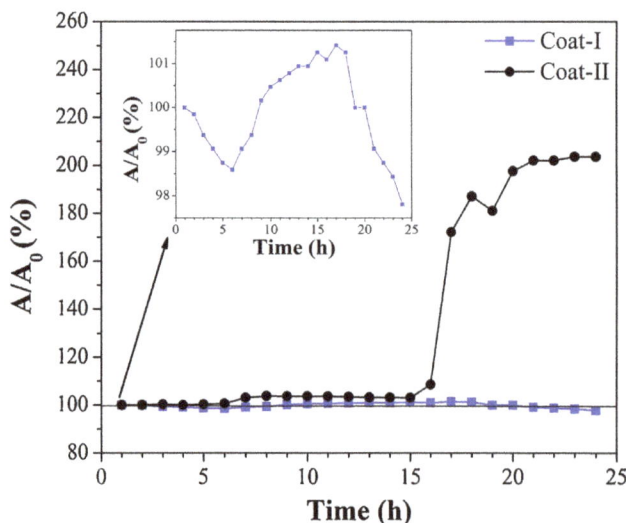

FIGURE 14.3 Ratio of maximum admittance value determined in a defect area at a particular duration (A) and the initial maximum admittance value (A_0), for Coat-I and Coat-II, during immersion in 0.005 mol/L NaCl. Inset: Coat-I (magnified scale). (Reproduced with permission from Nardeli (2020) © 2020 Elsevier Ltd.)

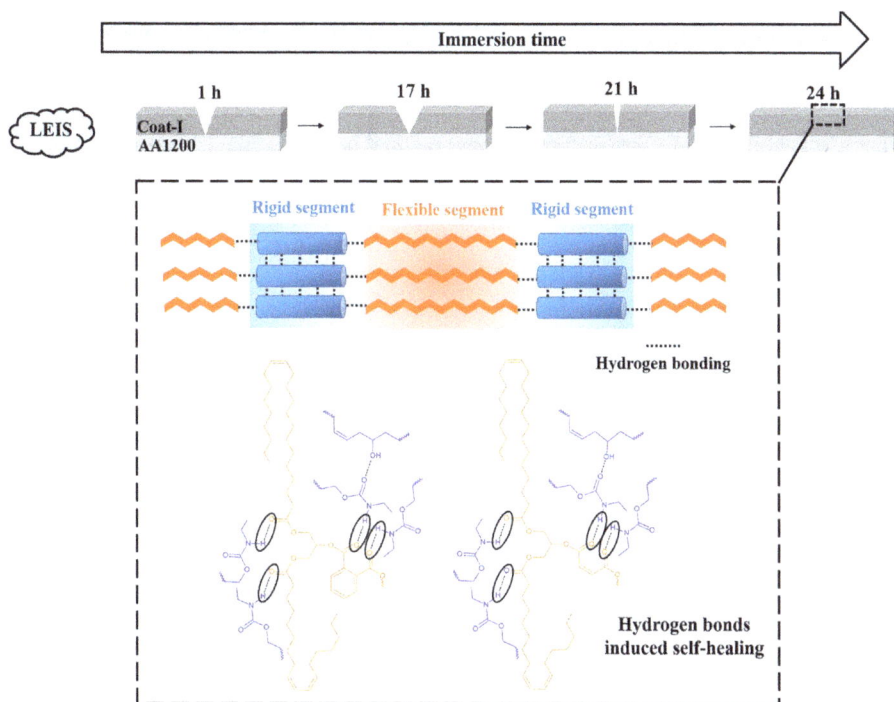

FIGURE 14.4 A schematic of the self-healing process of Coat-I in NaCl solution. (Reproduced with permission from Nardeli (2020) © 2020 Elsevier Ltd.)

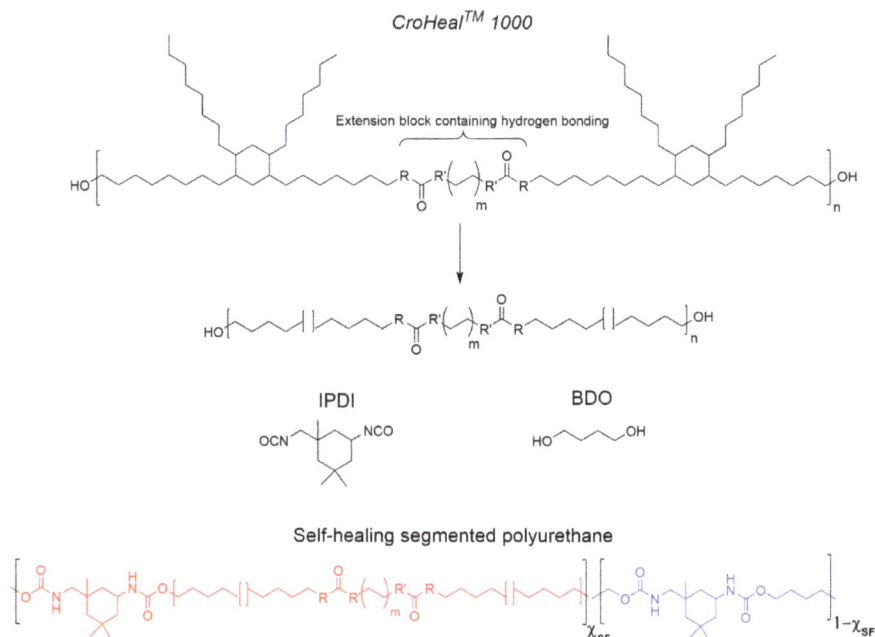

FIGURE 14.5 Monomers used (CroHeal 1000, IPDI, and BDO) and the segmented molecular structure of the synthesized PUs as a function of the soft/hard block fraction (χ_{SF}). The red structure represents the nominal soft blocks, and the blue structure represents the hard blocks. (Reproduced with permission from Montano (2021) (doi: 10.1021/acsapm.1c00323) © 2021 The Authors. For further permission related to the material excerpted should be directed to the American Chemical Society.)

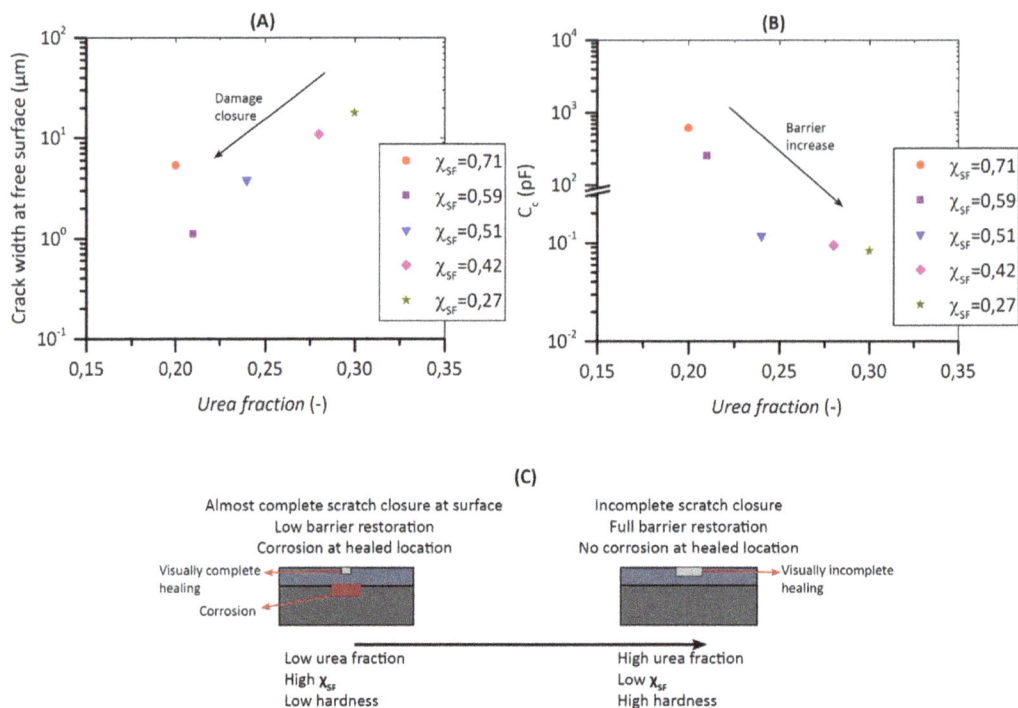

FIGURE 14.6 Effect of urea fraction on self-healing of PU coatings. (a) Residual crack width at the free surface after 60 min of healing at 65°C as measured by SEM, (b) Coating capacitance as a measure of barrier protection after 60 min of healing at 65°C obtained after five polarization cycles. (c) Schematic representation of the effects of the polymer architecture (χ_{SF}) and urea fraction on scratch closure and barrier restoration. (Reproduced with permission from Montano (2021) (doi: 10.1021/acsapm.1c00323) © 2021 The Authors. For further permission related to the material excerpted should be directed to the American Chemical Society.)

14.4   CONCLUSIONS AND PERSPECTIVES

The chapter provides a concise description of the effect of supramolecular architecture in designing anti-corrosion PU coatings with intrinsic self-healing properties. Only a few works are available on the self-healing properties of PUs in anti-corrosion applications, as discussed in this chapter. Among the different non-covalent interactions, the most explored one is hydrogen bonding. The reported outcomes clearly showed that with precise optimization, it is possible to develop PU coatings with enhanced self-healing properties without compromising the excellent barrier protection. An adequately designed supramolecular PU coating not only heals the scratches but also regains the coating barrier properties. In this regard, the critical point is to choose the proper fabrication approaches and use suitable monomers and coating additives. This type of PU coating can extend its service life to protect the metal structures from corrosion.

REFERENCES

Aguirresarobe, R. H., Nevejans, S., Reck, B., Irusta, L., Sardon, H., Asua, J. M., and Ballard, N. 2021. Healable and self-healing polyurethanes using dynamic chemistry. *Prog. Polym. Sci.* 114: 101362.
Chattopadhyay, D. K., and Raju, K. V. S. N. 2007. Structural engineering of polyurethane coatings for high performance applications. *Prog. Polym. Sci.* 32: 352–418.
Chattopadhyay, D. K., and Webster, D. C. 2009. Thermal stability and flame retardancy of polyurethanes. *Prog. Polym. Sci.* 34: 1068–1133.
Deflorian, F., Rossi, S., and Scrinzi, E. 2013. Self-healing supramolecular polyurethane coatings: preliminary study of the corrosion protective properties, *Corros. Eng. Sci. Technol.* 48: 147–154.

Habib, S., Shakoor, R. A., and Kahraman, R. 2021. A focused review on smart carriers tailored for corrosion protection: developments, applications, and challenges. *Prog. Org. Coat.* 154: 106218.

Montano, V., Vogel, W., Smits, A., Zwaag, S. V. D., and Garcia, S. J. 2021. From scratch closure to electrolyte barrier restoration in self-healing polyurethane coatings. *ACS Appl. Polym. Mater.* 3: 2802–2812.

Nardeli, J. V., Fujiwara, C. S., Taryba, M., Montemor, M. F., and Benedetti, A. V. 2020. Self-healing ability based on hydrogen bonds in organic coatings for corrosion protection of AA1200. *Corros. Sci.* 177: 108984.

Rahman, M. M. 2020. Polyurethane/zinc oxide (PU/ZnO) composite-synthesis, protective property and application. *Polymers* 12: 1535.

Rahman, M. M., Hasneen, A., Chung, I. D., Kim, H. D., Lee, W. K., and Chun, J. H. 2013. Synthesis and properties of polyurethane coatings: the effect of different types of soft segments and their ratios. *Compos. Interface* 20: 15–26.

Saji, V. S. 2019. Supramolecular concepts and approaches in corrosion and biofouling prevention. *Corr. Rev.* 37: 187–230.

Willocq, B., Odent, J., Dubois, P., and Raquez, J. M. 2020. Advances in intrinsic self-healing polyurethanes, and related composites. *RSC Adv.* 10: 13766–13782.

Xie, F., Zhang T., Bryant, P., Kurusingal, V., Colwell, J. M., and Laycock, B. 2019. Degradation and stabilization of polyurethane elastomers. *Prog. Polym. Sci.* 90: 211–268.

Yan, X. Liu, Z., Zhang, Q., Lopez, J., Wang, H., Wu, H. C., Niu, S. Yan, H., Wang, S., Lei, T., Li, J., Qi, D., Huang, P., Huang, J., Zhang, Y., Wang, Y., Li, G., Tok, J. B. H., Chen, X., and Bao, Z. 2018. Quadruple H-bonding cross-linked supramolecular polymeric materials as substrates for stretchable, antitearing, and self-healable thin film electrodes. *J. Am. Chem. Soc.* 140: 5280–5289.

15 Supramolecular Polyurethane Corrosion Inhibitors

Muthukumar Nagu
Research and Development Center

Viswanathan S. Saji
King Fahd University of Petroleum & Minerals

CONTENTS

15.1 INTRODUCTION

Chemical inhibitors are one of the most widely used corrosion control methods in various industrial sectors. They are essential additives in refinery processing units, oil and gas production units, gas oil separation units, petroleum products transporting pipelines, cooling water systems, steam generators and several other industries. Extensive research and development have been performed in developing corrosion inhibitors for various metals in different aggressive media (Nathan 1973; Sastri 1998; Saji 2010; Muthukumar 2014; Saji and Umoren 2020). Supramolecular compounds and approaches were also explored in this direction (Saji 2019). An industrially relevant corrosion inhibitor should be cheap, readily available, and environmentally friendly. It could function effectively under different process parameters of temperature, pH, H_2S, CO_2, chlorides, sulfates and microbes.

Polymers, in general, can form complex supramolecular structures with different functional groups attached to their backbone. The corrosion inhibitive power of several polymers is associated with the robust surface adsorption via cyclic rings with π electrons and heteroatoms. Through various functional groups, polymeric inhibitors can form coordination complexes with metal ions, providing a protective covering from corrosive species. They are also attractive due to their inherent stability, cost-effectiveness and superb adhesion properties. More details on different natural and synthetic polymeric corrosion inhibitors can be found elsewhere (Arthur et al. 2013; Umoren and Solomon 2014; Sabirneeza et al. 2015; Tiu and Advincula 2015; Umoren and Eduok 2016).

Polyurethanes (PUs) (Figure 15.1) constitute one of the most important and valuable classes of thermoplastic elastomers. They have an extensive range of applications in coatings, adhesives, foams, packaging materials and biomedical aids. They are known for their excellent biocompatibility (Wirpsza 1993; Mishra et al. 2010a, b; Banerjee et al. 2011a; Thomson 2005).

The principal factor limiting their use as corrosion inhibitors is low aqueous solubility. Incorporating hydrophilic units into the polymer backbone is often employed to overcome this drawback (Kumar et al. 2017). Banerjee et al. demonstrated the use of sulfonate functional groups to improve the solubility and conductivity of PUs, which also improved their corrosion inhibition efficiency (Banerjee et al. 2011a,b). This chapter provides a concise description of reported works on PU-based corrosion inhibitors (see Section 15.3 and Table 15.1).

DOI: 10.1201/9781003169130-17

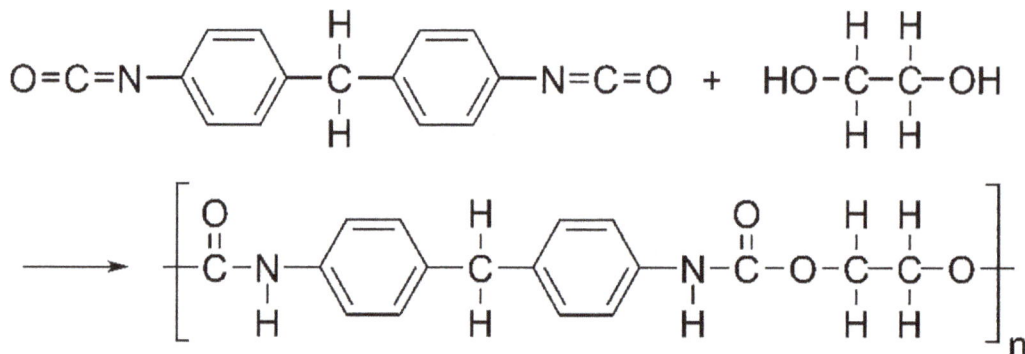

FIGURE 15.1 A typical synthesis of conventional PU. (Reproduced from Wikipedia under Creative Commons Attribution license.)

15.2 SUPRAMOLECULAR POLYURETHANES

Fabrication and assembling of PUs involve both dynamic covalent (Diels–Alder, alkoxyamine chemistry, etc.) and non-covalent (hydrogen bonding, π–π stacking, etc.) interactions, facilitating them with excellent self-healing properties (Willocq et al. 2020; Li et al. 2020). Typically, the PU structure consists of rigid and flexible segments bearing non-covalent hydrogen bonding between them. The non-covalent fraction can be increased by substituting covalent bonds with hydrogen bonds (Deflorian et al. 2013; Yan et al. 2018). Such partial replacement could enhance their intrinsic self-healing properties, as described in Chapter 14. A recent review by Willocq et al. has provided different self-healing mechanisms in PUs, namely, (i) self-healing through dynamic covalent chemistry, and (ii) self-healing through supramolecular interactions. The self-healing via supramolecular interactions includes (i) hydrogen bonding, (ii) metal–ligand coordination, (iii) π–π and (iv) ionic interactions (Willocq et al. 2020).

There are several reported works on supramolecular PUs. For example, Comí et al. reported the synthesis of supramolecular PU elastomers by post-synthetic non-covalent crosslinking based on ionic hydrogen-bonding interactions. The ionic networks were formed between the tertiary amino groups of castor oil-derived PU and small-molecular biological acids such as sebacic acid (Comí et al. 2017). A pyrenyl-functionalized telechelic PU was blended with polydiimide, forming π–π interactions by chain folding and complemented with hydrogen bonds of the PU backbone (Burattini et al. 2009). Feula et al. reported a well-defined self-healable supramolecular elastomer PU capable of self-assembling via π–π stacking and hydrogen bonding (Feula et al. 2015). Metal–ligand crosslinking in the PU matrix could be accomplished via suitable monomer selection (Wang et al. 2018).

Xiao et al. grafted 2-amino-4-hydroxy-6-methylpyrimidine (UPy) into the side chain of water-borne PU to provide self-healing capability (Xiao et al. 2020). Chen et al. reported novel zwitterionic PUs having excellent multi-shape-memory properties and self-healing effects where ionic interactions greatly influenced the properties (Chen et al. 2015). Wang et al., in a recent work, reported a robust self-healing material by integrating high-density non-covalent bonds at the interfaces between the dendritic tannic acid-modified tungsten disulfide nanosheets and PU matrix (Wang et al. 2021). More details on supramolecular PUs could be found elsewhere (Willocq et al. 2020).

15.3 POLYURETHANE CORROSION INHIBITORS

There are only a few works available on PUs-based corrosion inhibitors (Table 15.1). Banerjee et al. synthesized a series of PU ionomers with various degrees of sulfonation (DS) and explored for mild steel corrosion in an acid medium. For the synthesis, polytetramethylene glycol (PTMG)

TABLE 15.1
PU-Based Corrosion Inhibitors

Type of PU	Alloy/Medium	Efficiency (%)	Mechanism/ Adsorption	Ref.
Sulfonated PUs	Mild steel/0.5 M H$_2$SO$_4$	98	Mixed, chemisorption, Langmuir	Banerjee et al. (2011a)
Sulfonated PUs	Mild steel/0.5 M H$_2$SO$_4$	100	Mixed, chemisorption, Langmuir	Banerjee et al. (2011b)
PUs	Mild steel/1 M H$_2$SO$_4$	97	Mixed, physisorption & chemisorption, Langmuir	Eswaramoorthi et al. (2016)
Poly(N-vinyl-pyrrolidone)-b-PU-b-poly(N-vinyl-pyrrolidone), and poly(dimethyl-aminoethyl-methacrylate)-b-PU-b-poly(dimethyl-aminoethyl-methacrylate)	Mild steel/0.5 M H$_2$SO$_4$	98	Mixed, physisorption & chemisorption, Langmuir	Kumar et al. (2016)
Poly(N-isopropyl-acrylamide)-b-PU-b-poly(N-isopropyl-acrylamide), and poly(tert-butylacrylate)-b-PU-b-poly(tert-butylacrylate)	Mild steel/0.5 M H$_2$SO$_4$	99	Mixed, physisorption & chemisorption, Langmuir and El-Awady, respectively	Kumar et al. (2017)
PU functionalized with sulfonated graphene	Mild steel/0.5 M H$_2$SO$_4$	85	Mixed, Langmuir	Patel et al. (2017)
Trianiline PU	Mild steel/1 M HCl	97	Mixed, physisorption & chemisorption, Langmuir	Yang et al. (2017)
Cobaltocenium cation-based PU	Acid media	-	-	Yan et al. CN110577626
PU as microcapsule shell and triazole-based compound as core	-	-	-	Kim et al. KR1317645B1

and 4,4'-diphenylmethane diisocyanate (MDI) were first treated (70°C, 4 h) to form an isocyanate-terminated pre-polymer, and subsequently, the pre-polymer was treated with a chain extender (butane diol) in the presence dibutyltin dilaurate (DBTDL) catalyst (70°C, 24 h, inert atmosphere). Sulfonated PUs (SPUs) were then prepared via nucleophilic substitution of urethane hydrogen by reacting PU with NaH, which was then reacted with propane sultone. Figure 15.2a presents the synthesis scheme of SPUs. The corrosion inhibition efficiencies of the synthesized SPUs were examined by weight loss and electrochemical methods in 0.5 M H$_2$SO$_4$. Representative results obtained from the potentiodynamic polarization (PDP) studies are shown in Figure 15.2b. The inhibition efficiency increased with increasing DS from 48% to 97% with the following order for the synthesized SPUs: SPU-48 < SPU-60 < SPU-88 < SPU-97. The increase in inhibition efficiency with DS was attributed to the enhanced absorbability owing to the increased number of active centers (ionic sulfonate groups). Tafel plots showed that SPUs reduced both the anodic and the cathodic current densities, supporting mixed-type inhibition. The inhibitor adsorption followed the Langmuir isotherm (Figure 15.2c). Electrochemical impedance spectroscopy (EIS) results also confirmed the better inhibition efficiency of SPUs (Banerjee et al. 2011a).

In related work, the authors explored the efficiency of SPUs prepared from two different routes. In addition to propane sultone, H$_2$SO$_4$ was also used for the sulfonation. The results confirmed that SPUs are suitable inhibitors of mild steel in acidic media, regardless of the mode of sulfonation. The study also showed that incorporating nanoclay to SPU chains (prepared through ion-exchange

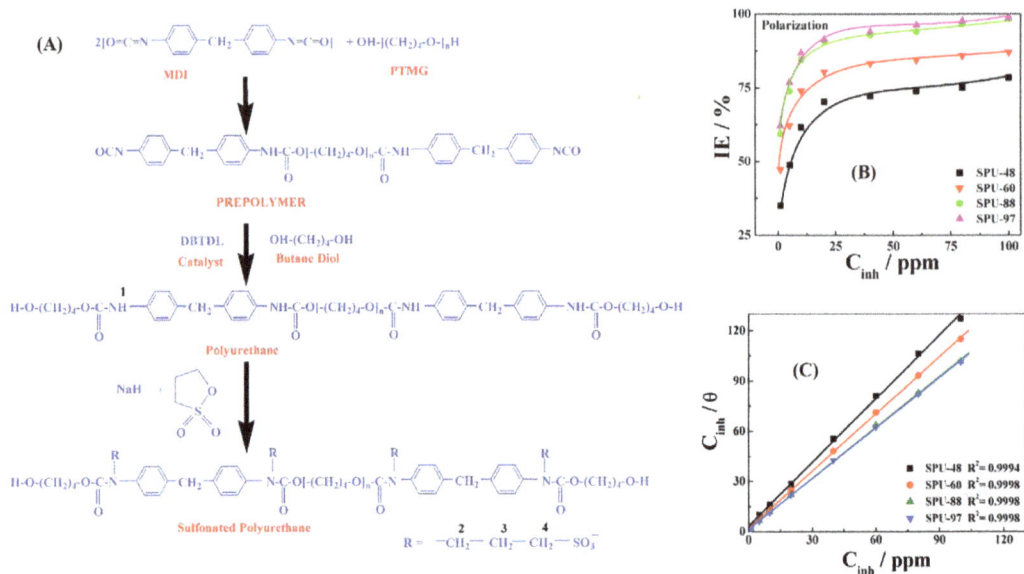

FIGURE 15.2 (a) Schematic showing the synthesis of pure and sulfonated polyurethanes. (b) Inhibition efficiency as a function of the inhibitor concentration and degree of sulfonation. (c) Langmuir adsorption isotherms. (Reproduced with permission from Banerjee et al. (2011a) © 2011 The Royal Society of Chemistry.)

reaction) dramatically enhanced the inhibition efficiency. The composite inhibitor with 30 ppm ionomer concentration and 2 ppm nanoclay exhibited nearly 100% efficiency (Banerjee et al. 2011b).

Eswaramoorthi et al. demonstrated PU synthesis from soya bean oil via three steps: (i) epoxidation of soya bean oil, (ii) synthesis of polyol, and (iii) synthesis of PU. The corrosion inhibition effect of PUs on mild steel in 1 M H_2SO_4 was studied using weight loss, PDP and EIS. The inhibition efficiency of PUs was directly correlated to the inhibitor concentration (Figure 15.3), which was attributed to the blocking effect by adsorption and film formation. Even though the PDP plots (Figure 15.3a) remained the same, the curves shifted to lower current densities in the presence of inhibitors. The change in the corrosion potential was only 29 mV, suggesting that the inhibitor retarded both the anodic and cathodic reactions (mixed-type). The Nyquist plots showed one single depressed semicircle, attributed to the charge transfer process, and the diameters of the semicircle increased with the increase in inhibitor concentration (Figure 15.3b). Their weight loss studies recorded a rise in efficiency from 95.15% to 97.69%, with increasing inhibitor concentration from 25 to 200 ppm. The presence of phosphate groups in the structure was beneficial in enhancing the efficiency by allowing the inhibitor to achieve a flat orientation on the metal surface. The inhibitor adsorption obeyed the Langmuir isotherm. The SEM surface morphology of the mild steel specimen in the absence and presence of the inhibitor is also shown in Figure 15.3. In the presence of PUs, the surface damage has diminished considerably. However, the efficiency decreased as the temperature increased to 333 K (Eswaramoorthi et al. 2016).

Kumar et al. reported two different PU-based tri-block copolymers, synthesized through atom transfer radical polymerization (ATRP), namely poly(N-vinyl-pyrrolidone)-*b*-PU-*b*-poly(N-vinyl-pyrrolidone), and poly(dimethyl-aminoethyl-methacrylate)-*b*-PU-*b*-poly(dimethyl-aminoethyl-methacrylate). Their inhibition performance was investigated for mild steel in 0.5 M H_2SO_4. Potentiostatic polarization studies showed that the first polymer displayed a surface passivating effect over the entire concentration range studied (400–1600 ppm). In contrast, with the second polymer, the passivation characteristics were significant at higher concentrations only (1200 and 1600 ppm). EIS studies supported polymer adsorption forming a pseudo-capacitive interface. The adsorption mechanism involved both physisorption and chemisorption. AFM images showed a

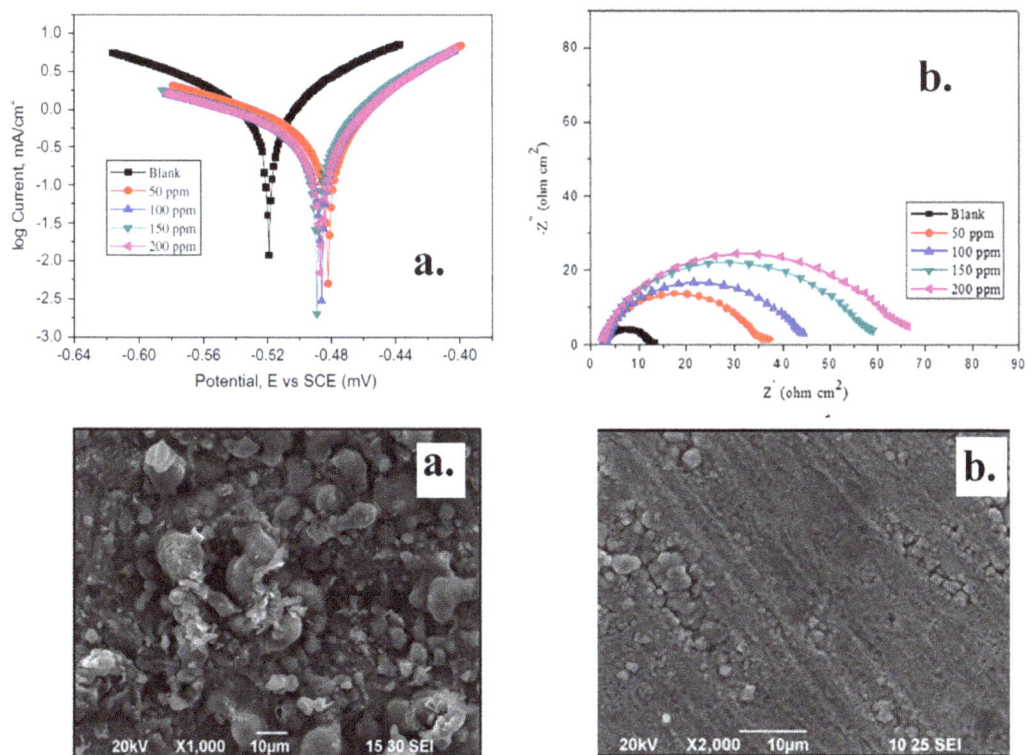

FIGURE 15.3   (a) Potentiodynamic polarization, and (b) Nyquist plots for mild steel in 1 M H$_2$SO$_4$ for different concentrations of PU. SEM images of mild steel in 1 M H$_2$SO$_4$ in the (a) absence and (b) presence of the inhibitor. (Reproduced with permission from Eswaramoorthi et al. (2016) © 2015 Taylor & Francis.)

smoother inhibited surface with significantly reduced surface roughness. Quantum chemical calculations and molecular dynamic simulations provided supportive evidence (Kumar et al. 2016).

In a subsequent work, the authors synthesized two new copolymers, namely, poly(N-isopropyl-acrylamide)-*b*-PU-*b*-poly(N-isopropyl-acrylamide) (PIA-PU-PIA), and poly(tert-butylacrylate)-*b*-PU-*b*-poly(tert-butylacrylate) (PtBA-PU-PtBA) by ATRP (Figure 15.4a). Both experimental and theoretical studies indicated that the inhibition efficiency of PIA-PU-PIA was better than PtBA-PU-PtBA. Efficiencies as high as 99% were obtained at 1600 ppm of the inhibitor concentration for mild steel in 0.5 M H$_2$SO$_4$. Electrochemical studies revealed that the polymers were mixed-type with passivating characteristics. Competitive physisorption and chemisorption were suggested as the adsorption mechanism. The adsorption of PIA-PU-PIA obeyed the Langmuir isotherm, while that of PtBA-PU-PtBA followed the El-Awady isotherm. The gas-phase optimized structures and the electron density distributions of the polymers are also shown in Figure 15.4. The HOMO and LUMO, respectively, provide information about the preferable sites of the inhibitor molecule for electron donation (to vacant orbitals of the metal atom) and retrodonation. The electron densities were distributed over the aromatic benzene ring and the heteroatoms. A higher E_{HOMO} (−6.161 eV) of PIA-PU-PIA than PtBA-PU-PtBA (−6.190 eV), suggested a greater tendency of the former to donate electrons to vacant d orbitals of metal and hence better surface adsorption. The inhibitor surface coverage was also shown to be directly related to the molar volume. The results also suggested that a lower dipole moment of PIA-PU-PIA has also enhanced the inhibition efficiency (Kumar et al. 2017).

PU was chemically tagged with sulfonated graphene using chain extension through an amine group. Even though the presence of graphene reduced the predominant hydrogen bonding in PU,

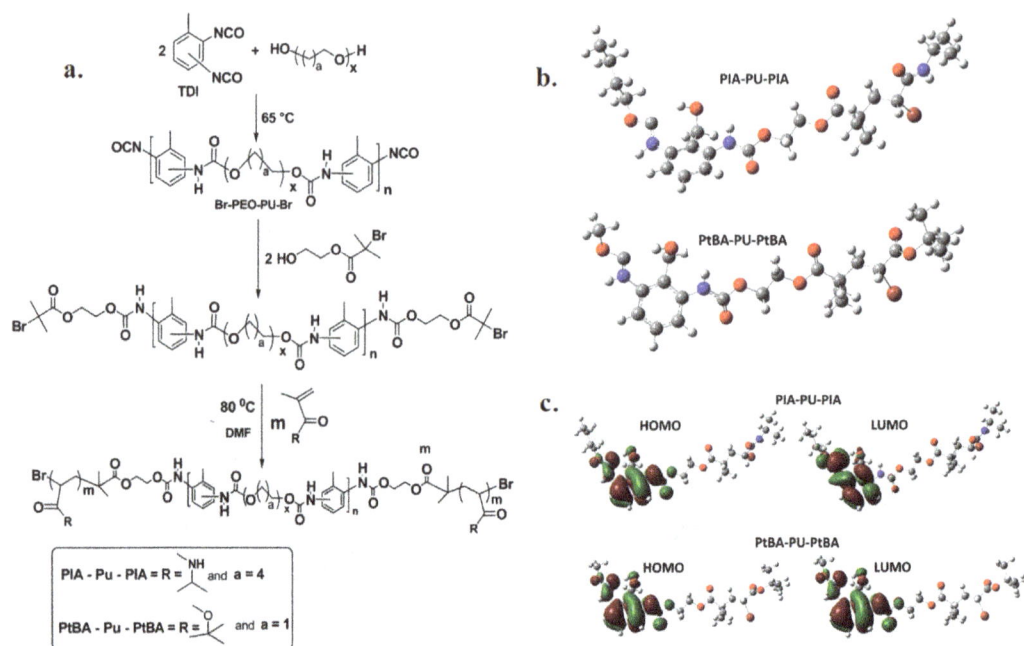

FIGURE 15.4 (a) Schematic of synthesis of PIA-PU-PIA and PtBA-PU-PtBA. (b) Optimized molecular structures obtained at B3LYP/6-31(G) level of theory. (c) HOMO and LUMO electron density surfaces. (Reproduced with permission from Kumar et al. (2017) © 2017, American Chemical Society.)

the nanohybrid exhibited better corrosion inhibition as compared to pure PU (Patel et al. 2017). Yang et al. synthesized a novel trianiline containing PU by polymerizing polyethylene glycol, toluene diisocyanate and amine-capped aniline trimer. The inhibition efficiency (mild steel, 1 M HCl) increased with PU concentration and reached ~97% at 200 mg/L (Yang et al. 2017).

Yan et al. patented a cobaltocenium cation-based water-based PU. The preparation steps consisted of adding an acetonitrile solution of trimethylsilyl-protected ethynylcobaltocenium hexafluorophosphate and diazido neopentyl glycol into a catalyst (CuI or CuBr) solution, adding anhydrous potassium carbonate, reacting at 25°C–90°C for 10–48 h under nitrogen gas, separating to obtain monomer cobaltocenium cation-based diol, and further allowing reaction with a diisocyanate in the presence of dibutyltin dilaurate and tetrabutylammonium salt, to obtain the final product. The inhibitor combined the advantages of inorganic, organic and polymeric corrosion inhibitors, and was shown to be suitable for steel pickling (Yan et al. 2019). Kim et al. patented a core–shell-type corrosion inhibitor containing PU microcapsule as shell and a triazole-based compound as core (Kim et al. 2013).

From Table 15.1, it is clear that PUs are effective corrosion inhibitors in acidic media. In aggressive media, it is expected that the adsorbed inhibitor film constantly gets challenged with the destructive ions. The surface film needs to have the ability to self-heal the initiated local damages at the nano/microscale. The better performance of PUs in acidic media supports that their intrinsic self-healing capability has a role in the inhibition performance.

The self-healing mechanism in supramolecular polymers can be extrinsic (incorporating healing agents) or intrinsic (see Chapter 8). In Section 15.2, we have briefly discussed representative examples of intrinsic approaches involving non-covalent interactions. These supramolecular interactions in the adsorbed inhibitor film indeed help in self-healing. The aqueous acidic pH can favor enhanced hydrogen-bonding interactions. Compared to dynamic covalent chemistry interactions, the supramolecular interactions are fast, specific and could allow multiple healing cycles with limited energy input (Willocq et al. 2020).

15.4 CONCLUSIONS AND PERSPECTIVES

This chapter provides a concise description of PUs-based corrosion inhibitors. As Table 15.1 reveals, all the available reports explored PUs for corrosion protection of mild steel in acidic media. All these studies supported the high inhibition efficiency of PUs and their potential suitability as corrosion inhibitors for mild steel in applications such as acid pickling, acid cleaning, and descaling. Even though most of the works discussed the supramolecular architecture of PUs, precise information on the role of supramolecular interactions in self-healing and inhibition efficiency is not available. The non-covalent interactions and the block structure of PUs could likely facilitate easy repair of the damages that can occur on an adsorbed inhibitor film at the nano/microscale.

REFERENCES

Arthur, D. E., Jonathan, A., Ameh, P. O., and Anya, C. 2013. A review on the assessment of polymeric materials used as corrosion inhibitor of metals and alloys. *Int. J. Ind. Chem.* 4: 2.

Banerjee, S., Mishra, A., Singh, M. M., Maiti, B., Ray, B., and Maiti, P. 2011a. Highly efficient polyurethane ionomer corrosion inhibitor: the effect of chain structure. *RSC Adv.* 1: 199–210.

Banerjee, S., Mishra, A., Singh, M. M., and Maiti. P. 2011b. Effects of nanoclay and polyurethanes on inhibition of mild steel corrosion, *J. Nanosci. Nanotechnol.* 11: 966–978.

Burattini, S., Colquhoun, H. M., Fox, J. D., Friedmann, D., Greenland, B. W., Harris, P. J. F., Hayes, W., Mackay, M. E., and Rowan, S. J. 2009. A self-repairing, supramolecular polymer system: healability as a consequence of donor-acceptor π-π stacking interactions. *Chem. Commun.* 6717–6719.

Chen, S., Mo, F., Yang, Y., Stadler, F. J., Chen, S., Yang, H., and Ge, Z. 2015. Development of zwitterionic polyurethanes with multi-shape memory effects and self-healing properties. *J. Mater. Chem. A* 3: 2924–2933.

Comí, M., Lligadas, G., Ronda, J. C., Galià, M., and Cádiz, V. 2017. Adaptive bio-based polyurethane elastomers engineered by ionic hydrogen bonding interactions. *Euro. Poly. J.* 91: 408–419.

Deflorian F., Rossi S., and Scrinzi E. 2013. Self-healing supramolecular polyurethane coatings: preliminary study of the corrosion protective properties. *Corros. Eng. Sci. Technol.* 48:147–154.

Eswaramoorthi, V., Jagadeesan, S., Palanisamy, S., Kandhasamy, P., and Chitra, S., 2016. Soya bean oil based polyurethanes for corrosion inhibition of mild steel in acid medium. *J. Adhes. Sci. Technol.* 30: 468–493.

Feula, A., Pethybridge, A., Giannakopoulos, I., Tang, X., Chippindale, A., Siviour, C. R., Buckley, C. P., Hamley, I. W., and Hayes, W. 2015. A thermoreversible supramolecular polyurethane with excellent healing ability at 45°C. *Macromolecules* 48: 6132–6141.

Kim, Y. U., Jung, G. U., Kim, N. G., and Ko, E. J. 2013. Polyurethane microcapsule corrosion inhibitor containing triazole based compound, its preparation method and multilayer coating having excellent corrosion resistance. Korean Patent, KR 1317645 B1 20131015.

Kumar, S., Vashisht, H., Olasunkanmi, L. O., Bahadur, I., Verma, H., Singh, G., Obot, I. B., and Ebenso, E. E. 2016. Experimental and theoretical studies on inhibition of mild steel corrosion by some synthesized polyurethane tri-block co-polymers. *Sci. Rep.* 6: 30937.

Kumar, S., Vashisht, H., Olasunkanmi, L. O., Bahadur, I., Verma, H., Goyal, M., Singh, G., and Ebenso, E. E. 2017. Polyurethane based tri-block-copolymers as corrosion inhibitors for mild steel in 0.5 M H_2SO_4. *Ind. Eng. Chem. Res.* 56: 441–456.

Li, M., Ding, H., Yang, X., Xu, L., Xia, J., and Li, S. 2020. Preparation and properties of self-healing polyurethane elastomer derived from tung-oil-based polyphenol, *ACS Omega* 5: 529–536.

Mishra, A., Aswal, V. K., and Maiti, P., 2010a. Nanostructure to microstructure self-assembly of aliphatic polyurethanes: the effect on mechanical properties. *J. Phys. Chem. B* 114: 5292–5300.

Mishra, A., Purkayastha, B. P. D., Roy, J. K., Aswal, V. K., and Maiti, P. 2010b. Tunable properties of self-assembled polyurethane using two-dimentional nanoparticles: potential nano-biohybrid. *Macromolecules* 43: 9928–9936.

Muthukumar, N. 2014. Petroleum products transporting pipeline corrosion- a review, Chapter 21, *The Role of colloidal systems in environmental protection.* 1st Edition, Elsevier (Fanun, M. Ed.), pp. 527–571.

Nathan, C. C. *Corrosion Inhibitors.* Houston: NACE, 1973, ISBN 9780915567898.

Patel, D. K., Senapati, S., Mourya, P., Singh, M. M., Aswal, V. K., Ray, B., and Maiti, P. 2017. Functionalized graphene tagged polyurethanes for corrosion inhibitor and sustained drug delivery. *ACS Biomater. Sci. Eng.* 3: 3351–3363.

Sabirneeza, A. A. F., Geethanjali, R., and Subhashini, S. 2015. Polymeric corrosion inhibitors for iron and its alloys: a review. *Chem. Eng. Commun.* 202: 232–244.

Saji, V. S. 2010. A review on recent patents in corrosion inhibitors. *Recent. Pat. Corros. Sci.* 2: 6–12.

Saji, V. S. 2019. Supramolecular concepts and approaches in corrosion and biofouling prevention. *Corros. Rev.* 37: 187–230.

Saji, V. S., and Umoren, S. A. 2020. *Corrosion Inhibitors in the Oil and Gas Industry.* Wiley-VCH Verlag GmbH & Co. KGaA. ISBN 978-3-527-34618-9.

Sastri, V. S. 1998. *Corrosion Inhibitors, Principles and Applications.* New Jersey: Wiley. ISBN 978–0471976080.

Thomson, T., 2005. *Polyurethanes as Specialty Chemicals: Principles and Applications.* Boca Raton, FL: CRC Press. ISBN 9780849318573.

Tiu, B. D. B., and Advincula, R. C. 2015. Polymeric corrosion inhibitors for the oil and gas industry: design principles and mechanism. *React. Funct. Polym.* 95: 25–45.

Umoren, S. A., and Eduok, U. M. 2016. Application of carbohydrate polymers as corrosion inhibitors for metal substrates in different media: a review. *Carbohydr. Polym.* 140: 314–341.

Umoren, S. A., and Solomon, M. M. 2014. Recent developments on the use of polymers as corrosion inhibitors – a review. *Open. Mater. Sci. J.* 8: 39–54.

Wang, Y., Huang, X. and Zhang, X. 2021. Ultrarobust, tough and highly stretchable self-healing materials based on cartilage-inspired noncovalent assembly nanostructure. *Nat. Commun.* 12: 1291.

Wang, Z., Xie, C., Yu, C., Fei, G., Wang, Z., and Xia, H. 2018. A facile strategy for self-healing polyurethanes containing multiple metal-ligand bonds. *Macromol. Rapid Commun.* 39(6): 1700678.

Willocq, B., Odent, J., Dubois, P., and Raquez, J. M. 2020. Advances in intrinsic self-healing polyurethanes and related composites. *RSC Adv.* 10: 13766–13782.

Wikipedia, https://en.wikipedia.org/wiki/Polyurethane#/media/File:Polyurethane_synthesis.svg., June 15, 2021.

Wirpsza, Z. 1993. *Polyurethanes: Chemistry, Technology and Application.* New York: Harwood Publishing.

Xiao, L., Shi, J., Wu, K., and Lu, M. 2020. Self-healing supramolecular waterborne polyurethane based on host–guest interactions and multiple hydrogen bonds. *React. Funct. Polym.* 148: 104482.

Yan, J., Yan, Y., Li, X., Zhang, J., and Guan, H. 2019. Cobaltocenium cation-based water-based polyurethane applied as corrosion inhibitor, and its corrosion inhibitor system and preparation method. Chinese Patent, CN 110577626 A 20191217.

Yan, X., Liu, Z., Zhang, Q., Lopez, J., Wang, H., Wu, H. C., Niu, S., Yan, Wang, S., Lei, T., Li, J., Qi, D., Huang, P., Huang, J., Zhang, Y., Wang, Y., Li, G., Tok, J. B. H., Chen, X., and Bao Z. 2018. Quadruple H-bonding cross-linked supramolecular polymeric materials as substrates for stretchable, antitearing, and self-healable thin film electrodes. *J. Am. Chem. Soc.* 140: 5280–5289.

Yang, F., Li, X., Qiu, S., Zheng, W., Zhao, H., and Wang, L. 2017. Water soluble trianiline containing polyurethane (TAPU) as an efficient corrosion inhibitor for mild steel. *Int. J. Electrochem. Sci.* 12: 5349–5362.

16 Crown Ethers, Dendrimers, Cyclodextrins, and Calixarene Derivatives as Corrosion Inhibitors

Vitalis I. Chukwuike and Rakesh C. Barik
CSIR-Central Electrochemical Research Institute
Academy of Scientific and Innovative Research (AcSIR)

CONTENTS

16.1 INTRODUCTION

Crown ethers, dendrimers, cyclodextrins, and calixarene derivatives are supramolecules by virtue of their structures. Based on the endowed superstructural arrangements, these compounds are capable of different applications, such as drug release (Chehardoli and Bahmani 2019; Hu et al. 2014), fabrication of nanocontainers (He et al. 2016; Amiri and Rahimi 2014), biological activities (Nimse and Kim 2013; Chen and Liu 2010), phase-transfer catalyses (Astruc et al. 2010), and corrosion inhibition (Chen and Fu 2012; Liu et al. 2018; Saji 2019). In the course of this chapter, we are going to limit our discussion to the corrosion inhibition performance of these four supramolecules.

It is well known that corrosion inhibitors are chemical compounds that, when added to a liquid or gaseous medium, decrease the corrosion rate of materials, usually metals or alloys, immersed in the environment (Chukwuike et al. 2019; Hou et al. 2017). The effectiveness of corrosion inhibitors depends on fluid composition, water quality, and flow regime (Chukwuike et al. 2021; Prasannakumar et al. 2020). Earlier, corrosion inhibition of metals has relied on the addition of toxic inorganic chemicals such as chromates (Son et al. 2016), vanadate (Iannuzzi and Frankel 2007), molybdates (Ilevbare and Burstein 2003), and tungstates (Qu et al. 2009). On the other hand, organic inhibitors, though not environmentally harmless, have been equally utilized, however, for a long time. Research was limited to small monomers such as benzotriazole, and other simple compounds, with little or no attention to complex macromolecules such as supramolecules. This could be attributed to challenges such as poor aqueous solubility, and difficulties in synthesis.

In recent times, macromolecules such as supramolecules have been shown to possess corrosion inhibition abilities completely different from smaller monomers and inorganic inhibitors. This stems from their fascinating characteristics, such as superstructure, selectivity, high availability, facile synthesis, functionalization, possible large-scale production, and less environmental harmfulness (Tang et al. 2020; Dehghani et al. 2020). These properties have made supramolecules very attractive in the area of corrosion inhibitor formulation. For example, several crown ether derivatives have been tested and found to be good corrosion inhibitors for different types of stainless steel in an acidic environment (Fouda et al. 2010). Dendrimers have been demonstrated by (Wang et al. 2018) as corrosion inhibitors for Cu and other metals. Cyclodextrins (CDs) and their derivatives are shown to be good corrosion inhibitors for carbon steel (de Souza et al. 2016), Al (Amiri and Rahimi 2016), and Zn (Altin et al. 2018). Several works have shown their capability as nanocontainers for inhibitor encapsulation (Dehghani et al. 2020) in the fabrication of intelligent release of corrosion inhibitors. Calixarene derivatives have been tested as corrosion inhibitors for steel (Kaddouri et al. 2013). This chapter, therefore, is focused on research work relating to the corrosion inhibition activities of these compounds on different metals in various corrosive environments. This work will deepen the understanding of macromolecules, their corrosion inhibition mechanisms, and improvement of metal protection through the development of supramolecular corrosion inhibitors.

16.2   SELECTION OF SUPRAMOLECULES AS CORROSION INHIBITORS

"Organic substances used for corrosion inhibition are selected essentially from empirical knowledge-based on their macroscopical physicochemical properties" (Abdallah 2008). Most supramolecules and organic inhibitors usually promote the formation of chelates at the surface as a point of bonding interaction, which helps in adsorption and film adherence for corrosion mitigation (Fouda et al. 2016). Under these conditions, for a given metal, the efficiency of a corrosion inhibitor depends on the stability of the formed chelate film. The inhibitor molecule should have centres capable of forming bonds with the metal surface by electron transfer or sharing. The four supramolecules discussed in this chapter possess electron dense atoms and heteroatoms such as O, N, and/or S with free electron pairs which are readily available for sharing. Thus, the metal substrate acts as an electrophile, whereas the supramolecules could act as Lewis bases. This phenomenon is obtainable in other heterocyclic organic compounds. Therefore, supramolecules equally share these properties with the additional steric advantage of large size for hindering the adsorption spaces for other species such as corrosive ions in the environment. The size of supramolecules presents both disadvantages and advantages. Increasing the density of the carbon chain may lead to challenges such as insolubility and difficulties in the isolation of molecules. However, the presence and positions of functional groups are higher in macromolecules than in small monomers. Molecular stability to thermal and other environmental factors may improve with the size of inhibitor molecules. Among these compounds, CDs have also been observed to behave as semi-hydrophilic/hydrophobic molecules due to the accumulation of -OH groups on the molecular exterior as well as the interior cavity, which is responsible for the hydrophobic property (de Souza et al. 2016; Yang et al. 2019), thus enhancing their corrosion inhibition abilities.

Encapsulation is another interesting property of supramolecules and one of the trending methods in the formulation of corrosion inhibitors. CDs are among the additives that are very good in the formation of inclusion compounds whereby some organic compounds are encapsulated as corrosion inhibitors with more endowed capabilities, such as self-healing and controlled release. The dendrimers, much like CDs, possess a hydrophobic core and a hydrophilic periphery to exhibit micelle-like behaviour and have container properties in solution as well, which is good for corrosion inhibitors and coating fabrication. The dendrimers typically involve conjugating other chemical species to the dendrimer surface, leading to multivalent systems. In other words, one dendrimer molecule has hundreds of possible sites for coupling to an active species.

Moreover, supramolecules are found to be environmentally benign. Searching for less toxic and environmentally-friendly corrosion inhibitors is becoming increasingly important due to the increased awareness of the importance of green chemistry applications. The less harmful nature of supramolecules makes them ideal materials for the present-day research in corrosion inhibition.

16.2.1 PHYSICAL PROPERTIES

Physical properties such as colour, solubility, melting, and boiling points of supramolecules are highly diverse depending on the molecular structure and size. However, these properties are very important as they define the interactions of the macro compounds with the surrounding environments, which affect the lifetime and overall corrosion inhibition of the materials. Thermal stability and aqueous solubility have thus become very crucial properties. Due to size and high functionality, supramolecules have been found to balance these properties, for example, 18-Crown-6 is an organic compound with the formula $[C_2H_4O]_6$ and the IUPAC name of 1,4,7,10,13,16-hexaoxacyclooctadecane. It is a white, hygroscopic crystalline solid with a low melting point (Steed 2001). The dipole moment of 18-crown-6 varies in different solvents and under different temperatures. Under 25°C, the dipole moment of 18-crown-6 is 2.76±0.06 D in cyclohexane and 2.73±0.02 D in benzene (Caswell and Savannunt 1988). It shows a boiling point of around 116°C with a fair solubility of 75 g/L. Other supramolecules have different sizes and functional groups that determine the way they physically react with the environment. All CDs are white, water-soluble solids with minimal toxicity. For example, α-CD is a hexasaccharide (six glucose units) derived from glucose. It is related to the β- and γ-types of CDs, which contain seven and eight glucose units, respectively. There is no clear marked melting and boiling points for most CDs as the compounds undergo decomposition in the presence of heat. However, they show good solubility, for example, α-CD is soluble up to 145 g/L, and thus, this type of macromolecule can function well as a room temperature corrosion inhibitor. The tuneable nature of supramolecules makes it possible to find an optimum condition for a selected material in a given environment.

16.2.2 CHEMICAL PROPERTIES

It is widely known that the chemistry of supramolecules involves molecular recognition and the selective binding to metals by natural as well as synthetic macrocyclic and macropolycyclic ligands. Non-covalent self-assembly and self-organization are the important chemistries of these compounds that facilitate their surface activity, metal coordination, hydrogen bonding, dipole-dipole interactions, and donor-acceptor interactions (Wipff et al. 1982; Oshovsky et al. 2007; Akhondi and Jamalizadeh 2020). Apart from these interactions, the factors that determine the resultant structures of the assembly include final thermodynamic stability of the self-assembly, unique structure of individual molecular units, exact stoichiometric ratio of the different molecular units, assembly medium, and concentration of the individual units. Therefore, supramolecular structure arises from the interactions of the discrete number of molecules leading to the formation of a complex structure.

16.2.3 STRUCTURAL PROPERTIES

16.2.3.1 Crown Ethers

Ethers generally are organic compounds characterized structurally by an oxygen atom bonded to two alkyl or aryl groups, thus replacing both hydrogen atoms in the water molecule. Crown ethers specifically are a group of cyclic ethers containing several (i.e., 4, 5, 6, or more) oxygen atoms

(Figure 16.1). Therefore, crown ethers are referred to as heterocyclic polyethers, in their simplest form, they are cyclic oligomers of dioxane. They are named by using the parent name *crown* in the form of X-crown-Y, where X is the number describing the size of the ring and Y is the number of oxygen atoms in the ring. The crown refers to the crown-like shape formed by the carbon chain between two oxygen atoms. Crown ethers were discovered by Charles Pedersen. In his initial works, he found that dibenzo[18]crown-6 could complex with alkali metal cations. Thereafter, he and his coworkers continued synthesizing a series of polyethers with different numbers and types of atoms. Although most of the crown ethers synthesized by their group were unable to form stable complexes, only polyether rings that contain 5–10 oxygen atoms with two carbons between each oxygen atom were able to form good complexation. The lone pair of electrons on the oxygen atoms facilitates the formation of complexes with specific metal cations which in turn depends on the number of atoms in the ring (Figure 16.2). As the size of the crown ether increases, the cavity accommodates larger metal ions. The selectivity for binding to alkali metal and alkali earth metal cations by crown ethers in the solution phase is also an added advantage for the use of these supramolecules as corrosion inhibitors. Their structure and binding properties can be improved by changing the number and type of donor atoms. Additional functional groups can equally be added to the macrocyclic platform to alter their geometry and affinity toward guests (Tan et al. 2017).

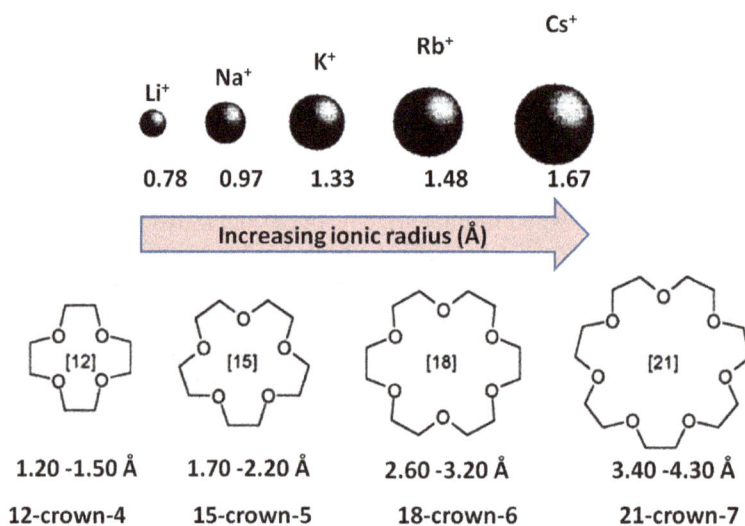

FIGURE 16.1 Structures of the different sizes of common crown ethers with a comparison of metallic ionic radii. (Reproduced with permission from Steed (2001) © 2001 Elsevier B.V.)

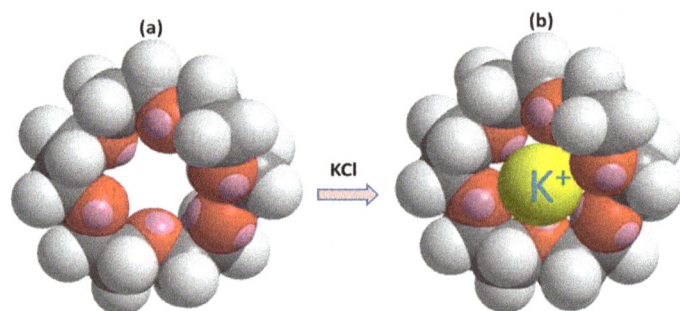

FIGURE 16.2 Structure of 18-crown-6 ether (a) without and (b) with potassium ion complexed in it after exposure to KCl solution. (Reproduced with permission from Gul et al. (2019) © 2019 MDPI, under Creative Commons Attribution License.)

Based on the ionic radius and the size of certain crown ethers as already shown, a selective complexing of a particular metal ion can take place as follows.

Over the decades, other types of crown ethers have been developed, including imino and aza crown ethers. Imino crown ethers contain Schiff base functional groups (N=C) while Aza type has nitrogen atoms within the macrocyclic ring structure. The simple form of aza crown ether is the cyclen, which is composed of four amine moieties separated by four ethylene units. Other advanced derivative structures are available, having additional rings or heteroatoms. For example, dibenzo-18-crown-6, 4,13-diacethyl, 1,7,10,16-tetraoxa-4,13-diazacyclo-octadecane, 4,7,13,21,24-hexa-oxa-1,10-diazo-bicyclo-(8,8,8)-hexacosane and dibenzo-24-crown-8 (Fouda et al. 2010).

16.2.3.2 Dendrimers

Dendrimers are molecules consisting of one or more dendrons. Dendrons are part of a molecule with only one free valence, comprising exclusively dendritic and terminal constitutional repeat units from different generations (G) (Figure 16.3). A dendrimer molecule comprising only one dendron is sometimes referred to as a dendron. Dendrimers and dendrons are monodisperse and usually highly symmetric, spherical compounds. They are in the form of highly branched star-shaped nano-sized macromolecules, which are associated with a large number of functional groups, and compact molecular structures (Figure 16.4).

16.2.3.3 Cyclodextrins

CDs are a series of cyclic oligosaccharides obtained from the enzymatic conversion of starch. CDs are composed of five or more sugar molecules linked at one and four carbons to form a ring. The five-sugar CD is less common due to ring strain; however, its synthesis is possible. The most frequently used CDs are α-CD, β-CD, and γ-CD which are composed of 6-, 7-, and 8-membered sugars, respectively (for structure, see Chapters 2 and 10). These common CDs can be produced through enzymatic conversion. In addition, the surface -OH groups on CDs are easily modified and functionalized with other species (Nikitenko and Prassolov 2013). CDs have a truncated cone structure and are known to have a hydrophobic interior and a hydrophilic exterior. Owing to this property, many hydrophobic guests are often used to form host–guest complexes with CDs as an advantage for corrosion inhibitor

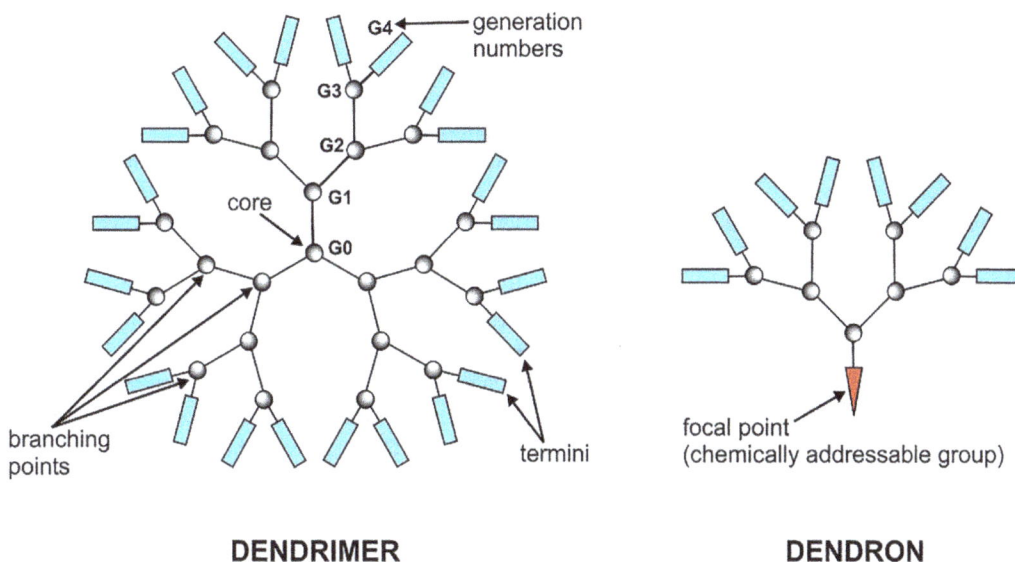

DENDRIMER   **DENDRON**

FIGURE 16.3   Schematic representation of dendrimer and the different generations of dendrons. (Reproduced with permission from Atav (2018) © 2018 Elsevier Ltd.)

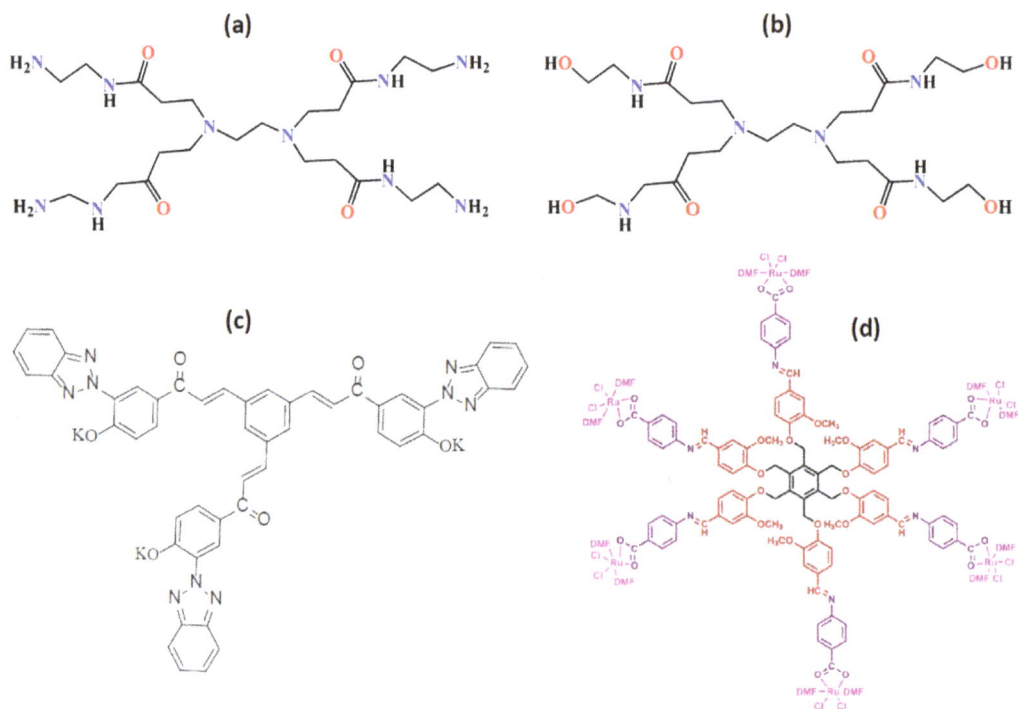

FIGURE 16.4 Structures of dendrimer derivatives with different generations of dendrons (a, b) polyamido-aminedendrimers with terminal NH$_2$ and OH groups, (c) 4,4′, 4″-((2E, 2′E, 2″E)-3,3′, 3″-(benzene-1,3,5-triyl)-tris-(acryloyl))-tris-(2-(2H-benzo[d][1,2,3]-triazol-2-yl)-phenyolate)-potassium, (d) Ru$_6$(hexbzaD)(DMF)12Cl12. (Reproduced with permission from Ren et al. (2021); Wang et al. (2018); Keshtiara et al. (2020) © 2020, 2018 and 2020 Elsevier.)

formulation. Moreover, as CDs are composed of sugar subunits, they are water-soluble and nontoxic; a requisite property for eco-friendly corrosion (Tan et al. 2017; Zou et al. 2013; Fan et al. 2014).

16.2.3.4 Calixarenes

Calixarenes are macrocyclic compounds derived from the coupling of phenol with an aldehyde. Calixarenes have a basket structure; they can be present in many conformations, such as cone, partial cone, 1,2-alternate, and 1,3-alternate, due to the rotation around the methylene groups, and this rotation can be avoided by adding a bulky substituent on the top of the ring to lock up the conformation. Calixarenes are difficult to produce because random polymerization occurs inside complex mixtures of linear and cyclic oligomers with different numbers of repeating units. However, with finely tuned starting materials and reaction conditions, synthesis can also be surprisingly facile. The nomenclature of calixarenes is as follows: calix[n]arenes, where n is the number of repeat units within the cyclic structures. In addition to the common structures of calixarenes such as calix[4] arene known as *p-tert*-Butylcalix[4]arene (Simaan and Biali 2003) (Figure 16.5a), there are other types of calixarenes such as those containing imidazolium units linked with a methylene moiety leading to a structure in the form of calix[n]imidazolium with a different number of imidazolium (Figure 16.5b and c). This way, calix[4]arene derivatives can be made by changing the number of 4-imidazolylethylcarboxamide substituents. The variation of 4-imidazolylethylcarboxamide substituents was reported by (Kaddouri et al. 2012). These molecules combine the natural amphiphilicity of calixarenes with the particular properties of an imidazole unit. Other types of calixarenes include thiacalixarene. This group possesses a larger cavity size than the conventional calixarenes with the same repeat units and has sulfur functionalities for better metal complexation, due to the strong affinity of thiols for various metals (Tan et al. 2017).

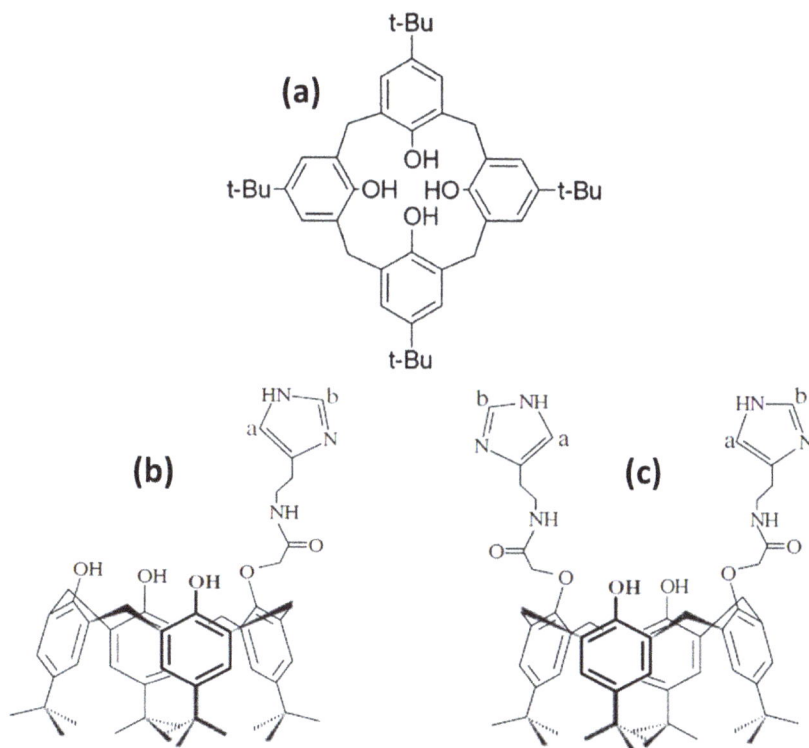

FIGURE 16.5   (a) Chemical structure of a cali[4]xarene, (b, c) Chemical structure of calixarene derivatives with the variation of 4-imidazolylethylcarboxamide substituents. ((a) Reproduced with permission
from Simaan and Biali (2003) © 2003 American Chemical Society; (b, c) Reproduced with permission from
Kaddouri et al. (2008) © 2008 Springer Science Business Media B.V.)

16.2.4   SOLUTION CHEMISTRY

Water molecules form a huge network of hydrogen bonds with large structured clusters (Dai
et al. 2019; Marczenko et al. 2018). Self-assembly and molecular interaction in water for the
four supramolecules in this chapter are considered to understand their functionalities and corrosion inhibition mechanisms. In solution, crown ethers can form stable inclusion complexes
with alkali, alkaline-earth, and primary ammonium cations. Therefore, complexation domi-
nates the solution chemistry of most supramolecules, however, with high selectivity for metal
ions to complex with. For example, 1H NMR titration studies of bis-crowns 8 and 9 show affinity
towards K+, Rb+, and Cs+ with preference for Cs+ (Steed 2001). 18-Crown-6 binds to a variety
of small cations, using all six oxygens as donor atoms (Steed 2001). Crown ethers can be used
in the laboratory as phase transfer catalysts. Salts that are normally insoluble in organic solvents
are made soluble by crown ether. For example, potassium permanganate dissolves in benzene
in the presence of 18-crown-6, giving the so-called "purple benzene", which can be used to
oxidize diverse organic compounds. Various substitution reactions are also accelerated in the
presence of 18-crown-6, which suppresses ion-pairing (Hadisaputra et al. 2014). The anions
thereby become naked nucleophiles. For example, 18-crown-6, potassium acetate is a more powerful nucleophile in organic solvents. Furthermore, the properties of crown ethers can easily be
tuned by modifying them into different derivatives through heteroatoms (O, N, S, and P), and by
adding π-electron groups such as benzene rings as multiple bonds to meet the criteria for corrosion inhibitors through chemisorption. The chemisorption mechanism is very important solution
chemistry of crown ethers. For example, crown-type polyethers (C-1 and C-2) adsorb onto steel

by protonation in acid media (Figure 16.6). Therefore, ligands C-1 and C-2 are usually in their protonated forms in acidic media with equilibrium between the neutral and protonated forms.

On the other hand, CDs tend to bind with other molecules in their quasi-cylindrical interiors. This inclusion (and release) behaviour is good for the fabrication of self-healing corrosion inhibitors (Akhondi and Jamalizadeh 2020; Liu et al. 2015). The formation of bridgehead (Zou et al. 2014a) is another very important solution chemistry of CDs, for example, chlorinated polyethylene glycol (PEG) can react with β-CD to yield β-CD–PEG under the influence of a certain amount of NaOH as a catalyst at 333 K as shown in Figure 16.7.

Dendrimers are highly soluble in aqueous solutions and can be used as solubilizing agents. These two characteristics favour the use of dendrimers as corrosion inhibitors as solubility in the test medium is a requisite property for inhibitor performance. In addition as a solubilizing agent, it acts as a surfactant and thus may interact with surface deposits or adsorbed corrosive agents to reduce the rate of corrosion. Its architecture with hydrophilic and hydrophobic core has made it a good candidate for host–guest chemistry similar to that of CDs. The entire solution chemistry of supramolecules can be summarized as ion–ion, ion–dipole, dipole–dipole, π–π, and cation–π interactions of different energies in addition to van der Waals and hydrogen bonding. Hydrophilic and hydrophobic ends are part of the influential forces, especially in the case of CDs and calixarenes, which both share the same abilities and similarities.

16.2.5 ADSORPTION PROPERTIES

The nature of inhibitor interactions on the metal surface is often deduced in terms of their adsorption or film formation characteristics. The different established isotherm models, such as Temkin, Freundlich, and Longmuir plots, can be employed to analyze the behaviour of inhibitor compounds (Solomon et al. 2017; Tian et al. 2015; Kaddouri et al. 2008). The relation between the

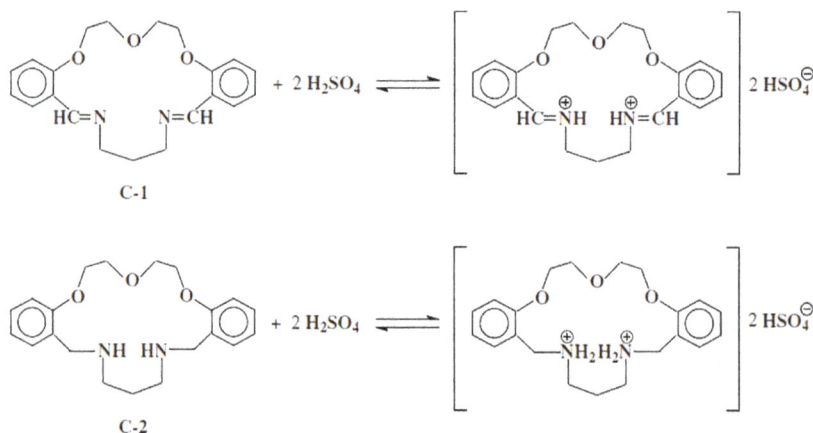

FIGURE 16.6 The equilibria between crown ethers and their protonated forms in acid solution. (Reproduced with permission from Hasanov et al. (2010) © 2009 Elsevier Ltd.)

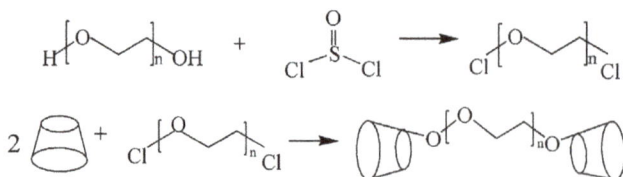

FIGURE 16.7 Schematic presentation of the formation of bridged β-CD–PEG in solution. (Reproduced with permission from Zou et al. (2014a) © 2014 Elsevier B.V.)

inhibitor concentration (C) divided by the surface coverage (θ), (C/θ) plotted against concentration C in mol or mmol/L at a certain temperature would give a straight line suggesting the validity of the model and indicating quality adsorption of the inhibitor on the metal surface. Reports on the studied supramolecules show that most of the macromolecules and their derivatives have good adsorption behaviour. For instance, calixarenes are amphiphilic molecules and their interfacial activity has been widely studied. Their adsorption on surfaces has been shown to offer users control of the surface reactivity (Kaddouri et al. 2008). Figure 16.8 shows the adsorption behaviour based on the Langmuir adsorption isotherm model of selected Crown ether, dendrimer, CD and calixarene derivatives, namely; (i) dibenzo-bis-imino crown ether (C-1), (ii) dendrimer known as 4,4′, 4″-((2E, 2′E, 2″E)-3,3′, 3″-(benzene-1,3,5-tryl)-tris-(acryloyl))-tris-(2-(2H-benzo[d] [1,2,3]-triazole -2-yl)-phenolate)-potassium, (iii) bridge type derivative of cyclodextrin (β-CD–PEG) and (iv) is the functionalized calixarenes termed as CA1, CA2, and CA3 containing one, two, and three 4-imidazolylethylamidocarbonyl groups, respectively, (Hasanov et al. 2010; Zou et al. 2014a; Kaddouri et al. 2008; Wang et al. 2018). The figures show straight-line plots indicating that all the selected supramolecules obey the Langmuir adsorption isotherm model.

However, besides the adsorption isotherm of an inhibitor, other complementary parameters must be considered, such as the number of adsorption active centers in the molecule and their charge densities, molecular size, delocalized π-electrons, mode of adsorption, the heat of hydrogenation, and the formation of metallic complexes to fully access a compound's interaction with the metal surface to achieve high inhibition efficiency (IE). Moreover, adsorption is further validated by a parameter such as the Gipps free energy (ΔG_{ads}). Generally, the value ΔG_{ads} around or below −20 kJ/mol is

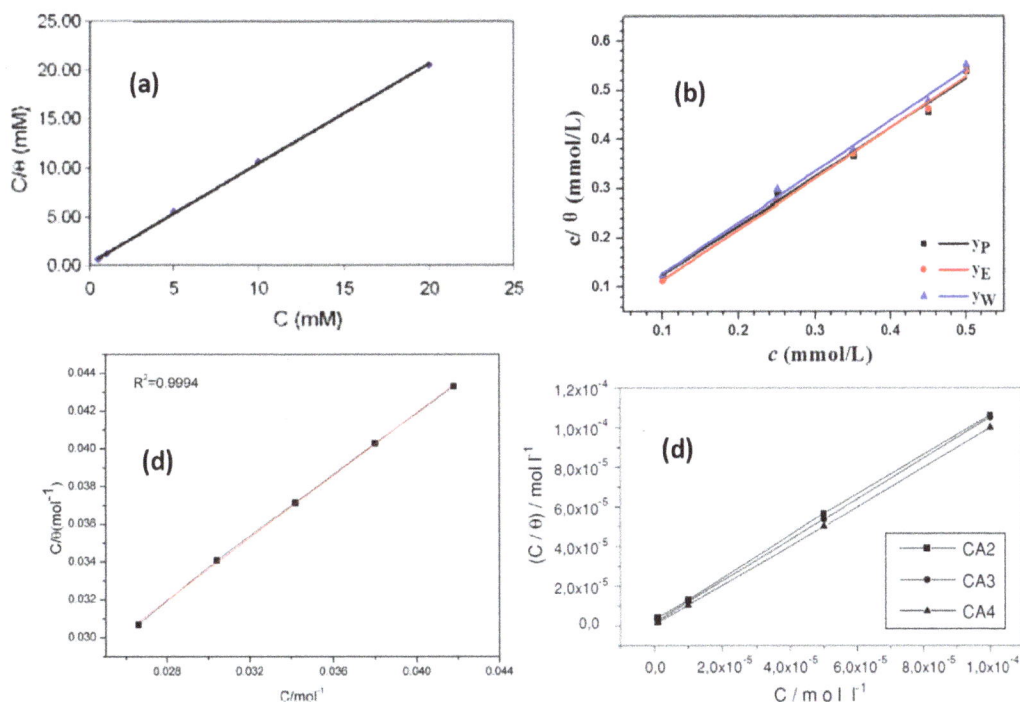

FIGURE 16.8 Langmuir adsorption isotherm model for selected derivatives of (a) Crown ether, (b) dendrimer (y_P, y_E and y_W correspond to data from potentiodynamic polarization, impedance spectroscopy, and weight loss, respectively) (c) cyclodextrins and (d) calixarene (CA1, CA2, and CA3 are calixarenes containing one, two, and three 4-imidazolylethylamidocarbonyl groups, respectively). (Reproduced with permission (a) from Hasanov et al. (2010) © 2009 Elsevier Ltd., (b) from Wang et al. (2018) © 2018 Elsevier B.V. (c) from Zou et al. (2014a), © 2014 Elsevier B.V., (d) from Kaddouri et al. (2008) © 2008 Springer Science Business Media B.V.)

attributed to physical adsorption (physisorption) whereas −40 kJ/mol or higher is correlated with chemical adsorption (chemisorption) (Donahue and Nobe 1965; Quartarone et al. 2012).

16.3   CROWN ETHER AND ITS DERIVATIVES AS CORROSION INHIBITORS

Researchers have shown that crown ethers as macrocyclic compounds could be a potential class of corrosion inhibitors. The effects of four crown ethers on the corrosion behaviour of 430 stainless steel in 2 M HCl at different temperatures were investigated by (Fouda et al. 2010). The crown ethers tested by the authors were: (i) 6,7,9,10,17,18,20,21-octahydro-5,8,11,16,19,22-hexaoxa-dibenzo[a, j]cyclooctadecene (dibenzo-18 crown-6), (ii) 4,13-diacetyl, 1, 7, 10, 16-tetraoxa-4, 13-diazacyclooctadecan (Kryptofix 22DD), (iii) 4,7,13,16,21,23-hexaoxa-1,10-diaza-bicyclo[8.8.7] pentacosane (Kryptofix 222) and (iv) 4, 7, 13, 16, 21, 24-hexa-oxa-1, 10-diazo-bicyclo-(8,8,8)-hexacosane (Dibenzo-24 crown-8). The inhibition efficiencies (%IE) of these compounds were found to increase with an increase in concentration according to Table 16.1. The %IE decreased in the following order: iv > iii > ii > i. The corrosion current densities (i_{corr}) show a decreasing trend as well; correlating to a reduction in corrosion rate with an increase in corrosion potential (E_{corr}). An increase in cathodic (β_c) and a decrease in anodic (β_a) Tafel slopes were also witnessed in Table 16.1, which indicates that the crown ether derivatives influenced both the cathodic and anodic

TABLE 16.1

Effects of the Concentrations of Crown Ethers on the Corrosion Parameters of 430 Stainless Steel in 2 M HCl from Potentiodynamic Polarization

Compound	Concentration 10^{-7} (M)	β_a mV/dec	β_c mV/dec	$-E_{corr}$ mV (SCE)	i_{corr} µA/cm^2	% IE
I	0	51	74	473	560	–
	1	48	79	468	528	15.9
	3	50	96	459	432	22.9
	5	47	98	453	343	38.7
	7	49	99	451	336	40.1
	9	47	93	447	244	56.5
	11	44	96	441	226	59.6
II	1	50	76	466	451	19.4
	3	43	103	445	378	32.5
	5	40	98	441	262	53.3
	7	37	100	434	235	58.0
	9	36	102	425	191	65.9
	11	28	106	411	152	72.9
III	1	58	95	449	277	50.6
	3	56	95	444	240	57.1
	5	56	93	443	195	65.1
	7	57	94	444	191	65.9
	9	59	94	444	167	70.2
	11	56	95	443	148	73.5
IV	1	64	100	449	237	57.6
	3	53	105	435	188	66.4
	5	50	103	431	134	76.1
	7	50	102	430	119	78.8
	9	51	102	429	97	82.7
	11	35	103	429	37	93.4

Source: Reproduced with permission from Fouda et al. (2010) © 2009 Elsevier B.V.

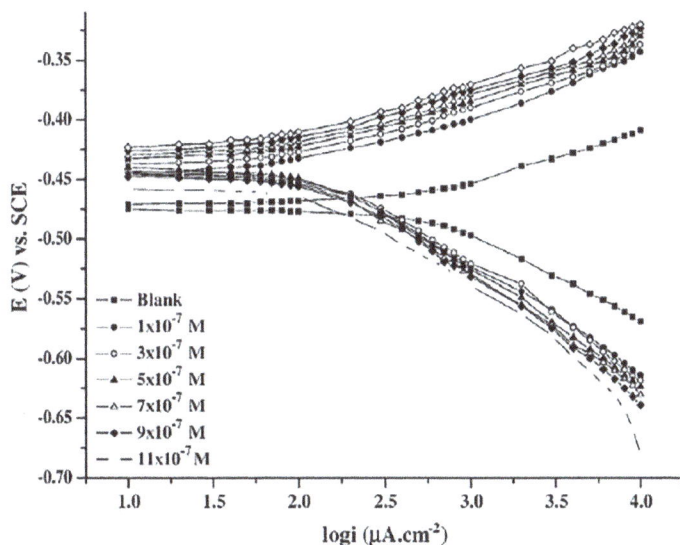

FIGURE 16.9 Representative potentiodynamic polarization studies of 430 stainless steel in 2 M HCl in the absence and presence of different concentrations of the compound IV (dibenzo-24 crown-8) at 30°C. (Reproduced with permission from Fouda et al. (2010) © 2009 Elsevier B.V.)

corrosion current densities. The potentiodynamic polarization (PDP) studies (Figure 16.9) thus, showed that crown ethers acted as mixed-type inhibitors, lowering the cathodic and anodic i_{corr} and shifting the E_{corr} towards the anodic region.

The inhibition effect was further proved by SEM surface morphologies (Figure 16.10). The micrographs in the absence of inhibitor (Figure 16.10a), show an extensive etching composed of green, dark, and white areas of corrosion products. The greenish areas reflect the part of chromium and its native protective oxide, while the white areas represent the ferrite phase, and the dark areas represent the pearlite, which is a mixture of ferrite and cementite. However, in the case of Figure 16.10b where $9 \times 10^{-7} M$ concentration of the compound (IV) was applied, the steel surface was protected as no evidence of corrosion damage was visible.

Other authors, such as Hasanov et al. (2010), equally showcased interesting data on the efficacy of different crown ether derivatives. The corrosion inhibition efficiencies of dibenzo-bis-imino

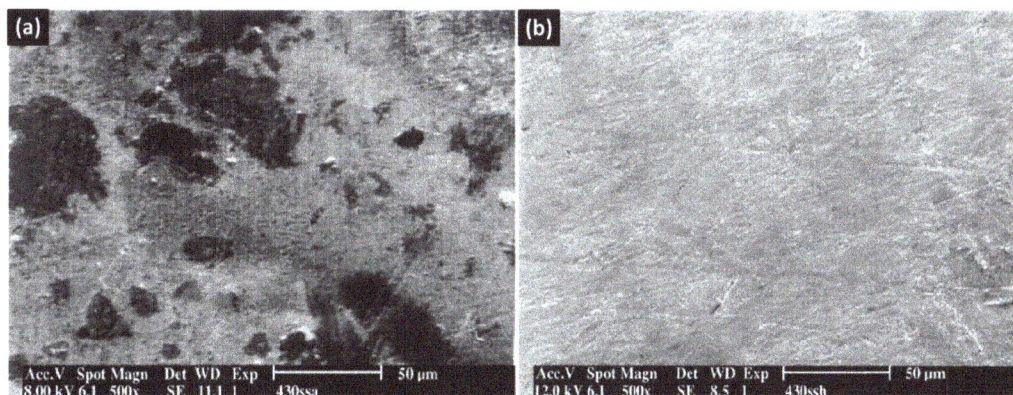

FIGURE 16.10 SEM micrographs of 430 stainless steel immersed in 2 M HCl at 30°C (a) in the absence and (b) presence of $9 \times 10^{-7} M$ of dibenzo-24 crown-8. (Reproduced with permission from Fouda et al. (2010) © 2009 Elsevier B.V.)

crown ether (termed as C-1) and dibenzo-diaza crown ether (C-2), which are crown type polyethers based on macrocyclic Schiff base and macrocyclic amine, respectively, were tested on steel metal in 1 M H_2SO_4. The experimental and theoretical data relating to these inhibitors are presented in Table 16.2. The IE of both the inhibitors shows maximum effects at 20 mM. Their inhibitive action can be explained based on the three oxygen atoms and the two nitrogen atoms present in the macrocyclic ring, which contribute to the donor-acceptor bond between the nonbonding electron pairs and the vacant orbitals of the metal surface. Moreover, the calculated ΔG_{ads} values for C-1 and C-2 were more negative than −40 kJ/mol, indicating that they undergo chemical adsorption on the steel surface. However, the authors described that the inhibitors equally have a contribution from barrier film formation, which might be made possible by the polymeric nature of these derivatives.

Several pieces of research have shown that the adsorption of an inhibitor on the metal surface can occur based on donor-acceptor interactions between the p-electrons of the heterocyclic compound and the vacant d-orbitals of the metal surface atoms (Hasanov et al. 2010; Obot, Obi-Egbedi, and Umoren 2009; Machnikova, Whitmire, and Hackerman 2008). Thus, according to the frontier molecular orbital theory, the formation of a transition state is due to an interaction between frontier orbitals (HOMO and LUMO) of reacting species (Behpour et al. 2008). This transition is facilitated by a high energy difference (ΔE) between the HOMO and LUMO orbitals. High E_{HOMO} values indicate that the molecule tends to donate electrons to appropriate acceptor molecules with vacant low energy molecular orbital while low E_{LUMO} indicates the ability of the molecules to accept electrons. The lower the E_{LUMO}, the higher the probability of accepting electrons. It also follows that the low absolute values of the ΔE give good inhibition efficiencies because the energy required to remove an electron from the last occupied orbital will be small. Therefore, high values of the E_{HOMO} facilitate the adsorption process and, consequently, the IE by influencing the transport process through the adsorbed layer. Table 16.2 shows the important parameters such as E_{HOMO}, E_{LUMO}, ΔE, dipole moment (μ), and the molecular surface area (Å^2) for C-1 and C-2. According to the results, the lowest E_{LUMO} (0.165 eV) and ΔE (9.517 eV) values are found for C-1.

16.4 DENDRIMER AND ITS DERIVATIVES AS CORROSION INHIBITORS

Dendrimers, due to their high effectiveness as organic compounds containing heteroatoms, have been reported as good corrosion inhibitors for mild steel in acidic media. The choice of these compounds as corrosion inhibitors is based on the fact that dendrimers can be easily and conventionally synthesized from commercially available chemicals. The higher solubility and reactivity of this compound is due to the presence of polar groups such as $-NH_2$ which facilitate the adsorption process. Dendrimer adsorbs over the metallic surface through unshared electron pair of nitrogen as well as pi-electrons of aromatic rings and amides (Verma et al. 2016). This is further facilitated

TABLE 16.2

The Experimental and Theoretical Data Relating to the Inhibitors (Dibenzo-bis-imino Crown Ether (C-1) and Dibenzo-diaza Crown Ether (C-2) on Steel in 1 M H_2SO_4

Compound	Spin	E_{HOMO} (eV)	E_{LUMO} (eV)	ΔE (eV)	μ (D)	Molecular Surface Area (Å^2)	Concentration (mM)	IE (%)
C-1	Alpha	−9.645	−0.165	9.480	0.803	439.98	20	98
	Beta	−9.595	−0.284	9.310				
C-2	Alpha	−9.295	0.222	9.517	1.320	460.52	20	98
	Beta	−9.295	0.221	9.516				

Source: Reproduced with permission from Hasanov et al. (2010) © 2009 Elsevier Ltd.

by its hyperbranched molecular nature. The branches of these molecules endow them with multiple functional groups needed for anchoring and adsorbing on the metal surface. This group investigated the inhibition properties of two NH_3-cored dendrimers on mild steel, in 1 M HCl. Their weight loss studies showed a maximum %IE of 96.95% at 50 mg/L concentration. The PDP plots showed that both the cathodic and anodic reactions were affected in the presence of the dendrimers, signifying a mixed mechanism. The IE of the dendrimer compounds in that study was dependent on their molecular size and the nature of the heteroatoms, which included O and N. From the HOMO electron distribution, it was suggested that the frontier electron density was mostly confined to the center N atom, suggesting that the N atom primarily participated in the electron donation process. Wang et al. demonstrated protection of Cu in 3 wt% NaCl solution by self-assembled dendric molecular aggregates based on strong π–π interactions. The results obtained from the molecular aggregates were better than those from the linear form of the molecules. This was correlated to the fact that the dendrimer aggregates could adsorb on the Cu surfaces efficiently to achieve anti-corrosion effects, unlike the linear form of the dendrimer inhibitor. The mechanism of this inhibitor was equally found to follow the Langmuir adsorption isotherm through chemisorption (Wang et al. 2018).

In a recent report, two dendrimer derivatives; 3,3′-((2-((3-((2-aminoethyl)amino)-3-oxopropyl)(4-((aminomethyl)amino)-3-oxobutyl)amino)ethyl)azanediyl)bis(N-(2-aminoethyl propanamide) termed as GON and 3,3′-((2-((3-((2hydroxyethyl)amino)-3-oxopropyl)(4-((hydroxylmethyl)amino)-3-oxobutyl)amino)ethyl)azanediyl)bis(N-(2-hydroxyethyl)propan- amide) (GOH) (see Figure 16.4 for structures) were tested on Al alloys in an aqueous NaCl solution (Ren et al. 2021). The results indicate that inhibition performances of GON and GOH are closely related and both are affected by the inhibitor concentrations, steric hindrance, and bonding effects. The authors reported that at the relatively low range of concentrations (0.03–0.10 mM) (see Table 16.3), GON displays better corrosion inhibition ability than GOH, owing to its possession of more anchoring functional groups. On the other hand, GOH shows better inhibition performance at higher concentrations from 0.50 to 1.00 mM.

SEM imaging was employed to view the surface morphologies after immersion in NaCl solution containing GON, (Figure 16.11), which confirmed the inhibitive effect. Panel (a) displays the attacked

TABLE 16.3
Electrochemical Data for Aluminium in NaCl Solution With and Without Inhibitor

Inhibitor	Concentration (mM)	R_{ct} ($\Omega.cm^2$)	%IE
Blank	0	6145	-
GON	0.03	8276	25.7
	0.05	10,710	42.6
	0.07	6151	85.5
	0.10	4340	79.1
	0.30	4535	70.3
	0.50	3294	53.6
	0.70	5861	48.9
GOH	0.07	6341	-
	0.30	6241	-
	0.50	17,280	64.4
	0.70	7099	81.8
	1.00	4644	67.5

Source: Reproduced with permission from Ren et al. (2021) © 2020 Elsevier B.V.

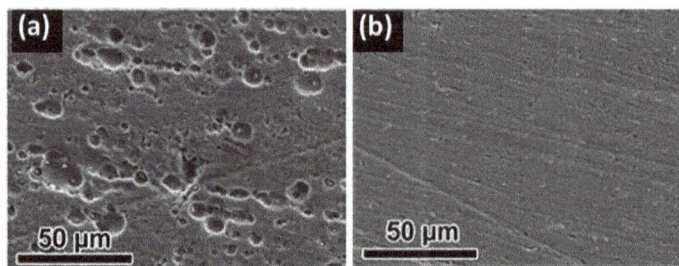

FIGURE 16.11 SEM micrographs of Al electrode after immersion in 3% NaCl solution for 12 h (a) without GON inhibitor, and (b) with GON. (Reproduced with permission from Ren et al. (2021) © 2020 Elsevier B.V.)

surface by Cl^- that caused typical pitting corrosion morphology. Nevertheless, the surface shown in panel (b) is protected by GON, as evident by the distinct nature without pitting. Hence, they concluded that GON and GOH have anti-corrosion performance on Al alloys in NaCl solution.

Further research shows that dendrimers can equally be used in combination with other compatible compounds for synergistic corrosion inhibition. Zhang et al. (2015) in their findings showcased the potential synergistic effect of polyamidoamine dendrimers and sodium silicate on the corrosion of carbon steel in soft water. The PDP result showed an IE of 82%; obtained at relatively low inhibitor dosages. The mechanism was attributed to chemisorption, which obeyed the Langmuir adsorption isotherm (Zhang et al. 2015).

16.5 CYCLODEXTRIN AND ITS DERIVATIVES AS CORROSION INHIBITORS

Although most CD corrosion inhibition is based on the derivatives used as inclusion complexes or polymerized forms, research has shown that these derivatives are good corrosion inhibitors. Here we discuss a few of these works where the authors prepared the derivatives of CDs and tested them by dissolving them in solution as corrosion inhibitors. For example, Zou et al. prepared a CD derivative by reacting 2-phosphonobu-tane-1,2,4-tricarboxylicacid (PBTCA) with β-CD and the formed β -CD-PBTCA complex was tested as a corrosion inhibitor for Q235 carbon steel in 0.1 M H_2SO_4. The result revealed that a maximum IE of 90% was achieved by introducing the product in the aggressive solution, which means that the inclusion complex can act as a corrosion inhibitor for carbon steel which was not possible in the case of PBTCA alone (Zou et al. 2013). Similarly, Fan et al. demonstrated corrosion inhibition properties of a supramolecular complex based on octadecylamine (ODA) and β-CD for protection of mild steel in condensate water. The mechanism of protection by the complex is based on the release of ODA molecules in the condensate water, which forms a monomolecular hydrophobic layer on the carbon steel to protect the metal by isolating it from dissolved oxygen and corrosive ions in solution (Fan et al. 2014).

Destabilizing the point defect region in the oxide layer by β-CD has been explored for corrosion inhibition of Zn (Altin et al. 2019). Monomolecular layer adsorption of β-CD was detected on the Zn oxide surface and this was responsible for blocking the defects (pores) which hitherto promote the propagation of pitting corrosion. Adsorption hence changes the defect density in the protective ZnO layer. This mechanism of corrosion inhibition reveals that influencing the defect chemistry of passivating films by molecular inhibitors may be a viable strategy to control corrosion of metals. Further, a new bridged CD derivative, where β-CD reacted with polyethylene glycol (PEG) to give β-CD–PEG was found to be an efficient corrosion inhibitor for Q235 carbon steel in 0.5 M HCl (Zou et al. 2014a). Electrochemical analysis revealed that β-CD–PEG acted as a mixed-type inhibitor. A high IE (92%) at an inhibitor concentration of 110 mg/L was achieved and the result was correlated to the adsorption of the compound's molecules on the steel surface followed by polymeric protective film formation. The performance of this CD derivative is shown in a representative Nyquist diagram

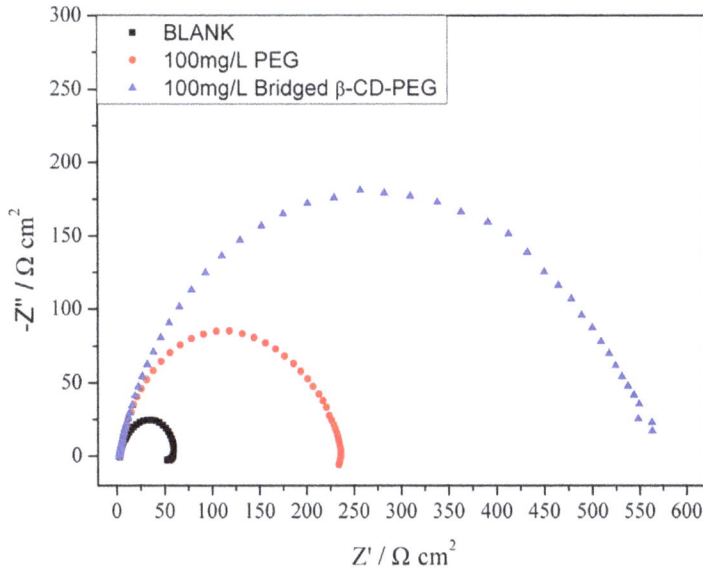

FIGURE 16.12 Nyquist plots of Q235 carbon steel immersed in 0.5 M HCl in the absence and presence of PEG and bridged β-CD–PEG inhibitor at a concentration of 100 mg/L. (Reproduced with permission from Zou et al. (2014a) © 2014 Elsevier B.V.)

of Q235 carbon steel in 0.5 M HCl (Figure 16.12). The bridged form shows the highest diameter of the capacitive loop semicircle, correlating to the highest inhibition performance of the three. The mechanism for the inhibition effect in HCl could be that the hydroxyl group of PEG and β-CD can be bridged between the surfaces of the electrode, coupled with a larger molecular size which ensures greater coverage of the metal, thus resulting in an inhibiting effect in the solution. The corrosion inhibition result was buttressed by SEM imaging. The SEM morphology of the system with and without the bridged β-CD–PEG inhibitor after immersion in 0.5 M HCl solution revealed a very rough surface for the blank system without the inhibitor, whereas in the presence of the bridged β-CD–PEG, the surface of the specimen was smooth, corresponding to a reduced rate of corrosion (Zou et al. 2014a). Other CDs tested as corrosion inhibitors are summarized in Table 16.4.

TABLE 16.4
Summary of Various CD Derivatives Reported as Corrosion Inhibitors

CD derivative	Concertation	Metal	Test Medium	Temp	%IE	Ref.
β-CD-modified acrylamide polymer	150 mg/L	X70 steel	0.5 M H_2SO_4	303 K	85	Zou et al. (2014b)
CD-based functionalized graphene oxide	15 (v/v %)	X60 steel	1 M HCl	Room temp.	81	Haruna et al. (2019)
Aniline trimer-carboxymethylated β-CDs	250 mg/L	Q235 steel	1 M HCl	Room temp.	99	Yang et al. (2019)
Dibenzylthiourea with hydroxypropylated β-CDs	10^{-4} mol/L	Carbon steel	1 M HCl	Room temp.	96	de Souza et al. (2016)
Dibenzylthiourea with hydroxypropylated α-CDs	10^{-4} mol/L	Carbon steel	1 M HCl	Room temp.	94	de Souza et al. (2016)
CD polymer and trans-cinnamaldehyde	125 mg/L	mild steel	3.5% NaCl	303 K	92	Ma et al. (2019)

16.6 CALIXARENE AND ITS DERIVATIVES AS CORROSION INHIBITORS

Calixarene derivatives containing 1–4 4-imidazolylethylamidocarbonyl groups termed as CA1 to CA4 were tested as corrosion inhibitors for mild steel in 1 M HCl at 308 K. It can be seen from Table 16.5 that all the calixarene compounds yield some degree of corrosion inhibition, though the extent of inhibition depends upon the concentration and the number of the substituent. Thus, IE increases with increasing inhibitor concentration for the different derivatives from 33, 94, and 95% to 100% at 10^{-4} M. Therefore, the most effective corrosion inhibitor out of these four derivatives is CA4, which was reported to provide complete protection throughout 6 h duration at a concentration as

TABLE 16.5
Weight-Loss Measurements for Mild Steel after 6 h Immersion in 1 M HCl at 308 K in the Absence and Presence of Various Concentrations of Calix[4]arene Derivatives

Inhibitor	Concentartion (M)	Weight loss (W) (mg cm^{-2}h^{-1})	%IE
No inhibitor (HCl)	1	4.406	–
CA 1	10^{-6}	4.338	1.5
	10^{-5}	3.787	14.0
	5×10^{-5}	3.509	20.3
	10^{-4}	2.938	33.3
CA 2	10^{-6}	3.354	23.8
	10^{-5}	1.011	77.1
	5×10^{-5}	0.563	87.2
	10^{-4}	0.274	93.7
CA 3	10^{-6}	2.699	38.7
	10^{-5}	0.803	81.7
	5×10^{-5}	0.303	93.1
	10^{-4}	0.286	93.5
CA 4	10^{-6}	1.751	60.2
	10^{-5}	0.156	96.4
	5×10^{-5}	0.0023	99.9
	10^{-4}	0.000	100

Source: Reproduced with permission from Kaddouri et al. (2008) © 2008 Springer Science Business Media B.V.

TABLE 16.6
Calixarene Derivatives Reported as Corrosion Inhibitors (test medium – 1 M HCl, temp. – 308 K)

Calixarene Derivatives	Concentration (M)	Metal	IE (%)	Ref.
Calix[6]arene	5×10^{-5}	C38 steel	96	Souane et al. (2009)
Calix[8]arenes	10^{-3}	Mild steel	95	Kaddouri et al. (2012)
Calix[8]arenes	10^{-3}	Mild steel	80	Kaddouri et al. (2012)
Sulfonated calix[8]arenes	10^{-3}	Mild steel	98	Kaddouri et al. (2013)
Sulfonated calix[8]arenes	10^{-3}	Mild steel	84	Kaddouri et al. (2013)
1,3,5-tri-carboxy 2,4,6-tri-methoxy di-nitro calix[6]arene	10^{-4}	Mild steel	87	Benabdellah et al. (2007)
1,3,5-tri-(2-ethyl imidazole acetamide) 2,4,6-trimethoxy calix[6]arene	10^{-4}	Mild steel	92	Benabdellah et al. (2007)

low as $\sim 5 \times 10^{-5}$ M. The IE is equally confirmed to be a linear function of the number of imidazole-containing substituents. This shows that as the number of substituents increases, the IE increases as well, irrespective of the concentration used. Other Calixarene derivatives equally showed impressive corrosion inhibition efficiencies on different metals in various aggressive media tested by different authors, as summarized in Table 16.6.

16.7 CONCLUSIONS AND PERSPECTIVES

Crown ethers, dendrimers, cyclodextrins, and calixarene derivatives are supramolecules, and these compounds recently received attention and have been tested as metal corrosion inhibitors in which the results show high inhibition efficiencies. During the past decades, they have served mainly as complexing agents, drug delivery materials, sensors, visualization agents, materials for phase transfer catalyses, and models for biological phenomena. With their discovery as corrosion inhibitors, reports related to the corrosion inhibition performance of different metals in various corrosive environments are on the increase. This chapter is therefore devoted to the theories and experimental analysis related to their potential as corrosion inhibitors. Factors that influence their corrosion inhibition performances, such as chemical structure, reactivity, solution chemistry, and adsorption properties, are the main focus of the chapter. The understanding of the corrosion inhibition mechanisms, major challenges, and developmental tendencies relating to these compounds is a major concern. Our survey shows that the corrosion inhibition performance of these macro compounds is based on adsorption and film formation through polymeric complexes. The structural nature of the macromolecules promises a higher corrosion inhibition performance than small monomers due to the presence of a large number of heteroatoms. However, the size should be optimized with a good balance of functional groups to retain solubility requirements. Poor solubility is a major challenge for large alkyl groups, which provide a hydrophobic advantage for corrosion mitigation at the optimum length. Overall, large adsorption sites provided by lone pairs of heteroatoms, delocalized electrons density of the benzene rings, and cavities for an easy host–guest relationship with large space for extra functionalization possibilities must make supramolecules a huge research opportunity for corrosion inhibition formulation.

REFERENCES

Abdallah, M. 2008. "Corrosion inhibiting properties of some crown ethers on corrosion of stainless steel (types 304 and 316) in hydrochloric acid." *Zaštita Materijala* 49 (3): 9–22.

Akhondi, M., and Jamalizadeh, E. 2020. "Fabrication of β-cyclodextrin modified halloysite nanocapsules for controlled release of corrosion inhibitors in self-healing epoxy coatings." *Progress in Organic Coatings* 145: 105676.

Altin, A., Krzywiecki, M., Sarfraz, A., Toparli, C., Laska, C., Kerger, P., Zeradjanin, A., Mayrhofer, K. J. J., Rohwerder, M., and Erbe, A. 2018. "Cyclodextrin inhibits zinc corrosion by destabilizing point defect formation in the oxide layer." *Beilstein Journal of Nanotechnology* 9 (1): 936–944.

Altin, A., Vimalanandan, A., Sarfraz, A., Rohwerder, M., and Erbe, A. 2019. "Pretreatment with a β-cyclodextrin-corrosion inhibitor complex stops an initiated corrosion process on zinc." *Langmuir* 35 (1): 70–77.

Amiri, S., and Rahimi, A., 2014. "Preparation of supramolecular corrosion-inhibiting nanocontainers for self-protective hybrid nanocomposite coatings." *Journal of Polymer Research* 21 (10): 566.

Amiri, S., and Rahimi, A. 2016. "Anticorrosion behavior of cyclodextrins/inhibitor nanocapsule-based self-healing coatings." *Journal of Coatings Technology and Research* 13 (6): 1095–1102.

Astruc, D., Boisselier, E., and Ornelas, C. 2010. "Dendrimers designed for functions: from physical, photophysical, and supramolecular properties to applications in sensing, catalysis, molecular electronics, photonics, and nanomedicine." *Chemical Reviews* 110 (4): 1857–1959.

Atav, R. 2018. Dendritic molecules and their use in water repellency treatments of textile materials, In Waterproof and Water Repellent Textiles and Clothing, The Textile Institute Book Series, 2018, PP 191–214. https://doi.org/10.1016/B978-0-08-101212-3.00007-1.

Behpour, M., Ghoreishi, S. M., Soltani, N., Salavati-Niasari, M., Hamadanian, M., and Gandomi, A. 2008. "Electrochemical and theoretical investigation on the corrosion inhibition of mild steel by thiosalicylaldehyde derivatives in hydrochloric acid solution." *Corrosion Science* 50 (8): 2172–81.

Benabdellah, M., Souane, R., Cheriaa, N., Abidi, R., Hammouti, B., and Vicens, J. 2007. "Synthesis of calixarene derivatives and their anticorrosive effect on steel in 1M HCl." *Pigment and Resin Technology* 36 (6): 373–81.

Caswell, L. R., and Savannunt, D. S. 1988. "Temperature and solvent effects on the experimental dipole moments of three crown ethers." *Journal of Heterocyclic Chemistry* 25 (1): 73–79.

Chehardoli, G., and Bahmani, A. 2019. "The role of crown ethers in drug delivery." *Supramolecular Chemistry* 31 (4): 221–38.

Chen, T., and Fu, J. 2012. "An intelligent anticorrosion coating based on PH-responsive supramolecular nanocontainers." *Nanotechnology* 23 (50): 505705.

Chen, Y., and Liu, Y. 2010. "Cyclodextrin-based bioactive supramolecular assemblies." *Chemical Society Reviews* 39 (2): 495–505.

Chukwuike, V. I., Prasannakumar, R. S., Gnanasekar, K., and Barik, R. C. 2021. "Copper corrosion mitigation: a new insight for fabricating a surface barrier film against chloride ion under hydrodynamic flow." *Applied Surface Science* 555: 149703.

Chukwuike, V. I., Sankar, S. S., Kundu, S., and Barik, R. C. 2019. "Capped and uncapped nickel tungstate (NiWO4) nanomaterials: a comparison study for anti-corrosion of copper metal in NaCl solution." *Corrosion Science* 158: 108101.

Dai, X., Kreyenschulte, C., Rabeah, J., Wang, H., Shi, F., Brückner, A., and Adomeit, S. 2019. "Sustainable Co-synthesis of glycolic acid, formamides and formates from 1,3-dihydroxyacetone by a Cu/Al2O3 catalyst with single active sites." *Angewandte Chemie International Edition* 58 (16): 5251–5255.

Dehghani, A., Bahlakeh, G., and Ramezanzadeh, B. 2020. "Beta-cyclodextrin-zinc acetylacetonate (β-CD@ZnA) inclusion complex formation as a sustainable/smart nanocarrier of corrosion inhibitors for a water-based siliconized composite film: integrated experimental analysis and fundamental computational electroni." *Composites Part B: Engineering* 197: 108152.

de Souza, T. M., Cordeiro, R. F. B., Viana, G. M., Aguiar, L. C. S., de Senna, L. F., Malta, L. F. B., and D'Elia, E. 2016. "Inclusion compounds of dibenzylthiourea with hydroxypropylated-cyclodextrins for corrosion protection of carbon steel in acidic medium." *Journal of Molecular Structure* 1125: 331–39.

Donahue, F. M., and Nobe, K. 1965. "Theory of organic corrosion inhibitors: adsorption and linear free energy relationships." *Theory of Organic Corrosion Inhibitors: Adsorption and Linear Free Energy Relationships* 112 (9): 886–91.

Fan, B., Wei, G., Zhang, Z., and Qiao, N. 2014. "Characterization of a supramolecular complex based on octadecylamine and β-cyclodextrin and its corrosion inhibition properties in condensate water." *Corrosion Science* 83: 75–85.

Fouda, A. E. A. S., Etaiw, S. E. D. H., El-Bendary, M. M., and Maher, M. M. 2016. "Metal-organic frameworks based on silver (I) and nitrogen donors as new corrosion inhibitors for copper in HCl solution." *Journal of Molecular Liquids* 213 (3): 228–34.

Fouda, A. S., Abdallah, M., Al-Ashrey, S. M., and Abdel-Fattah, A. A. 2010. "Some crown ethers as inhibitors for corrosion of stainless steel type 430 in aqueous solutions." *Desalination* 250 (2): 538–43.

Gul, S., O'neill, L., Cassidy, J., and Naydenova, I. 2019. "Modified surface relief layer created by holographic lithography: application to selective sodium and potassium sensing." *Sensors* 19 (5): 1026.

Hadisaputra, S., Canaval, L. R., Pranowo, H. D., and Armunanto, R. 2014. "Theoretical study on the extraction of alkaline earth salts by 18-crown-6: roles of counterions, solvent types and extraction temperatures." *Indonesian Journal of Chemistry* 14 (2): 199–208.

Haruna, K., Saleh, T. A., Obot, I. B., and Umoren, S. A. 2019. "Cyclodextrin-based functionalized graphene oxide as an effective corrosion inhibitor for carbon steel in acidic environment." *Progress in Organic Coatings* 128: 157–67.

Hasanov, R., Bilge, S., Bilgiç, S., Gece, G., and Kiliç, Z. 2010. "Experimental and theoretical calculations on corrosion inhibition of steel in 1 M H2SO4 by crown type polyethers." *Corrosion Science* 52 (3): 984–90.

He, Y., Zhang, C., Wu, F., and Xu, Z. 2016. "Fabrication study of a new anticorrosion coating based on supramolecular nanocontainer." *Synthetic Metals* 212: 186–94.

Hou, B., Li, X., Ma, X., Du, C., Zhang, D., Zheng, M., Xu, W., Lu, D., and Ma, F. 2017. "The cost of corrosion in China." *Npj Materials Degradation* 1 (1): 4.

Hu, Q., Tang, G., and Chu, P. K. 2014. "Cyclodextrin-based host-guest supramolecular nanoparticles for delivery: from design to applications." *Accounts of Chemical Research* 47 (7): 2017–25.

Iannuzzi, M., and Frankel, G. S. 2007. "Mechanisms of corrosion inhibition of AA2024-T3 by vanadates." *Corrosion Science* 49 (5): 2371–91.

Ilevbare, G. O., and Burstein, G. T. 2003. "The inhibition of pitting corrosion of stainless steels by chromate and molybdate ions." *Corrosion Science* 45 (7): 1545–69.

Kaddouri, M., Bouklah, M., Rekkab, S., Touzani, R., Al-Deyab, S. S., Hammouti, B., Aouniti, A., and Kabouche, Z. 2012. "Thermodynamic, chemical and electrochemical investigations of calixarene derivatives as corrosion inhibitor for mild steel in hydrochloric acid solution." *International Journal of Electrochemical Science* 7 (9): 9004–23.

Kaddouri, M., Cheriaa, N., Souane, R., Bouklah, M., Aouniti, A., Abidi, R., Hammouti, B., and Vicens, J. 2008. "Novel calixarene derivatives as inhibitors of mild C-38 steel corrosion in 1 M HCl." *Journal of Applied Electrochemistry* 38 (9): 1253–58.

Kaddouri, M., Rekkab, S., Bouklah, M., Hammouti, B., Aouniti, A., and Kabouche, Z. 2013. "Experimental study of inhibition of corrosion of mild steel in 1 M HCl solution by two newly synthesized calixarene derivatives." *Research on Chemical Intermediates* 39 (8): 3649–67.

Keshtiara, P., Hadadzadeh, H., Daryanavard, M., Mousavi, N., and Dinari, M. 2020. "New dendrimers containing ruthenium nanoparticles as catalysts for hydrogenation of citral to 3,7-dimethyloctanol." *Materials Chemistry and Physics* 249: 122962.

Liu, C., Zhao, H., Hou, P., Qian, B., Wang, X., Guo, C., and Wang, L. 2018. "Efficient graphene/cyclodextrin-based nanocontainer: synthesis and host-guest inclusion for self-healing anticorrosion application." Research-article. *ACS Applied Materials and Interfaces* 10 (42): 36229–39.

Liu, Y., Zou, C., Yan, X., Xiao, R., Wang, T., and Li, M. 2015. "B-cyclodextrin modified natural chitosan as a green inhibitor for carbon steel in acid solutions." *Industrial and Engineering Chemistry Research* 54 (21): 5664–5672.

Ma, Y., Fan, B., Zhou, T., Hao, H., Yang, B., and Sun, H. 2019. "Molecular assembly between weak crosslinking cyclodextrin polymer and trans-cinnamaldehyde for corrosion inhibition towards mild steel in 3.5% NaCl solution: experimental and theoretical studies." *Polymers* 11 (4): 635.

Machnikova, E., Whitmire, K. H., and Hackerman, N. 2008. "Corrosion inhibition of carbon steel in hydrochloric acid by furan derivatives." *Electrochimica Acta* 53 (20): 6024–32.

Marczenko, K. M., Mercier, H. P. A., and Schrobilgen, G. J. 2018. "A stable crown ether complex with a noble-gas compound." *Angewandte Chemie - International Edition* 57 (38): 12448–52.

Nikitenko, N. A., and Prassolov, V. S. 2013. "Non-viral delivery and therapeutic application of small interfering RNAs." *Acta Naturae* 5 (18): 35–53.

Nimse, S. B., and Kim, T. 2013. "Biological applications of functionalized calixarenes." *Chemical Society Reviews* 42 (1): 366–86.

Obot, I. B., Obi-Egbedi, N. O., and Umoren, S. A. 2009. "The synergistic inhibitive effect and some quantum chemical parameters of 2,3-diaminonaphthalene and iodide ions on the hydrochloric acid corrosion of aluminium." *Corrosion Science* 51 (2): 276–82.

Oshovsky, G. V., Reinhoudt, D. N., and Verboom, W. 2007. "Supramolecular chemistry in water." *Angewandte Chemie - International Edition* 46 (14): 2366–2393.

Prasannakumar, R. S., Chukwuike, V. I., Bhakyaraj, K., Mohan, S., and Barik, R. C. 2020. "Electrochemical and hydrodynamic flow characterization of corrosion protection persistence of nickel/multiwalled carbon nanotubes composite coating." *Applied Surface Science* 507: 145073.

Qu, Q., Li, L., Bai, W., Jiang, S., and Ding, Z. 2009. "Sodium tungstate as a corrosion inhibitor of cold rolled steel in peracetic acid solution." *Corrosion Science* 51 (10): 2423–28.

Quartarone, G., Ronchin, L., Vavasori, A., Tortato, C., and Bonaldo, L. 2012. "Inhibitive action of gramine towards corrosion of mild steel in deaerated 1.0 M hydrochloric acid solutions." *Corrosion Science* 64: 82–89.

Ren, X., Xu, S., Gu, X., Tan, B., Hao, J., Feng, L., Ren, W., Gao, F., Zhang, S., Xiao, Y., and Huang L. 2021. "Hyperbranched molecules having multiple functional groups as effective corrosion inhibitors for Al alloys in aqueous NaCl." *Journal of Colloid and Interface Science* 585: 614–626.

Saji, V. S. 2019. "Supramolecular concepts and approaches in corrosion and biofouling prevention." *Corrosion Reviews* 37 (3): 187–230.

Simaan, S., and Biali, S. E. 2003. "Synthesis of P-tert-butylcalix[4]arene derivatives with trans-alkyl substituents on opposite methylene bridges." *Journal of Organic Chemistry* 68 (9): 3634–39.

Solomon, M. M., Gerengi H., Kaya, T., and Umoren S. A. 2017. "Performance evaluation of a chitosan/silver nanoparticles composite on St37 steel corrosion in a 15% HCl solution." *ACS Sustainable Chemistry and Engineering* 5 (1): 809–20.

Son, A., Musiani, M., Orazem, M. E., Pébère, N., Tribollet, B., and Vivier, V. 2016. "Impedance study of the influence of chromates on the properties of waterborne coatings deposited on 2024 aluminium alloy." *Evaluation and Program Planning* 109: 174–81.

Souane, R., Kaddouri, M., Bouklah, M., Cheriaa, N., Hammouti, B., and Vicens, J. 2009. "Investigation of adsorption and inhibitive effect of calixarene derivative newly synthesized towards C38 steel in molar HCl." *Surface Review and Letters* 16 (3): 401–406.

Steed, J. W. 2001. "First- and second-sphere coordination chemistry of alkali metal crown ether complexes." *Coordination Chemistry Reviews* 215 (1): 171–221.

Tan, S. Y., Ang, C. Y., and Zhao, Y. 2017. Smart therapeutics achieved via host-guest assemblies. In *Comprehensive Supramolecular Chemistry II*. Second Edition. Vol. 5. pp. 391–420.

Tang, W., Zou, C., Da, C., Cao, Y., and Peng, H. 2020. "A review on the recent development of cyclodextrin-based materials used in oilfield applications." *Carbohydrate Polymers* 240 (8): 116321.

Tian, H., Cheng, Y. F., Li, W., and Hou, B. 2015. "Triazolyl-acylhydrazone derivatives as novel inhibitors for copper corrosion in chloride solutions." *Corrosion Science* 100: 341–352.

Verma, C., Ebenso, E. E., Vishal, Y., and Quraishi, M. A. 2016. "Dendrimers: a new class of corrosion inhibitors for mild steel in 1 M HCl: experimental and quantum chemical studies." *Journal of Molecular Liquids* 224: 1282–93.

Wang, Z., Gong, Y., Zhang, L., Jing, C., Gao, F., Zhang, S., and Li, H. 2018. "Self-assembly of new dendrimers basing on strong π-π intermolecular interaction for application to protect copper." *Chemical Engineering Journal* 342: 238–50.

Wipff, G., Weiner, P., and Kollman, P. 1982. "A molecular mechanics study of 18-crown-6 and its alkali complexes: an analysis of structural flexibility, ligand specificity, and the macrocyclic effect." *Journal of the American Chemical Society* 104 (12): 3249–58.

Yang, F., Liu, Y., Liu, T., Liu, S., and Zhao, H. 2019. "Aniline trimer-including carboxymethylated β-cyclodextrin as an efficient corrosion inhibitor for Q235 carbon steel in 1 M HCl solution." *RSC Advances* 9 (52): 30249–58.

Zhang, B., He, C., Chen, X., Tian, Z., and Li, F. 2015. "The synergistic effect of polyamidoamine dendrimers and sodium silicate on the corrosion of carbon steel in soft water." *Corrosion Science* 90: 585–96.

Zou, C., Liu, Y., Yan, X., Qin, Y., Wang, M., and Zhou, L. 2014a. "Synthesis of bridged β-cyclodextrin-polyethylene glycol and evaluation of its inhibition performance in oilfield wastewater." *Materials Chemistry and Physics* 147 (3): 521–27.

Zou, C., Yan, X., Qin, Y., Wang, M., and Liu, Y. 2014b. "Inhibiting evaluation of β-cyclodextrin-modified acrylamide polymer on alloy steel in sulfuric solution." *Corrosion Science* 85: 445–454.

Zou, C. J., Tang, Q. W., Zhao, P. W., Guan, E. D., Wu, X., and Ye, H. 2013. "Further study on the inclusion complex of 2-phosphonobutane-1,2,4-tricarboxylic acid with B-cyclodextrin: a new insight of high inhibition efficiency for protecting steel corrosion." *Journal of Petroleum Science and Engineering* 103: 29–35.

17 Coordination Polymers as Corrosion Inhibitors

Bhawna Chugh, Sheetal, and Sanjeeve Thakur
Netaji Subhas University of Technology

Balaram Pani
Bhaskaracharya College of Applied Science

Ashish Kumar Singh
Bharati Vidyapeeth's College of Engineering

CONTENTS

17.1 INTRODUCTION

It is well known that the destruction of metallic bodies is caused by a process named corrosion. It is a severe problem and is defined as a natural mechanism that transforms polished metals into more stable forms like oxides, and hydroxides by chemical or electrochemical reactions. The losses due to corrosion are so high that they have a significant impact on the world's economy. It is assumed that the cost of corrosion in industrialized nations is 3%–4% of their GDP. Economically, the losses due to corrosion in India are approximately stated as 25,000 crores per year (Chugh et al. 2020a).

Therefore, from an economic point of view, it is necessary to come up with ideas and concepts to prevent corrosion. There are several modes through which we can prevent corrosion like electrochemical methods, organic and inorganic coatings, usage of corrosion inhibitors, and so on. Corrosion inhibitors are chemicals incorporated into smaller quantities in the environment in which metal corrodes. They reduce the rate of corrosion or prevent corrosion. Among various types of inhibitors, coordination polymers constitute an important class. These are compounds in which a metal cation is surrounded by ligands. Utilizing the concepts of host and guest, these compounds are formed, and the interactions between them are mainly non-covalent, i.e., hydrogen bonding, electrostatic, closed-shell interactions, and so on.

Supramolecular chemistry has been a very interesting area of science that received the Nobel Prize in 1987. Significant research has been carried out on supramolecular compounds for numerous applications, such as sensors, biomedical, and stimuli frameworks (Atwood et al. 2017; Atwood and Steed 2004). Supramolecular compounds such as organic polymers, coordination polymers, metal–organic frameworks (MOFs), and crown ether types of moieties have been used significantly in the field of corrosion inhibition, as represented in Figure 17.1.

Coordination polymers are a kind of hybrid inorganic or organic structure formed mainly by transition metals that are linked either one, two, or three-dimensionally with the ligands. In particular, the metal cations involved are transition metals, as filled d orbitals can hybridize differently in comparison to lanthanoids and alkali metals (Nworie 2018). Owing to their wide range of properties like porosity, thermal stability, and solvent resistance, they are used in many areas as efficient applicants.

According to the research, some of the polymers may be employed as corrosion inhibitors due to their ligand functional groups and show complexation with the metal ions, occupying a large surface area and protecting the metal (Arthur et al. 2013). This chapter deals with the studies of coordination polymers having different metal cations, primarily focussing on Schiff's bases as ligands and their corrosion inhibitive characteristics. The complexing of metal ions with polydentate Schiff base ligands is considered a well-researched field of coordination chemistry science these days. It also describes other coordination complexes and MOFs as corrosion inhibitors.

17.2 SCHIFF'S BASE POLYMER COMPLEXES

It has been experimentally proven that there are several heterocyclic molecules possessing heteroatoms such as N, O, S, and so on, which exhibit excellent anti-corrosion performance (Yang et al. 2008). Such types of organic compounds are strongly adsorbed on the surface of mild steel (MS) or any other alloy due to the tendency of the bonding of lone pairs or π electrons with the metallic surface, thereby reducing corrosion (Sherif and Almajid 2010).

Schiff's bases are a class of organic compounds (–N=CH) that are formed from the reaction of carbonyl compounds with amines and have acquired much attention from the researchers in the field of corrosion inhibition. It has been observed that the ability to resist corrosion by inhibitors is significant if it has both N and S heteroatoms (Hamani et al. 2014). It has been stated that the protection mechanism of the inhibitor complexes widely depends on their electronic structures,

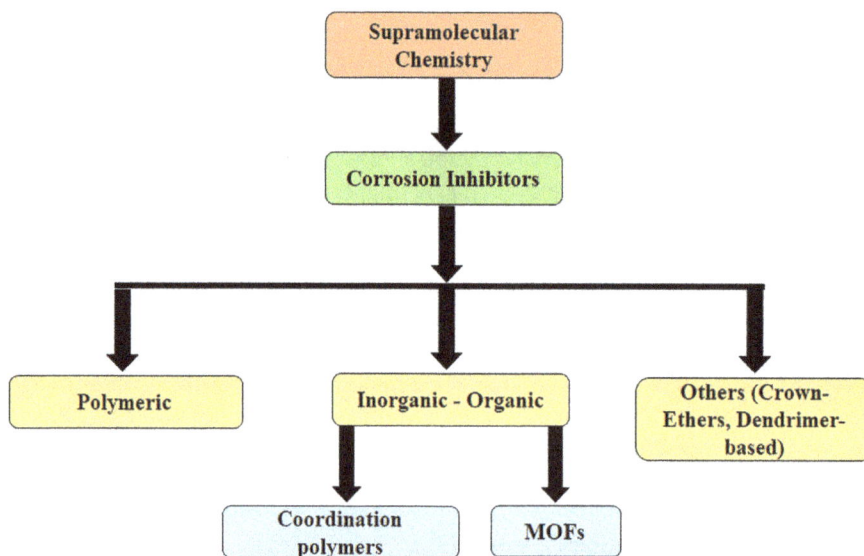

FIGURE 17.1 A schematic representation showing the application of supramolecular compounds in corrosion inhibition.

the electron density of donor atoms, aromaticity, and the existence of various functional groups like −C=NH, −N=N−, −CHO, R−OH, C=C, and so on. (Singh and Quraishi 2010). Schiff's bases monomers and their respective polymer complexes are potent molecules and exhibit a wide range of properties like anti-inflammatory, antipyretic, and anti-fungal. (Prakash and Adhikari 2011). Literature reveals that Schiff base monomers and polymers have been widely examined as corrosion inhibitors for acid medium (Saha et al. 2015; Chugh et al. 2020b; Achary et al. 2007; Safak et al. 2012; Hosseini et al. 2007). Such ligands have the ability to bond with numerous transition metal ions, resulting in stable and coloured metal complexes. The coordination complexes of metal ions with these Schiff's bases as ligands are elaborated, and further, their potential to prevent corrosion has been of great interest to researchers.

17.3   SYNTHESIS OF COORDINATING POLYMERS

The coordinating polymers are prepared by self-assembly. Self-assembly is the association of small molecules to produce a structurally well-defined aggregate. The principle by which self-assembly works is based on the concept of molecular recognition, which demands steric and electronic complementarity among the molecules forming the self-assembly. Thus, the molecular recognition-directed association of complementary molecules leads to self-assembly, in which the complementary components cannot be described as guests or hosts. The formation of such big aggregates is entropically disfavoured. However, the collective non-covalent interactions within the self-assembly produce an enthalpy favour that can overcome this entropic disfavour (Steed and Atwood 2009). Depending on the procedure, the self-assembly approaches have been clustered into two classes, given as follows:

   i. **Strict self-assembly**: It's a reversible mechanism that doesn't require any covalent alteration.
   ii. **Self-assembly with covalent modification**: In this procedure, during the occurrence of overall self-assembly of the starting molecules, at a certain step, they need covalent modification. This makes the overall process an irreversible one.

The synthesis of coordination polymers has received immense interest in coordination and crystal engineering by researchers because of their potential applications. El-Bindary et al. presented a review focussing on the coordination chemistry of supramolecular Schiff base polymer complexes, discussing their synthesis, structure, and characterization (El-bindary et al., 2016). El-Sonbati and coworkers presented many research papers on polychelates, acryloyl chloride, azo polymeric, and polymeric complexes of many aliphatic and aromatic amines (El-sonbati et al. 2011a, b; Diab and Mohamed 2010; Diab and Attallah 2012).

Poly(5-vinyl salicylidene anthranilic acid) (PHL5) homopolymer (1) and polymer complexes of 5-vinyl salicylidene anthranilic acid (HL5) with few transition metal salts (2–5) have been synthesized as shown in Figure 17.2 (Diab et al. 1990).

The reaction of the homopolymer with copper (II) salts such as Cl⁻ and CH_3COO^-, cobalt (II), nickel (II), cadmium, palladium (II) and dioxouranium(VI) salts resulted in the synthesis of mononuclear and binuclear complexes of poly (5-vinylsalicylidene2-aminopyridine) (PHL6) (6). Metal (II) (Cu, Co, Ni, Cd, and UO_2) acetates and palladium chloride have been identified to form mononuclear complexes (8–13) (Figure 17.3), whereas cupric chloride resulted in a binuclear complex (7). In determining the stereochemistry and the structure of the polymer complexes, the molar ratios of the reactants, the framework pH, and the anion's nature play an essential role. The homopolymer in all the polymer complexes is chelated to the metal ion through the azomethine atomic nitrogen and the phenolic oxygen atom (El-Sonbati and El-Bindary 1994). El-Sonbati et al. prepared and confirmed poly-(cinnamaldehyde-2-anthranilic acid) homopolymer and polymer complexes of cinnamaldehyde-2-anthranilic acid with Cu(II), Ni(II), Co(II), Zn(II), Cd(II) and Hg(II). The homopolymer was examined as a corrosion inhibitor for aluminium in HCl media (El-sonbati et al. 1993).

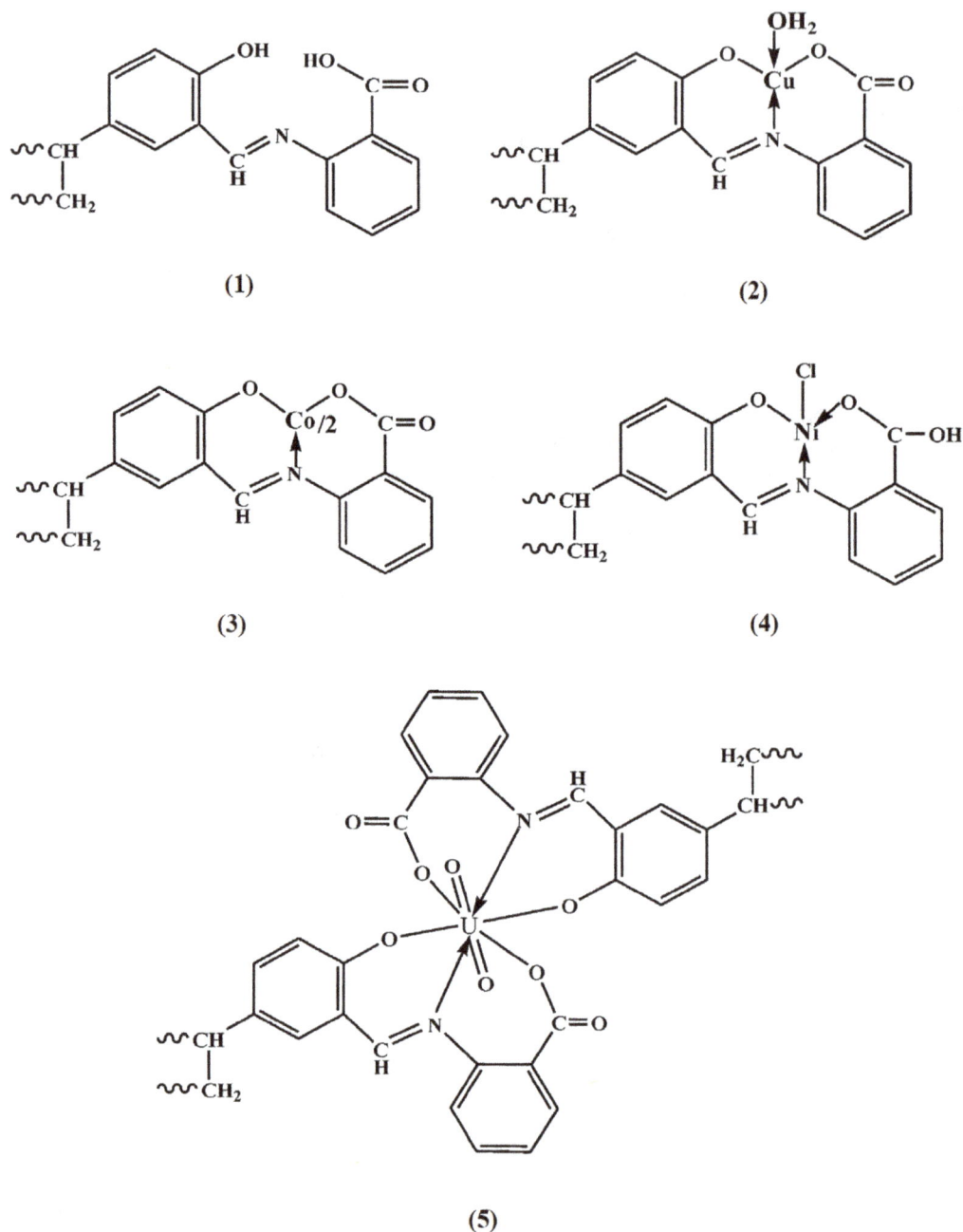

FIGURE 17.2 Chemical structure of coordination polymer complexes (1–5). (Reproduced with permission from El-Bindary et al. (2016) © 2015 Elsevier B.V.)

17.4 COORDINATION POLYMERS WITH CADMIUM AS METAL CENTRE

17.4.1 COORDINATION POLYMER OF CADMIUM WITH SCHIFF'S BASE LIGAND

Cadmium is a toxic and non-essential element for biological systems (Genchi et al. 2020). Still, besides its limitations, it can serve as an efficient corrosion inhibitor for different metals, such as nickel, steel, and so on. It gets complexed with the ligand (Schiff's base) and forms a coordination polymer.

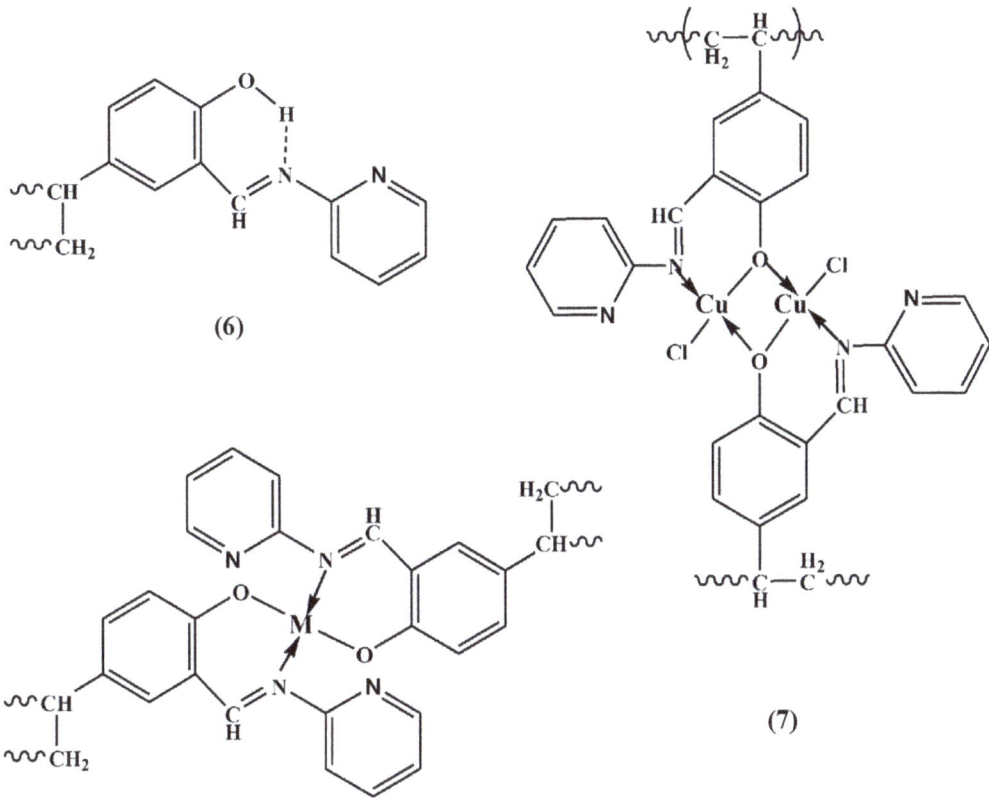

FIGURE 17.3 Chemical structures of coordination polymer complexes (6–13). (Reproduced with permission from El-Bindary et al. (2016) © 2015 Elsevier B.V.)

Schiff bases are selected as ligands as they form very stable bonds with the metal cations. Das et al. studied three ligands namely *L1* (N, N-dimethyl-N'(I-pyridin-2-yl-ethylidene)-ethane-1,2diamine], *L2* [2 morpholino-N-(1-pyridin-2-yl)ethylidene)ethanamine), and *L3* [2-(piperidin-1-yl)-N(1-pyridin-2-yl)ethylidene)ethanamine)] which formed five Schiff's base coordination complexes with Cd(II) namely [Cd(L1)$_2$](ClO$_4$)$_2$], [Cd(L1) (Cyanoacetate)(OAc)], [Cd$_2$(L1)$_2$(N$_3$)$_4$], [Cd(L$_2$)(N$_3$)$_2$]$_n$, and [Cd$_2$(L$_3$)$_2$(N$_3$)$_4$]$_n$ and the corrosion inhibitor performance of prepared complexes were studied for MS in 15% HCl solution (Figure 17.4). Additionally, Figure 17.4 provides representative coordinating configurations of the polymer unit as well as the supramolecular framework of complexes 3 and 4 (Das et al. 2017).

Complex 1 with the ligand *L1* didn't show significant corrosion prevention, and the later introduction of a co-ligand that is cyanoacetate also provided no considerable results. But further on, the introduction of azides with several nitrogen atoms as donor atoms interestingly lead to good results. Later complexes 4 and 5 with ligands *L2* and *L3* also provided efficient results, and the probable reason for this might be the greater availability of nitrogen as adsorbing sites in azide molecules. In addition, a positive deviation in corrosion potential was seen due to the addition of Cd (II) complexes. Weight loss measurement studies illustrated that complex 4 showed maximum deposition on the surface of the metal and hence prevented corrosion at the maximum rate. SEM micrographs of MS showed that in the presence of inhibitor molecules, the uniformity and the smooth texture of

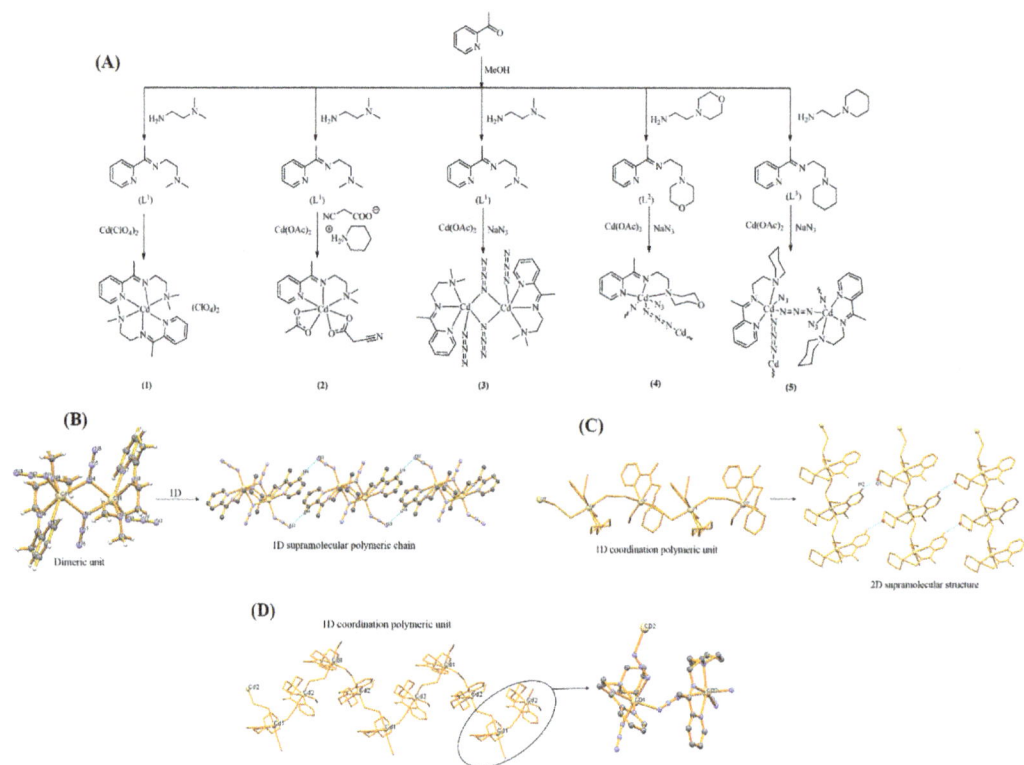

FIGURE 17.4 (a) Synthetic scheme of Cd (II) complexes (1–5). (b) Dimeric unit arrangement and supramolecular network structure of complex 3. (c) Coordination polymeric structure and the 2D supramolecular network of complex 4. (d) 1D coordination polymeric structure of complex 5. (Reproduced with permission from Das et al. (2017) Under Creative Commons Licence © The Royal Society of Chemistry.)

the metal were improved. The inhibitor molecules, when they get adsorbed over the surface of the metal, various physicochemical phenomena take place, such as electrostatic interactions between different functional groups of inhibitors and metal. The inhibitor species interact with the metallic surface through an electrostatic mode of interaction, coordinate bonding, covalent bonding, and so on. Co-ligands like azides with various donor sites and lone pairs adsorb with higher efficiency and thus result in negative values of Gibbs free energy. Experimental values of Gibbs free energy for complexes 3, 4, and 5 (−27.96, −28.35, and −28.20 kJ/mol) didn't differ much because all three possess azides as co-ligands. From these negative values, we can presume that the values of Gibb's free energy of adsorption will be more negative with an increase in the availability of heteroatoms. The presence of these heteroatoms facilitates their adsorption with the metal through electrostatic- and coordination-type bonding. Thus, conclusively, we can define that both physisorption and chemisorption take place in such systems (Das et al. 2017).

17.4.2 COORDINATION POLYMER OF CADMIUM WITH NICOTINATE LIGAND

The selection of ligands and the metal ions play a vital role in constructing a coordination polymer with a potential range of applications (Wang et al. 2004). Furthermore, if the ligands selected are confined to those molecules with nitrogen and oxygen as donor atoms, they would construct interesting structures (Tao et al. 2000; Konar et al. 2003).

This coordination polymer $[CdN_3(en)(n)]$, where en stands for ethylnicotinate and n stands for nictotinate, having cadmium as a metal cation, has been prepared by reacting $CdSO_4.5H_2O$ with ethyl nicotinate (en) in combination with sodium azide. This polymer is synthesized in the presence of water or CH_3CN as a medium. The sole purpose behind the addition of sodium azide is to bridge the metal centre from either end to end or end on. From a structural point of view, one molecule of this polymer constitutes the asymmetric unit of the structure, consisting of one portion having one cadmium atom and two crystallographically different units of ethylnicotinate and nictotinate ligands, and one azide moiety (Fouda et al. 2018).

The molecules having heteroatoms like oxygen and nitrogen as their donor atoms have a certain tendency to prevent corrosion due to their capability to provide electrons very easily. Potentiodynamic polarization (PDP) and impedance spectroscopy (EIS) methods are employed to test the corrosion inhibition efficiency. We know that a corrosion inhibitor controls the process of corrosion either by retarding anodic reaction or by cathodic dissolution, or it could be a mixed-type inhibitor. Here, polarization studies reveal that this coordination polymer works by slowing the rate of both anodic as well as cathodic reactions (El Rehim et al. 2001; Bockris and Yang 1991). Tafel plots showed that there are parallel anodic as well as cathodic lines, which indicate that the coordination polymer is first adsorbed over the surface of the metal merely by restricting the active reaction sites of the metal and thus preventing corrosion without decreasing the anodic and cathodic reaction processes (Asan et al. 2005).

Furthermore, it is a fact that the adsorption of inhibitors over the metallic surface is the most crucial step in this process of corrosion prevention (Chaieb et al. 2005). The mechanism involved suggests that the organic molecules that adsorb over the surface replace the water molecules from the metallic surface. Adsorption here involves the donation of electrons from the heteroatom like nitrogen of the inhibitor to the copper surface, and the adsorption can also happen in other ways by the interaction of π electrons between the three-dimensional structure of the polymer and the copper metal surface.

Negative values of Gibbs free energy, i.e. (–23.60 kJ/mol), indicate a strong association of the metal with the polymer. A rise in temperature results in a decline in the inhibition efficiency of the polymer as there is an increase in the value of activation energy, which provides a demarcation of an increase in columbic and electrostatic interactions. Negative and large values of entropy state that this process involves association rather than dissociation.

The mechanism of inhibition mainly includes the adsorption of coordination polymers over the metal surface, where the nature of the polymer and its functional groups play a vital role. This polymer is seen to chemisorb over the metallic surface with the chemical bond between the positively charged copper surface and positively charged nitrogen atoms (Zhang et al. 2004).

17.5   COORDINATION POLYMER OF COPPER AS METAL CENTRE

Copper is one of those transition metals which has a number of potential applications in various fields such as electronics, production of sheets, wires, and so on, and also in the preparation of alloys. This metal is highly susceptible to corrosion if placed in an acidic medium or alkaline solution. $[Cu(C_6H_4N_3)]_n$ is the metal coordination polymer that has been prepared by the chemical reaction between cupric nitrate, 1H-benzotriazole (BTAH), and aqueous ammonia in a hydrothermal environment. The structural unit of this complex comprises three copper atoms in Cu(I) oxidation state and two H-benzotriazole ligands. Two copper atoms, one having a two-coordination number and the other having tetrahedral geometry, and the third Cu(I) is having a generalized planar geometry (coordination number 3). The two copper cations are bridged with the help of two BTAH ligands to produce a subunit $[Cu_2(BTA)_2]$, and these two subunits are attached to the same ligands in an anti-parallel way to generate a secondary building unit. Further, these subunits are linked with the help of copper atoms and Cu-N bonding and form a one-dimensional chain, and these one-dimensional chains are linked through van der Walls interactions and generate a supramolecular network. The corrosion inhibition mechanism of this polymer involves the interaction of the heteroatoms such as nitrogen with the metallic surface by donation of the electrons, thereby decreasing the corrosion rate (Liu et al. 2014).

17.6 CERIUM (III)–MELAMINE COORDINATION POLYMER AS CORROSION INHIBITOR

Cerium, being a lanthanoid, is considered a potential corrosion inhibitor because it tends to get precipitated over the cathodic site as insoluble oxides and hydroxides, thereby blocking the cathodic mechanism (Hu et al. 2019). Melamine (2,4,6-triamino-s-triazine), however, is recognized to be a poor ligand as it has a network due to hydrogen bonding, and because of this, it has quite low solubility in common solvents. This ligand has nitrogen as a donor atom, so ligating it with the cerium atom produced a polymer CMCP that was further tested to prevent corrosion (Zorainy et al., 2021). Here, the corrosion protection of this polymer was tested for AA2024 Al alloy in an aerated 3.5% NaCl medium. PDP and EIS studies were employed to test and compare the potential of CMCP to prevent corrosion with that of blank, ligand, and the metal cerium. The AA2024 specimen was taken and dipped for 1 h in the different solutions containing blank, polymer, and the metal ion at a temperature of 25°C. Anodic current density elevated so sharply, indicating a pitting type of corrosion. Nyquist plots further confirmed the efficiency of the inhibitor, which could be seen from the enlargement of the diameter of the inhibitor compared to blank. In addition, the results with cerium and polymer showed superior inhibition tendency (Figure 17.5). SEM micrographs of the specimen after immersion for 72 h in 3.5% NaCl solution containing 10 ppm of CMCP revealed no sign of corrosion (pitting). It has been proven that the coordination

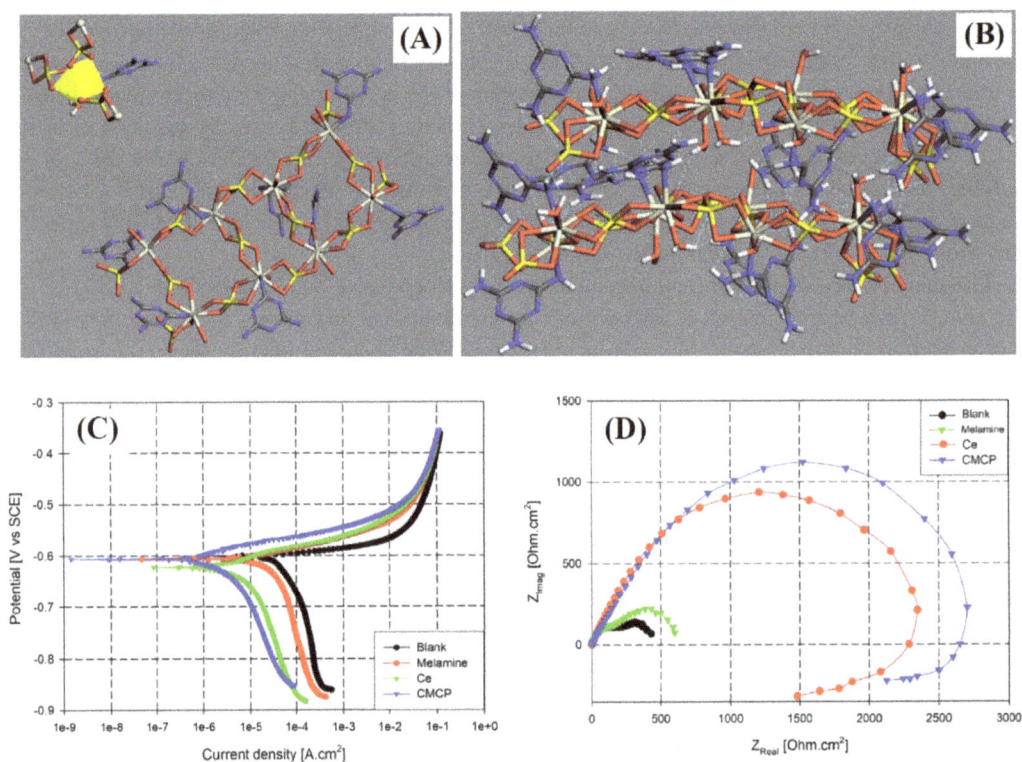

FIGURE 17.5 (a) Optimized chemical structure of Ce (III) coordination sphere (top left) and double-strand chain (centre). (b) Optimized geometry and hydrogen bonding representation of two adjacent double-strand chains. (c) Tafel curves of AA2024 in 3.5% NaCl with various inhibitors. D) Nyquist curves of AA2024 in 3.5% NaCl solution with varied inhibitors (Ce=Ce(III) sulfate, CMCP=cerium(III)–melamine coordination polymer). (Reproduced with permission from Zorainy et al. (2021) Under Creative Commons Licence © The Royal Society of Chemistry.)

occurred through the -NH$_2$ group instead of the N-triazine. The melanin, being a ligand here, plays several roles in the prevention of corrosion, such as:

- It is adsorbed over the metal surface with the help of nitrogen atoms through lone pairs.
- It improves the interaction of the polymer and acts as a structure director.
- It helps in reducing cerium from +4 to +3 oxidation state, which in turn oxidizes it.
- And more specifically, it helps to regenerate cerium during the process of immersion, and this ultimately leads to an increase in corrosion inhibition by CMCP (Zorainy et al., 2021).

17.7   METAL–ORGANIC FRAMEWORKS

MOFs are considered to be a type of coordination polymers, although they are quite different from the usual coordination polymers. Elaborating it further, one- and two-dimensional extended structures are best defined as coordination polymers, while three-dimensional structures can be classified as MOFs (Steed and Atwood 2009). The MOFs have excellent properties in comparison to normal coordination polymers as they possess high thermal stability, permanent porosity, and structural robustness (Silva et al. 2015). MOFs are observed to accommodate various metal ions, including alkali and lanthanoids (Mueller et al. 2006). Several ligands bridge between these metal ions and have heteroatoms like oxygen and nitrogen as their donor atoms mainly. The main ligands include pyridyl and cyano groups, carboxylates and crown ethers, and so on. The dimensionality of the final framework is overall linked to the coordination number, and the electronic structure, coordination modes, and the hardness of the metal cation greatly affect the topology of the final metal framework (Cheetham et al. 2006).

MOFs are stated to have a range of applicability, like catalytic activity, magnetism, luminescence, and corrosion inhibition of metal surfaces (Morozan and Jaouen 2012). Some examples where MOFs are used as corrosion inhibitors have been discussed in further sections. In one of the studies, MOF having Cadmium as the metal centre and thiocyanate as the ligand was prepared, and its potential to prevent corrosion for copper in 1 M HCl was tested (Fouda et al. 2010). The MOF 1 has been prepared by the self-assembly of Cd^{2+} cation, 6-methyl quinoline, and KSCN in H$_2$O/CH$_3$CN mixture at ambient conditions (Etaiw and El-bendary 2017). The MOF 1's structure is made up of cyclic (CdSCN)$_n$ building blocks with chair conformation that forms a 1D chain of coordinated 6-mquin on both sides. The 3D structure having cavities of the MOF 1 has been achieved through strong hydrogen bonds and face to face π⋯π associations.

Looking at the adsorption isotherms, it is clear that the adsorption of inhibitors over the surface of the metal is one of the most crucial steps (Abdul Azim et al. 1974). In addition, from the values obtained in the case of Gibbs free energy, i.e., −31.332 kJ/mol, it can be concluded that there is a sort of strong bonding between the metal and the adsorption (Li and Mu 2005). From the literature, it has been assumed that if the values of Gibb's free energy are in the range of 20 kJ/mol then, it is considered to be physisorption, but if it comes in the range of 40 kJ/mol, then it is said to be chemisorption (Aljourani et al. 2009). Here the value is in between both ranges; thus the mode is combined physisorption and chemisorption. More details of MOF-based corrosion inhibitors are described in Chapter 18.

17.8   CONCLUSIONS AND PERSPECTIVES

This chapter provides a comprehensive outlook on coordination polymers and their potential applications in the corrosion field. The different coordination complexes having potential ligands and metal atoms have been illustrated, showing great potential to inhibit corrosion. Various techniques like the weight loss method, electrochemical impedance spectroscopy, pH change, and so on, and their mechanistic view of inhibiting corrosion have been listed. Polymers having hetero atoms as donor atoms in their ligands interact with the metal ions through their heteroatoms, particularly the unshared pair of electrons. In this way, they transfer the electrons to the vacant d-orbitals of the metal iron. Such a mechanism further slows down the process of corrosion. In addition, there are a few other possible corrosion inhibition mechanisms by metal complexes involving their bonding with metal or alloy with the respective supramolecular interactions (electrostatic, hydrogen bonding, etc.).

Hence, it is concluded that the coordination polymers of different complexes of metals with ligands, particularly those having the heteroatoms and the atoms with unshared pairs of electrons, interact with the metals, and this way helps to reduce the corrosion and therefore act as efficient corrosion inhibitors.

Further, looking at the future perspective of this research, it can be analysed that there is a bit less literature available regarding this topic compared to others, so this chapter can promote researchers to work on this topic owing to the wide range of applications of these polymers as well as supra-molecular interactions. The literature review of this topic provides ample evidence and approaches which assure us that in the near future, if work is done on coordination polymers as corrosion inhibitors, then undoubtedly relatable and much efficient outcomes will be obtained. Therefore, researchers in this field should head in this direction and synthesize coordination polymers with different metal ions as well as different ligands and must check their corrosion prevention ability. Further, these could be tested for different metals and alloys.

REFERENCES

Abdul Azim, A. A., Shalaby, L. A., and Abbas, H. 1974. Mechanism of the corrosion inhibition of Zn anode in NaOH by gelatine and some inorganic anions. *Corros. Sci.* 14: 21–24.

Achary, G., Sachin, H. P., Shivakumara, S., Naik, Y. A., and Venkatesha, T. V. 2007. Surface treatment of zinc by Schiff's bases and its corrosion study. *Russ. J. Electrochem.* 43: 844–849.

Aljourani, J., Raeissi, K., and Golozar, M. A. 2009. Benzimidazole and its derivatives as corrosion inhibitors for mild steel in 1 M HCl solution. *Corros. Sci.* 51: 1836–1843.

Arthur, D. E., Jonathan, A., Ameh, P. O., and Anya, C. 2013. A review on the assessment of polymeric materials used as corrosion inhibitor of metals and alloys. *Int. J. Ind. Chem.* 4: 1–9. https://doi.org/10.1186/2228-5547-4-2.

Asan, A., Kabasakaloglu, M., Işıklan, M., and Kılıç, Z. 2005. Corrosion inhibition of brass in presence of terdentate ligands in chloride solution. *Corros. Sci.* 47: 1534–1544.

Atwood, J., Gokel, G., and Barbour, L. 2017. *Comprehensive Supramolecular Chemistry II.* 2nd ed., Amsterdam: Elsevier.

Atwood, J., and Steed, J. 2004. *Encyclopedia of Supramolecular Chemistry.* Florida: CRC Press.

Bockris, J. O. M., and Yang, B. 1991. The mechanism of corrosion inhibition of iron in acid solution by acetylenic alcohols. *J. Electrochem. Soc.* 138: 2237.

Chaieb, E., Bouyanzer, A., Hammouti, B., and Benkaddour, M. 2005. Inhibition of the corrosion of steel in 1 M HCl by eugenol derivatives. *Appl. Surf. Sci.* 246: 199–206.

Cheetham, A. K., Rao, C. N. R., and Feller, R. K. 2006. Structural diversity and chemical trends in hybrid inorganic–organic framework materials. *Chem. Commun.* 4780–4795.

Chugh, B., Singh, A. K., Chaouiki, A., Salghi, R., Thakur, S., and Pani, B. 2020a. A comprehensive study about anti-corrosion behaviour of pyrazine carbohydrazide: gravimetric, electrochemical, surface and theoretical study. *J. Mol. Liq.* 299: 112160.

Chugh, B., Singh, A. K., Poddar, D., Thakur, S., and Pani, B. 2020b. Relation of degree of substitution and metal protecting ability of cinnamaldehyde modi Fi Ed chitosan. *Carbohydr. Polym.* 234: 115945.

Das, M., Biswas, A., Kundu, B. K., Mobin, S. M., Udayabhanu, G., and Mukhopadhyay, S. 2017. Targeted synthesis of cadmium(II) schiff base complexes towards corrosion inhibition on mild steel. *RSC Adv.* 7: 48569–48585.

Diab, M. A., El-Sonbati, A. Z., and Hilali, A. S. 1990. Polymer complexes : part XIV thermal stability of poly (5-vinyl salicylidene anthranilic acid) homopolymer and polymer complexes of 5-vinyl salicylidene anthranilic acid with some transition metal salts. *Polym. Degrad. Stabil.* 29: 165–173.

Diab, M. A., and Mohamed, R. H. 2010. Polymer complexes. LIII. Supramolecular coordination modes and structural of novel sulphadrug complexes. *Spectrochim. Acta A* 77: 795–801.

Diab, M. A., and Attallah, M. E. 2012. Polymer complexes. LV. Spectroscopic, thermal studies, and coordination triazolo)] acrylamide polymer complexes. *J. Coord. Chem.* 65: 539–549.

El-sonbati, A. Z., El-bindary, A. A., and Diab, M. A. 1993. Polymer complexes : XXIV. Physico-chemical studies on coordination and stability in relation to IR data for poly (cinnamaidehyde- anthranilic acid) complexes of d-block elements. *Polym. Degrad. Stabil.* 42: 1–11.

El-sonbati, A. Z., and Ei-bindary, A. A. 1994. Polymer complexes : 25. Complexing ability of poly (5-vinylsalicylidene-2- aminopyridine) towards different metal (n) salts. *Polymer* 35: 647–652.

El-sonbati, A. Z., Belal, A. A. M., Diab, M. A., and Balboula, M. Z. 2011a. Polymer complexes. LVIII. Structures of supramolecular assemblies of vanadium with chelating groups. *Spectrochim. Acta A* 78: 1119–1125.

El-sonbati, A. Z., Belal, A. A. M., Diab, M. A., and Mohamed, R. H. 2011b. Polymer complexes. LIV. Structural and spectral studies of supramolecular coordination polymers built from Ni (II), Fe (II) and Pd (II) with sulphadrug. *J. Mol. Struct.* 990: 26–31.

El-bindary, A. A., El-sonbati, A. Z., Diab, M. A., Ghoneim, M. M., and Serag, L. S. 2016. Polymeric complexes - LXII. Coordination chemistry of supramolecular schiff base polymer complexes - a review. *J. Mol. Liq.* 216: 318–329.

Etaiw, S. E. H., and El-bendary, M. M. 2017. A new metal-organic framework based on cadmium thiocyanate and 6-methylequinoline as corrosion inhibitor for copper in 1 M HCl solution. *Prot. Met. Phys. Chem. Surf.* 53: 937–949.

Fouda, A. S., Abdallah, M., Al-ashrey, S. M., Abdel-fattah, A.A. 2010. Some crown ethers as inhibitors for corrosion of stainless steel type 430 in aqueous solutions. *Desalination* 250: 538–543.

Fouda, A. S., El-bendary, M. M., Etaiw, S. E. H., and Maher, M. M. 2018. Structure, characterizations and corrosion inhibition of new coordination polymer based on cadmium azide and nicotinate ligand. *Prot. Met. Phys. Chem. Surf.* 54: 689–699.

Genchi, G., Sinicropi, M.S., Lauria, G., Carocci, A., and Catalano, A. 2020. The effects of cadmium toxicity. *Int. J. Environ. Res. Public Health* 17(11): 1–24.

Hamani, H., Douadi, T., Al-noaimi, M., Issaadi, S., Daoud, D., and Chafaa, S. 2014. Electrochemical and quantum chemical studies of some azomethine compounds as corrosion inhibitors for mild steel in 1 M hydrochloric acid. *Corros. Sci.* 88: 234–245.

Hosseini, M. G., Ehteshamzadeh, M., and Shahrabi, T. 2007. Protection of mild steel corrosion with schiff bases in 0.5 M H_2SO_4 solution. *Electrochim. Acta* 52: 3680–3685.

Hu, T., Shi, H., Hou, D., Wei, T., Fan, S., Liu, F., and Han, E. 2019. A localized approach to study corrosion inhibition of intermetallic phases of AA 2024-T3 by cerium malate. *Appl. Surf. Sci.* 467–468: 1011–1032.

Konar, S., Mukherjee, P. S., Drew, M. G. B., Ribas, J., and Chaudhuri, N.R. 2003. Syntheses of two new 1D and 3D networks of Cu (II) and Co (II) using malonate and urotropine as bridging ligands: crystal structures and magnetic studies. *Inorg. Chem.* 42: 2545–2552.

Li, X., and Mu, G. 2005. Tween-40 as corrosion inhibitor for cold rolled steel in sulphuric acid: weight loss study, electrochemical characterization, and AFM. *Appl. Surf. Sci.* 252: 1254–1265.

Liu, J., Li, Z., Yuan, X., and Wang, Y. 2014. A copper(I) coordination polymer incorporation the corrosion inhibitor 1H-benzotriazole: poly[M3-Benzotriazolato-κ(3)N(1):N(2):N(3)-copper(I)]. *Acta Crystallogr. C Struct. Chem.* 70: 599–602.

Morozan, A., and Jaouen, F. 2012. Metal organic frameworks for electrochemical applications. *Energy Environ. Sci.* 5: 9269–9290.

Mueller, U., Schubert, M., Teich, F., Puetter, H., and Pastre, J. 2006. Metal – organic frameworks - prospective industrial applications. *J. Mater. Chem.* 16: 626–636.

Nworie, F. S. 2018. Emerging trends in coordination polymers and metal-organic frameworks: perspectives, synthesis, properties and applications. *Arc. Org. Inorg. Chem. Sci.* 1: 39–51.

Prakash, A., and Adhikari, D. 2011. Application of schiff bases and their metal complexes-a review. *Int. J. Chem. Tech. Res.* 3(4): 1891–1896.

El Rehim, S. S. A., Ibrahim, M. A. M., and Khalid, K. F. 2001. The inhibition of 4- (2-Amino-5-methylphenylazo) antipyrine on corrosion of mild steel in HCl solution. *Mater. Chem. Phys.* 70: 268–273.

Şafak, S., Duran, B., Yurt, A., and Türkoğlu, G. 2012. Schiff bases as corrosion inhibitor for aluminium in HCl solution. *Corros. Sci.* 54: 251–259.

Saha, S. K., Dutta, A., Ghosh, P., Sukul, D., and Banerjee, P. 2015. Adsorption and corrosion inhibition effect of schiff base molecules on the mild steel surface in 1 M HCl medium: a combined experimental and theoretical approach. *Phys. Chem. Chem. Phys.* 17: 5679–5690.

Sherif, E. S. M., and Almajid, A. A. 2010. Surface protection of copper in aerated 3. 5% sodium chloride solutions by 3-amino-5-mercapto-1, 2, 4-triazole as a copper corrosion inhibitor. *J Appl Electrochem* 40: 1555–1562.

Silva, P., Vilela, S. M. F., Tome´, J. P. C., Paz Almeida, F. A. 2015. Multifunctional metal – organic frameworks: from academia to industrial applications. *Chem. Soc. Rev.* 44: 6774–6803.

Singh, A. K., and Quraishi, M. A. 2010. The effect of some bis-thiadiazole derivatives on the corrosion of mild steel in hydrochloric acid. *Corros. Sci.* 52: 1373–1385.

Steed, J., and Atwood, J. L. 2009. *Supramolecular Chemistry.* 2nd ed., Chichester UK: Wiley.

Tao, J., Tong, M., and Chen, X. 2000. Hydrothermal synthesis and crystal structures of three-dimensional co-ordination frameworks constructed with mixed terephthalate (Tp) and 4,4'-bipyridine (4,4'-Bipy) ligands: [M(Tp)(4,4'-Bipy)] (M=CoII, CdII or ZnII). *J. Chem. Soc. Dalton Trans.* 20: 3669–3674.

Wang, X., Qin, C., Wang, E., Li, Y., Hao, N., Hu, C., and Xu, L. 2004. Syntheses, structures, and photoluminescence of a novel class of d 10 metal complexes constructed from pyridine-3, 4-dicarboxylic acid with different coordination architectures. *Inorg. Chem.* 43: 1850–1856.

Yang, H., Meng, X., Liu, Y., Hou, H., Fan, Y., and Shen, X. 2008. Syntheses, crystal structures and properties of two 1-D cadmium (II) coordination polymers based on 1, 1 0- (1,3-propanediyl) bis-1H-benzimidazole. *J. Solid State Chem.* 181: 2178–2184.

Zhang, D., Gao, L., and Zhou, G. 2004. Inhibition of copper corrosion in aerated hydrochloric acid solution by heterocyclic compounds containing a mercapto group. *Corros. Sci.* 46: 3031–3040.

Zorainy, M., Boffito, D. C., Gobara, M., Baraka, A., Naeema, I., and Hesham, T. 2021. Synthesis of a novel Ce(III)/melamine coordination polymer and its application for corrosion protection of AA2024 in NaCl solution. *RSC Adv.* 11: 6330–6345.

18 MOFs-Based Corrosion Inhibitors

Elyor Berdimurodov
Karshi State University

Lei Guo
Tongren University

Abduvali Kholikov and Khamdam Akbarov
National University of Uzbekistan

Mengyue Zhu
East China Jiaotong University

CONTENTS

18.1 INTRODUCTION

The corrosion of metallic materials causes significant economic problems around the world. According to NACE, the annual cost of corrosion in the United States equals \$276 billion, approximately 3.1% of GDP of the world (2002). This amount dramatically rose to \$2.2 trillion in 2011. The cost of corrosion is high in developed countries, such as India, Russia, Brazil and European countries. Recently, NACE confirmed that the global annual cost of corrosion is \$2.5 trillion (2016), equaling 3.4% of the worldwide GDP (Prabhu et al. 2020).

The chloride, sulfate and hydroxyl anions and hydrogen cations are mainly responsible for metallic corrosion. These ions could quickly react with metallic materials to form corrosion products such as $FeCl_2$, $FeCl_3$, $Fe(OH)_3$ and $Fe_2(SO_4)_3$, and get deposited on the metallic surface. One of the major parameters influencing the rate of corrosion is temperature. Typically, the rate of electrochemical anodic and cathodic reactions rises with an increase in temperature as the aggressive ions become more active at higher solution temperatures. The corrosion behaviors indeed change with the nature of the metal. Metals like iron and magnesium are more prone to corrosion. The surface metal oxides formed on such metals are typically less stable or porous (Galai et al. 2020; Dai et al. 2020).

When the corrosion inhibitors are added to the aggressive solutions, they get adsorbed onto the metal surface through physical and chemical adsorption. One of the most used corrosion inhibitors are organic heterocyclic compounds having functional groups such as $-COOH$, $-CONH_2$, $-COOC_2H_5$, $-NH_2$, $-HS$, $-H_2PO_4$, $-CH_3$, $-OCH_3$, $-OH$ and $-NO_2$. The unshared electron pairs of

DOI: 10.1201/9781003169130-20

heteroatoms, conjugated π-electrons of multiple bonds and the π-electron systems of heterocycles are the adsorption centers. The inhibitor adsorption depends on various factors, such as solution temperature, concentration, solution pH, solution type, the pressure of natural gases (oxygen, hydrogen sulfide and carbon dioxide), exposure time, electronic structure of inhibitor molecules, nature of the electrolyte, and nature and magnitude of charge present on the metal (Phadagi et al. 2021; Verma et al. 2020; El Ibrahimi et al. 2020).

18.2   MOFs AS GREEN AND EFFICIENT CORROSION INHIBITORS FOR STEEL

This section presents work reported on MOFs-based corrosion inhibitors for steel (Table 18.1). The table also shows the chemical structures and names of MOFs, the type of steel used, electrolytic media and protection degree. For example, [Ag(qox)(4-ab)] MOF was reported as a corrosion inhibitor for carbon steel in 1 M HCl (Etaiw et al. 2011a). The MOF was synthesized from the reaction of $AgNO_3$ with 4-amino benzoic acid (4-aba) and quinoxaline (qox). The corrosion inhibition efficiency was evaluated by potentiodynamic polarization (PDP) and electrochemical impedance spectroscopy (EIS). The observed results confirmed that this MOF is a mixed-type inhibitor, which means that it can effectively influence both anodic and cathodic electrochemical reactions. The inhibitor adsorption was characterized by the Freundlich adsorption isotherm, indicating physical and chemical adsorption mechanisms. The maximum inhibition efficiency was 84.3% at 1×10^{-4} M at 25°C. The protection efficiency of this inhibitor was nearly stable at high temperatures (55°C). The obtained large amount of enthalpies of activation showed the endothermic nature of the inhibition process (Etaiw et al. 2011a). Chen et al. suggested that Cu-based MOFs (Cu-MOFs) are excellent corrosion inhibitors for carbon steel in 1 M HCl. The maximum inhibition efficiency of 82.42% was achieved at 50 mg/L of the inhibitor. The results suggested that the MOF was firmly adsorbed onto the metal surface, resulting in a dense protective film. The inhibitor was mixed-type and simultaneously inhibited both the cathodic and anodic processes (Chen et al. 2020).

Lashgari and co-workers researched the anti-corrosion properties of newly made MOFs (Lashgari et al. 2020a, b), which are based on the zeolite–imidazole frameworks, ZIF-67 and ZIF-67@APS (Table 18.1). Electrochemical studies showed that the ZIF-67 and ZIF-67@APS corrosion inhibitors are mixed type, with inhibition efficiencies of 65% and 81% (at 2 g/L), respectively. The 2-methylimidazole functional groups of the ZIF-67@APS were chemically linked with Fe^{2+} in the anodic zones. In the process, the electron pairs in the 2-methylimidazole functional groups were shared with free d-orbitals of Fe. In contrast, the positively charged N atoms of 2-methylimidazole molecules are electrostatically connected to the negatively charged metal surface. The metal surface was negatively charged with the adsorbed chloride ions in the saline medium (Lashgari et al. 2020b). ZIF-67/graphene oxide was reported as a green and excellent inhibitor for mild steel in 3.5 wt% NaCl (Lashgari et al. 2020a). In another research work, $[Ag_6(CN)_4(qox)_2]$ was introduced as an effective inhibitor for carbon steel in an acidic medium. It was synthesized from the reaction of $AgNO_3$, quinoxaline (qox) and KCN. The inhibition properties of $[Ag_6(CN)_4(qox)_2]$ were studied using PDP and EIS. This inhibitor is also a mixed type. According to Langmuir, the adsorption may be a combination of both physisorption and chemisorption. The inhibitor formed a stable-protective film on the metal surface. As a result, the charge transfer rose and the double layer capacitance decreased (Etaiw et al. 2011b). The MOF inhibitors have donor–acceptor interactions with positively charged metal surfaces, and they can replace adsorbed water molecules on the metal surface. The electron pairs of N atoms are transferred to vacant d-orbitals of iron. Some heteroatoms are negatively charged in acidic solution. These atoms are electrostatically linked to the metal surface. Theoretical investigations suggest that the MOF molecules are parallelly adsorbed on the metal surface. This orientation is associated with the interaction of the p-electrons of unshared pairs of N atoms with the vacant d-orbitals of the surface iron atoms. Therefore, the p-electron system and lone pairs of electrons of N atoms are mainly responsible for the inhibition efficiency of MOF corrosion inhibitors.

TABLE 18.1

MOFs as Corrosion Inhibitors for Steel Materials

Chemical Structure and Name of MOF	Steel Type	Corrosive Medium	Maximum Protection (%)	Inhibitor Type	Ref.
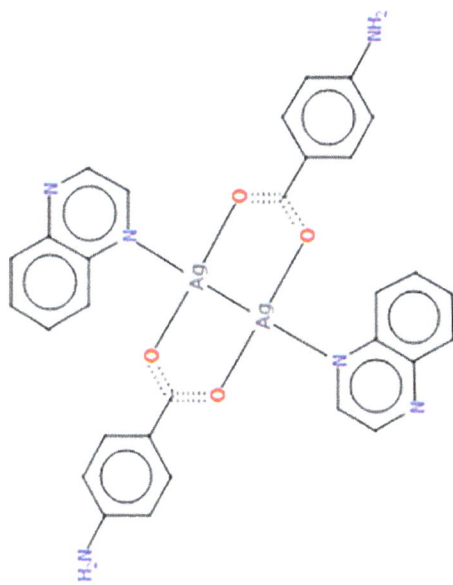 [Ag(qox)(4-ab)]	Carbon steel	1 M HCl	84.3% at $1 \times 10^{-4 M}$ at 25°C	Freundlich, mixed-type	Etaiw et al. (2011a)
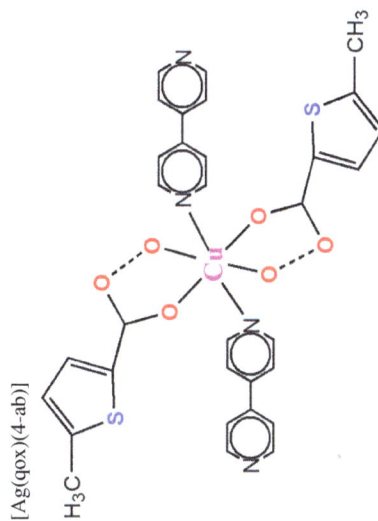 Cu–MOFs	Carbon steel	1 M HCl	79.01%±0.43% at 100 mg/L	Mixed-type	Chen et al. (2020)

(Continued)

TABLE 18.1 (Continued)

MOFs as Corrosion Inhibitors for Steel Materials

Chemical Structure and Name of MOF	Steel Type	Corrosive Medium	Maximum Protection (%)	Inhibitor Type	Ref.
Co²⁺ (as central atom) and 2-methylimidazole MOF	Bare Steel	3.5 wt% NaCl	65% (ZIF-67)	Mixed-type	Lashgari et al. (2020b, 2021)
Co²⁺ (as central atom) and 2-methylimidazole assembled to fabricate ZIF-67@APS (aminopropyl triethoxysilane) MOFs			81% (ZIF-67@ APS) at 2 g/L		

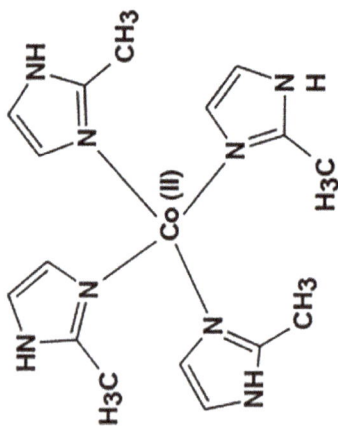

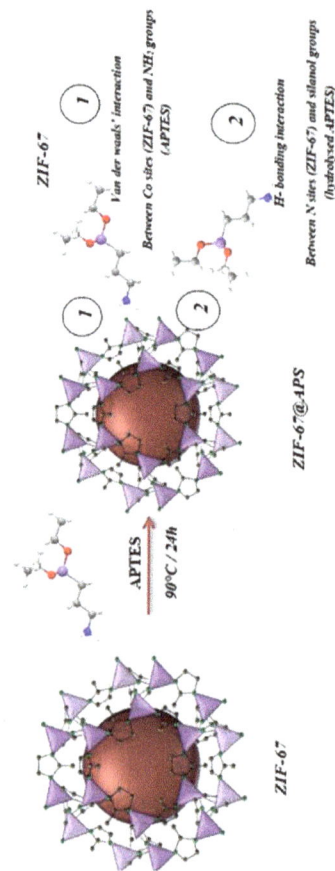

(Continued)

TABLE 18.1 (Continued)

MOFs as Corrosion Inhibitors for Steel Materials

Chemical Structure and Name of MOF	Steel Type	Corrosive Medium	Maximum Protection (%)	Inhibitor Type	Ref.
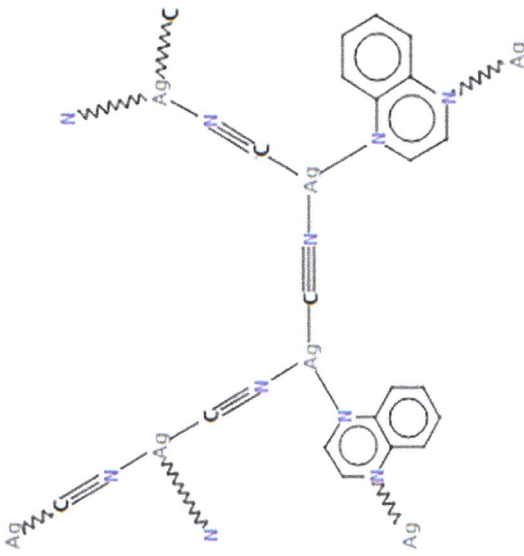 [Ag₆(CN)₄(qox)₂]	Carbon steel	1 M HCl	75.9% at 1×10^{-4} M	Langmuir, mixed-type	Etaiw et al. (2011b)
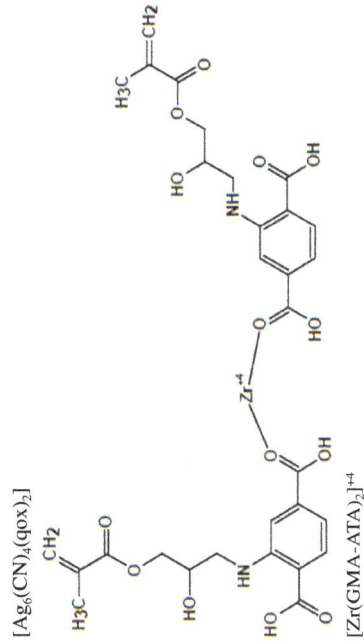 [Zr(GMA-ATA)₂]⁺⁴	Mild steel	pH 2, HCl	85.93% at 200 mg/L	Mixed-type	Ramezanzadeh et al. (2021)

(Continued)

TABLE 18.1 (Continued)

MOFs as Corrosion Inhibitors for Steel Materials

Chemical Structure and Name of MOF	Steel Type	Corrosive Medium	Maximum Protection (%)	Inhibitor Type	Ref.
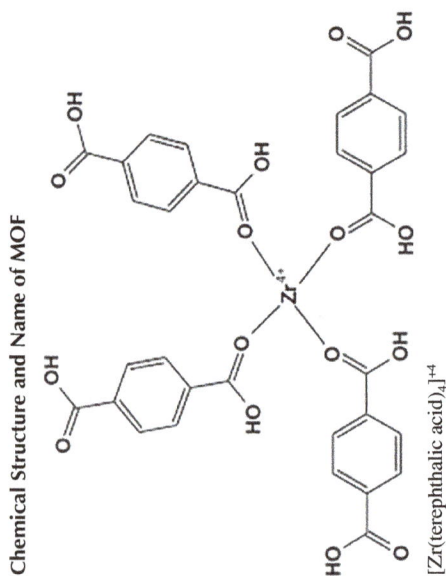 [Zr(terephthalic acid)₄]⁺⁴	Mild steel	3.5 wt% NaCl	69.31% at 200 mg/L	Mixed-type	Ramezanzadeh et al. (2021)
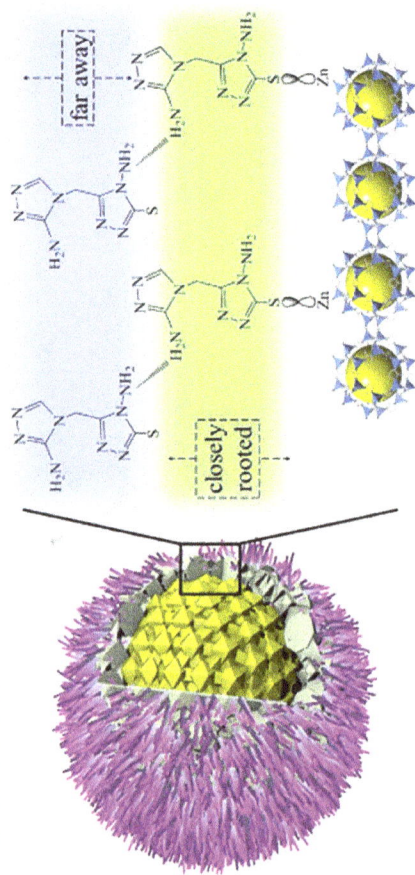 4-amino-5-(3-amino-1,2,4-triazol-4-ylmethyl)-1,2,4-triazole-3-thiol (ATT) + ZIF-8.	Mild steel	0.5 M NaCl	97% at 1.3×10^{-4} M	Mixed-type	Tian et al. (2018)

(Continued)

TABLE 18.1 (Continued)
MOFs as Corrosion Inhibitors for Steel Materials

Chemical Structure and Name of MOF	Steel Type	Corrosive Medium	Maximum Protection (%)	Inhibitor Type	Ref.
ZIF-8	Carbon steel	1 M HCl	87.4% at 700 ppm	Langmuir, mixed type	Zafari et al. (2019)
ZIF-8@{Mo$_{132}$}			92.3% at 700 ppm		Ghahramaninezhad et al. (2018)

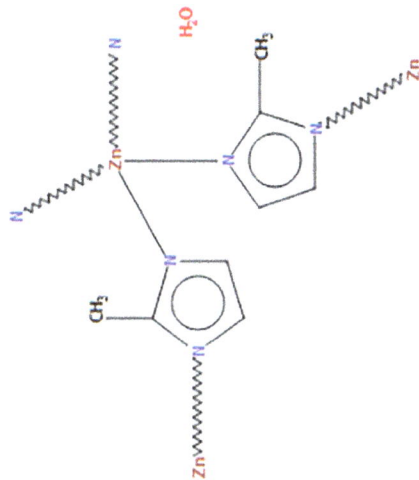

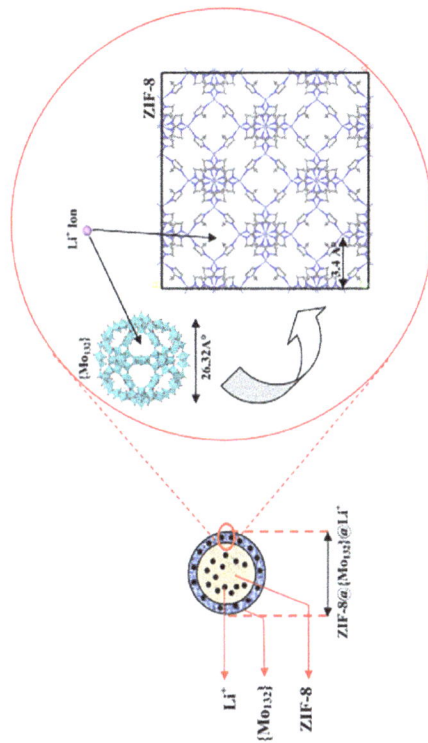

(Continued)

TABLE 18.1 (*Continued*)
MOFs as Corrosion Inhibitors for Steel Materials

Chemical Structure and Name of MOF	Steel Type	Corrosive Medium	Maximum Protection (%)	Inhibitor Type	Ref.
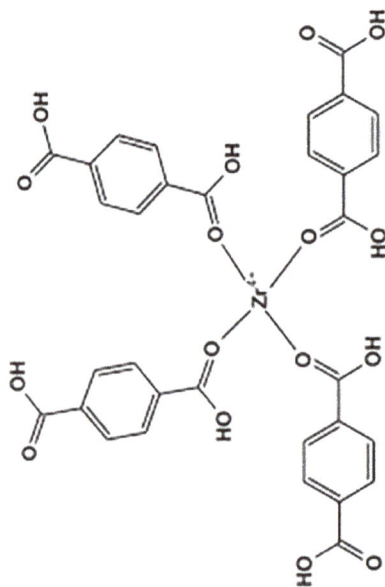 [Zr(Terephthalic acid)$_4$]$^{+4}$	Mild steel	3.5 wt% NaCl (pH 7.5 and pH 12)	87.65% at 200 mg/L	Mixed-type	Ramezanzadeh et al. (2021)
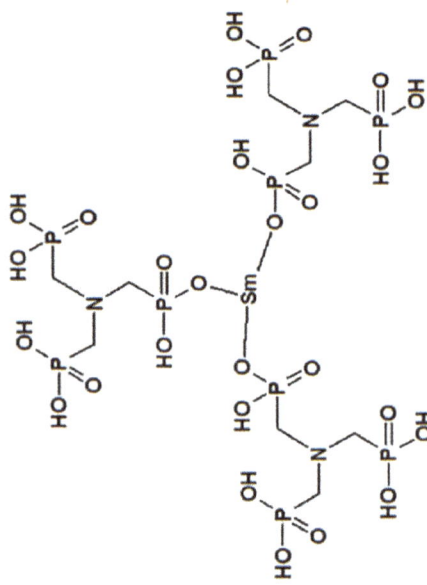 Combination of Sm (III) nitrate with [bis (phosphonomethyl) amino] methylphosphonic acid (ATMP).	Mild steel	3.5 wt% NaCl	98% at 200:600ppm of ATMP: Sm(III)	Dominant anodic inhibition.	Dehghani et al. (2020)

(Continued)

TABLE 18.1 (*Continued*)

MOFs as Corrosion Inhibitors for Steel Materials

Chemical Structure and Name of MOF	Steel Type	Corrosive Medium	Maximum Protection (%)	Inhibitor Type	Ref.
$Co(C_2H_3PO_3) \times (H_2O)_n$ $Zn(C_2H_3PO_3) \times (H_2O)_n$	Mild steel	0.014 M HNO_3 with 3% NaCl	96.8% at 2 mM 87.3% at 2 mM	Mixed type	Maranescu et al. (2019)

(Continued)

TABLE 18.1 (Continued)

MOFs as Corrosion Inhibitors for Steel Materials

Chemical Structure and Name of MOF	Steel Type	Corrosive Medium	Maximum Protection (%)	Inhibitor Type	Ref.
	Mild steel	3.5 wt% NaCl	73% (ZIF-67/ GO) and	Mixed type	Lashgari et al. (2020a)
			83% (ZIF-67/ GO@APS) at 2 g/L		

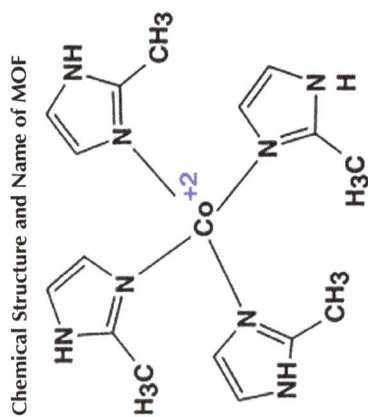

Cobalt (II) cation/2-methyl imidazole complexes (ZIF-67/GO)

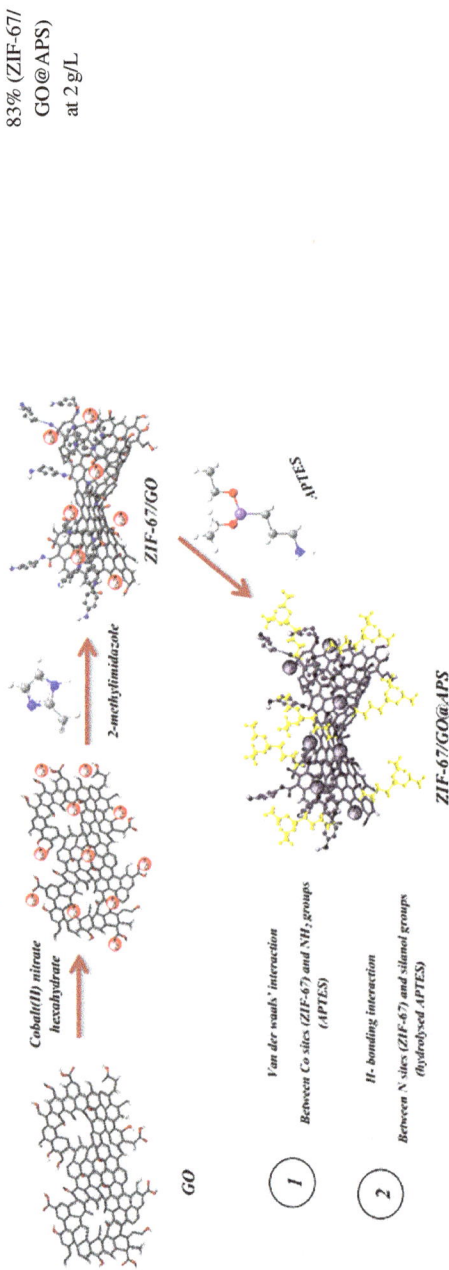

*All the images in the table are reproduced with permission from the source references. Some of them are redrawn based on the original reference.

Ramezanzadeh et al. explored MOFs-based corrosion inhibitors for steel in acidic and alkaline solutions. They have selected $[Zr(GMA-ATA)_2]^{+4}$ and $[Zr(terephthalic\ acid)_4]^{+4}$. $[Zr(GMA-ATA)_2]^{+4}$ was synthesized through Zr^{+4} cation, glycidyl methacrylate (GMA) and 2-aminoterephthalic acid (ATA), whereas $[Zr(terephthalic\ acid)_4]^{+4}$ was synthesized from Zr^{+4} cation and terephthalic acid. The study confirmed that $[Zr(GMA-ATA)_2]^{+4}$ and $[Zr(terephthalic\ acid)_4]^{+4}$ were more effective corrosion inhibitors in pH 2 (HCl) and 3.5 wt% NaCl solutions with their effective mixed-type inhibition. Electrochemical findings indicated significantly enhanced charge–transfer resistance in the presence of the inhibitors. The hydroxyl, carboxyl and amine functional groups can support physical and chemical interactions between the inhibitor and the metal surface (Ramezanzadeh et al. 2021). Tian et al. introduced a multi-substituted triazole MOF (ATT+ZIF-8) as an efficient inhibitor for mild steel in 0.5 M NaCl. The results of PDP and EIS analyses showed that the cathodic and anodic electrochemical-corrosion reactions were excellently blocked by the formation of a protective film, with a maximum inhibition efficiency of 98% at 5.5×10^{-4} M. Figure 18.1 reveals the adsorption characteristics of (ATT+ZIF-8) MOF on the on mild steel. There are two adsorption layers on the metal surface. In the first layer, the inhibitor is firmly adsorbed on the metal surface through chemical interaction. The electron pairs of N atoms are transferred to vacant d-orbitals. As a result, covalent bonds are formed between the metal surface and the corrosion inhibitor. In the second layer, the triazole molecules support the adsorption of inhibitors and insulate the first layer from water dipoles (Tian et al. 2018).

Zafari et al. introduced a novel class of MOF inhibitors which are named as ZIF-8@{Mo132} for carbon steel in 1 M HCl. The obtained data showed that the inhibition efficiency depends on the inhibitor concentration and the solution temperature. The maximum efficiencies recorded for ZIF-8 and ZIF-8@{Mo132} were 87.4% and 92.3% (at 700 ppm), respectively. These MOFs are more stable in the acidic medium at high temperatures. Their thermodynamic investigation showed that chemisorption occurred due to the high tendency of anionic ZIF-8@{Mo132} to absorb over the steel surface, resulting in a stable-protective film (Zafari et al. 2019).

Dehghani et al. suggested samarium-based MOFs as an effective corrosion inhibitor for mild steel in 3.5 wt% NaCl. It was synthesized from the reaction between Sm(III) nitrate and [bis(phosphonomethyl)amino] methylphosphonic acid (ATMP). The maximum inhibition efficiency (98%) was obtained at 200:600 ppm of ATMP:Sm (III) during 120 h of mild steel exposure to the saline media. The inhibitor was anodic, indicating that the inhibitor neutralized iron ions forming an electrostatic barrier. The inhibitor film enhanced the hydrophobicity of the metal surface. XPS confirmed the uniform deposition of the protective film, which contained ATMP-Sm^{3+} and

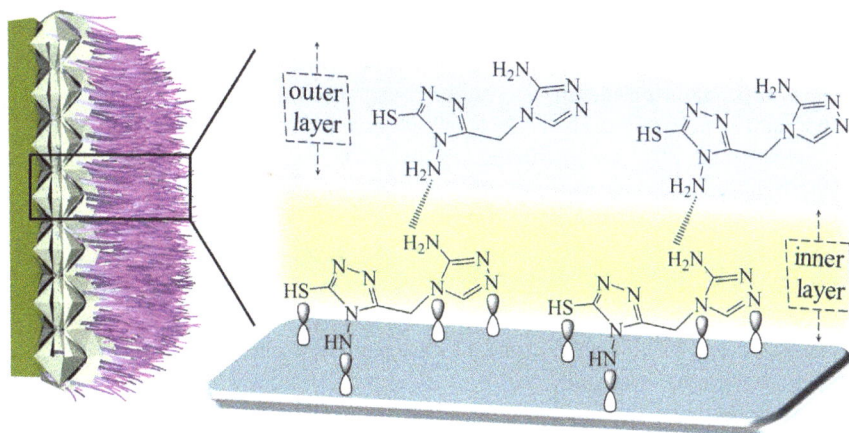

FIGURE 18.1 Adsorption characteristics of (ATT+ZIF-8) MOF on mild steel. (Reproduced with permission from Tian et al. (2018) © 2017 Elsevier Ltd.)

ATMP-Fe^{2+} complexes, Sm oxides and hydroxides (Dehghani et al. 2020). $Co(C_2H_3PO_3) \times (H_2O)_n$ and $Zn(C_2H_3PO_3) \times (H_2O)_n$ were introduced as effective corrosion inhibitors for mild steel in saline media (Maranescu et al. 2019).

18.3 MOFs AS CORROSION INHIBITORS FOR COPPER

The corrosion of Cu continues to be a serious issue in several industries. Table 18.2 illustrates recently investigated MOF inhibitors for Cu in acidic, alkaline, saline and other corrosive media. For instance, Li et al. introduced CTAB@HKUST-1 as an excellent corrosion inhibitor for Cu in 1 M NaCl. CTAB@HKUST-1 contained cetyltrimethyl ammonium bromide (CTAB), which was responsible for inhibition performance. The results confirmed that the protective film of CTAB@HKUST-1 had excellent anti-corrosion efficiency for copper. Figure 18.2 shows the EIS Bode and phase angle plots and adsorption mechanism of CTAB@HKUST-1. The inhibition performance improved as the immersion time increased. The maximum protection was achieved at 96 h. The phase angle results suggest that CTAB@HKUST-1 can control the corrosion processes through the charge transfer mechanism. The charge transfer between the anode and cathode regions on the Cu surface was greatly blocked by the adsorption. The protective film of CTAB@HKUST-1 effectively insulated the Cu surface from the attacks of chloride ions and water dipoles (Li et al. 2018).

Etaiw et al. suggested new cadmium thiocyanate and 6-methylequinoline [Cd(SCN)$_2$(6-mquin)$_2$] MOF as a corrosion inhibitor for Cu in 1 M HCl solution. [Cd(SCN)$_2$(6-mquin)$_2$] is an environmentally friendly inhibitor. It was synthesized by the reaction of CdSO$_4$·5H$_2$O with 6-methylequinoline (6-mquin) in the presence of KSCN. The inhibition performance was characterized by PDP and EIS methods. The difference in corrosion potentials of samples in the corrosive and inhibited medium was 78 mV, indicating that the [Cd(SCN)$_2$(6-mquin)$_2$] is a mixed-type inhibitor. The PDP plots showed that the anodic slope was more affected than the cathodic one, revealing anodic dominance inhibition. The inhibitor adsorption obeyed the Langmuir isotherm. The inhibition performance closely depends on the inhibitor concentration and the solution temperatures (Etaiw et al. 2017).

Fouda et al. synthesized [Ag(qox)NO$_3$] and [Ag(pyzca)] from the reaction of AgNO$_3$ with quinoxaline (qox) and pyrazine carboxylic acid (pyzca). The inhibition efficiency of Cu in acidic medium was explored by PDP and EIS. It was found that the maximum efficiency of [Ag(qox NO$_3$] and [Ag(pyzca)] were respectively, 90.9% and 85.8% at $21 \times 10^{-6\,M}$ at 25°C. The maximum corrosion potential displacement between the corrosive and inhibited samples was 75 mV for [Ag(qox)NO$_3$] and 64 mV for [Ag(pyzca)], confirming that both are mixed-type inhibitors. The Langmuir adsorption isotherm recommended a physical adsorption mechanism. The inhibition mechanism of these MOFs on the Cu surface depends on their molecular structure. The π-electrons of benzoyl rings and p-electron pairs of N atoms are shared with vacant d-orbitals of Cu (chemisorption). In contrast, the positively charged Ag atoms are responsible for physisorption. In this action, the positively charged Ag atoms are electrostatically linked with the negatively charged (due to adsorbed chloride ions) Cu surface. Therefore, the protective film on the Cu surface was formed by chemisorption and physisorption (Fouda et al. 2016).

Wei et al. investigated the inhibition properties of {[Cu(2,3-Hpyza)]·2(H$_2$O)} and {[Nd$_2$(2,3-Hpyza)$_3$]·4(H$_2$O)}$_n$ for Cu in 1 mol/L HCl. It was found that both the inhibitors obeyed the Langmuir isotherm and are mixed–type inhibitors. The highest inhibition efficiency achieved with the two inhibitors was 91.4% and 87.5% (at 0.5 mmol/L), respectively. Figure 18.3 shows SEM micrographs of Cu surfaces after immersion studies in corrosive and inhibited solutions. The images clearly show that the MOF inhibitors significantly improved the metal surface by the formation of a protective film (Wei et al. 2020).

TABLE 18.2

MOFs as Corrosion Inhibitors for Copper

Chemical Structure and Name of MOF	Corrosive Media	Maximum Protection (%)	Inhibitor Type	Ref.
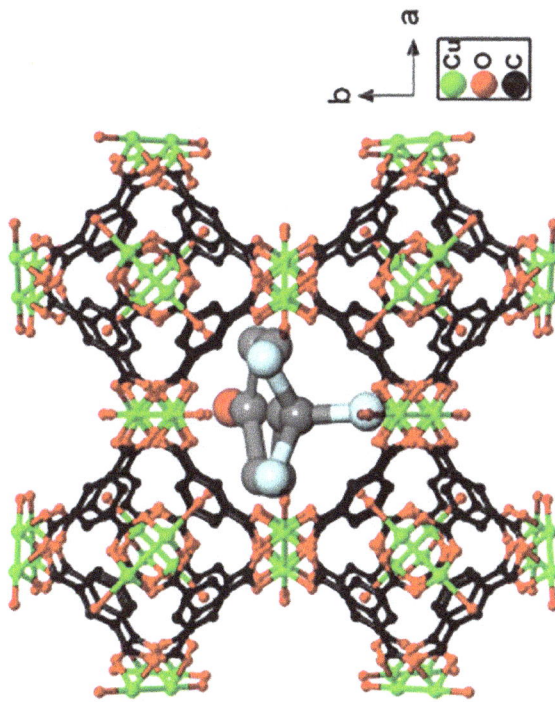 CTAB@HKUST-1	1 M NaCl	–	–	Li et al. (2018)

(Continued)

TABLE 18.2 (Continued)

MOFs as Corrosion Inhibitors for Copper

Chemical Structure and Name of MOF	Corrosive Media	Maximum Protection (%)	Inhibitor Type	Ref.
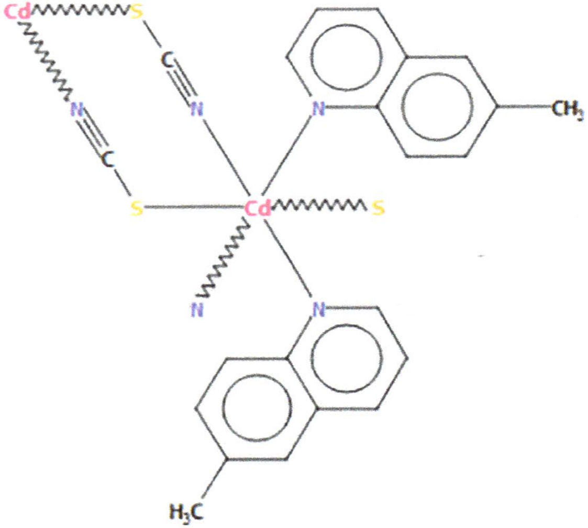 [Cd(SCN)$_2$(6-mquin)$_2$]	1 M HCl	88.3% at 21×10^{-6} M at 25°C	Mixed-type	Etaiw et al. (2017)
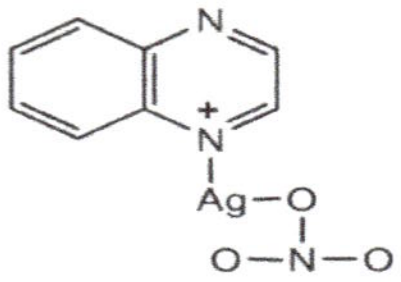 [Ag (qox) NO$_3$]	1 M HCl	90.9% at 21×10^{-6} M at 25°C	Langmuir, mixed type, physical adsorption	Fouda et al. (2016)
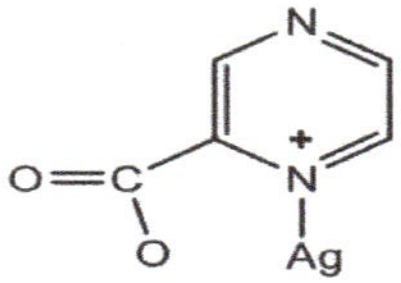 [Ag(pyzca)]		85.8% at 21×10^{-6} M at 25°C		

(Continued)

TABLE 18.2 (Continued)

MOFs as Corrosion Inhibitors for Copper

Chemical Structure and Name of MOF	Corrosive Media	Maximum Protection (%)	Inhibitor Type	Ref.
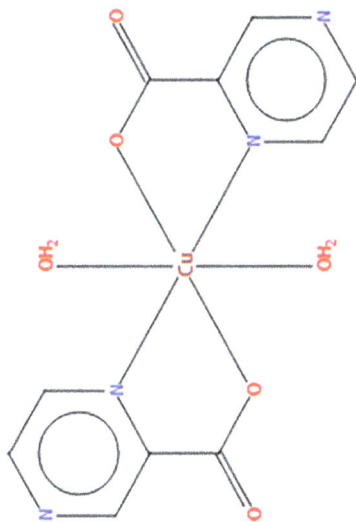 {[Cu(2,3-Hpyza)] 2(H₂O)}	1 mol/L HCl	91.4% at 0.5 mmol/L	Langmuir, mixed type, chemisorption and physisorption	Wei et al. (2020)
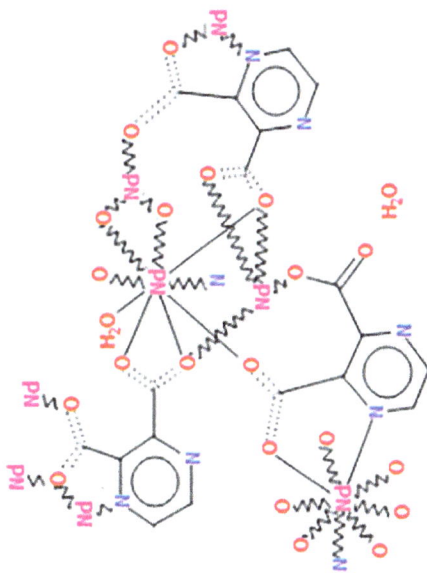 {[Nd₂(2,3-Hpyza)₃] 4(H₂O)}ₙ		87.5% at 0.5 mmol/L		

*All the images in the table are reproduced with permission from the source references. Some of them are redrawn based on the original references.

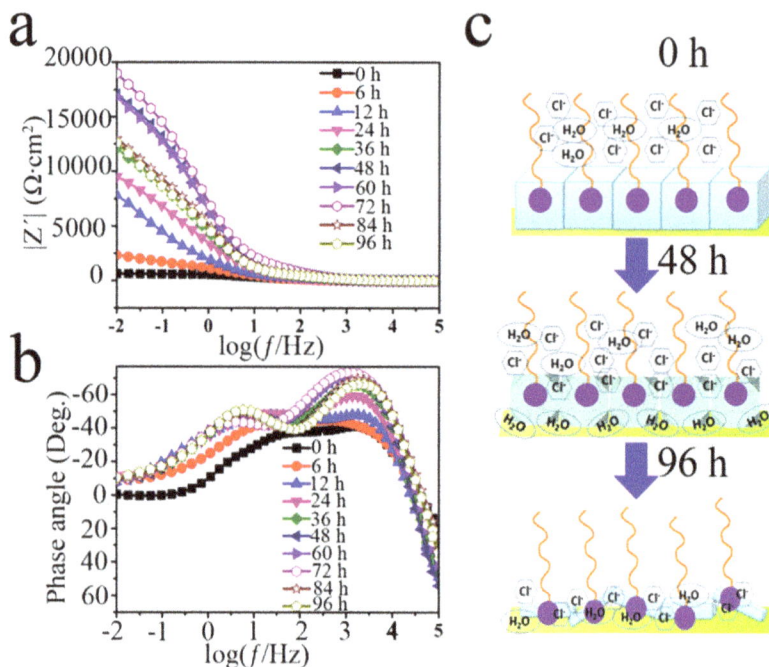

FIGURE 18.2 (a) Bode, (b) phase angle plots, and (c) adsorption mechanism of CTAB@HKUST-1 on copper in 1 M NaCl. (Reproduced with permission from Li et al. (2018) © 2018 American Chemical Society.)

FIGURE 18.3 SEM images of the copper surface after immersion in 1 mol/L HCl (50°C, 72 h) (a) in the absence, and (b, c) the presence of 50 μmol/L of (b) {[Cu(2,3-Hpyza)]2(H$_2$O)}, and (c) {[Nd$_2$(2,3-Hpyza)$_3$]·4(H$_2$O)}$_n$. (Reproduced with permission from Wei et al. (2020) © 2020 Springer Nature.)

18.4 MOFs AS CORROSION INHIBITORS FOR MAGNESIUM

Mg and its alloys are attractive due to their desirable mechanical properties, light density, and high strength-to-weight ratio. They are widely researched as alternatives for steel and aluminum for electronic, aerospace and automotive applications. However, they are prone to corrosion (Saji 2019). Recently, MOFs are shown to be good inhibitors for Mg alloys (Table 18.3). For example, Mesbah et al. researched the inhibition performance of Mg(C$_7$H$_{13}$O$_2$)$_2$×3H$_2$O and Mg(C$_{10}$H$_{19}$O$_2$)$_2$×(H$_2$O)$_3$ in ASTM D1384–87 standard water. It was found that Mg(C$_7$H$_{13}$O$_2$)$_2$×3H$_2$O was not efficient whereas Mg(C$_{10}$H$_{19}$O$_2$)$_2$×(H$_2$O)$_3$ had 70% efficiency at 5×10^{-2} M. The inhibition performance was correlated to their hydrophobic ability. The Mg(C$_7$H$_{13}$O$_2$)$_2$×3H$_2$O and Mg(C$_{10}$H$_{19}$O$_2$)$_2$×(H$_2$O)$_3$ effectively

TABLE 18.3

MOFs as Corrosion Inhibitors for Magnesium

Chemical Structure and Name of MOF	Corrosive Media	Maximum Protection (%)	Inhibitor Type	Ref.
$Mg(C_7H_{13}O_2)_2 \times 3H_2O$ $Mg(C_{10}H_{19}O_2)_2 \times (H_2O)_3$	ASTM D1384-87 standard water, pH adjusted to 8.3 with NaOH.	Not efficient 70% at 5×10^{-2} M.	-	Mesbah et al. (2007)
$Al(OH)(O_2C-CH=CH-CH-CO_2)$	30% ethylene glycol solution containing 0.5 M NaCl.	88.35% at 400 ppm.	Langmuir, mixed type, physisorption and chemisorption	Mohamadian-Kalhor et al. (2021)

deposited on the surface of the Mg sample and significantly insulated the metal surface from the corrosion solution. The surface view images of Mg samples after immersion studies in the absence and presence of the inhibitor showed that the Mg surface was seriously destroyed in the inhibitor-free corrosive medium. In comparison, the addition of $Mg(C_{10}H_{19}O_2)_2 \times (H_2O)_3$ to the corrosive solution resulted in the formation of a protective surface film. The deposited Mg salts are thermodynamically stable in the aggressive acidic medium (Mesbah et al. 2007).

Mohamadian-Kalhor et al. introduced aluminum fumarate ($Al(OH)(O_2C-CH=CH-CH-CO_2)$) as an inhibitor (50–400 ppm) for AM60B Mg alloy in 30% ethylene glycol solution containing 0.5 M NaCl. The maximum protection (88.35%) was achieved at 400 ppm. The obtained values of Gibbs free energy confirmed that the $Al(OH)(O_2C-CH=CH-CH-CO_2)$ adsorbed on the Mg surface by both electrostatic interactions (physisorption), and charge sharing between the inhibitor and the metal (chemisorption). The adsorption of this inhibitor obeyed the Langmuir isotherm. The p-electron pairs of oxygen atoms were responsible for the good inhibition performance (Mohamadian-Kalhor et al. 2021).

18.5 INHIBITION AND ADSORPTION MECHANISM

These results confirm that MOFs are excellent corrosion inhibitors for steel, copper and magnesium alloys. The inhibitors improved the surface morphology by the formation of a protective film on the metal surface. The heteroatoms and functional groups were mainly responsible for the adsorption, which means that the heteroatoms and functional groups formed the adsorption centers. The value of Gibbs free energy recommends that MOF inhibitors are mainly mixed types. The PDP studies also confirmed that MOFs are mixed-type inhibitors, revealing that the inhibitors influence both the cathodic and anodic electrochemical reactions. The enthalpy variation showed the endothermic nature of the adsorption process. Figure 18.4 represents the corrosion and inhibition mechanism of the ATMP:Sm MOF corrosion inhibitor as discussed in Section 18.2 (Dehghani et al. 2020). The Sm ions formed Sm oxide/hydroxide on the metal surface to protect the cathodic sites of the steel sample. The inhibitor forms complexes with iron ions on the metal surface, which also promotes the adsorption performance. The inhibition and adsorption mechanism of ATMP: Sm MOF corrosion inhibitor on steel surface can be described as follows:

i. The free p-electron pairs of N and O atoms are transferred to vacant d-orbitals of steel to adsorb on the metal surface.
ii. The ATMP molecules electrostatically interact with the metal surface. This interaction promoted to enhance the adsorption performance.
iii. ATMP molecules could adsorb onto the metal surface through hydrogen bonds between the -OH groups and the steel hydrated oxide layer. This observation signifies the importance of non-covalent interactions in the formation of protective inhibitor films.

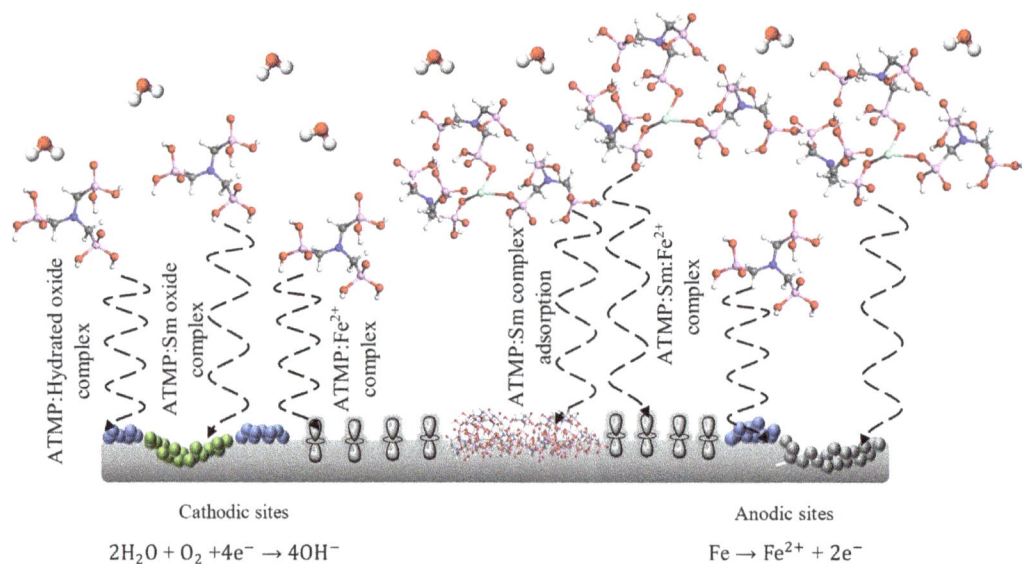

FIGURE 18.4 Corrosion and inhibition mechanism of ATMP:Sm MOF for mild steel in 3.5 wt% NaCl solution. (Reproduced with permission from Dehghani et al. (2020) © 2019 Elsevier Ltd.)

18.6 CONCLUSIONS AND PERSPECTIVES

MOFs-based corrosion inhibitors reported for steel, copper, and magnesium for various corrosive media are presented. Our analysis suggested that MOFs are excellent inhibitors, effective at low concentrations. They are highly stable and eco-friendly. The surface adsorbed protective film is thermodynamically stable with better stability at higher solution temperatures. These inhibitors are generally mixed-type and obey Langmuir adsorption. MOFs-based compounds could be excellent alternatives to the traditionally used organic and inorganic corrosion inhibitors. This is credited to their high chemical stability, less volatility, high chemical stability, non-flammability, non-toxicity and high solubility. These advantages make the MOF compounds green and excellent corrosion inhibitors.

REFERENCES

Chen, M, Cen, H, Guo, C, Guo, X, Chen, Z. 2020. Preparation of Cu-MOFs and its corrosion inhibition effect for carbon steel in hydrochloric acid solution. *J. Mol. Liq.* 318, 114328.

Dai, J, Feng, H, Li, H-B, Jiang, Z-H, Li, H, Zhang, S-C, Zhou, P, Zhang, T. 2020. Nitrogen significantly enhances corrosion resistance of 316L stainless steel in thiosulfate-chloride solution. *Corros. Sci.* 174, 108792.

Dehghani, A, Poshtiban, F, Bahlakeh, G, Ramezanzadeh, B. 2020. Fabrication of metal-organic based complex film based on three-valent samarium ions-[bis (phosphonomethyl) amino] methylphosphonic acid (ATMP) for effective corrosion inhibition of mild steel in simulated seawater. *Constr. Build. Mater.* 239, 117812.

El Ibrahimi, B, Bazzi, L, El Issami, S. 2020. The role of pH in corrosion inhibition of tin using the proline amino acid: theoretical and experimental investigations. *RSC Adv.* 10, 29696–29704.

Etaiw, SEH, El-bendary, MM, Fouda, AE-AS, Maher, MM. 2017. A new metal-organic framework based on cadmium thiocyanate and 6-methylequinoline as corrosion inhibitor for copper in 1 M HCl solution. *Prot. Met. Phys. Chem. Surf.* 53, 937–949.

Etaiw, SEH, Fouda, AES, Amer, SA, El-Bendary, MM. 2011a, Structure, Characterization and anti-corrosion activity of the new metal-organic framework [Ag(qox)(4-ab)]. *J. Inorg. Organometal. Polym. Mater.* 21, 327–335.

Etaiw, SEH, Fouda, AES, Abdou, SN, El-bendary MM. 2011b. Structure, characterization and inhibition activity of new metal-organic framework. *Corros. Sci.* 53, 3657–3665.

Fouda, A, Etaiw, S, El-bendary, MM, Maher, MM. 2016. Metal-organic frameworks based on silver (I) and nitrogen donors as new corrosion inhibitors for copper in HCl solution. *J. Mol. Liq.* 213, 228–234.

Galai, M, Rbaa, M, Ouakki, M, Abousalem, AS, Ech-chihbi, E, Dahmani, K, Dkhireche, N, Lakhrissi, B, EbnTouhami, M. 2020. Chemically functionalized of 8-hydroxyquinoline derivatives as efficient corrosion inhibition for steel in 1.0 M HCl solution: experimental and theoretical studies. *Surf. Interface* 21, 100695.

Ghahramaninezhad, M, Soleimani, B, Niknam Shahrak, M. 2018. A simple and novel protocol for Li-trapping with a POM/MOF nano-composite as a new adsorbent for CO_2 uptake. *New J. Chem.* 42, 4639–4645.

Lashgari, SM, Yari, H, Mahdavian, M, Ramezanzadeh, B, Bahlakeh, G, Ramezanzadeh, M. 2020a. Synthesis of graphene oxide nanosheets decorated by nanoporous zeolite-imidazole (ZIF-67) based metal-organic framework with controlled-release corrosion inhibitor performance: Experimental and detailed DFT-D theoretical explorations. *J. Hazard Mater.* 404, 124068.

Lashgari, SM, Yari, H, Mahdavian, M, Ramezanzadeh, B, Bahlakeh, G, Ramezanzadeh, M. 2020b. Unique 2-methylimidazole based inorganic building brick nano-particles (NPs) functionalized with 3-aminopropyltriethoxysilane with excellent controlled corrosion inhibitors delivery performance; experimental coupled with molecular/DFT-D simulations. *J. Taiwan Inst. Chem. Eng.* 117, 209–222.

Lashgari, SM, Yari, H, Mahdavian, M, Ramezanzadeh, B, Bahlakeh, G, Ramezanzadeh, M. 2021. Application of nanoporous cobalt-based ZIF-67 metal-organic framework (MOF) for construction of an epoxy-composite coating with superior anti-corrosion properties. *Corros. Sci.* 178, 109099.

Li, W, Ren, B, Chen, Y, Wang, X, Cao, R. 2018. Excellent efficacy of MOF films for bronze artwork conservation: The key role of HKUST-1 film nanocontainers in selectively positioning and protecting inhibitors. *ACS Appl. Mater. Interfaces* 10, 37529–37534.

Maranescu, B, Plesu, N, Visa, A. 2019. Phosphonic acid vs phosphonate metal organic framework influence on mild steel corrosion protection. *Appl. Surf. Sci.* 497, 143734.

Mesbah, A, Juers, C, Lacouture, F, Mathieu, S, Rocca, E, François, M, Steinmetz, J. 2007. Inhibitors for magnesium corrosion: metal organic frameworks. *Solid State Sci.* 9, 322–328.

Mohamadian-Kalhor, S, Edjlali, L, Basharnavaz, H, Es'haghi, M. 2021. Aluminum fumarate metal–organic framework: synthesis, characterization, andapplication as a novel inhibitor against corrosion of AM60B magnesium alloy in ethylene glycol solution. *J. Mater. Eng. Perform.* 30, 720–726.

Phadagi, R, Singh, S, Hashemi, H, Kaya, S, Venkatesu, P, Ramjugernath, D, Ebenso, EE, Bahadur, I. 2021. Understanding the role of dimethylformamide as co-solvents in the dissolution of cellulose in ionic liquids: experimental and theoretical approach. *J. Mol. Liq.* 328, 115392.

Prabhu, PR, Prabhu, D, Rao, P. 2020. Analysis of Garcinia indica Choisy extract as eco-friendly corrosion inhibitor for aluminum in phosphoric acid using the design of experiment. *J. Mater. Res. Technol.* 9, 3622–3631.

Ramezanzadeh, M, Ramezanzadeh, B, Bahlakeh, G, Tati, A, Mahdavian, M. 2021. Development of an active/barrier bi-functional anti-corrosion system based on the epoxy nanocomposite loaded with highly-coordinated functionalized zirconium-based nanoporous metal-organic framework (Zr-MOF). *Chem. Eng. J.* 408, 127361.

Saji, VS. 2019. Review of rare-earth-based conversion coatings for magnesium and its alloys, *J. Mater. Res. Technol.* 8, 5012–5035.

Tian, H, Li, W, Liu, A, Gao, X, Han, P, Ding, R, Yang, C, Wang, D. 2018. Controlled delivery of multi-substituted triazole by metal-organic framework for efficient inhibition of mild steel corrosion in neutral chloride solution. *Corros. Sci.* 131, 1–16.

Verma, C, Olasunkanmi, LO, Akpan, ED, Quraishi, MA, Dagdag, O, El Gouri, M, Sherif, E-SM, Ebenso, EE. 2020. Epoxy resins as anticorrosive polymeric materials: a review. *React. Funct. Polym.* 156, 104741.

Wei, W, Liu, Z, Liang, C, Han, G-C, Li, Y, Li, Q, Han, J, Zhang, S. 2020. Corrosion inhibition performance of nitrogen-containing metal organic framework compounds oncopper flakes in dilute HCl medium. *Prot. Met. Phys. Chem. Surf.* 56, 638–650.

Zafari, S, Niknam Shahrak, M, Ghahramaninezhad, M. 2019. New MOF-based corrosion inhibitor for carbon steel in acidic media. *Met. Mater. Int.* 26, 25–38.

19 Self-Assembled Monolayers in Corrosion Inhibition

Sunder Ramachandran, Carlos Menendez,
and Tracey Jackson
Baker Hughes Company, Inc.

CONTENTS

19.1 INTRODUCTION

Corrosion inhibitors (CIs) are surface-active molecules used in small concentrations to prevent corrosion. At low surface concentrations, a monolayer of molecules is formed, which may at higher concentrations form more complex surface structures such as bilayers/multilayers/semi-micelles. The structure at low surface concentration is important as users of CIs often apply the minimum possible concentration to optimize cost and performance for their systems. In this chapter, the occurrence of CIs as monolayers and bilayers are considered.

Self-assembled monolayers (SAMs) form spontaneously upon immersion of a substrate in a dilute solution of adsorbate. They are more stable than Langmuir–Blodgett films (Bain and Whitesides 1989). SAMs are created from an adsorption process that builds the layer, while Langmuir-Blodgett films are a quick way of transferring air–liquid monolayers onto a substrate. Monolayer formation is characterized by strong specific interactions between the head group and the surface of the substrate (Bain and Whitesides 1989). The stronger this interaction, the better the molecule performs as a CI. Recent coarse-grained molecular modeling studies have illustrated the effect of changing head group size and surface affinity on the number of molecules adsorbed (Ko and Sharma 2017, 2020).

There is an important distinction between conventional SAMs and SAMs that offer corrosion protection at low concentrations. Not all surface-active molecules can protect metals from

DOI: 10.1201/9781003169130-21

the electrochemical actions of corrosion. This chapter describes the unique features of some molecules that allow them to bind more strongly to the metal surface, which is one of the attributes of a SAM that can prevent corrosion. Molecules with this attribute slow or prevent corrosion at relatively low concentrations.

The SAM mechanism of corrosion inhibition is a theory that describes the low concentration inhibition of CIs such as oleic imidazoline to decrease the corrosion of carbon dioxide on mild steel (Ramachandran et al. 1996). The mechanism for oleic imidazoline shows several important features regarding strong, ordered binding to the surface and the formation of a hydrophobic barrier. This strong, ordered binding is what makes CI SAMs unique in the world of self-assembly.

19.2 ADSORPTION OF CIs

The adsorption of ionic surfactants on metal oxide surfaces has been well studied (Zhang and Somasundaran 2006). Figure 19.1 shows the main features of these types of isotherms based on results obtained for dodecyl sulfate on alumina (Zhang and Somasundaran 2006).

Surfactant adsorption can be characterized into four regions (Zhang and Somasundaran 2006).

- **Region I**: At low surfactant concentrations, individual molecules in the solution adsorb onto the solid surface.
- **Region II**: Surface aggregate structures begin to form.
- **Region III**: Lateral interactions between the surface aggregates become important.
- **Region IV**: At sufficient aqueous surfactant concentration, critical micelle concentration (CMC) is attained, the surfactant monomer concentration becomes constant and has little or no impact on the adsorption density.

Linear polarization resistance (LPR), potentiodynamic scans, and electrochemical impedance spectroscopy (EIS) measurements were conducted on a mixture of nbenzalkonium chloride (BAC) with different alkyl chain lengths on a carbon steel substrate to illustrate the effect of the surface aggregation concentration as the concentration of the CI was increased at various temperatures (40°C, 50°C and 60°C) (Zhu et al. 2016). EIS results of this work are shown in Figure 19.2.

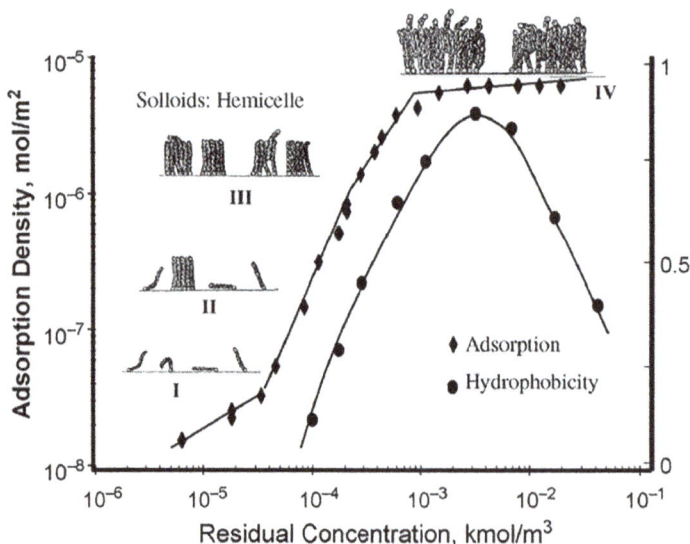

FIGURE 19.1 Schematic representation of surface concentration versus residual concentration of dodecyl sulfate on alumina. (Reproduced with permission from Zhang and Somasundaran (2006) © 2006 Elsevier B.V.)

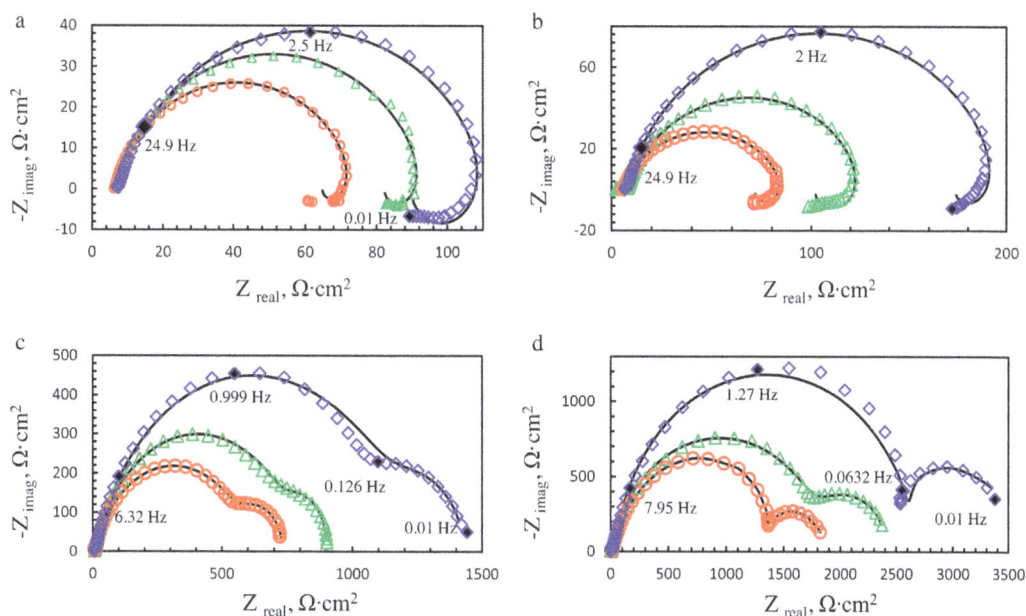

FIGURE 19.2 Nyquist plot of X65 steel electrode exposed in CO_2 –saturated 0.171 M NaCl aqueous solution containing (a) 0 μM (Blank), (b) 9 μM, (c) 72 μM and (d) 360 μM mixed alkyl length BAC surfactants at different temperatures (diamond: 40°C; triangle: 50°C; circle: 60°C); Solid lines indicate equivalent circuit curve fitting. The number next to the solid symbols in Nyquist plots represents the corresponding frequency. (Reproduced with permission from Zhu et al. (2016) © 2015 Elsevier Ltd.)

At a concentration of 9 μM, the formation of a monolayer is not complete and the inhibitor films still show inductive behavior at low frequencies. At 72 μM, the formation of a monolayer is complete and it is in transition to a bilayer or other types of aggregates. At a concentration of 360 μM, a second semicircle is seen, indicating the surface coverage above the monolayer level as a bilayer/multilayer/semi-micelles. This second semicircle could also be indicative of a porous monolayer film. Concentrations of 9 and 72 μM are in the region where SAMs are formed (Zhu et al. 2016). In the following sections, these regions of low surface concentration of CI are discussed in greater detail in relation to SAMs. Bilayer systems are also briefly considered.

19.3 SELF-ASSEMBLED MONOLAYERS

SAMs are regular systems characterized by a strong specific interaction between an adsorbate and a substrate. Among the first systems to be systematically studied were thiols on gold surfaces (Bain and Whitesides 1989). SAMs in the form of Langmuir–Blodgett films have been used in systematic electrochemical studies to investigate the electrochemical nature of thin films (Lee and Bard 1988; Bard and Faulkner 2001). SAMs have been observed with organosilicon derivatives of silicon oxide, aluminum oxide, mica, zinc selenide, germanium oxide, and gold (Ulman 1996). Fatty acids on aluminum oxide, silver oxide, and copper oxide (Ulman 1996) systems have also been characterized.

SAMs bind with the surface forming ordered structures with varying lattice spacing (Ulman 1996). For example, $CH_3(CH_2)_{20}COOH$ monolayers on AgO adsorb in a p (2×2) structure. The alkyl tails in this arrangement are in the all *trans* configuration and are tilted 26.7° from the surface normal (Ulman 1996). The phenomenon of tight binding to the surface, embodied by a regular structure with organic chain groups oriented in a way to preserve close packing of the chains, is a characteristic of many CI SAM systems.

19.4   SAM MECHANISM OF CORROSION INHIBITION

The corrosion inhibition of iron by imidazolines was proposed to follow a SAM mechanism (Ramachandran et al. 1996). This mechanism is applicable to systems where inhibition is achieved at very low concentrations. The important features of the model are the following:

1. Ordered adsorption to the native oxide of the metal. In the case of imidazoline, a $\sqrt{3}\times\sqrt{3}$ ordered structure.
2. Strong binding to the surface (for imidazolines, Lewis base binding to Lewis acid sites on the metal oxide surface). It was found that the sp^2 ligand imidazoline has an extremely strong binding with iron, consistent with the excellent corrosion inhibition of oleic imidazoline compounds (Ramachandran et al. 1997).
3. SAM formation on the native oxide of the metal acts as a hydrophobic barrier. For imidazolines, the hydrocarbon must have over 12 or more carbon atoms.
4. Tilting of the hydrocarbon tail to form a tightly packed hydrophobic monolayer.
5. Sufficient partitioning of the organic compound to the aqueous phase to allow fast enough diffusive transfer to the surface.

This mechanism applies to surface-active CIs for metals that work at dosages below the critical micelle concentration.

19.4.1   METAL SURFACE DURING CORROSION

Anodic and cathodic surface reaction sites make metal corrosion complex and heterogeneous. Inhibitors interact with cathodic or anodic sites. They either retard electrochemical reactions or limit the transport of reactive corrosive species from the solution.

Substantial concentrations of Fe^{2+} are found on the oxide film near the iron metal surface (Jovancicevic and Bockris 1986). $FeCO_3$, Fe_3O_4, and Fe_2O_3 have been found as corrosion products during the corrosion process of mild steel due to CO_2. Which species are present depends upon temperature and the partial pressure of CO_2 in the environment (John et al. 1998). Oxide surface films are likely to be present on most metals and are a significant factor in the CI SAM mechanism.

19.4.2   INHIBITOR BINDING

CIs are well known to bind strongly to metal surfaces. Inhibitor studies using surface-enhanced Raman spectroscopy (SERS) of a quaternary amine on mild steel have shown that the inhibitor helped form Fe_3O_4 and was strongly adsorbed on the oxide (Oblonsky et al. 1995). Second-harmonic generation at iron surfaces has been used to study the adsorption kinetics of oleic imidazoline onto the surface of a mild steel electrode (Joseph and Klenerman 1992). Early studies of oleic imidazoline as a CI involved both *ab initio* studies of clusters of iron hydroxide with water and oleic imidazoline (Ramachandran et al. 1997). Figure 19.3 shows the atomic spacing and bond lengths in an iron hydroxide cluster. Section 19.4.3 will discuss the relevance of this spacing to inhibitor binding in more detail.

The *ab initio* cluster results were used to develop an appropriate force field for iron oxides (Ramachandran et al. 1997). Strong binding of the oleic imidazoline molecule was obtained in this work (Ramachandran et al. 1997). In this study, binding energy was assessed from a variety of cluster systems involving snap and adiabatic calculations of the bond energy (Ramachandran et al. 1997). More recent work has used *ab initio* techniques that recreate the effect of the solvent (Rodriguez-Valdez et al. 2006). The calculations have been used to compute the HOMO and LUMO energies. The HOMO and LUMO energies should be close to the Fermi level of the metal surface.

As aforementioned, strong binding to specific surface sites is an important part of the SAM mechanism of corrosion inhibition and, as seen below, this spacing is optimal for binding.

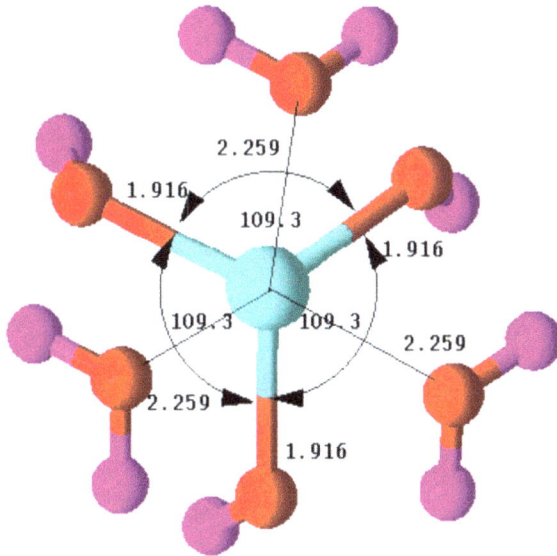

FIGURE 19.3 Iron hydroxide cluster studied in earlier work. (Reproduced with permission from Ramachandran et al. (1997) © 1997 American Chemical Society.)

19.4.3 SURFACE STRUCTURE OF BINDING GROUPS

Strong adsorption to the surface and the steric nature of an inhibitor dictates the coverage of an inhibitor to the surface. The available surface sites are spaced differently on various surfaces. In Figure 19.4, the spacing of Fe^{2+} on the 111 Miller plane of Fe_3O_4 is shown (Ramachandran and Jovancicevic 1998).

Different functional groups have different sizes. For a bulky molecule like imidazoline, its steric bulk restricts adsorption to the next nearest neighbor sites as shown in Figure 19.5 (Ramachandran et al. 1996).

Similarly, adsorption at next nearest neighbor sites on the $\sqrt{3} \times \sqrt{3}$ unit cell of adsorbed oleic imidazoline also provides coverage of ⅓ of the iron sites on the (001) surface of hematite (Ramachandran et al. 1996). This adsorption is one molecule per $65.94 \, \text{Å}^2$. This corresponds well to the value of $65 \, \text{Å}^2$ per molecule obtained from ^{14}C tracer studies (McMahon 1991).

Molecules that have smaller sizes or shapes, such as carboxylic acids or alkyl amines, would have larger surface coverage (Ramachandran et al. 1996). This is in accordance with adsorption data on alkyl amines (Yao 1963) and experiments with stearic acid Langmuir-Blodgett monolayers deposited on iron (Xing et al. 1995).

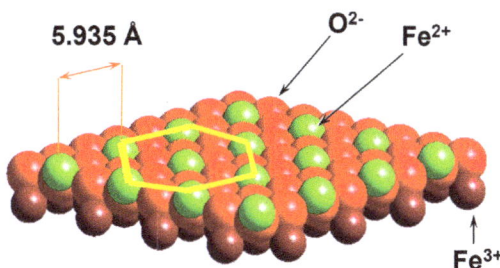

FIGURE 19.4 Spacing of Fe^{2+} sites on the 111 Miller plane of Fe_3O_4. (Modified after Ramachandran and Jovancicevic (1998).)

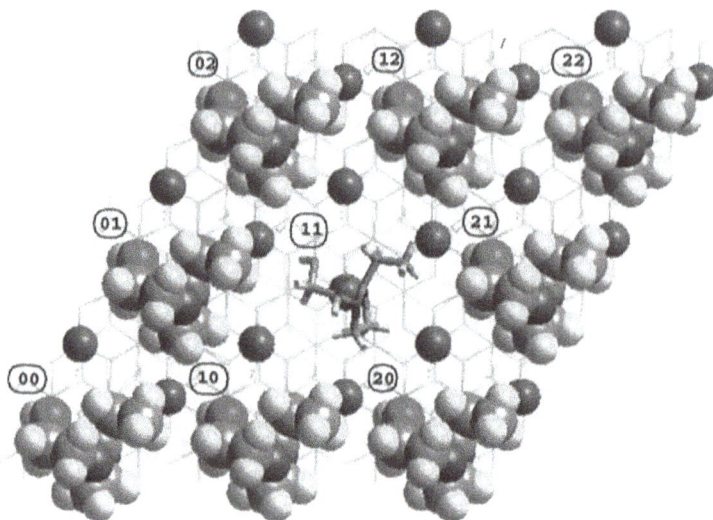

FIGURE 19.5 Surface Structure of 1-Ethyl 2-Aminoethyl imidazoline on hematite (α-Fe$_2$O$_3$) The ethyl head group was chosen to avoid obscuring the head group. Binding of the bulky imidazoline molecule precludes adsorption at the nearest neighbor sites, but binding occurs at the six next nearest sites. (Reproduced with permission from Ramachandran et al. (1996) © 1996 American Chemical Society.)

19.4.4 HYDROPHOBIC TAIL

The hydrophobic tail is an important aspect of CI performance at concentrations sufficient to create monolayers. There are several factors to consider. Recent molecular dynamics studies have indicated that the tails orient parallel or at an angle with the metal surface depending on concentration (Singh and Sharma 2019; Ko and Sharma 2017). At low concentration, the tails lay flat on the surface. This has two benefits for the hydrocarbon tail, (i) it maximizes the surface contact, and (ii) it is a net positive entropy mechanism due to the loss of solvent molecules on the portion of the tail in contact with the metal surface. At higher concentrations, the molecules tend to orient at an angle close to 90° to the surface, so they can be arranged into a close-packed SAM structure. However, this close packing is not the lowest energy state for inhibitor molecules with head groups strongly bound to the surface. This causes the structure to tilt. The hydrophobic tail must block the surface sites to maintain inhibition. This means that bulky binding groups must have sufficient tail length to block the surface sites. Longer tail groups have stronger cohesive energies that keep the monolayer intact. They are also less soluble in water, which prevents them from being solubilized once they are on the surface. The conformations of the alkyl tail are restricted on the surface while they are less restricted in solution. This means that there is an entropy loss with monolayer formation.

19.4.4.1 Tilt Angle and Surface Coverage

While inhibitor spacing on the surface is defined by the spacing of the metal surface sites and the steric bulk of the binding group, the hydrophobic tail wants to remain in a close-packed structure to minimize its free energy profile. For alkyl chains, the spacing will be similar to the spacing seen in polyethylene, which is a spacing of 4.53 Å (Ramachandran et al. 1996). The hydrophobic tails tilt at angles defined by eq. (19.1) (Ramachandran et al. 1996).

$$\cos(\theta) = \frac{d}{l\sqrt{3}} \tag{19.1}$$

In eq. (19.1) d is the close-spaced packing of the hydrocarbon chains (4.53 Å for alkyl chains) while l is the spacing of the binding group on the surface. The value of l is 10.28 Å for imidazoline on the

(111) surface of Fe_3O_4 (Ramachandran and Jovancicevic 1998). It is 8.725 Å for imidazoline on (001) the surface of hematite (Ramachandran et al. 1996).

When the steric bulk of the binding group prevents full coverage of the surface sites. It is important that the alkyl chain be long enough to cover the surface. In the case of imidazolines on a hematite surface, this only occurs for an alkyl chain length that has 12 carbon atoms or longer (Ramachandran et al. 1996). The bulkier the binding group, the smaller the surface coverage will be. Hence, longer hydrocarbon tail groups are needed for complete coverage. Conversely, binding groups with less steric bulk can achieve higher surface coverage and may not need long alkyl chain lengths to form a SAM and prevent corrosion.

19.4.4.2   Cohesive Energy and Entropy Changes from Solution to Surface

In the formation of monolayers on a hematite surface, the cohesive energy, E_{coh} attracting the tails was found to have a bi-linear dependence on chain length dependent upon when the alkyl chain interacts with other molecules bound to the surface. When the alkyl chain length is 6 or greater, 4 neighbors interact with the alkyl chain while interactions with all 6 neighbors occur when the alkyl chain is 12 or greater (Ramachandran et al. 1996).

When an imidazoline molecule is bound to the surface. The alkyl chains have restricted entropy as the extended system (all trans) will have the highest cohesive energy. In solution, the alkyl chains can have many conformations and greater entropy. Rotational isomeric states theory can be used to estimate the entropy of the alkyl chain in solution (Ramachandran et al. 1996).

19.4.4.3   Solubility, Partitioning, and Diffusive Transport to the Surface

CIs need to be present in sufficient concentrations in the water to prevent corrosion. Many inhibitors with long alkyl tails may have very low concentrations in the water. If the concentration is extremely low, the time for the inhibitor to adsorb on the surface may be restricted by mass transfer (Ramachandran et al. 1996). A model to assess solubility and diffusive transport to the surface was proposed for imidazolines (Ramachandran et al. 1996).

19.5   BILAYER FORMATION

At higher concentrations, SAMs can rearrange into bilayers and other more supramolecular complex structures may form on the surface. The formation of a bilayer will take place in Region IV of Figure 19.1. The local bilayer areas may be in the form of cylindrical or rod-like micelles, in which the surfactant molecules adsorb with a reverse, 180°, orientation (Zhu et al. 2017; Ko and Sharma 2020). The nanostructure of surface aggregates for a single component solution can be predicted using a packing parameter (Israelachvili et al. 1976) defined as:

$$P = \frac{V}{a_0^l} \tag{19.2}$$

Where V is the volume of the surfactant molecule, a_o is the optimal cross-sectional area of the hydrophilic group and l is the hydrocarbon chain length. This kind of prediction can assist in understanding where the SAM will no longer exist to study the benefits of a SAM versus more complex structures. A packing parameter of 1/3 indicates a transition from a sphere to a cylinder, whereas a packing parameter of 1/2 indicates the transition from a cylinder to a bilayer and 1 the transition to inverted micelles (Israelachvili et al. 1976). In the case of surface aggregates consisting of more than one component, a mixed packing parameter was introduced as shown in eq. (19.3) (Zhang and Somasundaran 2006).

$$P_{mix} = \sum x_i V_i / (a_0 \times l_i) \tag{19.3}$$

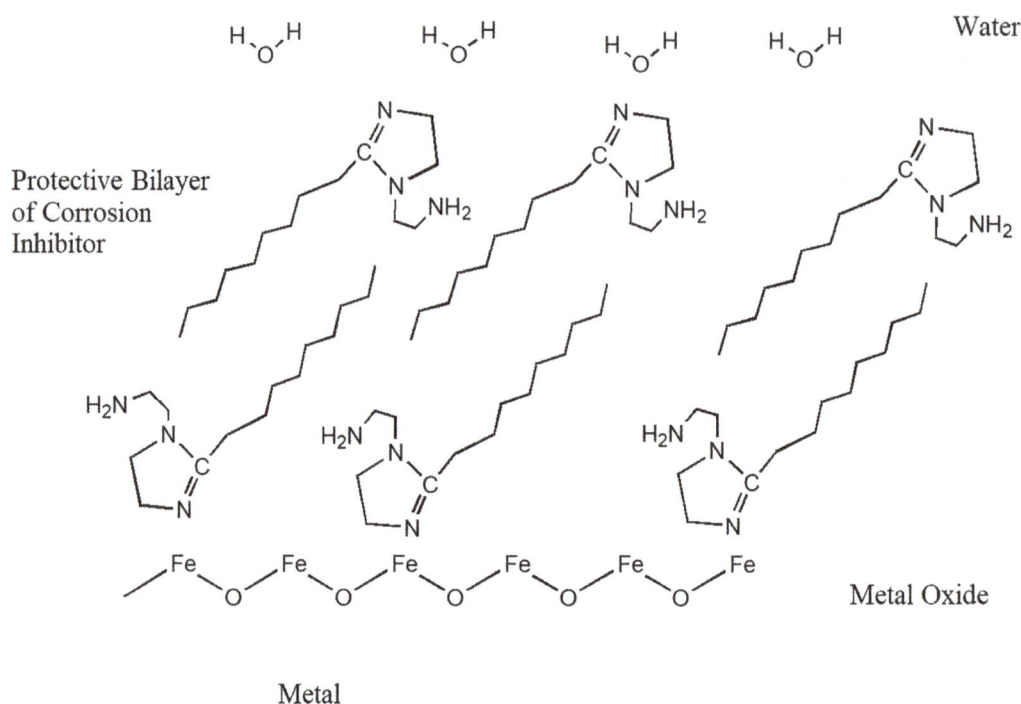

FIGURE 19.6 Schematic diagram of the bilayer film of imidazolines on the iron oxide surface. (Modified after Jovancicevic et al. (1998).)

where V_i and l_i are the volume and length of the hydrocarbon chain for component i, respectively. The a_0 in eq. (19.3) is the average packing area at a composition obtained purely from adsorption data using eq. (19.4), where N is Avogadro's constant and Γ_{max} is the maximum adsorption density at the interface (mol/m²):

$$a_0 = \frac{10^{20}}{N\Gamma_{max}} \tag{19.4}$$

While more complex than SAMs, bilayers share many features with SAMs, such as a monolayer arranged around the metal oxide spacing as well as the tilted packing required for the underlying monolayer. A schematic diagram of the bilayer formation of oleic imidazolines on a metal oxide surface is shown in Figure 19.6. This schematic illustrates the forced tilt from the strong binding of the imidazoline head group (Jovancicevic et al. 1998).

The cohesive energy of bilayer films increases with alkyl chain length, as shown in Figure 19.7 (Ramachandran and Jovancicevic 1998). This increased cohesive strength increases the stability of inhibitor films with longer chain lengths. This phenomenon also allows long-chain hydrocarbons to integrate into the monolayer to enhance the stability of the film as shown experimentally (Wang and Guang-Ling Song 2019) and in molecular modeling studies (Ko and Sharma 2020).

19.6 EXPERIMENTAL STUDIES ON CHAIN LENGTH INVOLVEMENT IN THE SAM MECHANISM

A study of the mechanism of corrosion inhibition by imidazolines and their precursors in the CO_2 environment was carried out using the rotating cylinder electrode using LPR. The minimum effective concentration to retard CO_2 corrosion at these conditions is shown in Figure 19.8. This study

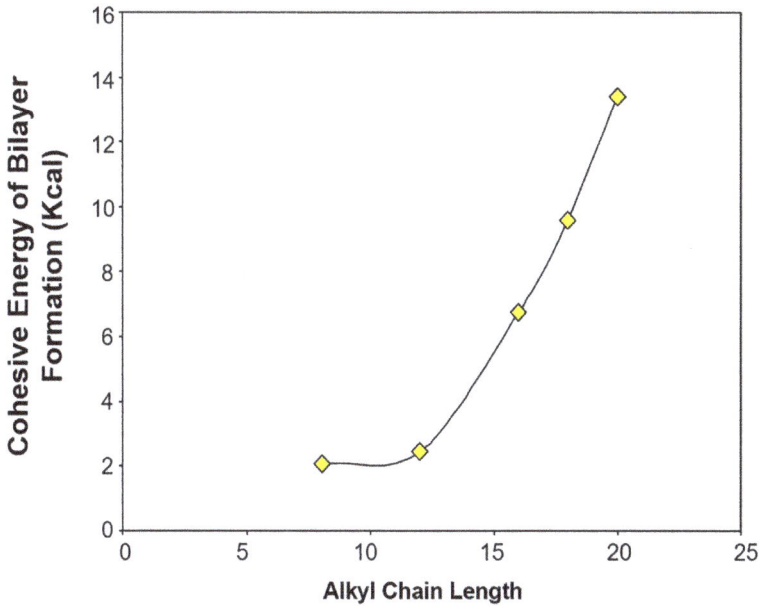

FIGURE 19.7 Cohesive energy of bilayer formation of imidazolines as a function of alkyl chain length. (Modified after Ramachandran and Jovancicevic (1998).)

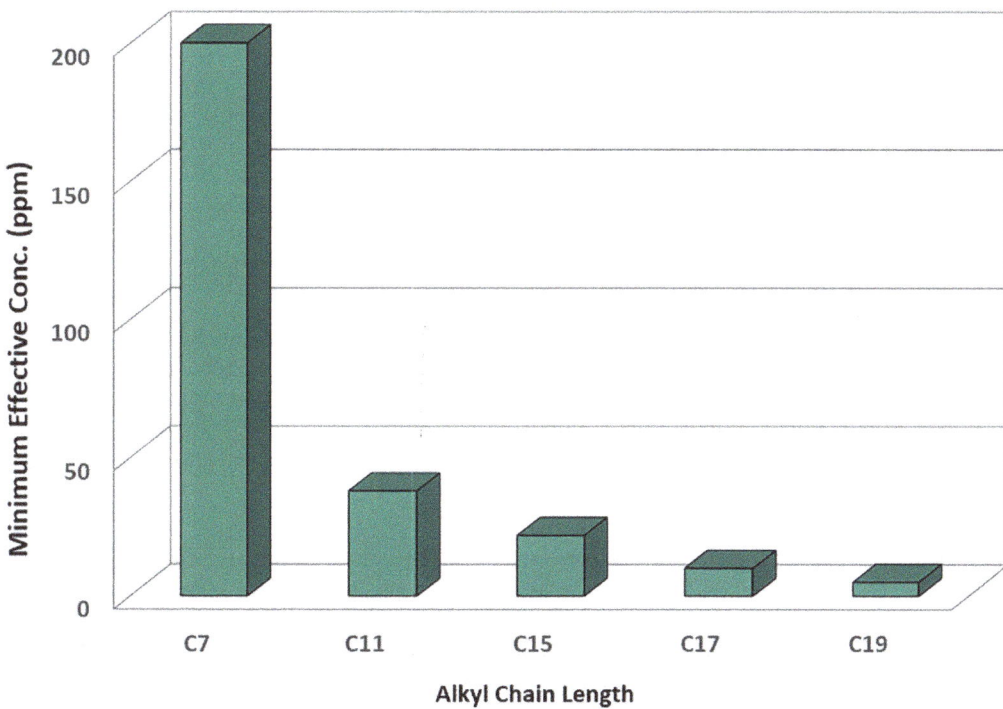

FIGURE 19.8 Minimum effective concentrations to retard CO_2 corrosion for various straight-chain imidazolines from rotating cylinder electrode experiments in a CO_2 saturated brine (1 atm) at 66°C, pH 6.3, with a rotation speed of 6000 rpm. (Modified after Jovancicevic et al. (1998).)

shows that longer chain lengths are more effective at lower concentrations, but these data are also a bit misleading because, at chain lengths higher than C19, the inhibitor becomes so insoluble that it begins to form a second phase even at low concentrations (Jovancicevic et al. 1998).

19.7   LIMITATIONS AND DEFICIENCIES OF THE SAM CI MECHANISM

Metal corrosion is complex and heterogeneous due to the presence of various anodic and cathodic surface reaction sites. The mechanism shown here applies to an idealized representation of the surfaces. Below are a few aspects not covered by this theory, most of which are related to dosage higher than the CMC.

- This mechanism does not consider the role played by cementite, which could be an important factor in the inhibition of CO_2 corrosion of carbon steel (Guldbransen et al. 2000; Cao et al. 2017).
- The model also does not consider the role of surface aggregation that manifests at higher concentrations. This model does not explore detrimental increases in corrosion rate at concentrations higher than the CMC (Smith et al. 2009).
- In formulated CI products, many components interact with one another, each changing the activity coefficients of the others. These complexities can invalidate some of the assumptions of this model.
- This model does not deal with the complex behavior interference from conventional surfactants or other oil field chemicals which could alter the Lewis base functionality of the CI.

Despite these limitations, the SAM CI theory has applicability across many known inhibitors and deviations can often be explained in simple terms.

19.8   CONCLUSIONS AND PERSPECTIVES

A SAM mechanism for corrosion inhibition for systems is described above, where inhibition is achieved with very low inhibitor concentrations. The mechanism has the following features:

1. Ordered adsorption of the CI molecules to the native oxide of metals
2. Strong CI binding to the surface which is unique for CI SAMs
3. SAM alignment on the native oxide of the metal where the inhibitor head aligns with the spacing of the oxide lattice.
4. Tilting of the hydrocarbon tail to form a tightly packed hydrophobic monolayer. This feature is sensitive to tail length with greater than C12 having the highest cohesive energy and greater than C17 showing the lowest minimum effective concentration for inhibition.
5. Sufficient partitioning of the organic molecules to the aqueous phase is required to allow fast enough diffusive transfer to the metal surface. This is also related to alkyl chain length and is optimal between C12 and C19.

The mechanism is valid for metal CIs that perform at dosages below the critical micelle concentration.
    At higher CI concentrations, the corrosion protection mechanism is more complex and film morphology can change with CI concentration compounding the complexity. Multilayer films could be present in the process of batch treating, which may involve complex aggregates. The film persistency on the surface will still be a function of the aggregate binding force to the surface and cohesiveness of the aggregates to each other to prevent water penetration into the metal surface. The formation of aggregate or supramolecular structures on metal surfaces (gold, mica, graphite, MoS_2) has been extensively studied since the 1990s (Manne and Gaub 1996). Creating supramolecular surface structures could result in more persistent CI films, but this area of research is still

in its infancy. However, morphology changes to supramolecular structures have also been shown to decrease CI performance in some instances (Smith et al. 2009). Unfortunately, little research has been conducted on the morphology of high-performing CI films in an aqueous solution on iron. Molecular modeling could provide insight into this area, but sufficient force fields for iron and iron oxides are not yet available as of this writing.

ACKNOWLEDGMENTS

The authors would like to thank Baker Hughes and the many internal and external collaborators that made this work possible.

REFERENCES

Bain, C. D., and Whitesides, G. M. 1989. Modeling organic surfaces with self-assembled monolayers, *Angew. Chem.* 101(4), 522–528.

Bard, A., and Faulkner, L. R. 2001. *Electrochemical Methods: Fundamentals and Applications*, 2nd ed., John Wiley & Sons, Inc., Hoboken, NJ.

Cao, F., Mennito, A., Schmatz, D., Luo, S., Dierolf, M., Xiong, Y. and Ling, S. 2017. A Mechanistic Study of Corrosion Inhibitor Partitioning and Performance in Sweet Corrosion Environments, Paper No. 9399 Presented at CORROSION/17, NACE International: Houston, TX.

Gulbrandsen, E., Nyborg, R., Loland, T., and Nisancioglu, K. 2000. Effect of Steel Microstructure and Composition on Inhibition of CO_2 Corrosion, Paper No. 23 Presented at CORROSION/00, NACE International: Houston, TX.

Israelachvili, J. N., Mitchell, D., and Ninham, B. W. 1976. Theory of self-assembly of hydrocarbon amphiphiles into micelles and bilayers, *J. Chem. Soc. Faraday Trans.* 2(72), 1525–68.

John, R. C., Jordan, K. G., Young, A. L., Kapusta, S. D., Thompson, W. T. 1998. Sweetcor: An Information System for The Analysis of Corrosion of Steels by Water and Carbon Dioxide, Paper No. 20 Presented at CORROSION/98, NACE International: Houston, TX.

Joseph, M., and Klenerman, D. 1992. A second-harmonic generation study of a corrosion inhibitor on a mild steel electrode, *J. Electroanal. Chem.* 340, 301–313.

Jovancicevic, V., and Bockris, J. O. M. 1986. The mechanism of oxygen reduction on iron in neutral solutions, *J. Electrochem. Soc.* 133, 1797.

Jovancicevic, V., Ramachandran, S., and Prince, P. 1998. Inhibition of CO_2 Corrosion of Mild Steel by Imidazolines and their Precursors, Paper No. 18, Presented at CORROSION/98, NACE International: Houston, TX.

Ko, X., and Sharma, S. 2017. Adsorption and self-assembly of surfactants on metal–water interfaces. *J. Phys. Chem. B* 121(45), 10364–10370.

Ko, X., and Sharma, S. 2020. Adsorption and Self-Assembly of Corrosion Inhibitors on Metallic Surfaces Studied Using Molecular Simulations, Paper No. 15056 at CORROSION/20, NACE International: Houston, TX.

Lee, C.-W., and Bard, A. 1988. Comparative electrochemical studies of N-methyl-N'hexadecyl viologen monomolecular films formed by irreversible adsorption and the Langmuir-Blodgett method, *J. Electroanal. Chem. Interfacial Electrochem.* 239, 441–446.

Manne, S., and Gaub, H. E. 1995. Molecular organization of surfactants at solid-liquid interfaces, *Science* 270, 1480.

McMahon, A. J. 1991. The Mechanism of action of an oleic imidazoline for oilfield use, *Colloids and Surfaces* 59. 187–208.

Oblonsky, L. J., Chesnut, G. R., and Devine, T. M. 1995. Adsorption of octadecyldimethylbenzylammonium chloride to two carbon steel microstructures as observed with surface-enhanced Raman spectroscopy, *Corrosion* 51, 891–899.

Ramachandran, S., and Jovancicevic, V. 1998. Molecular Modeling of the Inhibition of Mild Steel CO_2 Corrosion by Imidazolines, Paper No. 17, Presented at CORROSION/98, NACE International: Houston, TX.

Ramachandran, S., Tsai, B. L., Blanco, M., Chen, H., Tang, Y., and Goddard, W. A. 1996. Self-assembled monolayer mechanism for corrosion inhibition of iron by imidazolines, *Langmuir* 12, 6419–6428.

Ramachandran, S., Tsai, B. L., Blanco, M., Chen, H., Tang, Y., and Goddard, W. A. 1997. Atomistic simulations of oleic imidazolines bound to ferric clusters, *J. Phys. Chem. A* 101, 83–89.

Rodriguez-Valdez, L. M., Villiamsar, W., Casales, M., Gonzales- Rodriguez, J. G., Martinez-Villafane, A., Martinez, L., and Glossman-Mitinik, D. 2006. Computational simulations of the molecular structure and corrosion properties of amidoethyl, aminoethyl and hydroxyethyl imidazoline inhibitors, *Corros. Sci.* 48, 4053–4046.

Singh, H. and Sharma, S. 2019. Aggregation and Adsorption Behavior of Organic Corrosion Inhibitors studied using Molecular Simulations, CORROSION/19, paper 12953, NACE International: Houston, TX.

Smith, C., Li, C., Bedard, T.C. and Kimler, J. 2009. March. A new MIC control strategy in low velocity gas gathering pipelines, Paper No. 09402 Presented at CORROSION/09, NACE International: Houston, TX.

Ulman, A. 1996. Formation and structure of self-assembled monolayers, *Chem. Rev.* 96, 1533–1554.

Wang, Z. M., and Guang-Ling Song, G.-L. 2019. Fast Evaluation of Corrosion Inhibitors Used in Oil/Water Mixed Fluids, Paper No. 12887 Presented at CORROSION/19, NACE International: Houston, TX.

Xing, W., Shan, Y., Guo, D., Lu, T., and Xi, S. 1995. Mechanism of iron inhibition by stearic acid Langmuir-Blodgett monolayers, *Corrosion* 51, 45–49.

Yao, Y. 1963. Heats of adsorption of amines, hydrocarbons, and water vapor on reduced and oxidized iron surfaces, *J. Phys. Chem.* 67, 2055–2061.

Zhang, R., and Somasundaran, P. 2006. Advances in adsorption of surfactants and their mixtures at solid/solution interfaces, *Adv. Colloid Interface Sci.* 123, 213–229.

Zhu, Y., Free, M. L., Woollam, R., and Durnie, W. 2017. A review of surfactants as corrosion inhibitors and associated modeling, *Prog. Mat. Sci.* 90, 159–223.

Zhu, Y., Free M. L., and Yi, G. 2016. The effects of surfactant concentration, adsorption, aggregation, and solution conditions on steel corrosion inhibition and associated modeling in aqueous media, *Corros. Sci.* 102, 233–250.

Section III

Applications in Biofouling Protection

20 Supramolecular Polymers/Gels/Self-Assemblies for Anti-Biofouling Surfaces

Drishya Elizebath and Akhil Padmakumar
CSIR-National Institute for Interdisciplinary
Science and Technology (CSIR-NIIST)
Academy of Scientific and Innovative Research (AcSIR)

Rakesh K. Mishra
National Institute of Technology, Uttarakhand (NITUK)

Vakayil K. Praveen
CSIR-National Institute for Interdisciplinary
Science and Technology (CSIR-NIIST)
Academy of Scientific and Innovative Research (AcSIR)

CONTENTS

20.1 INTRODUCTION

Supramolecular chemistry, commonly defined as "chemistry beyond the molecule," focuses on developing complex systems through the association of molecular species bound by synergistic interaction of various non-covalent forces (Lehn, 1988). From the humble beginning as an area of chemistry, this relatively young field has emerged as an inter-disciplinary subject, spreading its realm to physics, materials and biology. Supramolecular soft materials such as molecular aggregates, supramolecular polymers and gels (Aida and Meijer, 2020; Mishra et al., 2018), which link supramolecular chemistry and materials, have lured research around the globe due to their superior properties such as dynamic nature, self-healing, stimuli-responsive behavior, and processability and have been utilized for various applications such as functional soft materials, biomaterials, sensing and organic electronics (Amabilino et al., 2017; Du et al., 2015; Anees et al., 2017; Ghosh et al., 2016). In recent years, supramolecular materials have started to find themselves useful in new avenues, such as hybrid materials, art conservation, marine oil-spill recovery, corrosion prevention, three-dimensional printing, soft-robotics, anti-counterfeiting, and so on (Vedhanarayanan et al., 2018; Carretti et al. 2010; Prathap and Sureshan, 2017; Saji, 2019;

DOI: 10.1201/9781003169130-23

Chivers and Smith, 2019; Whitesides, 2018; Praveen et al., 2020). The modular and reversible properties of supramolecular systems useful for various applications in modern-day human life make these materials an exciting topic of research worldwide. This chapter discusses an emerging application of supramolecular polymers/gels/self-assemblies in the development of anti-biofouling surfaces with the help of selected illustrative examples.

20.2   BIOFOULING

Biofouling is the undesired accumulation of various microorganisms, plants, animals, and their by-products on unprotected surfaces, causing health, financial, and environmental risks in the medical, food packing and storage, water purification systems, marine and industrial fields (Kirschner and Brennan, 2012; Banerjee et al., 2011; Bixler and Bhushan, 2012). Medical biofouling occurs in implants, sensors, catheters and various medical equipment. Biofouling of the implants results in thrombosis or infections such as *Staphylococcus aureus* bacteria that also causes methicillin resistance. The formation of biofilms on medical devices increases the risk of antibody resistance, leading to medical conditions like cystic fibrosis. Fouling can also cause bio-incompatibility and malfunction of the sensors and rejection of implants.

Food processing and packaging is another sector seriously affected by biofouling. Biofilm formation by various pathogens has become problematic in a wide range of food industries, including seafood, meat, poultry, and dairy processing, causing quality issues such as foul smell, bad odor, and food spoilage (Verran, 2002; Singh et al., 2019; González-Rivas et al., 2018; Carrascosa et al., 2021). The complexity of the food-processing units, long production periods, and the nutrients present in the food materials such as fats, proteins, and carbohydrates provide a suitable environment for the growth of biofilm and microorganisms. Various biofilm-forming pathogens are *Bacillus cereus*, *Escherichia coli*, *Salmonella enterica*, *Staphylococcus aureus*, *Geobacillus stearothermophilus*, and so on. The intake of fouled food products can cause sickness and diarrhea, hemolytic uremic syndrome, listeriosis, antibiotic resistance, and so on.

Biofouling of membranes is a major challenge faced in the sustainable purification of potable water, industrial wastewater treatment and desalination, hindering the nanofiltration and reverse osmosis membrane filtration processes (Zhao et al., 2018; Zhang et al., 2016; Nguyen, et al., 2012). Removal of biofoulants from water is quite challenging since a minimal leftover of the pathogens can continue the growth of biofoulants utilizing different biodegradable substances in the feed water. ‖‖‖Biofouling can have several adverse effects on membrane systems, such as membrane flux decline, increased differential pressure and feed pressure, membrane biodegradation, increased salt passage through the membrane, reduced quality of the product water and increased energy consumption.

Industrial fouling associated with the accumulation of microorganisms from flowing water on the surfaces of processing equipment affects various industries, including nuclear reactors, power plants, paper production, textile industry, and so on (Bott, 2011; Thomason, 2009; Flemming et al., 2009). Biofouling in nuclear reactors and power plants can cause energy losses due to increased fluid frictional resistance in pipelines and propellers, increased heat transfer resistance in condensers and process heat exchangers, which creates safety concerns during accidents or other emergencies, decreased pipe pressures, increased operational cost due to capital losses for replacement of equipment due to corrosion and excess downtime for cleaning the components. In paper manufacture, the water recycled during the process is ideal for biological growth in terms of temperature and nutrients, fouling the heat exchange equipment or rolled steel, causing quality issues in the manufactured paper.

The most prevalent form of biofouling occurs in the marine environment (Callow and Callow, 2011; Cao et al., 2011). More than 4000 kinds of marine biofouling species have been reported globally, with sizes ranging from micrometer to centimeter, classified as microfoulants such as

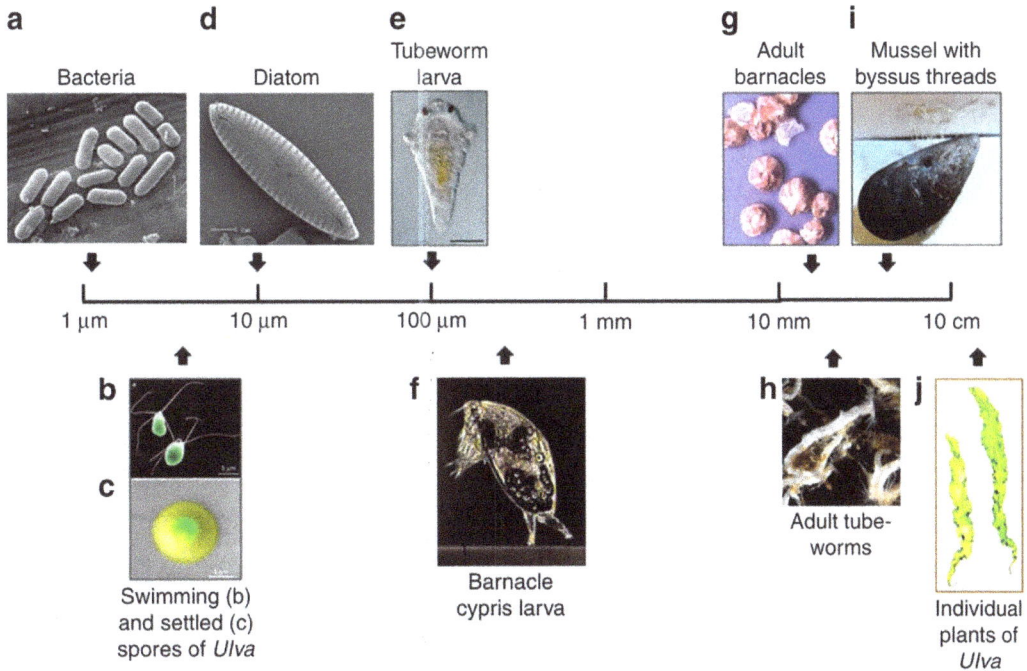

FIGURE 20.1 Diversity and size scales of biofouling organisms. (Reproduced with permission from Callow and Callow (2011) © 2011, Nature Publishing Group.)

bacteria, diatoms, algae spores and macrofoulants such as barnacles, tube-worms, oysters, and mussels, respectively (Figure 20.1). The extent of marine biofouling is dependent on several factors such as temperature, presence of nutrients and pH of the aquatic environment, and surface properties, such as surface energy and wettability. Marine biofouling affects unprotected surfaces of ships such as waterline, propeller and rudder blades, buoys, offshore rigs, bridge pillars, pipelines, underwater cables and instruments such as sonar, heat exchangers, aquaculture cages, fishing nets and other submerged structures in aquatic environments. Marine biofouling causes hull roughness, hull drag, engine stress, corrosion and reduced speed, leading to higher maintenance costs and safety concerns for shipping vessels. In addition, the various cleaning processes to remove biofoulants generate a large number of toxic substances that are discharged into the ocean, affecting the aquatic ecosystem.

20.3 MECHANISM OF BIOFOULING

Biofouling is a dynamic process with a timeline of seconds to days or months. It is often considered to progress following a linear successional model as given in Figure 20.2a, occurring in four steps: adsorption of organic molecules, primary colonization, soft macrofouling and hard macrofouling (Wahl and Lafargue, 1990; Yang et al., 2014). In the first step, organic molecules, such as protein, polysaccharides and proteoglycans, attach to the unprotected substrate surface through a physical reaction. In the next step, a biofilm is formed as bacteria and diatoms adsorb to the surface, followed by microalgae and protozoans. The depth of the biofilm depends on the season, geographic location and physical effects such as electrostatic interactions, gravity, water flow and van der Waals forces. Biofilms provide increased adhesion and protect the attached microorganisms by providing enhanced adhesion and controlling the diffusion of nutrients within the colony. In the last step, macroorganisms such as spores of macroalgae, barnacle larvae, bryozoans, mollusks,

a

SURFACE CONDITIONING FILM

⇐ BACTERIA ⇐ ALGAE ⇐ INVERTEBRATE LARVAE

b

SURFACE CONDITIONING FILM

BACTERIA

ALGAE

INVERTEBRATE LARVAE

PARTICULATES

MARINE SNOW

SINK | SOURCE | SINK

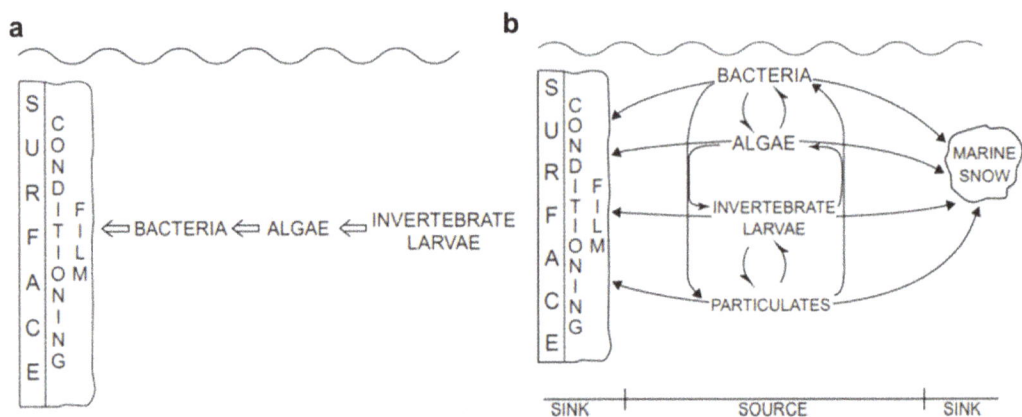

FIGURE 20.2 (a) Linear successional model and (b) dynamic/availability model of biofouling. (Reproduced with permission from Clare et al. (1992) © 1992 Taylor & Francis.)

polychaetes, tunicates, and coelenterates attach to the surface, leading to their growth. However, it is interesting to note that the sequence of biofouling does not always strictly follow this successional model. The formation of biofilm on the surface is not an essential criterion and different marine species may settle onto the surface simultaneously. Hence, controlling the initial stages of biofouling does not result in anti-biofouling. Thus, another model called the dynamic model is considered, as depicted in Figure 20.2b; as per this, the major driving force in fouling is the relative amount of each foulant in the aquatic environment (Clare et al., 1992). This model suggests that the initial biofouling events may influence subsequent events but are not necessarily precursors to the progress of the fouling process.

20.4   ANTI-BIOFOULING STRATEGIES

The term antifouling was considered synonymous with biocidal in the early twentieth century (Omae, 2003; Yebra et al., 2004). The most common coatings used in the 1960s contained tin-based biocides such as tributyltin and tributyltin oxide. However, the adverse effects of the accumulation of these compounds in aqueous environments resulted in the voluntary withdrawal of organotin compounds from the market in 2003 and they were banned by the International Maritime Organization in 2008. Since then, antifouling research has focused on the development of non-toxic or green antifouling coatings, considering strategies such as preventing attachment of the biofoulants to the surfaces by reducing adhesion of the foulants to the surfaces and killing of biofoulants, with more emphasis on utilizing antifouling mechanisms found in nature (Li et al., 2021; Banerjee et al., 2011; Callow and Callow, 2011; Yang et al., 2014).

20.5   SUPRAMOLECULAR SELF-ASSEMBLED MATERIALS AS ANTI-BIOFOULING COATINGS

Supramolecular assemblies and hydrogels derived from peptide-based scaffolds (Du et al., 2015; Draper and Adams, 2019) find promising applications in tissue regeneration techniques due to their ability to create a hydrated environment near its vicinity, which is found to be a suitable environment for the functioning of the host cell. However, this specific condition could also accelerate microbial growth, making the post-implantation processes cumbersome. In this scenario, a hydrogel capable of showing the intended tissue regeneration purpose along with the antibacterial property would be ideal.

Schneider and co-workers were the pioneers in disclosing the inherent antibacterial properties associated with peptide-based hydrogels (Salick et al., 2007). For this purpose, they chose a 20-residue peptide, VKVKVKVKV^{D}PPTKVKVKVKV-NH$_2$ (MAX1), which is rich in valine and lysine amino acids. When the peptides get dissolved in water, the lysine chains get protonated and this prevents the peptide from folding and ends up in a randomly coiled conformation. However, effective charge screening of the protonated lysine-side chain would help the peptide to form an amphiphilic β-hairpin conformation, which rapidly self-assembles into a mechanically rigid hydrogel. Various microscopic and rheological studies provide evidence for interconnected fibrils rich in β-sheet conformation. On a closer look, we can see that the interior of each fibril is constituted by hydrophobic valine residues and their exteriors by hydrophilic lysine residues, which are known to have antibacterial activity (Shima et al., 1984).

The MAX1 gels exhibited broad-spectrum antibacterial activity for solutions ranging from concentrations of 2×10^3 colony forming units (CFUs)/dm^2 to 2×10^9 CFUs/dm^2. Detailed studies on this particular hydrogel surface showed MAX1 hydrogel is active against Gram-positive and Gram-negative bacteria found in hospital environments. Though the mechanism of the bactericidal properties is not yet clear, live-dead assays using laser scanning confocal microscopy and α-galactosidase leakage experiments indicate that direct cellular interaction with the surface may be required for membrane disruption and eventually cell death. In addition to the bactericidal property, the hydrogel surface allowed mammalian cell adhesion and proliferation. Most importantly, the gel surfaces were found to be nonhemolytic toward human red blood cells. So based on these observations, this particular peptide-based β-hairpin hydrogel can be considered an ideal candidate for tissue regeneration purposes with the added advantage of bactericidal properties.

Later, Liu, Li, Yang and co-workers introduced an environmentally benign antimicrobial coating derived from naturally occurring lysozyme, an enzyme popularly known for its antimicrobial activity in the natural defense system (Gu et al., 2017). However, the usage of the lysozyme on Gram-negative bacteria and fungi is limited due to its poor activity and instability under ambient conditions. Through this work, the group developed a phase-transited lysozyme from hen egg-white without any complex chemical and physical treatment (Figure 20.3). When the lysozyme is mixed with tris(2-carboxyethyl)phosphine buffer at neutral pH, the disulfide bonds in lysozyme reduces, resulting in a partially unfolded monomer aggregation as two-dimensional (2D) nanofilm. The antimicrobial activity of this nanofilm can be explained by the enrichment of positive charges on the surface. The phase-transited lysozyme nanofilm coated on glass slides showed an effectiveness of 95, 92, and 94% toward *E. coli*, *S. aureus*, and *C. albicans*, respectively. Even though the anti-Gram-positive bacterial activity was reasonably expected from the conventional lysozyme, its activity toward the Gram-negative bacteria was quite surprising. This result can be explained with the added advantage of the positively charged surface of 2D nanofilm obtained by the polymeric aggregation processes. On the other hand, the bactericidal activity toward Gram-positive bacteria could be explained based on the ability of lysozyme to cleave the β-1,4-glycosidic linkages in peptidoglycans present in Gram-positive bacteria cell wall.

The next challenge was then to incorporate the nanofilm into medical devices with various requirements. The system showed a significantly easy and effective coating on materials of any shape or surface composition using simple methods. In the first method, a material was soaked into the phase transition buffer and the assembled lysozyme was transferred onto the immersed material surface to form the phase-transited lysozyme nanofilm coating. In the next method, the film was generated on the phase transfer buffer, which could be lifted by the substrate. Apart from these two wetting techniques, they also tried the printing method where phase-transited lysozyme nanofilm was performed on an agarose gel that could be transferred to the substrate through stamping or peeling. An *in vivo* antimicrobial test on a coated catheter implanted in a rat showed excellent infection inhibition. Hence, hen egg-white lysozyme offers lower environmental impact and can be prepared on a large scale from low-cost renewable feedstock without physical doping or chemical synthesis compared to other antimicrobials such as silver nanoparticles, synthetic polymers, or peptides.

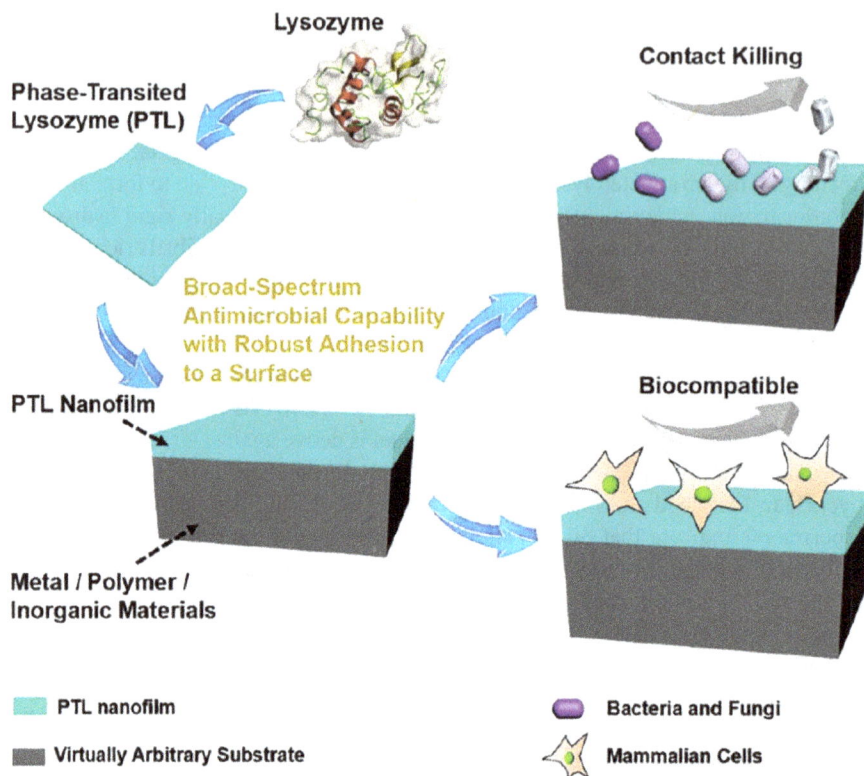

FIGURE 20.3 Strategy to develop phase-transited lysozyme nanofilm toward a broad-spectrum antimicrobial coating on virtually arbitrary materials. (Reproduced with permission from Gu et al. (2017) © 2016 American Chemical Society.)

In 2014, Reches and co-workers introduced a supramolecular anti-biofouling coating based on a peptide unit with improved biocompatibility (Maity et al., 2014). Rather than using a polypeptide, as seen in the earlier case, the research group utilized a relatively small tripeptide unit (Figure 20.4). Here, the design strategy of the self-assembling monomer consists of a conceptual merging of a fluorinated derivative of diphenyl alanine along with a third peptide unit, 3,4-dihydroxy-L-phenylalanine (Figure 20.4). The latter amino acid moiety is best known for its adhesion properties in

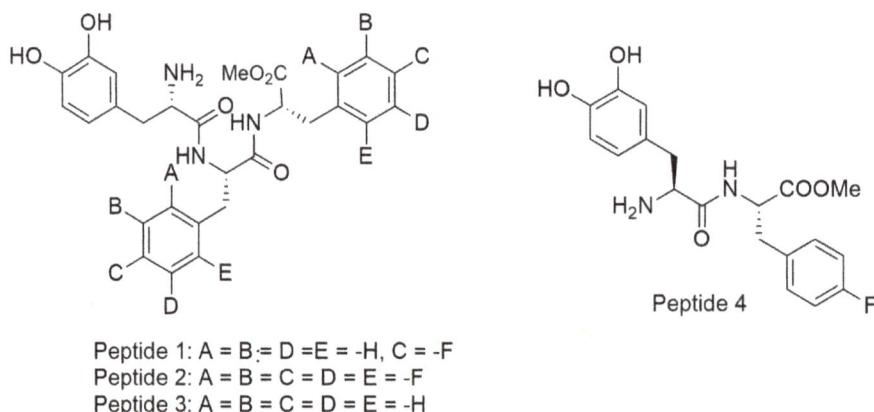

Peptide 1: A = B = D =E = -H, C = -F
Peptide 2: A = B = C = D = E = -F
Peptide 3: A = B = C = D = E = -H

FIGURE 20.4 Molecular structures of self-assembling short peptides used for the anti-biofouling studies. (Reproduced with permission from Maity et al. (2014) © 2014 The Royal Society of Chemistry.)

marine mussels. However, the former unit rich in fluorine atoms is supposed to prevent proteins from attaching to the surface and therefore acts as the anti-biofouling motif. The short peptides (peptides 1–4) were dissolved in various polar solvents and coated onto surfaces like gold, silicon, titanium, glass, stainless steel, and so on. As anticipated, the authors could find a significant improvement in the hydrophobicity of the system with an increase in the number of fluorine atoms in the monomer unit, which can be quantified from the contact angle measurement of the prepared coating.

The anti-biofouling activity of the peptide systems was studied with bovine serum albumin (BSA) and lysosome. Peptide 1 with the highest number of fluorine atoms showed the best anti-biofouling activity upon coating. These results were quantified by the test kit available for BSA and lysosome (Figure 20.5a). Similarly, the antibacterial property of the coating was also evaluated for *P. aeruginosa and E. coli*. The staining of the substrate with crystal violet and its optical quantification is given in Figure 20.5b. The extracted crystal violet dye from the substrate is proportional to the number of bacteria attached to the surface. As expected, peptide 1 showed significant antibacterial properties owing to its hydrophobicity. The anti-biofouling activity of such tripeptide units opens a wide possibility of their utility as functional materials with less synthetic effort.

On a comparable basis, the dual functionality of the peptides was introduced by the research group of Reches (Saha et al., 2020). For this purpose, the authors used a self-assembling amphiphilic peptide (Figure 20.6a), which is capable of both anti-microbial and antifouling properties due to the lysine amino acid residue and the fluorinated hydrophobic amino acids, respectively. As in the previous study, the 3,4-dihydroxy-L-phenylalanine (L-DOPA) portion enhances the adhesion property of the system. The antifouling properties of the self-assembling peptide-coated titanium surfaces were evaluated by various optical microscopy analyses. Fluorescent microscopy technique with staining substances like SYTO9 and propidium iodide showed a clear distinction between the dead and live bacteria upon treating the peptide material under study (Figure 20.6b). The fluorescence-activated cell sorting technique provides more quantitative measures in the result mentioned above. Moreover, the resistance to the protein adsorption of the system was confirmed by a quartz crystal microbalance technique using BSA.

Apart from this sort of biologically derived system, there were reports of antimicrobial hydrogels based on synthetic polymers like polyelectrolyte complexes, polycationic chitosan-graft-poly(ethylene

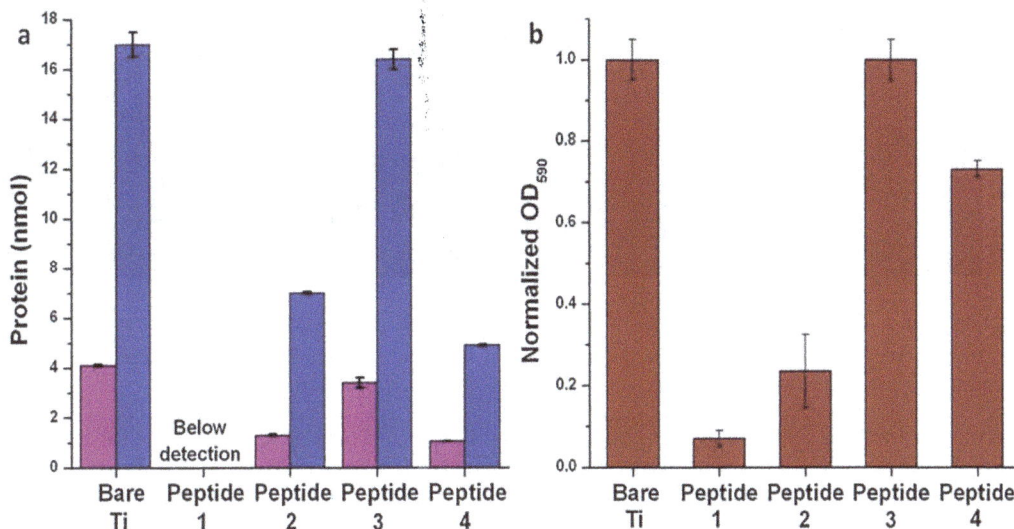

FIGURE 20.5 Adsorbed amounts of (a) bovine serum albumin and lysozyme, represented in violet and blue, respectively, on bare and peptide coated titanium substrates. (b) Optical density quantification of P. aeruginosa on bare and peptide coated titanium substrates. (Reproduced with permission from Maity et al. (2014) © 2014 The Royal Society of Chemistry.)

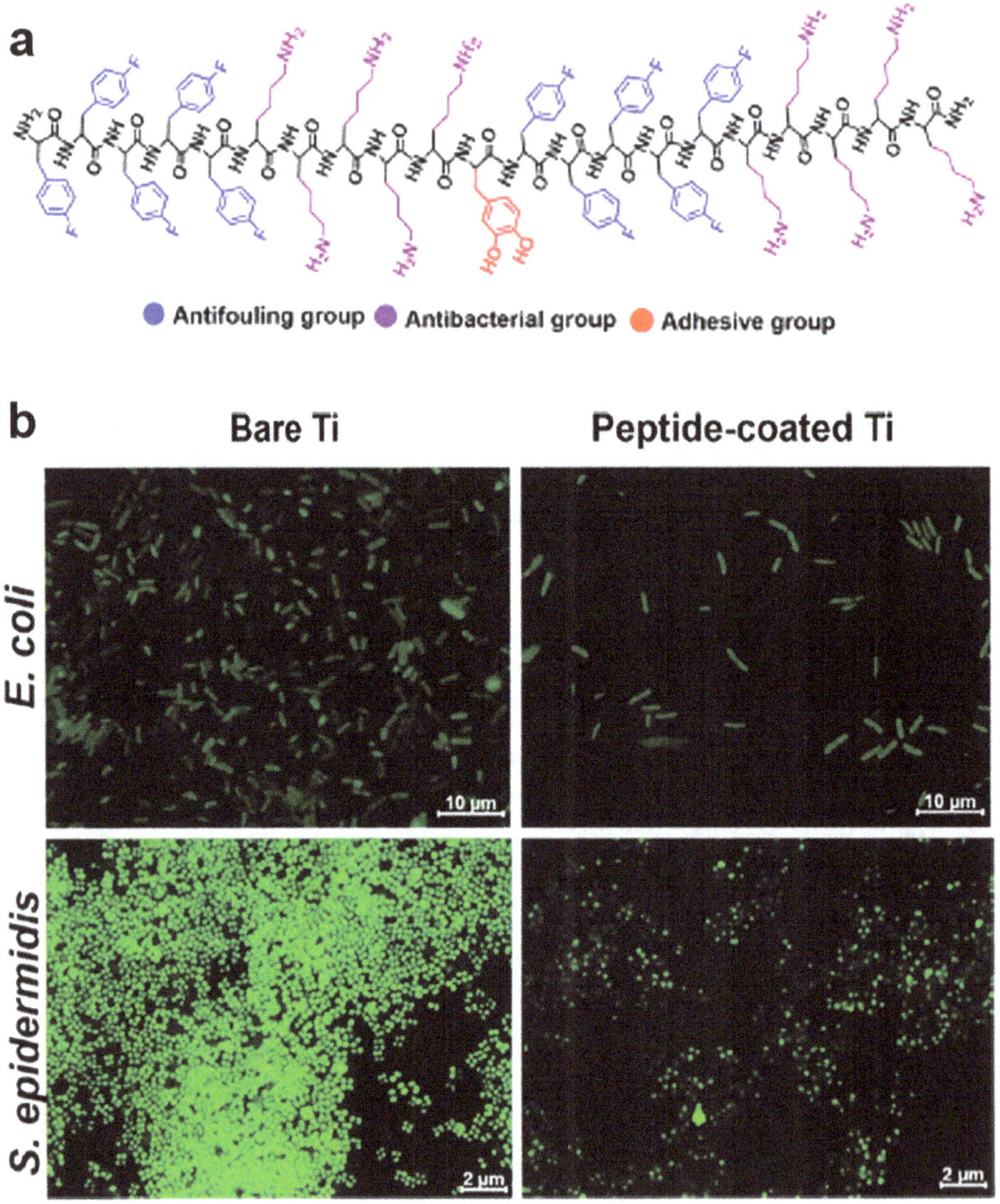

a

● Antifouling group   ● Antibacterial group   ● Adhesive group

b

FIGURE 20.6 (a) Molecular structure of the self-assembling amphiphilic peptide. (b) Fluorescence microscopy images of bare and peptide-coated titanium surfaces demonstrating the antibacterial activity. (Reproduced with permission from Saha et al. (2020) © 2020 American Chemical Society.)

glycol) methacrylate, ε-poly-l-lysine-graft-methacrylamide, and peptide self-assembly (Tsao et al., 2010; Li et al., 2011; Zhou et al., 2011; Salick et al., 2007). However, there is significant limitations associated with these sorts of systems, like immunogenicity and material variance or the cost of the peptides. Therefore, there is a pressing demand for hydrogels derived from synthetic variants of the peptides more cost-effectively. In this aspect, Yang, Hedrick and co-workers developed a stimuli-responsive antimicrobial gel formed from stereocomplexation of biodegradable poly(l-lactide)-*b*-poly(ethylene glycol)-*b*-poly(l-lactide) (PLLA-PEG-PLLA) and a charged biodegradable polycarbonate triblock polymer (PDLA-CPC-PDLA) (Figure 20.7a) (Li et al., 2013). The

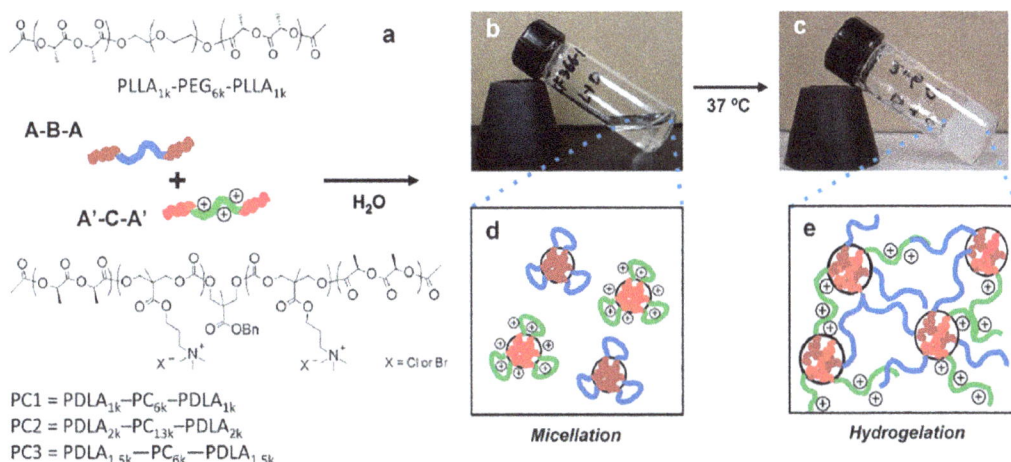

PLLA$_{1k}$-PEG$_{6k}$-PLLA$_{1k}$

A-B-A

+

A'-C-A'

H$_2$O

O⁻OBn

X = Cl or Br

PC1 = PDLA$_{1k}$–PC$_{6k}$–PDLA$_{1k}$
PC2 = PDLA$_{2k}$–PC$_{13k}$–PDLA$_{2k}$
PC3 = PDLA$_{1.5k}$—PC$_{6k}$—PDLA$_{1.5k}$

a

b

c

37 °C

d

Micellation

e

Hydrogelation

FIGURE 20.7 (a) Schematic representation and pictures of 10% w/v solution of PLLA-PEG-PLLA and PDLA-CPC-PDLA. (b) at 25°C and (c) at 37°C (d) flower-type micelle formation of the polymer at 25°C (e) stereocomplex formation between the enantiomeric pure polylactide segments in the micelle cores. (Reproduced with permission from Li et al. (2013) © 2013 Wiley-VCH Verlag GmbH & Co. KGaA, Weinheim.)

stereocomplexes existed as soluble micelles in the ambient aqueous medium at 25 °C; however, the system significantly improved the material properties upon heating to the physiological temperature, 37 °C (Figure 20.7b and c). The hydrogel formation under physiological conditions with significant improvement in the antimicrobial properties makes this system a promising candidate in this area.

The antimicrobial activities of stereocomplex were evaluated against various pathogens like *S. aureus* (Gram-positive), *E. coli* (Gram-negative), and *C. albicans* (fungus). The gels prepared from PLLA-PEG-PLLA, PDLA-CPC-PDLA, and PDLA-PEG-PDLA in a molar ratio of 1:0.15:0.85 were found to suppress the growth of bacteria with 100% efficiency. However, a gel with an increased amount of PDLA-CPC-PDLA was needed for the fungicidal activity.

As we discussed in the previous case, the mechanism of antimicrobial action is the same, i.e., cell wall/membrane lysis. However, here the authors supported the plausible explanation with morphological evidence from SEM studies (Figure 20.8). Cellular deformation and surface roughness on the cell after treatment with the hydrogel indicate the catastrophic membrane failure mechanism of the process involved. The noncovalent interactions involved in the system provide an added advantage of processability to the system, like the shear thinning effect. Moreover, the hydrogels showed no signs of cytotoxicity nor hemolysis.

So far, we have discussed anti-biofouling hydrogel materials, which are either derived from naturally derived polymers or synthetic polymers. Recently, Cai, Gu and co-workers introduced an inorganic-organic hybrid material capable of showing anti-biofouling activity with enhanced material properties (Lian et al., 2018). The ionic interaction between the negatively charged sodium polyacrylate entwined clay nanosheets and the positive-charged polyhedral oligomeric silsesquioxane (POSS) core-based generation one (*L*-Arginine) dendrimer (POSS-R) stabilized the hydrogel even at low concentrations of POSS-R. Moreover, they were able to improve the material properties by increasing the POSS-R concentration. The hydrogel with the enhanced material property was then used to test the anti-biofouling property with MC3T3 cells. The cell death was monitored by staining the cell with a dye, which can stain selectively metabolically active cells and be visualized with green fluorescence. The mechanistic details of this anti-biofouling property can be explained by the electrostatic repulsion exerted by the sodium polyacrylate surface and cell bearing the same surface charge. This particular system can be considered an excellent example of an inorganic–organic hybrid soft material with enhanced material strength and anti-biofouling property.

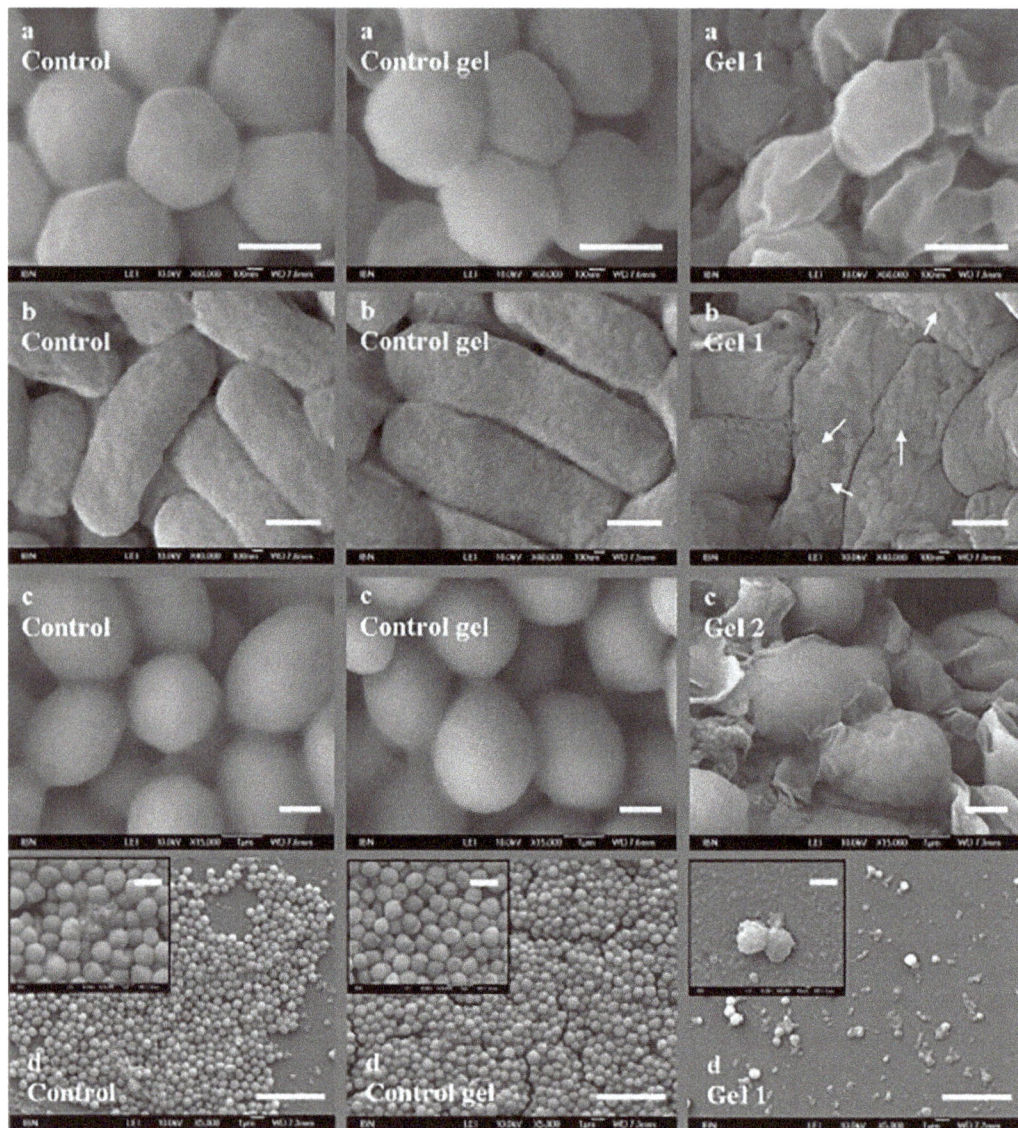

FIGURE 20.8 SEM images of (a) S. aureus, (b) E. coli (scale bar for a, b: 500 nm), (c) C. albicans (scale bar: 1 mm), and (d) Methicillin-resistant S. aureus biofilm before and after treatment with control gel, gel 1 (S. aureus, E. coli and Methicillin-resistant S. aureus), and gel 2 (C. albicans) for 2 h (scale bar: 5 mm). Scale bar for control and control gel samples (inset): 1 mm; gel 1 sample (inset): 500 nm. (Reproduced with permission from Li et al. (2013) © 2013 Wiley-VCH Verlag GmbH & Co. KGaA, Weinheim.)

Dankers and co-workers came up with a synthetic supramolecular polymeric system that is capable of showing the anti-biofouling activity based on a self-complementary 2-ureido-4[1H]-pyrimidinone (UPy) moiety (Pape et al. 2017). The multicomponent supramolecular system consists of a base self-assembling unit (PCLdiUPy) and an additive to regulate hydrophilicity (Figure 20.9). The additives are rich in poly(ethylene glycol) (PEG) chains, and that could bring more hydrophilicity, thereby achieving the anti-biofouling property. Moreover, the dodecyl spacer between UPy and the PEG acts as a barrier to protect the interaction of the urea group from the hydrated water of the PEG chain, which can enhance the anchoring in the base unit (PCLdiUPy). The results were quantified based on contact angle measurement on the film prepared by

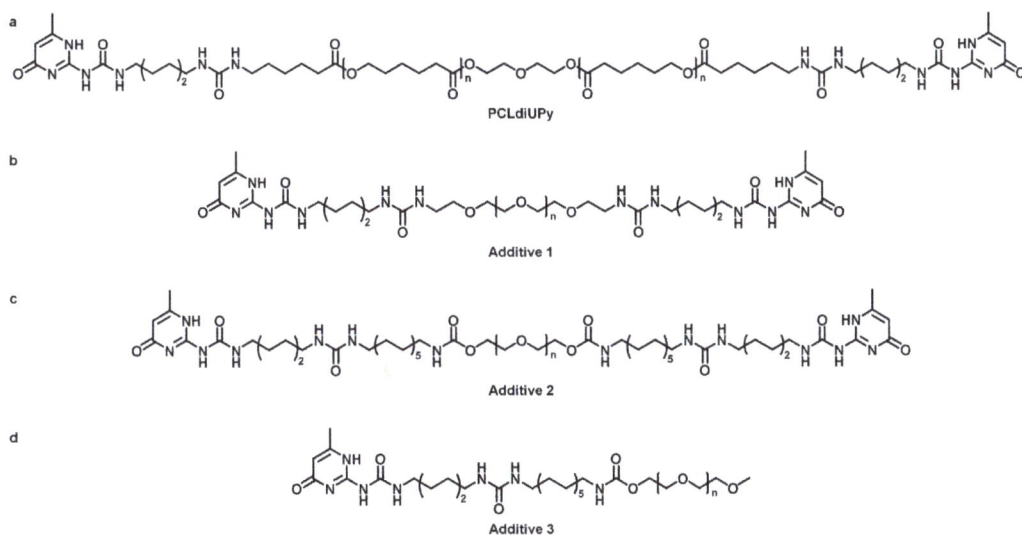

FIGURE 20.9 Chemical structures of the base self-assembling unit and additives. (a) UPy-modified poly-caprolactone (b) Bifunctional UPy poly(ethylene glycol) (c) Bifunctional PEG$_{10k}$diUPy (d) Monofunctional UPyPEG. (Reproduced with permission from Pape et al. (2017) © 2017 American Chemical Society.)

spin-casting of the self-assembled materials from the PCLdiUPy base unit and with different additives 1–3. Pristine PCLdiUPy exhibited a water contact angle of 74.7°C±0.2°C while the addition of hydrophilic substances decreased the contact angle to 59.8°C±1.3°C for PCLdiUPy with additive 1, 56.7°C±0.9°C with additive 2, and 44°C±14°C with additive 3. The cell adhesion property studies carried out with Human Umbilical Vein Endothelial cells (HUVECs) and cells from Human Kidney (HK-2) showed an apparent decrease in adhesion and spreading on mixtures with additive 2 and additive 3 compared to additive 1, where HK-2 cells showed a well-spread morphology though fewer HUVECs adhered. This result indicates that the length of the glycol chain and the improved anchoring imparted into the polymer due to the additional alkyl spacer are crucial factors for non-cell adhesive properties.

Later, the same group reported more interesting anti-biofouling materials based on the covalent modification of supramolecular materials (Goor, Brouns et al., 2017; Goor, Keizer et al., 2017). Like in the previous report, the authors used the PCLdiUPy as a base self-assembling unit and incorporated UPy-Tz (Figure 20.10) as an additive with end group functionalities capable of post-covalent modification through an inverse electron demand Diels-Alder cycloaddition. Interestingly, the anti-biofouling properties of the material were found to be dependent on the polymers functionalized with bicyclononyne moieties (PEG-BCN) being used (Figure 20.10). Among the three varieties opted, the star-shaped PEG-BCN showed the highest anti-biofouling property owing to its ability to form a cyclic loop at the material interface along with the statistical factor of enhanced surface interaction with the material surface.

The protein and cell adhesion of the material were carried out using various known standards. The supramolecular PCLdiUPy with 10% UPy-Tz and the star-PEG-BCN showed a significant reduction in protein adhesion toward different proteins like BSA, γ-gloubulin and fibrinogen. On the other hand, both bi-functional and star-PEG-BCN were capable of showing a significant reduction in cell adhesion toward human kidney cells (HK-2). Altogether the system shows the possibility of the supramolecular functional materials toward the intended bioactive materials in regenerative medicine application (Catoira et al., 2019). Similarly, varying the base material to the one with more traditional hydrogen-bonded systems like bisurea hard blocks was also evaluated for the anti-biofouling properties (Ippel et al., 2019).

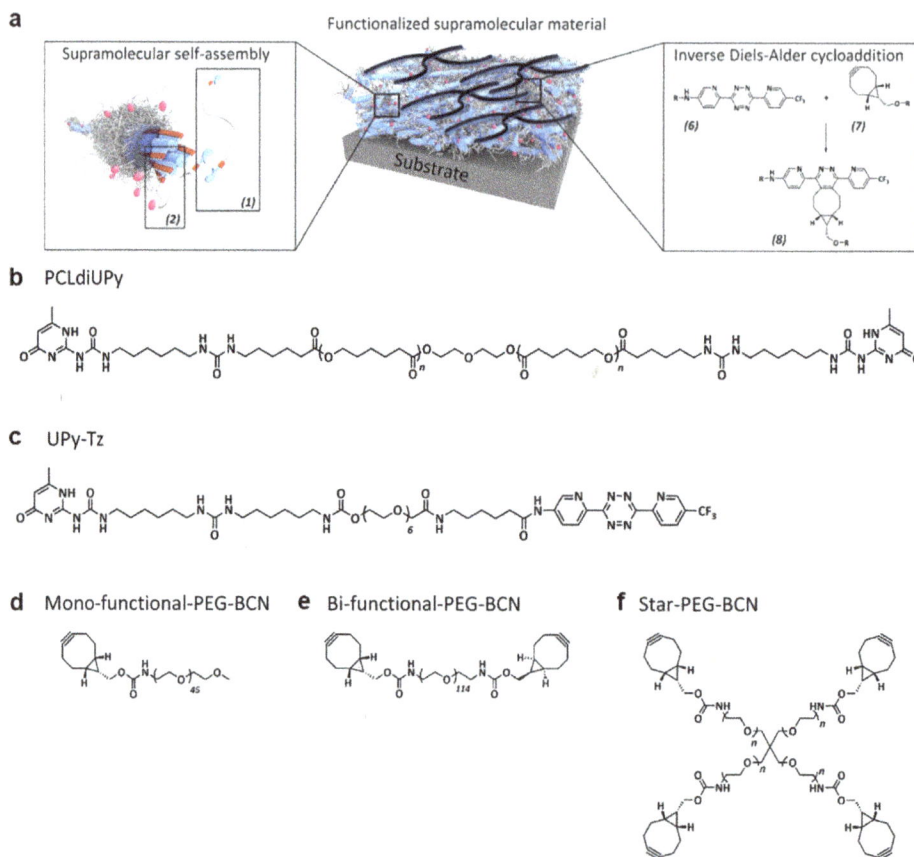

a Functionalized supramolecular material

Supramolecular self-assembly

Inverse Diels-Alder cycloaddition

Substrate

b   PCLdiUPy

c   UPy-Tz

d Mono-functional-PEG-BCN   **e** Bi-functional-PEG-BCN   **f** Star-PEG-BCN

FIGURE 20.10   (a) Schematic representation of the supramolecular functionalization, with self-assembly of the UPy-functionalized polymer (PCLdiUPy) with the UPy-additive (UPy-Tz) on the left and inverse Diels–Alder cycloaddition on the right. Molecular structures of (b) $PCL_{2k}diUPy$, (c) UPy-Tz additive, (d) mono-functional PEG_{2k}-BCN, (e) bi-functional-PEG_{5k}-BCN and (f) star-PEG_{10k}-BCN. (Reproduced with permission from Goor et al. (2017) under Creative Commons Attribution License © 2017 The Royal Society of Chemistry.)

20.6   CONCLUSIONS AND PERSPECTIVES

In summary, we have reviewed the potential of supramolecular self-assembled materials as anti-biofouling coatings on various surfaces. Although a lot of research has been undertaken on anti-biofouling materials, issues such as biocompatibility, low sensitivity, low stability and limited activity on specific surfaces often hamper their widespread applicability. These can be tackled by designing coatings for biomolecules or by combining the required functions with biocompatible molecules. The zwitterion formation of the peptides often leads to the heavy hydration shell around the charged amino acid species, providing an excellent fouling resistant coating driven by electrostatic interactions. Succeeding developments in this area have provided novel materials with enhanced hydrophobicity and immobilization capacity by chemical modification. However, the immunogenicity associated with this sort of expensive peptide-based scaffold made a pressing demand for a more cost-effective, biodegradable material resistant to unfavorable proteolysis. In this aspect, the PEGylated derivatives that can heavily hydrate the surface act as a promising candidate. However, there is a pressing demand for the novel design of supramolecular materials that can provide improved properties such as hydrophobicity, roughness, wettability, and so on, for the development of anti-biofouling surfaces. Active research to develop new strategies to overcome these limitations will achieve 100% efficient anti-biofouling materials in the future.

ACKNOWLEDGMENTS

D.E. is grateful to the Council of Scientific and Industrial Research (CSIR), Government of India for a research fellowship. A.P. acknowledges the University Grants Commission (UGC), Government of India for a research fellowship. V.K.P. is grateful to CSIR fast-track projects (MLP0027 and MLP0040) for the financial support.

REFERENCES

Aida, T., Meijer, E.W. 2020. Supramolecular polymers – we've come full circle. *Isr. J. Chem.* 60: 33–47.

Amabilino, D.B., Smith, D.K., Steed, J.W. 2017. Supramolecular materials. *Chem. Soc. Rev.* 46: 2404–2420.

Anees, P., Praveen, V.K., Kartha, K.K., Ajayaghosh, A. 2017. *Self-Assembly in Sensor Nanotechnology in Comprehensive Supramolecular Chemistry II*. Vol. 9 (Raston, C. ed.), Elsevier Ltd. doi: 10.1016/B978-0-12-409547-2.12644-7.

Banerjee, I., Pangule, R.C., Kane, R.S. 2011. Antifouling coatings: recent developments in the design of surfaces that prevent fouling by proteins, bacteria, and marine organisms. *Adv. Mater.* 23: 690–718.

Bixler, G.D., Bhushan, B. 2012. Biofouling: lessons from Nature. *Philos. Trans. R. Soc., A* 370: 2381–2417.

Bott, T.R. 2011. *Industrial Biofouling*. Elsevier Ltd., Amsterdam, The Netherlands. doi: 10.1016/C2009-0-22884-4.

Callow, J.A., Callow, M.E. 2011. Trends in the development of environmentally friendly fouling-resistant marine coatings. *Nat. Commun.* 2: 244.

Cao, S., Wang, J.D., Chen, H.S., Chen, D.R. 2011. Progress of marine biofouling and antifouling technologies. *Chin. Sci. Bull.* 56: 598–612.

Carrascosa, C., Raheem, D., Ramos, F., Saraiva, A., Raposo, A. 2021. Microbial biofilms in the food industry-a comprehensive review. *Int. J. Environ. Res. Public Health* 18: 2014.

Carretti, E., Bonini, M., Dei, L., Berrie, B.H., Angelova, L.V., Baglioni, P., Weiss, R.G. 2010. New frontiers in materials science for art conservation: responsive gels and beyond. *Acc. Chem. Res.* 43: 751–60.

Catoira, M.C., Fusaro, L., Francesco, D.D., Ramella, M., Boccafoschi, F. 2019. Overview of natural hydrogels for regenerative medicine applications. *J. Mater. Sci.: Mater. Med.* 30:115.

Chivers, P.R.A., Smith, D.K. 2019. Shaping and structuring supramolecular gels. *Nat. Rev. Mater.* 4: 463–478.

Clare, A.S., Rittschof, D., Gerhart, D.J., Maki, J.S. 1992. Molecular approaches to nontoxic antifouling. *Invertebr. Reprod. Dev.* 22: 67–76.

Du, X., Zhou, J., Shi, J., Xu, B. 2015. Supramolecular hydrogelators and hydrogels: from soft matter to molecular biomaterials. *Chem. Rev.* 115: 13165–13307.

Draper, E.R., Adams, D.J. 2019. Controlling the assembly and properties of low-molecular-weight hydrogelators. *Langmuir* 35:6506–6521.

Flemming, H.-C., Murthy, P.S., Venkatesan, R., Cooksey, K. 2009. *Marine and Industrial Biofouling*. Springer Series on Biofilms. Vol. 4, Berlin, Heidelberg. doi: 10.1007/978-3-540-69796-1.

Ghosh, S., Praveen, V.K., Ajayaghosh, A. 2016. The chemistry and applications of π-gels. *Annu. Rev. Mater. Res.* 46: 235–262.

González-Rivas, F., Ripolles-Avila, C., Fontecha-Umaña, F., Ríos-Castillo, A.G., Rodríguez-Jerez, J.J. 2018. Biofilms in the spotlight: detection, quantification, and removal methods. *Compr. Rev. Food Sci. Food Saf.* 17: 1261–76.

Goor, O.J.G.M., Brouns, J.E.P., Dankers, P.Y.W. 2017. Introduction of anti-fouling coatings at the surface of supramolecular elastomeric materials via post-modification of reactive supramolecular additives. *Polym. Chem.* 8: 5228–5238.

Goor, O.J.G.M., Keizer, H.M., Bruinen, A.L., Schmitz, M.G.J., Versteegen, R.M., Janssen, H.M., Heeren, R.M.A., Dankers, P.Y.W. 2017. Efficient Functionalization of additives at supramolecular material surfaces. *Adv. Mater.* 29: 1604652.

Gu, J., Su, Y., Liu, P., Li, P., Yang, P. 2017. An environmentally benign antimicrobial coating based on a protein supramolecular assembly. *ACS Appl. Mater. Interfaces* 9: 198–210.

Ippel, B.D., Keizer, H.M., Dankers, P. Y. W. 2019. Supramolecular antifouling additives for robust and efficient functionalization of elastomeric materials: molecular design matters. *Adv. Funct. Mater.* 29: 1805375.

Kirschner, C.M., Brennan, A.B. 2012. Bio-inspired antifouling strategies. *Annu. Rev. Mater. Res.* 42: 211–229.

Lehn, J.-M. 1988. Supramolecular chemistry-scope and perspectives molecules, supermolecules, and molecular devices (Nobel lecture). *Angew. Chem., Int. Ed.* 27: 89–112.

Li, P., Poon, Y.F. Li, W., Zhu, H.Y., Yeap, S.H., Cao, Y., Qi, X., Zhou, C., Lamrani, M., Beuerman, R.W., Kang, E.T., Mu, Y., Li, C.M., Chang, M.W., Leong, S.S., Chan-Park, M.B. 2011. A polycationic antimicrobial and biocompatible hydrogel with microbe membrane suctioning ability. *Nat. Mater.* 10: 149–156.

Li, S., Chen, J., Wang, J., Zeng, H. 2021. Anti-biofouling materials and surfaces based on mussel-inspired chemistry. *Mater. Adv.* 2: 2216–2230.

Li, Y., Fukushima, K., Coady, D.J., Engler, A.C., Liu, S., Huang, Y., Cho, J.S., Guo, Y., Miller, L.S., Tan, J.P.K., Lai, P., Ee, R., Fan, W., Yang, Y.Y., Hedrick, J.L. 2013. Broad-spectrum antimicrobial and bio-film-disrupting hydrogels: stereocomplex-driven supramolecular assemblies. *Angew. Chem., Int. Ed.* 52: 674–678.

Lian, X., Shi, D., Ma, J., Cai, X., Gu, Z. 2018. Peptide dendrimer-crosslinked inorganic-organic hybrid supra-molecular hydrogel for efficient anti-biofouling. *Chin. Chem. Lett.* 29: 501–504.

Maity, S., Nir, S., Zada, T., Reches, M. 2014. Self-assembly of a tripeptide into a functional coating that resists fouling. *Chem. Commun.* 50: 11154–11157.

Mishra, R.K., Das, S., Vedhanarayanan, B., Das, G., Praveen, V.K., Ajayaghosh, A. 2018. *Stimuli-Responsive Supramolecular Gels, in Molecular Gels: Structure and Dynamics.* Monographs in Supramolecular Chemistry Series (Weiss, R. G. ed.), RSC, pp. 190–226. doi: 10.1039/9781788013147-00190.

Nguyen, T., Roddick, F.A., Fan, L. 2012. Biofouling of water treatment membranes: a review of the underlying causes, monitoring techniques and control measures. *Membranes* 2: 804–840.

Omae, I. 2003. Organotin antifouling paints and their alternatives. *Appl. Organomet. Chem.* 17: 81–105.

Pape, A.C.H., Ippel, B.D., Dankers, P.Y.W. 2017. Cell and protein fouling properties of polymeric mixtures containing supramolecular poly(ethylene glycol) additives. *Langmuir* 33: 4076–4082.

Prathap, A., Sureshan, K.M. 2017. Organogelator-cellulose composite for practical and eco-friendly marine oil-spill recovery. *Angew. Chem., Int. Ed.* 56: 9405–9409.

Praveen, V.K., Vedhanarayanan, B., Mal, A., Mishra, R.K., Ajayaghosh, A. 2020. Self-Assembled extended π-systems for sensing and security applications. *Acc. Chem. Res.* 53: 496–507.

Saha, A., Nir, S., Reches, M. 2020. Amphiphilic peptide with dual functionality resists biofouling. *Langmuir* 36: 4201–4206.

Saji, V.S. 2019. Supramolecular concepts and approaches in corrosion and biofouling prevention. *Corros. Rev.* 37: 187–230.

Salick, D.A., Kretsinger, J.K., Pochan, D.J., Schneider, J.P. 2007. Inherent antibacterial activity of a peptide-based β-hairpin hydrogel. *J. Am. Chem. Soc.* 129: 14793–14799.

Shima, S., Matsuoka, H., Iwamoto, T., Sakai, H. 1984. Antimicrobial action of ε-poly-l-lysine. *J. Antibiot.* 37: 1449–1455.

Singh, P., Basak, G., Sharma, B., Jain, U., Mishra, R. 2019. Biofilm: an alarming niche in dairy industry. *Int. J. Livest. Res.* 9: 10–24.

Thomason, J.C. 2009. Fouling on shipping: data-mining the world's largest antifouling archive. In *Biofouling* (Dürr, S., and Thomason, J.C. eds.), Wiley-Blackwell, Oxford, UK, pp. 207–216. doi: 10.1002/9781444315462.ch14.

Tsao, C.T., Chang, C.H., Lin, Y.Y., Wu, M.F., Wang, J.L., Han, J.L., Hsieh, K.H. 2010. Antibacterial activ-ity and biocompatibility of a chitosan-gamma-poly(glutamic acid) polyelectrolyte complex hydrogel. *Carbohydr. Res.* 345: 1774–1780.

Vedhanarayanan, B., Praveen, V.K., Das, G., Ajayaghosh, A. 2018. Hybrid materials of 1D and 2D carbon allotropes and synthetic π-systems. *NPG Asia Mater.* 10: 107–126.

Verran, J. 2002. Biofouling in food processing: biofilm or biotransfer potential? *Food Bioprod. Process.* 80: 292–298.

Wahl, M., Lafargue, F. 1990. Marine epibiosis. *Oecologia* 82: 275–282.

Whitesides, G.M. 2018. Soft robotics. *Angew. Chem., Int. Ed.* 57: 4258–4273.

Yang, W.J., Neoh, K.-G., Kang, E.-T., Teo, S.L.-M., Rittschof, D. 2014. Polymer brush coatings for combating marine biofouling. *Prog. Polym. Sci.* 39: 1017–1042.

Yebra, D.M., Kiil, S., Dam-Johansen, K. 2004. Antifouling technology - past, present and future steps towards efficient and environmentally friendly antifouling coatings. *Prog. Org. Coat.* 50: 75–104.

Zhang, R., Liu, Y., He, M., Su, Y., Zhao, X., Elimelech, M., Jiang, Z. 2016. Antifouling membranes for sustain-able water purification: strategies and mechanisms. *Chem. Soc. Rev.* 45: 5888–5924.

Zhao, X., Zhang, R., Liu, Y., He, M., Su, Y., Gao, C., Jiang, Z. 2018. Antifouling membrane surface construc-tion: chemistry plays a critical role. *J. Membr. Sci.* 551: 145–171.

Zhou, C., Li, P., Qi, X., Sharif, A.R., Poon, Y.F., Cao, Y., Chang, M.W., Leong, S.S., Chan-Park, M.B. 2011. A photopolymerized antimicrobial hydrogel coating derived from epsilon-poly-L-lysine. *Biomaterials* 32: 2704–2712.

21 Cyclodextrins in Anti-Biofouling Applications

Jyotirmayee Mohanty and Achikanath C. Bhasikuttan
Bhabha Atomic Research Centre
Homi Bhabha National Institute

CONTENTS

21.1 INTRODUCTION

Biofouling, a cumulative process, is defined as the uncontrolled accumulation/non-specific adhesion of unwanted biological contaminants (biofoulants) such as microorganisms/microbes, plants, algae or small animals on a wetted material surface, mostly at an aqueous liquid/solid interface, triggering structural or other functional deficiencies (Bixler and Bhushan 2012). Biofouling poses major challenges in health care (Busscher et al. 2012), food processing (Verran 2002), water treatment (Flemming 2002), and marine industries (Callow and Callow 2011; Schultz et al. 2011; Fitridge et al. 2012; Learn et al. 2020). The nature and degree of biofouling depend on the local environment and organisms; that differs considerably among health care, marine and industrial environments. In medical applications, biofouling due to microbial contamination leads to significant health threats such as the spread of contagious diseases, implant rejection, and malfunction of biosensors (Sakala and Reches 2018; Maan et al. 2020). Industrial fouling causes detrimental effects such as additional energy needs, pipeline blockage, reduced efficacy in power plants, and contamination in membrane systems for water treatment and in the food industry (Maan et al. 2020). In marine environments, ship hull biofouling increases drag, corrosion, added fuel consumption and engine stress (Callow and Callow 2011; Bixler and Bhushan 2012). Generally, medical biofouling includes the biofilm generated by the microorganisms, whereas marine and industrial biofouling include a combination of biofilm and macrofouling (macro-scale biofouling created by macro-organisms) (Bixler and Bhushan 2012).

DOI: 10.1201/9781003169130-24

21.1.1 Anti-Biofouling Materials

Anti-biofouling (ABF) materials prevent the deposition of proteins, cells, and/or micro- and macro-organisms (Hoffman 1999; Zhang and Chiao 2015), which reduces the consequences of unnecessary biological material accumulation on critical interfaces. Since biofouling occurs on the surface, it largely depends on the surface properties, e.g. surface energy, surface contact, hydrophilicity, wettability, and surface topology. Modification in the surface property and structure by treating the surface with an anti-fouling coating is the most common method to combat and control biofouling. There are two major ABF coating strategies: (i) active ABF coating and (ii) passive ABF coating, which are mainly based on the degradation of adherent biofoulants or resistance to the accumulation of biofoulants or release of biofoulants (Banerjee et al. 2011; Harding and Reynolds 2014; Learn et al. 2020a).

Active ABF coating methods mainly involve the use of biocidal agents such as a chemical substance (Yebra et al. 2004; Learn et al. 2020a), or thermal treatment (Learn et al. 2020b) or pulses of energy on the material surface to destroy the organisms responsible for biofouling. Silver compounds in the form of Ag nanoparticles/nanocomposites and antibiotic drugs are usually used on medical implant surfaces (Punitha et al. 2017; Learn et al. 2020a). These substances are amalgamated into an anti-fouling surface coating, either through physical adsorption or through chemical modification including both covalent and noncovalent modifications. Biofouling happens on surfaces with the initial formation of a biofilm that readies a surface on which consecutive layers of other microorganisms grow. Tactfully, the microorganisms responsible for creating the initial biofilm are targeted with coatings using biocides (Damodaran and Murthy 2016; Sakala and Reches 2018; Huang et al. 2020; Maan et al. 2020). Once the initial layer is destroyed, the biomass is unable to accumulate and grow and get detached from the surface (Damodaran and Murthy 2016; Sakala and Reches 2018; Huang et al. 2020; Maan et al. 2020).

Passive ABF coating methods employ anti-adhesive materials as surface coating to diminish the chances of biological contaminants/organisms to adhere to the material surface (Bixler and Bhushan 2012; Learn et al. 2020a). Biofouling can also be avoided by reducing contact or attractive forces, especially electrostatic forces, by coating with a slippery surface. Such ultra-low fouling surface can be obtained with the use of zwitterions or polymer coating, thereby creating nanoscale surface topologies, which offer poor anchoring points (Sakala and Reches 2018; Learn et al. 2020a; Maan et al. 2020). Natural coatings containing anti-fouling materials extracted from organisms are difficult to commercialize due to the limited source/availability/supply, high cost, short-term efficacy and specificity of natural anti-fouling materials (Sakala and Reches 2018; Maan et al. 2020). Over the years, efforts have been made for polymer-based coatings, which are low cost, nontoxic, biocompatible, easy to process with wide-range efficacy, and above all, the functionalities and the topology can be modified/altered easily to tune their anti-fouling properties (Maan et al. 2020). The polymer-based coatings are mainly used as membranes in water treatment applications. However, the hydrophobic surface of the polymer reduces the ABF capability in membranes by increasing the adsorption of proteins on the surface and thereby blocking the surface pores and reduces the membrane durability/capability (Yu et al. 2015). Various studies show that electrically neutral hydrophilic surfaces, especially those containing ample H-bond acceptors and without H-bond donors have the utmost resistance towards protein adsorption (Ostuni et al. 2001; Learn et al. 2020a). In general, charge neutralization diminishes electrostatic interactions that could otherwise attract proteins (or cells) to a surface (Holmlin et al. 2001). Glycocalyx-mimetic carbohydrate surfaces which contain many H-bond donors also efficiently resist non-specific protein adsorption (Holland et al. 1998). It is understood that water held at the surface through H-bonding is critical towards the resistance of protein adsorption. As a result, neutrally charged, H-bond-acceptor-rich, hydrophilic polymers such as polyzwitterions, poly(ethylene glycol) (PEG), poly(2-hydroxyethyl methacrylate) (pHEMA), cyclodextrin polymers (pCD) and many polysaccharides are used as polymer coatings on the surface of ABF (Minett et al. 1984; Ekblad et al. 2008; Schlenoff 2014; Learn et al. 2020a). In this context, the supramolecular approach through host–guest complexation using macrocyclic hosts with hydrophilic portals is one of the approaches to increase the hydrophilicity/wettability of the surface which in turn reduces the protein adsorption and the attachment of microorganisms and hence, is expected to enhance the membrane durability (Yu et al. 2015). Further, hydrophilic cyclo-dextrins (CDs) (used as excipients in drug formulation) are used in the polymer coatings to increase the wettability of the polymer coatings.

21.1.2 CYCLODEXTRIN MACROCYCLIC HOST

Supramolecular functional assemblies through covalently-linked or noncovalently modified by macrocyclic hosts have received growing interest in different fields of chemistry, material science and bioscience (Lehn 1988; Busseron et al. 2013; Prochowicz et al. 2017). Host–guest complexation is one of the approaches to construct supramolecular assemblies where macrocyclic organic mol-ecules such as crown ethers (Pedersen 1988), CDs (van de Manakker 2009; Kandoth et al. 2010; Shinde et al. 2015a; Kalyani et al. 2017; Mohanty et al. 2017; Prochowicz et al. 2017; Khurana et al. 2019b), calixarenes (Shinde et al. 2016a, b, 2017a; Mehra et al. 2019), cucurbiturils (Sayed et al. 2015; Shinde et al. 2015b; Goel et al. 2016; Barooah et al. 2017; Bhasikuttan and Mohanty 2017; Khurana et al. 2017; Mohanty et al. 2017, 2019; El-Sheshtawy et al. 2018; Ruz et al. 2021), etc. act as host molecules and are bound to guest molecules reversibly. Among these macrocycles, CDs, cyclic oligosaccharides, having a hydrophobic central cavity and hydrophilic exterior rims, are extensively studied with a variety of important guest molecules such as drugs, fluorescent dyes, etc. for creating supramolecular assemblies with different functionalities (Prochowicz et al. 2017). CDs comprise of D-(+)-glucopyranosyl units linked by α-1,4-glycosidic bonds in a cyclic manner and depending upon the six, seven and eight numbers of glucopyranose unit, α-, β- and γ-CDs are known (Manakker 2009; Kandoth et al. 2010; Shinde et al. 2015a; Khurana et al. 2019b). Due to the chair conformation of the glucopyranose units and the presence of secondary hydroxyl groups in the wider rim and the primary hydroxyl groups in the narrow rim, CD molecules are shaped like truncated cones (for the structure, see Chapters 2 and 10) (Prochowicz et al. 2017). For the first time, CDs were discovered in 1891 by Villiers from the enzymatic degradation of cellulose (Mohanty et al. 2017) and the cyclic oligosaccharide forms were identified by Schardinger in 1904 (Prochowicz et al. 2017). The native CDs exhibit moderate water solubility which has been increased by functionalizing the hydroxyl groups (Shinde et al. 2015a; Khurana et al. 2019a). The hydrophilic CDs are used as excipients in drug formulation as they are nontoxic at low to mod-erate oral dosages (Shinde et al. 2015a, 2017b; Khurana et al. 2019a). In recent years, CDs have been incorporated in the polymer coatings by covalent or noncovalent-linking to increase the ABF properties by increasing the wettability/hydrophilicity of the polymer surface (Saji 2019). These CD-functionalized polymer coatings find potential applications in health care (Learn et al. 2020a), water treatment (Yu et al. 2015), marine industries (Punitha et al. 2017) by preventing the uncon-trolled accumulation of micro- and macro-organisms.

In this chapter, the CD-functionalized polymer-based coatings for the ABF performance have been discussed. In addition, the use of CD-capped Ag nanoparticles (Ag NPs) incorpo-rated into PEG polymer as biocides has been described. The promising applications of these CD-functionalized polymer coatings in health care, membrane separation process, nanofiltration composite membrane and marine biofouling have been deliberated. Besides, stimuli-responsive β-CD-based anti-fouling coatings promising for biomedical and biotechnological applications have also been elaborated.

21.2 CYCLODEXTRIN-FUNCTIONALIZED POLYMER-BASED COATINGS

21.2.1 CYCLODEXTRIN POLYMER COATINGS: DETERRENCE OF PROTEIN FOULING, MAMMALIAN CELL ADHESION, AND BACTERIAL ATTACHMENT

CD-based materials are useful as coatings in industrial or medical applications where biofouling-resistant and/or drug-delivering surfaces are required. The major limitation of surface coatings based on carbohydrates is the lack of ability to form inclusion complexes with small drug molecules for drug delivery applications (Holland et al. 1998; Luk et al. 2000). PEG-based surface coatings are prone to oxidation in the presence of oxygen and metal ions in physiological solutions (Ulbricht et al. 2014), whereas zwitterionic-based polymers are difficult and costly to synthesize (Leng et al. 2018). On the other hand, the wettability of the material surface plays a vital role in the anti-fouling behaviour/activity (Learn et al. 2020a). The combination of macrocyclic hosts and hydrophilic portals such as CDs with hydroxyl groups and their derivatives increases surface wettability. In this regard, Learn et al. have synthesized polymer coatings based on CD subunits and evaluated their anti-fouling performance by measuring the non-specific adsorption of protein, adhesion of mammalian cells (NIH/3T3) and attachment of bacterial cell (*S. aureus and E. coli*) (Learn et al. 2020a). In this work, polymerized β-CD (pCD) has been applied as a pre-coating for polypropylene (PP) layer and is made to undergo nonthermal plasma treatment to enhance the uniformity and adherence of pCD coating (Learn et al. 2020a, b). It has been observed that lightly crosslinked pCD have excellent passive resistance towards the attachment of protein, cells and bacteria, which can be attributed to the hydrophilic and electrically neutral surface properties of these coatings (Learn et al. 2020a). However, the increased hexamethylene diisocyanate (HDI) cross-linking of pCD restricts the chain mobility and the existence of polar moieties at the network polymer surface. As a result, the decrease in hydrophilicity facilitates protein adsorption and the increase in substrate rigidity promotes mammalian cell adhesion, spreading and bacterial attachment and hence, subsequent biofouling (Learn et al. 2020a). This study envisages that pCD can be applied as coatings for medical devices and can prevent post-surgical adhesions as low cross-linked pCD are capable of mitigating non-specific adsorption, denaturation of protein and adhesion of fibroblasts. In addition, pCD coatings may be useful for application onto PP water treatment membranes as these coatings resist bacterial biofouling and remove small-molecule pollutants from water (Learn et al. 2020a).

21.2.2 NOVEL ANTI-FOULING β-CYCLODEXTRIN–PVDF MEMBRANE

In the last few decades, membrane separation technology has become more fascinating and attracted more research interest than conventional separation technology, owing to its high efficacy, high selectivity, consistency in separation, ease of operation, and low energy intake (Pendergast and Hoek 2011). Polyvinylidene fluoride (PVDF) is widely used in membrane preparation due to its distinct mechanical properties, chemical, thermal and radiation stability (Liu et al. 2011). However, the application of hydrophobic PVDF-based membrane in water treatment is restricted as the natural organic substances such as proteins are prone to adsorb on the hydrophobic surface of the membrane, block the membrane pores and reduce the membrane flux which leads to foul the solution (Yu et al. 2015). In recent times, numerous methods have been explored to increase the hydrophilicity of PVDF membranes to reduce the adsorption and deposition of proteins (Yu et al. 2015). A novel β-CD-based PVDF membrane through an interfacial reaction has been developed by Yu et al. by using trimesoyl chloride (TMC) and β-CD as respective cross-linking and modification agents (Figure 21.1A) (Yu et al. 2015). The presence of hydroxyl groups in β-CD grafted on the surface of the PVDF membrane improves the hydrophilicity of the membrane by lowering the water contact angle of modified PVDF than the unmodified PVDF. The roughness along with the permeate flux and hydrophilicity of β-CD grafted PVDF membranes increase as compared to those of pristine PVDF membranes (Yu et al. 2015).

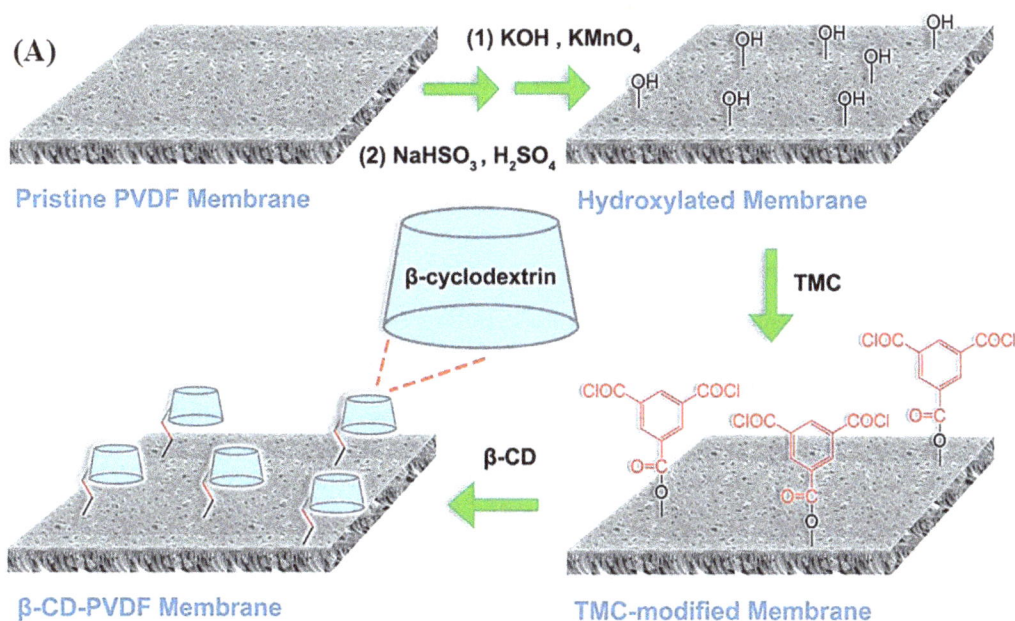

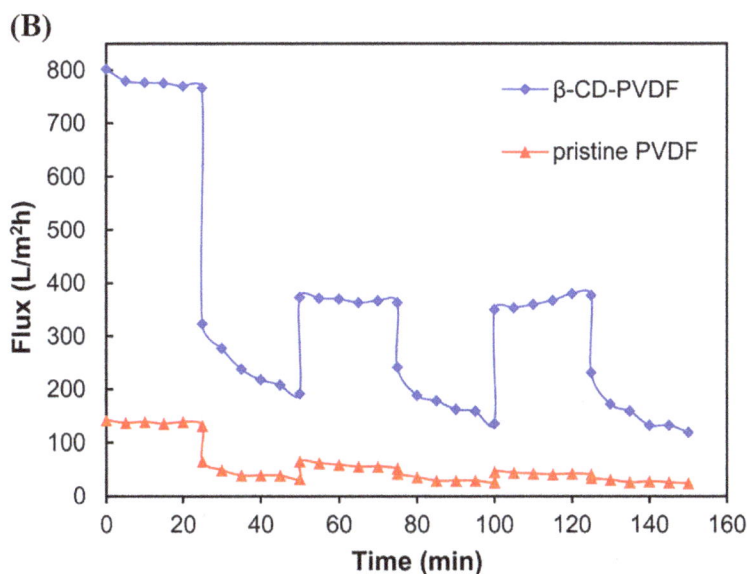

FIGURE 21.1 (A) Schematic diagram of the β-CD–PVDF membrane preparation. (B) Water flux recovery of pristine PVDF membrane and β-CD-PVDF membrane (1.8%) during two cycles. (Reproduced with permission from Yu et al. (2015) © The Royal Society of Chemistry.)

Figure 21.1B shows the anti-fouling properties of both β-CD modified and unmodified PVDF membrane, during two cycles of flux recovery and the results of relative flux reduction (RFR) and the flux recovery ratio (FRR) are given in Table 21.1 (Yu et al. 2015). In the first cycle, the initial water flux of β-CD-PVDF membrane is considerably higher than that of the pristine membrane, however, the fast flux reduction was observed in both pristine and modified PVDF, when bovine serum albumin (BSA) protein solution was filtered through the membrane. Though the RFR of

TABLE 21.1

The Relative Flux Reduction (RFR) and the Flux Recovery Ratio (FRR) of the Membranes in BSA Filtration Experiments

	First Cycle		Second Cycle	
Sample	RFR (%)	FRR (%)	RFR (%)	FRR (%)
PVDF membrane	75.0	40.0	80.0	31.0
β-CD-PVDF membrane	75.0	47.4	82.3	49.1

Source: Reproduced with permission from Yu (2015) © The Royal Society of Chemistry.

the two membranes is 75.0%, the β-CD-PVDF membrane has higher BSA flux than that of the pristine membrane. In the last stage, after the membrane was cleaned by water, the pristine PVDF membrane water flux recovered up to only 40.0% (FRR) of the initial flux, while β-CD–PVDF membrane water flux recovered up to 47.4% (FRR). The FRR of the modified membrane is better than that of the pristine membrane (Yu et al. 2015). The superior anti-fouling performance of β-CD-based PVDF membrane holds promise for their practical application in membrane separation areas.

21.2.3 Anti-Fouling β-Cyclodextrin/Polyester Thin Film Nanofiltration Composite Membrane

Nanofiltration (NF) membrane is a particular type of pressure-driven membrane with separation characteristics between reverse osmosis (RO) and ultrafiltration (UF) membrane. NF membranes are generally thin film composite (TFC) based membranes synthesized by interfacial polymerization (Wu et al. 2013). The major issue lies with the anti-fouling property of the membrane for long-term practical operation. With time the fouling degrades the membrane, which subsequently reduces the membrane performance and increases the maintenance cost (Boussu et al. 2007). As a result, numerous research efforts have been made for the development of anti-fouling TFC membrane by employing different strategies including the modification of hydrophilicity by introducing hydrophilic organic molecules or macrocyclic hosts with hydrophilic portals (Lin et al. 2010; Wagner et al. 2011). In this context, Wu et al. have synthesized β-CD/polyester thin film nanofiltration composite membranes through *in situ* interfacial polymerization of trimesoyl-chloride (TMC) and triethanolamine (TEOA) in the presence of β-CD (Figure 21.2A) (Wu et al. 2013). In addition, they have also investigated the effects of β-CD concentration and pendant group/portal groups on the separation efficiency, morphology and anti-fouling performance of NF composite membrane. The rise in water flux and the decrease in the rejection of Na_2SO_4 salt with an increase in the concentration of β-CD have been observed (Figure 21.2B) (Wu et al. 2013). When the concentration of β-CD is 1.8% (w/v) in the aqueous phase, the water flux of the NF composite membrane reaches a value that is almost two times that of the bare polyester membrane. It is also found that the contact angle, inversely proportional to the hydrophilicity, decreases with increasing concentration of β-CD. But at a higher concentration of β-CD (3.0%, w/v), the contact angle increases due to the severe agglomeration of β-CD (Wu 2013). This result indicates that at lower β-CD concentration, the hydrophilicity of the NF composite membrane increases due to the presence of multi-hydroxyl groups of β-CD in the membrane surface which shows enhanced hydrogen bond interaction with water molecules and thereby facilitates the water transport through the NF composite membrane (Wu et al. 2013).

On the other hand, the incorporation of sulphated β-CD (sβ-CD), in which the hydroxyl groups are substituted by sulphonic acid groups at the pendant region of β-CD, the NF composite membrane exhibits a remarkable increase in negative charge density (measured in terms of streaming

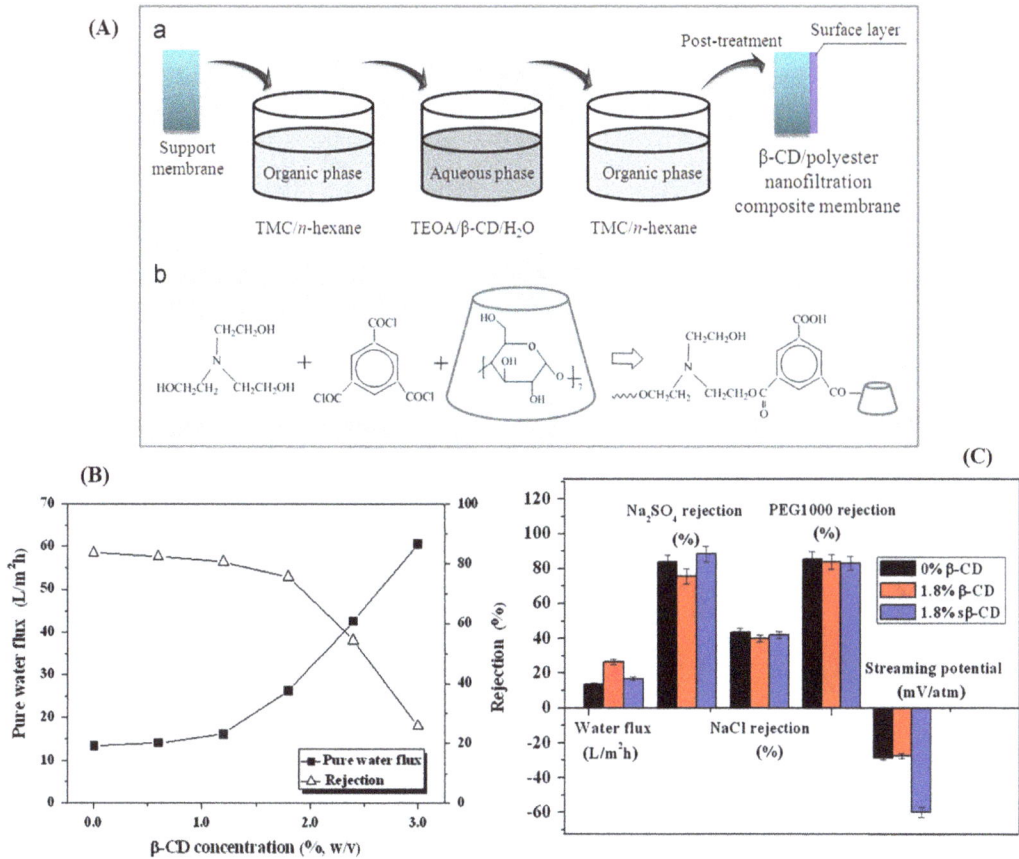

FIGURE 21.2   (A) (a) Schematic of the preparation procedure of the β-CD/polyester nanofiltration compos-ite membrane; (b) Interfacial polymerization formulas between TMC and TEOA in the presence of β-CD. (B) Effect of β-CD concentration on the pure water flux and Na₂SO₄ rejection of NF composite membranes at 0.6 MPa operating pressure. (C) Effect of the pendant group of β-CD on pure water flux, Na₂SO₄ rejection, NaCl rejection, PEG1000 rejection, and streaming potential of NF composite membranes. (Reproduced with permission from Wu et al. (2013) © Elsevier.)

potential), hydrophilicity and salt rejection (Figure 21.2C) (Wu et al. 2013). sβ-CD/polyester nano-filtration composite membrane can also improve the permeability with an enhanced separation performance, especially when applied in separating multi-valent anion.

The anti-fouling performance of β-CD/polyester NF composite membrane has been examined in both Na₂SO₄ and Na₂SO₄/BSA mixtures. It is reported that the rate and the extent of fouling are most significantly influenced by the roughness of the membrane surface. The existence of multi-hydroxyl groups of β-CD retards the cross-linking reaction and forms strong intermolecular hydrogen bond-ing, leading to a smoother membrane surface. Thus β-CD/polyester NF composite membrane exhib-its an excellent improvement in the anti-fouling ability, which is attributed to its smoother surface and higher hydrophilicity (Wu et al. 2013). Therefore, this study shows that β-CD can be considered as a potential additive to improve the property of nanofiltration composite membrane.

21.2.4   β-Cyclodextrin Stabilized PEG-Based Silver Nanocomposites

Self-polishing polymer composites have been extensively used in the design and development of antimicrobial agents and anti-fouling coatings through a controlled release of the metal biocide.

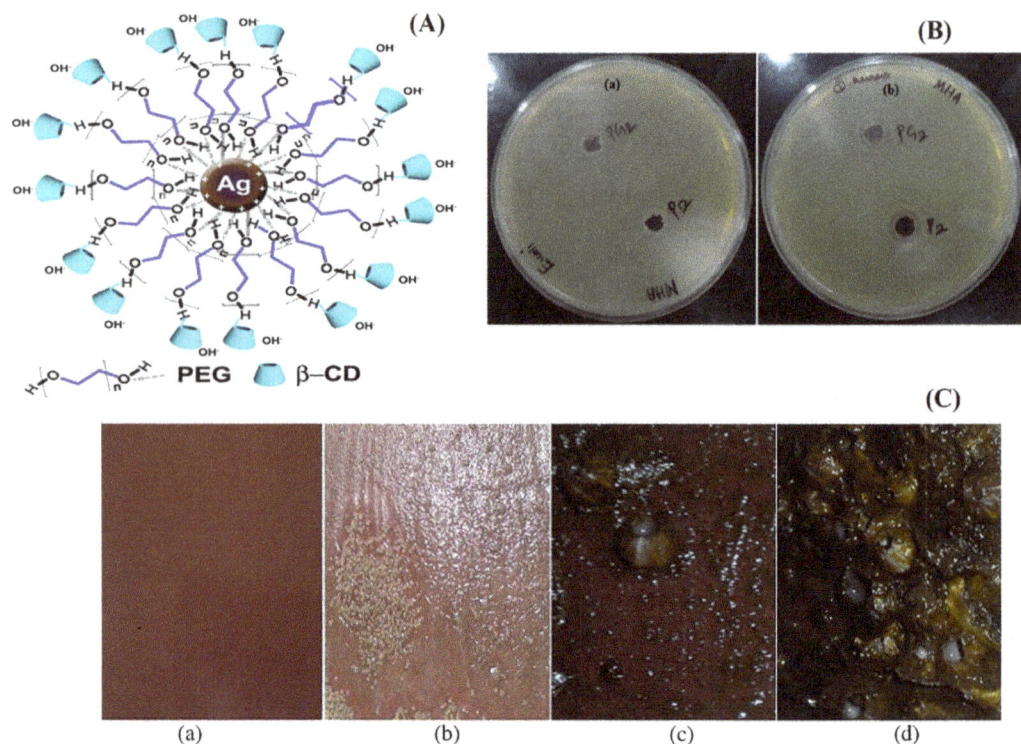

FIGURE 21.3 (A) Proposed model of Ag-loaded biopolymer complex. (B) Zone of inhibition of β-CD stabilized Ag-PEG NCs. (a) *E. coli* and (b) *S. aureus* using the well diffusion method. (C) Photograph of the coating panels (a) before immersion, (b) after 6 months (c) after 12 months and (d) Control panel without biocide. (Reproduced with permission from Punitha et al. (2017) © 2017 Elsevier.)

In this regard, Ramesh et al. have established an environment-friendly green synthesis of functionalized Ag nanocomposites (SNCs) by using adhesive resistance PEG and biocompatible β-CD (Figure 21.3A) and investigated their antibacterial and ABF activity for marine biofouling (Punitha et al. 2017). The prepared SNCs display excellent micro-fouling activities in comparison to the normal Ag NPs.

The antibacterial activity of Ag-PEG and β-CD stabilized SNCs against Gram-negative (*E. coli*) and Gram-positive (*S. aureus*) bacteria have been examined by using the zone of inhibition test (Figure 21.3B) (Punitha et al. 2017). The minimum inhibitory concentration (MIC) of SNCs (10 βg/mL) was sufficient to inhibit bacterial growth. This antibacterial activity originated due to Ag NPs and the Ag$^+$ ions on the surface interact with sulphur and phosphorus comprising bio-molecules such as DNA or other biological moieties, thereby triggering cell damage and is promising for ABF applications (Punitha et al. 2017).

The anti-fouling studies were carried out by recording the photographs of the test panels (polyethylene sheet coated with SNCs infused epoxy resins) before and after immersion in the sea for 1 year (Figure 21.3C) (Punitha et al. 2017). The presence of very few bacterial colonies and a very thin biofilm on the surface of SNCs epoxy resin coated polyethylene sheet after 1 year suggests superior fouling resistance of SNCs. This behaviour is mainly controlled by hydrophobic–hydrophilic equilibrium of the PEGlyation of Ag NPs (Punitha et al. 2017). The slower rate of fouling occurrence on the surfaces could be attributed to the presence of self-cleaning nature of a brush-like lengthy chain of PEG which has greatly reduced barnacle attachment while also preventing blistering and deterioration of the surfaces. The low surface energy of SNCs and combined effect of PEG and β-CD demonstrate the synergistic effect which is accountable for the superior fouling resistance of the system containing Aradur HY951 triethylenetetramine curing agent and PEG-based SNCs (Punitha et al. 2017). Immersion tests in the marine environment confirmed that the epoxy nano-hybrid anti-fouling coating provides better protection by killing both air-borne and water-borne bacteria, due to its self-cleaning ability. Above all the epoxy nano-hybrid coating system exhibits superior barrier properties, visual characteristics and anti-fouling ability under accelerated and natural ageing conditions (Punitha et al. 2017).

21.2.5 Stimuli-Responsive Supramolecular Anti-Biofouling Coatings

Smart biointerfaces having ability to control cell-surface interactions against external stimuli such as pH, temperature, light, ionic strength of the solutions, etc. are of immense importance to both fundamental research and practical applications (Maan et al. 2020). The combination of stimuli-responsiveness with anti-fouling properties of materials can become advantageous for killing and releasing microorganisms from the surface and also for the regeneration of material surface with anti-fouling properties on an external trigger (Maan et al. 2020). Wei et al. have developed a smart and light-responsive supramolecular anti-fouling coating (Wei et al. 2017) where the surface comprised of an azobenzene (Azo)-containing self-assembled monolayer (SAM) and a biocidal β-CD derivative, attached with seven quaternary ammonium salt moieties (CD-QAS). The surface-immobilized Azo groups in the *trans* form interact with CD-QAS to form host–guest inclusion complexes and result in a strongly bactericidal surface that eradicates more than 90% attached bacteria (Wei et al. 2017). The inclusion complex formation between Azo groups and CD-QAS is weak and reversible. The density of bacteria attached to the surface before and after treatment is determined by using fluorescence microscopy. Upon UV light irradiation, the Azo moieties switch over to *cis* form, as a result, Azo/CD-QAS complex dissociates and releases 90.6%±3.9% inactivated bacteria from the surface. The release of dead bacteria is further improved to ≥95% by using a strong competitive binder with higher binding affinity such as adamantadine hydrochloride (Ada) towards β-CD under UV irradiation (Figure 21.4A) (Wei et al. 2017).

The regeneration of clean and active surface by following the inactivate-and-release cycle can be utilized for further use by re-inclusion of fresh CD-QAS with *trans* Azo group generated during visible light irradiation (Figure 21.4B) (Wei et al. 2017). The use of supramolecular host–guest interaction to construct smart biointerfaces with stimuli-responsive switchable functions for the bacterial killing and release cycle provides a promising strategy for the design and development of multifunctional surfaces, which find prospective applications in the biomedical and biotechnology fields (Wei et al. 2017).

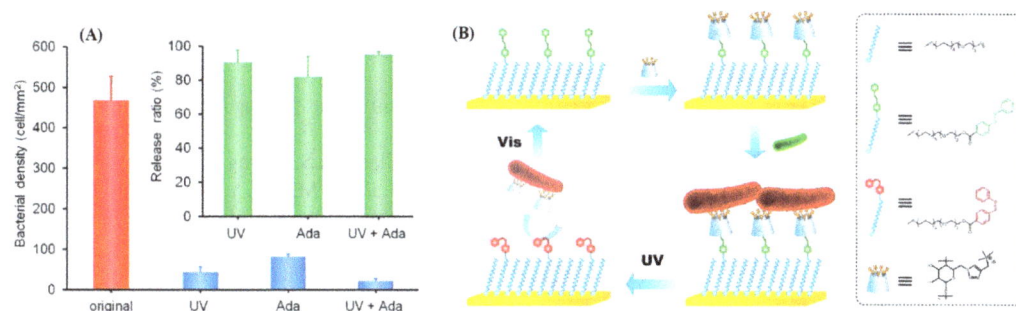

FIGURE 21.4 (A) Bacterial density on Au-Mix$_{10}$/CD-QAS surface before and after different treatments. Inset shows corresponding bacterial release ratios. Error bars represent standard deviation ($n = 3$). Three replicates were chosen for each surface type and images of 15 randomly chosen fields were captured for each replicate. (B) Smart antibacterial surface with photo-switchable biocidal activity and bacteria releasing ability. (Reproduced with permission from Wei et al. (2017) © 2017 American Chemical Society.)

21.3 CONCLUSIONS AND PERSPECTIVES

In this chapter, we have summarized most of the reports on CD-based polymer coatings which includes β-CD polymer coating on polypropylene surface, β-CD-linked polyvinylidene fluoride polymer/polyester TFC, and β-CD stabilized PEG-based Ag nanocomposites, their anti-fouling behaviour and their potential applications in sustained drug release, membrane separation technology, water treatment and marine industry and so forth. In addition, light-responsive supramolecular anti-fouling coatings containing biocidal β-CD derivative for kill-release antibacterial surface offer a favourable platform for the fabrication of multifunctional surfaces, promising for biomedical and biotechnological applications. It is interesting to synthesize CD-linked polymers containing other homologues of β-CD (α-and γ-CDs) and to examine the effect of homologues on the polymer affinity, rigidity, wettability, swellability and the subsequent ABF performance, as the CD homologues have different cavity sizes and water solubilities. Currently, it is established that there is substantial amelioration in the antibacterial activity of drugs in the presence of macrocyclic hosts such as cucurbiturils, calixarenes and pillarenes/calixarene-functionalized nanoparticles. Thus, it is challenging to synthesize polymer coatings using these macrocyclic hosts/nanocomposites as linking agents or subunits and to assess the modification in the prime features such as affinity, hydrophilicity as well as anti-fouling capacity. Furthermore, such stimuli-responsive self-assembled monolayers containing desired hydrophilic macrocycles will be promising for the construction of smart biointerfaces for applications in water treatment, healthcare and biomedical fields.

REFERENCES

Banerjee, I., Pangule, R.C., Kane, R.S. 2011. Antifouling coatings: recent developments in the design of surfaces that prevent fouling by proteins, bacteria, and marine organisms. *Adv. Mater.* 23:690–718.

Barooah, N., Kunwar, A., Khurana, R., Bhasikuttan, A.C., Mohanty, J. 2017. Stimuli-responsive cucurbit[7]uril-mediated BSA nanoassembly for uptake and release of doxorubicin. *Chem. Asian J.* 12:122–129.

Bhasikuttan, A.C., Mohanty, J. 2017. Detection, inhibition and disintegration of amyloid fibrils: role of optical probes and macrocyclic receptors. *Chem. Commun.* 53:2789–2809.

Bixler, G.D., Bhushan, B. 2012. Biofouling: lessons from nature. *Philos. Trans R. Soc. A* 370:2381–2417.

Boussu, K., Belpaire, A., Volodin, A., van Haesendonck, C., Van der Meeren, P., Vandecasteele, C., Van der Bruggen, B. 2007. Influence of membrane and colloid characteristics on fouling of nanofiltration membranes. *J. Membr. Sci.* 289:220–230.

Busseron, E., Ruff, Y., Moulin, E., Giuseppone, N. 2013. Supramolecular self-assemblies as functional nanomaterials. *Nanoscale* 5:7098–7140.

Busscher, H.J., van der Mei, H.C., Subbiahdoss, G., Jutte, P.C., van den Dungen, J.J.A.M., Zaat, S.A.J., Schultz, M.J., Grainger, D.W. 2012. Biomaterial-associated infection: locating the finish line in the race for the surface. *Sci. Trans. Med.* 4:153rv10.

Callow, J.A., Callow, M.E. 2011. Trends in the development of environmentally friendly fouling-resistant marine coatings. *Nat. Commun.* 2:244.

Damodaran, V.B., Murthy, N.S. 2016. Bio-inspired strategies for designing antifouling biomaterials. *Biomater. Res.* 20:18.

Ekblad, T., Bergström, G., Ederth, T., Conlan, S.L., Mutton, R., Clare, A.S., Wang, S., Liu, Y., Zhao, Q., D'Souza, F., Donnelly, G.T., Willemsen, P.R., Pettitt, M.E., Callow, M.E., Callow, J.A., Liedberg, B. 2008. Poly(ethylene glycol)-containing hydrogel surfaces for antifouling applications in marine and freshwater environments. *Biomacromolecules* 9:2775–2783.

El-Sheshtawy, H.S., Chatterjee, S., Assaf, K.I., Shinde, M.N., Nau, W.M., Mohanty, J. 2018. A supramolecular approach for enhanced antibacterial activity and extended shelf-life of fluoroquinolone drugs with cucurbit[7]uril. *Sci. Rep.* 8:13925.

Fitridge, I., Dempster, T., Guenther, J., de Nys R. 2012. The impact and control of biofouling in marine aquaculture: a review. *Biofouling* 28:649–669.

Flemming, H.C. 2002. Biofouling in water systems-cases, causes and counter measures. *Appl. Microbiol. Biotechnol.* 59:629–640.

Goel, T., Barooah, N., Mallia, M.B., Bhasikuttan, A.C., Mohanty, J. 2016. Recognition-mediated cucurbit[7] uril-heptamolybdate hybrid material: a facile supramolecular strategy for ^{99m}Tc separation. *Chem. Commun.* 52:7306–7309.

Harding, J.L., Reynolds, M.M. 2014. Combating medical device fouling. *Trends Biotechnol.* 32:140–146.

Hoffman A. 1999. Non-fouling surface technologies. *J. Biomater. Sci. Polym. Ed.* 10:1011–1014.

Holland, N.B., Qiu, Y., Ruegsegger, M., Marchant, R.E. 1998. Biomimetic engineering of non-adhesive glyco-calyx-like surfaces using oligosaccharide surfactant polymers. *Nat. Chem.* 392:799–801.

Holmlin, R.E., Chen, X., Chapman, R.G., Takayama, S., Whitesides, G.M. 2001. Zwitterionic SAMs that resist nonspecific adsorption of protein from aqueous buffer. *Langmuir* 17:2841–2850.

Huang, D., Wang, J., Ren, K., Ji, J. 2020. Functionalized biomaterials to combat biofilms *Biomater. Sci.* 8:4052–4066.

Kalyani, V.S., Malkhede, D.D., Mohanty, J. 2017. Cyclodextrin-assisted modulation of the photophysical properties and acidity constant of pyrene-armed calix[4]arene. *Phys. Chem. Chem. Phys.* 19:21382–21389.

Kandoth, N., Dutta Choudhury S., Mohanty, J., Bhasikuttan, A.C., Pal, H. 2010. Inhibiting intramolecular electron transfer in flavin adenine dinucleotide by host-guest interaction: a fluorescence study. *J. Phys. Chem. B* 114:2617–2626.

Khurana, R., Barooah, N., Bhasikuttan, A.C., Mohanty, J. 2017. Modulation in the acidity constant of acridine dye with cucurbiturils: stimuli-responsive pK_a tuning and dye relocation into live cells. *Org. Biomol. Chem.* 15:8448–8457.

Khurana, R., Kakatkar, A.S., Chatterjee, S., Barooah, N., Kunwar, A., Bhasikuttan, A.C., Mohanty, J. 2019a. Supramolecular nanorods of (N-methylpyridyl) porphyrin with captisol: effective photosensitizer for anti-bacterial and anti-tumor activities. *Front. Chem.* 7:452.

Khurana, R., Mohanty, J., Padma, N., Barooah, N., Bhasikuttan, A.C. 2019b. Redox-mediated negative differential resistance (NDR) behavior in perylenediimide derivative: a supramolecular approach. *Chem. Eur. J.* 25:13939–13944.

Learn, G.D., Lai, E.J., von Recum, H.A. 2020a. Cyclodextrin polymer coatings resist protein fouling, mammalian cell adhesion, and bacterial attachment. *bioRxiv*:doi:10.1101/2020.01.16.909564.

Learn, G.D., Lai, E.J., von Recum, H.A. 2020b. Nonthermal plasma treatment Improves uniformity and adherence of cyclodextrin-based coatings on hydrophobic polymer substrates. *Coatings* 10:1056.

Lehn, J.-M. 1988. Supramolecular chemistry-scope and perspectives molecules, supermolecules, and molecular devices. *Angew. Chem., Int. Ed.* 27:89–112.

Leng, C., Huang, H., Zhang, K., Hung, H.-C., Xu, Y., Li, Y., Jiang, S., Chen, Z. 2018. Effect of surface hydration on antifouling properties of mixed charged polymers. *Langmuir* 34:6538–6545.

Lin, N.H., Kim, M.M., Lewis, G.T., Cohen, Y. 2010. Polymer surface nano-structuring of reverse osmosis membranes for fouling resistance and improved flux performance. *J. Mater. Chem.* 20:4642–4652.

Liu, F., Hashim, N.A., Liu, Y., Abed, M.R.M., Li, K. 2011. Progress in the production and modification of PVDF membranes. *J. Membr. Sci.* 375:1–27.

Luk, Y.-Y, Kato, M., Mrksich, M. 2000. Self-assembled monolayers of alkanethiolates presenting mannitol groups are inert to protein adsorption and cell attachment. *Langmuir* 16:9604–9608.

Maan, A.M.C., Hofman, A.H., de Vos, W.M., Kamperman, M. 2020. Recent developments and practical feasibility of polymer-based antifouling coatings. *Adv. Funct. Mater.* 2000936.

Mehra, C., Gala, R., Kakatkar, A., Kumar, V., Khurana, R., Chatterjee, S., Kumar, N.N., Barooah, N., Bhasikuttan, A.C., Mohanty, J. 2019. Cooperative enhancement of antibacterial activity of sanguinarine drug through *p*-sulfonatocalix[6]arene functionalized silver nanoparticles. *Chem. Commun.* 55:14275–14278.

Minett, T.W., Tighe, B.J., Lydon, M.J., Rees, D.A. 1984. Requirements for cell spreading on polyHEMA coated culture substrates. *Cell. Biol. Int. Rep.* 8:151–159.

Mohanty, J., Dutta Choudhury, S., Barooah, N., Pal, H., Bhasikuttan, A.C. 2017. Mechanistic aspects of host-guest binding in cucurbiturils: physicochemical properties. In (G.W., Gokel, J. L., Atwood, eds.), *General Principles of Supramolecular Chemistry and Molecular Recognition, Comprehensive Supramolecular Chemistry II*, Elsevier. pp. 435–457.

Mohanty, J., Khurana, R., Barooah, N., Bhasikuttan, A.C. 2019. Cucurbituril-functionalized supramolecular assemblies: gateways to diverse applications. In (D., Tuncel, ed.), *Cucurbituril-Based Functional Materials*, Royal Society of Chemistry. pp. 235–257.

Ostuni, E., Chapman, R.G., Holmlin, R.E., Takayama, S., Whitesides, G.M. 2001. A survey of structure-property relationships of surfaces that resist the adsorption of protein. *Langmuir* 17:5605–5620.

Pedersen, C.J. 1988. The discovery of crown ethers. *Angew. Chem., Int. Ed.* 27:1021–1027.

Pendergast, M.M., Hoek, E.M.V. 2011. A review of water treatment membrane nanotechnologies. *Energy Environ. Sci.* 4:1946–1971.

Prochowicz, D., Kornowicz, A., Lewiński, J. 2017. Interactions of native cyclodextrins with metal ions and inorganic nanoparticles: fertile landscape for Chemistry and materials science. *Chem. Rev.* 117:13461–13501.

Punitha, N., Saravanan, P., Mohan, R., Ramesh, P.S. 2017. Antifouling activities of b-cyclodextrin stabilized peg based silver nanocomposites. *Appl. Surf. Sci.* 392:126–134.

Ruz, P., Banerjee, S., Khurana, R., Barooah, N., Sudarsan, V., Bhasikuttan, A.C., Mohanty, J. 2021. Metal-free supramolecular catalytic hydrolysis of ammonia borane through cucurbituril nanocavitands. *ACS Appl. Mater. Interface* 13:16218–16226.

Saji, V.S. 2019. Supramolecular concepts and approaches in corrosion and biofouling prevention. *Corros. Rev.* 37:187–230.

Sakala, G.P., Reches, M. 2018. Peptide-based approaches to fight biofouling. *Adv. Mater. Interfaces* 5:1800073.

Sayed, M., Biedermann, F., Uzunova, V.D., Assaf, K.I., Bhasikuttan, A.C., Pal, H., Nau, W.M., Mohanty, J. 2015. Triple emission from *p*-dimethylaminobenzonitrile-cucurbit[8]uril triggers the elusive excimer emission. *Chem. Eur. J.* 21:691–696.

Schlenoff, J.B. 2014. Zwitteration: coating surfaces with zwitterionic functionality to reduce nonspecific adsorption. *Langmuir* 30:9625–9636.

Schultz, M.P., Bendick, J.A., Holm, E.R., Hertel, W.M. 2011. Economic impact of biofouling on a naval surface ship. *Biofouling* 27:87–98.

Shinde, M.N., Barooah, N., Bhasikuttan, A.C., Mohanty, J. 2016a. Inhibition and disintegration of insulin amyloid fibrils: a facile supramolecular strategy with *p*-sulfonatocalixarenes. *Chem. Commun.* 52:2992–2995.

Shinde, M.N., Bhasikuttan, A.C., Mohanty, J. 2015a. The contrasting recognition behavior of b-cyclodextrin and its sulfobutylether derivative towards 4′, 6-diamidino-2-phenylindole. *ChemPhysChem* 16:3425–3432.

Shinde, M.N., Bhasikuttan, A.C., Mohanty, J. 2016b. Recognition-mediated contrasting fluorescence behaviour of 4′, 6-diamidino-2-phenylindole (DAPI): probing the pK_a of *p*-sulfonatocalix[4/6]arenes. *Supramol. Chem.* 28:517–525.

Shinde, M.N., Dutta Choudhury, S., Barooah, N., Pal, H., Bhasikuttan, A.C., Mohanty, J. 2015b. Metal-ion-mediated assemblies of thiazole orange with cucurbit[7]uril: a photophysical study. *J. Phys. Chem. B* 119:3815–3823.

Shinde, M.N., Khurana, R., Barooah, N., Bhasikuttan, A.C., Mohanty, J. 2017a. Metal ion-induced supramolecular pK_a tuning and fluorescence regeneration of *p*-sulfonatocalixarene encapsulated neutral red dye. *Org. Biomol. Chem.* 15:3975–3984.

Shinde, M.N., Khurana, R., Barooah, N., Bhasikuttan, A.C., Mohanty, J. 2017b. Sulfobutylether-β-cyclodextrin for inhibition and rupture of amyloid fibrils. *J. Phys. Chem. C* 121:20057–20065.

Ulbricht, J., Jordan, R., Luxenhofer, R. 2014. On the biodegradability of polyethylene glycol, polypeptoids and poly(2-oxazoline)s. *Biomaterials* 35:4848–4861.

van de Manakker, F., Vermonden, T., van Nostrum, C.F., Hennink, W.E. 2009. Cyclodextrin-based polymeric materials: synthesis, properties, and pharmaceutical/biomedical applications. *Biomacromolecules* 10:3157–3175.

Verran, J. 2002. Biofouling in food processing: Biofilm or biotransfer potential? *Food Bioprod. Process.* 80:292–298.

Wagner, E.M.V., Sagle, A.C., Sharma, M.M., La, Y.H., Freeman, B.D. 2011. Surface modification of commercial polyamide desalination membranes using poly(ethylene glycol)diglycidyl ether to enhance membrane fouling resistance. *J. Membr. Sci.* 367:273–287.

Wei, T., Zhan, W., Yu, Q., Chen, H. 2017. A smart biointerface with photoswitched functions between bactericidal activity and bacteria-releasing ability. *ACS Appl. Mater. Interfaces* 9:25767–25774.

Wu, H., Tang, B., Wu, P. 2013. Preparation and characterization of anti-fouling b-cyclodextrin/polyester thin film nanofiltration composite membrane. *J. Membr. Sci.* 428:301–308.

Yebra, D.M., Kiil, S., Dam-Johansen. K. 2004. Antifouling technology - past, present and future steps towards efficient and environmentally friendly antifouling coatings. *Prog. Org. Coat.* 50:75–104.

Yu, Z., Pan, Y., He, Y., Zeng, G., Shiab, H., Di, H. 2015. Preparation of a novel anti-fouling b-cyclodextrin–PVDF membrane. *RSC Adv.* 5:51364–51370.

Zhang, H., Chiao, M. 2015. Anti-fouling coatings of poly(dimethylsiloxane) devices for biological and biomedical applications. *J. Med. Biol. Eng.* 35:143–155.

22 Mesoporous Silica-Based Systems for Anti-Biofouling Applications

Silvia Rosane S. Rodrigues
Instituto Federal de Educação, Ciência e Tecnologia Sul-rio-grandense

Viviane Dalmoro
Universidade Federal do Rio Grande do Sul

João Henrique Z. dos Santos
Instituto Federal de Educação, Ciência e Tecnologia Sul-rio-grandense

CONTENTS

22.1 INTRODUCTION

Biofouling is the term used for the adhesion of marine/aquatic microorganisms to all kinds of underwater surfaces. To look for solutions to prevent fouling escalation, it is mandatory to understand the mechanism of the formation of biofouling (Hadfield 2011; Cavalcanti et al. 2020). The first step of adhesion is the formation of a very tiny layer composed of many microbial species, including bacteria, diatoms, fungi, and protozoa, at the microscopic level that works as a substrate for the attachment of other larger organisms and microorganisms (Cavalcanti et al. 2020). Some researchers have dedicated their studies to EPS (extracellular polymer substance) once the microbial origin of EPS was realized and that it acts as a glue to attach microorganisms to the surfaces. This forms the biofilm system that is responsible for both adhesion and cohesion, binding together cells and particulate matter onto the substrate. After contact with a solid surface, the physiologically sessile bacterium secretes a cementing substance (glycocalyx). Based on scanning electron microscopy evidence, the presence of glycocalyx appears to be able to withstand enormous shear forces, which prevents the leaching out of biofoulants (Flemming 2016). This can explain why, in places with strong waves, those waves are not able to remove or prevent biofouling. Furthermore, the extracellular matrix protects the biofilm, forming water-filled channels as circulatory systems to spread nutrients until planktonic bacterial cells are released (Jayaprakashvel et al. 2020). The next phase of biofouling is the development of a multicellular species community containing microalgae, debris, and sediments on the surface, followed by attachment of larger marine invertebrates such as barnacles, mussels and macroalgae (Nurioglu et al. 2015).

DOI: 10.1201/9781003169130-25

The biofouling problem has worsened with the spread of invasive species around the world. For example, mussels can be moved from one country to another through ballast water from ships and vessels. The greatest problem of invasive species occurs in regions where there are no predators for those species, such as the zebra mussel, native to the Caspian and Black Seas (Depew et al. 2020). These species increase the biofouling community, affecting any underwater structures in all aquatic environments, and thus become a plague.

On the other hand, biofouling is a very expensive problem because it can form on any kind of surface, including pipelines, ducts, and marine litter. Biofouling can increase surface corrosion and necessitate shutdowns in underwater equipment, increase the frequency of cleaning marine marker buoys, among other issues (Póvoa et al. 2020; Rajala et al. 2016). Biofouling in ships and ocean vessels reduces their speed and increases fuel consumption. An example is that the US Navy costs related to biofouling are between \$180 and \$260 million per year only in terms of oil consumption (WHOI 2012). In addition, there is a negative ecological impact resulting from this increase in fuel consumption since it contributes to greenhouse gas emissions (CO_2, SO_x, and NO_x) (Selim et al. 2020). Severe impairment to the performance of hydroelectric power plants results in an economic impact due to the additional maintenance costs of biofouling removal or biocide expenses (chlorine and others). In addition, shutdowns are sometimes necessary to remove biofouling (Flemming 2020). Intensive efforts and innovative strategies to improve the prevention of fouling of ship hulls are of particular interest and have resulted in numerous advanced and innovative anti-biofouling strategies, including fouling-release, fouling-degrading, and fouling-resistant coatings.

There is no single solution to solve the problems related to biofouling, due to its complexity, as many different organisms are living in a biofouling environment (Bhoj et al. 2021). Since marine transport is part of human history, it is difficult to know exactly when people started using anti-biofouling substances. The anti-biofouling evolution is summarized in Figure 22.1.

The local and global environmental impacts associated with biocides, added to the international prohibition of the use of several biocides, have boosted the search for anti-biofouling alternatives.

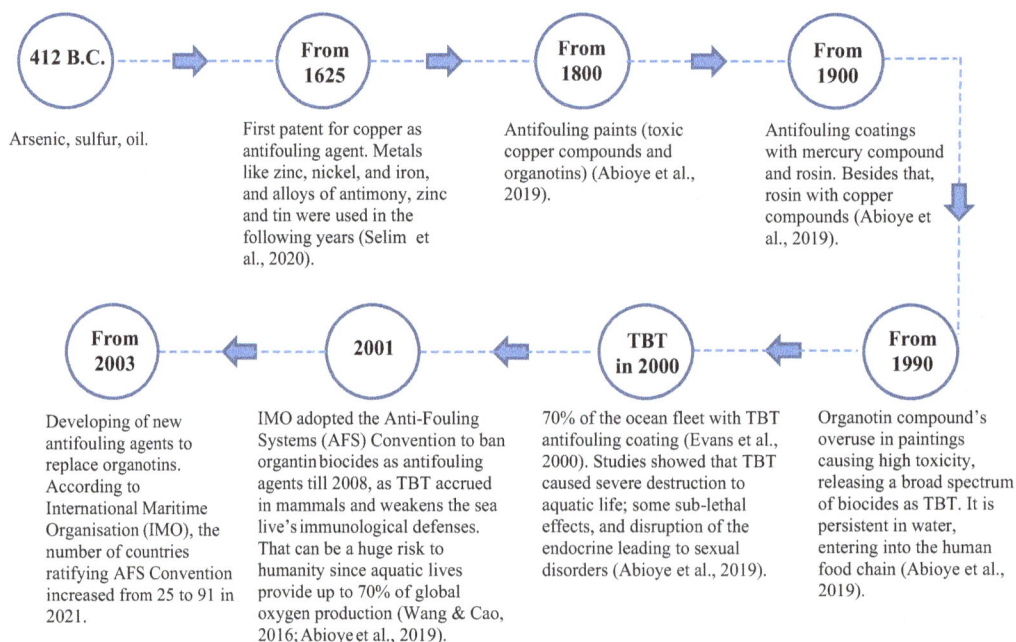

FIGURE 22.1   A scheme showing anti-biofouling evolution.

Thus, considering the need to avoid the formation of fouling and simultaneously protecting the environment, it is understandable that there remains the challenge of finding anti-biofouling agents that meet both of these conditions. Anti-adhesive materials that prevent or limit biofouling are desirable, and their development includes modification in surface chemistry, topography, and architecture. Numerous studies have revealed the importance of surface roughness and topography; these paramount physical traits play an important role in surface interactions with various biomolecules. Changes in surface roughness can therefore affect the hydrodynamic performance of anti-biofouling coatings. Thus, these physical traits must be seriously considered when designing novel approaches for biofouling prevention (Maan et al. 2020).

Surface roughness and topography impact surface interactions with various biomolecules. Air trapped within grooves and ridges that occur naturally on rough material surfaces prevents water penetration, enabling superhydrophobic surfaces to obtain anti-biofouling properties. Although physical alteration of the microtopography is a nontoxic, versatile, and easy way to reduce biofouling, its anti-biofouling action is often too weak, species-dependent, and efficient for only a short time. Consequently, control of microtopography alone is unable to prevent maritime biofouling (Maan et al. 2020).

Another approach for fouling prevention is chemical surface modification designed by hydrophobic and hydrophilic coatings. The anti-biofouling mechanism of hydrophilic polymers is based on retaining a layer of tightly bound water acting as an energetic and physical barrier preventing the attachment of biofouling since protein and initial bacterial attachment promotes subsequent biofilm formation (Huang and Ghasemi 2020; Ulaeto et al. 2017; Selim et al. 2017a; Nurioglu et al. 2015). Several anti-biofouling hydrophilic polymers have attracted attention, such as poly(ethylene glycol) (PEG), hydroxyethyl methacrylate (HEMA), zwitterion-containing polymers and their derivatives, hyperbranched polymers, and hydrogels (Selim et al. 2017a; Yang et al. 2014). These polymers can form hydration layers through hydrogen bonding and electrostatic attraction (Huang and Ghasemi 2020). The drawbacks of these systems include their water solubility, poor mechanical durability, complicated fabrication process, and the susceptibility of the hydration layers to disruption over prolonged service time, blemishing their application as unique coatings (Huang and Ghasemi 2020).

Zwitterionic (ZI) polymer (macromolecules with both positive and negative charges) coatings can be mentioned as an alternative for safe anti-biofouling coatings. Taking into consideration that on a clean surface, like hulls, for instance, an electrical double layer is formed at the liquid/substrate interface, promoting fouling attachment by its charge and the proteins present in fouling microorganisms. In this sense, those coatings can passivate the surface, therefore reducing microorganism adhesion and fouling attachment. Examples such as copolymers of [2-(methacryloyloxy)ethyl]trimethyl ammonium chloride (TM) and 3-sulfopropyl methacrylate (SA) monomers, or [2-(acryloyloxy)ethyl]trimethyl ammonium chloride (TMA) and 2-carboxyethyl acrylate (CAA) can be cited. Considering this passivation phenomenon, it was possible to achieve ultralow-fouling properties with cationic TM and anionic SA in a ratio of 1:1 in relation to the investigated proteins. Polybetaines, which are polymers containing both cationic and anionic groups on the same unit, have been described to have equivalent or better anti-biofouling properties to PEG (Blackman et al. 2018). Due to their highly charged nature, polybetaines present a high degree of hydration. These polymers have shown good resistance, avoiding nonspecific protein attachment. According to some reports, the levels of protein attachment have been lower than the detection limit (<0.3 ng/cm^{-1}) (Jiang and Cao 2010).

Hydrophobic surfaces allow the detachment of biofoulant when hydrodynamic forces occur at the ship's hull due to its low friction and nonstick surfaces (Ulaeto et al. 2017; Nurioglu et al. 2015). These surfaces are good for preventing macrofouling but still suffer from microbial biofilm adhesion. High solid–liquid interface properties can increase both surface roughness and hydrophobicity. In this regard, non-biocide-releasing anti-fouling coatings or fouling release (FR) coatings are composed of a layer of low-surface-energy chemical molecules resulting in nonstick and nontoxic surfaces with fouling anti-adhesion mechanisms. FR resins, including polysiloxane and fluoro-based polymers, have demonstrated effective functions as easily cleaned systems through the physical hindering behavior of settled biofouling. Fluoropolymer structures inherit high stiffness from the strong binding of fluorine atoms, making structural rotation along the matrix backbone more difficult, and thus preventing the easy release of biofouling attachments (Selim et al. 2020). High price and production difficulty of fluoropolymers have led to their replacement with those derived from silicone. Several commercial anti-biofouling formulations contain polydimethylsiloxane (PDMS) (Nurioglu et al. 2015), but its biodegradability has been little investigated (Selim et al. 2017a). The disadvantage of these hydrophobic coatings is their poor adhesion to the hull surface. Thus, intermediate coatings are necessary to enhance the adhesion of the coatings, but such an approach may increase the cost of their use (Gu et al. 2020; Selim et al. 2020).

Nature can also inspire researchers. It is possible to develop new solutions mimicking natural anti-fouling; this is known as biomimicry (Bhushan 2009). Natural surfaces can be biomimetic in terms of minimized free energy, micro/nanoroughness, anisotropic heterogeneity, superior hydrophobicity, tunable nonwettability, antibacterial efficiency, and mechanical robustness. One such surface was prepared by mimicking *Laminaria japonica* using an elastomer and polyelectrolyte layer composed of sodium alginate with guanidine-hexamethylenediamine-polyethylenimine (poly(GHPEI)) (Zhao et al. 2020). Chapman, inspired by *Saccharina latissima* and *Fucus guiryi* macroalgae, developed a surface doped with 3-bromo-5-diphenylene-2(5H)-furanone mimicking the anti-biofouling action of the algae surface with good results (Chapman et al. 2014).

There are some commercial and nonmetallic anti-biofouling products, such as selektope, a medetomidine-based product of the Swedish SME I-Tech, which stimulates barnacle larvae to maintain the swimming mode, thus avoiding their attaching to hulls. Nippon developed a coating based on a polymer used to construct artificial heart and blood vessels on which no organisms can stick. Jansen launched Econea with an anti-biofouling potential similar to copper. In Econea technology, tralopyril and arylpyrrole compounds are included, and AkzoNobel is developing a UV technology linked to a coating to avoid fouling formation to be launched in 2023 (Zainzinger 2019).

On this innovative maritime anti-biofouling scene, approaches have been reported involving mesoporous silica. In the following sections, the application of this material to anti-biofouling coatings is discussed. These particles allow the modification of its properties, as well as the more controlled release of the biocide and the modification of the physical and chemical properties of the coating.

22.2 MESOPOROUS SILICA

Mesoporous silica nanoparticles (MSNs) are composed of an amorphous silica matrix with ordered and periodic arrangements of uniformly sized pores (diameter between 2 and 50 nm), presenting large surface area, high pore volume, tunable pore diameter, and size (Popat et al. 2012). Other interesting properties are their biocompatibility, thermal and chemical stability, and inexpensive synthesis. In addition to their porosity, surface, and morphological properties, it is possible to tailor them to match the physicochemical properties according to the intended guest compounds and applications (Chan et al. 2017). For example, the surface energy and wettability were modified by using organofunctional silanes, improving the dispersity and release profile by stimuli-sensitive molecules and carrier interactions (Chan et al. 2017; Lakshmi and Pola 2020).

There are multiple applications of MSNs, such as in wastewater remediation, indoor air cleaning, biocatalysis, CO_2 capture, bioanalytical sample preparation, catalytic chemistry, biosensing, bioimaging (Lakshmi and Pola 2020), carriers for pharmaceutical drugs, corrosion inhibitors, and biocidal agent delivery (Chan et al. 2017). Based on extensive physicochemical, ecotoxicological safety, and epidemiological data, it is evident that there are no ecological or health hazards associated with those materials. More basic details of MSNs could be found in Chapter 11.

22.3   SURFACE MODIFICATION BY MESOPOROUS SILICA PARTICLES

An effective coating technique for biofouling prevention has been the insertion of different nanofillers (including metal oxides, noble metals, and graphene-based materials) into coatings, which can increase the resistance against microorganism attack (Selim et al. 2017b, 2020; Maan et al. 2020). Control over the surface functionality and morphology, nanoscale size, and active site function of nanocomposites can be intrinsically improved by embedding and distributing nanofillers in homopolymer and copolymer matrices. They can impart distinct topographies leading to the artificial production of lotus surfaces, for example (Elbourne et al. 2019). In the present scenario, MSNs are being utilized in several applications as antibacterial and anti-biofilm agents, some selected studies are highlighted in Table 22.1. In this sense, the use of MSNs in anti-biofouling compounds is an opportunity to have a safer material considering that silica is an inert and nontoxic material. It is also necessary to mention that the ability of bare silica nanoparticles to prevent biofouling by hydrophobicity/hydrophilicity degree modification is practically unexplored (Boguslavsky et al. 2018). In terms of the degree of silica nanoparticle modification, it must be considered that inorganic-organic surfaces can improve the superhydrophobic and anti-biofouling abilities by reducing the free energy of the surface. On the other hand, the nanofiller's amount needs to be controlled to avoid agglomerations, cracking, and brittleness that occur at high nanofiller concentrations (Selim et al. 2020).

PDMS coatings are known for their high molecular mobility, ultrasmooth surface, and low friction, which lead to weakening of the fouling adhesion strength. Selim et al. (2018) improved the anti-fouling properties of the PDMS coating by insertion of an Ag@SiO_2 core–shell nanocomposite, which can be prepared by the Stöber-modified method. Water contact angle (WCA) measurements obtained on a virgin PDMS coating and PDMS coating with insertion of 0.5% Ag@SiO_2 core–shell nanoparticles were 107.1° and 156.3°, respectively. Hysteresis in the CA values was observed in their reduction from 19.4° to 6° after the insertion of 0.5 wt% Ag@SiO_2 core–shell nanoparticles into the PDMS matrix. High WCA values (>150°) and low CA hysteresis (<10°) are fundamental for ultrahydrophobic self-cleaning properties. The addition of 0.5% Ag@SiO_2 core–shell nanoparticles resulted in high resistance against fouling attachments. The same research group reported similar results with PDMS filled with ZnO@SiO_2 particles (Selim et al. 2017b).

It has been emphasized that particle size and filler percentage control are fundamental for good anti-biofouling performance. Large particles provide large pores, less roughness, and less capillary air pressure, allowing the cells to contact the material surface, resulting in a high degree of biofouling (Ee et al. 2019). An increase in the surface free energy was observed by enhancing the nanoparticles loading caused by particles agglomeration and clustering (Selim et al. 2018). The surface smoothness, the lotus effect, was reduced by particle clustering, reducing the self-cleaning ability and favoring the settlement of fouling organisms. Nanoparticle aggregation impacts matrix–nanoparticle interfacial bonding reduction and brittleness by destabilizing due to the high nanoparticle concentration (Selim et al. 2018, 2017b, 2020). High concentrations of up to 3% Ag@SiO_2 core–shell nanoparticles (Selim et al. 2018) and 5% ZnO@SiO_2 particles (Selim et al. 2017b) in silicone composites cause agglomeration and topological heterogeneity, inducing minimum water repellency and biofouling attachment increments.

A zwitterionic core-shell nanocarrier based on poly(sulfobetaine methacrylate) (pSBMA)-coated MSNs was obtained by grafting pSBMA brushes onto the MSNs surface using atom transfer radical polymerization (ATRP). A well-shaped monodispersed spherical MSN (95±15 nm particle diameter, 2.8 nm pore diameter) was surface-functionalized with an ATRP initiator, 2-bromo-2-methyl-N-3-[(triethoxysilyl)propyl]-propanamide (BrTEOS). The study allowed understanding that there is a correlation between polymer chain density and length with the protein adhesion. The layers with very short chains and low graft densities or, the opposite, well-packed layers but with long chains were the shell morphologies with the highest adhesion percentage in the protein test. Better anti-biofouling performance was obtained with high packing density and intermediate molecular weight. The results suggested that the functionalization time of 6–8 h and SBMA/CuBr (catalyst) ratio of

TABLE 22.1

Different Strategies Involving Silica Nanoparticles to Biofouling Prevention

Type of Silica	Biocide*	Fouling Agent Used to Evaluate the Anti-fouling Efficacy	Ref.
Mesoporous silica nanocapsules	MBT and DCOIT	Bioluminescent E. coli.	Maia et al. (2015)
Silica nanocapsules (Si-NCs) and silica nanoparticles (Si-MNPs)	Usnic acid and zosteric acid	Biodeteriogens growing on outdoor surfaces of Roman monuments (filamentous and coccoids cyanobacteria of the genera, Gloeocapsa and Chroococcus and green algae	Ruggiero et al. (2020a)
Mesoporous silica (MCM-48 SiO$_2$) with chemically attached quarternary ammonium salts on the surface	DCOIT	Gram-negative E. Coli, Gram-positive S. aureus. Panels exposed to the northern Red Sea, Eilat, Israel for 5 months	Michailidis et al. (2017)
Mesoporous silica (MCM-48 SiO$_2$)	DCOIT	E. Coli (Gram-negative), S. aureus (Gram-positive), Red Sea Brachidontes pharaonis mussels. Panels exposed to the northern Red Sea, Eilat, Israel for 6 months	Michailidis et al. (2020)
Spherical particles	Econea®	S. aureus	Loureiro et al. (2018)
Mesoporous silica nanocapsules	DCOIT	Bivalve P. perna	dos Santos et al. (2020)
Silica mesoporous nanocapsules	DCOIT and Ag	Bacteria V. fischeri and diatom P. tricornutum	Figueiredo et al. (2020)
Silica nanoparticles (SiNP)	ZnPT and CuPT	Diatom P. tricornutum and mussel M. edulis	Avelelas et al. (2017)
Spherical mesoporous silica nanocapsules (SiNC)	ZnPT and CuPT	Bryozoan B. neritina, from the Mediterranean and the Red Sea, mussel M. galloprovincialis from Atlantic coast and mussel B. pharaonis from the Mediterranean and the Red Sea	Gutner-Hoch et al. (2018)
Spherical mesoporous silica nanocapsules (SiNC)	ZnPT and CuPT	B. pharaonis mussel and B. neritina larvae, Artemia salina, Paracentrotus lividus sea urchins	Gutner-Hoch et al. (2019)
Silica mesoporous nanocapsules	DCOIT and Ag	V. fischeri bacteria, P. tricornutum diatoms, and M. galloprovincialis mussels	Figueiredo et al. (2019)
Silica nanoparticles	Allyl isothiocyanate and cinnamaldehyde	E. coli and P. aeruginosa	Chan et al. (2017)
Polydimethylsiloxane/Ag@SiO$_2$	Ag	B. subtilis and S. aureus (gram-positive), P. aeruginosa and E. coli (Gram-negative), C. albicans (yeast) and A. niger (fungus)	Selim et al. (2018)
Monodispersed silica nanoparticles (SNPs)	Biocide-free	B. licheniformis and P. aeruginosa	Boguslavsky et al. (2018)
PDMS/spherical SiO$_2$-doped ZnO	Without biocides	Micrococcus sp. (Gram-positive) and P. putida (Gram-negative) and A. niger fungi	Selim et al. (2017)

*CuPT (Copper pyrithione), DCOIT (4,5-Dichloro-2-octyl-4-isothiazolin-3-one), MBT (2-Mercaptobenzothiazole), PDMS (Polydimethylsiloxane) and ZnPT (Zinc pyrithione)

*Most investigation were based on biocide encapsulation in silica nanoparticles and some (Selim et al. 2018; Boguslavsky et al. 2018; Selim et al. 2017) through surface modification by silica nanoparticles.

*Generally for anti-fouling evaluation, the silica nanoparticles containing the biocide are exposed to seawater with fouling agents. Nevertheless, Michailidis et al. (2017) and Michailidis et al. (2020) assessed the polyvinyl chloride (PVC) panels coated with paint containing biocide silica nanoparticles and Maia et al. (2015) used carbon steel coated with hydroxyfunctional polyacrylic dispersion combined with aliphatic polyisocyanates or epoxy-based resin cured with polyamide in the anti-fouling assays.

*Only Ruggiero et al. (2020a) employed a natural biocide

100, were the condition/variables to achieve the best anti-biofouling performance under the studied conditions (Beltrán-Osuna et al. 2017, 2018). The existence of the upper critical solution temperature (UCST) was confirmed by the polymer chain conformational change in the evaluation. The researchers observed using light scattering (DLS) that in the solution with free zwitterionic polymer chains, there was a UCST once the polymer underwent a hydrophobic–hydrophilic phase transition, which was explained by the dipole–dipole interactions between the betaine groups. According to them, as UCST is strictly defined for free polymer chains in solution, higher energy requirements seem to be necessary for pSBMA attached to MSNs particles. The particle size diameter close to the native MSNs was attributed to a conformational change of the polymer chain by collapsing on its surface at low temperature. Therefore, by increasing the temperature, the transition state could be reached, and the energy requirements could be overcome, breaking the electrostatic interaction between zwitterion molecules and allowing chain solvation. The higher transition temperature results observed in that study were related to the fact that grafted chains on silica allowed higher pSBMA concentrations. It had already been reported that UCST increased from 15°C to 27°C as the polymer concentration was increased from 0.8 to 5.5 wt%. While varying the pH medium, the zwitterionic range was established for pH values of 5–9, allowing both charges (from quaternary ammonium and sulfonate groups) to become balanced. In terms of the zeta potential (ζ-potential) of pSBMA-MSN dispersions, a positive ζ-potential was measured for acidic media, while a negative ζ-potential was obtained for basic media, displaying an isoelectric point at a pH range of 5.5–6.5. ζ-potential values were found to have a directly proportional trend with temperature. Since the results were independent of graft density or polymer layer thickness, this trend was considered evidence that, for ζ-potential variations, polymer brushes with the different morphologies tested were all effective in particle surface modification. It was possible to modulate the particle behavior in solution by adjusting the pH, temperature, and ionic concentration of the medium. In this sense, those results should be applied to achieve a specific response for pSBMA-MSNs to be used as carriers for controlled delivery applications (Beltrán-Osuna et al. 2020).

22.4 MESOPOROUS SILICA ENCAPSULATION AS ANTI-BIOFOULING ALTERNATIVE

Several investigations have demonstrated that massive amounts of biocides released into seawater lead to marine environment contamination with its associated biocide toxicity to nontarget species (Tornero and Hanke 2016; Figueiredo et al. 2020). In the coating formulation, the biocides are embedded and dispersed molecularly in the polymer matrix and self-diffuse through the coating matrix. Considering that the biocide is immediately available to the surrounding environment, a very high biocide concentration is thus necessary to maintain the coating anti-biofouling activity for longer periods once the anti-biofouling activity decreases during its lifetime (Ruggiero et al. 2020a).

One strategy reported to overcome these drawbacks is biocide encapsulation. This technology is well known for controlled delivery systems of pharmaceuticals and corrosion inhibitors. The encapsulation process using MSNs is used as a drug nanocarrier. It was first adopted by the pharmaceutical industry once the high thermal properties, biocompatibility, and nontoxicity of these particles were recognized. The same strategy may be applied for anti-biofouling design applications considering its properties, such as the pore volume and pore size, as well as its surface area (Amolegbe et al. 2018). It is possible to prevent the direct interaction between biocide and coating components by using suitable carriers for encapsulation, decreasing the leaching rate of biocides (Maia et al. 2015; Avelelas et al. 2017; Ruggiero et al. 2020b), and maximizing the coating lifetime and anti-biofouling efficiency, even at lower doses, reducing the environmental impact related to the anti-fouling biocides.

In silica nanoparticle (Si-NP) evaluation using infra-red (FTIR) spectroscopy, a shift to higher wavenumbers was observed for the band assigned to the asymmetric ν(Si–O–H) stretching mode after encapsulating 2-mercaptobenzothiazole (MBT) into Si-NP (Ruggiero et al. 2020b). This result indicated that the silica network deformed to accommodate the MBT molecules, forming larger siloxane rings with greater Si–O–Si angles and longer Si–O bond lengths. Distinctly different results were obtained for mesoporous silica when synthesized with MBT, revealing the different spatial confinement of the biocide. In Si-MNPs, the biocide is embedded within the sponge-like structure, while the MBT is contained within the internal core volume of the Si-NP. In the synthesis procedure of both containers, a cationic surfactant (CTAB) was employed. Raman analysis showed that the CTAB was present even after several washes with water, along with the interaction of MBT through the nitrogen atom with the residual CTAB, and the tendency of the MBT to dimerize. These findings imply different release rates for the two containers, as is also verified by monitoring the release profiles obtained in water over 120 days. Although the total amount of biocide released after 4 months is approximately the same, the release kinetics of the two systems were different (Ruggiero et al. 2019). For Si-NC, linear and slow release of MBT was observed for 70 days, followed by a faster rate at longer times. For the Si-MNP-MBT system, an initial linear release of 0.00092 mM/day was observed, faster than that of 0.00046 mM/day for the Si-NP-MBTs, followed by saturation at approximately 0.12 mM. The authors attributed the faster initial release for the Si-MNP-MBT system to more direct contact between the biocide and water, due to the sponge structure, and higher BET specific area and average pore volume than those of the Si-NP-MBT sample, considering that both nanosystems exhibit mesoporosity. Thus, it is evident that the high surface area and the distribution of mesopores are important in controlling the kinetics of biocide release.

Analysis of nitrogen adsorption–desorption isotherms, according to the BET method, demonstrated the textural property modification of silica nanocapsules (Si-NCs) with the encapsulation of MBT and 4,5-dichloro-2-octyl-4-isothiazolin-3-one (DCOIT) (Maia et al., 2015). The researchers verified an increase in the surface area and in the pore size distribution with biocide encapsulation, with the greatest effect observed on Si-NCs with DCOIT. The increase in the pore size distribution and surface area was attributed to the increase in the size of the encapsulated molecules and the meso- and macro-porosity in the nanocapsule shell, respectively. In addition, DCOIT molecules are larger than MBT molecules and are solvated, resulting in larger pores for DCOIT@Si-NC. The solvent employed for Si-NCs synthesis contributes to pore formation. For MBT encapsulation, only ethyl ether was used, and channels and gradual porosity were created during vaporization (37°C). Xylene was mixed with ethyl ether for DCOIT encapsulation; thus, ethyl ether, with a lower boiling point, vaporized first, generating micro- and small mesopores, but due to its higher boiling point (140°C), xylene remained in the initial structure formed during the condensation step. After the vaporization of xylene, larger mesopores were formed in comparison to empty capsules, and those were prepared with MBT. The small pores of MBT@Si-NC affected its release, and for DCOIT@ Si-NC, the release was limited by the solubility once those pores were larger. Carbon steel panels were coated with two different formulations (one aqueous and one non-aqueous solvent-based formulation) containing MBT@Si-NC and DCOIT@Si-NC. The inactivation of recombinant bioluminescent strains of *E. coli* was measured by real-time monitoring of bacterial death from the biocidal release of Si-NCs. The antibacterial efficacy was dependent on the formulation. Increasing the DCOIT@Si-NC and MBT@Si-NC concentrations in the aqueous-based coating resulted in a high inactivation level. Conversely, for systems with the solvent-based coating, no biocide and no bacterial inactivation were detected. These results were correlated with the matrix permeability of the solvent-based coating, which was smaller than that of the aqueous-based coating.

Several investigations have been conducted screening the efficacy, toxicity, and deleterious effects on marine organisms (target and nontarget species) of encapsulated/immobilized biocides, as well as the influencing effects of different geographical regions on anti-biofouling action (dos Santos et al. 2020; Figueiredo et al. 2019, 2020; Avelelas et al. 2017; Gutner-Hoch et al. 2018, 2019). Anti-biofouling assays with spherical mesoporous Si-NCs and Zn-Al layered double hydroxides (LDHs) loaded with two commercial biocides, zinc pyrithione (ZnPT) and copper pyrithione (CuPT), were performed using adult mussels from three geographic regions: the eastern Mediterranean, Eilat, the Gulf of Aqaba, northern Red Sea, and the Atlantic coast, presenting different annual seawater salinities and annual temperature ranges (Gutner-Hoch et al. 2018). Mussels and larvae of bryozoan Bugula neritina have been used as a model system to evaluate the anti-biofouling properties of biocidal compounds by determining both half-maximal effective concentrations (EC50) and median lethal concentrations (LC50) values. The researchers verified the efficacy of these anti-biofouling systems for all geographic regions exposed to different environmental conditions; nevertheless, the Red Sea mussels and bryozoans yielded the highest efficacy.

As nontarget species, the green microalgae *Tetraselmis chuii*, *Isochrysis galbana* and *Nannochloropsis gaditana*, the rotifer *Brachionus plicatilis*, the bivalves *Cerastoderma edule*, the polychaete *Hediste diversicolor*, the crustaceans *Artemia salina* and *Palaemon varians* and the echinoderm *Paracentrotus lividus* have been evaluated (Avelelas et al. 2017; Figueiredo et al. 2019). The brine shrimp *Artemia salina* and sea urchin *Paracentrotus lividus* are successful model species in marine ecotoxicology and acute toxicity tests, respectively (Gutner-Hoch et al. 2019). The toxicity shown is dependent on the nontarget species, engineered micro/nanomaterials (the LDHs being evaluated), mesoporous Si-NCs, polyurea (PU) microcapsules, biocide type DCOIT, ZnPT, CuPT, and Ag (dos Santos et al. 2020; Figueiredo et al. 2019, 2020; Avelelas et al. 2017; Gutner-Hoch et al. 2018, 2019). Biocide encapsulation reduces toxicity to nontarget species without compromising anti-biofouling efficacy against target species (dos Santos et al. 2020; Figueiredo et al. 2019, 2020; Gutner-Hoch et al. 2019).

The use of secondary metabolites from marine bioresources is effective against biofouling (Wang et al. 2017), and the positive aspects are the nontoxicity or low toxicity to the marine ecological environment, and their degradability (Prasad et al. 2020). Silica nanocontainers were loaded with natural products derived from marine plants, such as Zostera marina, which produced and continuously released zosteric acid (Ruggiero et al. 2018, 2020a), and with saxicolous lichens as usnic acid (Ruggiero et al. 2020a) to reduce the environmental impact of traditional biocides. In vitro tests and optical microscopy observations under visible light showed that the anti-biofouling nanosystems are effective against photosynthetic microorganisms (algae and cyanobacteria), lacking in situ tests to better understand the anti-fouling and biocidal power of those substances. Interestingly, mesoporous nanoparticles loaded with usnic acid and zosteric salt changed the shape from spherical to hexagonal (Ruggiero et al. 2020a). According to the authors, during the micellization process, the biocides acted as cosurfactants and enhanced the amount of effective surfactant, leading to micelle modification in the self-assembled array from spherical to cylindrical micelles.

Michailidis et al. investigated the synergic action of surface modification and biocide encapsulation on mesoporous silica (MCM-48 SiO_2) (Michailidis et al. 2017). First, the surface MCM-48 SiO_2 nanoparticles were functionalized with dimethyloctadecyl [3-(trimethoxysilyl) propyl] ammonium chloride or dimethyltetradecyl [3-(triethoxysilyl) propyl] ammonium chloride) quaternary ammonium salts (QASs), as confirmed by FTIR, thermogravimetric analysis (TGA), elemental analysis, and ζ-potential measurements. In a second step, it was loaded with a ecofriendly liquid biocide (Parmetol), whose active component was DCOIT. The isoelectric point of QAS-modified silica particles shifted to a lower pH from 2.0 to 1.5. The surface was functionalized with positively charged quaternary ammonium groups instead of negatively charged hydroxyl groups present on pristine silica. Moreover, the high ζ-potential is evidence of its good stability in colloidal suspensions. PVC panels were coated with pristine paint and paint with 5 wt% modified MCM-48 SiO_2. Afterward, those panels were submitted to a 5-month field trial in the northern Red Sea, Eilat (Israel). The combination of surface modification and encapsulation improved the antibacterial and anti-biofouling performance. It was observed that biofouling coverage followed the order: the samples painted with QAS-modified silica loaded with DCOIT > painted with QAS-modified nanoparticles > pristine paint. The authors supposed that the QAS groups on the surface of the silica particles afforded anti-fouling properties even after DCOIT was released from the paint as a synergistic effect. The QAS's action to kill the bacteria can be explained by their adsorption in the bacteria through electrostatic interactions between the negative surfaces of the bacterial cells and the positively charged cationic compounds. After that, diffusion occurs through the cell wall binding to the cytoplasmic membrane, causing its rupture and cell death. Further study by this group (Michailidis et al. 2020) investigated the concentration of the modified silica in the paint. The addition of 5% QAS-modified silica loaded with DCOIT to the coating formulation was more effective than the addition of 2% DCOIT to mitigate fouling.

To control Econea® biocide release, Loureiro et al. proposed physical and chemical impregnation on silica particles (Loureiro et al. 2018). They performed chemical grafting between the oxirane ring of the glycidoxypropyltrimethoxysilane (GPTMS) microscaffold's silica precursor and the secondary amino groups of the biocide. Beyond physical entrapment within the "worm-like" morphology by scaffolds with different pore sizes, this configuration can contribute to controlling biocide release. The FTIR analysis performed after 3 days of immersion in saline conditions confirmed that the biocide was maintained on silica; furthermore, that system was effective against *S. aureus* bacterial growth.

22.5   CONCLUSIONS AND PERSPECTIVES

The unique properties of mesoporous silica, such as ordered pore structures, very high specific surface areas, and morphological diversity, render it a promising material for anti-biofouling design applications. From this point of view, biocide encapsulation in mesoporous silica improves the biocide release profile, avoiding both premature release and the employment of large quantities of biocides. The negative consequences of biocide impacts are minimized, contributing to the generation of efficient anti-biofouling coatings with lower environmental hazards. Nevertheless, there is still room to improve the knowledge related to the chemical interactions of biocides with silica functionalization compounds, as well as the release kinetics that are not yet completely understood. Furthermore, superhydrophobic properties and lower surface energy are obtained by the incorporation of mesoporous silica particles into the coating, leading to a reduction in fouling organism attachment. The control of particle size, surface functionality, geometrical morphology, and dispersion into the coating are critical to the success of the tuning of the surface properties as well as obtaining the desired anti-biofouling functionality. Considering the diversity of fouling organisms, the production of anti-biofouling coatings is complex and inhospitable; thus, intensive investigation in this area remains mandatory, and mesoporous silica represents a considerable potential to control biofouling.

ACKNOWLEDGMENTS

The authors thank the support of INCT MIDAS, CNPq (301408/2019-9) and FAPERGS (19/2551-0001869-0))

REFERENCES

Abioye, O.P., Loto, C.A., Fayomi, O.S.I. 2019. Evaluation of anti-biofouling progresses in marine application. *J. Bio. Tribo. Corros.* 5: 1–8.
Amolegbe, S.A., Hirano, Y., Adebayo, J.O., Ademowo, O.G., Balogun, E.A., Obaleye, J.A., Krettli, A.U., Yu, C., Hayami, S., 2018. Mesoporous silica nanocarriers encapsulated antimalarials with high therapeutic performance. *Sci. Rep.* 8: 3078.
Avelelas, F., Martins, R., Oliveira, T., Maia, F., Malheiro, E., Soares, A.M.V.M., Loureiro, S., Tedim, J. 2017. Efficacy and ecotoxicity of novel anti-fouling nanomaterials in target and non-target marine species. *Mar. Biotechnol.* 19: 164–174.
Beltrán-Osuna, Á.A., Gómez-Ribelles, J.L., Perilla, J.E. 2020. Temperature and pH responsive behavior of antifouling zwitterionic mesoporous silica nanoparticles. *J. Appl. Phys.* 127: 135106.
Beltrán-Osuna, A.A., Ribelles, J.L.G., Perilla, J.E. 2017. A study of some fundamental physicochemical variables on the morphology of mesoporous silica nanoparticles MCM-41 type. *J. Nanopart. Res.* 19: 381.
Beltrán-Osuna, Á.A., Ródenas-Rochina, J., Gómez-Ribelles, J.L., Perilla, J.E. 2018. Antifouling zwitterionic pSBMA-MSN particles for biomedical applications. *Polym. Adv. Technol.* 2018: 1–10.
Bhoj, Y., Tharmavaram, M., Rawtani, D. 2021. A comprehensive approach to antifouling strategies in desalination, marine environment, and wastewater treatment. *Chem. Phys. Impact.* 2: 100008.
Bhushan, B. 2009. Biomimetics: lessons from nature – an overview. *Philos. Trans. R. Soc. A* 367: 1445–1486.

Blackman, L.D., Gunatillake, P.A., Cass, P., Locock, K.E.S. 2018. An introduction to zwitterionic polymer behavior and applications in solution and at surfaces. *Chem. Soc. Rev.* 48: 757–770.

Boguslavsky, Y., Shemesh, M., Friedlander, A., Rutenberg, R., Filossof, A.M., Buslovich, A., Poverenov, E. 2018. Eliminating the need for biocidal agents in anti-biofouling polymers by applying grafted nano-silica instead. *ACS Omega* 3: 12437–12445.

Cavalcanti, G.S., Alker, A.T., Delherbe, N., Malter, K.E., Shikuma, N.J. 2020. The influence of bacteria on animal metamorphosis. *Annu. Rev. Microbiol.* 74: 137–158.

Chan A.C., Cadena, M.B., Townley, H.E., Fricker, M.D., Thompson, I.P. 2017. Effective delivery of volatile biocides employing mesoporous silicates for treating biofilms. *J. R. Soc. Interface* 14: 20160650.

Chapman, J., Hellio, C., Sullivan, T., Brown, R., Russell, S., Kiterringham, E., Le Nor, L., Regan, F. 2014. Bioinspired synthetic macroalgae: Examples from nature for antifouling applications. *Int. Biodeterior. Biodegrad.* 86: 6–13

Depew D.C., Krutzelmann, E., Watchorn, K.E., Caskenette, A., Enders, E.C. 2021. The distribution, density, and biomass of the zebra mussel (Dreissena polymorpha) on natural substrates in Lake Winnipeg 2017–2019. *J. Great Lakes Res.* 47: 556–566.

dos Santos, J.V.N., Martins, R., Fontes, M.K., de Campos, B.G., Prado e Silva, M.B.M., Maia, F., Abessa, D.M.S., Perina, F.C. 2020. Can encapsulation of the biocide DCOIT affect the anti-fouling efficacy and toxicity on tropical bivalves? *Appl. Sci.* 10: 8579–8591.

Ee, U., Udo, G.J., Koffi, U.S., Alswafy, O.B. 2019. The Role of Particle Size on Bio-Fouling Properties of Oil-impregnated Nano-porous Silica Coatings. *J. Appl. Sci. Environ. Manage.* 33: 1153–1157.

Elbourne, A., Chapman, J., Gelmi, A., Cozzolino, D., Crawford, R.J., Truong, V.K. 2019. Bacterial-nanostructure interactions: The role of cell elasticity and adhesion forces. *J. Colloid Interface Sci.* 546: 192–210.

Evans, S.M., Birchenough, A.C., Brancato, M.S. 2000. The TBT ban: Out of the frying pan into the fire? *Mar. Pollut. Bull.* 40: 204–211.

Figueiredo, J., Loureiro, S., Martins, R. 2020 Hazard of novel anti-fouling nanomaterials and biocides DCOIT and silver to marine organisms. *Environ. Sci. Nano* 7: 1670–1680.

Figueiredo, J., Oliveira, T., Ferreira, V., Sushkova, A., Silva, S., Carneiro, D., Cardoso, D.N., Gonçalves, S.F., Maia, F., Rocha, C., Tedim, J., Loureiro S., Martins, R. 2019. Toxicity of innovative anti-fouling nano-based solutions in marine species. *Environ. Sci. Nano* 6: 1418–1429.

Flemming, H.C. 2016. EPS - then and now. *Microorganisms* 4: 1–18.

Flemming, H.C. 2020. Biofouling and me: my Stockholm syndrome with biofilms. *Water Res.* 173: 115576.

Gu, Y., Yu, L., Mou, J., Wu, D., Xu, M., Zhou, P., Ren, Y. 2020. Research strategies to develop environmentally friendly marine antifouling coatings. *Mar. Drugs* 18: 1–22.

Gutner-Hoch, E., Martins, R., Maia, F., Oliveira, T., Shpigel, M., Weis, M., Tedim, J., Benayahu, Y. 2019. Toxicity of engineered micro- and nanomaterials with antifouling properties to the brine shrimp Artemia salina and embryonic stages of the sea urchin Paracentrotus lividus. *Environ. Pollut.* 251: 530–537.

Gutner-Hoch, E., Martins, R., Oliveira, T., Maia, F., Soares, A.M.V.M., Loureiro, S., Piller, C., Preiss, I., Weis, M., Larroze, S.B., Texeira, T., Tedim, J., Benayahu, Y. 2018. Antimacrofouling efficacy of innovative inorganic nanomaterials loaded with booster biocides. *J. Mar. Sci. Eng.* 6: 6–17.

Hadfield, M.G. 2011. Biofilms and marine invertebrate larvae: What bacteria produce that larvae use to choose settlement sites. *Ann. Rev. Mar. Sci.* 3: 453–470.

Huang, Y., Callahan, S., Hadfield, M.G. 2012. Recruitment in the sea: bacterial genes required for inducing larval settlement in a polychaete worm. *Sci. Rep.* 2: 228.

Huang, Z., Ghasemi, H. 2020. Hydrophilic polymer-based anti-biofouling coatings: preparation, mechanism, and durability. *Adv. Colloid Interface Sci.* 284: 102264.

Jayaprakashvel M., Sami M., Subramani R. 2020. Antibiofilm, antifouling, and anticorrosive biomaterials and nanomaterials for marine applications. In *Nanostructures for Antimicrobial and Antibiofilm Applications* (Prasad, R., Siddhardha, B., Dyavaiah, M., eds.), Nanotechnology in the Life Sciences. Springer, Cham.

Jiang, S., Cao, Z. 2010. Ultralow-fouling, functionalizable, and hydrolyzable zwitterionic materials and their derivatives for biological applications. *Adv. Mater.* 22: 920–932.

Lakshmi P., Pola S. (2020) Mesoporous silica nanomaterials as antibacterial and antibiofilm agents. In *Nanostructures for Antimicrobial and Antibiofilm Applications* (Prasad, R., Siddhardha, B., Dyavaiah, M., eds.), Nanotechnology in the Life Sciences. Springer, Cham.

Loureiro, M.V., Vale, M., De Schrijver, A., Bordado, J.C., Silva, E., Marques, A.C. 2018. Hybrid custom-tailored sol-gel derived microscaffold for biocides immobilization. *Microporous Mesoporous Mater.* 261: 252–258.

Maan, A.M.C., Hofman, A.H., de Vos, W.M., Kamperman, M. 2020. Recent developments and practical feasibility of polymer-based antifouling coatings. *Adv. Funct. Mater.* 30: 2000936.

Maia, F., Silva, A.P., Fernandes, S., Cunha, A., Almeida, A., Tedim, J., Zheludkevicha, M.L., Ferreira, M.G.S. 2015. Incorporation of biocides in nanocapsules for protective coatings used in maritime applications. *Chem. Eng. J.* 270: 150–157.

Michailidis, M., Gutner-Hoch, E., Wengier, R., Onderwater, R., D'Sa, R.A., Benayahu, Y., Semenov, A., Vinokurov, V, Shchukin, D.G. 2020. Highly effective functionalized coatings with antibacterial and antifouling properties. *ACS Sustainable Chem. Eng.* 8: 8928–8937.

Michailidis, M., Sorzabal-Bellido, I., Adamidou, E.A., Diaz-Fernandez, Y.A., Aveyard, J., Wengier, R., Grigoriev, D., Raval, R., Benayahu, Y., D'Sa, R.A., Shchukin, D. 2017. Modified mesoporous silica nanoparticles with dual synergetic antibacterial effect. *ACS Appl. Mater. Interfaces* 9: 38364–38372.

Nurioglu, A.G., Catarina, A., Esteves, C., With, G. 2015. Non-toxic, non-biocide-release antifouling coatings based on molecular structure design for marine applications. *J. Mater. Chem. B* 3: 6547–6570.

Popat, A., Liu, J., Hu, Q., Kennedy, M., Peters, B., Lu, G.Q., Qiao, S.Z. 2012. Adsorption and release of biocides with mesoporous silica nanoparticles. *Nanoscale* 4: 970–975.

Póvoa, A.A., Skinner, L.F., Araújo, F.V. Fouling organisms in marine litter (rafting on abiogenic substrates): A global review of literature. *Mar. Pollut. Bull.* 166: 112189.

Prasad, R., Siddhardha, B., Dyavaiah, M. (Eds.), *Nanostructures for Antimicrobial and Antibiofilm Applications.* Springer Nature, Switzerland, Cham, 2020. ISBN: 978-3-030-40337-9.

Rajala, P., Sohlberg, E., Priha, O., Tsitko, I., Väisänen, H., Tausa, M., Carpén, L. 2016. Biofouling on coated carbon steel in cooling water cycles using brackish seawater. *J. Mar. Sci. Eng.* 4: 1–15.

Ruggiero, L., Bartoli, F., Fidanza, M.R., Zurlo, F., Marconi, E., Gasperi, T., Tuti, S., Crociani, L., Di Bartolomeo, E., Caneva, G., Ricci, M.A., Sodo, A. 2020a. Encapsulation of environmentally-friendly biocides in silica nanosystems for multifunctional coatings. *Appl. Surf. Sci.* 514: 145908.

Ruggiero, L., Crociani, L., Zendri, E., El Habra, N., Guerriero, P. 2018. Incorporation of the zosteric sodium salt in silica nanocapsules: synthesis and characterization of new fillers for antifouling coatings. *Appl. Surf. Sci.* 439: 705–711.

Ruggiero, L., Di Bartolomeo, E., Gasperi, T., Luisetto, I., Talone, A., Zurlo, F., Peddis, D., Ricci, M.A., Sodo, A. 2019. Silica nanosystems for active antifouling protection: nanocapsules and mesoporous nanoparticles in controlled release applications. *J. Alloys Compd.* 798: 144–148.

Ruggiero, L., Sodo, A., Cestelli-Guidi, M., Romani, M., Sarra, A., Postorino, P., Ricci, N.A. 2020b. Raman and ATR FT-IR investigations of innovative silica nanocontainers loaded with a biocide for stone conservation treatments. *Microchem. J.* 155: 104766.

Selim, M.S., El-Safty, S.A., Shenashen, M.A., Higazy, S.A., Elmarakbi, A. 2020. Progress in biomimetic leverages for marine antifouling using nanocomposite coatings. *J. Mater. Chem. B* 8: 3701–3732.

Selim, M.S., Yang, H., Wang, F., Li, X., Huang, Y., Fatthallah, N.A. 2018. Silicone/Ag@SiO_2 core–shell nanocomposite as a self-cleaning antifouling coating material. *RSC Adv.* 8: 9910–9921.

Selim, M.S., Shenashen, M.A., Elmarakbi, A., Fatthallah, N.A., Hasegawa, S., El-Safty, S.A. 2017b. Synthesis of ultrahydrophobic and thermally stable inorganic–organic nanocomposites for self-cleaning foul release coatings. *Chem. Eng. J.* 320: 653–666.

Selim, M.S., Shenashen, M.A., El-Safty, S.A., Higazy, S.A., Selim, M.M., Isago, H., Elmarakbi, A. 2017a. Recent progress in marine foul-release polymeric nanocomposite coatings. *Prog. Mater. Sci.* 87: 1–32.

Tornero, V., Hanke, G. 2016. Chemical contaminants entering the marine environment from sea-based sources: A review with a focus on European seas. *Mar. Pollut. Bull.* 112: 17–38.

Ulaeto, S.B., Rajan, R., Pancrecious, J.K., Rajan, T.P.D., Pai, B.C. 2017. Developments in smart anticorrosive coatings with multifunctional characteristics. *Prog. Org. Coat.* 111: 294–314.

Wang, W., Cao, Z. 2016. Opinion on the recent development of environmentally friendly marine anti-fouling coating. *Sci. China Tech. Sci.* 59: 1968–1970.

Wang, K.L., Wu, Z.H., Wang, Y., Wang, C.Y., Xu, Y. 2017. Mini-review: Antifouling natural products from marine microorganisms and their synthetic analogs. *Mar. Drugs* 15: 266.

WHOI, Woods Hole Oceanographic Institution. (2012). Barnacles and Biofilms: Could tiny predators help banish barnacles? (Publication of Winner, C. in December 5, 2012). Retrieved from https://www.whoi.edu/oceanus/feature/barnacles-and-biofilms.

Yang, W.J., Neoh, K.G., Kang, E.T., Teo, S.L.M., Rittschof, D. 2014. Polymer brush coatings for combating marine biofouling. *Prog. Polym. Sci.* 39: 1017–1042.

Zhao, L., Chen, R., Lou, L., Jing, X., Liu, Q., Liu, J., Yu, J., Liu, P., Wang, J. 2020. Layer-by-layer-assembled antifouling films with surface microtopography inspired by Laminaria japonica. *Appl. Surf. Sci.* 511: 145564.

Zainzinger, V. 2019. Antifouling coatings cling to copper. Chemistry World (RSC): https://www.chemistryworld.com/news/antifouling-coatings-cling-to-copper/3010011.article.

23 MOFs-Based Systems for Anti-Biofouling Applications

Ubong Eduok
University of Saskatchewan

CONTENTS

23.1 INTRODUCTION

Exposed and submerged surfaces are readily fouled when their exteriors are outgrown by unwanted substances, even living organisms. Foulants may also be environmentally influenced surface-growths, and their presence would impede or outrightly interfere with the working functions of these surfaces. This surface phenomenon is known as fouling and may be diverse, depending on the types and scales of adhering foulants (and fouling organisms). Fouling substances may be found on heat-transfer components (e.g., coolants), marine vessels (e.g., ship hulls), oral cavity (e.g., tooth-based plaques), filtration devices (e.g., membrane), and so on. (Tian et al. 2020).

DOI: 10.1201/9781003169130-26

23.1.1 Classification of Fouling

There may be several classifications of fouling; however, they could be macro or macrofouling, depending on the size of the foulants (Martín-Rodríguez et al. 2015; Tian et al. 2020). Fouling is mainly caused by either biological and/or inorganic coarse matter. The main lines of cooling systems, including the pumps, are readily fouled due to substances from water sources (e.g., open seawater). These fouling substances can easily detach and be transferred into water circuitry, thereby leading to deteriorating cooling efficiency and reduced performance. In heat exchangers, these substances alter heat transfer coefficient by blocking channels, leading to the redistribution of heat flow. When the foulants are on the microscale, the fouling phenomenon is classified as microfouling. Fouling substances could be precipitating scales (e.g., from salts and oxides deposits) and even biofoulants (e.g., algae) (Martín-Rodríguez et al. 2015; Tian et al. 2020). For the sake of this chapter, attention will be focused on the latter. Otherwise referred to as biological fouling, biofouling is the stepwise assemblage of unwanted microorganisms (e.g., green algae and diatoms) and macro-organisms (e.g., barnacles and polychaetas), even plants (e.g., lichens) and bacteria (e.g., *Escherichia coli*) on submerged surfaces. Biofouling is common on surfaces of hulls of ships, storage reservoirs of sewage and filtration membranes. This phenomenon could lead to microbial induced corrosion (MIC) in the later stages if the resident bacteria have sufficient carbon sources to support their growth. Here, the host metal substrate should also have the requisite electrode potential within appropriate pH (Martín-Rodríguez et al. 2015, Tian et al. 2020). The consequences of biofouling may lead to the anodic dissolution of the metal surfaces that they are attached. A case in point being the MIC of ferrous metals caused by sulfate-reducing bacteria (SRB, e.g., *Desulfovibrio desulfuricans*) and sulfide-oxidizing bacteria (SOB, e.g., *Acidithiobacillus ferrooxidans*). The effects of produced hydrogen sulfide in SRB and sulfuric acid from SOB contribute significantly to the gross corrosion process (Choudhury et al. 2018).

23.1.2 Stages of Biofouling

The stages of the biofouling process are depicted in Figure 23.1. The first stage involves the formation of conditioning films on the surface of the solid substrate immediately after exposure. This is

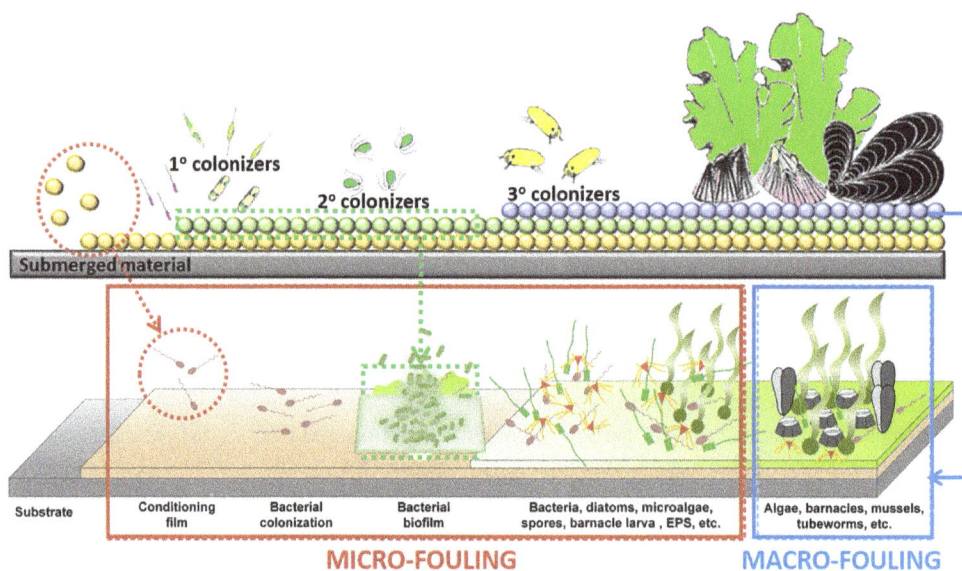

FIGURE 23.1   A schematic showing the stages of biofouling process. The upper part of the figure is reproduced with permission from Martín-Rodríguez et al. (2015) under Creative Commons Attribution License. (The lower part is reproduced with permission from Tian et al. (2020) © 2019 Elsevier B.V.)

possible through the absorption of dissolved organic and inorganic matters. Within several minutes to few hours, surface conditioning is followed by bacterial colonization (Crouzet et al. 2014). When bacterial cells settle on surfaces, the formation of biofilms may be spontaneous. After cellular attachment, the formation of colonies is followed by maturation of biofilms and even occasional removal and formation of new colonies (Costerton et al. 1995; Landini et al. 2010). These slimy biofilms are a complex multicellular three-dimensional community of growing bacterial cells within extracellular polymeric substances (EPSs). EPS is a high molecular-weight polymer secreted from the bacterial colony to uphold the structural integrity of the biofilm network. EPS also keeps the nutrients that maintain the bacterial living process. EPS comprises of polysaccharides, lipids, nucleic acids, and so on. It is also important for the maintenance of the growth process of bacterial biofilm formation while also maintaining intercellular communication via quorum sensing (Costerton et al. 1995). The biofilm houses sessile bacterial cells and those in the dormant growth phase; they may also be exhibiting planktonic phenotypes (Stoodley et al. 2002; Vilain et al. 2004). EPS also upholds the resistance of the biofilm against environmental influences, including pressure, temperature, and even antibiotics (Høiby et al. 2010). It is the structural biofilms that make bacterial infections very transmissible (Costerton et al. 1999; Lewis 2001). EPS makes the understanding of biofilms overly complex as bacterial cells shift between planktonic to sessile and vice versa (O'Toole et al. 2000). These bacteria could be aerobic (needing oxygen for survival in open systems) or anaerobic (without the need for oxygen in closed systems) within their growth media containing carbon sources.

Once the growth of a stable bacterial biofilm is attained, the population of other microorganisms gradually takes over. This may occur between days and weeks, concurrently with the previous stage. The micro-fouling stage is a successive progression of different colonizers; it starts with the primary colonizers (complex multispecies of bacteria). Before the tertiary colonizers with invertebrate larva develop into macroscopic organisms, one other set (i.e., secondary colonizers) arrive (Crouzet et al. 2014). They are also attached to these surfaces via EPS produced by the bacterial biofilms. At this stage, most of the fouling organisms are diatoms, microalgae, spores, and larva of other common unicellular lower animals. The next stage involves macrofouling organisms. These are much larger and more distinct organisms to the human eye. Depending on the climate and locality, the macrofouling stage may consist of multicellular organisms like barnacles, reefs, tubeworms, and mussels and even marine plant populations (Crouzet et al. 2014). Figure 23.2a–c depicts some photographic images of the extent of biofouling on coated panels after marine field tests in oceanic bays in China and Saudi Arabia. A typical reported case is the antibiofilm activities of Ni-MOFs against Methicillin-resistant *Staphylococcus aureus* (*S. aureus*, MRSA ATCC 33591) and N7 clinical disease-causing fouling strains are depicted in the microscopic images shown in Figure 23.2d (Zhao et al. 2019; Guo et al. 2019; Eduok et al. 2015; Raju et al. 2020).

Beyond submerged surfaces as mentioned above, biofouling also readily occurs on filtration membranes in cooling systems. The growth of these organisms is heavily fostered by changes in physical conditions, of which temperature is a major factor (Choudhury et al. 2018). One of the key challenges surrounding membrane filtration efficiency is the abundance of foulant variants in wastewater capable of covering their inherent micropores. This also hinders separation processes and gross membrane performance. The presence of nutrients within the media also supports the growth of cellular deposits on membrane surfaces and within their pore walls (Rana and Matsuura 2010). At any stage in the growth process in microfiltration (MF) and ultrafiltration (UF) membranes, these organisms are capable of pore blocking while also facilitating scaling of non-biological inorganic salts (e.g., $CaCO_3$). When certain organisms (e.g., bacteria) are involved, biofouling is accompanied by the formation of stable biofilm layers. The presence of bacterial EPS makes way for firm attachment of their colonies to membrane surfaces. The genera of common bacteria involved in membrane biofouling include *Corynebacterium, Pseudomonas, Bacillus, Arthrobacter, Flavobacterium, and Aeromonas*. Apart from bacteria, fungi of *Penicillium* and *Trichoderma* genera are also among the most common eukaryotic microorganisms on membranes (Matin et al. 2011; She et al. 2016; Mansouri et al. 2010).

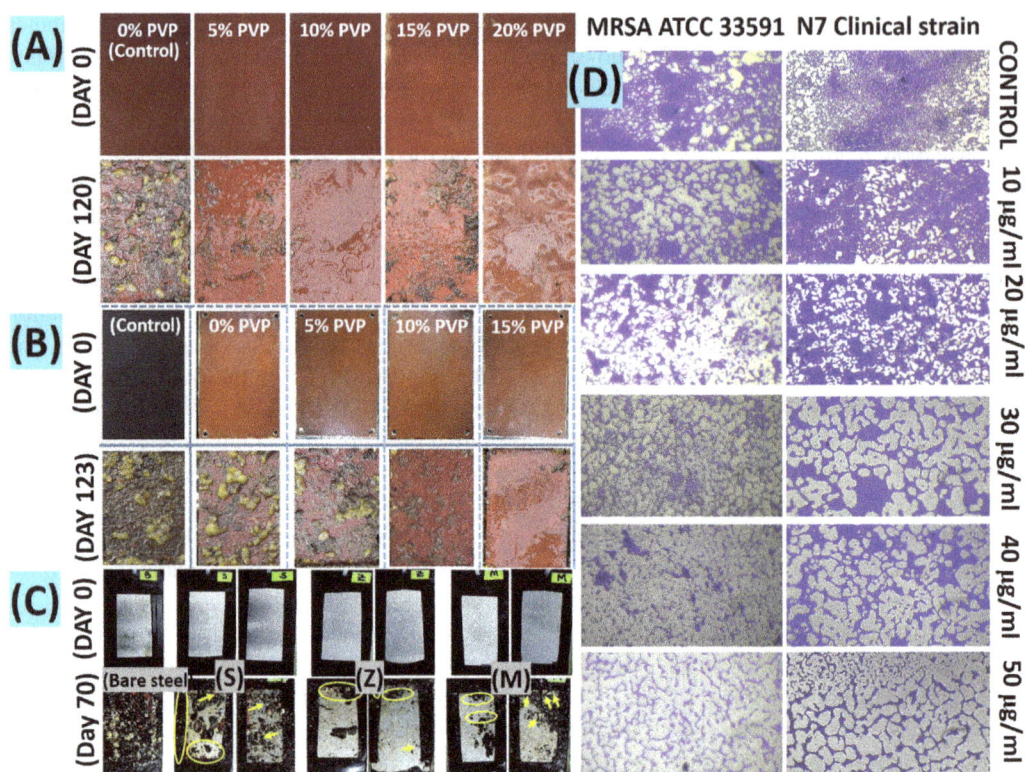

FIGURE 23.2 Photographic images of bare and uncoated panels showing the extent of biofouling after marine field test: (a) PVC control plate and those coated with PDMS, and polyvinyl pyrrolidone/poly(methyl methacrylate)/poly(butyl acrylate) graft polymer coatings with varying polyvinyl pyrrolidone (PVP) contents before and after immersion in seawater for 4 months at Qingdao bay (near Yellow sea), China. (Reproduced with permission from Zhao et al. (2019) © 2019 Elsevier B.V. (b) PDMS and PDMS-PVP coatings with different PVP contents before and after immersion in seawater for 4 months at Qingdao bay (near Yellow sea), China. (Reproduced with permission from Guo et al. (2019) © 2019 Elsevier B.V.) (c) Siloxane/PDMS hybrid coatings with two zinc-based pigments (zinc aluminium polyphosphate (Z) and zinc molybdate (M)) before and after immersion in hypersaline seawater for 70 days at Half-Moon Bay, Al Khobar, Saudi Arabia. (Reproduced with permission from Eduok et al. (2015) under Creative Commons Attribution License.) (d) Microscopic images showing the antibiofilm activity of Ni-MOFs against Methicillin-resistant S. aureus and clinical strain N7. (Reproduced with permission from Raju et al. (2020) © 2020 Elsevier B.V.)

23.1.3 Cost of Marine Biofouling

On deep-sea marine vessels, the accumulation of biofouling organisms contributes to an increased economic cost of fuel consumption and maintenance. Their presence on the hulls of ships causes hydrodynamic drag, thereby reducing sailing speed and maneuverability due to high frictional resistance. The increasing surface roughness due to adhering marine organisms also increases the gross vessel weights. This leads to increased emissions of greenhouse gas since more fuel needs to be burnt to compensate for weight-related drag. The US war vessels (e.g., Navy destroyers) spend about 10.3% more fuel per year due to fouling (Schultz et al. 2011). About $5 billion USD is expended yearly to combat marine biofouling in this industry, excluding the $1 million USD linked with ship's hull biofouling per ship per year (Yebra et al. 2004; Schultz et al. 2011). Larger modern marine vessels record 15% speed loss due to an 80% increase in associated hydraulic friction drag for every 1-mm biofouling thickness (Gordon and Mawatari 1992). Apart from increasing the tendency for biosecurity risk, biofouling leads to associated corrosion damage on ships and offshore structures (Adland et al. 2018; Davidson et al. 2016). Biofouling increases shipping cost by 77% for every 40% of fuel consumption (Yebra et al. 2004). A few common macrofouling organisms on the hulls of ships are fungi, large seaweeds, barnacles, and tubeworms (e.g., *Hydroides elegans*).

23.1.4 BIOFOULING CONTROL TECHNIQUES

The increasing costs associated with frequent dry docking for periodic cleaning have adversely affected the economic growth of shipping companies. Therefore, this calls for efficient and reliable scientific techniques for removing and/or preventing biofouling organisms on marine vessels (Galhenage et al. 2016). The historical development of antifouling systems has witnessed successes in a number of antifouling techniques. From lead and copper sheathing to the use of biocidal and non-biocidal antifouling paints, these techniques have been deployed in recent pasts to control marine fouling. With the ban on biocidal tributyltin (TBT, $(C_4H_9)_3Sn$) compounds by international agencies due to their toxicity, there are calls for other effective alternative antifouling techniques. Recently, Tian et al have tabulated biofouling control techniques as depicted in the flow diagram in Figure 23.3. For purposes of the present discussion, only the use of antifouling systems based on metal–organic frameworks (MOFs) will be discussed. This preventive technique focuses on the use of specially designed chemical MOF additives to prevent the adsorption of biofouling organisms on submerged surfaces. Most MOFs act as either biological or surface control additives (Tian et al. 2020). With MOFS, coated surfaces that made of these substances possess antimicrobial functions. Their presence on these surfaces retards biofouling by preventing the preliminary attachment of biofilm-forming microbes and/or discouraging the formation of conditioning films.

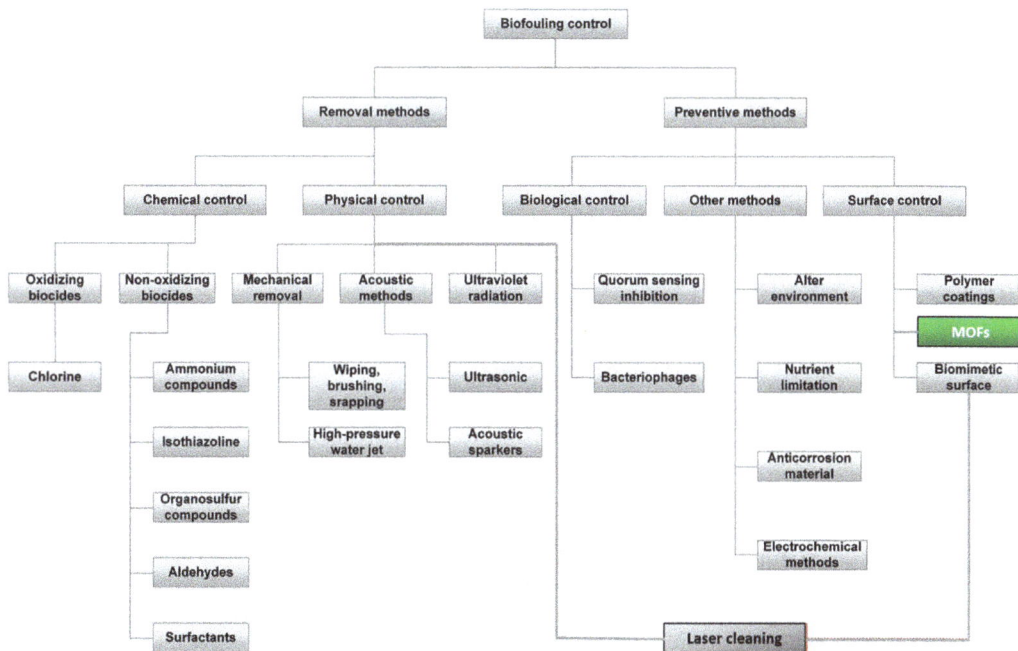

FIGURE 23.3 A flow diagram depicting an overview of biofouling control techniques; attention is focused only on the use of MOFs in this chapter. (Reproduced with permission from Tian et al. (2020) © 2019 Elsevier B.V.)

23.2 MOFs: WHAT ARE THEY?

MOFs are a large class of extremely ordered crystalline one-, two-, or three-dimensional hierarchical structured and ultra-porous hybrid materials. Normally coordinated with inorganic-organic hybrid groups within their molecular units, MOFs consist of metal ion/clusters coordinated to organic binding ligands (Nadar et al. 2018). MOFs possess these units as repeating coordination entities, extending into the above-named dimensional network structures with flexible skeletons as presented in Figure 23.4. Their ease toward functionalization makes their tunable properties incredibly unique for many applications in various domains. They may have a large surface area up to 2000–8000 m²/g within their crystalline structures (Farha et al. 2012; Furukawa et al. 2010; Chae et al. 2004).

Depending on the choice of application, desired physicochemical properties of MOFs can be achieved by regulating reaction conditions. The regulated conditions may include either of the following: synthesis parameters, types of metal cluster centers and even organic ligands (e.g., with appropriate size, geometry, and connectivity). These changes normally result in MOF products with unique pore capacity, structured crystallinity, and spatial cavity arrangements (Lee et al. 2013; Stoeck et al. 2012; Morris et al. 2012). For instance, zeolite-like MOFs are readily synthesized via milder conditions that lead to enhanced crystallinity and porosity (Eddaoudi et al. 2015; Zhou et al. 2012). However, this is not the case for regular mesoporous materials (e.g., metal oxides). Avenues for structural modification are enormous in MOF synthesis and most of them are geared toward improving properties and applicability. The presence of mixed linkers in modern MOF designs has been known to improve their gross porosity and stability. This and other research concerns

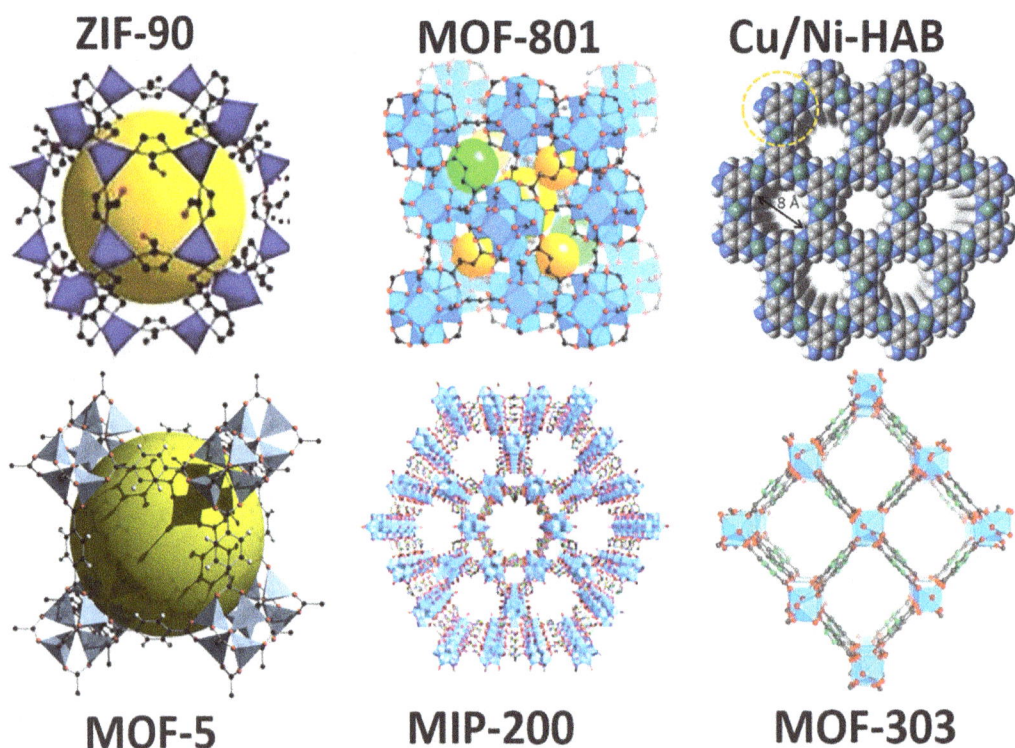

FIGURE 23.4 3D structures of some MOFs with repeated units of metal centers and organic ligands. MOFs possess these units as repeating coordination entities extending into the 3D dimensional network structures with flexible skeletons. N/B: These structures are only presented as examples of MOFs in this chapter; they may not have been utilized as antibiofouling additives. (Reproduced with permission from Tan et al. (2021) © 2021 Elsevier B.V.; (Zu et al. 2020) © 2020 Elsevier B.V.; and (Ma et al. 2021) © 2021 Elsevier B.V.)

associated with biocompatibility have been addressed with the use of bio-ligands (e.g., peptides). Physical properties like porosity and conductivity have also been improved by incorporating missing ligands into MOF structures. More than 20,000 MOFs have been synthesized and studied in several applications within the last decade (Isaeva et al. 2010; Li et al. 2016; Wang et al. 2012; Zhao et al. 2015; Huo and Yan 2012). Most of these applications cover research domains in chemistry (e.g., catalysis), model engineering (e.g., hydrogen storage), biomedicine (e.g., drug delivery), separation science (e.g., extraction of toxic substances), environmental monitoring (e.g., extraction sensor devices), fouling control (antibacterial MOFs), and so on. However, the focus of this chapter is centered on the use of MOFs-based systems for anti-biofouling applications within selected domains. It should be noted that the applicability of MOFs is vastly enormous in the true sense of the word as the applications accompanying their usage are limitless.

23.3 MOFs-BASED SYSTEMS FOR ANTI-BIOFOULING APPLICATIONS

Antifouling agents of this type are chemical additives that work in one of these three basic mechanisms or, in most cases, act concurrently. First, their presence disallows the initial adsorption on biofouling organisms on surfaces and may also inhibit their accumulation after surface adhesion in a second mechanism. Thirdly, antimicrobial MOFs are also capable of killing adhering microbes upon reaching submerged surfaces, with the goal of adverting biofilm formation and subsequent growth. Most MOFs are both antimicrobial and antibacterial agents, since they may exhibit biocidal capacities in most applications, including growth of bacteria and marine organisms in filtration membranes, under-water marine equipment, and biomedical implants (Nadar et al. 2018).

23.3.1 INHIBITING BACTERIAL GROWTH WITH MOFs AND MOFs-BASED NANOCOMPOSITES: HIGHLIGHTING TYPES OF ANTIBACTERIAL MECHANISMS

Bacteria readily form biofilms within a few seconds to a couple of minutes of surface adhesion, and they continue to grow into large colonies. Bacterial biofilms have also been identified as a major threat to public health since they foster infection spread and drug resistance. Where traditional medicine administration (e.g., antibiotic therapy) has failed to arrest bacterial infection and even contributed to resistance against potent antibiotics, more drastic approaches have been deployed. In the past decade, nanotechnology has provided the necessary scientific platform for designing new systems to curb antibiotic overuse and dependency (Nadar et al. 2018). From their enormous potential and usage in combatting bacterial infections, the advent of MOFs has been an emerging tool in this success story. Between food-bound pathogens within the food industry and hospital-acquired bacterial infections, modern advances in MOF chemistry have rare potentials as superior antimicrobial nanomaterials. With MOFs possessing chemical and physical properties that make them effective against bacterial growth, it is pertinent to mention that their working principles may not be consistent with regular antibiotic mechanisms. The antibacterial capacities of MOFs are linked with the following mechanisms: by physical contact, metal ions and ligands, oxidative stress, photothermal effect and a synergy of any of them combined (Nadar et al. 2018). Let's briefly explain these with a few reported examples.

23.3.1.1 Physical Interactions

With physical contact with antibacterial MOFs, there is an established means of interaction that may result in the death of bacterial cells and their colonies. This mechanism is in many ways similar to those exhibited by any other nanomaterial with antibacterial ability. There must be an interaction or a combination of several interactions; namely, van der Waals, electrostatic interaction, hydrophobic and receptor-ligand interactions. Here, the most affected part of the bacterial cell is the membrane where targeted disruptions occur. This may also lead to the exposure of cytoplasmic components, cellular inactivation of cellular physiology and eventual bacterial death. In reported studies by Yuan and Zhang (2017) and Yuan et al. (2019), the authors investigated the antibacterial activities of a zeo-litic imidazolate framework (ZIF) with bactericidal surfaces. They termed it "nano-dagger" due to its tendency for membrane rupture by its unique morphologies. This nanocomposite demonstrated unique bactericidal activities against *S. aureus, Candida albicans* (*C. albicans*) and *Escherichia coli* (*E. coli*). A magnitude of *log* reduction of more than seven units was recorded as a difference in the results between the control and test products. The authors also speculated that this nano-dagger may have enhanced cellular adhesion while also facilitating the needed contacts that resulted in bacterial death. The daggers must have punctured the cell membranes by direct contact with their sharp tips with positive polarity. This material (within coated gauze) demonstrated improved bio-compatibility with significantly lowered cytotoxicity and hemolysis compared to normal Ag gauze.

23.3.1.2 Contributions of Released MOF Components

With metal ions and ligands, MOFs' antibacterial capacity is a bit different compared to the first mechanism above. The two main components of MOF, metal cluster centers and organic linkers, also have high antibacterial abilities. Here, since MOFs are hybrids of both inorganic and organic components, they may be deployed against bacterial growth by remotely releasing them. It is also possible that a combination of both components can also be released from disintegrated MOF skeletons, thereby acting in synergy. Any of these components, especially the metal ions, can penetrate the cell membranes to disrupt normal cellular functions upon destroying the cytoplas-mic components (Nadar et al. 2018). It is also worthy of note to mention that each ion (within the metal cluster) and organic linker may have varying antibacterial responses. Modern advances in this field have also witnessed the development of MOF systems capable of controlled release of metal ions and ligands. This is an important factor for regulating health tissues related to inherent toxicity of both components. In a study by Zhuang et al., the author investigated the antibacterial activity of Co^{2+}-based MOF against *E. coli* growth. Here, they reported the least bactericidal level at 10–15 ppm. The observed activity was also linked with inherent active ligand sites on this mate-rial that initiated the disruption of bacterial cellular membrane after oxidative degradation of lipids (Zhuang et al. 2012). Rodríguez et al. investigated the antibacterial activity of cellulose-MOF-199 against *E. coli*. The authors reported 100% dead bacterial cells on both agar plates and liquid cultures. This result was attributed to the contributions of active copper sides of distinct oxidative state of the MOF material. The fabrication technique reported within this study ensured a strong interaction between MOF and the cellulosic membrane. This allows for the reuse of MOF coated fabrics (Rodríguez et al. 2014).

The antibacterial activity of another ZIF-8 modified with graphene oxide (GO) has been reported for thin film nanocomposite (TFN) membranes (Wang et al. 2016). The test was conducted for the membrane alone and in combination with different contents of ZIF-8/G nanocomposites. The photo-graphic images showing the extent of antibacterial activities against *E. coli* growth after exposure to blank/control solution, TFC, TFN-GO and TFN-ZIF/GO membranes are presented in Figure 23.5. The authors observed an optimal antibacterial activity for the hybrid nanocomposites up to 128 μg/mL (MIC, minimal inhibition concentration) compared to individual component considered alone. The best ZIF-8/GO nanocomposite system was TFN-ZG3, with 84.3% antibacterial potency against *E. coli* growth. This observed activity was attributed to the disruption of liposomes by zinc ions and the ZIF-8 bound imidazole groups. A similar test was also conducted by Ahmad et al. against *E. coli* and *S. aureus* growths. Related results are presented in Figure 23.5g–j. The SEM micrographs show changes in the morphologies of the cells of both bacteria, before and after exposure to ZIF-8/GO nanocomposite: (upper row) *E. coli* and (lower row) *S. aureus*. The authors speculated that the observed differences between the systems could be due to the inherently different antibacterial mechanisms. They concluded that bacterial growth inhibition was due to the combined effects of both antibacterial ZIF-8 and GO additives within the nanocomposite. According to these authors, the GO tips could also have contributed to the physical damage of cellular membranes. After bac-terial exposure to the nanocomposite, their membrane morphologies changed significantly due to

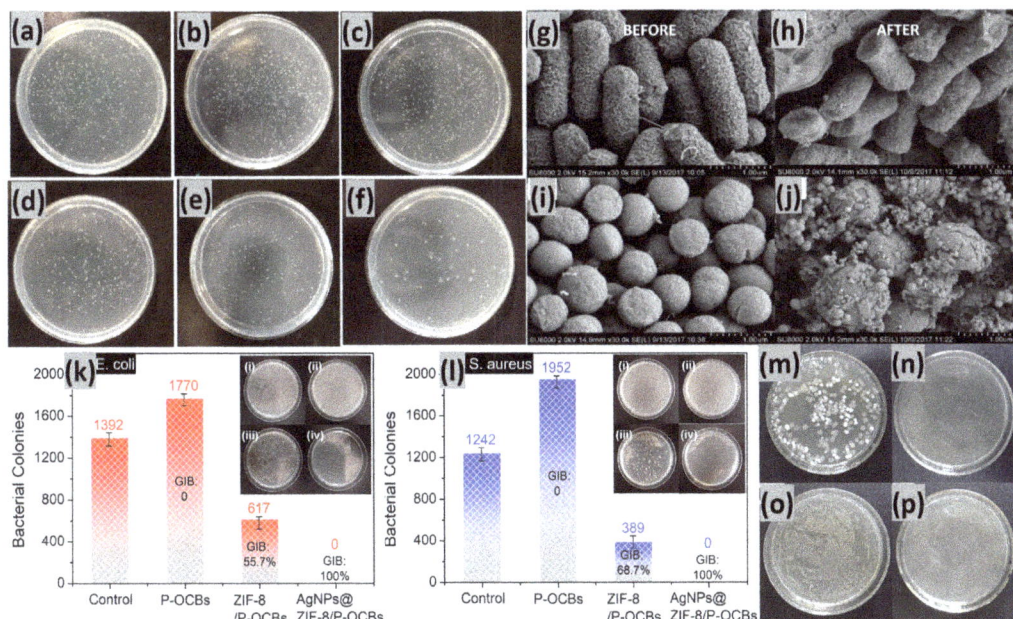

FIGURE 23.5 (a–f) Photographic images showing the extent of antibacterial activities against E. coli growth after exposure to (a) blank/control solution, (b) TFC membrane only, (c) TFN-GO), (d) TFN-ZG1, (e) TFN-ZG2, and TFN-ZG3 membranes (f). (Reproduced with permission from (Wang et al. 2016) © 2016 Elsevier B.V.) (g–j) SEM micrographs of biofilms housing bacterial cells, before and after exposure to ZIF-8/ GO nanocomposite: (upper row) E. coli and (lower row) S. aureus. (Reproduced with permission from Ahmad et al. (2020) © 2019 Chinese Society of Particuology and Institute of Process Engineering, Chinese Academy of Sciences. Published by Elsevier B.V.) (k–l) Trend of bacterial growth in colonies of (k) E. coli and (l) S. aureus after exposure. (Reproduced with permission from Meng et al. (2020) © 2020 Elsevier B.V.) (m–p) The extent of B. subtilis (upper row) and E. coli (lower row) growth on agar plates (m, o) without and (n, p) with 10 mg/mL Ag@ZIF-8, respectively. (Reproduced with permission from Guo et al. (2018) © 2017 Elsevier B.V.)

substantial damage induced by the nanocomposites (Ahmad et al. 2020). In another report by Meng et al., the authors were investigating the efficacy of corncob (P-OCBs) particles modified with Ag/ ZIF-8 nanocomposite. It is later referred to as Ag NPs@ ZIF-8 nanohybrid MOF. Figure 23.5 also shows the bacterial growth trends for *E. coli* (k) and *S. aureus* (l) colonies after exposure to (i) blank solution, (ii) P-OCBs, (iii) corncob with ZIF-8 MOF (ZIF-8s/P-OCBs) and (iv) a combination of both Ag and ZIF-8 additives (Ag NPs@ ZIF-8s/P-OCBs). Evidently, the nanohybrid showed excellent antibacterial potency compared to the rest of the test materials due to the synergistic antimicrobial activities of ZIF-8s (both zinc ions and imidazole groups) and Ag NPs additives. There were more colonies on P-OCBs corncob for both bacteria, as there was no growth inhibition (Meng et al. 2020). A similar study was reported by Guo et al. using varying concentrations of Ag@ZIF-8 hybrid MOF. In this study, 10 mg/mL Ag@ZIF-8 exhibited the most antibacterial effect against *B. subtilis* and *E. coli* growths as presented in the agar plates in Figure 23.5m–p (Guo et al. 2018).

23.3.1.3 Production of Toxic ROS

Inherent production of toxic reactive oxygen species (ROS) by MOFs is known for being detrimental to a range of bacteria (Vatansever et al. 2013). This is because ROS disrupts bacterial cells, their membranes as well as their cellular metabolism. ROS can be generated during metal ion release catalyzed by MOFs during photochemotherapy. However, the use of photosensitizers is also required. The integration of photosensitizers is also possible with MOFs via adsorption, entrapment and covalent attachment, and in most cases, this is accomplished by incorporating porphyrins (Felgentrager et al. 2014). Porphyrins-MOFs are widely generated during photodynamic therapies. Zhou et al. (2018) investigated the antibacterial activities of porphyrin-based MOFs against *E. coli*. At the end of the test, the authors reported a more than 98% dead rate after ROS generation. ROSs were activated using visible light. Another group of authors also reported the inhibition of cellular viability reaching values between 45%–63% and 18%–38% for three MOFs, without and with indocyanine green. This study was against *Enterococcus faecalis* growth (Golmohamadpour et al. 2018).

23.3.1.4   Synergy of Antibacterial Mechanisms (Including Photothermal Effects)

Photothermal disinfection is known for its adverse effects on bacterial growth. However, its effectiveness is also greatly influenced by the types of photothermal transducing agents deployed (Abbas et al. 2017). For efficient activity, these transducing agents are designed to demonstrate significantly reduced cytotoxicity and reliable biocompatibility (Cheng et al. 2014). Emerging development in MOF research has brought about new directions in the use of these transducing agents since they are also developed and utilized as light-absorbing agents. In this antibacterial mechanism, MOFs are biodegradable and also possess inherent photothermal effects. The synergy of antibacterial activities between MOFs and other transducing agents have been widely reported (Liu et al. 2021). Actually, photothermal disinfection is regularly utilized with other systems of antibacterial mechanisms. The use of a single mechanism is rare, and if higher bactericidal efficiency must be achieved at exceptionally low MOF limits, a combination of mechanisms is required. This is the reason why the synergy between several mechanisms is gaining more attention. For instance, Liu et al. (2020) have recently investigated the use of ZIF-8-humic acid nanocomposite against *E. coli* and *S. aureus* growths. The dead rate stood at 99.37% and 99.59% for these bacteria after a few minutes of irradiation. This MOF material exhibited enhanced photothermal effect, and it was linked with the adsorptive capacity of humic acid in conjunction with the release of zinc ions upon ZIF-8 degradation. Liu et al. (2021) have enlisted suitable examples centering on the use of dual bactericidal systems of many MOFs, including those with synergistic effects with other antibacterial mechanisms. Table 23.1 shows published examples for individual MOFs and MOF-based nanocomposites as antibacterial agents.

23.3.2   Inhibiting Surface Adhesion of Marine Organisms using MOFs and MOFs-Based Nanocomposites

As already presented in the earlier subsections, the progression in the life cycles of biofouling communities continues after the formation of conditioning films on submerged surfaces. The size and ages of successive sub-groups gradually change from one population of colonizers to the other, depending on the niches and species of organisms available within them. This in turn detects the types of biofouling removal and/or preventive techniques. In this subsection, attention will be focused on anti-biofouling applications against core marine organisms, including few single-celled marine bacteria. A few recent related examples are drawn directly from the literature, between the types of organisms and the MOF utilized in controlling their growth processes. Since film coatings are the first barriers against the adhesion of species of surfaces, the characteristics of their inhibition layers are also controlled by factors related to stimulus-responsive chemistries.

Most of these responsive techniques depend solely on the interactions between the coating additives and the organisms. However, since the TBT ban in 2001 by the International Maritime Organization, the choice of the right antifouling additives in coatings has become increasingly challenging. The use of highly porous and hybrid coordinated MOF polymer material has been considered a useful alternative antibiofouling tool (An et al. 2009; Horcajada et al. 2008). The antimicrobial activities of their central metal clusters and organic linkers, when released, play vital roles in this surface inhibition against the adsorption of marine organisms. A summarized result from

TABLE 23.1
Summarized Antibacterial Activities of Few MOFs and MOFs-Based Nanocomposites

S/N	MOF/MOFs-Based Nanocomposites	Mechanism for Antibacterial Activity/Test Bacteria Studied	Antibacterial Efficiency (%)	Ref.
1.	ZIF-L nano-dagger	Direct physical contact/*C. albicans, E. coli* and *S. aureus.*	*Log* reduction of 4, 7 and 8 for *C. albicans, E. coli* and *S. aureus*, respectively	Yuan and Zhang (2017)
2.	HKUST-1	Cu^{2+} ions released from MOF/*S. aureus* and *E. coli*	Unspecified	Ren et al. (2019)
3.	ZIF-8	ROS generation/*E. coli*	Approx. 100%	Li et al. (2019)
4	ZIF-L nano-dagger	Direct physical contact/*E. coli* and *S. aureus*	*Log* reduction > 7 units	Yuan et al. (2019)
5.	HKUST-1	Cu^{2+} ions released from MOF/Methicillin-resistant *S. aureus* and *P. aeruginosa*	*100µg/mL/#50–200µg/mL (*S. aureus*); *50µg/mL/#200µg/mL (*P. aeruginosa*)	Azad et al. (2016)
6.	Ag/Zr-MOF	Ag^+ ions released from MOF/*E. coli*	*6.5µg/mL; #6.5µg/mL	Mortada et al. (2017)
7.	Ag-MOF/GO	Significantly reduced electrostatic charge/*E. coli* and *B. subtilis*	95%	Firouzjaei et al. (2018)
8.	Co-MOF/GO	Co^{2+} ions released from MOF and effects of GO/*E. coli* and *S. aureus*	>99%	Hatamie et al. (2019)
9.	Ag-based MOFs	Ag^+ ions released from MOF/*E. coli* and *S. aureus*	>90%–96%	Zirehpour et al. (2017)
10.	ZIF-8	Synergy between Zn^{2+} ions released from MOF/*S. mutans*	Unspecified	Chen et al. (2017)
11.	HKUST-1	Cu^{2+} ions released from MOF/*E. coli* and *S. aureus*	Approx. 100%	Ren et al. (2019)
12.	ZIF-8	Release of organic ligand and Zn^{2+} ions from MOF; ROS generation / *E. coli*	Approx. 99%	(Wang et al. 2020)

*Minimum inhibitory concentration (MIC); #Minimum Bactericidal Concentration (MBC)

recent studies showing the applications of MOFs-based systems for anti-biofouling applications against marine organisms are presented in Table 23.2.

In a study by Sancet et al. (2013), the authors had synthesized new MOFs (SURMOF), named after their surface-anchorage and crystalline nature with Cu-bound microstructure (Cu-SURMOF 2). Here, these materials were found to be biocidal against one distinct marine bacterium, *Cobetia marina* (*C. marina*). After allowing surface settlement, these MOFs immediately attacked the bacterial cells, thereby preventing them from forming biofilms. How was this possible? Aqueous degradation of these MOF materials prompted the release of their consistent antimicrobial components that acted against this bacterium. In this case, the released Cu^{+2} ions impeded bacterial growth within the culture media. Before the test, selected bacterial colonies from preculture agar plate were inoculated with 20 mL Marine Broth. This broth was then left overnight with gentle stirring (65 rpm) at 33°C–35°C. The authors ensured that this sterile medium was diluted to a 1:100 ratio. Appropriate assays were utilized

TABLE 23.2

Antibiofouling Activities of MOFs and MOFs-Based Nanocomposites

S/N	MOF/ MOFs- Nanocomposites	Mechanism of Antibacterial Activity/Test Organism(s) under Study	Culture Medium/Marine Environment	Antibiofouling Efficiency	Ref.
1.	Cu-containing MOF (Cu-SURMOF 2)	Cu⁺² ions release from degraded MOF acted as antimicrobial agents/marine bacterium (*C. marina*)	Selected bacterial colonies from agar plate were inoculated with 20mL Marine Broth and left overnight with gentle stirring (65 rpm) at 33°C–35°C.	100% (approx.)	Sancet et al. (2013)
2.	Ag@TA-SiO₂	Ag⁺ ions release from degraded MOF/2 microalgae (*N. closterium* and *D. zhanjiangensis*) and 2 bacteria (*E. coli* and *S. aureus*)	The bacteria were incubated at 37°C for 24h in the dark. Both algae were cultured in artificial seawater for 8 days at 23°C within an incubator in 12 equal hours of light and dark cycles.	Antibiofouling activity up 93.5 and 97.6% for microalgae *N. Closterium* and *Dicrateria zhanjiangensis*, respectively; Antibacterial efficiency of 99.1% and 82.7% for *E. coli* and *S. aureus*, respectively.	Deng et al. (2021)
3.	ZIF-8	Release of Zn²⁺ ions upon MOF degradation; bactericidal potency of MOF-bound imidazole groups; MOF also prevented the attachment of diatom cells/marine diatom (*N. closterium*)	Algae were cultured in artificial seawater in a 12-h light and dark cycle, illuminated within an incubator that was stirred between 20–23°C.	Biofouling attachment rate of 0.51% was recorded for best coating variant compared to the control (92%).	Yang et al. (2021)
4.	Co, Zn and Ag-MOFs	Algal growth was inhibited by Ag-MOF; antibacterial activity was initiated by Co²⁺, Zn²⁺, Ag⁺ ions released after MOF degradation/2 cyanobacterial strains, one filamentous, *Anabaena sp.* and one unicellular, *Synechococcus sp.*; and one unicellular green alga strain, *C. reinhardtii*.	Both organisms were cultured in specialized media on rotatory shakers set under controlled conditions (28°C at 60 μmol photons m²/s light intensity).	Approx. 95% antifouling efficiency	Martín-Betancor et al. (2017)

(Continued)

TABLE 23.2 (*Continued*)

Antibiofouling Activities of MOFs and MOFs-Based Nanocomposites

S/N	MOF/ MOFs- Nanocomposites	Mechanism of Antibacterial Activity/Test Organism(s) under Study	Culture Medium/Marine Environment	Antibiofouling Efficiency	Ref.
5.	Ni-MOFs	Ni-MOFs acted as efficient cargos of pesticide and antibiotics. The larvicidal activity was dependent on MOF dosage/larva of viral vector (*A. aegypti*) mosquito and an aquatic crustacean, *A. salina*.	Introducing defined dosage of Ni-MOFs (20–100 μg/mL) into sterile water with 20 mosquito larva; this medium was subjected to 16 h of light and 8 h of dark cycles.	LC50 and LC90 values of 43.62±1.25 and 74.72±1.97 μg/mL, respectively, for mosquito (*Aedes aegypti*) larva. Ni-MOFs was non-toxic at lower doses toward the aquatic crustacean; LC50 value up to 138.33±3.72 μg/mL was recorded for highest dose under study, denoting mild toxicity.	Raju et al. (2020)
6.	Ag doped chitosan-GO/ ZIF-8	Antimicrobial activities of Ag⁺ ions and GO release within the MOF, ZIF-8 also inhibited surface cellular adhesion/ marine diatom (*Nitzschia*).	Uranium solutions and nature seawater (Bohai Sea, China) were deployed as test media.	Inhibited algal growth (70% cell mortality)	Guo et al. (2020)

to study the marine bacterial viability and microfluidic bacterial detachments. Figure 23.6a and b depicts the viability test of adhering *C. marina* cells in the presence of Cu-SURMOF 2. These are fluorescence microscopy images of growing cells after culture. The bright green patches show live cells within the control media and the unviable ones with Cu-SURMOF 2, confirming its antibiofouling ability. Figure 23.6c–f shows an annotated schematic depicting of degradation of Cu-SURMOF 2 with approaching bacterial cells. In Figure 23.6a, a viable bacterial cell is in contact with SURMOF (c). The settled cell begins to secrete EPS around SURMOF as shown on the provided SEM micrograph, depicting cells with unaffected growth rate (d). After the degradation of SURMOF, Cu^{2+} ions are released (e). The accompanied SEM image depicts cellular contact with Cu-SURMOF 2 within the first 2 h of incubation. As presented, the deformed bacterial cell has disrupted the cell membrane as a consequence of MOF's antibacterial activity. This dead bacterial cell losses its viability and detaches from the surface due to inherently reduced adhesion strength (f). The accompanied SEM image shows weakened surface adhesion due to the impact of SURMOF. The observed black spot is that of a detached bacteria cell (Sancet et al. 2013).

In another study by Deng et al. (2021), new hybrid Ag@TA-SiO$_2$ MOF structured nanoparticles were fabricated as presented in the schematic in Figure 23.6g. The reaction following the formation of this highly crosslinked material involved a free radical co-polymerization process with glycidyl methacrylate, 2-hydroxyethyl methacrylate and 3-(methacryloxypropyl) trimethoxysilane. Ag was immobilized to provide the needed antimicrobial property. These authors were investigating the antifouling activity of Ag@TA-SiO$_2$ nanoparticles against the growth of two microalgae and bacteria. The model bacteria were *E. coli* and *S. aureus*. Two model microalgae were also studied, (i) *Nitzschia Closterium* (*N. closterium*) with a siliceous cell wall and the other (ii) *Dicrateria zhanjiangensis* (*D. zhanjiangensis*) with no distinct cell wall. Both algae were cultured for 8 days at 23°C within an incubator in 12 equal light and dark hour cycles. These MOF nanoparticles served as antifouling barriers due to the embedded Ag. They acted as functional biocides to inhibit growth and formation of attachment on the solid surface. In this study, antifouling and bactericidal capacities

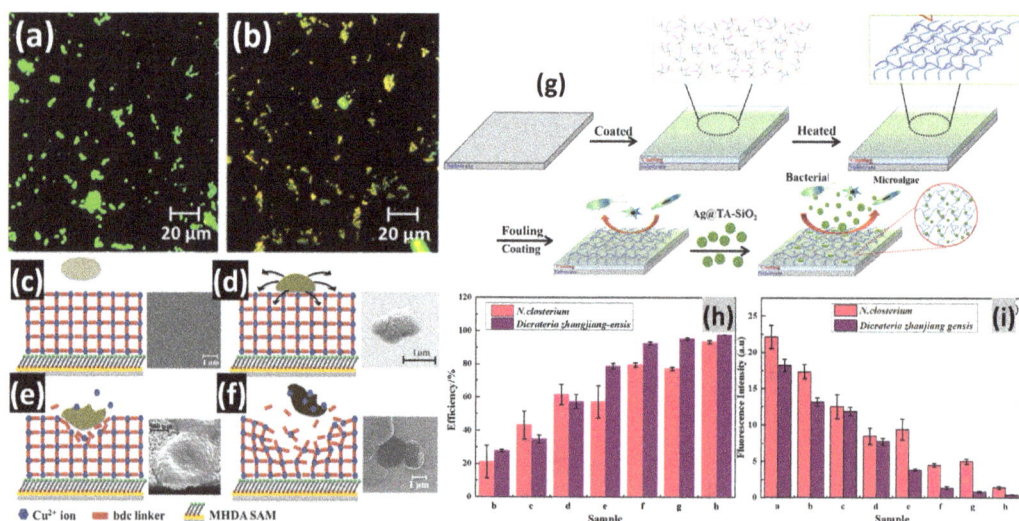

FIGURE 23.6 Fluorescence images showing the viability of C. marina without (a) and with (b) Cu-SURMOF-2. (c–f) An illustrative schematic showing the antibiofouling mechanism (stepwise, from c–f) against C. marina growth upon Cu-SURMOF 2 aqueous degradation. Reproduced with permission from Sancet et al. (2013) under Creative Commons Attribution License. (g) Schematic showing an Ag@TA-SiO$_2$ MOF modified coating design. (h) Antibiofouling efficiencies and (i) fluorescence intensities of MOF modified coatings after exposure to microalgae N. closterium and Dicrateria zhanjiangensis for 8 days. The letterings a-h are designated for the various coating variants. (Reproduced with permission from Deng et al. (2021) © 2020 Elsevier B.V.)

of Ag@TA-SiO$_2$ nanoparticles were investigated by observing bacterial growth and microalgae attachment. The authors recorded an efficient antibiofouling activity up to 93.5% and 97.6% for microalgae *N. Closterium* and *D. zhanjiangensis,* respectively (Figure 23.6h). These parameters were computed from the magnitudes of florescence intensity presented in Figure 23.6i for the variant coatings (PGHMK0–1600, a–h) under study. The authors also recorded values of antibacterial efficiency up to 99.1% and 82.7% for *E. coli* and *S. aureus,* respectively. These antibacterial and biocidal activities were possible through Ag+ ion released from degraded MOFs that also acted as antimicrobial agents. The authors ascribed the observed biocidal activity to the formation of ROS and weakened adhesion of approaching microalgae toward the surface of the modified coatings. Upon the release of Ag@TA-SiO$_2$ nanoparticles, photosynthesis was also inhibited due to reduced chlorophyll activities, hence, impacting the microalgal growth (Deng et al. 2021).

Yang et al (2021) investigated the antibiofouling capacity of polyurethane (PU) coating modified with ZIF-8 MOF (PHZ1–3) against the growth and surface settlement of marine diatom (*N. closterium*). These marine algae were cultured in artificial seawater in a 12-h light and dark cycle, illuminated within an incubator stirred at 100–500 rpm between 20–23°C. The authors concluded that ZIF-8 prevented the surface attachment of *N. closterium* cells by disallowing the initial adhesion. They recorded a biofouling attachment rate of 0.51% for the best coating variant compared to the control (92%). The authors also reported that the size of the ZIF-8 particle sizes also affected their distribution densities within the coating. Here, larger MOF sizes weakened the gross particulate cohesion, leading to lowered mechanical and thermal stabilities. However, these changes had a limited effect on the surface hydrophobicity of the modified coating. The authors also drew a correlation between MOF concentration, particle size, hydrophobicity, and antifouling performance. In all, ZIF-8 reinforced the gross mechanism strength of this hybrid PU/ZIF-8 coating while also contributing to the inhibition of algal biofouling rate. The algal cells were cultured in a medium containing these modified coating variants. In the end, their viability in each coating was examined by means of Florescence microscopy. Results showing varying degrees of cellular attachment of the coating are presented as florescence images in Figure 23.7. These images depict changing surface adhesion behaviors at different culture's durations, from 0 to 30 days. Live cells are observed with bright red florescence. As expected, the control coating (without ZIF-8) displayed enhanced florescence due to extensive algal growth. However, the ZIF-8 modified coatings (PHZ1–3) disrupted algal growth, hence, biofouling inhibition within the culture durations. The mean area of algal growth and surface attachment for each coating variant were computed from Figure 23.7. These authors recorded 92% surface coverage by living algal cells after 30 days for the pristine coating compared to the coating with superior performance. The lowest coverage (0.51%) was recorded for PHZ1, hence, the matrix was the best anti-algal character. In all, inherent ZIF-8 NPs within the coating significantly enhanced the anti-biofouling property of the coatings due to Zn^{2+} ion release upon ZIF-8 degradation. Apart from the contribution of Zn^{2+}, the authors also attributed the observed antibiofouling property of these modified coatings to the bactericidal potency of the MOF-bound imidazole groups. The imidazole groups are known for their adverse impacts on cell membranes and cytoplasmic contents of living algal cells (Yang et al. 2021). A similar study to that reported by Yang et al. is the investigation featuring the antibiofouling capacity of Ag-doped chitosan–GO/ZIF MOF. In this later study, the authors ascribed the antifouling performance of their MOF material to the antimicrobial activities of the Ag$^+$ ion and GO released. Conducted against the growth of marine diatom (*Nitzschia*) in seawater, the authors recorded a 70% biocidal rate. ZIF-8 inhibited the growth rate of this diatom and also significantly reduced its surface adhesion (Guo et al. 2020).

Martín-Betancor et al. have reported the antibiofouling activity of three MOFs (Co, Zn and Ag-MOFs) against the growth of two cyanobacterial strains, *Anabaena* sp. PCC 7120 (filamentous) and *Synechococcus* sp. PCC 7942 (unicellular). The authors also conducted similar tests for a green unicellular alga (*Chlamydomonas reinhardtii*). These tests were carried out in specialized media with 5 mM nitrate and tris-minimal phosphate at pH 7.5 (as well as other undisclosed minerals), depending on the test organism. The culture media were illuminated and stirred at 28°C for

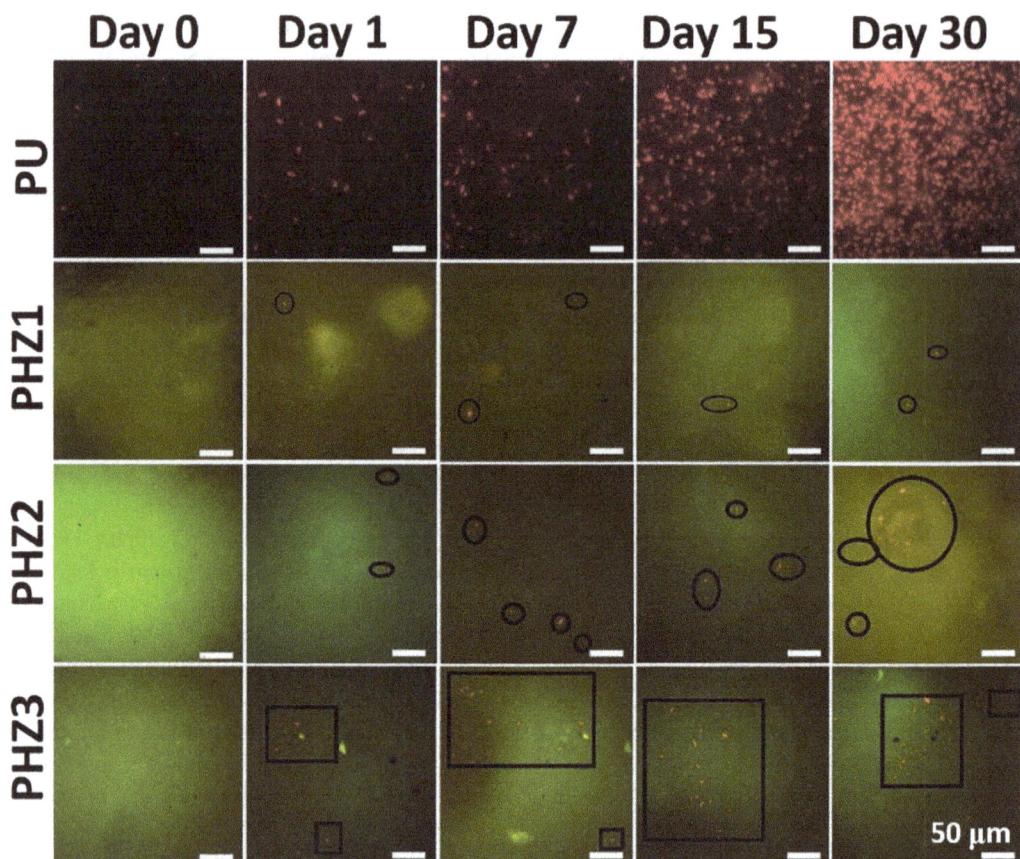

FIGURE 23.7 Fluorescence microscopic images showing the extent of algal cellular attachment (in circles) on polyurethane/ZIF-8 MOF coating variants (PHZ 1–3) at different culture durations. The labels PHZ 1–3 were provided based on the different particle sizes of ZIF-8 NPs within the polyurethane coatings. (Reproduced with permission from Yang et al. (2021) © 2020 Elsevier B.V.)

a few days. These MOFs significantly reduced the growth rates of cyanobacteria left within the media after 24 h. However, algal growth only impacted by Ag-MOF. Growth patterns were unaffected by Co-MOF and Zn-MOF. The authors concluded that the observed antibiofouling activities were initiated by the released Co^{2+}, Zn^{2+}, Ag^+ ions after MOF degradation. These released ions later acted as antimicrobial and biocidal agents against bacterial and algal growths, respectively. Apart from the above-provided reasons, the authors also opined that the anti-biofouling capacity of these MOFs must also have depended on a few other factors. These factors were related to the bioavailability and toxicity of these free ions as well as their release rates from the three MOFs. Other key factors that they pointed out were the kinds of target organisms and the sizes of their colonies (Martín-Betancor et al. 2017).

In another study, Raju et al. designed a combined antimicrobial-larvicidal strategy against targeted organisms. In this study, the authors utilized Ni-bound MOFs capable of multifunctional performances against the larva of a disease-causing mosquito (*Aedes aegypti*). Larvicidal studies were conducted by introducing defined dosage of Ni-MOFs (20–100 µg/mL) into sterile water samples with 20 mosquito (*A. aegypti*) larva. This medium was subjected to 16 h of light and 8 h of dark cycles. In the end, the authors observed a distinct larvicidal performance against this viral vector. Ni-MOFs acted as efficient cargos of pesticide and antibiotics in way that indicated that larvicidal activity was dependent on MOF dosage. The authors also recorded LC50 and LC90 values

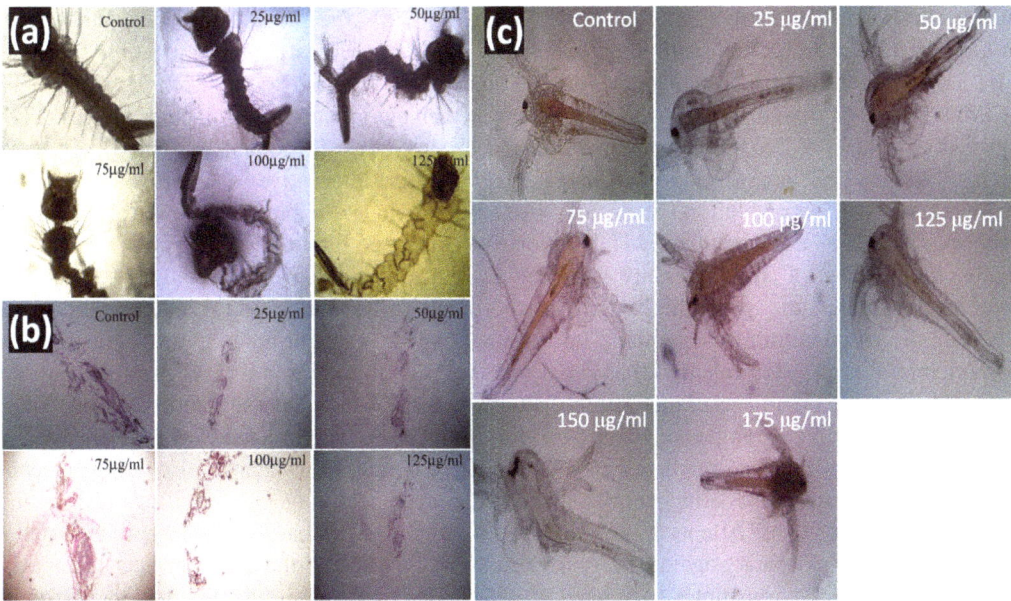

FIGURE 23.8 (a) Phase contrast microscopic images showing larvicidal activity of Ni-MOFs against A. aegypti mosquito larvae; (b) Captured histopathology images of larvae after exposure to Ni-MOFs (magnification 10×); (c) Microscopic images showing the assessment of cytotoxic effect of Ni-MOFs under in vivo condition using A. salina nauplii as animal model system. (Reproduced with permission from Raju et al. (2020) © 2020 Elsevier B.V.)

of 43.62 ± 1.25 and $74.72 \pm 1.97 \, \mu g/mL$, respectively. Figure 23.8a shows histopathological changes after exposing the mosquito larva to Ni-MOFs. The presence of this hybrid material impacted adversely on the larval morphology. These microscopic images reveal inherent damage to the head and thorax features, even at $25 \, \mu g/mL$ Ni-MOF concentration. Distinct growth inhibitions were observed at 75 and $100 \, \mu g/mL$. The captured histopathology images also reveal few cellular damages and fragmentation at lower MOF concentrations. There was also organ crumpling or outright organ loss at higher concentrations, as displayed in Figure 23.8b. The observed organ distortion (including papillae and around the cuticle) and the eventual larval death could have been due to the antimicrobial Ni^{2+} ion transport as well as the released imidazole groups. Could the use of Ni-MOF as biocide against mosquitoes affect other aquatic lives? To answer this question, these authors also deployed a model aquatic crustacean, *Artemia salina* (*A. salina*), a brine shrimp, to monitor its capacity for safety. They observed that Ni-MOFs were non-toxic toxic at lower doses. However, an LC50 value up to $138.33 \pm 3.72 \, \mu g/mL$ was recorded for highest dose under study, denoting mild toxicity. Figure 23.8c reveals inherent changes in the morphology of *A. salina* due to MOF impact at higher concentration. Higher MOF level contributed adversely to its growth process, leading to mild mortality within the duration of the test (Raju et al. 2020).

23.4 CONCLUSIONS AND PERSPECTIVES

The fouling of submerged surfaces of marine vessels is due to the adhesion of marine organisms. Their presence on the hulls of ships causes hydrodynamic drag, thereby reducing sailing speed and maneuverability due to induced frictional resistance. Between the unreasonably high costs of maintenance, periodic repairs, and fuel consumption, biofouling has impacted adversely on global maritime economies in the past several decades. The development of antifouling paints with organotin tributylin (TBT) in the 1960s was effective in combating this scourge until their toxicity subsequently led to their eventual ban. Even at that, it took a while! Before the 1990s, the global journey to complete ban of harmful TBT substances in antifouling paints was far from successful. The initial call to reduce TBT-led pollution by the International Maritime Organization was initiated in the 1992s Chapter 17 of IMO Agenda 21 developed at Rio. A resolution was then put forward to end the use of these TBT biocides on January 1, 2003. It was not until 5 years later that they were completely banned.

Just like the TBT ban, the journey to reestablish effective alternatives also witnessed positive change and scientific revolution. There are so many techniques of biofouling control. So many new ones have been added and are widely reported to date (see Figure 23.3). As presented in this figure, fouling control techniques are classified as removal and preventive methods. Alongside presented examples, the use of highly porous coordinated MOFs as a useful antibiofouling tool has been discussed within this chapter. The antimicrobial activities of their central metal clusters and organic linkers, when released, play vital roles in growth inhibition and adsorption of micro and macro-marine organisms. This chapter features recent developments on antibiofouling MOF materials design in nanocomposite coating systems with protective capacities and potency against surface adhesion by marine organisms. Beyond enlisted examples, this chapter also outlines recent research areas in aspect of coating technology as well as accompanying antibiofouling mechanisms of selected MOF systems. Attention is focused on MOF and MOFs-based nanocomposite systems with anti-biofouling capacity against preventive bacterial growth and those of other marine organisms. A few case-study examples have been provided. However, there may be a few loopholes in this field. Most of them range from random questions and unresolved concerns to actual need for further studies. Below are some future perspectives on this topic as well as suggested aspects of antibiofouling that still need to be addressed:

23.4.1 Synergy of Antibiofouling Mechanisms

While discussing the inhibition of bacterial growth with MOF and MOFs-based nanocomposites, an emerging development in MOF research was mentioned. This development opened new directions to the use of MOF materials capable of more than one antibiofouling mechanism. A number of MOFs may act against the adhesion of bio-foulants, either by (i) release of antibacterial MOF components, (ii) photothermal disinfection, or (iii) inherent production of toxic ROS by MOFs, or (iv) those that act by direct physical contact with marine organisms. However, live examples are not restricted to the aforementioned examples. More studies still need to be conducted in order to establish a synergy of antibacterial activities between MOFs and other transducing agents. Antibiofouling systems with single mechanism are rare in nature, if higher bactericidal efficiency must be achieved at reduced MOF concentrations, a synergy between mechanisms should be further investigated (Liu et al. 2020, 2021).

23.4.2 From MOFs to MOF-on-MOFs

The use of MOFs with highly efficient antibiofouling potency has been highlighted within this chapter. Discussions are also accompanied with lists of organisms from defined stages of biofouling. MOF's antibacterial capacity is dependent on their compositions, and architectural structures. So, if simple or common MOFs are capable of these functions, imagine if their structural units can be increased for enhanced robust potentials. MOF-on-MOFs are a new class of hybrid materials that are normally constructed from more than one or two MOFs units (Liu et al. 2021). Their conjugation chemistries endow them with properties and capabilities that are a combination of their individual constituent units. As a recommendation, more intensive research should be carried out of the antibiofouling potentials of these new materials. Attention should be focused on their conjugated MOF units and even those from non-MOF derivatives. Figure 23.9 shows the architectural morphologies of some common MOF-on-MOFs.

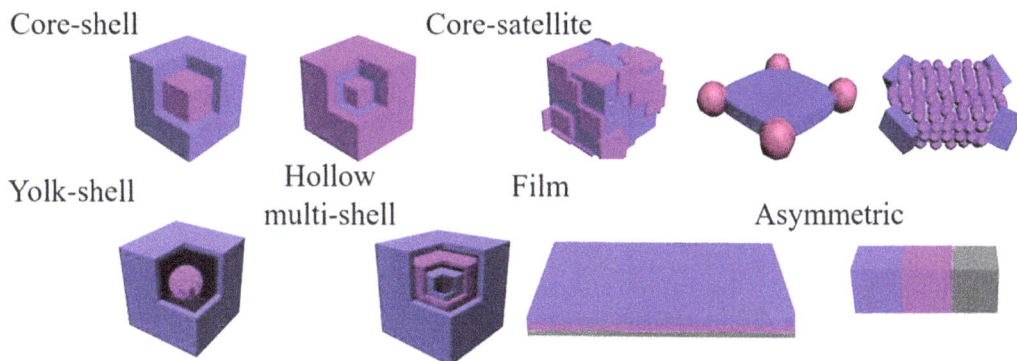

FIGURE 23.9 Schematics of some common MOF-on-MOF architectural morphologies. (Reproduced with permission from Liu et al. (2021) © 2021 Elsevier B.V.)

23.4.3 BEYOND COMMON DISEASE-CAUSING BACTERIA

The antibacterial abilities of most MOFs are brought about by their ability to inhibit the growth processes of certain common bacterial targets. However, successful trials on specific axenic colonies do not automatically translate to enhanced potency against other bacteria of the same genera. For instance, ZIF-L nano-daggers have been reported to inhibit the growth of *E. coli* with more than 90% efficiency (Yuan and Zhang 2017). However, can this MOF also act beyond this Gram-negative anaerobic bacterium? It is not enough to assume that this research was conducted with the goal of combating *E. coli* poisoning in food industries. How about food toxicity caused by other common bacteria? This is where further research is needed, and this remains a future perspective into the efficient use of MOFs in mitigating bacterial growth.

23.4.4 ANTIMICROBIAL MOF SYSTEMS WITH CONTROLLED RELEASE MECHANISMS ARE NEEDED

The antibiofouling capacities of MOFs enlisted in Tables 23.1 and 23.2 are linked with their tendency to release their metal cluster centers and organic linkers components. This happens during aqueous degradation of MOF in a way that also depends on their structural stability. However, since there are several microbes inherent within a defined service environment, there is a call for targeted fouling inhibition strategies capable of controlled release of these MOF components. This can be addressed during syntheses. Desired properties of MOFs can also be achieved by regulating reaction conditions, synthesis parameters as well as ensuring the proper choice of metal cluster centers and their organic ligand counterparts.

23.4.5 MORE MOFS WITH ROBUST CAPACITIES FOR ANTIBIOFOULING ARE REQUIRED

At the later stages of biofouling, higher multicellular organisms completely take over submerged surfaces, thereby replacing the micro-foulers. Since most macro-foulers have different life forms within their relatively short metamorphic lifecycles, it also means that each organism is capable of adhesion characteristics that are distinctly different from the next. For example, the adhesion patterns of barnacles on submerged surfaces depend on which lifeform is considered. Knowing that all lifeforms (e.g., pelagic suspension-feeding nauplius larva, non-feeding cypris larva, and benthic suspension-feeding adult) may be capable of surface attachment, makes the design of antifouling systems challenging. So, if a MOF material can inhibit the attachment of barnacles, which lifeform is considered? This is the reason more MOFs with robust capacities for antibiofouling are required for a wide range of growth stages of various marine organisms (e.g., dormant and free-swimming eggs, larva, cyprids, young and full adults, etc.).

REFERENCES

Abbasloo, F., Khosravani, S.A., Ghaedi, M., Dashtian, K., Hosseini, E., and Manzouri, L. Sonochemical-solvothermal synthesis of guanine embedded copper based metal-organic framework (MOF) and its effect on oprD gene expression in clinical and standard strains of Pseudomonas aeruginosa. *Ultrason Sonochem.* 2018;42:237–243.

Adland, R., Cariou, P., Jia, H., and Wolff, F.C. The energy efficiency effects of periodic ship hull cleaning. *J. Clean Prod.* 2018;178:1–13.

Ahmad, N., Md Nordin, N.A., Jaafar, J., Malek, N.A.N.N., Ismail, A.F., Yahya, M.N.F., Hanim, S.A.M., and Abdullah, M.S. Eco-friendly method for synthesis of zeolitic imidazolate framework 8 decorated graphene oxide for antibacterial activity enhancement. *Particuology* 2020;49:24–32.

An, J., Geib, S.J., Rosi, N.L. Cation-triggered drug release from a porous zinc-adeninate metal-organic framework. *J. Am. Chem. Soc.* 2009;131:8376.

Azad, F.N., Ghaedi, M., Dashtian, K., Hajati, S., and Pezeshkpour, V. Ultrasonically assisted hydrothermal synthesis of activated carbon–HKUST-1-MOF hybrid for effcient simultaneous ultrasound-assisted removal of ternary organic dyes and antibacterial investigation: Taguchi optimization. *Ultrason. Sonochem.* 2016;31:383–393.

Chae, H.K., Siberio-Pérez, D.Y., Kim, J., Go, Y., Eddaoudi, M., Matzger, A.J., O'Keeffe, M., and Yaghi, O.M. A route to high surface area, porosity and inclusion of large molecules in crystals. *Nature* 2004;427:523–527.

Chen, J., Zhang, X., Huang, C., Cai, H., Hu, S., and Wan, Q. Osteogenic activity and antibacterial effect of porous titanium modifed with metal-organic framework flms. *J. Biomed. Mater. Res. A* 2017;105:834–846.

Cheng, L., Wang, C., Feng, L., Yang, K., and Liu, Z. Functional nanomaterials for phototherapies of cancer. *Chem. Rev.* 2014;114:10869–10939.

Choudhury, R.R., Gohil, J.M., Mohantya, S., and Nayak, S.K. Antifouling, fouling release and antimicrobial materials for surface modification of reverse osmosis and nanofiltration membranes. *J. Mater. Chem. A* 2018;6:313–333.

Costerton, J.W., Lewandowski, Z., Caldwell, D.E., Korber, D.R., and Lappin-Scott, H.M. Microbial biofilms. *Annu. Rev. Microbiol.* 1995;49:711–745.

Costerton, J.W., Stewart, P.S., and Greenberg, E.P. Bacterial biofilms: a common cause of persistent infections. *Science* 1999;284:1318–1322.

Crouzet, M., LeSenechal, C., Brözel, V.S, Costaglioli, P., Barthe, C., Bonneu, M., Garbay, B., and Vilain, S. Exploring early steps in biofilm formation: set-up of an experimental system for molecular studies. *BMC Microbiol.* 2014;14:1–12.

Davidson, I., Scianni, C., Hewitt, C., Everett, R., Holm, E., Tamburri, M., and Ruiz, G. Mini-review: Assessing the drivers of ship biofouling management – aligning industry and biosecurity goals. *Biofouling* 2016;32:411–428.

Deng, Y., Song, G.L., Zheng, D., and Zhang, Y. Fabrication and synergistic antibacterial and antifouling effect of an organic/inorganic hybrid coating embedded with nanocomposite Ag@TA-SiO2 particles. *Colloids Surf., A* 2021;613:126085.

Eddaoudi, M., Sava, D.F., Eubank, J.F., Adil, K., and Guillerm, V. Zeolite-like metalorganic frameworks (ZMOFs): design, synthesis, and properties. *Chem. Soc. Rev.* 2015;44:228–249.

Eduok, U., Suleiman, R., Gittens, J., Khaled, M., Smith, T.J., Akid, R., ElAli, B., and Khalil, A. Anticorrosion/antifouling properties of bacterial spore-loaded sol–gel type coating for mild steel in saline marine condition: a case of thermophilic strain of Bacillus licheniformis. *RSC Adv.* 2015;5:93818–93830.

Farha, O.K., Eryazici, I, Jeong, N.C., Hauser, B.G., Wilmer, C.E., Sarjeant, A.A., Snurr, R.Q., Nguyen, S.T., Yazaydın, A.Ö., and Hupp, J.T. Metal-organic framework materials with ultrahigh surface areas: is the sky the limit? *J. Am. Chem. Soc.* 2012;134:15016–15021.

Felgentrager, A., Maisch, T., Spath, A., Schroder, J. A., and Baumler, W. Singlet oxygen generation in porphyrin-doped polymeric surface coating enables antimicrobial effects on Staphylococcus aureus. *Phys. Chem. Chem. Phys.* 2014;16:20598–20607.

Firouzjaei, M.D., Shamsabadi, A.A., Sharifan Gh, M., Rahimpour, A., Soroush, M. A novel nanocomposite with superior antibacterial activity: a silver-based metal organic framework embellished with graphene oxide. *Adv. Mater. Interfaces* 2018;5:1701365.

Furukawa, H., Ko, N., Go, Y.B., Aratani, N., Choi, S.B., Choi, E., Yazaydin, A.Ö., Snurr, R.Q., O'Keeffe, M., Kim, J., Yaghi, O.M. Ultrahigh porosity in metal-organic frameworks. *Science* 2010;329(80):424–428.

Galhenage, T.P., Hoffman, D., Silbert, S.D., Stafslien, S.J., Daniels, J., Miljkovic, T., Finlay, J.A., Franco, S.C., Clare, A.S., Nedved, B.T., Hadfield, M.G., Wendt, D.E., Waltz, G., Brewer, L., Teo, S.L.M., Lim, C.S., Webster, D.C. Fouling-release performance of silicone oil-modified siloxane-polyurethane coatings. ACS Appl Mater Interfaces. 2016;8:29025–29036.

Golmohamadpour, A., Bahramian, B., Khoobi, M., Pourhajibagher, M., Barikani, H.R., Bahador, A. (2018). Antimicrobial photodynamic therapy assessment of three indocyanine green-loaded metal-organic frameworks against *Enterococcus faecalis*. Photodiagnosis Photodyn Ther. 2018;23:331–338.

Gordon, D.P., Mawatari, S.F. (1992) Atlas of Marine-Fouling Bryozoa of New Zealand Ports and Harbours. Miscellaneous Publishers, New Zealand.

Guo, H., Yang, J., Zhao, W., Xua, T., Lin, C., Zhang, J., Zhang, L.Direct formation of amphiphilic crosslinked networks based on PVP as a marine anti-biofouling coating, Chem Eng J. 2019;374:1353–1363

Guo, X., Yang, H., Liu, Q., Liu, J., Chen, R., Zhang, H., Yu, J., Zhang, M., Li, R., Wang, J. A chitosan-graphene oxide/ZIF foam with anti-biofouling ability for uranium recovery from seawater. Chem Eng J. 2020;382:122850.

Guo, Y.F., Fang, W.J., Fu, J.R., Wu, Y., Zheng, J., Gao, G.Q., Chen, C., Yan, R.W., Huang, S.G., Wang, C.C. Facile synthesis of Ag@ZIF-8 core-shell heterostructure nanowires for improved antibacterial activities. Appl Surf Sci. 2018;435:149–155

Hatamie, S., Ahadian, M.M., Zomorod, S, Torabi, S., Babaie, A., Hosseinzadeh, S. Antibacterial properties of nanoporous graphene oxide/cobalt metal organic framework. Mater Sci Eng C Mater Biol Appl. 2019;104:109862.

Høiby, N., Bjarnsholt, T., Givskov, M., Molin, S., Ciofu, O. Antibiotic resistance of bacterial biofilms. Int J Antimicrob Agents. 2010;35:322–332.

Horcajada, P., Serre, C., Maurin G., Ramsahye N.A., Balas, F., Vallet-Regi, M., Sebban, M., Taulelle, F., Ferey, G. Flexible porous metal-organic frameworks for a controlled drug delivery. J Am Chem Soc. 2008;130:6774–6780.

Huo, S.H, Yan, X.P. Facile magnetization of metal–organic framework MIL101 for magnetic solid-phase extraction of polycyclic aromatic hydrocarbons in environmental water samples. Analyst. 2012;137:3445–3450.

Isaeva, V.I., Kustov, L.M. The application of metal-organic frameworks in catalysis (Review). Pet Chem 2010;50:167–180.

Landini, P., Antoniani, D., Burgess, J.G., Nijland, R. Molecular mechanisms of compounds affecting bacterial biofilm formation and dispersal. Appl Microbiol Biotechnol. 2010;86:813–823.

Lee, Y.R., Kim, J., Ahn, W.S. Synthesis of metal-organic frameworks: a mini review. Korean J Chem Eng. 2013;30:1667–1680.

Lewis, K. Riddle of biofilm resistance. Antimicrob Agents Chemother. 2001;45:999–1007.

Li, P., Li, J., Feng, X., Li, J., Hao, Y., Zhang, J. Metal-organic frameworks with photocatalytic bactericidal activity for integrated air cleaning. Nat Commu. 2019;10:2177–2180.

Li, S., Yang, K., Tan, C., Huang, X., Huang, W., Zhang, H. Preparation and applications of novel composites composed of metal–organic frameworks and two-dimensional materials. Chem Commun. 2016;52:1555–1562.

Liu J., Wu D., Zhu N., Wu Y., Li G., Antibacterial mechanisms and applications of metal-organic frameworks and their derived nanomaterials. Trends Food Sci Tech. 2021;109:413–434.

Liu, C., Wang, J., Wan, J., Yu, C. MOF-on-MOF hybrids: Synthesis and applications, Coord Chem Rev. 2021;432:213743.

Liu, Z., Tan, L., Liu, X., Liang, Y., Zheng, Y., Yeung, K. W. K. Zn^{2+}-assisted photothermal therapy for rapid bacteria-killing using biodegradable humic acid encapsulated MOFs. Colloid Surfaces B. 2020;188:11078–11083.

Mansouri, J., Harrisson, S., Chen, V. Strategies for controlling biofouling in membranefiltration systems: challenges and opportunities. J Mater Chem. 2010;20:4567–4586.

Martín-Betancor, K., Aguado, S., Rodea-Palomares, I., Tamayo-Belda, M., Leganés, F., Rosal, R., Fernández-Piñas, F. Co, Zn and Ag-MOFs evaluation as biocidal materials towards photosynthetic organisms, Sci Total Environ. 2017;595:547–555.

Martín-Rodríguez, A.J., Babarro, J.M.F., Lahoz, F., Sansón, M., Martín, V.S., Norte, M. From Broad-Spectrum Biocides to Quorum Sensing Disruptors and Mussel Repellents: Antifouling Profile of Alkyl Triphenylphosphonium Salts. PLoS ONE. 2015;10:e0123652.

Matin, A., Khan, Z., Zaidi, S.M.J., Boyce, M.C. Biofouling in reverse osmosis membranes for seawater desalination: Phenomena and prevention. Desalination. 2011;281:1–16.

Meng, X., Duan, C., Zhang, Y., Lu, W., Wang, W., Ni, Y. Corncob-supported Ag NPs@ ZIF-8 nanohybrids as multifunction biosorbents for wastewater remediation: Robust adsorption, catalysis and antibacterial activity. Compos Sci Technol. 2020;200:108384.

Morris, W., Volosskiy, B., Demir, S., Gándara, F., McGrier, P.L., Furukawa, H., Cascio. D., Stoddart, J.F., Yaghi, O.M. Synthesis, structure, and metalation of two new highly porous zirconium metal–organic frameworks. *Inorg. Chem.* 2012;51:6443–6445.

Mortada, B., Matar, T.A., Sakaya, A., Atallah, H., Kara, A, Z., Karam, P. Post metalated zirconium metal organic frameworks as a highly potent bactericide. *Inorg. Chem.* 2017;56:4740–4745.

Nadar, S.S., Varadan, N.O., Suresh, S., Rao, P., Ahirrao, D.J., and Adsare, S. Recent progress in nanostructured magnetic framework composites (MFCs): Synthesis and applications. *J. Taiwan Inst. Chem. Eng.* 2018;91:653–677.

O'Toole, G.A., Gibbs, K.A., Hager, P.W., Phibbs, Jr., P.V., and Kolter, R. The global carbon metabolism regulator Crc is a component of a signal transduction pathway required for biofilm development by Pseudomonas aeruginosa. *J. Bacteriol.* 2000;182:425–431.

Raju, P., Ramalingam, T., Nooruddin, T., and Natarajana, S. In vitro assessment of antimicrobial, antibiofilm and larvicidal activities of bioactive nickel metal organic framework. *J. Drug Deliv. Sci. Technol.* 2020;56:101560.

Rana, D., and Matsuura, T. Surface modifications for antifouling membrane. *Chem. Rev.* 2010;110:2448–2471.

Ren, X., Yang, C., Zhang, L., Li, S., Shi, S., Wang, R. Copper metal-organic frameworks loaded on chitosan flm for the effcient inhibition of bacteria and local infection therapy. *Nanoscale* 2019;11:11830–11838.

Rodríguez, H. S., Hinestroza, J. P., Ochoa-Puentes, C., Sierra, C. A., Soto, C.Y. Antibacterial activity against Escherichia coliof Cu-BTC (MOF-199) metal-organic framework immobilized onto cellulosic fbers. *J. Appl. Polym. Sci.* 2014;131:1–5.

Sancet, M.P.A., Hanke, M., Wang, Z., Bauer, S., Azucena, C., Arslan, H.K, Heinle, M., Gliemann, H., Wöll, C., Rosenhahn, A. Surface anchored metal-organic frameworks as stimulus responsive antifouling coatings. *Biointerphases.* 2013;8:1–8.

Schultz, M.P., Bendick, J.A., Holm, E.R., Hertel, W.M. Economic impact of biofouling on a naval surface ship. *Biofouling* 2011;27:87–98.

She, Q., Wang, R., Fane, A.G., Tang, C.Y. Membrane fouling in osmotically driven membrane processes: a review. *J. Membr. Sci.* 2016;499:201–233.

Stoeck, U., Krause, S., Bon, V., Senkovska, I., and Kaskel, S.A. highly porous metal-organic framework, constructed from a cuboctahedral super-molecular building block, with exceptionally high methane uptake. *Chem. Commun.* 2012;48:10841–10849.

Stoodley, P., Sauer, K., Davies, D.G., and Costerton, J.W. Biofilms as complex differentiated communities. *Annu. Rev. Microbiol.* 2002;56:187–209.

Tian, Z., Lei, Z., Chen, X., Chen, Y., Zhang, L.C., Bi, J., and Liang, J. Nanosecond pulsed fiber laser cleaning of natural marine microbiofoulings from the surface of aluminum alloy. *J. Clean. Prod.* 2020;244:118724.

Vatansever, F., de Melo, W. C., Avci, P., Vecchio, D., Sadasivam, M., and Gupta, A. Antimicrobial strategies centered around reactive oxygen species–bactericidal antibiotics, photodynamic therapy, and beyond. *FEMS Microbiol. Rev.* 2013;37:955–989.

Vilain, S., Cosette, P., Zimmerlin, I., Dupont, J.P., Junter, G.A., and Jouenne, T. Biofilm proteome: homogeneity or versatility? *J Proteome Res.* 2004;3:132–136.

Wang, C., Zhang, T, and Lin, W. Rational synthesis of noncentrosymmetric metal–organic frameworks for second-order nonlinear optics. *Chem. Rev.* 2012;112:1084–1089.

Wang, D., Jana, D., and Zhao, Y. Metal-organic framework derived nanozymes in biomedicine. *Acc. Chem. Res.* 2020;53:1389–1400.

Wang, J., Wang, Y., Zhang, Y., Uliana, A., Zhu, J., Liu, J., and Van der Bruggen, B. Zeolitic imidazolate framework/graphene oxide hybrid nanosheets functionalized thin film nanocomposite membrane for enhanced antimicrobial performance. *ACS Appl. Mater. Interfaces* 2016;8:25508–25519.

Yang, H., Guo, X., Chen, R., Liu, Q., Liu, J., Yu, J., Lin, C., Wang, J., Zhang, M. Enhanced anti-biofouling ability of polyurethane anti-cavitation coating with ZIF-8: a comparative study of various sizes of ZIF-8 on coating. *Eur. Polym. J.* 2021;144:110212.

Yebra, D.M., Kiil, S., Dam-Johansen, K. Antifouling technology—past, present and future steps towards efficient and environmentally friendly antifouling coatings. *Prog. Org. Coat.* 2004;50:75–104.

Yuan, Y., and Zhang, Y. Enhanced biomimic bactericidal surfaces by coating with positively-charged ZIF nano-dagger arrays. *Nanomed.: Nanotechnol. Biol. Med.* 2017;13:2199–2207.

Yuan, Y., Wu, H., Lu, H., Zheng, Y., Ying, J. Y., and Zhang, Y. ZIF nano-dagger coated gauze for antibiotic-free wound dressing. *Chem. Commun.* 2019;55:699–702.

Zhao, W., Yang, J., Guo, H., Xu, T., Li, Q., Wen, C., Sui, X., Lin, J., Zhang, J., and Zhang, L. Slime-resistant marine anti-biofouling coating with PVP-based copolymer in PDMS matrix. *Chem. Eng. Sci.* 2019;20:790–798.

Zhao, X., Liu, S., Tang, Z., Niu, H., Cai, Y., Meng, W., Wu, F., and Giesy. J.P. Synthesis of magnetic metal-organic framework (MOF) for efficient removal of organic dyes from water. *Sci. Rep.* 2015;5:11849–11851.

Zhou, H.C., Long, J.R., and Yaghi, O.M. Introduction to metal-organic frameworks. *Chem. Rev.* 2012;112:673–674.

Zhou, W., Begum, S., Wang, Z., Krolla, P., Wagner, D., and Brase, S. High antimicrobial activity of metal-organic framework-templated porphyrin polymer thin films. *ACS Appl. Mater. Interfaces* 2018;10:1528–1533.

Zhuang, W., Yuan, D., Li, J. R., Luo, Z., Zhou, H. C., and Bashir, S. Highly potent bactericidal activity of porous metal-organic frameworks. *Adv. Healthc. Mater.* 2012;1:225–238.

Zirehpour, A., Rahimpour, A., Arabi Shamsabadi, A., Sharifan Gh, M., and Soroush, M. Mitigation of thin-film composite membrane biofouling via immobilizing nano-sized biocidal reservoirs in the membrane active layer. *Environ. Sci. Technol.* 2017;51:5511–5522.

24 Nanocontainers-Based Anti-Biofouling Coatings – A Pilot Study

George Kordas
Peter the Great St. Petersburg Polytechnic University

CONTENTS

24.1 INTRODUCTION

An object immersed in seawater is first coated by a biofilm consisting of microorganisms and then colonized by algae and invertebrates. In the end, the process culminates with corrosion which results in a reduction in its lifespan in the sea (Armstrong et al. 2000). This process is called biofouling and is one of the most serious problems in modern shipping industries. Its control is one of the major attractions of marine biotechnology. The adhesion of marine invertebrates to the hulls of ships increases corrosion, reduces the speed of ships, upsurges fuel consumption resulting in increased air pollution. The International Maritime Organization (IMO) estimates in a recent study that the global commercial fleet in 2020 burns about half a billion tons of fuel a year (Schultz 2007). This translates to a boat with heavy hull fouls requiring an additional 70% propulsion power compared to a boat with clean hulls. In 2020, effective antifouling protection will save the world over USD 120 billion. In addition to the appalling economic costs, it is estimated that a 5% increase in bio-pollution is causing a 14% increase in CO_2, NO_x, and SO_2 gas emissions (Townsin 2003). You must not forget that shipping companies spend over 5 billion €/year for ship protection. It's not just ships, but also structures like buoys, offshore oil rigs, wind farms, membrane bioreactors and desalination plants, refrigeration systems, aquaculture of marine fish, in particular aquaculture based on marine cages and oil pipelines of power plants falling within the same biofouling category at significant economic cost.

Antifouling paints are currently based on copper compounds and commemorative biocides that, when submerged, release toxic compounds causing adverse environmental impacts. The total concentration of copper oxide in colors ranges from 20% to 76% wt. (Brooks et al. 2009). With the ever-increasing cost of copper, incorporating high amounts of copper oxide into antifouling marine paints makes this technology a costly exercise. In the past, tributyltin (TBT) was used as an antifouling agent but proved destructive to the marine environment. A study with two bacteria, *Bacillus stearothermophilus* and *Bacillus subtilis* demonstrated that TBT produces alterations in growth, and respiratory activity with oxygen absorption reduction (Martins et al. 2005). Scientists are concerned about the toxicity of TBT, which acts as an endocrine disrupter e.g., mussels (Morcillo and Porte 2000) or causes masculinity by increasing testosterone levels in female gasteropods (Fent 1996) leading to population deterioration. Today, we use Sea-Nine™ 211 (SN), which acts against non-target phytoplankton, bivalves, echinoderms, and ascidians by causing embryotoxicity, larval mortality, and immunotoxicity. Similar damage results from the use of copper oxide, which is also very harmful to marine organisms, although aquatic species are greatly different in their sensitivity to the heavy metal. Some organisms withstand high concentrations of heavy metals, but others are destroyed by very low concentrations of copper. In general, copper toxicity mechanisms involve the inhibition of some key respiratory enzymes, the generation of oxygen species, and changes in gene transcription especially related to oxidative stress (Kiaune 2011).

The new technology must solve all these problems and demonstrate novel antifouling multifunctional systems that will offer radically improved performance over currently used marine paints on vessels, offshore oil platforms, wind farms, pipelines, and aquaculture facilities. This study shows the way to solve these problems using flu-like molecules from the sea such as bromosphaerol, anticorrosion molecules harmless to the marine environment such as 8-hydroxyquinoline (8-HQ), and nanotechnology which will trap these molecules and release them when required to protect the ship hulls. Nanotechnology is based on copper oxide, zinc oxide, and cerium-molybdenum (CeMo) nanocontainers where they are safe in small quantities in the marine environment.

24.2 PRODUCTION OF NANOCONTAINERS

We used cerium(III) acetylacetonate, zinc acetylacetonate hydrate, sodium molybdate, polyvinylpyrrolidone (average molecular weight 555000), potassium persulfate, sodium dodecyl sulfate, sodium chloride, 8-HQ, bromosphaerol, SeaNine™ 211, and bovine serum albumin in this study. The manufacturing procedures of the nanocontainers were described in our previous publications (Kordas 2019, 2020).

Because the use of copper is not so safe and there is a discussion about reducing or completely removing it in its antifouling paints, other pigments have been proposed to replace it. Among them, Zn (II) was proposed (Kordas 2019). Considering this perspective, copper and zinc nanocontainers were developed and synthesized through the sol–gel method in the entrapment of bromosphaerol. CeMo nanocontainers were also added which were used to entrap 8-HQ to protect the metal surfaces when incorporated into the dyes. The following reactions summarize the sequence of chemical reactions in the formulation of copper nanocontainers:

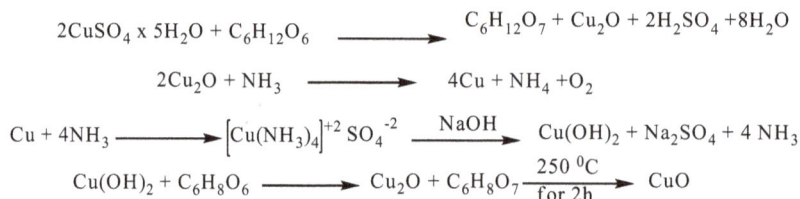

$$2CuSO_4 \times 5H_2O + C_6H_{12}O_6 \longrightarrow C_6H_{12}O_7 + Cu_2O + 2H_2SO_4 + 8H_2O$$

$$2Cu_2O + NH_3 \longrightarrow 4Cu + NH_4 + O_2$$

$$Cu + 4NH_3 \longrightarrow \left[Cu(NH_3)_4\right]^{+2} SO_4^{-2} \xrightarrow{NaOH} Cu(OH)_2 + Na_2SO_4 + 4NH_3$$

$$Cu(OH)_2 + C_6H_8O_6 \longrightarrow Cu_2O + C_6H_8O_7 \xrightarrow[\text{for 2h}]{250\ ^0C} CuO$$

The manufacture of CeMo nanocontainers was based on the method of coating polystyrene nanosphere templates following heat treatment at 600°C to burn off the template.

Figure 24.1 shows the SEM micrographs of the CeMo, ZnO, and CuO nanocontainers. One can perceive from these micrographs a homogeneous distribution of their size. After the synthesis of the nanocontainers, the active compounds bromosphaerol and 8-HQ were encased in the spheres of copper oxide, zinc, and CeMo, respectively. The calculation of the entrapment rate was made through spectroscopy reference curve (Kordas 2019). The results showed that for bromosphaerol

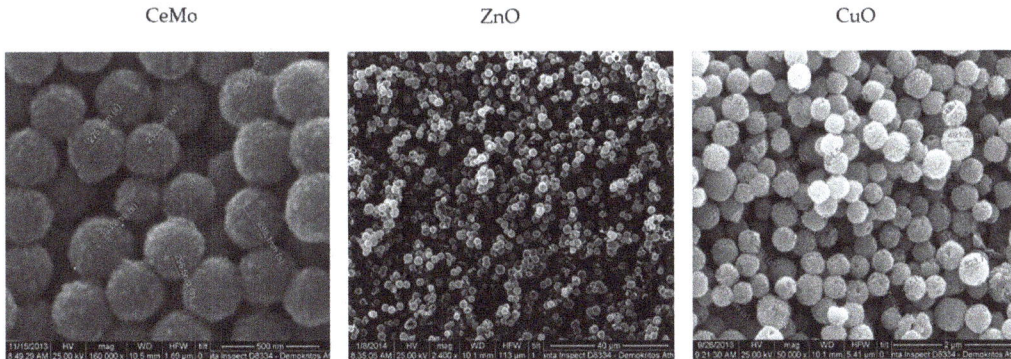

FIGURE 24.1   SEM micrographs of the CeMo, ZnO, and CuO spheres.

in CuO, the % encapsulation efficacy (EE) was 50.38 and % loading capacity was 25.2. In the case of ZnO, these values were 67.85% and 33.94%, respectively. Correspondingly, the values for CeMo nanocontainers were 80% and 40.2%.

The nanocontainers were then introduced into the paints by a procedure detailed below. Biocides are usually small molecules that, if freely scattered in antifouling paint, are released from the coating very quickly by relatively thin coatings ~100 μm. Premature loss is a serious problem because we don't get lifetime in a marine paint of about 5 years when embedded in the seawater environment. Antifoulant immobilization is achieved in this work by trapping bromospherol, SeaNineTM, and 8-HQ into nanocontainers where the actual diffusion coefficient is controlled by the porosity of the nanocontainers. Diffusion control of molecules incorporated into nanocontainers was proven in previous studies by our group (Kartsonakis et al. 2008).

In order to ensure uniform coating and avoidance of materials that will not allow the development of sufficient coherence between metal and paint, the shipbuilding sheet was cleaned with the sandblasting process.

24.3   PAINTING PROCESS

The painting was carried out with an air pistol with a nozzle diameter of 1 mm and a pressure of 2 bar (5, 10, and 15 layers). For the finished product, 15 layers were used and the paint speed for each layer was approximately 5 samples/s. The quantities described in Table 24.1 were used. After the first layer (primer), the samples were painted with anticorrosion paint. At first, the pistol was used with a nozzle of 1.0 mm. The paint contained the appropriate percentage of CeMo nanospheres with 8-HQ (CeMo-8-HQ) inside Wilckens' anticorrosion paint.

TABLE 24.1
Inggredients and Quantities Used

Ingredients	Quantity			
	A	**B**	**C**	**D**
Eposist HBS[1]	150 g	150 g	75 g	75 g
ERX 1755[2] hardener	40.2 mL/37.5 g	40.2 mL/37.5 g	23.4 mL/12.5 g	23.4 mL/12.5 g
Xylene	51.7 mL/45 g	51.7 mL/45 g	25.8 mL/22.5 g	25.8 mL/22.5 g

[1] Egotist HBS: black color Παρ002543 | BN | 130606 | ERP 9005 | Wilckens paints
[2] ERX 1755: 52161 | Batch No. 130605 | Production 2548 | Wilckens paints

For the third paint layer, similar conditions were used with the two previous layers. The materials and colors used were (i) Wilckens Ecomar AF 2000 with Cu, (ii) Wilckens Ecomar AF 2000 without Cu, (iii) copper oxide nanocontainers, (iv) zinc oxide nanocontainers, (v) bromosphaerol (BrSp), and (vi) SeaNine-211™. The quantities of microspheres used are indicated in the related figures. After the paint was diluted with xylene, the microspheres were added, and then the suspension was stirred with a mechanical stirrer for 5 min (moderate to strong stirring). The nanospheres exhibited sufficient stability in the paint. The coating thickness and morphology were determined using scanning microscopy. Figure 24.2 shows typical examples.

24.4 PERFORMANCE EVALUATION

Electrochemical Impendence Spectroscopy (EIS) was used to evaluate the performance of the paints. To simulate real conditions, metal samples (1 by 2 cm) were first cleaned, painted, and placed in a tube in the seawater in the Mikrolimano harbor. To simulate motion, a propeller was placed at one end of the tube, moving the water at a speed corresponding to 14 knots. After 3 months of exposure under this condition, samples were taken out every month for corrosion test using the EIS spectroscopy.

Figure 24.3a shows the EIS spectrum before and after exposure of the sample painted with the primer only in the seawater for 3 months. |Z| is 50×10^8 and 30 Ω for the coating with the primer only before and after seawater exposure. The primer doesn't provide corrosion protection as expected, but the binding ability of the paints above. Figure 24.3b shows the EIS spectrum before and after exposure of the sample painted with the primer and the anticorrosion paint including the CeMo(8-HQ) nanocontainers. For this coating, |Z| is greater than 5×10^8 Ω and equal to 10^8 Ω before and after exposure to the seawater for 3 months, respectively. Figure 24.3c shows the EIS spectrum before and after exposure of the sample painted with the primer, the anticorrosion paint including the CeMo(8-HQ) nanocontainers and top paint including $Cu_2O(BrSp)$. For the two coatings |Z| is about 10^{10} Ω before and after exposure in the seawater for the months, indicating additional protection due to the top coating with $Cu_2O(BrSp)$. This unchanged impedance value is an indication of "self-healing". One will need scanning vibrating electrode technique (SVET) analysis for unequivocal proof of the phenomenon, but we leave this for studies in the future.

Figure 24.4a shows the EIS spectra of the paint, including the primer, anticorrosion paint with CeMo(8-HQ) and antifouling paint with equal amounts of $Cu_2O(SN)$ and $ZnO(SN)$. |Z| is about 10^{10} Ω of this paint and becomes better than this value after exposure of this sample to the seawater for 3 months. This improvement can be attributed to the "self-healing" effect achieved by the coating with nanocontainers. If we remove $ZnO(SN)$ from the top coating and add the same amount of $Cu_2O(SN)$, as shown in Figure 24.4b, |Z| is better than 10^{10} Ω before and after exposure to the

substrate + primer (~67-81μm)   + 2nd layer with CeMo(8-HQ) (~67-81μm)   + 3rd layers CuO(BR)+ZnO(BR) (~50μm)

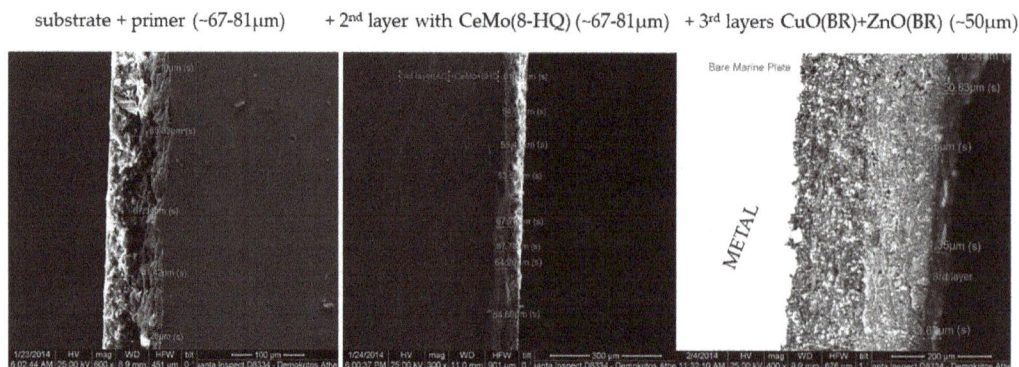

FIGURE 24.2 Indicative paint thicknesses: first layer, first and second layer, and final layer thickness.

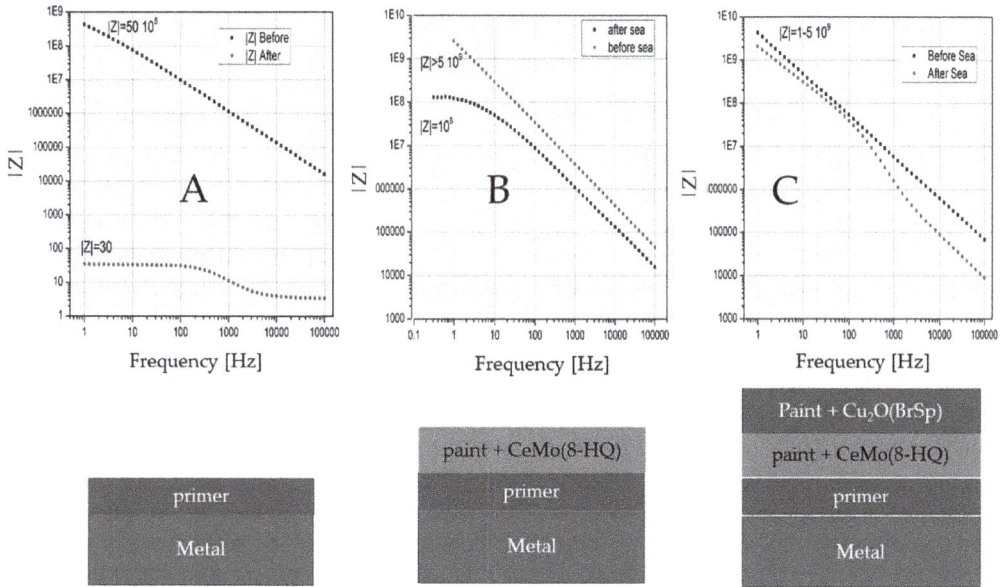

FIGURE 24.3 EIS measurements of the painted samples with (a) primer, (b) anticorrosion paint with CeMo(8-HQ) nanocontainers, and (b) antifouling paint with Cu$_2$O.

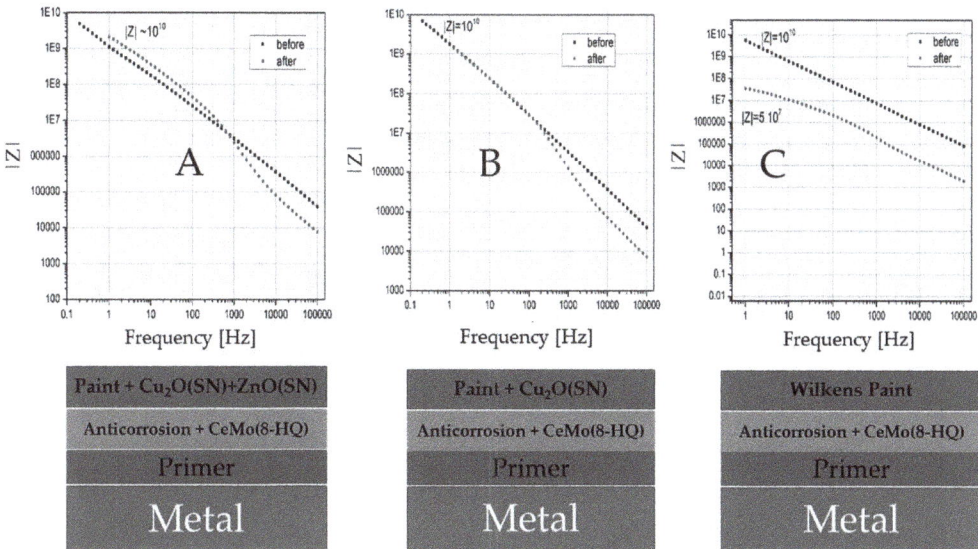

FIGURE 24.4 EIS measurements of the painted samples with (a) primer, anticorrosion paint with CeMo(8HQ), and Cu$_2$O(SN)+ZnO(SN); (b) with primer, CeMo(8-HQ), and Cu$_2$O(SN); and (c) primer, anticorrosion paint with CeMo(8-HQ) and a commercial antifouling top coat.

seawater, an additional proof that nanocontainers loaded with inhibitors and antifouling agents offer the phenomenon called "self-healing". If we remove the nanocontainers from the top coating and put the antifouling coating of Wilkens, then |Z| before and after exposure to seawater for 3 months were in the order of 10^{10} and 10^7 Ω, respectively. |Z| drops significantly after the exposure to seawater for 3 months in this last configuration of paint, in contrast to the paints containing all the nanocontainers with additives. Our technology is revolutionary and we showed that it can be applied to a multitude of coatings and substrates. All these nanocontainer-loaded coatings exhibit the "self-healing" phenomenon (Kartsonakis and Kordas 2010, 2011; Kartsonakis et al. 2012; Montemor et al. 2012; Balaskas et al. 2012; Harbers et al. 2007).

24.5 PILOT TEST

The technology developed in this work that turned out to be better than commercial paints had to be demonstrated in the field. To this end, a passenger ship traveling between Greece and Italy in the Adriatic was made available, which was painted on a strip and traveled for a year at a speed of 14 knots. During this period of experimental time, the temperature of the Adriatic Sea changes. The long distance also offers a change in the salinity of the seawater (Kordas 2019). Figure 24.5 shows the boat that after a year went out for inspection in a Greek shipyard and showed that the strip of the ship painted with our own technology is in better condition than the rest of the ship painted with commercial paint. This demonstration is in line with the laboratory results presented in the above text. Similar results were received on a second ship painted in Norway with paints of Re-Turn which after more than a year was a better condition on the strip painted with our paint. At this point, we report that extensive studies have been done on the changes made to the surface of our coatings. The evaluation was carried out with Wilkens and Ru-Turn paints in marine waters of Greece and Thailand where the results are very good and will be published later in another paper.

24.6 ZERO LIGHT ANTIBACTERIAL PROPERTIES OF CEMO

The antibacterial properties of CeMo were studied using sodium ampicillin salt, yeast extract, bacto tryptone, and bacteria agar, sodium hydroxide, and sodium chloride. *E. coli* (DH size 5 Å) was grown anaerobically in a 15-mL conical glass bottle containing 2 mL of Liquid Broth nutrient (LB) at 37°C in a controlled rotating stirrer for 16 h. The speed of the stirrer was set to 230 rpm. The ingredients of 1 L of LB are shown in Table 24.2. Then, the cell culture is diluted 4.3 orders of size in sterile distilled water. The final cell concentration was determined by the measurement of colony-forming units (CFU) in LB agar petri plates.

Fraction of the ship hull painted with the nanotechnology

Observation of the nanotechnology paint after a year traveling with 14 knots across Adriatic Sea

FIGURE 24.5 Evaluation of the nanotechnology paints using a strip of a ship painted with our nanocontainers loaded with inhibitors and antifouling agents. The ship was traveling across the Adriatic Sea for a year with a speed of 14 knots.

TABLE 24.2
Ingredients for Preparing 1 L of Liquid Broth Nutrient

Material	Quantity (g)
Yeast extract	5
Bacto tryptone	10
Sodium chloride	10
Sodium hydroxide 1N	3.5
Bacto agar	17
Distilled water	1000

For photocatalysis experiments, stock nanosphere aqueous suspenders are always prepared before the photocatalysis reaction and kept in the dark. Multiple times, the stock suspension of 1 mL is added to a 50-mL boiling glass containing 8 mL of sterile distilled water and 1 mL of cell solution. The nanosphere-cell suspension is stirred continuously in a magnetic stirrer and irradiated with two 8-W blue-black fluorescent lamps (Sylvania 8 W/BLB), one above the boiling glass and one below the boiling glass. Loss of cell viability was examined by the process of measuring living organisms. An *E. coli* suspension without nanocontainers was irradiated as a control suspension. The reaction of nano-cell suspension to darkness was also examined. For both experiments, samples were taken at intervals of 10 min for 60 min in triads. The viability measurement was performed on LB agar petri plates. All petri plates were incubated at 30°C for 16 h. The viability of cells processed with CeMo nanocontainers was determined by CFU measurements after 16 h of incubation. The initial cell concentration was 524 CFU/0.5 mL. For the experiment with cells processed with CeMo, the viability curve in Figure 24.6 shows that when *E. coli* cells are irradiated for 10 min in the presence of 10 mg/mL CeMo, only 0% of cells maintain their viability. The CeMo nanocontainers are "zero light" antibacterial materials reducing the bacteria down to zero concentration in 10 min without any light exposure.

24.7 ROLE OF NANOCONTAINERS

This study focuses on describing results obtained from an industrial multi-year development of technology to protect ships in the marine environment. The achievement of this technology is the development and application of cost-effective and ecologically compatible antifouling and anticorrosion systems where we use quantities of CuO nanocontainers of the order of 5% w/w instead of 20%–76% w/w, replacing chemicals dangerous to the marine environment and minimizing the maintenance costs of these structures. By extension, this technology can also be used in incorporating paints

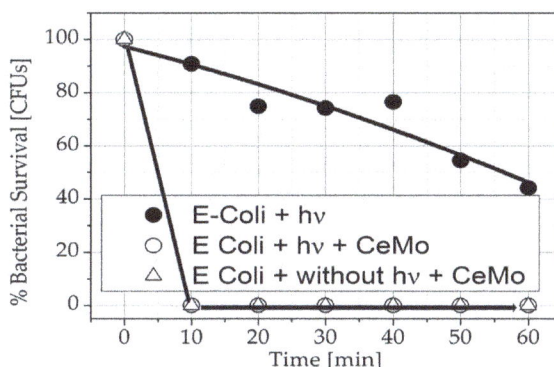

FIGURE 24.6 Photocatalytic effect of CeMo nanocontainers.

with nanocontainers through dyeing of threads and braiding of aquaculture nets. Anticorrosion protection is achieved by producing CeMo nanocontainers filled with 8-HQ integrated into the metal surface paint intermediate layer. The technology has led to the development of three categories of new antifouling systems with a significant impact on: (i) marine paints which, thanks to their new composition, will protect ship's vessels and conventional aquaculture nets and structures; (ii) more efficient coatings for offshore oil rigs, wind farms, and pipelines; (iii) durable yarns incorporating active agents for the manufacture of aquaculture nets. The new paints offer multifunctional properties: nanocontainers loaded with active monomers and corrosion inhibitors that offer self-healing, self-polishing, and anticorrosion action. In addition, new surfaces improve hydrophobia and the morphology of coating surfaces by reducing water friction (Kordas 2019). The new technology increases the life of paints thanks to the controlled release of antifouling agents from nano-systems and the extension of maintenance intervals. The new products are environmentally friendly, with reduced raw material costs which can be reduced by ten, as in the case of reduced quantities of copper oxides to be used as nano combinations. New technology ensures the development of a sustainable solution to the ongoing and costly problem of biofouling (Kordas 2019).

Biofouling begins by changing the state of microorganisms from freely suspended to attached to the metal surface. When this attachment is made, the new condition is no longer reversible associated with the surface chemical bonding. This is avoided if the adhesion is disconnected at the beginning. Such a situation prevails on the skin of the shark because it consists of small triangular surfaces of geometric length ~350 μm, with gaps between them of the order of 70 μm. This structure does not allow the permanent installation of microorganisms on the skin of the shark because the movement is rinsed. This hypothesis in our case, like others, must prove more convincing, but one is the result that this surface prevents the establishment of algae, barnacles, and bacteria.

Polymers that their surface is modified with different molecular structures that determine their length can lead to a new surface with a wide variety of biosystems. An example is the addition of a polyethylene glycol (PEG) layer with a co-polar connection to a polymer substrate, leading to bacterial adhesion reduction (Kartsonakis and Kordas 2011). Another work showed that a PEG-based coating inhibits the adhesion of proteins, bacteria, and mammalian cells (Harris et al. 2004). A sequence poly(L-lysine) - PEG creates a coating that prevents cell adhesion, such as fibroblast and osteoblast cells (Mcgary 1960). The disadvantage of PEG-based systems is the lack of stability that is often observed over time (Wach et al. 2008) and is the main reason it is not used. A successful coating involved attaching vancomycin to aneseline chromosomes through a PEG connector (Broxton et al. 1983). The aneseline chromophore allows a strong connection to TiO_2 surfaces and vancomycin provides antimicrobial action. Polyexamethylene biguanides can also be used as a potential antimicrobial agent (Zeng et al. 2019).

We know from previous studies the behavior of Cu and Zn elements in their antifouling ability (Kordas 2019; Krzak et al. 2012). Here should be added the discovery, that CeMo nanocontainers also offer antimicrobial properties added to the anticorrosion properties. These nanocontainers work without irradiation of light and we call them zero light antifouling components. We also note that the anticorrosion property is close to 10^{10} Ω in the CeMo(8-HQ) coating and 10^8 Ω after 3 months of exposure to the marine environment. The performance of the commercial paint was inferior to this.

Lately, there has been a great effort in the literature on the development of supramolecular chemistry, which is one of the exciting branches of chemistry where we get non-covalent interactions between molecules and subsequent supermolecular structures for corrosion and prevention of biofouling (Saji 2019). We note that different strategies that are not based on industrial coatings currently used will fail because the ship painting industry is very conservative and the penetration of foreign technologies is very difficult. New technology will be adopted more easily if it is based on current commercial paints, as we did in this study.

Our technology was validated with the collaboration of two commercial companies (Wilkens and Re-Turn) where we painted two ships using their paints. One traveled to the Adriatic Sea for a year at speeds of 14 knots and the other to the North Sea again for a year (Kordas 2019, 2020). The ships were withdrawn after a year and it appeared that the paint based on our nanotechnology was many times better than the commercial paints.

24.8   CONCLUSIONS AND PERSPECTIVES

In our laboratory, we developed an anticorrosion and antifouling technology that we believe will not go unnoticed by the major industries in the field. This nanotechnology is based on materials that are friendly to the marine environment and demand the solution of the problem. The quantities of materials are small and make this technology economically viable. This new technology will not require licenses in different countries because the materials come from the marine environment like bromosphaerol and can be easily obtained on an industrial scale. We note here that our technology uses one-eleventh of the CuO quantity used by today's technology with better results with regard to the antifouling protection of metals in marine water.

Other possibilities include: to replace copper oxide, known for its impact on the marine environment, one can think about replacing zinc with another form of nanocontainer to dose the solution, as in the future, other types of nanocontainers may also be developed more friendly to the environment. The nanocontainer technology presents a vital solution to the problem that will be part of the fourth industrial revolution in a number of different applications in medicine, self-healing, energy storage, energy transport, textiles, home appliances, and more other thinkable applications.

ACKNOWLEDGMENTS

Supported by the Ministry of Education and Science of the Russian Federation on application No. 2020–220–08–6558 for state support of scientific research conducted under the guidance of leading scientists in Russian educational institutions of higher education, scientific institutions, and state scientific centers of the Russian Federation.

REFERENCES

Armstrong E., Boyd K.G., Burgess J.G. Prevention of marine biofouling using natural compounds from marine organisms. *Biotechnol. Annu. Rev.* 6 (2000) 221–241.

Balaskas A.C., Kartsonakis I.A., Tziveleka L.A., Kordas G. Improvement of anti-corrosive properties of epoxy-coated AA 2024-T3 with TiO_2 nanocontainers loaded with 8-hydroxyquinoline. *Prog. Org. Coat.* 74 (3) (2012) 418–426.

Brooks N.S., Waldock M., Helio C., Yebra D. *Advances in Marine Antifouling Coatings and Technologies*, Woodhead Publishing Ltd., Cambridge (2009), p. 495.

Broxton P., Woodcock P.M., Gilbert P. A study of the antibacterial activity of some polyhexamethylene biguanides towards Escherichia coli ATCC 8739. *J. Appl. Bacteriol.* 54 (3) (1983) 345–353.

Fent K. Organotin compounds in municipal wastewater and sewage sludge: contamination, fate in treatment process and ecotoxicological consequences. *Sci. Total Environ.* 185 (1–3) (1996) 151–159.

Fresenius Environmental Bulletin (FRESEN ENVIRON BULL). 19, 2297–2302.

Harbers G.M., Emoto K., Greef C., Metzger S.W., Woodward H.N., Mascali J.J., Grainger D.W., Lochhead M.J. A functionalized poly (ethylene glycol)-based bioassay surface chemistry that facilitates bio-immobilization and inhibits non-specific protein, bacterial, and mammalian cell adhesion. *Chem. Mater.* 19 (18) (2007) 4405–4414.

Harris L.G., Tosatti S., Wieland M., Textor M., Richards R.G. Staphylococcus aureus adhesion to titanium oxide surfaces coated with non-functionalized and peptide-functionalized poly(l-lysine)-grafted-poly (ethylene glycol) copolymers. *Biomaterials* 25 (18) (2004) 4135–4148.

Kartsonakis I.A., Danilidis I.A., Kordas G. Encapsulation of the corrosion inhibitor 8-hydroxyquinoline into Ceria nanocontainers. *J. Sol. Gel Sci. Technol.* 48 (1–2) (2008) 2431.

Kartsonakis I.A., Kordas G. Synthesis and characterization of cerium molybdate nanocontainers and their inhibitor complexes. *J. Am. Ceram. Soc.* 93 (1) (2010) 65–73.

Kartsonakis I.A., Balaskas A.C., Kordas G. Influence of cerium molybdate containers on the corrosion perfor-
    mance of epoxy coated aluminium alloys 2024-T3. *Corros. Sci.* 53, (2011) 3771–3779.
Kartsonakis I.A., Balaskas A.C., Koumoulos E.P., Charitidis C.A., Kordas G. Incorporation of ceramic nano-
    containers into epoxy coatings for the corrosion protection of hot dip galvanized steel. *Corros. Sci.* 57
    (2012) 50–53.
Kartsonakis I.A., Balaskas A.C., Kordas G. Influence of cerium molybdate containers on the corrosion perfor-
    mance of epoxy coated aluminium alloys 2024-T3. *Corros. Sci.* 53 (11) (2011) 3771–3779.
Kordas G. CuO (Bromosphaerol) and CeMo (8 hydroxyquinoline) microcontainers incorporated into com-
    mercial marine paints. *J. Am. Ceram. Soc.* 103 (4) (2020) 2340–2350.
Kordas G. Nanotechnology to improve the biofouling and corrosion performance of marine paints: from lab
    experiments to real tests in sea. *Int. J. Phys. Res. Appl.* 2 (2019) 033–037.
Kordas G. Novel antifouling and self-healing eco-friendly coatings for marine applications enhancing the
    performance of commercial marine paints. In *Engineering Failure Analysis* (2019), pp. 161–169,
    Intech Open.
Kiaune L. Pesticidal copper (I) oxide: environmental fate and aquatic toxicity. *Rev. Environ. Contam. Toxicol.*
    213 (2011) 1–26.
Krzak M., Tabor Z., Nowak P., Warszyński P., Karatzas A., Kartsonakis I.A., Kordas G. Water diffusion in
    polymer coatings containing water-trapping particles. Part 2. Experimental verification of the math-
    ematical model. *Prog. Org. Coat.* 75 (3) (2012) 207–214.
Martins J., Jurado A.S., Moreno A.J.M., Madeira V.M.C. Comparative study of tributyltin toxicity on two
    bacteria of the genus Bacillus. *Toxicol. Vitro* 19 (7) (2005) 943–949.
Mcgary C.W. Degradation of poly (ethylene oxide). *J. Polym. Sci.* 46 (147) (1960) 51–57.
Montemor M.F., Snihirova D.V., Taryba M.G., Lamaka S.V., Kartsonakis I.A., Balaskas A.C., Kordas G.,
    Tedim J., Kuznetsova A., Zheludkevich M.L., Ferreira M.G.S. Evaluation of self-healing ability in
    protective coatings modified with combinations of layered double hydroxides and cerium molybdate
    nanocontainers filled with corrosion inhibitors, *Electrochim. Acta.* 60 (2012) 31–40.
Morcillo Y., Porte C. Evidence of endocrine disruption in clams-Ruditapes decussata--transplanted to a
    tributyltin-polluted environment. *Environ. Pollut.* 107 (1) (2000) 47–52.
Saji V.S. Supramolecular concepts and approaches in corrosion and biofouling prevention. *Corros. Rev.* 37 (3)
    (2019) 187–230.
Schultz M.P. Effects of coating roughness and biofouling on ship resistance and powering. *Biofouling* 23
    (2007) 331–341.
Townsin R. The ship hull fouling penalty. *Biofouling* 19 (2003) 9–15.
Wach J.Y., Bonazzi S., Gademann K. Antimicrobial surfaces through natural product hybrids. *Angew. Chem.,
    Int. Ed.* 47 (37) (2008) 7123–7126.
Zeng L., Wu Y., Xu J.F., Wang S., Zhang X. Supramolecular switching surface for antifouling and bactericidal
    activities. *ACS Appl. Bio Mater.* 2 (2019) 638–643.

25 Supramolecular Surface Modifications of Titanium Implants

Karan Gulati
The University of Queensland

CONTENTS

25.1 INTRODUCTION

Owing to its appropriate mechanics, biocompatibility, and corrosion resistance, titanium (Ti) is an ideal choice to replace diseased/damaged tissues, hence widely applied as an implant in dentistry and orthopaedics, and as cranioplasty supports and coronary stents (Guo et al. 2021a). A very thin titanium dioxide (TiO_2) surface layer is responsible for the biocompatibility of Ti (Choi et al. 2010). The structural and functional connection between bone tissue and implant (osseointegration) is critical to the long-term success of Ti implants. Besides, in a dental implant setting, early establishment and long-term maintenance of soft-tissue integration at the transmucosal region of dental implants is crucial to prevent the ingress of oral microbes (Guo et al. 2021a). Further, modulation of immune responses to a more reparative (wound healing) phase rather than inflammatory phase is also equally important and correlates with osseointegration (Gulati et al. 2018). In summary, of the host cells, increased bioactivity of osteoblasts (osseointegration) and fibroblasts/epithelial cells (soft-tissue integration), and modulation of macrophages (immunomodulation) are important determinants of implant acceptance and long-term success. However, implant placement surgery is a trauma, which results in a '*race to invade*' between host cells and bacteria (Guo et al. 2021b). The bacteria from the patient's skin or inside the body and local operating surfaces may be the source of pathogens in an orthopaedic setting, while for dental implants, bacteria are always present in the oral cavity (Chevalier and Gremillard 2009; Narendrakumar et al. 2015). The *race* between host cells and bacteria dictates the fate of the implant. While Ti-based implants favour high success rates in healthy patient conditions, in compromised scenarios like patients with poor bone quality/

DOI: 10.1201/9781003169130-28

quantity, age, and diabetes; early stability and long-term success may be challenged (Guo et al. 2021b). This is also attributed to the biological inertness of bare Ti.

Bacterial infection represents a major challenge towards the long-term success of Ti-based dental and orthopaedic implants. Briefly, bacterial adhesion onto the implant surface can be described in two stages: (i) rapid and reversible initial interaction, and (ii) slowly reversible (or irreversible) interaction between proteins on bacterial fimbriae/pili and implant (Chouirfa et al. 2019). It is noteworthy that upon the establishment of a mature biofilm, bacterial eradication becomes very cumbersome. Furthermore, due to poor penetration of antibiotics and compromised vascularization, simple irrigation or debridement are often unsuccessful. Repeated antibiotic administration can exacerbate the situation through the emergence of resistant strains. Clearly, for long-term success, Ti-based implants must enable local bactericidal functions, along with ensuring osseo- and soft-tissue integration and immunomodulation (Gulati et al. 2015).

Antibacterial strategies can be either anti-adhesion, anti-colonization, anti-biofilm, or anti-proliferation (Zhang et al. 2021). To understand and optimize the bactericidal performance of Ti implants, the influence of surface topography (micro- to nanoscale) on cell adhesion and biofilm formation has been studied (Lee et al. 2021). Furthermore, to achieve effective contact-killing of bacteria, various surface modifications at the nanoscale, including bioinspired nanostructures, nanopillars, and nanotextures, have been reported (Lee et al. 2021; Tripathy et al. 2017). Such 'static' antibacterial surfaces mechanically damage bacteria and prevent initial adhesion. However, dead and attached bacteria may protect other microbes from the underlying antibacterial surface. Hence, 'dynamic' antibacterial topographies have been proposed which can enable effective antibacterial functions using triggers like voltage, pressure, or magnetic field (Jayasree et al. 2021). More recently, the research has been focussed on the generation of anti-biofouling surfaces using local release of antibacterial agents (Chopra et al. 2021b). This chapter focusses on the next-generation of anti-biofouling surface modifications of Ti-based implants using advanced nano-engineering, with a focus on supramolecular surface modifications. This will inform the readers of the latest developments in the domain of anti-biofouling supramolecular modification of Ti implants and enable easy understanding of the key fabrication techniques, anti-biofouling efficacy, and bioactivity assessments, towards enabling future research bridging the gap to clinical translation.

25.2 NANOSCALE SURFACE MODIFICATIONS OF TITANIUM IMPLANTS

Surface modifications of Ti implants have been performed at micro- and nanoscales, with a special focus on reducing/inhibiting biofilm formation, towards the generation of bactericidal or bacteriostatic implant surfaces (Chouirfa et al. 2019; Feng et al. 2016). It is noteworthy that the nanoscale implant surface offers increased surface area and anchoring points, which can augment both bacterial and osteoblast adhesion, and hence the inclusion of anti-biofouling feature in nano-engineered implants is very crucial. In addition, the influence of surface roughness on bacterial colonization for orthopaedic/dental implants revealed that a Ra value less than or equal to 0.088 μm had minimal bacterial adhesion (Rimondini et al. 1997). Ideally, surface characteristics including nanotopography, roughness, and wettability must be controlled to reduce bacterial adhesion on implants (Müller et al. 2006). As a result, surface modification of Ti-based implants has been performed using various physical, chemical, and biological means to address *three Is*: integration, inflammation, and infection (Dhaliwal et al. 2019; Goodman et al. 2013; Gulati and Ivanovski 2017; Neoh et al. 2012; Raphel et al. 2016). Various techniques have been employed to achieve such implant surface nano-engineering, including plasma treatment, micro-machining, polishing/grinding, particle blasting, chemical etching, and electrochemical anodization (Chopra et al. 2021b). In addition, bioinspired nanotexturing, grafting of anchoring molecules, and immobilization of bioactive molecules have been explored on Ti implants (Córdoba et al. 2016; Gulati et al. 2015; Kunrath et al. 2020). Figure 25.1a summarizes the various surface modifications performed on Ti implants to enable anti-biofouling functions.

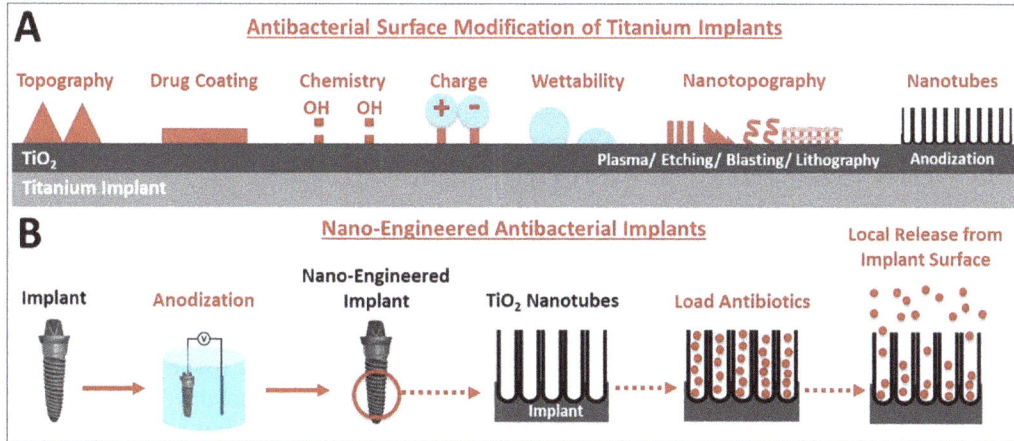

FIGURE 25.1   Antibacterial titanium implants: (a) various surface modifications and (b) electrochemically anodized titanium with TiO$_2$ nanotubes towards local elution of antibiotics. (Reproduced with permission from Chopra et al. (2021b) © 2021 Acta Materialia Inc. Published by Elsevier Ltd.)

One of the most popular and widely researched Ti implant surface modification strategies is electrochemical anodization (EA) (Guo et al. 2021). EA is a simple, cost-effective, and scalable strategy to enable the fabrication of controlled TiO$_2$ nanostructures including nanotubes and nanopores on the surface of Ti implants (Figure 25.1b) (Gulati et al. 2018). Briefly, it involves self-ordering of nanostructures, when constant voltage/current is applied to an electrochemical cell with target implant as anode and non-target Ti as cathode, immersed in water and fluoride-containing electrolyte (Gulati et al. 2015). The more popular, titania/TiO$_2$ nanotubes (TNTs) are like empty test tubes (length 0.5–300 mm and diameter 10–300 nm), open at the top and closed at the bottom (Gulati et al. 2012). TNTs have been extensively explored to modify Ti-based orthopaedic and dental implants towards enhanced bioactivity and local drug release (Gulati et al. 2013, 2018). The empty tubular structure of TNTs can be loaded with a variety of therapeutics, including antibiotics, antibacterial metal ions/nanoparticles, bone-forming proteins, growth factors, anti-inflammatory and anti-cancer drugs, to cater to various conditions (Gulati et al. 2011, 2012; Kaur et al. 2016). Further, owing to the ease of further chemical/physical functionalization and modification of clinical implants, EA-based TNTs have been one of the most researched implant modifications (Martinez-Marquez et al. 2020; Li et al. 2018). It is also worth noting that EA is a versatile technique and controlled metal oxide nanostructures can be fabricated on Ti and its alloys, Zr, Al, and other biomedically relevant metals and alloys (Chopra et al. 2021a; Gulati et al. 2017; Saji et al. 2015). Towards clinical translation, more recently, controlled TiO$_2$ nanostructures with preserved underlying micro-roughness (dual micro-nano) have been fabricated on commercial dental implants (Gulati et al. 2018; Li et al. 2018).

25.3   ANODIZED NANO-ENGINEERED TI IMPLANTS

TNTs have also been widely applied towards achieving bactericidal and bacteriostatic functions, as reviewed recently (Chopra et al. 2021b). Various groups have investigated the influence of TNTs diameter on bacterial functions *in vitro* and have reported that higher nanotube diameters of 80/100 nm demonstrate the most robust antibacterial effect, which can be further enhanced through heat treatment (Narendrakumar et al. 2015; Puckett et al. 2010). While TNTs have been used to enhance the bioactivity of various cells including osteoblasts and fibroblasts (Gulati et al. 2020; Gulati et al. 2018), bare TNTs can enhance bacterial adhesion/attachment due to nanoscale roughness, incorporation of fluoride (from EA), amorphous nature, and increased number of dead bacteria. Hence, to ensure the long-term success of such nano-engineered implants, coupling of bioactive TiO$_2$ nanotubes with local elution of antimicrobials may be ideal. The influence of TNTs diameter, wettability, and crystallinity on bacterial adhesion and attachment can be found in a recent review by Chopra et al. (2021b). This section primarily focusses on antibacterial TiO$_2$ nanotube modified Ti implants through loading and release of potent antibacterial agents. A summary of tailoring drug loading and achieving controlled sustained release patterns from TNTs is shown in Figure 25.2. The figure also shows several supra-molecular interactions that can be achieved on nano-engineered Ti implants.

25.3.1 RELEASE OF ANTIBACTERIAL AGENTS

Based on simple diffusion, drugs/proteins/growth factors can be incorporated inside nanotubes through drop-casting, immersion, or vacuum-assisted loading, and further dependent on diffu-sion gradient the loaded drugs can be eluted into the surrounding media or tissue (Rahman et al. 2016). The idea is to achieve substantial loading (which depends on nanotube dimensions and

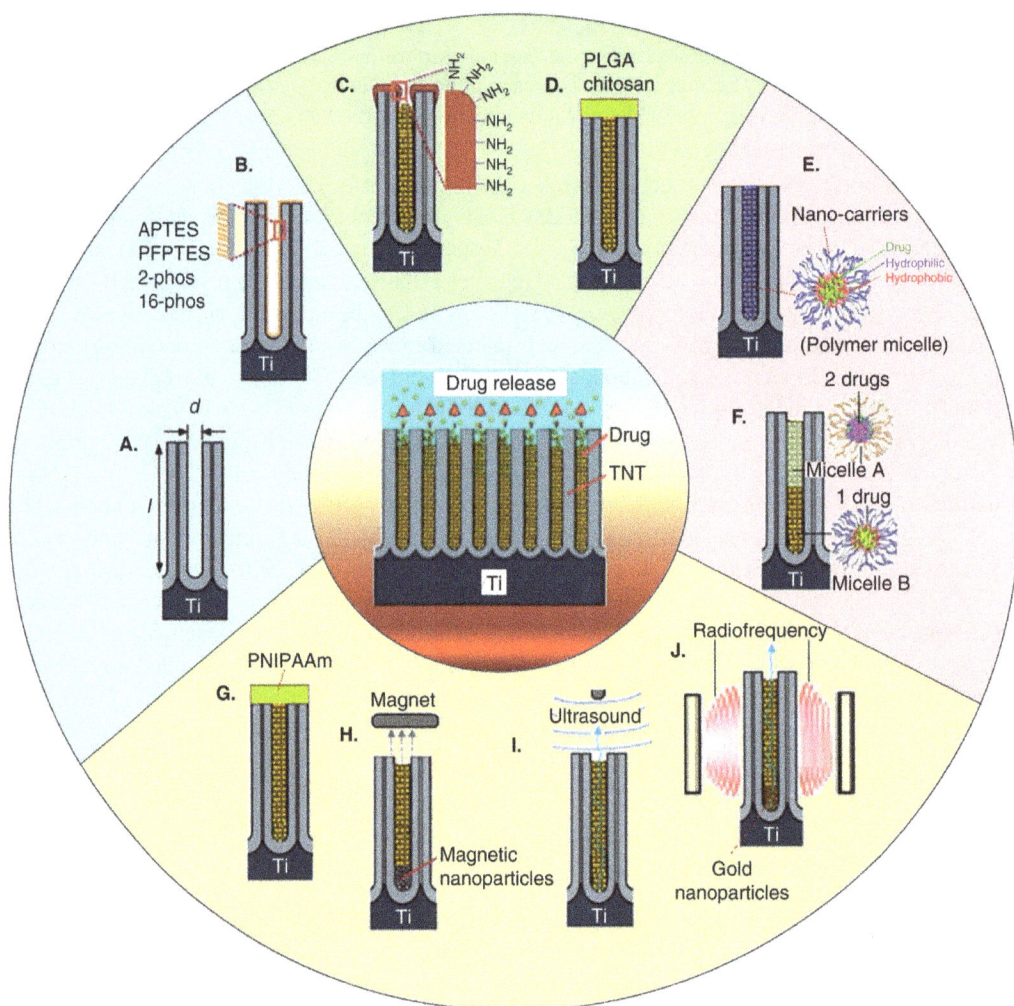

FIGURE 25.2 Scheme showing surface coating and functionalization of titania nanotubes (TNTs) to achieve superior loading, controlled and triggered release of therapeutics. PLGA: Poly(lactic-co-glycolic acid); Ti: Titanium; TNTs: Titania nanotubes. (Reproduced with permission from Losic et al. (2015) © 2014 Taylor & Francis.)

loading protocol), reduced initial burst release (IBR) which occurs due to a very high diffusion gradient, and sustained release over the long term. Early attempts at utilizing TNTs towards implant therapy involved loading and release of clinically used antibiotics including gentamicin and vancomycin (Popat et al. 2007; Zhang et al. 2013). Alternatively, minocycline, amoxicillin, cephalothin, and cefuroxime have also been loaded and released from TNTs modified Ti implants (Chennell et al. 2013; Park et al. 2014). While favourable outcomes have been obtained with drug release while maintaining the functions of osteoblasts, simple local antibiotic administration may be ineffective against certain resistant bacteria, especially the methicillin-resistant *Staphylococcus aureus* (MRSA). In a pioneering attempt, Ma et al. incorporated antimicrobial peptides (HHC-36) inside nanotubes and obtained a 99.9% bactericidal effect against MRSA (Ma et al. 2012). Besides the use of clinical antibiotics, metallic/semi-metallic ions or nanoparticles (NPs) including Au, Ag, Cu, B, P, Ca, Zn, and so on have also been loaded inside TNTs. Of these, Ag NPs have been most commonly explored to impart superior bactericidal efficacy, while maintaining favourable cytocompatibility (and limited cytotoxicity, which is often dose-dependent) (Das et al. 2008; Gunputh et al. 2018; Nandi et al. 2018; Shivaram et al. 2016). Further, incorporated B, P, Ca, F, Ga, and Cu ions/NPs inside TNTs have also shown promising outcomes in imparting antibacterial efficacy against common implant-infection-related bacteria (Arenas et al. 2013; Sopchenski et al. 2018; X. Zhang et al. 2018). In addition, (poly-DL-lactic acid) with Ga (III), Folic acid conjugated ZnO quantum dots and Ag-doped hydroxyapatite (Ag-HAp) have been immobilized on TNTs/Ti implants to demonstrate excellent antibacterial functions (Dong et al. 2019; Mirzaee et al. 2016; Xiang et al. 2018). It is noteworthy that all NPs (and some ions) pose a cytotoxicity risk, which is related to the size, shape, type, and concentration of NPs. Future studies aimed at long-term (several months) effects of such ion/NPs releasing implant coatings will be needed to prove the biosafety of such potent bactericidal systems.

25.3.2 CONTROLLED RELEASE USING BIOPOLYMERS

Post-implantation, a very high locally eluted drug concentration (IBR) from nanotubes can be cytotoxic to surrounding tissue. Further, it can lead to consumption of high drug amounts (40%–70%), which reduces long-term release, risking retriggered bacterial attachment and invasion. The local need for therapy is difficult to quantitate. While a few attempts have been made to qualitatively or quantitatively calculate drug release inside the tissue *ex vivo* or *in vivo* (Aw et al. 2012; Kaur et al. 2016; Rahman et al. 2016), it is noteworthy that a post-surgery traumatized tissue may behave entirely differently with restricted perfusion and diffusion (Daish et al. 2017). Further, tuning nanotube diameters and lengths (using EA voltage and time) gives easy control over release kinetics, whereby wider/longer nanotubes allow for higher loading amounts but high IBR (Gulati et al. 2016). A controlled release can be achieved if open pores of the TNTs can be covered with a biocompatible coating that would degrade with time, allowing for drug release. In that light, biopolymers chitosan, polydopamine, and PLGA [poly(lactic-co-glycolic acid)] have been used to cover drug-loaded nanotubes to receive degradation-dependent controlled local release with minimal IBR (Gulati et al. 2012; Kumeria et al. 2015). In an interesting attempt, drug (indomethacin) was encapsulated inside polymeric micelles (Pluronic F127) before loading inside TNTs, which were then coated with either PLGA or chitosan, which enabled controlled release of drugs, with the chitosan coating also enabling antibacterial and anti-biofilm formation abilities (Kumeria et al. 2015). In addition, polydopamine, Zn, and Ag have been immobilized on TNTs, which exhibited good antibacterial activity, with lower cytotoxicity and favourable biocompatibility (Ding et al. 2018). Further, torularhodin and Ag NPs have been incorporated inside TNTs with polydopamine to delay the release of potent antibiotic agents and achieve superior killing of implant-relevant bacteria (Jia et al. 2016; Ungureanu et al. 2016). Electrospun silk fibroin nanofibers have also been used to cover vancomycin-loaded TNTs and the findings revealed enhanced antibacterial activity against *S. aureus* (Fathi et al. 2019). While most of these studies report degradation of biopolymers when used as a drug release control, in 2016, Gulati et al. explored the fate of chitosan coatings on TNTs for varied times when immersed in PBS *in vitro* (Gulati et al. 2016). Interestingly, chitosan microtubules were formed on TNTs, providing a dual micro-nano architecture.

25.3.3 TRIGGERED RELEASE

Jayasree et al. recently reviewed the use of various intrinsic (pH or enzymatic reaction) and extrinsic (UV, magnetic/electric field, or electromagnetic radiation) triggers to turn the release of payloads from nanotubes on or off (Jayasree et al. 2021). Triggered release ensures that a premature IBR is minimized while the drugs are released when needed, which will be triggered internally (upon sensing a change) or externally (via clinician/patient). In two separate enzyme-based trigger studies, defroxamine and vancomycin were loaded inside 70 nm TNTs using surface modification through layer-by-layer (LBL) coating of chitosan and HA-gentamicin conjugate and LBL coating of catechol-modified HA and chitosan, respectively (Yu et al. 2020; Yuan et al. 2018). With trigger 'on', the release of deferoxamine was increased by 30%–60% (12 h), while that of vancomycin was increased from <10% to 44% (4 h), with both demonstrating reduced bacterial adhesion and high antibacterial efficacy (Yu et al. 2020; Yuan et al. 2018). Similarly, when the pH was reduced to 5.4–5.5, Ag NPs (bound to TNTs with a pH-liable acetal linker) and vancomycin (coated on TNTs with coordination polymers) were released, resulting in high antibacterial activity against *S. aureus* and *E. coli* (Dong et al. 2017; Wang et al. 2017). While internal triggers deliver high drug concentrations in response to locally detected change (microbial load), turning the release 'on/off' cannot be achieved, and upon trigger, entire payloads may be released. Stimulating the release externally can address these shortcomings. Electrical stimulation therapy (EST) has been utilized to trigger the release of Penicillin/Streptomycin (carbon nanotube modified TNTs electrodeposited with polymers) and vancomycin (using electrodeposition of chitosan) from TNTs, upon application of appropriate voltage pulses (Shi et al. 2013; Sirivisoot et al. 2011). In a pioneering attempt, conversion of semi-conducting TiO_2 nanotubes into electrically-conducting Ti nanotubes was shown using magnesiothermic conversion, to achieve enhanced EST ability (Gulati et al. 2016). Other external triggers, including magnetic fields, radiofrequency, and near-infra-red light, have also been used to deliver therapeutics with high IBR % achieved *in vitro* (Aw et al. 2012; Bariana et al. 2014; Zhao et al. 2020). Alternatively, triggered release of tetracycline and ampicillin using visible light and UV irradiation has also been reported (Zhou et al. 2018). It is worth noting that while triggers have worked successfully in proof-of-concept *in vitro* studies, there remain research gaps to enable their translation, including the effect of the very high release of therapeutics, accidental triggers, and cytotoxicity concerns.

25.4 SUPRAMOLECULAR SURFACE MODIFICATIONS

Supramolecular assembly (SA) is recognized as a promising surface modification strategy for metals (Gu et al. 2017). To augment the bioactivity of Ti-based implants, surface modification employing bioactive molecules using adsorption, silicate coupling, LBL, self-assembled monolayers, and Langmuir–Blodgett membranes holds great promise (Liu et al. 2002; Nanci et al. 1998; Salditt and Schubert 2002). In the following sections, we explore representative examples of SA coatings on Ti implants to enhance anti-biofouling efficacy while maintaining the functions of host cells towards bioactivity.

25.4.1 LYSOZYME-BASED COATINGS

SA of a phase-transited lysozyme (PTL) nanofilm is an emerging technology that can be used to modify metals, oxides, semiconductors, and polymers (Wu and Yang 2015). Derived from the structural phase of the natural lysozyme, the disulphide bond is reduced through tris(2-carboxy-ethyl)phosphine (TCEP) buffer under physiological conditions, which yields a supramolecular structure that can be used to form a 2D nanofilm (Ding et al. 2020). Hen egg white lysozyme (HEWL) possesses antimicrobial activity attributed to hydrophobic amino acid residues and positive charge on the lysozyme nanofilm. However, the use of wild-type lysozymes like HEWL can have limitations, including poor effectiveness against Gram-negative bacteria/fungi, instability, and difficulty incorporating them as coatings on various materials (Hu et al. 2020). Bypassing and complex chemical or physical treatments, Gu et al. utilized SA proteinaceous material to enable broad-spectrum antimicrobial properties, with excellent blood and mammalian cell biocompatibility (Gu et al. 2017). The transparent coating can easily be performed on a variety of material surfaces (including Ti) through simple aqueous soaking or contact printing (Figure 25.3). The antifouling and anti-biofilm therapy (against Gram-positive *S. aureus*, Gram-negative *E. coli*, and fungus *C. albicans*) was attributed to triple-combination of positive charge, hydrophobicity, and surface hydration effect of the protein nanofilm (also demonstrated for modified Ti screw). Further, cytotoxicity assays with murine osteoblast cells *in vitro* confirmed favourable cytocompatibility. In addition, medical catheters treated with the protein nanofilm and pre-seeded with *E. Coli* were implanted in Sprague–Dawley female rats subcutaneously *in vivo*. After 3 days, implants were harvested and characterized. The findings confirmed that modified catheters presented normal tissue without inflammation, as compared to controls where serious inflammation was found. Fabricating a 'green' antimicrobial modification using SA of lysozyme, extracted from daily food, is the novelty of the study.

Ha et al. have reported the use of PTL film on Ti plates to induce nucleation and growth of hydroxyapatite (HAp) (Ha et al. 2018). The multi-step functionalization approach included: (i) surface modification of Ti through PTL nanofilm (containing amyloid-like assembly nanostructures), (ii) surface binding of Ca^{2+} and (iii) growth of Hap in simulated body fluid (SBF)The abundance of carboxyl and hydroxyl functional groups on the PTL surface attributed to the nucleation and growth of Hap crystals, correlating with natural mineral growth. Mechanically robust HAp@PTL (Hap on PTL-coated Ti) showed favourable biocompatibility with rat bone marrow mesenchymal stem cells (rBMSCs) *in vitro*. Furthermore, using an ectopic osteogenesis animal model, the implants were placed under the skin of the rats *in vivo* for 4 and 8 weeks. Compared with bare Ti substrates and Ti-PTL surfaces, Hap@PTL demonstrated excellent osteoconductivity through ectopic bone formation (4 weeks) and bone matrix formation (8 weeks) (Figure 25.4).

Furthermore, in 2018, Xu et al. reported coating of Ti with a lysozyme-polyphosphate composite coating towards antibacterial and bioactivity functions (Xu et al. 2018). The coating was formed through sequential soaking of Ti substrate in reduced lysozyme and polyphosphate solutions. Further, effective antibacterial efficacy against Gram-negative *E. coli* and upregulated osteoblast functions (adhesion, proliferation, and differentiation) *in vitro* were confirmed.

The use of Strontium (Sr) in bone remodelling is well established and clinically approved (Pilmane et al. 2017). As a result, attempts have been made to incorporate Ti implants with Sr to enhance their osteogenic potential, especially applicable to compromised patient conditions including poor bone quality/quantity (Ding et al. 2020). Various strategies have been employed to incorporate Sr onto Ti, including micro-arc oxidation, alkaline heat treatment, hydrothermal, and ion exchange (Liu et al. 2014; Okuzu et al. 2017; Xu et al. 2016; Zhao et al. 2013). More recently, Ding et al. reported the fabrication of Sr-doped protein nanofilm on Ti towards orchestrating osteogenesis (Ding et al. 2020). Briefly, Sr-containing phase change lysozyme coating on Ti was performed through supramolecular self-assembly. *In vitro* release of Sr^{2+} ions in PBS confirmed a linear release profile, with fast release in the first 3 days. Furthermore, the culture of bone marrow stromal cells (BMSCs) *in vitro* revealed enhanced cellular pseudopodia on Sr modified Ti, increased cell viability, and high levels of osteogenic genes. Next, modified implants were surgically placed in the femur of male SD rats for 4 weeks *in vivo* and the findings confirmed new bone formation on Sr modified Ti.

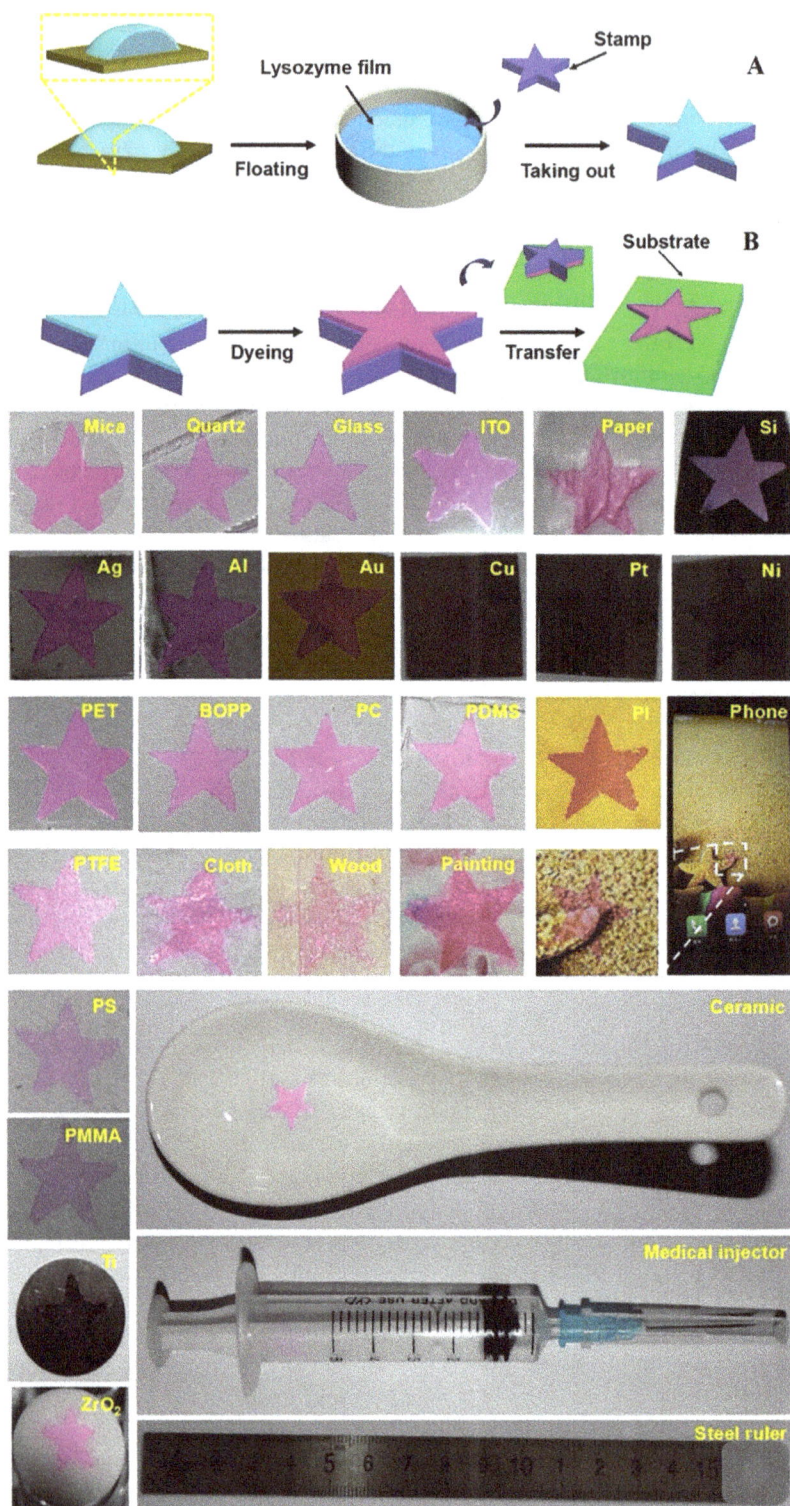

FIGURE 25.3   Demonstration of implantation of phase-transited lysozyme (PTL) nanofilm on a variety of materials through contact printing. (Reproduced with permission from Gu et al. (2017) © 2016 American Chemical Society.)

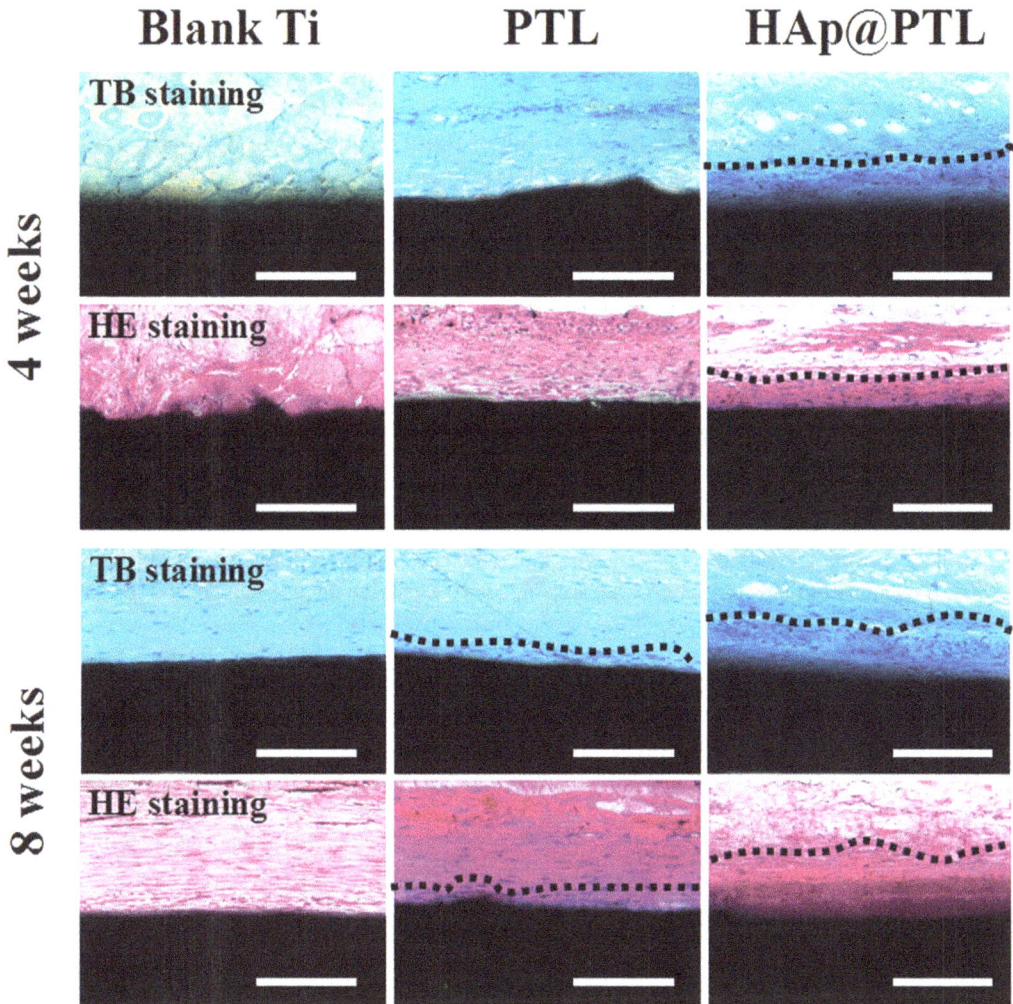

FIGURE 25.4 Phase-transited lysozyme (PTL) nanofilm modification towards enhancing osseointegration of Ti implants. Hematoxylin and eosin (HE, bone matrix) and Toluidine blue (TB, new bone formation) staining of blank Ti, original PTL-coated Ti, and HAp@PTL-coated Ti after 4/8 weeks in a rat model in vivo. (Reproduced with permission from Ha et al. (2018) © 2017 WILEY-VCH Verlag GmbH & Co. KGaA, Weinheim.)

25.4.2 BIOPOLYMER SELF-ASSEMBLY

In a pioneering attempt, Cai et al. reported the use of LBL self-assembly of chitosan and gelatin on Ti driven by polyelectrolyte-mediated electrostatic adsorption (Cai et al. 2005). Followed by the deposition of one layer of positively charged poly(ethylene imine) (PEI) on Ti, alternately negatively charged Gelatin and positively charged chitosan were coated. Further, increased proliferation and viability of osteoblasts *in vitro* was confirmed for chitosan/gelatin-modified Ti surfaces, which paved the way for further research advances in this domain. Next, in two different studies, the same group used chitosan with poly (styrene sulfonate) (PSS) or silk fibroin to augment osteoblast functions (Cai et al. 2007. Further, Hu et al. performed gene functionalization of Ti using chitosan and plasmid DNA (pEGFP–hBMP2, pGB) multilayered structures and evaluated its influence on mesenchymal stem cells *in vitro* (Cai et al. 2005). The results indicated that modified Ti surface-enhanced production of alkaline phosphatase and osteocalcin (7 and 14 days) and could induce MSCs differentiation. In 2011, Shu et al. employed heparin-chitosan LBL self-assembly to modify Ti implants (Shu et al. 2011). First, the Ti surface was treated with NaOH, followed by single-layer deposition of positively charged poly-L-lysine (PLL). Next, alternate deposition of negatively charged Heparin and positive-charged chitosan was performed, with the outermost layer of chitosan. The uniform coating of Hep-Chi significantly augmented the adhesion, proliferation, and differentiation of osteoblasts *in vitro*.

As mentioned previously, soft-tissue integration (STI) at the transmucosal component of dental implants is crucial to implant success, as it maintains a physical barrier against the ingress of oral pathogens (Guo et al. 2021b). To achieve timely STI from dental implants, Zhang et al. performed a multi-layer sequential coating (LBL assembly) using chitosan, hyaluronic acid, and collagen on polished Ti surfaces (J. Zhang et al. 2018). Next, the gene transduction coating of the laminin α3 gene (LAMA3) was cross-linked onto the modified Ti. Biological assessments with human keratinocytes *in vitro* and implantation in male Wistar rats *in vivo* (4 weeks) confirmed favourable biocompatibility, upregulated expression of laminin α3 in epithelial cells, and augmented biological sealing between the implant and soft tissue. Figure 25.5 shows the histological observation of laminin α3 deposition along the tooth-junctional epithelium interface, with strong laminin α3 immunostaining visible for CS/(HA/COL)5-LAMA3 group.

In 2012, Calliess et al. performed copolymer surface modification through dip coating on TiAl6V4 surfaces towards antimicrobial and bioactivity enhancements, which was proposed for maintenance of skin sealing at the skin-implant interface (Calliess et al. 2012). Briefly, different ratios of 4-vinylpyridine (VP) and phosphonate monomers were synthesized through free radical copolymerization, and the spin coating of copolymer solutions in methanol was performed on polished Ti alloy. The findings confirmed that a suitable implant surface coating can be developed based on specifically designed copolymers that show superior antibacterial efficacy against *S. epidermidis* and *S. aureus* while maintaining good biocompatibility.

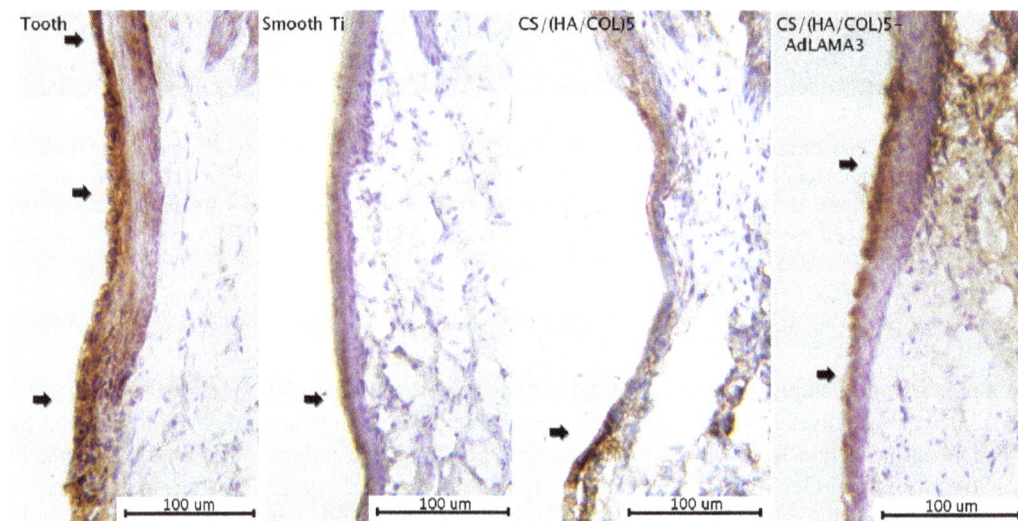

FIGURE 25.5 Enhanced soft-tissue integration ability of laminin α3 gene (LAMA3) modified Ti. Immunohistochemical detection of laminin α3 expression at the implant-tissue interface at 4 weeks. CS: Chitosan, HA: Hyaluronic acid; COL: Collagen. (Reproduced with permission from Zhang et al. (2018) © 2017 Elsevier B.V.)

25.4.3 HOST-GUEST CHEMISTRY-BASED SURFACE MODIFICATION

Host-guest chemistry has also been utilized to achieve effective anti-biofouling and stimuli-responsive biointerfaces (Cai et al. 2016; Yang et al. 2014). In general, host molecules contain large cavity volume [e.g. cyclodextrins (CD), cucurbiturils (CB), and calixarenes]; while guests have complementary shape and interaction with the host. The host–guest interaction is attributed to non-covalent interactions including H-bonding, electrostatic, van der Waals, and hydrophobic interactions (Yang et al. 2014). In 2016, Cai et al. reported the host-guest functionalization of Ti implants towards antifouling applications (Cai et al. 2016). Briefly, simple reactive ester-amine reaction and amidation were used to synthesize two new catecholic derivatives [dopamine 4-(phenylazo)benzamide (AZODopa) and dopamine 1-adamantanecarboxamide (AdaDopa)]. Next, anchoring of AZODopa and AdaDopa on Ti–6Al–4V alloy (guest) was performed for interaction with antifouling zwitterionics (host) [heptakis[6-deoxy-6-(N-3-sulfopropyl-N, N-dimethylammonium ethyl sulfanyl)]-β-cyclodextrin (SBCD) and hydrophilic β-CD polymer (CDP). *In vitro* evaluations confirmed that zwitterionic SBCD- and hydrophilic CDP-functionalized Ti can decrease protein adsorption (shown using bovine plasma fibrinogen) and adhesion of *E. coli*. The technique described is very versatile as the functionalization can be performed on other metal/metal oxide substrates and the zwitterionics can be replaced with bioactive agents to achieve tailored local therapy.

25.5 CONCLUSION AND PERSPECTIVES

Antibacterial surface modification of titanium implants shows great promise in achieving potent bactericidal functions without initiating any adverse cytotoxicity. It is noteworthy that most studies are *in vitro* and without mechanical loading, and these proof-of-concept advances may be distant from clinical translation. The following summarizes key clinical translation challenges and considerations for anti-biofouling titanium surface modifications (both anodized and supramolecular):

- Testing the mechanical stability of surface modification including nanotubes and biopolymer modifications under mechanical loading conditions
- Implantation of drug-releasing antibacterial implants in infection models *in vivo* for long durations (several months or years)
- Exploring the biosafety and cytotoxicity concerns associated with use of metal ions/nanoparticles and triggers
- Investigating the influence of local infection, coagulation, and related trauma in the tissue microenvironment on drug-releasing implant performance and functioning
- Achieving reduced IBR and sustained release for long duration and considering removal of dead bacteria or inclusion of surface decontamination

Further, supramolecular modifications of Ti-based implants discussed in this chapter including lysozyme and biopolymer self-assemblies and host-guest modifications have shown favourable outcomes in terms of achieving effective anti-biofouling and maintaining bioactivity. However, only a few studies have been reported with a direct focus on supramolecular interactions, and future focussed research is needed addressing the above-mentioned research challenges to advance supramolecular surface modifications of titanium towards customized dual anti-biofouling and bioactivity enhancements, in comparison with established clinical standards.

ACKNOWLEDGEMENTS

KG is supported by the National Health and Medical Research Council (NHMRC) Early Career Fellowship (APP1140699).

REFERENCES

Arenas, M. A., Pérez-Jorge, C., Conde, A., Matykina, E., Hernández-López, J. M., Pérez-Tanoira, R., de Damborenea, J. J., Gómez-Barrena, E., & Esteba, J. (2013). Doped TiO2 anodic layers of enhanced antibacterial properties. *Colloids and Surfaces B: Biointerfaces*, *105*, 106–112. https://doi.org/https://doi.org/10.1016/j.colsurfb.2012.12.051

Aw, M. S., Addai-Mensah, J., & Losic, D. (2012). Magnetic-responsive delivery of drug-carriers using titania nanotube arrays [10.1039/C2JM16819G]. *Journal of Materials Chemistry*, *22*(14), 6561–6563. https://doi.org/10.1039/C2JM16819G

Aw, M. S., Khalid, K. A., Gulati, K., Atkins, G. J., Pivonka, P., Findlay, D. M., & Losic, D. (2012). Characterization of drug-release kinetics in trabecular bone from titania nanotube implants. *International journal of nanomedicine*, *7*, 4883–4892. https://doi.org/10.2147/ijn.S33655

Bariana, M., Aw, M. S., Moore, E., Voelcker, N. H., & Losic, D. (2014). Radiofrequency-triggered release for on-demand delivery of therapeutics from titania nanotube drug-eluting implants. *Nanomedicine*, *9*(8), 1263–1275. https://doi.org/10.2217/nnm.13.93

Cai, K., Hu, Y., & Jandt, K. D. (2007). Surface engineering of titanium thin films with silk fibroin via layer-by-layer technique and its effects on osteoblast growth behavior. *Journal of Biomedical Materials Research Part A*, *82A*(4), 927–935. https://doi.org/https://doi.org/10.1002/jbm.a.31233

Cai, K., Hu, Y., Jandt, K. D., & Wang, Y. (2007). Surface modification of titanium thin film with chitosan via electrostatic self-assembly technique and its influence on osteoblast growth behavior. *Journal of Materials Science: Materials in Medicine*, *19*(2), 499. https://doi.org/10.1007/s10856-007-3184-5

Cai, K., Rechtenbach, A., Hao, J., Bossert, J., & Jandt, K. D. (2005). Polysaccharide-protein surface modification of titanium via a layer-by-layer technique: Characterization and cell behaviour aspects. *Biomaterials*, *26*(30), 5960–5971. https://doi.org/https://doi.org/10.1016/j.biomaterials.2005.03.020

Cai, X. Y., Li, N. N., Chen, J. C., Kang, E.-T., & Xu, L. Q. (2016). Biomimetic anchors applied to the host-guest antifouling functionalization of titanium substrates. *Journal of Colloid and Interface Science*, *475*, 8–16. https://doi.org/https://doi.org/10.1016/j.jcis.2016.04.034

Calliess, T., Sluszniak, M., Winkel, A., Pfaffenroth, C., Dempwolf, W., Heuer, W., Menzel, H., Windhagen, H., & Stiesch, M. (2012). Antimicrobial surface coatings for a permanent percutaneous passage in the concept of osseointegrated extremity prosthesis. *Biomedizinische Technik/Biomedical Engineering*, *57*(6), 467–471. https://doi.org/doi:10.1515/bmt-2011-0041

Chennell, P., Feschet-Chassot, E., Devers, T., Awitor, K. O., Descamps, S., & Sautou, V. (2013). In vitro evaluation of TiO2 nanotubes as cefuroxime carriers on orthopaedic implants for the prevention of periprosthetic joint infections. *International Journal of Pharmaceutics*, *455*(1), 298–305. https://doi.org/https://doi.org/10.1016/j.ijpharm.2013.07.014

Chevalier, J., & Gremillard, L. (2009). Ceramics for medical applications: A picture for the next 20 years. *Journal of the European Ceramic Society*, *29*(7), 1245–1255. https://doi.org/https://doi.org/10.1016/j.jeurceramsoc.2008.08.025

Choi, B.-H., Choi, Y. S., Kang, D. G., Kim, B. J., Song, Y. H., & Cha, H. J. (2010). Cell behavior on extracellular matrix mimic materials based on mussel adhesive protein fused with functional peptides. *Biomaterials*, *31*(34), 8980–8988. https://doi.org/https://doi.org/10.1016/j.biomaterials.2010.08.027

Chopra, D., Gulati, K., & Ivanovski, S. (2021a). Towards Clinical Translation: Optimized Fabrication of Controlled Nanostructures on Implant-Relevant Curved Zirconium Surfaces. *Nanomaterials*, *11*(4), 868. https://www.mdpi.com/2079-4991/11/4/868

Chopra, D., Gulati, K., & Ivanovski, S. (2021b). Understanding and optimizing the antibacterial functions of anodized nano-engineered titanium implants. *Acta Biomaterialia*. https://doi.org/https://doi.org/10.1016/j.actbio.2021.03.027

Chouirfa, H., Bouloussa, H., Migonney, V., & Falentin-Daudré, C. (2019). Review of titanium surface modification techniques and coatings for antibacterial applications. *Acta Biomaterialia*, *83*, 37–54. https://doi.org/https://doi.org/10.1016/j.actbio.2018.10.036

Córdoba, A., Hierro-Oliva, M., Pacha-Olivenza, M. Á., Fernández-Calderón, M. C., Perelló, J., Isern, B., González-Martín, M. L., Monjo, M., & Ramis, J. M. (2016). Direct Covalent Grafting of Phytate to Titanium Surfaces through Ti-O-P Bonding Shows Bone Stimulating Surface Properties and Decreased Bacterial Adhesion. *ACS Applied Materials & Interfaces*, *8*(18), 11326–11335. https://doi.org/10.1021/acsami.6b02533

Daish, C., Blanchard, R., Gulati, K., Losic, D., Findlay, D., Harvie, D. J. E., & Pivonka, P. (2017). Estimation of anisotropic permeability in trabecular bone based on microCT imaging and pore-scale fluid dynamics simulations. *Bone Reports*, *6*, 129–139. https://doi.org/https://doi.org/10.1016/j.bonr.2016.12.002

Das, K., Bose, S., Bandyopadhyay, A., Karandikar, B., & Gibbins, B. L. (2008). Surface coatings for improvement of bone cell materials and antimicrobial activities of Ti implants. *Journal of Biomedical Materials Research Part B: Applied Biomaterials*, *87B*(2), 455–460. https://doi.org/https://doi.org/10.1002/jbm.b.31125

Dhaliwal, J. S., Rahman, N. A., Knights, J., Ghani, H., & de Albuquerque Junior, R. F. (2019). The effect of different surface topographies of titanium implants on bacterial biofilm: a systematic review. *SN Applied Sciences*, *1*(6), 615. https://doi.org/10.1007/s42452-019-0638-6

Ding, X., Zhang, Y., Ling, J., & Lin, C. (2018). Rapid mussel-inspired synthesis of PDA-Zn-Ag nanofilms on TiO2 nanotubes for optimizing the antibacterial activity and biocompatibility by doping polydopamine with zinc at a higher temperature. *Colloids and Surfaces B: Biointerfaces*, *171*, 101–109. https://doi.org/https://doi.org/10.1016/j.colsurfb.2018.07.014

Ding, Y., Yuan, Z., Liu, P., Cai, K., & Liu, R. (2020). Fabrication of strontium-incorporated protein supramolecular nanofilm on titanium substrates for promoting osteogenesis. *Materials Science and Engineering: C*, *111*, 110851. https://doi.org/https://doi.org/10.1016/j.msec.2020.110851

Dong, J., Fang, D., Zhang, L., Shan, Q., & Huang, Y. (2019). Gallium-doped titania nanotubes elicit anti-bacterial efficacy in vivo against Escherichia coli and Staphylococcus aureus biofilm. *Materialia*, *5*, 100209. https://doi.org/https://doi.org/10.1016/j.mtla.2019.100209

Dong, Y., Ye, H., Liu, Y., Xu, L., Wu, Z., Hu, X., Ma, J., Pathak, J. L., Liu, J., & Wu, G. (2017). pH dependent silver nanoparticles releasing titanium implant: A novel therapeutic approach to control peri-implant infection. *Colloids and Surfaces B: Biointerfaces*, *158*, 127–136. https://doi.org/https://doi.org/10.1016/j.colsurfb.2017.06.034

Fathi, M., Akbari, B., & Taheriazam, A. (2019). Antibiotics drug release controlling and osteoblast adhesion from Titania nanotubes arrays using silk fibroin coating. *Materials Science and Engineering: C*, *103*, 109743. https://doi.org/https://doi.org/10.1016/j.msec.2019.109743

Feng, W., Geng, Z., Li, Z., Cui, Z., Zhu, S., Liang, Y., Liu, Y., Wang, R., & Yang, X. (2016). Controlled release behaviour and antibacterial effects of antibiotic-loaded titania nanotubes. *Materials Science and Engineering: C*, *62*, 105–112. https://doi.org/https://doi.org/10.1016/j.msec.2016.01.046

Goodman, S. B., Yao, Z., Keeney, M., & Yang, F. (2013). The future of biologic coatings for orthopaedic implants. *Biomaterials*, *34*(13), 3174–3183. https://doi.org/https://doi.org/10.1016/j.biomaterials.2013.01.074

Gu, J., Su, Y., Liu, P., Li, P., & Yang, P. (2017). An Environmentally Benign Antimicrobial Coating Based on a Protein Supramolecular Assembly. *ACS Applied Materials & Interfaces*, *9*(1), 198–210. https://doi.org/10.1021/acsami.6b13552

Gulati, K., Atkins, G., Findlay, D., & Losic, D. (2013). *Nano-engineered titanium for enhanced bone therapy* (Vol. 8812). SPIE. https://doi.org/10.1117/12.2027151

Gulati, K., Aw, M. S., Findlay, D., & Losic, D. (2012). Local drug delivery to the bone by drug-releasing implants: perspectives of nano-engineered titania nanotube arrays. *Therapeutic Delivery*, *3*(7), 857–873. https://doi.org/10.4155/tde.12.66

Gulati, K., Aw, M. S., & Losic, D. (2011). Drug-eluting Ti wires with titania nanotube arrays for bone fixation and reduced bone infection. *Nanoscale Research Letters*, *6*(1), 571. https://doi.org/10.1186/1556-276X-6-571

Gulati, K., Aw, M. S., & Losic, D. (2012). Nanoengineered drug-releasing Ti wires as an alternative for local delivery of chemotherapeutics in the brain. *International journal of nanomedicine*, *7*, 2069–2076. https://doi.org/10.2147/IJN.S29917

Gulati, K., Hamlet, S. M., & Ivanovski, S. (2018). Tailoring the immuno-responsiveness of anodized nano-engineered titanium implants [10.1039/C8TB00450A]. *Journal of Materials Chemistry B*, *6*(18), 2677–2689. https://doi.org/10.1039/C8TB00450A

Gulati, K., & Ivanovski, S. (2017). Dental implants modified with drug releasing titania nanotubes: therapeutic potential and developmental challenges. *Expert Opinion on Drug Delivery*, *14*(8), 1009–1024. https://doi.org/10.1080/17425247.2017.1266332

Gulati, K., Johnson, L., Karunagaran, R., Findlay, D., & Losic, D. (2016). In Situ Transformation of Chitosan Films into Microtubular Structures on the Surface of Nanoengineered Titanium Implants. *Biomacromolecules*, *17*(4), 1261–1271. https://doi.org/10.1021/acs.biomac.5b01037

Gulati, K., Kogawa, M., Maher, S., Atkins, G., Findlay, D., & Losic, D. (2015). Titania Nanotubes for Local Drug Delivery from Implant Surfaces. In D. Losic & A. Santos (Eds.), *Electrochemically Engineered Nanoporous Materials: Methods, Properties and Applications* (pp. 307–355). Springer International Publishing. https://doi.org/10.1007/978-3-319-20346-1_10

Gulati, K., Kogawa, M., Prideaux, M., Findlay, D. M., Atkins, G. J., & Losic, D. (2016). Drug-releasing nano-engineered titanium implants: therapeutic efficacy in 3D cell culture model, controlled release and stability. *Materials Science and Engineering: C*, *69*, 831–840. https://doi.org/https://doi.org/10.1016/j.msec.2016.07.047

Gulati, K., Li, T., & Ivanovski, S. (2018). Consume or Conserve: Microroughness of Titanium Implants toward Fabrication of Dual Micro–Nanotopography. *ACS Biomaterials Science & Engineering*, *4*(9), 3125–3131. https://doi.org/10.1021/acsbiomaterials.8b00829

Gulati, K., Maher, S., Chandrasekaran, S., Findlay, D. M., & Losic, D. (2016). Conversion of titania (TiO2) into conductive titanium (Ti) nanotube arrays for combined drug-delivery and electrical stimulation therapy [10.1039/C5TB02108A]. *Journal of Materials Chemistry B*, *4*(3), 371–375. https://doi.org/10.1039/C5TB02108A

Gulati, K., Moon, H.-J., Kumar, P. T. S., Han, P., & Ivanovski, S. (2020). Anodized anisotropic titanium surfaces for enhanced guidance of gingival fibroblasts. *Materials Science and Engineering: C*, *112*, 110860. https://doi.org/https://doi.org/10.1016/j.msec.2020.110860

Gulati, K., Moon, H.-J., Li, T., Sudheesh Kumar, P. T., & Ivanovski, S. (2018). Titania nanopores with dual micro-/nano-topography for selective cellular bioactivity. *Materials Science and Engineering: C*, *91*, 624–630. https://doi.org/https://doi.org/10.1016/j.msec.2018.05.075

Gulati, K., Prideaux, M., Kogawa, M., Lima-Marques, L., Atkins, G. J., Findlay, D. M., & Losic, D. (2017). Anodized 3D–printed titanium implants with dual micro- and nano-scale topography promote interaction with human osteoblasts and osteocyte-like cells. *Journal of Tissue Engineering and Regenerative Medicine*, *11*(12), 3313–3325. https://doi.org/https://doi.org/10.1002/term.2239

Gulati, K., Ramakrishnan, S., Aw, M. S., Atkins, G. J., Findlay, D. M., & Losic, D. (2012). Biocompatible polymer coating of titania nanotube arrays for improved drug elution and osteoblast adhesion. *Acta Biomaterialia*, *8*(1), 449–456. https://doi.org/https://doi.org/10.1016/j.actbio.2011.09.004

Gulati, K., Santos, A., Findlay, D., & Losic, D. (2015). Optimizing Anodization Conditions for the Growth of Titania Nanotubes on Curved Surfaces. *The Journal of Physical Chemistry C*, *119*(28), 16033–16045. https://doi.org/10.1021/acs.jpcc.5b03383

Gunputh, U. F., Le, H., Handy, R. D., & Tredwin, C. (2018). Anodised TiO2 nanotubes as a scaffold for antibacterial silver nanoparticles on titanium implants. *Materials Science and Engineering: C*, *91*, 638–644. https://doi.org/https://doi.org/10.1016/j.msec.2018.05.074

Guo, T., Gulati, K., Arora, H., Han, P., Fournier, B., & Ivanovski, S. (2021a). Orchestrating soft tissue integration at the transmucosal region of titanium implants. *Acta Biomaterialia*, *124*, 33–49. https://doi.org/https://doi.org/10.1016/j.actbio.2021.01.001

Guo, T., Gulati, K., Arora, H., Han, P., Fournier, B., & Ivanovski, S. (2021b). Race to invade: Understanding soft tissue integration at the transmucosal region of titanium dental implants. *Dental Materials*, *37*(5), 816–831. https://doi.org/https://doi.org/10.1016/j.dental.2021.02.005

Guo, T., Oztug, N. A. K., Han, P., Ivanovski, S., & Gulati, K. (2021). Old is Gold: Electrolyte Aging Influences the Topography, Chemistry, and Bioactivity of Anodized TiO2 Nanopores. *ACS Applied Materials & Interfaces*, *13*(7), 7897–7912. https://doi.org/10.1021/acsami.0c19569

Ha, Y., Yang, J., Tao, F., Wu, Q., Song, Y., Wang, H., Zhang, X., & Yang, P. (2018). Phase-Transited Lysozyme as a Universal Route to Bioactive Hydroxyapatite Crystalline Film. *Advanced Functional Materials*, *28*(4), 1704476. https://doi.org/https://doi.org/10.1002/adfm.201704476

Hu, X., Tian, J., Li, C., Su, H., Qin, R., Wang, Y., Cao, X., & Yang, P. (2020). Amyloid-Like Protein Aggregates: A New Class of Bioinspired Materials Merging an Interfacial Anchor with Antifouling. *Advanced Materials*, *32*(23), 2000128. https://doi.org/https://doi.org/10.1002/adma.202000128

Jayasree, A., Ivanovski, S., & Gulati, K. (2021). ON or OFF: Triggered therapies from anodized nano-engineered titanium implants. *Journal of Controlled Release*, *333*, 521–535. https://doi.org/https://doi.org/10.1016/j.jconrel.2021.03.020

Jia, Z., Xiu, P., Li, M., Xu, X., Shi, Y., Cheng, Y., Wei, S., Zheng, Y., Xi, T., Cai, H., & Liu, Z. (2016). Bioinspired anchoring AgNPs onto micro-nanoporous TiO2 orthopedic coatings: Trap-killing of bacteria, surface-regulated osteoblast functions and host responses. *Biomaterials*, *75*, 203–222. https://doi.org/https://doi.org/10.1016/j.biomaterials.2015.10.035

Kaur, G., Willsmore, T., Gulati, K., Zinonos, I., Wang, Y., Kurian, M., Hay, S., Losic, D., & Evdokiou, A. (2016). Titanium wire implants with nanotube arrays: A study model for localized cancer treatment. *Biomaterials*, *101*, 176–188. https://doi.org/https://doi.org/10.1016/j.biomaterials.2016.05.048

Kumeria, T., Mon, H., Aw, M. S., Gulati, K., Santos, A., Griesser, H. J., & Losic, D. (2015). Advanced biopolymer-coated drug-releasing titania nanotubes (TNTs) implants with simultaneously enhanced osteoblast adhesion and antibacterial properties. *Colloids and Surfaces B: Biointerfaces*, *130*, 255–263. https://doi.org/https://doi.org/10.1016/j.colsurfb.2015.04.021

Kunrath, M. F., Diz, F. M., Magini, R., & Galárraga-Vinueza, M. E. (2020). Nanointeraction: The profound influence of nanostructured and nano-drug delivery biomedical implant surfaces on cell behavior. *Advances in Colloid and Interface Science*, *284*, 102265. https://doi.org/10.1016/j.cis.2020.102265

Lee, S. W., Phillips, K. S., Gu, H., Kazemzadeh-Narbat, M., & Ren, D. (2021). How microbes read the map: Effects of implant topography on bacterial adhesion and biofilm formation. *Biomaterials*, *268*, 120595. https://doi.org/https://doi.org/10.1016/j.biomaterials.2020.120595

Li, T., Gulati, K., Wang, N., Zhang, Z., & Ivanovski, S. (2018). Bridging the gap: Optimized fabrication of robust titania nanostructures on complex implant geometries towards clinical translation. *Journal of Colloid and Interface Science*, *529*, 452–463. https://doi.org/https://doi.org/10.1016/j.jcis.2018.06.004

Liu, Q., Ding, J., Mante, F. K., Wunder, S. L., & Baran, G. R. (2002). The role of surface functional groups in calcium phosphate nucleation on titanium foil: a self-assembled monolayer technique. *Biomaterials*, *23*(15), 3103–3111. https://doi.org/https://doi.org/10.1016/S0142-9612(02)00050-9

Liu, Y.-T., Kung, K.-C., Yang, C.-Y., Lee, T.-M., & Lui, T.-S. (2014). Engineering three-dimensional structures using bio-inspired dopamine and strontium on titanium for biomedical application [10.1039/C4TB00822G]. *Journal of Materials Chemistry B*, *2*(45), 7927–7935. https://doi.org/10.1039/C4TB00822G

Losic, D., Aw, M. S., Santos, A., Gulati, K., & Bariana, M. (2015). Titania nanotube arrays for local drug delivery: recent advances and perspectives. *Expert Opinion on Drug Delivery*, *12*(1), 103–127. https://doi.org/10.1517/17425247.2014.945418

Ma, M., Kazemzadeh-Narbat, M., Hui, Y., Lu, S., Ding, C., Chen, D. D. Y., Hancock, R. E. W., & Wang, R. (2012). Local delivery of antimicrobial peptides using self-organized TiO2 nanotube arrays for peri-implant infections. *Journal of Biomedical Materials Research Part A*, *100A*(2), 278–285. https://doi.org/https://doi.org/10.1002/jbm.a.33251

Martinez-Marquez, D., Gulati, K., Carty, C. P., Stewart, R. A., & Ivanovski, S. (2020). Determining the relative importance of titania nanotubes characteristics on bone implant surface performance: A quality by design study with a fuzzy approach. *Materials Science and Engineering: C*, *114*, 110995. https://doi.org/https://doi.org/10.1016/j.msec.2020.110995

Mirzaee, M., Vaezi, M., & Palizdar, Y. (2016). Synthesis and characterization of silver doped hydroxyapatite nanocomposite coatings and evaluation of their antibacterial and corrosion resistance properties in simulated body fluid. *Materials Science and Engineering: C*, *69*, 675–684. https://doi.org/https://doi.org/10.1016/j.msec.2016.07.057

Müller, R., Abke, J., Schnell, E., Scharnweber, D., Kujat, R., Englert, C., Taheri, D., Nerlich, M., & Angele, P. (2006). Influence of surface pretreatment of titanium- and cobalt-based biomaterials on covalent immobilization of fibrillar collagen. *Biomaterials*, *27*(22), 4059–4068. https://doi.org/https://doi.org/10.1016/j.biomaterials.2006.03.019

Nanci, A., Wuest, J. D., Peru, L., Brunet, P., Sharma, V., Zalzal, S., & McKee, M. D. (1998). Chemical modification of titanium surfaces for covalent attachment of biological molecules. *J Biomed Mater Res*, *40*(2), 324–335. https://doi.org/10.1002/(SICI)1097-4636(199805)40:2<324::AID-JBM18>3.0.CO;2-L

Nandi, S. K., Shivaram, A., Bose, S., & Bandyopadhyay, A. (2018). Silver nanoparticle deposited implants to treat osteomyelitis. *Journal of Biomedical Materials Research Part B: Applied Biomaterials*, *106*(3), 1073–1083. https://doi.org/https://doi.org/10.1002/jbm.b.33910

Narendrakumar, K., Kulkarni, M., Addison, O., Mazare, A., Junkar, I., Schmuki, P., Sammons, R., & Iglič, A. (2015). Adherence of oral streptococci to nanostructured titanium surfaces. *Dental Materials*, *31*(12), 1460–1468. https://doi.org/https://doi.org/10.1016/j.dental.2015.09.011

Neoh, K. G., Hu, X., Zheng, D., & Kang, E. T. (2012). Balancing osteoblast functions and bacterial adhesion on functionalized titanium surfaces. *Biomaterials*, *33*(10), 2813–2822. https://doi.org/https://doi.org/10.1016/j.biomaterials.2012.01.018

Okuzu, Y., Fujibayashi, S., Yamaguchi, S., Yamamoto, K., Shimizu, T., Sono, T., Goto, K., Otsuki, B., Matsushita, T., Kokubo, T., & Matsuda, S. (2017). Strontium and magnesium ions released from bioactive titanium metal promote early bone bonding in a rabbit implant model. *Acta Biomaterialia, 63,* 383–392. https:// doi.org/https://doi.org/10.1016/j.actbio.2017.09.019

Park, S. W., Lee, D., Choi, Y. S., Jeon, H. B., Lee, C.-H., Moon, J.-H., & Kwon, I. K. (2014). Mesoporous TiO2 implants for loading high dosage of antibacterial agent. *Applied Surface Science, 303,* 140–146. https:// doi.org/https://doi.org/10.1016/j.apsusc.2014.02.111

Pilmane, M., Salma-Ancane, K., Loca, D., Locs, J., & Berzina-Cimdina, L. (2017). Strontium and strontium ranelate: Historical review of some of their functions. *Materials Science and Engineering: C, 78,* 1222–1230. https://doi.org/https://doi.org/10.1016/j.msec.2017.05.042

Popat, K. C., Eltgroth, M., LaTempa, T. J., Grimes, C. A., & Desai, T. A. (2007). Decreased Staphylococcus epidermis adhesion and increased osteoblast functionality on antibiotic-loaded titania nanotubes. *Biomaterials, 28*(32), 4880–4888. https://doi.org/https://doi.org/10.1016/j.biomaterials.2007.07.037

Puckett, S. D., Taylor, E., Raimondo, T., & Webster, T. J. (2010). The relationship between the nanostructure of titanium surfaces and bacterial attachment. *Biomaterials, 31*(4), 706–713. https://doi.org/https://doi.org/10.1016/j.biomaterials.2009.09.081

Rahman, S., Gulati, K., Kogawa, M., Atkins, G. J., Pivonka, P., Findlay, D. M., & Losic, D. (2016). Drug diffusion, integration, and stability of nanoengineered drug-releasing implants in bone ex-vivo. *Journal of Biomedical Materials Research Part A, 104*(3), 714–725. https://doi.org/https://doi.org/10.1002/jbm.a.35595

Raphel, J., Holodniy, M., Goodman, S. B., & Heilshorn, S. C. (2016). Multifunctional coatings to simultaneously promote osseointegration and prevent infection of orthopaedic implants. *Biomaterials, 84,* 301–314. https://doi.org/https://doi.org/10.1016/j.biomaterials.2016.01.016

Rimondini, L., Faré, S., Brambilla, E., Felloni, A., Consonni, C., Brossa, F., & Carrassi, A. (1997). The Effect of Surface Roughness on Early In Vivo Plaque Colonization on Titanium. *Journal of Periodontology, 68*(6), 556–562. https://doi.org/https://doi.org/10.1902/jop.1997.68.6.556

Saji, V. S., Kumeria, T., Gulati, K., Prideaux, M., Rahman, S., Alsawat, M., Santos, A., Atkins, G. J., & Losic, D. (2015). Localized drug delivery of selenium (Se) using nanoporous anodic aluminium oxide for bone implants [10.1039/C5TB00125K]. *Journal of Materials Chemistry B, 3*(35), 7090–7098. https://doi.org/10.1039/C5TB00125K

Salditt, T., & Schubert, U. S. (2002). Layer-by-layer self-assembly of supramolecular and biomolecular films. *Reviews in Molecular Biotechnology, 90*(1), 55–70. https://doi.org/https://doi.org/10.1016/S1389-0352(01)00049-6

Shi, X., Wu, H., Li, Y., Wei, X., & Du, Y. (2013). Electrical signals guided entrapment and controlled release of antibiotics on titanium surface. *Journal of Biomedical Materials Research Part A, 101A*(5), 1373–1378. https://doi.org/https://doi.org/10.1002/jbm.a.34432

Shivaram, A., Bose, S., & Bandyopadhyay, A. (2016). Mechanical degradation of TiO2 nanotubes with and without nanoparticulate silver coating. *Journal of the Mechanical Behavior of Biomedical Materials, 59,* 508–518. https://doi.org/https://doi.org/10.1016/j.jmbbm.2016.02.028

Shu, Y., Ou, G., Wang, L., Zou, J., & Li, Q. (2011). Surface Modification of Titanium with Heparin-Chitosan Multilayers via Layer-by-Layer Self-Assembly Technique. *Journal of Nanomaterials, 2011,* 423686. https://doi.org/10.1155/2011/423686

Sirivisoot, S., Pareta, R., & Webster, T. J. (2011). Electrically controlled drug release from nanostructured polypyrrole coated on titanium. *Nanotechnology, 22*(8), 085101. https://doi.org/10.1088/0957-4484/22/8/085101

Sopchenski, L., Cogo, S., Dias-Ntipanyj, M. F., Elifio-Espósito, S., Popat, K. C., & Soares, P. (2018). Bioactive and antibacterial boron doped TiO2 coating obtained by PEO. *Applied Surface Science, 458,* 49–58. https://doi.org/https://doi.org/10.1016/j.apsusc.2018.07.049

Tripathy, A., Sen, P., Su, B., & Briscoe, W. H. (2017). Natural and bioinspired nanostructured bactericidal surfaces. *Advances in Colloid and Interface Science, 248,* 85–104. https://doi.org/https://doi.org/10.1016/j.cis.2017.07.030

Ungureanu, C., Dumitriu, C., Popescu, S., Enculescu, M., Tofan, V., Popescu, M., & Pirvu, C. (2016). Enhancing antimicrobial activity of TiO2/Ti by torularhodin bioinspired surface modification. *Bioelectrochemistry, 107,* 14–24. https://doi.org/https://doi.org/10.1016/j.bioelechem.2015.09.001

Wang, T., Liu, X., Zhu, Y., Cui, Z. D., Yang, X. J., Pan, H., Yeung, K. W. K., & Wu, S. (2017). Metal Ion Coordination Polymer-Capped pH-Triggered Drug Release System on Titania Nanotubes for Enhancing Self-antibacterial Capability of Ti Implants. *ACS Biomaterials Science & Engineering, 3*(5), 816–825. https://doi.org/10.1021/acsbiomaterials.7b00103

Wu, Z., & Yang, P. (2015). Simple Multipurpose Surface Functionalization by Phase Transited Protein Adhesion. *Advanced Materials Interfaces, 2*(2), 1400401. https://doi.org/https://doi.org/10.1002/admi.201400401

Xiang, Y., Liu, X., Mao, C., Liu, X., Cui, Z., Yang, X., Yeung, K. W. K., Zheng, Y., & Wu, S. (2018). Infection-prevention on Ti implants by controlled drug release from folic acid/ZnO quantum dots sealed titania nanotubes. *Materials Science and Engineering: C*, *85*, 214–224. https://doi.org/https://doi.org/10.1016/j. msec.2017.12.034

Xu, K., Chen, W., Hu, Y., Shen, X., Xu, G., Ran, Q., Yu, Y., Mu, C., & Cai, K. (2016). Influence of strontium ions incorporated into nanosheet-pore topographical titanium substrates on osteogenic differentiation of mesenchymal stem cells in vitro and on osseointegration in vivo [10.1039/C6TB00724D]. *Journal of Materials Chemistry B*, *4*(26), 4549–4564. https://doi.org/10.1039/C6TB00724D

Xu, X., Zhang, D., Gao, S., Shiba, T., Yuan, Q., Cheng, K., Tan, H., & Li, J. (2018). Multifunctional Biomaterial Coating Based on Bio-Inspired Polyphosphate and Lysozyme Supramolecular Nanofilm. *Biomacromolecules*, *19*(6), 1979–1989. https://doi.org/10.1021/acs.biomac.8b00002

Yang, H., Yuan, B., Zhang, X., & Scherman, O. A. (2014). Supramolecular Chemistry at Interfaces: Host–Guest Interactions for Fabricating Multifunctional Biointerfaces. *Accounts of Chemical Research*, *47*(7), 2106–2115. https://doi.org/10.1021/ar500105t

Yu, Y., Ran, Q., Shen, X., Zheng, H., & Cai, K. (2020). Enzyme responsive titanium substrates with antibacterial property and osteo/angio-genic differentiation potentials. *Colloids and Surfaces B: Biointerfaces*, *185*, 110592. https://doi.org/https://doi.org/10.1016/j.colsurfb.2019.110592

Yuan, Z., Huang, S., Lan, S., Xiong, H., Tao, B., Ding, Y., Liu, Y., Liu, P., & Cai, K. (2018). Surface engineering of titanium implants with enzyme-triggered antibacterial properties and enhanced osseointegration in vivo [10.1039/C8TB01918E]. *Journal of Materials Chemistry B*, *6*(48), 8090–8104. https://doi.org/10.1039/C8TB01918E

Zhang, E., Zhao, X., Hu, J., Wang, R., Fu, S., & Qin, G. (2021). Antibacterial metals and alloys for potential biomedical implants. *Bioactive Materials*, *6*(8), 2569–2612. https://doi.org/https://doi.org/10.1016/j. bioactmat.2021.01.030

Zhang, H., Sun, Y., Tian, A., Xue, X. X., Wang, L., Alquhali, A., & Bai, X. (2013). Improved antibacterial activity and biocompatibility on vancomycin-loaded TiO2 nanotubes: in vivo and in vitro studies. *International Journal of Nanomedicine*, *8*, 4379–4389. https://doi.org/10.2147/ijn.S53221

Zhang, J., Wang, H., Wang, Y., Dong, W., Jiang, Z., & Yang, G. (2018). Substrate-mediated gene transduction of LAMA3 for promoting biological sealing between titanium surface and gingival epithelium. *Colloids and Surfaces B: Biointerfaces*, *161*, 314–323. https://doi.org/https://doi.org/10.1016/j.colsurfb.2017.10.030

Zhang, X., Li, J., Wang, X., Wang, Y., Hang, R., Huang, X., Tang, B., & Chu, P. K. (2018). Effects of copper nanoparticles in porous TiO2 coatings on bacterial resistance and cytocompatibility of osteoblasts and endothelial cells. *Materials Science and Engineering: C*, *82*, 110–120. https://doi.org/https://doi. org/10.1016/j.msec.2017.08.061

Zhao, J., Xu, J., Jian, X., Xu, J., Gao, Z., & Song, Y.-Y. (2020). NIR Light-Driven Photocatalysis on Amphiphilic TiO2 Nanotubes for Controllable Drug Release. *ACS Applied Materials & Interfaces*, *12*(20), 23606–23616. https://doi.org/10.1021/acsami.0c04260

Zhao, L., Wang, H., Huo, K., Zhang, X., Wang, W., Zhang, Y., Wu, Z., & Chu, P. K. (2013). The osteogenic activity of strontium loaded titania nanotube arrays on titanium substrates. *Biomaterials*, *34*(1), 19–29. https://doi.org/https://doi.org/10.1016/j.biomaterials.2012.09.041

Zhou, J., Frank, M. A., Yang, Y., Boccaccini, A. R., & Virtanen, S. (2018). A novel local drug delivery system: Superhydrophobic titanium oxide nanotube arrays serve as the drug reservoir and ultrasonication functions as the drug release trigger. *Materials Science and Engineering: C*, *82*, 277–283. https://doi.org/ https://doi.org/10.1016/j.msec.2017.08.066

Index

Note: **Bold** page numbers refer to tables and *italic* page numbers refer to figures.

For Product Safety Concerns and Information please contact our EU
representative GPSR@taylorandfrancis.com
Taylor & Francis Verlag GmbH, Kaufingerstraße 24, 80331 München, Germany

www.ingramcontent.com/pod-product-compliance
Lightning Source LLC
Chambersburg PA
CBHW080137220326
41598CB00032B/5097

* 9 7 8 0 3 6 7 7 6 9 6 2 8 *